Flavours and Fragrances

Chemistry, Bioprocessing and Sustainability

R. G. Berger (Ed.)

Flavours and Fragrances

Chemistry, Bioprocessing and Sustainability

With 231 Figures and 61 Tables

Springer

Prof. Dr. Ralf Günter Berger
Universität Hannover
FB Chemie, Institut für Lebensmittelchemie
Wunstorferstraße 14
30453 Hannover, Germany
rg.berger@lci.uni-hannover.de

Library of Congress Control Number: 2006939012
ISBN 978-3-540-49338-9 Springer Berlin Heidelberg New York
DOI 10.1007/b136889

Springer is a part of Springer Science + Business Media
springeronline.com

Printed in Germany

Coverdesign: WMXDesign GmbH, Heidelberg
Typesetting & production: LE-TeX Jelonek, Schmidt & Vöckler GbR, Leipzig, Germany
Printed on acid-freepaper 2/3141/YL - 543210

Preface

Our ancestors lived in intimacy with nature and knew well that their survival depended on a safe and fertile environment. The introduction of three-field rotation in the eighth century bc, for example, counteracted the depletion of soil and increased crop yields without negative side effects. The first definition of the modern term "sustainability" is usually ascribed to forest chief captain H. C. von Carlowitz, who in 1713 in his *Sylvicultura Oeconomica* formulated principles for a sensible economy of wood. From J. S. Mill (*Of the Stationary State*) to modern academic representatives, such as K. Boulding, D. E. Meadows (*The Limits to Growth*), R. Easterlin and H. E. Daly, the "ecological economists" have remained a concerned but rather ignored minority. The situation started to change after the famous Brundtland report (*Our Common Future*) of the UN defined sustainability as a desirable characteristic of development, which will not only meet current needs of people, but also will not jeopardise the ability of future generations to meet their demands and to choose their style of life. This definition includes a social dimension and was also adopted by Agenda 21 of the UNCED in 1992 in Rio de Janeiro.

A set of rules may aid in assessing the sustainable quality of a process:

- Consumption and regeneration of the raw materials should be balanced.
- Non-regenerative goods should be replaced.
- Generation of waste and its biological elimination should be balanced.
- Technical processes should match biological processes on the time scale.

A merely growth oriented economy must violate these rules. According to the first law of thermodynamics, energy in a closed system like the planet earth is finite (if we neglect the solar photon flux). Today mankind secures its survival by exploiting low-entropy resources, such as fossil fuels, concentrated minerals and higher plants, and by converting them to high-entropy products, such as carbon dioxide, cars and fine chemicals. However, as proven by our office desks, high entropy levels can only be lowered by energy input. Here the first and the second law of thermodynamics collide, and we apparently encounter the inner core of the conflict.

With the world running out of crude oil, species dying out at an alarming rate and political leaders seemingly little concerned about the predicted disasters, scientists should feel challenged to suggest solutions. A sustainable production of natural flavours, like wood, fats and oils, saccharides, phytomedicines, bioethanol, biopolymers and natural colours, mainly depends on the existence of reliable plant sources. But how long will the traditional sources of flavours last? Quality of soil, unfavourable weather conditions, insect infestations and sociopolitical instabilities may all adversely affect classical agricultural production. Are there new biosources that could replace exhausted ones? Will, as with vanillin production, the exploitation of waste streams of the agricultural and food industries gain importance? "White biotechnology" is propagated as an alternative option, but will bioprocesses possess stability, specificity, up-scalability and profitability? Will the recent advances in biotechnology be successfully transferred to industrial scales? How can the aspired match of economy and ecology be achieved?

In an attempt to compile the current status of sustainability in the flavour industry and the developments in the foreseeable future of flavour production, the present volume discusses consumer trends and preferences, legal and safety aspects; it describes renewable resources of flavours, such as spice plants, fruits, vegetables, fermented and heated plants, and natural building blocks; it presents analytical methods, such as gas chromatography coupled to human or electronic noses or to mass spectrometers; it deals with the isolation, quality control and formulation of flavours for liquid or dry products, with biotechnology to provide novel renewable resources, with enzymes, microbial and fungal cells to biotransform cheap substrates or to produce flavours de novo, and with plant cells as a resource of genes coding for metabolic activities in transgenic producers.

The manufacturers of flavours and fragrances and their scientists are working at the leading edge of research, they look back on a long history of using natural resources, and are profitable on the basis of renewables. A wealth of experience has been gathered on issues such as provenance and quality, safety, authenticity and on problems of isolation, processing and shelf life. On the basis of this fundament of knowledge, we should start to deal with sustainability now, before the looming problems start to deal with us.

Finally, I should like to express my sincere thanks to the contributors for their thoughts and writing efforts, and to the publishers for their continuing support and patience.

Hanover, Summer 2006 Ralf Günter Berger

Contents

1 The Flavour and Fragrance Industry—Past, Present, and Future

Matthias Guentert
Symrise Inc., 300 North Street, Teterboro, NJ 07608, USA

The origin of using odorous substances simply for enjoyment or medicinal reasons is as old as mankind. People have used perfume oils, and unguents on their bodies for thousands of years in lesser or greater amounts dependent on fashion whims. The early Egyptians used perfumed balms as part of religious ceremonies and later as part of pre-love-making preparations. Myrrh and frankincense were exuded gums from trees used to scent the atmosphere in rituals. Other plants such as rose and peppermint were steeped in oils until a perfumed unguent formed. The unguent was then rubbed into the skin. It is interesting to note that perfume has come full circle today as more and more of us seek out high-quality aroma therapy perfumed oils to use in exactly the same way as our ancestors did. Perfume fell out of use during early Christianity, but was revived in the medieval period. By the 1600s scents were applied to objects such as furniture, gloves, and fans. In the Georgian era non-greasy eau de Cologne was developed and it had many uses from bath essence to mouthwash [1].

People have always been interested in the odour and use of essential oils (from herbs and spices). This is probably attributable to their aromas, and also to the bacteriostatic and antiseptic properties of many of the aroma chemicals they contain. While the use of essential oils is associated with mankind's history, the beginning of perfumery is more difficult to define.

The late nineteenth century was the first real era of perfume as we know it when new scents were created because of advances in organic chemistry knowledge. Synthetic perfume products were used in place of certain hard-to-find or expensive ingredients. At the same time a similar chemical knowledge development happened in textile printing dyes. The small town of Grasse in Provence, France, became a centre for flower and herb growing for the perfume industry. The men who treated leathers in the same area found the smells so bad they perfumed themselves and the leathers. They were knowledgeable about making the botanical essences and were the early perfume noses. But it was only in the twentieth century that scents and designer perfumes were really mass-produced. Before that, the few trade names that existed were Coty and Yardley, who made fairly light scents with familiar smells.

Obviously, these first perfumes were all natural, since the introduction of synthetic aroma chemicals happened only at the end of the nineteenth century. Along with the invention of certain aroma chemicals, the flavour and fragrance industry originated more than 150 years ago, at a time that is in general characterised by significant technological breakthroughs, largely in chemistry. At that time, the first flavour and fragrance companies were founded by entrepreneurial scientists or business people, and many still exist, either as such or as the nucleus of larger firms that evolved during the subsequent decades.

Over the years this industry has developed into a very profitable niche market. It serves retail companies in the food and beverage, cosmetics, toiletry, and household products markets, as well as the fragrance industries. Food service and private label companies play an increasingly significant role in this business. The total market for flavours, fragrances, and cosmetic ingredients is estimated at €15 billion. The market shares between the flavour and the fragrance parts are almost equal (€6.5 billion for flavours, €6.5 billion for fragrances). The largest markets are in the Europe, Africa, and Middle East region (36%) and North America (32%), followed by Asia-Pacific (26%) and South America (6%). Interesting emerging markets are, in particular, China, India, Russia, and Central America. There are eight major global companies that share about 60% of the world market. Aside from these multinationals—well-known names to insiders—there are virtually hundreds of smaller companies specialising in certain segments of this business covering the other half of the market. The two largest flavour and fragrance companies are Givaudan and International Flavors & Fragrances, followed by Firmenich and Symrise, Quest International, Takasago, T. Hasegawa, Sensient Technologies, Mastertaste, Danisco, and Mane. The top two companies have a turnover in excess of $2 billion, the next three companies have a turnover in excess of $1 billion each. For respective information on the flavour and fragrance business and the companies, readers are referred to [2]. Another good and recently published source of information on flavours and fragrances is [3]. The latest development of the industry happened just at the end of 2006 when Givaudan announced that they will acquire Quest International. In the following, the history and achievements of some of the companies are described.

In 1993, Bell Flavors & Fragrances acquired the operations of the former firm Schimmel & Co. in Leipzig, Germany. This company, originally founded in 1829, is considered the founding firm of the flavour and fragrance industry. The scientific accomplishments developed at Schimmel formed the basis for the technology still used in the industry today. Works such as *The Encyclopedia of Essential Oils*, published by doctors Gildemeister and Hoffmann in 1899 and *The Theory of the Extraction and Separation of Essential Oils by way of Distillation*, published by Carl V. Rechenberg in 1908, became the standards for the production and use of these products. Outstanding achievement in the field of terpene chemistry was recognised when Otto Wallach received the Nobel Prize in Chemistry in 1910 [4].

In 1874, Holzminden chemists Ferdinand Tiemann and Wilhelm Haarmann first succeeded in synthesising vanillin from coniferin. Holzminden became the site where vanillin was first produced industrially. Haarmann & Reimer was the world's first factory in which synthetic scents and flavourings were produced [5].

Milestones Haarmann & Reimer:

1874 Haarmann & Reimer founded in Holzminden; industrial production of aroma chemicals.

1953 Acquisition by Bayer AG in Leverkusen, Germany.

1990 Company expands via acquisitions, including Creations Aromatiques and Florasynth.

1998 Improvement of margin and performance; focus on major clients and emerging markets; key account management; centres of expertise created; regionalisation and market-oriented innovation management.

2003 Merger of Holzminden companies Haarmann & Reimer and Dragoco. A new corporation is formed: Symrise.

Milestones Dragoco:

1919 Family business founded in Holzminden, Germany, by Carl-Wilhelm Gerberding.

1930 Flavourings first produced.

1949 Aroma chemicals first produced.

1955 Carl-Heinz Gerberding becomes CEO; company expands internationally and focuses on independence and profitability.

1981 Horst-Otto Gerberding becomes CEO; extensive investment programme for regional centres; implementation of a global divisional organisation.

1993 The Dragoco group is restructured and the parent company is turned into a joint-stock company.

2003 Merger of Holzminden companies Haarmann & Reimer and Dragoco. A new corporation is formed: Symrise. The latest milestone of the new company Symrise has been reached at the end of 2006 when the company became publicly traded.

Firmenich was founded in 1895 in Geneva, by Philippe Chuit, a talented Swiss chemist, in association with Martin Naef, a shrewd businessman. They were joined shortly after by Fred Firmenich, who soon became the majority partner. Since then, Firmenich has remained a family-owned business, building on a solid foundation of pioneering and entrepreneurial vigour. Today, it is the world's largest private company in the fragrance and flavour industry worldwide. Since 1895, Firmenich has built its business on innovative research. Leopold Ruzicka, professor at ETH-Zurich and Nobel Prize winner in 1939, was Firmenich's first research director and a life-long consultant [6].

The history of Givaudan [7)], International Flavors & Fragrances [8] and Quest International [9] can be looked up at their respectively cited Web sites.

The flavour and fragrance business has always been very research driven and innovative. All larger companies spend about 7–8% of their total sales per annum on research and development. They all have large research centres, usually centred in their headquarters, as well as development and innovation centres around the globe. The general focus of their research is on new products, offering better performance at the lowest cost. This can be new molecules but also a new technique to concentrate (fold) a citrus oil, or a new way to encapsulate a flavour or fragrance. New products must be innovative, environmentally friendly, and safe. The key to success nowadays is to bring these new research results to market as quickly as possible; therefore, the major companies all use the concept of the innovation funnel to ensure proper project management and efficient commercialisation of innovative ideas. Strong research only pays off in combination with innovative flavour and fragrance chemists (flavourists and perfumers) as well as strong application teams in combination with technical marketing. In addition, all major companies use worldwide IT systems to enable their product developers and regulatory people to work with a consistent set of raw materials and product formulas on a worldwide basis.

Flavourists and perfumers are professionals engaged in the study and exploitation of materials capable of impacting the human senses of taste, smell, and chemesthesis. Flavourists work primarily with substances that are either derived (directly or indirectly) from plant or animal sources or chemically synthesised from petrochemicals to develop products intended for use in foods and beverages. Perfumers work mostly with materials of plant, animal, or petrochemical origin to create perfumes, fragranced personal care products, and scented household goods.

Research carried out by flavour and fragrance companies is generally for the purpose of understanding, designing, or improving upon the sensory characteristics and/or the functionality of existing or new products. This often starts with the detailed chemical analysis of a specific target: a finished product or raw materials used in its manufacture. Creative flavourists or perfumers, respectively, with the help of product technologists, may then try to reconstitute flavours or fragrances that match or improve upon the sensory properties of the target. In the case of flavourists, matching a specific natural or processed food or beverage is usually the objective, while a perfumer often has more latitude in cases where the target fine perfume or household air freshener, for example, may be little more than a marketing concept. Product technologists help ensure that flavours and fragrances are stable in products and are released effectively and are therefore perceivable at the time of consumption or use. Results of chemical analysis may alternatively be used. For example, to design better flavour or fragrance molecules; to make improvements in ingredient formulations or manufacturing processes. It can be mentioned here that the instrumental analysis part in the major flavour and fragrance companies is very sophisticated and remarkable. The typical instrumentation ranges from capillary gas chromatography (GC) to high-performance liquid chromatography (HPLC), Fourier transform IR spectroscopy (FTIR), and nuclear magnetic resonance (NMR) to coupled techniques like capillary GC–mass spectrometry (MS), HPLC-MS, and GC-FTIR. More recent advances are the coupled techniques GC-MS-MS as well as HPLC-NMR. This enables industry to separate virtually all kinds of complex product mixes analytically and also to elucidate the chemical structures of unknown components. Needless to say that a lot of year-long experience and know-how is involved when it comes to research and development in flavour and fragrance companies (personal communication within Symrise).

Although the industry is about 150 years old, in particular the use of synthetic materials started only about 60 years ago, after World War II. In 1954, the flavour use of coumarin was banned by the FDA, television screens were small and round and only showed black-and-white pictures, and a fine house could be purchased for less than $17,000. It would be 4 years until Congress enacted the Food Additives Amendment of 1958 and both the FDA and the Flavour and Extract Manufacturer's Association began developing the generally recognised as safe (GRAS) lists. The first commercial production of synthetic linalool, geraniol, and derivatives had not yet started. Fewer than 500 volatile components had been found in foodstuffs. In 1955 the first primitive commercial gas chromatograph was introduced and it would be about 15 years later before the full power of capillary GC-MS became practical and another 15 years with the use of computerised data bases. Only four of the five basic tastes were generally accepted and theories of olfaction were extremely theoretical. Chirality was rarely considered as important in the synthesis of flavour or fragrance chemicals. Much has occurred in the last 50 years [10]. In a slightly different way, the development of the last about 50 years is shown in Fig. 1.1. The flavours used in the 1950s were mostly liquid. They consisted of natural extracts and essential oils. The first big paradigm shift happened in the 1960s when several developments happened at the same time. The first synthetic components started being used, while instrumental analysis and information technology began influencing the flavour and fragrance industry. Spray-dried flavours were developed and the food market started to embrace convenience food. The big era of analytical flavour research started at that time, characterised by many scientific publications and patents in subsequent years. The next big change happened in the 1990s when research started to become a lot more applications-driven. Flavour release and integrated product concepts played a role, and food-on-the-go was developed. In the new century the term "productivity" came up, a clear sign that shrinking margins led to the consolidation of the food companies and the search for more cost-effective ingredients and flavours. Taste and taste modifications as well as mouth sensations became prevalent. Sensory started becoming consumer research, and health aspects played into product development. This is the phase the industry is still in, and we will see when the next paradigm shift is going to happen. Nowadays the palettes of a perfumer and flavourists are fully developed. There are still new aroma chemicals entering the market every year but the number is certainly smaller compared with that 10–20 years ago, and the organoleptical differences of these new molecules from known ones are typically smaller, which means business success is usually and primarily not built anymore solely on new molecules. At the same time the typical analytical research from the end of the last century that was going on in all large flavour and fragrance companies with the goal to analyse natural materials (preferably foods, essential oils, and flower scents) and find new molecules that could be synthesised and used as nature-identical materials in new compositions is not the main focus anymore. Nowadays research is a lot more applications-driven. Innovation happens foremost at the finished-product level; hence, flavour and fragrance companies work a lot closer together with their large consumer-goods customers and in many cases have taken on a part of the work that used to be done in their laboratories. On the flavour side, research on taste and taste modification has become a lot more important than the work on volatile materials. Topics like salt taste enhancement or sweetness enhancement prevail. Mastering the flavour release in various applications and encapsulating liquid flavours with different matrices to keep even critical ingredients (e.g. citrus) stable for up to 4 years have opened the door for different food and beverage concepts and have also helped to make the food business more global and more convenience-driven than it ever was before. The ideal scenario today in flavour and fragrance research is to find a new molecule whose structure can be patented and used in a new formulation that helps to improve an application for a consumer-goods company significantly. The application can be everything from a cosmetic product to a household article to seasoning for a potato chip or a canned coffee product (personal communication within Symrise).

"Sustainable development" describes and stands for the policy of a company of how it conducts business, treats its employees and resources, and interacts with society and the environment. It is basically the corporate philosophy around the pillars ecology, economy, and society. There are many other phrases and acronyms for more or less the same type of activity used. The most common one is corporate social responsibility (CSR). Sustainable development has become an important initiative for many industries and companies over the last few years. Many chemical companies have started a sustainable development initiative over the last few years. Strong points in there are the environmental/ecological aspects as well as the workers' safety programmes. It is a distinct sign that the industry has learned to deal with its weaknesses in a much more offensive way than in the first decades after World War II when major environmental crises represented for the public how the industry operated. One example is the little town of Seveso in the industrialised north of Italy. It was heavily affected in 1976 when a major chemical accident led to the outbreak of chlorine gas and dioxins into the environment. Since those years, the chemical industry has invested a lot and has learned significantly more about how to manufacture even hazardous materials in such a way that this type of crisis is prevented from happening. In addition, chemical waste is treated differently, energy is used a lot more economically, and odours are prevented from being released.

The flavour and fragrance industry's weak points from an environmental/ecological point of view evidently are, in particular, odour emissions, the handling of chemicals and chemical reactions in manufacturing, and the handling of wastewater. Every company that has started sustainable development activities has looked at its weak (vulnerable) points. Their statements show that the sustainable development programme is used to turn weaknesses into strengths or at least show work being done continuously on these weak points. By looking at the pillar "society", another challenge becomes apparent. While it seems to be obvious for most consumers why pharmaceuticals are needed and beneficial, the use of flavours for foods and beverages as well as fragrances for various applications is not so easily understood by a certain part of the population. Unfortunately, this is sometimes abused by certain authors in common publications when flavours are described as potential risks for humans and fragrances are classified as luxury goods or simply unnecessary and annoying. Therefore, it is important to educate the population about the safety of flavour and fragrances and the benefits for their use in consumer products. Obviously, this attempt is complicated by the fact that the flavour and fragrance industry does not usually deal directly with consumers.

In the following a few activities are listed that can be measured by a flavour and fragrance company in a sustainable development programme:

- Measurable reduction of energy (water, electricity)
- Measurable reduction of odour emissions
- Improvement of manufacturing processes
- Financial support for charities, aid organisations, and local cultural activities
- Consistent and transparent equal rights and compensation policies throughout the company
- No child labour throughout the company

An important part of such an initiative is the search for sustainable raw materials. There are virtually thousands of different raw materials used in the flavour and fragrance industry. They typically comprise a combination of chemicals, essential oils, extracts, distillates, and others. Many essential oils and other ingredients come from tropical countries and/or parts of the world that are (still) outside of the mainstream business countries, e.g. China, Vietnam, Indonesia, Côte d'Ivoire (cocoa). The supplier companies of these raw materials for the flavour and fragrance industry need to make sure that the supply is sustainable, i.e. specific business practices need to be applied by those companies to maintain and secure the supply. The Chiquita company may serve as a good example in the food industry [11]. Chiquita is by far the most popular banana in the world. The company is number 1 in Europe and number 2 in the USA. The total sales of the Chiquita Company are about US $4 billion. It has been working together for many years with the Rainforest Alliance [12] in order to guarantee the consumers in nine European countries the certified requirements of an independent environmental organisation. The nucleus of these requirements covers social, legal, and ecological conditions that the banana farmers in the respective countries of origin (such as Costa Rica) have to fulfil. Although the Chiquita bananas cost about twice as much as non-certified ones, the concept seems to be being well received by consumers.

One of the important tasks of a marketing department in the flavour and fragrance industry is to study consumer and lifestyle trends to help research and development departments to work on the appropriate long-term projects and the sales force to target the right customers and product categories. In particular, the fragrance and cosmetics part of the flavour and fragrance business is dependent on interpreting these consumer trends ahead of time and correctly.

At the moment the following trends are observable (communication from Symrise's Marketing departmenrs):

1. Consumer segmentation:
   (a) Traditional family continues to alter:
   - Single parent homes
   - Same-sex families
   - Communal living
   - Fewer children
   - Nestlings/"boomerang" kids
   - Longer lifespan
   - Multi-cultural families

   (b) Breakdown in traditional demographic categories:
   - A redefinition of youth:
     - Young—tween, teen, early 20s
     - Super youths—25–39, refuse to get "old"
     - Hip-hops—the new parent, home-owning, trend-setting
     - New seniors—trendier more active
   - A redefinition of all-American: a global population on the move: city to city; country to country

   (c) Shift in ethnicity of USA:
   - Latina population continues to grow (67.5% between 1990 and 2002 vs. 8.1% non-Hispanic)
   - Increasing affluence
   - Very appearance oriented
     - Spend 27% more on cosmetics
     - Spend 43% more on fragrance
     - Spend $1.6 billion annually on personal care

2. Well-being.
   (a) Holistic trend responsible for considerable launch activity:
   - Aromatherapy
   - Aromachology
   - Spa
   - Deng-shui
   - Ayurveda
   - Ki

   (b) Satisfying the consumer's need for feeling restored, rejuvenated, repaired

3. Sensorial branding.
   (a) Products that offer a multisensory experience via:
   - Unique fabrications

- Translucency
- Organic tactility
- Colour infusion
- Light diffusion
- Thermal reaction
- Enticing aromas

(b) Satisfying the consumer's need for feeling stimulated, intoxicated, involved

4. New luxury.
   (a) A quality-of-life approach available to the masses:
   - Masstige
   - Time-sensitive
   - Limited editions
   - Artisan approach

   (b) Satisfying the consumer's need for feeling pampered, special, extraordinary

5. New simplicity.
   (a)Subtle means of self-expression versus bold and blatant branding:
   - Designer labels inside not outside

   (b) Invisible branding/whisper campaigns/viral marketing/underground communication:
   - Flyers, stickers, creating a buzz
   - Street-based promotion
   - Calvin Klein's CRAVE approach to launch

   (c) Satisfying the consumer's need for feeling edgy, unique, "in-the-know"

6. Return to the classics
   (a) Glamour has found its way to centre stage:
   - Tiffany is opening an Iridesse pearl boutique
   - Ladylike designs return to fashion
   - Warmth and character returns to home décor
   - Elegance, grace, style are en vogue

   (b) Move toward:
   - Authenticity
   - Quality
   - Yesteryear

   (c) Satisfying the consumer's need for feeling refined, elegant, glamorous, pampered

But product marketing is getting more and more important for the flavour business as well. Similar to the fragrance side of the business, the early recognition of consumer trends and of course the understanding of the major food and beverage brands is one of the keys for success. At the moment the following ten global trends are observable (communication from Symrise's Marketing departments):

1. Age nullification. Manufacturers need to break away from traditional stereotyping of age groups and explore opportunities of targeting other age groups with their products.
   Food solutions:
   (a) Cool
   (b) Fashionable
   (c) Healthy
   (d) Targeted

2. Gender complexity. Strong cross-over of product usage and behaviour from men to women and vice versa.
   Food solutions:
   (a) Distinguished
   (b) Stylish
   (c) Fashionable
   (d) Personality

3. Life stage complexity. Marketers need to categorise groups by attitudinal and behavioural rather than by traditional demographics.
   Food solutions:
   (a) Convenience
   (b) Small portions
   (c) Meal replacements
   (d) On-line shopping
   (e) Virtual communities

4. Hypertasking. Consumers are becoming more aware of time. It has become an essential part of life.
   Food solutions:
   (a) Convenient
   (b) Bite-size portions
   (c) Accessible
   (d) Easy to use
   (e) Portable
   (f) Back to basics
   (g) Resealable

5. Spending complexity. Understanding the complex mindset of consumers regarding spending money and quality of life.

Food solutions:
(a) Quality
(b) Benefit
(c) Value
(d) Indulgence

6. Health and wellness. Modern consumers are increasingly focused on personal well-being (physical and mental health, beauty).
Food solutions:
(a) Functional
(b) Low and light
(c) Nutritious
(d) Organic
(e) Botanicals
(f) Holistic

7. Sensory sensations. Growing stress, rising affluence, and availability of global foods are driving consumer demands for new and more intense taste sensations.
Food solutions:
(a) Ethnic
(b) Fresh
(c) Gourmet
(d) Novelty
(e) New sensation
(f) Textured
(g) Tryvertising [13]

8. Individualism. Being yourself and having personal needs recognised rather than being part of the mass market.
Food solutions:
(a) Customised
(b) Personalised
(c) Self-expressing
(d) Single-serving
(e) Premium
(f) Trendy and unique
(g) Exclusive

9. Comfort space. Building a secure environment wherever you are is an eminent part of developing a stable and close relationship within your environment.
Food solutions:
(a) Nostalgic
(b) Traditional

(c) Comfort foods
(d) Ethnic

10. Connectivity. Developing a lifestyle that is invigorating and prosperous based on information and opinion as well as life experiences.
Food solutions:
(a) Customised
(b) Personalised
(c) Tryvertising [13]
(d) Health and wellness

Figure 1.2 shows the cuisines of the world with the most potential for growth. In a way, this development is not very different from what is going on in the business and the technology sectors around the world. In a recently published book by Thomas L. Friedman, a phrase was used that describes the world as becoming flat, i.e. growing more and more together [14].

It is interesting to compare the trends on the fragrance and flavour side with each other and see that there are certainly communalities between the two. Figure 1.3 emphasises this point by showing emerging tastes and fragrance trends side by side. In general it is worth mentioning that although the technical elements of the two businesses are very different, i.e. the raw materials used in

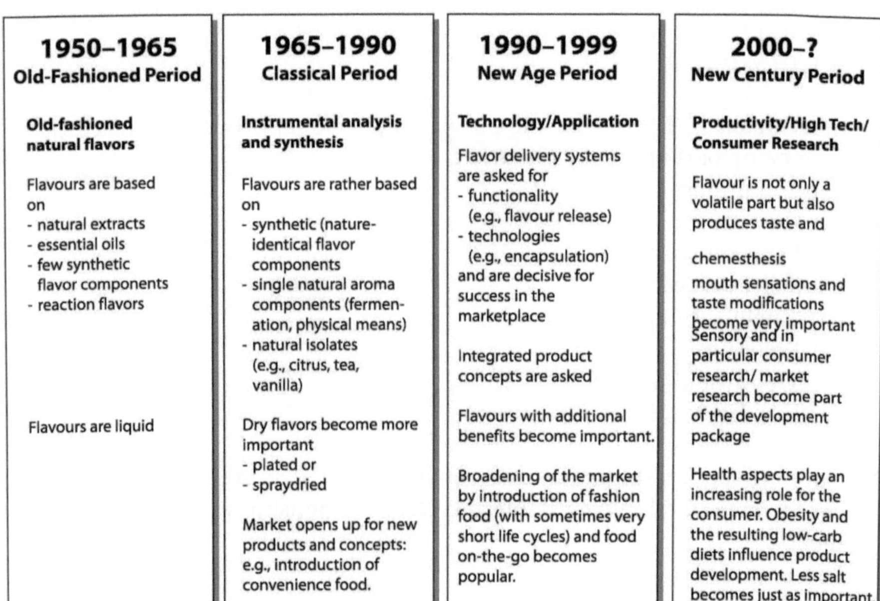

Fig. 1.1 The history of flavour development and taste

North America | **Europe** | **South America** | **Asia - Pacific**

North America

USA:
- **Asian Influence:** (sesame, ginger wasabi, noodle and Asian cabbage)
- **Indian Influence:** (fruit, spice and toasted nut chutney -quince, pear, roasted coriander, pistachio, peanut, almond and walnut)
- **Blue Cheese and Goat Cheese**

Mexico:
Tamarind, Squash Flowers, Huitlacoche (corn mushroom), Portobello Mushroom, Duck Meat

Europe
- **Mediterranean Influence**
- **Indian Influence**
- **Middle East Influence**
- **Slow Food**

South America
- **Fusion Style:** (Thai, Indonesian, Vietnamese influences)
- **Contemporary Cuisine**
- **Mediterranean Influence**
- **Exotic Combinations**

Asia - Pacific
- **Fusion Style:** (Thai/Chinese, Western/Chinese, Indonesian/Thai, American/Mediterranean)
- **Italian**
- **French**

Fig. 1.2 The cuisines with most potential for growth

- **Exotic Infusions**
  - A spicy kick of lemongrass, curcuma, pepper, coriander, ginger, basil, cardamom, cinnamon, oregano
- **Red Pleasures**
  - Strawberry, cranberry, pomegranate, rooibos, greengage, rhubarb, plum, blood orange, cherry variants, black currant, huckleberry
- **Black Health**
  - Black tea, black vinegar, black sesame seeds, black soybeans, black rice, black sugar, malt
- **Botanical Power**
  - Honeysuckle, lavender blossom, elderflower, hibiscus, sunflower blossom, rose

Attracting Opposites
  - Spicy/mild, sweet/sour, hot/cold, fire/ice
- **Ethnic Revival**
  - Traditional tastes & flavors are rediscovered, African hibiscus, Japanese cherry blossom or Maroccan kumquat
- **Flavor Migration**
  - Different categories start to mingle, dessert drinks, coffee cocktails

- **Sophisticated Red Fruit**
  - Pomegranate, redcurrant and raspberry leaves
- **Red Fruit will "go darker"**
  - Blackcurrant, blackberry, black rose and black plum
- **Gourmand notes**
  - Chocolate is big in replacing vanilla as a base, brown sugar
- **Milky notes**
  - milk, milky coconut
- **More specific exotic fruit**
  - Passion fruit, star fruit, kiwi, guava and litchi sorbet instead of pineapple and coconut
- **Pink Pepper**
  - New spicy note
- **Oriental Influences**
  - Tea (red tea and green tea), ginger and bamboo for herbal notes based on oriental influences

Fig. 1.3 Emerging flavours vs. fragrance trends

fragrances are different with a few exceptions, it happens quite frequently that perfumers seek unique bouquet notes on the flavour side. A typical example is the use of fruity notes, in particular tropical fruit notes, over the last few years (communication from marketing departments within Symrise).

Acknowledgements

I would like to thank the respective colleagues at Symrise who helped me write this contribution. I am in particular grateful to my colleague and friend Emmanuel Laroche for supporting me with his expertise in the world of marketing.

References

1. http://www.fashion-era.com/perfume_history.htm
2. http://www.leffingwell.com
3. Rowe DJ (ed) (2005) Chemistry and technology of flavors and fragrances. CRC, Boca Raton
4. http://www.bellff.com
5. http://www.symrise.com
6. http://www.firmenich.com
7. http://www.givaudan.com
8. http://www.iff.com
9. http://www.ici.com/ICIPLC/divisions/Quest.jsp
10. Leffingwell JC (2004) Reflections on a half century of flavor chemistry. Speech at the 50th anniversary flavor symposium in October 2004 of the Society of Flavor Chemists. http://www.flavorchemist.org
11 Article published in Frankfurter Rundschau (in German) on 13 October 2005 \CEnote{Please provide the title of the article and the page number}
12. http://www.rainforest-alliance.org
13. http://www.trendwatching.com
14. Friedman TL (2005) The world is flat—a brief history of the twenty-first century. Farrar, Strauss and Giroux, New York

2 Flavours: the Legal Framework

Dirk A. Müller
Takasago Europe GmbH, Postfach 1329, 53905 Zülpich, Germany

2.1 Definitions

Flavourings are a major category of ingredients intentionally added to food and feeding stuff. Flavourings are concentrated preparations with the primary purpose to impart flavour except for substances that have an exclusively sweet, sour or salty taste. They are added in small amounts to food or feeding stuff but are not intended to be consumed as such.

Flavourings may contain flavouring substances, flavouring preparations, process flavourings, smoke flavourings and flavouring adjuvants.

Flavouring substances are chemically defined substances with flavouring properties. There are three different categories of flavouring substances defined in the definitions of the IOFI Code of Practice and EU Flavour Directive 88/388/EEC [1, 2]:

1. Natural flavouring substances
2. Nature-identical flavouring substances
3. Artificial flavouring substances

Flavouring preparations are natural complexes used because of their flavouring properties. They contain flavouring constituents and they are obtained by appropriate physical, microbiological or enzymatic processes from foodstuffs or other material of vegetable or animal origin, either in the raw state or after processing for human consumption by traditional food-preparation processes (including drying, torrefaction and fermentation).

Process flavourings means products which are obtained according to good manufacturing practices by heating a mixture of ingredients to a temperature not exceeding 180 °C for a period not exceeding 15 min , the ingredients themselves not necessarily having flavouring properties, and at least one of which contains nitrogen (amino) and another is a reducing sugar.

Smoke flavourings means smoke extracts used in traditional foodstuff smoking processes. The EU Regulation on smoke flavourings subdivides them into four categories:

1. 'Primary smoke condensate' shall refer to the purified water-based part of condensed smoke and shall fall within the definition of 'smoke flavourings'.

2. 'Primary tar fraction' shall refer to the purified fraction of the water-insoluble high-density tar phase of condensed smoke and shall fall within the definition of 'smoke flavourings'.
3. 'Primary products' shall refer to primary smoke condensates and primary tar fractions.
4. 'Derived smoke flavourings' shall refer to flavourings produced as a result of the further processing of primary products and which are used or intended to be used in or on foods in order to impart smoke flavour to those foods.

Flavouring adjuvants are foodstuffs, food additives, other food ingredients or processing aids which are necessary to ensure the safety and quality of flavourings and to facilitate the production, storage and intended use of flavourings.
Flavouring adjuvants may also include flavour modifiers.

2.2 Legal Positions

In the following, the regulations on flavourings of three major regions are presented. Several other countries have similar legal regulations or accept flavourings produced according to these regulations. One major difference is the general classification of flavourings. In some countries, flavourings are classified as food additives, like in the USA or Japan. In other regions, flavourings are considered to be a special type of foodstuff, like in the EU.

2.2.1 Current Situation in the EU

In 1988 the "Council Directive of 22 June 1988 on the approximation of the laws of the Member States relating to flavourings for use in foodstuffs and to source materials for their production" was published. Together with the amending Directive 91/71/EEC regulating the labelling of flavourings for end consumers, this Directive defined the categories of flavouring ingredients, purity criteria and maximum levels for certain "biological active principles" (BAPs). With this Directive the frame for following specific regulations was established [2, 3].

Two specific Regulations mentioned in the indent of EU Flavour Directive 88/388/EEC have been established.

1. The EU Regulation on smoke flavourings which was published in 2003 [4]. The major subject of this Regulation is to establish:
   (a) A Community procedure for the evaluation and authorisation of primary smoke condensates and primary tar fractions for use as such in or on foods or in the production of derived smoke flavourings for use in or on foods.

(b) A Community procedure for the establishment of a list of primary smoke condensates and primary tar fractions authorised to the exclusion of all others in the Community and their conditions of use in or on foods.
The evaluation of those primary products will be carried out by the respective panels of the European Food Safety Authority (EFSA).
After finishing the evaluation procedures a positive list of primary smoke condensates, including purity criteria and maximum levels for contaminants, will be established in the EU.

2. The EU Regulation on food additives necessary for storage and use of flavourings, including respective conditions for their use, has been established. Following several years of intensive discussion and several drafts, Directive 2003/114/EC amending Miscellaneous Directive 95/2/EC was been published on 22 December 2003 [5, 6].

The Directive states that the levels of additives present in flavourings should be the minimum required to achieve the intended purpose. Flavouring adjuvants should not have a remaining technological function in the final foodstuff. With regard to this requirement, the possibility of the "carryover" of additives used in flavourings is especially mentioned in the Directive.

If the additive still has its technological function in the final food, labelling of this additive will be necessary for the final foodstuff as well.

In addition, the sixth indent of this Directive mentions that in accordance with the provisions of the EU Flavour Directive, quantitative labelling of each component which is subject to quantitative limitation, expressed either numerically or by *quantum satis* principles, is required for the flavour.

Following article 5 of the EU Flavour Directive, EU Regulation 2232/96 defined the basic rules for the use of flavouring substances for foodstuffs in the EU. In addition, it lays down a procedure for establishing a positive list for flavouring substances in the EU [7].

The procedure for evaluation was published as Commission Regulation 1565/2000 [10].

In 1998 the EU Commission within the Commission Decision 199/217/EEC published an inventory of flavouring substances used in the EU. This inventory (including its amendments) lists most of the flavouring substances which are subject to evaluation, leading to a positive list of flavouring substances to be used in foodstuffs in the EU [8].

The agreed timetable according to Commission Regulation 622/2002 mentioning the finalisation by 2005 has been postponed to 2007/2008 because the evaluation could not been finalised within the expected period [9].

In the meantime the existing national regulations of EU member states regarding flavouring substances are still in force. These existing national regulations show an unlimited permission of use for natural and nature-identical flavouring substances as defined in the EU Flavour Directive in all EU member states except Italy. Italy has kept a specific limitation for seven nature-identical flavouring substances.

Regarding artificial flavouring substances, four EU member states (Germany, Italy, Spain and the Netherlands) have specific positive lists with use levels, whereas all other EU member states permit all artificial flavouring substances suitable for human consumption.

With regard to the flavouring preparations, some EU countries have negative lists for plant materials which should not be used for production of flavouring preparations.

The labelling requirements for flavourings in the EU are laid down in EU Flavour Directive 88/388/EEC for the flavourings themselves and in EU Directive 91/72/EEC concerning the designation of flavourings in the list of ingredients of the final foodstuff.

It is required to use the word "flavouring" or a more specific name or description of the flavouring.

The word "natural" or a word of similar meaning may only be used if the flavouring ingredients are exclusively natural flavouring substances or flavouring preparations. Mentioning the flavouring source together with the word "natural" is only permitted if the flavouring ingredients have been isolated solely or almost solely from this source [2, 3].

2.2.2 Expected Regulations on Flavourings in the EU in the Future

The EU Commission is currently preparing a revision of the EU Flavour Directive. The publication of the finalised version for submission to the Council and the European Parliament is expected for 2006. In order to reduce the number of Regulations and Directives, the EU Commission will present a Regulation combining the Additive Directive, the revision of the Flavour Directive and a new Enzyme Regulation in one framework regulation of those "food improving agents".

Summarising the previous discussion and drafts, the new EU Flavour Regulation will show some new definitions of flavouring ingredients, like "flavour precursors" and "other flavourings". Within the definition of flavouring ingredients, it shall be distinguished between flavouring ingredients derived from food and material of vegetable or animal origin not consumed as food (non-food). Such flavouring ingredients derived from non-food material will be evaluated and will need explicit authorisation. The respective principles for authorisation and the procedures will be implemented in the new regulation. Also the existing regulations for genetically modified material will be implemented for flavourings as well.

The new regulation will define the permitted processes for production of natural flavouring ingredients. Definitions and provisions for the use of ingredients containing BAPs have been renewed. Some additional restrictions for source materials for production of flavouring ingredients will be added. The sales descriptions for flavourings will be revised and some definitions will be added [11].

2.2.3
Current Situation in the USA

Under the terms of the US Food Regulations, flavourings fall under the definition of food additives. The respective definition was implemented in the Federal Food Drug and Cosmetic Act by the Food Additives Amendment of 1958. With this amendment, the general requirement of "safety" became the major topic for food additives.

Flavouring substances which were not covered by one of the two grandfather clauses of the Food Additives Amendment needed either an approval or an evaluation as "generally recognised as safe" (GRAS). Under the supervision of the FDA, several flavouring materials have been evaluated. The permitted components are listed in the Code of Federal Regulation (CFR) Title 21, parts 170–180. Components approved as GRAS are listed in parts 182–184 of the same CFR Title [12].

Later, the US Food and Drug Adminstration (FDA) passed the responsibility for evaluation of the GRAS status for flavouring materials to the Flavour Expert Panel (FEXPAN). This panel of scientists from different related scientific areas evaluates new flavouring substances which are applied for notification. The FEXPAN is not affiliated with the flavour industry but is organised by the US Flavour and Extract Manufacturer's Association (FEMA). In publications currently up to GRAS 22, the positively evaluated flavouring substances are published with name, synonyms, identification number and the average maximum-use levels.

The US Regulations only distinguish between natural and artificial flavourings. The European category "nature-identical" is unknown in the legal definitions. If such substances are synthetically produced, they are classified as artificial flavouring substances in the USA.

For labelling of the final foodstuff, use of the term "natural" is divided into two subcategories. First, flavourings which contain only flavouring ingredients from the named source, the so-called from the named fruit flavourings (FTNF). In this case the name of the source can be used together with the word "natural" and followed by the word "flavoured". If the food contains a flavouring where the flavouring ingredients are natural but not solely from the named source, the additional words "with other natural flavour" (WONF) are required. Smoke flavourings derived from smoked wooden or plant materials are natural in the USA. The same applies for process flavourings prepared with natural raw materials. Only if synthetically produced substances were used for the production of a process flavour, it would be artificial, unless these are non-flavouring substances which are declared separately [13].

2.2.4 Current Situation in Japan

The Japanese Food Regulations are based on the Food Sanitation Law (FSL). The FSL was first enacted in 1947 by the Ministry of Health and Welfare, now the Ministry of Health, Labour and Welfare (MHLW) of Japan [14]. The purpose of the FSL is to prevent the occurrence of health hazards arising from human consumption of food, by making necessary regulations and taking any measure sfor the protection of the health of the people. It enables the MHLW to establish detailed regulations to manage immediately diverse issues related to international food distribution and the need for international harmonisation of food regulations [15].

Flavourings are considered to be food additives according to the principle definition of the FSL. Food additives need an authorisation for use. Article 6 of the FSL mentions the general terms under which the use of a food additive is not permitted in and for food. In the FSL Enforcement Regulations, tables with the list of existing and permitted food additives, including the synthetically derived flavouring substances, are mentioned. The food additives appearing on this list are not subject to article 6 of the FSL [16].

Natural flavouring agents and substances generally provided as food and used as food additives are also not subject to the provisions of article 6 of the FSL. Under the terms of the FSL, a list of "origin of natural flavouring agents" and a list of substances generally provided for eating and drinking as food and used as food additives have been compiled and published by the MHLW. In the list of existing and permitted flavouring substances, only about 84 flavouring substances are mentioned by individual name; the other flavouring ingredients are mentioned only by chemical groups.

These chemical groups are:

- Isothiocyanates (except those generally recognised as highly toxic)
- Indoles and its derivatives
- Ethers
- Esters
- Ketones
- Fatty acids
- Aliphatic higher[1] alcohols
- Aliphatic higher aldehydes (except those generally recognised as highly toxic)
- Aliphatic higher hydrocarbons (except those generally recognised as highly toxic)
- Thioethers
- Thiols (thioalcohols)
- Terpene hydrocarbons
- Phenol ethers
- Phenols

[1] "Higher" means C6 or more.

- Furfurals and its derivatives (except those generally recognised as highly toxic)
- Aromatic alcohols
- Aromatic aldehydes (except those generally recognised as highly toxic)
- Lactones (except those generally recognised as highly toxic)

Not listed and therefore not permitted are substances from chemical groups like pyrazines, pyridines, amines, amides, or aliphatic lower alcohols, aldehydes and hydrocarbons (C5 and lower) if not mentioned by individual name [17].

The MHLW is currently evaluating several individual flavouring substances not covered by the aforementioned groups but that are of commercial interest. Most of these substances are lower alcohols, aldehydes and pyrazines. As soon as the evaluation has finished, the result will be published and in positive cases the substances will by added to the list of permitted substances.

2.2.5 Global Approach

A major lack in global comparability of flavourings is the difference between the permitted flavouring ingredients . Owing to the difference in the regulations mentioned, a broad range of flavouring substances are only permitted in one region or country. In 2000, the Japan Flavour and Fragrance Materials Association (JFFMA) started a survey with the objective to create a list of all flavouring substances marketed in Japan and to compare them with the EU Register and the US FEMA listed substances. On the basis of the figures for 2001, the FEMA list covered 1,578 substances, the EU Register contained 2,702 substances, whereas 2,577 flavouring substances were reported as being used in Japan.

Comparison with the EU Register showed that 1,800 substances are covered in both lists. Of those, 777 substances were only used in Japan, whereas 902 substances were only mentioned in the EU Register, and of those 640 were not used in Japan and 181 were not permitted for use in Japan. Further, 81 substances from the EU Register are not classified as flavouring substances in Japan [18].

Compared with the FEMA-listed substances, similar results were obtained. A total of 1,182 substances were mentioned in the Japanese survey and the FEMA list. Of those, 1,342 substances were only reported in Japan and 396 substances were only on the FEMA list. In addition, 216 of them had no reported use, 73 were not permitted in Japan and 107 were not classified as flavouring substances [18]. Owing to the fact that the EU Register covered the substances from the FEMA list up to GRAS 21, only a few new substances from the GRAS 22 publication and some substances that were deleted because of no reported use are not implemented in the EU Register in the amended version.

But the comparison with the Japanese survey showed that 749 flavouring substances had reported use in Japan but were neither listed in the FEMA list nor in the EU Register [18].

Such figures indicate the necessity for a global regulatory approach for flavouring material.

The Codex Committee on Food Additives and Contaminants (CCFAC) of the Codex Alimentarius Commission agreed to propose work on the elaboration of a "Codex Guideline for the Use of Flavourings" that establishes safe conditions of use for such substances in foodstuffs. This should lead to globally accepted general requirements for flavourings, including definition, safe use, labelling and specifications. In addition it should provide a reference to the safety evaluations completed by JECFA as a global approach for evaluation and authorisation procedures [19].

2.3 Legal Situation and Natural Flavourings, a Brief Reflection

Most of the regulations on flavourings distinguish between natural derived flavouring components and substances produced synthetically. There are still some differences between the national rules regarding source materials and accepted techniques.

All flavourings considered natural in the EU should also be considered natural in the USA; however, the reverse is not necessarily true. Smoke flavourings and process flavourings are separate categories in the EU, and cannot be used in natural flavours, whereas smoke flavourings as well as process flavourings prepared with natural raw materials are considered natural in the USA. Another important difference between the EU and the USA is the methods allowed to obtain "natural flavouring substances". Under the legal terms of the USA, the naturalness of the starting material defines the status of the resulting product [13]. The EU Flavour Directive defines the staring materials as well as the permitted processes to obtain natural substances or natural preparations [2]. This excludes any type of chemical processing or chemically catalysed process. These differences in definition and handling of natural flavouring materials will be explained using α-ionone as an example.

α-Ionone is an important flavouring substance for a range of fruit flavour systems. In various fruit and plant species α-ionone was found as an almost optically pure *R* enantiomer, whereas the chemical synthesis will lead to a racemic mixture of both enantiomers. The chemical synthesis of α-ionone uses citral, which is condensed with acetone in basic media to the respective pseudo-ionone, followed by cyclisation in acidic media. If now, with this way of production, solely natural citral and natural acetone, derived from fermentation processes, together with natural pH adjusting materials are used, the resulting α-ionone fulfils the US requirements for natural flavouring substances. However the process is still a chemical reaction leading to a catalysed formation of a covalent C–C binding. Therefore the α-ionone derived from such a process will not fulfil the EU definitions for a natural flavouring substance. In the EU this α-ionone is still considered to be a "nature-identical flavouring substance" according to the EU Flavour Directive [2].

But the general goal to differentiate between natural and synthetically derived materials is obvious. This main issues of the flavouring regulations are resumed in the respective rules of the single categories of food products. Many of these regulations permit only the use of natural flavourings for specific types of foodstuffs. Often such products are more highly qualified or many contain the depicting or labelling of a respective source.

Not the general focus on "green chemistry" nor the move towards "natural" or even "organic" sources of food products observed over the last decade in industrial nations has led to this focus on natural flavourings. Most of these regulations focussing on natural flavourings were established long ago. With this long-term history of the respective regulations on flavourings focussing on "renewable" resources, the governments have emphasised the importance of the sensorial impression from the natural sources. This demand led to the current situation that the flavour industry and the respective research institutes have become one of the driving forces in development of new methods using "renewable resources" for the generation of flavouring materials.

References

1. IOFI code of practice. International Organization of the Flavour Industry, Brussels
2. Council Directive of 22 June 1988 on the approximation of the laws of the Member States relating to flavourings for use in foodstuffs and to source materials for their production (88/388/EEC). Official Journal of the European Communities no L 184, 15 July 1988
3. Commission Directive 91/71/EEC of 16 January 1991 completing Council Directive 88/388/EEC on the approximation of the laws of the Member States relating to flavourings for use in foodstuffs and to source materials for their production. *Official Journal* of the European Communities no *L 42, 15 February 1991*
4. Regulation (EC) no 2065/2003 of the European Parliament and of the Council of 10 November 2003 on smoke flavourings used or intended for use in or on foods. Official Journal of the European Communities no L 309, 26 November 2003
5. Directive 2003/114/EC of the European Parliament and of the Council of 22 December 2003 amending Directive 95/2/EC on food additives other than colours and sweeteners. *Official Journal* of the European Communities no *L 24, 29 January 2004*
6. Directive 95/2/EC of the European Parliament and of the Council of 20 February 1995 on food additives other than colours and sweeteners. *Official Journal* of the European Communities no *L 61, 18 March 1995*
7. Regulation (EC) no 2232/1996 of the European Parliament and of the Council of 28 October 1996 laying down a Community procedure for flavouring substances used or intended for use in or on foodstuffs. Official Journal of the European Communities no L 299, 23 November 1996
8. Commission Decision 1999/217/EC of 23 February 1999 adopting a register of flavourings substances used in or on foodstuffs drawn up in application of Regulation (EC) no 2232/96 of the European Parliament and of the Council of 28 October 1996. *Official Journal* of the European Communities no *L 84, 27 March 1999*

9. Commission Regulation (EC) no 622/2002 of 11 April 2002 establishing deadlines for the submission of information for the evaluation of chemically defined flavouring substances used in or on foodstuffs. Official Journal of the European Communities no L 95, 12 April 2002
10. Commission Regulation (EC) no 1565/2000 of 18 July 2000 laying down the measures necessary for the adoption of an evaluation programme in application of Regulation (EC) No 2232/1996 of the European Parliament and of the Council. Official Journal of the European Communities no. L 180, 19 July 2000
11. Proposal for a Regulation of the European Parliament and of the Council on flavourings and certain food ingredients with flavouring properties for use in and on foods. SANCO/2004/rev 22, WGF/002/02 rev3
12. Code of Federal Regulations, Title 21, Food and drugs, parts 170, 172, 173, 178, 182, 184, 189
13. Code of Federal Regulations, Title 21, Food and drugs, part 101, sect 102.22(a)–(i)
14. Food Sanitation Law of Japan, law no. 233, 24 December 24 1947; last amendment law no 55, 30 May 2003
15. Food Sanitation Law Enforcement Ordinances, Cabinet Order no 229, 31 August 1953; last amendment Cabinet Order no 511, 12 December 2003
16. Food Sanitation Law Enforcement Regulations, Ministry of Health and Welfare Ordinance no. 23, 13 July 1941; last amendment 31 March 2004, Ministry of Health, Labour and Welfare Ordinance no 78
17. Specifications and standards of foods, food Additives, etc. Under the Food Sanitation Law (abstracts). Japan External Trade Organisation, Tokyo
18. Survey on flavouring substances currently marketed or used in Japan (summary), March 2001. Flavor Committee, Japan Flavor and Fragrance Materials Association
19. Discussion paper on the development of a Codex Guideline that establishes safe conditions of use for flavourings in foods with a reference to the evaluations completed by JECFA (2005). Codex Alimentarius Commission, Joint FAO/WHO Food Standards Programme, Rome

3 Olfaction, where Nutrition, Memory and Immunity Intersect

J. Bruce German
Department of Food Science and Technology,
University of California, 1 Shields Avenue, Davis, CA 95616, USA

Chahan Yeritzian, Vladimir B. Tolstoguzov
Nestlé Research Centre,
P.O. Box 44, Vers-chez-les-Blanc,
1000 Lausanne 26, Switzerland

3.1 Introduction

The hypothesis of memory consolidation was first proposed 100 years ago by Müller and Pilzecker [1]. According to this hypothesis, "new" memories are initially labile and require additional (biochemical) reinforcement to be consolidated into long-term memories. Gradually, biochemical research is unravelling the time-dependent processes forming long-term memories. A century of studying memory, including the genetic determinants and molecular structures forming the basis of memory—where in the brain memory storage occurs and the mechanisms whereby memories are stored and maintained—was intensively reviewed [2–4]. The use of genetic and molecular approaches has led to the identification and characterisation of genes and molecules that play a fundamental role in the biological mechanisms underlying learning [2, 4].

This chapter considers memory development in terms of recent ideas about molecular mimicry and symbiosis [5–7], and proposes hypotheses and new approaches to the biological basis of memorisation. From the prebiotic history, nutrition and immunity of the cell are mutually interacting processes aiming at the same goal—namely to survive [5]. Immunity must resist the attack of exogenous invaders such as foreign macromolecules, viruses and bacteria. Nonetheless, food is the main source of consumed foreign macromolecular materials. Nutrition can thus be regarded as two contradictory, yet mutually interactive processes—the feeding and protection of the individual organism. Both nutrition and immunity aim at rendering food components useful and harmless. Both profit from acquiring memories of prior experiences. Both use denaturation, phase separation and hydrolysis of biopolymers to achieve extremely exacting definitions of structure for recognising foreign molecules. Both also employ epithelial membranes in which a mucosal layer provides exclusion and specific absorption of nutrients/molecules. For nutrition and immunity, the first mucosal surface of recognition and defence is the olfactory system. Recently, it was proposed that the olfactory G-proteins underlying odour perception evolved as a defence mechanism against dangerous foreign substances and their originators.

G-proteins and immune-related molecules can be regarded as the first lines of defence against dangerous components in the surroundings [8]. The effectiveness of such first lines of defence requires both identification and memorisation of the molecular signature of these dangers. Interestingly, in neither odour nor immunity are intact biopolymer macromolecules the basic signatures of structure that are recognised and memorised. Instead, smaller products of their hydrolysis liberated prior to or during consumption are the language of olfaction and immunity.

From the very simple to the highly complex, organisms are ostensibly the combination of molecular structures and metabolic processes. Each is capable of encoding information and profiting from previous events. The simplifying predominance of the two extreme conformations—globular and rod-like—typical of biopolymers has been used to define a principle of molecular interaction and symbiosis. That is, macromolecular biopolymers are not all mutually compatible, and this success or failure of compatibility that derives from the mutual influence of dissimilar macromolecules—symbionts—has been proposed to be important to the molecular evolution of the structures of the key biopolymers of living organisms [5]. Proteins and polysaccharides are thermodynamically incompatible, and their spontaneous separation provides a mechanism to build a thermodynamic barrier around virtually all living organisms, just as the lipid biomembrane provides a solubility barrier. This symbiotic macromolecular, thermodynamically driven interaction is beneficial to the activity of the biological system as a whole. One expression of molecular symbiosis is the increasingly well-characterised properties of excluded-volume effects of biopolymers, thermodynamic incompatibility of polymers and interbiopolymer complexation [8–10]. Symbiotic interactions of dissimilar macromolecules are not only based on differences in size and shape—excluded-volume effects—of macromolecules, but also include their mutually coordinated synthesis, modifications and transportation. The formation of memory can also be considered in this context, i.e. the development of a templated set of coordinated chemical reactions of modification, synthesis of biopolymers, their conformational changes, diffusion, complexation and coprecipitation is of principal importance. The development of "memorised" systems of chemical and physicochemical processes would underlie the adaptations providing appropriate and sufficiently rapid reactions to the environment. Memorisation processes were important throughout prebiotic evolution of biological structures as the means to develop an adaptation to the new surroundings of any organism.

One objective of this chapter is to discuss the biological basis of memory and the possible progress in the understanding of memory development, using recent ideas about thermodynamic features—symbiotic interactions—of biopolymers. The main applications of such basic knowledge are to gain insights into the development and memorisation of symbiotic biochemical processes in the scope of nutrition. Nutrition in the modern sense includes everything from basic essential nutrients to non-essential ingested substances, as well as the intestinal microflora and its symbiosis with the host. Another objective is to consider the directions in which it is now possible to evolve foods and flavours. With the rapidly expanding knowledge of genomics and the potential to expand agriculture beyond traditional commodities and surroundings, what principles should guide that expansion? The final objective is to highlight the concept that one application of understanding olfactory memorisation and flavour preference learning is to consider how to potentially guide formulations, e.g. new generations of foods as a means of "flavour education" strategy. Because the quality of diets is dictated in part by choices based on preference, knowing how preferences are developed should allow individuals to guide preferences to more suitable food choices. The multifaceted mechanisms of memory development, fixation and storage dictate a multidisciplinary approach. For this reason, the three thermodynamic, biochemical and evolutionary aspects of memory development are considered. Thermodynamic approaches are applied as the most general analytical technique for interactions within and between multicomponent systems, whereas evolutionary approaches are used to gain insight into the interactions of a biological system with its environment.

3.2 Memory Consolidation—Short-Term, Long-Term and Permanent Memories

In the most general sense, memory means the processes of accumulation, storage and reuse of information about the environment. Transformation of short-term (from minutes to days) memory into long-term (from days to years) memory extends from olfactory preference to acquired immunity. Environmental information flow (mainly through intestinal and nasal mucous membranes) in the form of different compounds selectively activates effectors (cells and organs, such as glands and muscles) that respond to a corresponding stimulus. The context, reciprocal relations between different sources of information (e.g. between the appearance, texture, flavour of a food, and satiety, pleasure and physiological well-being during and after eating) are important contributory factors for long-time memory. The variety of biochemical inputs that integrate to form the context of aroma perception to achieve a flavour preference as more permanent memory are not yet known; however, from *Drosophila* to humans, the complexity of the process has been noted [11]. It is also not known how widely the odour preference phenomenon can be generalised to other sensory memory (e.g. texture). Formation of concepts in terms of sensations makes memory about a past event or an experience more likely to be stored long term.

A biochemical approach to memory covers various aspects of perception, performance, learning, motor skill, thinking and problem-solving. It is assumed that the basic principle of memory underlies the construction of various increasingly successful (practised) responses as structural—in space and time—blocks of coordinated biochemical reactions. Each memorised biochemical block is a structured system of chemical and physicochemical processes, which are organised in a logical manner in response to changes to the surroundings. Each memorised block is formed by a memorisation mechanism and stored in accessible form to facilitate reuse, to adapt (metabolic memory) and protect the organism. In other words, memory can be considered conceptually as responsible for acquiring symbiosis on molecular, supramolecular, cellular and organ levels. Protective responses that are a set of corresponding mutually coordinated chemical and physical processes have not always been regarded as a long-term memory, but are nonetheless a form of semipermanent memory. Two types of persistent memories can thus be distinguished. The first is a stimulated expression of a particular subset of genetic elements whose functional response is established and fixed. The second is the establishment of a pattern of responses based on a decision taken at an early stage and the decision is memorised. Odour preference appears to be the latter form of memory. Similarly, allergy appears to be another. Exposure of an otherwise appropriate or benign antigen at the wrong moment and in an inappropriate context appears to establish a persistent wrong decision—allergy to that antigen. The development of preferences for foods that on balance constitute unhealthy diets can be considered another form of inappropriate decision with respect to, in this case, olfactory preference. How and when these decisions are made are thus critical. The majority of learning-induced persistent lifelong memories as olfactory preferences may be formed by puberty. The ability to alter these "memories" after adulthood is not known. For example, it is not clear in humans if the original odour preferences are undone and redecided or whether higher-order processes overwhelm the original pathway. Learning-induced additions to and changes in the permanent lifelong memory presumably must be initially induced as a short-term memory, and after time are converted into a long-term memory. For instance, similarly to a difference between a mother language studied in childhood and a foreign language studied later, both the formation rate and the stability of permanent memory decrease with age. The obvious implication of decreasing abilities to acquire new olfactory preferences is to limit substantially the ability both within an individual and across populations to redirect food preferences towards more desirable, i.e. healthy, food choices. By understanding the basis of the biochemical processes, it may be possible, however, to reconstitute the ability to form memories in a more pliable, i.e. adolescent manner.

Another potential consequence of a decreased ability to acquire persistent memory in adulthood, more precisely its negative consequences, relates to other compositional aspects of nutrition. The feeling of hunger in humans appears to relate primarily to macronutrient and calorie content and not to quality of food, though animals demonstrate nutrient-specific hunger [12]. Humans feel hunger for energy and apparently do not feel hunger for the essential nutrients except for water—thirst is a special "hunger" for water. Because energy governs the sensation of hunger, an unbalanced diet can be selected using foods differing in their content of essential nutrients, with satisfaction occurring only when the energy hunger has been overcome. It has been proposed that the proliferation of high-energy foods has resulted in an intake of high-energy diets—caloric overconsumption at the expense of vitamin-rich and mineral-rich diets. Apart from the individual nutritional history (i.e. our permanent and long-term nutritional memories), real nutritional requirements vary depending on individual physiological and psychological behaviour features, including functioning under normal and stress conditions and adaptation to new surroundings. Unfortunately, hunger as a sensation does not provide input beyond that of energy requirements.

Nutrition is the bridge between the physiology and the immediate environment as food choices by an individual. Ideally, persistent biochemical memories and olfactory preferences serve to coordinate a habitual physiological state with a successful set of food choices in an environment. As the availability of foods and the variation in composition of foods have increased dramatically in the past century, the basic sensory preference development processes may actually contribute to nutritional problems. Furthermore, agriculture itself is inadvertently being designed to uncouple composition from sensory cues. The quantity-based agricultural model, where the relative content of food energy per acre corresponds to a driving force for genetic breeding and agricultural practices, is not necessarily consistent with the content of nutrients that underlie food quality. Furthermore, processing that disassembles commodities into component biomolecule classes (proteins, carbohydrates and oils) serves to further dissociate sensory cues from the composition and quality of foods. Individually recommended consumption of specific nutrient compositions is even further from being differentiated in food-commodity planning. It is clear that it is not optimal if the food supply is to change simply to more homogeneous food without considering variable consumption. An important question is whether the processes of sensory preference development and biochemical memories can be considered as an asset in the future design of foods, diets and individual health.

Nutritionally deficient environments (e.g. low-protein and low-energy diets) are well known to affect memory development. Whether specific nutrients are able to augment the speed and persistence of memories has not been established. Nevertheless, the variation in ability to acquire memories at specific periods during growth suggests that the biochemical context during memory formation varies and that it is theoretically possible to recover this context via exogenous means. Short-term memory studies are mainly based on observations of electrical brain activity (electroencephalography) and use the electric circuit model. However, the nature of memorisation is not well understood [2–4, 13–17]. Food habits for at least one obvious example—lactase expression during adulthood—can cause short-term or long-term modifications of the gene pool. That is, food composition is a Darwinian selective factor. Are preference mechanisms selectable factors as well? Different brain structures (hippocampus, thalamus and amygdala) may be involved in the formation and storage of long-term memories, which is accompanied by chemical and structural changes. Both short-term and long-term memories presumably use the same synapses, but long-term memory requires synthesis of some special proteins at least. Reactivation of labile memories requires de novo protein synthesis for reconsolidation [15]. The formation of long-term and permanent memories requires gene expression, involves the formation and modification of particular synapses in the brain and the synthesis of new messenger RNAs and new proteins that control synaptic activity. Infusion of the protein synthesis inhibitor anisomycin into the lateral and basal nuclei of the amygdala shortly after training prevents consolidation of fear memories. This may reflect the development of a nutritional strategy for preventing the effects of early malnutrition on long-term memory development [16]. It was also shown that the initial percentage of body fat provides an individual metabolic memory (imprinting), e.g. energy efficiency and the extent to which the body's protein and fat—i.e. both energy reserves—are mobilised for fuel during starvation. This memory means an individual strategy ensuring maximum length of survival during long-term starvation contributes to human variability in energy partitioning [17]. The influence of diet on genetic development and the permanent hereditary memory is multifactorial. The dietetic management of such metabolic errors as phenylketonuria and galactosemia shows that nutrition can influence the exploitation of the genetic program (permanent memory). An insufficient adaptation of the human genome to the new surroundings could result in overfeeding, atherosclerosis and diabetes.

When thinking about the various non-genetic forms of "memorisation", it will be important to consider the diversity and biological information content of biomolecules themselves. For example, the most universal technique used for permanent memory formation in biological systems and their ingredients is separation of water in order to limit macromolecular mobility and to decrease the biological access and activity. The glassy state of densely packed globules of storage proteins—the interior of which is not accessible to water—spores, pollen and dry seeds preserves not only the biomolecules themselves, but in simple terms also represents a form of bioinformation [18, 19].

3.3
Multidimensional Biomemory

The most obvious goal of memory development is for defence in competitive organisms. Memorisation is needed to fix a negative (hostile, toxic) experience for defence and also to fix a positive experience concerning the measures and tools successfully used by an organism for nutrition. In other words, nutrition and immunity require a repetitive accomplishment of sets of biochemical defensive actions (operations), which, if templated and memorised, can be rapidly reused in the next similar situation. Memorisation could be regarded as a guide (programme) of actions to optimise the influence of the surroundings, or in other words, as a specific biological reaction to variability and uncertainty of the surroundings. Such surroundings would logically include competition or even symbiosis from other organisms. It is well recognised that the toxicity of products of pathogenic bacteria if experienced coincident with a novel aroma leads to memorisation of and avoidance (negative preference) of the odour. Thus, it is certainly possible that the beneficial products of symbiotic bacteria could provide a physiological "context" in experiencing a novel odour that leads to the memorisation of and subsequent positive preference for that odour.

In principle, information from all five senses as to the surroundings and information about all kinds of activity and regulatory and feedback systems would form the integrated "context" in which olfactory preference decisions were made. Memorisation is, presumably, responsible for functional control, integration and adaptive, purposive responses to (protective reactions against) important environmental changes. Faster consolidation of memory by an individual means faster adaptation to the new surroundings to protect and to nourish itself. In other words, successful adaptation is based upon memorised sequences of interacting biochemical and biophysical commands for execution of a templated series of coordinated chemical transformations that are consolidated as memories. Because memories are apparently sorted and consolidated largely during sleep, the processes of sleep become part of the success of preference development. Thus, adaptation in terms of olfactory preference development requires a large set of specific chemical reactions as symbiotic interactions, e.g. with variables as disparate as microflora and sleep as factors in the processes leading to preference. To date, multidimensional biomemorisation and formation of food preferences are among the least studied aspects of human nutrition.

3.4 Flavour Sensation as a Part of Personal Dietary Choices

In the modern, affluent food marketplace, the key to success is delivering food that pleases the consumer's palate [18, 19]. Although throughout history—or today in less-affluent cultures—cost and availability dictated food choices, today preference rules. Even the most nutritious foods are not routinely accepted and regularly consumed if they have poor sensory properties for the individual consumer choosing them. Therefore, in building a knowledge base of food choices and particularly the role of flavour, it is necessary to study and understand olfactory preferences. Food perception is more than the simple volatile compounds in food biomaterials capable of binding to olfactory receptors. Understanding flavour means first understanding the individual responses to olfactory stimuli and subsequently building an understanding of how those responses lead to preferences.

3.5 Measuring Flavour Perception Is Influenced by Several Factors

The flavour a consumer perceives has been described as the result of interactions among three factors that impact the overall flavour perception of foods (Fig. 3.1). The first factor includes all physical, chemical and biological aspects related to the isolated food or food material itself. Traditionally, analytical flavour research was concerned with extracting, identifying and quantifying the literally hundreds of different aroma and taste-active compounds in foods [20]. However, the molecules themselves are not the only factors that dictate flavour perception. The second factor concerns the various immediate processes of eating and all the aspects related to the physiology, anatomy and physicochemistry of the oral space. Whereas the first factor, the food, is independent of the consumer, the processes of eating are different among individuals. Important variables are all those that lead to the liberation of aroma and taste compounds and their transport from the oral cavity to either the nasal cavity, where the olfactory receptors are located, or to the taste buds on the tongue. When eating food, flavour compounds interact with the entire oral environment, including salivary and mucous layers. As these are all part of a person's perception of food flavour, novel approaches are needed that place the individual consumer inside the process of flavour analysis. In essence if the variation among individuals is key to olfactory perception and preference, it is necessary to move to individualised flavour science. Finally, perception itself is affected by the myriad memorisation processes discussed already that extend to psychosocial and cognitive factors such as culture, education and even mood [21–23]. In building a more individualised view of flavour perception, measuring aroma perception—the smell of food—as one central element of the overall sensory experience of food is becoming possible.

The aroma of foods is initiated when volatile aromatic compounds reach the olfactory epithelium in the upper part of the nose. One of the more obvious protective benefits of olfaction to the protection of the organism is the fact that aroma perception from food begins before eating is initiated. When volatiles emanating from the food are inhaled, they enter the nose through the orthonasal route. Decisions as to the risk benefits of consuming the food can be made prior to touching it. Once food is in the mouth, volatiles are released into the oral cavity and transported via the retronasal route to the throat (pharynx) and nose. The two specific "types of aromas" are the orthonasal and the retronasal aromas.

Because both the orthonasal and the retronasal aromas are dynamic, evolving over time, these dimensions must be captured analytically; hence, techniques are needed that are capable of analysing aroma profiles with the high time resolution appropriate to aroma perception in humans, and capturing the time–intensity patterns of the volatile compounds sweeping over the olfactory receptors. Furthermore, odour perception is an inherently non-equilibrium situation. Foods on a plate and in the mouth are open systems, and volatiles continuously escape into the air. The most effective way to measure the release of aroma during eating is to monitor the breath air as close as possible to the olfactory receptors in the nose. One approach to collecting the exhaled air at the nostril breath by breath is termed nosespace or in vivo aroma analysis [24, 25].

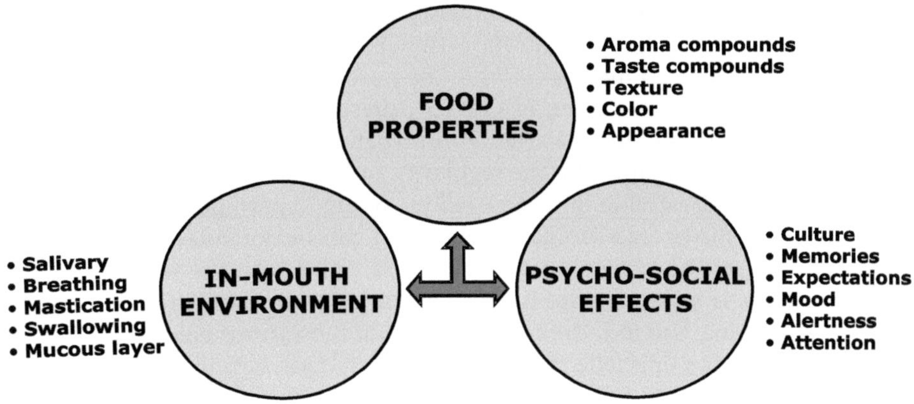

Fig. 3.1 Three factors influence flavour perception. The first includes all aspects that are related solely to the food, such as the aroma-active compounds present and interactions between the food matrix and aroma compounds. The second comprises all aspects related to the in-mouth situation. This makes the person eating the food an integral part of the system being analysed, and takes account of interactions between food and consumer. Finally, psychosocial and cognitive effects modulate aroma perception

3.6 The "Melody" of Coffee

Coffee is an interesting example of olfactory preference. The remarkable preference that some consumers develop for the volatile aroma molecules liberated from mature, partially fermented, dried, roasted and ground beans of the plant are testament to the integrative nature of the olfactory preference development process. The positive preferences that are developed are presumably a consequence, in part, of neurophysiological inputs from caffeine rather than nutritive or even microfloral inputs. Coffee aroma evolves in the mouth during drinking and finishes over several minutes after swallowing, with a typical afterodour in the mouth. The nosespace technique is able to capture many of these dynamic processes analytically, and gives a vivid insight into aroma release and its temporal evolution in the mouth. An abbreviated selection of 11 compounds that were simultaneously measured in the air exhaled through the nose during drinking of espresso coffee is shown (Fig. 3.2) using a technique termed proton-transfer reaction mass spectrometry (PTR-MS) [26, 27]. The top-left frame of Fig. 3.2 shows the in-mouth temperature measured with a tiny thermocouple in the coffee assessor's mouth. Prior to taking coffee into the mouth, the temperature was about 35 °C. As the individual sipped the coffee (at 50 s), the temperature rose immediately to 46 °C, and then decreased owing to the thermal conduction in the oral cavity. After keeping the coffee in the mouth for 10 s, the temperature dropped below 40 °C. In this specific aroma evaluation, the individual assessor was instructed to keep the coffee within the mouth for a relatively long time prior to swallowing in order to extend the measurements during the basic processes that occur in mouth. The concentration-time plots are those of compounds appearing at the indicated masses (m/z) in the nosespace air: m/z 37 corresponds to the protonated water cluster, $H_2O{\cdot}H_3O^+$, present in the breath air, whether or not the person had coffee in the mouth (natural humidity in breath). This water signal acts as a marker for the regularity and stability of breathing rhythm, another important variable in the overall in-mouth experience. Though somewhat arbitrary, the overall aroma development can be considered sequentially in stages. First, at the first contact of the liquid coffee with the in-mouth environment, there is an initial rise in the concentration of aromatic compounds, the first-sip aroma. Second, the concentration of the various compounds available to the olfactory epithelia peaks and then decreases rapidly. Breaking these individual compounds into discrete temporal curves of concentration versus time reveals that the rate of decrease is not the same for all compounds; hence, the overall profile of the coffee aroma exposed to the olfactory epithelia, again, changes with time. The rapid decrease of the concentrations of volatile compounds from coffee in the breath air is believed to be a combination of various phenomena: (1) temperature dependence of the air–water partition coefficient, (2) dilution of coffee with saliva, (3) interaction with saliva constituents and (4) adsorption and diffusion into the mucous layer. Third, when coffee is swallowed, coffee volatiles are released during the passage through the throat. The subsequent exhalation, the swallow-breath, entrains these volatiles through the nose and out through the nostrils. Accordingly, the corresponding aroma profile is called the swallow-breath aroma. For a series of compounds, high concentrations of volatiles are measured in the breath air just after swallowing. Fourth, when coffee is swallowed, the breath air continues to contain some of the coffee volatiles for several more minutes. This effect is known as the finishing or after-odour aroma. The persistence of various coffee aroma compounds in the breath air is reminiscent of coffee aroma, although it has a composition quite different from the aroma in the first sip, or the swallow breath.

The breath-by-breath observations of the retronasal aroma transport of a wide variety of subjects revealed inter- and intraindividual differences and documented the need to go beyond a static aroma description. Simply describing the odorant exposure experience requires that the various compounds be measured as an integrated and dynamic process, but the differences among subjects imply that additionally an individualised view be brought into the very first stages of flavour research—measurement of aroma exposure. The breakthroughs in methodologies that bring such analytical precision to studying olfactory exposure can now be brought to address a more concrete understanding of the customer's perception of food aroma in general. The analytical approaches described must now be coupled to means to evaluate the subjective aspects of flavour preference simultaneous with odorant exposure.

Ultimately, studies such as the evaluation presented will enable research to acquire a better understanding of how aromas lead to preferences for specific foods. The example of coffee aroma measurement revealed interindividual dif-

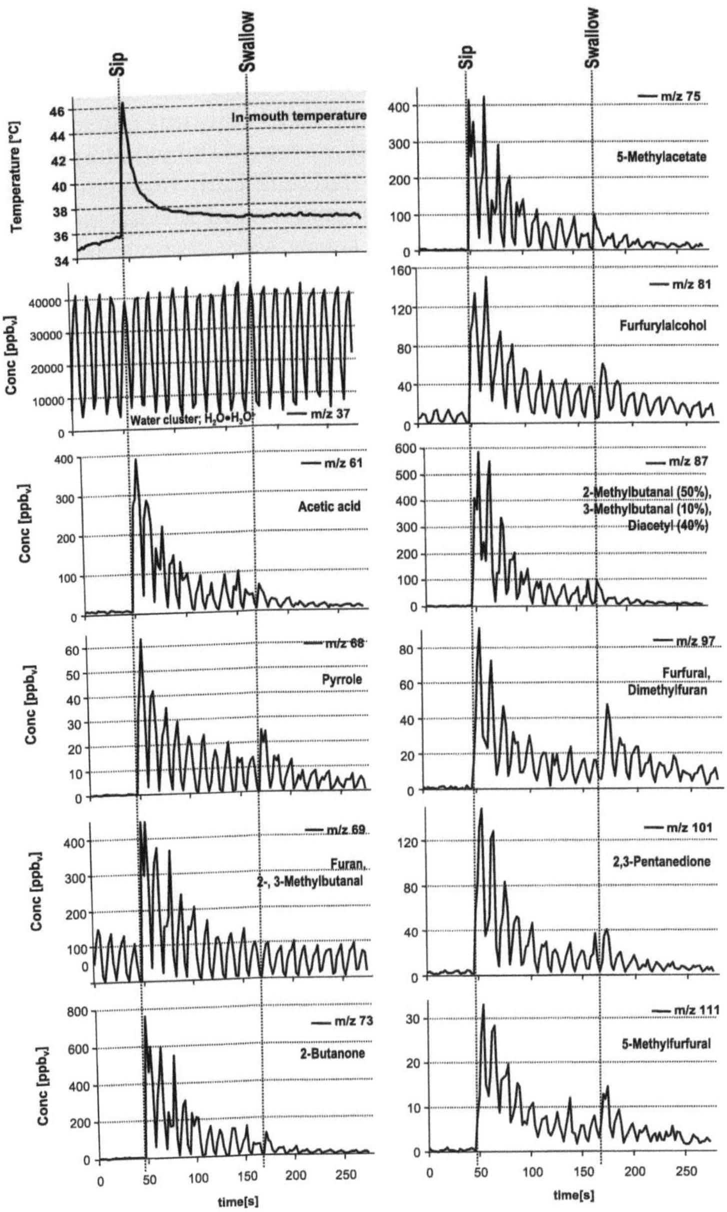

Fig. 3.2 Nosespace spectra during drinking of espresso coffee by an experienced coffee taster. The *top-left frame* shows the in-mouth temperature profile during drinking

ferences in the manipulation of the odorant exposure related to flavour preferences. That is, individuals who reported a greater preference for coffee manipulated the concentration of aromas to increase the net concentration and duration of exposure relative to individuals who did not regularly consume coffee. These approaches were thus capable of resolving novel aspects of the variation in individual consumers. For example, coffee is prepared differently from country to country. Individual preferences, modes of preparations and serving temperatures vary within a country and even within a family. Very accurate measures are necessary to resolve these subtle differences that are nonetheless critical to preference development. Recent studies have investigated the retronasal aroma from other foods such as ice cream or banana [28]. In all of these studies, a dynamic evolution was observed that was characteristic for the type of food and consumption temperature, and that revealed interindividual differences.

With methods in place to measure volatile aroma compounds within the olfactory space of individuals in real time, and to couple these to subjective reports of preference, it then becomes possible to combine these with more comprehensive measures of acute metabolism and physiology within an individual during the period when a novel food is being first perceived and olfactory preferences are being developed.

3.7 Metabolomics and the Metabolic Response to Foods

Metabolism in human and animal biofluids and tissues is the quantitative interaction of metabolic pathways with physiological demands and the consequences of eating. Metabolite concentrations are the direct reflection of metabolism. Measurements of metabolite concentrations, when comprehensive and accurate, reflect the range of biochemical effects induced by a condition or intervention. Metabolomics is emerging as a postgenomic science that seeks to measure all of the metabolites in a tissue, biofluid or cell. Metabolomics is seen as a field with substantial applications to biotechnology and medicine. Although the tools of metabolomics are still in the process of development, they are already being used to identify the functions of genes, describe the effects of toxicological, pharmaceutical, nutritional and environmental interventions, and to build integrated databases of metabolite concentrations across human and research animal populations [29]. When these measures are considered to be a reflection of the entire metabolite pool, i.e. metabolomics, data can be used to diagnose or predict disease, to stratify populations by an individual's specific metabolism or to determine the safety or efficacy of a therapeutic intervention [30]. Metabolomics can also be used to directly quantify and assess the consequences of eating. The only additional consideration is to include measurements of metabolites as a function of time after eating a standardised meal (ensemble of components).

Metabolite measurements have been used to assess health for decades, and so metabolomics is not a revolutionary approach to medicine, toxicity or nutrition. Measurements of fasting plasma glucose concentrations to assess diabetes or of cholesterol concentrations to predict the likelihood of cardiovascular disease are common, metabolite-based tools used to assess health. More recently, the idea of assessing postprandial metabolism has gained acceptance, with the ability of an individual's response to a standard glucose challenge to predict insulin sensitivity prior to the development of metabolic diseases of insulin failure [31]. Nonetheless, the use of metabolic measurements for assessing health status has, to date, been approached as an application of single biomarkers designed for diagnosis or prognosis of disease. Metabolomics offers a fresh perspective on this approach because of the scope of measurements that can now be made with modern analytical equipment. Because the products, intermediates and substrates for virtually all endogenous biochemical pathways can be measured by various analytical platforms, it is now possible to assemble a picture of individual health in its full context. This is already providing advantages for both discovery and clinical work in disease, but could be equally powerful in developing an understanding of the relations between metabolism and sensation and preference development.

Metabolism itself can be described comprehensively in breadth and depth and time. Rather than single metabolites, highly comprehensive sets of metabolite measurements are obtained by multiparallel analyses. Rather than averaging over large populations or trials, measurements of the metabolic profile of single individuals become discretely targeted information, and rather than attempting to identify a key point in time, measurements are taken as a function of time after various challenges, including diet, to reflect the true dynamics of metabolism. As the result, it is possible to approach the problem of assessing health and flavour preference development scientifically.

3.8 Profiling of Postprandial Plasma Lipid Composition

The structural and energetic lipids present in blood have proven to be a particularly informative class of metabolites for diagnosing and understanding changes in energy balance and transport, such as in disorders like atherosclerosis, type 2 diabetes and the metabolic syndrome. The concentration and composition of lipid metabolites in whole plasma, including the different phospholipids, sphingolipids, sterols, sterol esters, glycerides and free fatty acids, represent the integrated metabolism of key tissues exchanged continuously with the blood compartment. In particular, plasma lipid metabolites directly reflect metabolism from organs serving as biosynthetic sites for lipids, including liver, adipose tissue and intestine. A key advantage of profiling these metabolites is that most of the biochemical pathways responsible for their synthesis, metabolism and catabolism are known. Thus, the quantitative lipid metabolite profiles can be mapped against pathway knowledge to determine biological bases for metabolic changes in the profile. Lipomics's TrueMass® analytical platform (http://www.

lipomics.com/services) produces quantitative data on approximately 300 individual lipid metabolites from a single human serum or plasma sample.

Whole plasma can also be fractionated into specific lipoprotein size classes to further resolve the underlying biochemistry and metabolism of tissues that deliver these lipids to blood and selectively remove them. Thus, TrueMass® analysis can be used to measure the lipid profiles of very-low-density lipoprotein, quantify the lipid pathways responsible for metabolic changes in the liver and measure profiles of high-density lipoprotein to quantify the flux of lipids in reverse cholesterol transport.

3.9 Profiling Signalling Lipids

Polyunsaturated fatty acids and their derivatives (i.e. monoacylglycerols, amides and oxylipins) function as effectors of biological activities. Free fatty acids modulate the activity of phospholipases, ionic channels, ATPases, G-proteins and protein kinases; they also regulate the phosphoinositide and sphingomyelin cycle, hormonal signal transduction and gene transcription. Furthermore, enzymatic oxygenation of unsaturated fatty acids gives rise to a wide range of highly active oxylipins, which function as signalling molecules. As the oxylipins are synthesised from polyenoic fatty acids in response to different biological stimuli, their measurement provides a quantitative reflection of the state of cells and tissues being measured. A large part of the oxidised lipids present in biofluids and tissues is specifically biosynthesised from polyunsaturated fatty acids by action of acutely regulated enzyme(s). The amount and types of oxylipins in biofluids have been used to indicate inflammatory, damaged and explicitly diseased states.

In animals, the fatty acid arachidonic acid is considered to be the most important precursor of oxygenated derivatives—compounds commonly referred to as eicosanoids, i.e. derived from 20-carbon-chain-length fatty acids. Studies have documented that most of the primary oxygenation of arachidonic acid and other fatty acids in animal tissue is catalysed by cyclooxygenases (prostaglandin endoperoxide synthases) and lipoxygenases that are coordinated in a series of time-dependent events to control various processes of response to stress and infection [32]. These multiple enzymatically catalysed reactions lead to a number of oxygenated derivatives such as prostaglandin H2, leukotriene A4 and various fatty acid hydroperoxides, which can be further modified by secondary enzymes, including prostaglandin E, D and F synthases, thromboxane A synthase, prostacyclin synthase, leukotriene A4 hydrolase and leukotriene C4 synthase, to generate members of the prostaglandin, leukotriene and thromboxane families [33]. Cytochrome P450 monooxygenase activity can also lead to the formation of epoxy, hydroxy and dihydroxy derivatives, whereas non-enzymatic oxygenation of arachidonic acid and other polyunsaturated fatty acids can lead to the formation of the isoprostane group of compounds [34]. Taken together, the oxylipins exert remarkably diverse biological effects from acute cellular processes, from promoting the aggregation of blood platelets to curtail bleeding to muscle contraction to physiological processes such as reproduction. Importantly, the quantities and distributions of oxylipins are, in part, responsive to diet because all are derived from fatty acids that cannot be synthesised de novo by humans and must be ingested as components of foods. These compounds, by playing such diverse roles in the various processes of responses to stress constitute a valuable and measurable index of both stress detection and stress response and its effective (or potentially ineffective) resolution.

Much as measuring lipid metabolites comprehensively as opposed to a single biomarker provides a quantum leap in understanding structural lipid metabolism and status, quantitatively measuring lipid-signalling oxylipins has been shown to provide unique insight into shifts in phenotype in response to environmental stressors both systemically and in various tissues including the brain [35]. The 5-hydroxyeicosatrieneoic acid (5-HETE) from 5-LOX and PGF2a from COX are early markers of inflammatory progression, the epoxy eicosatrienoic acids (EETs) are potent anti-inflammatory and vasodilatory agents, and 12-HETE and 15-HETE are reported markers of cellular proliferation [36].

By measuring all of these signalling, energetic and structural metabolites during the period after consuming a meal, it is possible to generate a comprehensive perspective of the metabolic response to a meal, the stress responses occurring during this period and the overall state of physiological context in which the olfactory decision must take place. This type of database, once constructed, would represent the physiological context in which the initial experience to and the decision as to preferences of a specific olfactory molecule(s) could be established.

3.10 Conclusion

The development of mechanisms that simultaneously protect and nourish an organism within a particular environment is key to survival, and these mechanisms represent an important Darwinian selective pressure. The ability of organisms to learn from their surroundings and to improve their biochemical responses to that environment is becoming increasingly well established as forms of imprinting and metabolic memory. Within this context, the development of olfactory preferences is a vivid example of acquired memories. Food is not only a source of nutrients, but also the chemicals that elicit characteristic volatile aromas and lead to preferences for particular food choices. Ideally, the memories formed in response to exposures to diets enhance an individual's ability to succeed in a particular environment, including the available foods. However, the failure of modern diets to deliver increasing health to the entire population is testament to the inability of all humans to match food choices to optimal nutritional requirements in all environments and lifestyles. The ability to reformulate food commodities and foods with widely varying nutritional and flavour properties has the potential to both confound and enhance the processes of flavour preference and food choice. Enhancing food choices based on flavour preferences will require an understanding of precisely how flavour preferences are developed. The tools to simultaneously measure aroma exposure, aroma perception and metabolic responses to foods are at hand. Bringing these tools to practice and joining the fields of flavour science, nutrition and metabolic assessment into a new era of personalised diet and health is an attractive possibility.

References

1. Müller GE, Pilzecker A (1900) Z Psychol Ergänzungsbd 1:1
2. McGaugh JL (2000) Science 287:248
3. Ache BW, Young JM (2005) Neuron 48:417
4. Alberini CM (1999) J Exp Biol 202:2887
5. Tolstoguzov V (2000) Nahrung 44:89
6. Tolstoguzov V (1999) FEBS Lett 444:145
7. Tolstoguzov VB (2000) In: Walter H, Brooks DE, Srere PA (eds) Microcompartmentation and phase separation in cytoplasm. International review of cytology, vol 192. Academic, San Diego, p 3
8. Tolstoguzov V (1999) In: Roos YH, Leslie RB, Lillford PJ (eds) Water management in the design and distribution of quality foods. Technomic, Basel, p 199
9. Tolstoguzov VB (1997) In: Damodaran S, Paraf A (eds) Food proteins and their applications in foods. Dekker, New York, p 171
10. Tolstoguzov V (1998) In: Mitchell JR, Ledward DA, Hill S (eds) Functional properties of food macromolecules. Blackie, London, p 252
11. Yu D, Keene AC, Srivatsan A, Waddell S, Davis RL (2005) Cell 123:945
12. Leshem M, Del Canho S, Schulkin J (1999) Physiol Behav 67:555
13. Schooler J, Loftus EF (1997) In: McGraw-Hill encyclopedia of science and technology, 8th edn, vol 4. McGraw-Hill, New York, p 671
14. Garcia JA, Zhang D, Estill SJ, Michnoff C, Rutter J, Reick M, Scott K, Diaz-Arrastia R, McKnight SL (2000) Science 288:2226
15. Nader K, Schafe GE, Le Doux JE (2000) Nature 406:722
16. Dauncey MJ, Bicknell RJ (1999) Nutr Res Rev 12:231
17. Dulloo AG, Jacquet J (1999) Br J Nutr 82:339
18. Lindeman M, Stark K (2000). Appetite 35:263
19. Griep M, Mets T, Massart D, (2000) Br J Nutr 83:105
20. Spanier AM, Shahidi F, Parliament TH, Mussinan CJ, Ho C-T, Tratras Contis E (2001) Food flavors and chemistry: advances of the new millennium. Royal Society of Chemistry, Cambridge
21. Capaldi ED (2001) In: Capaldi ED (ed) Why we eat what we eat: the psychology of eating. American Psychological Association, Washington, p 53
22. Rozin P (1998) Towards a psychology of food choice. Danone Chair monograph. Institut Danone, Brussels
23. Mennella JA, Beauchamp G (2002) Early Hum Dev 68:71

24. Taylor AJ (2003) Compr Rev Food Sci Food Safety 1:45
25. Taylor AJ, Linforth R (2000) In: Roberts DD, Taylor AJ (eds) Flavour release. American Chemical Society, Washington, p 8
26. Lindinger W, Hansel A, Jordan A (1998) Int J Mass Spectrom Ion Process 173:191
27. Yeretzian C, Jordan A, Brevard H, Lindinger W (2000) In: Roberts DD, Taylor AJ (eds) Flavour release. American Chemical Society, Washington, p 58
28. Mayr D, Tilmann M, Lindinger W, Brevard H, Yeretzian C (2003) Int J Mass Spectrom 223:743
29. Watkins SM, German JB (2002) Curr Opin Biotechnol 13:512
30. Watkins SM, German JB (2002) Curr Opin Mol Ther 4:224
31. Ceriello A, Hanefeld M, Leiter L, Monnier L, Moses A, Owens D, Tajima N, Tuomilehto J (2004) Arch Intern Med 164:2090
32. Levy BD, Clish CB, Schmidt B, Gronert K, Serhan CN. (2001) Nat Immunol 2:612
33. Needleman P, Turk J, Jakschik BA, Morrison AR, Lefkowith JB (1986) Annu Rev Biochem 55:69
34. Lawson JA, Rokach J, Fitzgerald GA (1999) J Biol Chem 274:24441
35. Bazan NB (2005) Mol Neurobiol 32:89
36. Sharma GD, Ottino P, Bazan NG, Bazan HE (2005) J Biol Chem 280:7917

4 Chemistry of Essential Oils

K. Hüsnü Can Başer, Fatih Demirci
Department of Pharmacognosy,
Faculty of Pharmacy,
Anadolu University, 26470 Eskişehir, Turkey

4.1 What Is an Essential Oil?

Essential oil, also defined as essence, volatile oil, etheric oil or aetheroleum, is a complex mixture of volatile constituents biosynthesised by living organisms. Essential oils can be liberated from their matrix by water, steam and dry distillation, or expression in the case of citrus fruits [1–5]. Their occurrence and function in nature is still a question and the subject of ongoing research. However, there is evidence that organisms produce essential oils for defence, signalling or as part of their secondary metabolism. As a consequence essential oils comprise an important bioresource for renewable natural products [1–25].

Extracts of aromatic plant or animal materials obtained using organic solvents or fluidised gasses *are not* considered as essential oils [1, 23, 25–28]. Concretes, absolutes, spice oleoresins, *etc.* which can be classified as aromatic extracts are not covered in this chapter.

Essential oils, their fractions and their isolates are utilised in flavour and fragrance, food, perfumery, cosmetics and toiletries, fine chemicals, pharmaceutical industries and therapy. They are used as such or in diluted forms in the budding aromatherapy sector [1, 3, 5, 6, 8–14, 16–19, 21–35].

Essential oils may comprise volatile compounds of terpenoid or non-terpenoid origin. All of them are hydrocarbons and their oxygenated derivatives. Some may also contain nitrogen or sulphur derivatives. They may exist in the form of alcohols, acids, esters, epoxides, aldehydes, ketones, amines, sulphides, *etc.* Monoterpenes, sesquiterpenes and even diterpenes constitute the composition of many essential oils. In addition, phenylpropanoids, fatty acids and their esters, or their decomposition products are also encountered as volatiles [1–16, 21–33, 36–38].

Owing to their liquid nature at room temperature, essential oils are called as such. They should not be confused with fixed oils or fatty oils, which are composed of a naturally occurring mixture of lipids which may not necessarily be volatile. Therefore, essential oils differ entirely both in chemical and in physical properties from fatty oils. Essential oil evaporates completely when dropped on filter paper; however, fixed oil leaves a permanent stain which does not evaporate even when heated.

Essential oils occur mainly in aromatic plants. A few of them are found in animal sources, e.g. musk, civet and sperm whale, or are produced by microorganisms [1, 3, 6, 23, 25, 26, 29–33]. The Council of Europe describes "essential oil" as a product obtained from "vegetable raw material" [27]. Owing to a ban on animal-based flavour and fragrance materials, essential oils of trade are entirely of plant origin

Among many others, well-known families rich in essential oil bearing species are Apiaceae, Asteraceae, Cupressaceae, Hypericaceae, Lamiaceae, Lauraceae, Myrtaceae, Pinaceae, Piperaceae, Rutaceae, Santalaceae, Zingiberaceae and Zygophyllaceae [1–4, 8–11, 39].

In plants, essential oils occur in oil cells, secretory ducts or cavities, or in glandular hairs. In some cases, they are bound with carbohydrates in the form of glycosides [1–4, 8–14]. In such cases, they must be liberated by hydrolysis of the glycosidic bond. This is done by allowing enzymatic reactions to take place during wilting prior to distillation of fresh plant materials. Mosses, liverworts, seaweeds, sponges and fungi have also been shown to contain essential oils. Besides higher plants, some terrestrial and marine animals, insects, fungi and microorganisms are also known to biosynthesise volatile compounds [6–14, 30–33, 40, 41].

Essential oils are frequently associated with gums and/or resins. They are freed from such products by distillation.

Essential oil constituents can be classified as terpenoids and non-terpenoid hydrocarbons.

4.1.1 Non-terpenoid Hydrocarbons

Non-terpenoid hydrocarbons found in essential oils such as short chain alcohols and aldehydes are formed by metabolic conversion or degradation of phospholipids and fatty acids [12].

Hydrocarbons consist of carbon and hydrogen. They may also contain oxygen, nitrogen or sulphur. The simplest hydrocarbon is methane (CH_4), which is a colourless and odourless highly flammable gas. The carbons are connected by single, double or triple bonds to form higher molecular weight hydrocarbons. Saturated homologous straight-chain structures are called **alkanes**, while their unsaturated forms are called **alkenes**. Alkenes show isomerism owing to the

1 2

Structure 4.1

position of the hydrogen atoms attached to the double bond as in the examples *trans*-but-2-ene **1** and *cis*-but-2-ene **2** (Structure 4.1).

Molecules with three carbon atoms can only form a straight chain; however, with four carbons or more they can form both straight and branched chains. Then, their naming also changes accordingly. Isoprene is one such molecule; it is chemically 2-methyl-1,3-butadiene **3** (Structure 4.2).

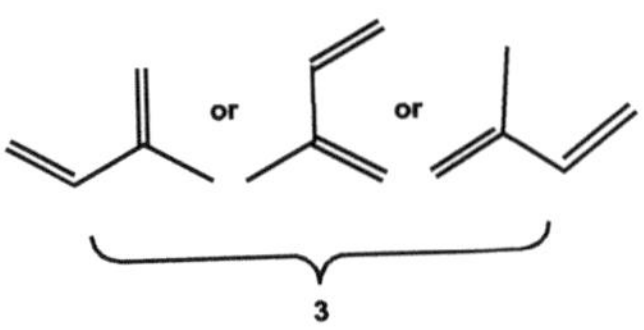

Structure 4.2

4.1.2 Terpenoids

Terpenes, also called isoprenoids, are one of the largest classes of natural chemicals formed by head-to-tail rearrangement of two or more isoprene molecules. More than 30,000 terpenoids have been isolated from plants, microorganisms and animals [3, 7, 11, 37, 42]. They are important constituents of essential oils. Molecules formed from two isoprene **3** molecules are called **monoterpenes** ($C_{10}H_{16}$). C_5H_8 compounds are **hemiterpenes. Sesquiterpenes** contain three isoprene units; hence they have the formula $C_{15}H_{24}$. $C_{20}H_{32}$ compounds formed from four isoprene units are **diterpenes**. Heavier terpenes like diterpenes are generally not found in essential oils. Isoprene itself is considered the only hemiterpene, but oxygen-containing derivatives such as prenol and isovaleric acid are hemiterpenoids, too [3, 7–14].

Kekulé, in 1880, was the first scientist to name $C_{10}H_{16}$ compounds as "terpenes", because of their existence in turpentine. His assistant Wallach (1910 recipient of the Nobel Prize in Chemistry) hypothetically proposed in 1887 that terpenes were constructed via two or more isoprene units. Three decades later, Robinson (1947 recipient of the Nobel Prize in Chemistry) perfected Wallach's "isoprene rule" by suggesting that the isoprene units should be connected in a head-to-tail fashion. A few years later, Ruzicka (1939 recipient of the Nobel Prize in Chemistry) proposed in 1950 the "biogenetic isoprene rule", further developing Wallach's hypothesis. The rule reiterated the formation of terpenes by head-to-tail rearrangement of two or more isoprene units. This rule stipulates that the terpenoids are derived from aliphatic precursors such as geraniol for the formation of monoterpenes, farnesol for the sesquiterpenes, geranylgeraniol for the diterpenes and squalene for triterpenes. It is interesting to note that three terpene scientists received the Nobel Prize in Chemistry within a span of 37 years [3, 7–14, 38, 39, 42].

4.1.2.1 Biosynthesis of Terpenes

Terpenes, biogenetically, arise from two simple five-carbon moieties. Isoprenyl-diphosphate (IPP) and dimethylallyldiphosphate (DMAPP) serve as universal precursors for the biosynthesis of terpenes. They are biosynthesised from three acetylcoenzyme A moieties through mevalonic acid (MVA) *via* the so-called *mevalonate pathway*. About 10 years ago, the existence of a second pathway leading to IPP and DMAPP was discovered involving 1-deoxy-*D*-xylulose-5-phosphate (DXP) and 2C-methyl-*D*-erythritol-4-phosphate (MEP). This so-called *non-mevalonate* or *deoxyxylulose phosphate pathway* starts off with the condensation of glyceraldehyde phosphate and pyruvate affording DXP. Through a series of reactions as shown in Fig. 4.1, IPP and DMAPP are formed, respectively [3, 7, 42, 43].

IPP and DMAPP lead to geranylpyrophosphate (GPP), which is an immediate precursor of monoterpenes. The formation of nerylpyrophosphate (NPP) from GPP gives rise to a wide range of acyclic, cyclic, bicyclic or tricyclic skeletons. Reactions like rearrangement, oxidation, reduction and hydration *via* various terpene cyclases result in the formation of numerous terpene derivatives. Condensation of GPP and IPP leads to farnesylpyrophosphate (FPP), the immediate precursor of sesquiterpenoids. Likewise, FPP and IPP are conducive to diterpenoids.

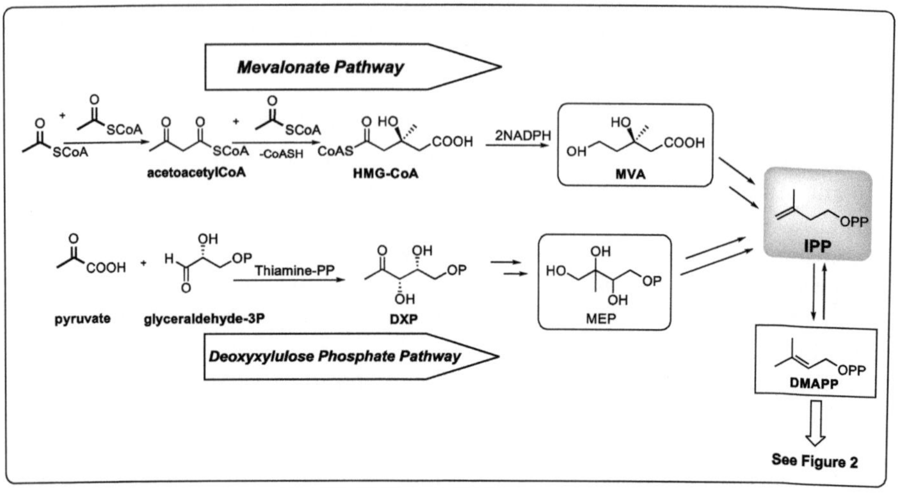

Fig. 4.1 Terpenoid biosynthesis: two independent pathways

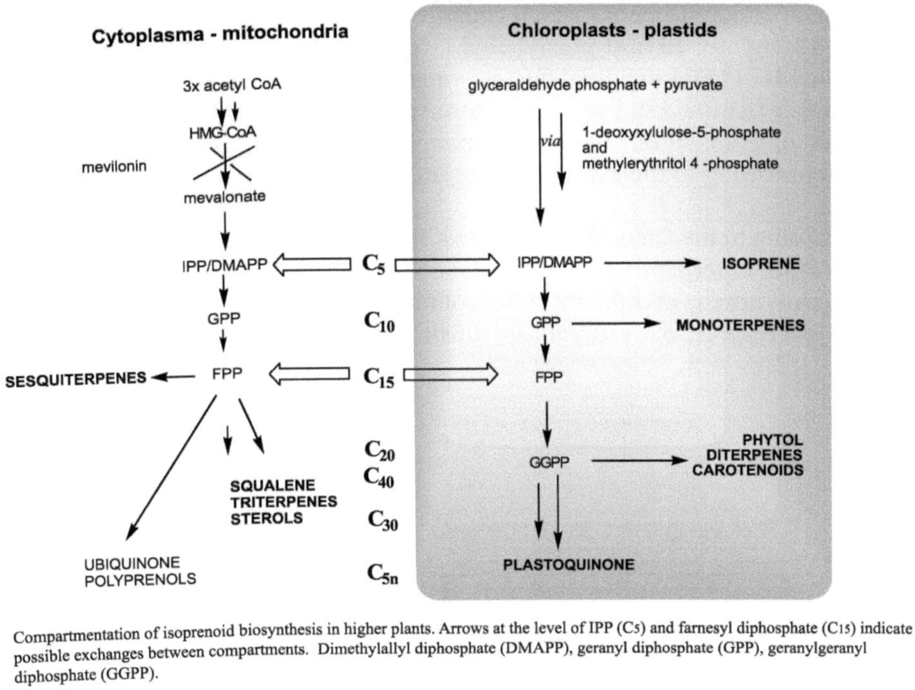

Fig. 4.2 Terpenoid biosynthesis sites and products (metabolites) (reprinted from Rohmer [46], copyright 2006, with kind permission from Elsevier)

The mevalonate-independent pathway is present in most bacteria and all phototropic organisms. In higher plants and most algae both pathways run independently. The mevalonate pathway is located in the cytoplasm and is responsible for the biosynthesis of most sesquiterpenoids. The mevalonate-independent pathway, in contrast, is restricted to the chloroplasts where plastid-related isoprenoids such as monoterpenes and diterpenes are biosynthesised via this pathway [43–45]. Figure 4.2 illustrates the interrelationships of both biosynthetic pathways connected to Fig. 4.1 [46].

Monoterpenes

Monoterpenes are formed from two attached isoprene **3** units: 2,6-dimethyloctane as the simplest skeleton. Thus, they can be acyclic or linear like *β*-myrcene **4**, (*E*)-*β*-ocimene **5**, (*Z*)-*β*-ocimene **6**, and *allo*-ocimene **7** (Structure 4.3). Or they can be cyclic, meaning ring-forming, such as in the simplest form like *p*-menthane **8** or *p*-cymene **9**. Monocyclic **8**, **9**, bicyclic *δ*-3-carene **10** and tricyclic tricyclene **11** type monoterpenes are found in essential oils [1–4, 6–14, 16–23, 38, 39, 42, 47, 48].

Aromatic monoterpenes which contain a benzene ring like *p*-cymene **9**, carvacrol **12**, thymol **13** and phenylethyl alcohol **14** (Structure 4.4) are common constituents of many essential oils, e.g. oregano (*Origanum* sp.), thyme (*Thymus* sp.), savory (*Satureja* sp.) and rose (*Rosa* sp.) oils. Another important constituent class of essential oils is phenypropanoids [36]. They are not considered as terpenoids owing to their different biogenetic origins, which will be mentioned later.

According to the *Dictionary of Natural Products* (DNP), there are 25 different classes of monoterpenes [37].

The biosynthesis of different classes of monoterpenes formed from α-terpinyl cation and respective precursors are illustrated in Fig. 4.3.

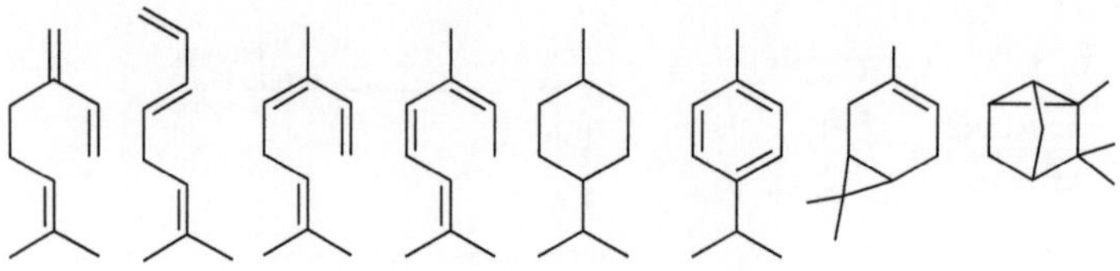

4 5 6 7 8 9 10 11

Structure 4.3

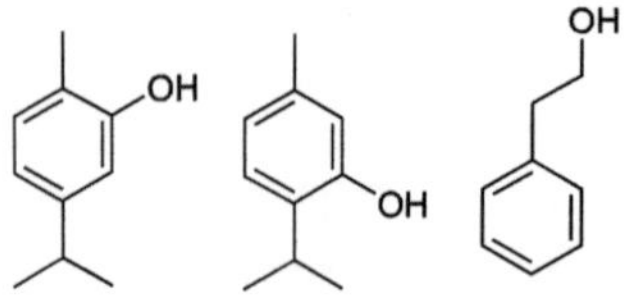

12 13 14

Structure 4.4

4.1.2.1.1.1
Acyclic Monoterpenes

These regular monoterpenes constitute a small class which includes the trienes myrcene **4** and ocimenes (**5–7**) and the alcohols geraniol **15**, nerol **16**, citronellol **17**, linalool **18**, *etc* (Structure 4.5).

Citral is the naturally occurring mixture of the aldehydes geranial **19** and neral **20** (Structure 4.6). Citronellal **21** is another acyclic aldehyde within this grouping. Variation of the 2,6-dimethyloctane skeleton is easily noticeable.

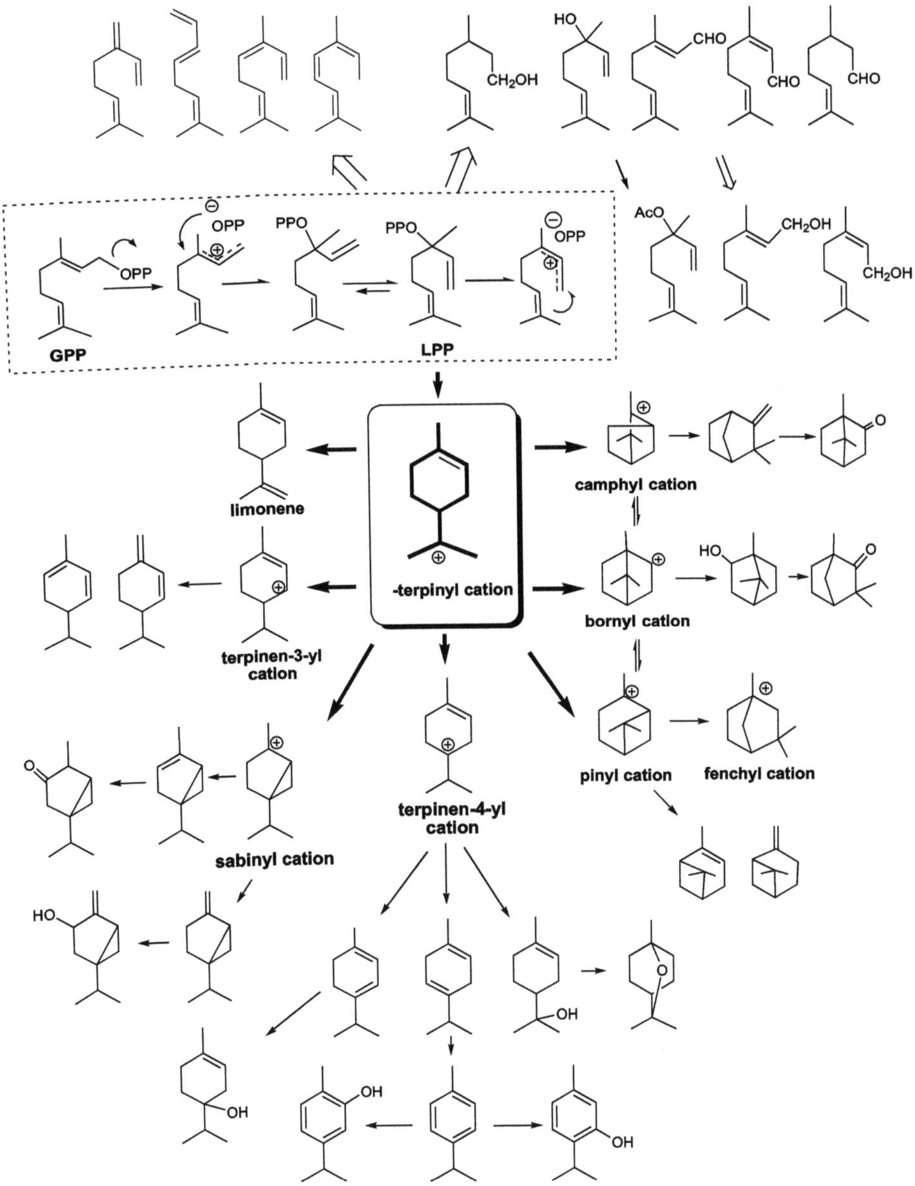

Fig. 4.3 Monoterpene (re)arrangements and important intermediates

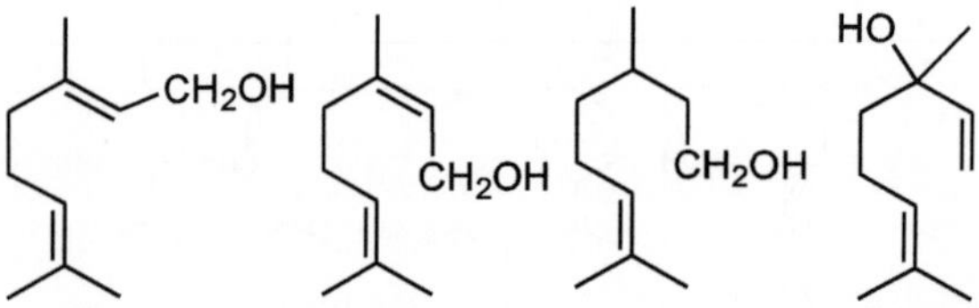

Structure 4.5

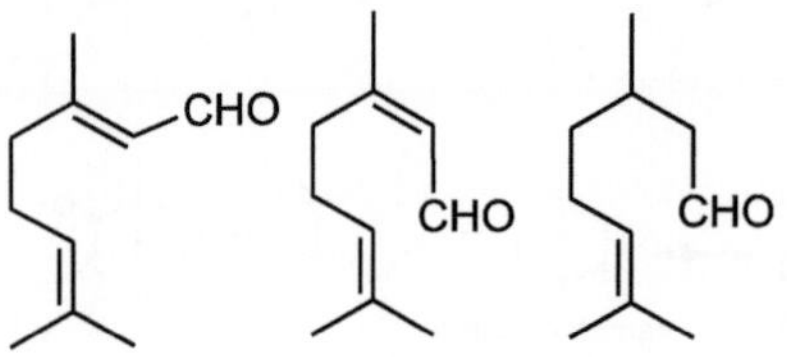

Structure 4.6

Cyclic Monoterpenes
Cyclic monoterpenes can be classified in three subgroups according their ring size such as:
1. Monocyclic monoterpenes
2. Bicyclic monoterpenes
3. Tricyclic monoterpenes

Monocyclic Monoterpenes
p-Menthane monoterpenes which possess the 1-methyl-4-isopropyl-cyclohexane **8** skeleton comprise the largest group of naturally occurring monoterpenes.

p-Menthadienes are limonene **22**, *α*-terpinene **23**, *β*-terpinene **24**, γ-terpinene **25**, terpinolene **26**, *α*-phellandrene **27** and *β*-phellandrene **28** (Structure 4.7) resulting from different rearrangements of the *α*-terpinyl cation (Fig. 4.3). This group is also classified among the monoterpene hydrocarbons.

Aromatic monoterpenes such as *p*-cymene **9** and its hydroxylated derivatives thymol **12** and its isomer carvacrol **13** always occur along with *α*-terpinene **23**, γ-terpinene **25** and terpinen-4-ol **29** (Structure 4.8). Metabolites, like *p*-cymene-8-ol **30** and cuminyl alcohol **31** may also be derived from *p*-cymene (Fig. 4.4).

Other important members of this class include oxygenated derivatives such as *α*-terpineol **32**, menthol **33**, isopulegol **34** and *cis*-hexahydrocuminyl alcohol **35**, also classified as monoterpene alcohols.

Aldehydes in this group are as follows: carvone **36**, dihydrocarvone **37**, isomenthone **38**, piperitone **39**, pulegone (piperitenone) **40**, isopulegone **41** (Structure 4.9).

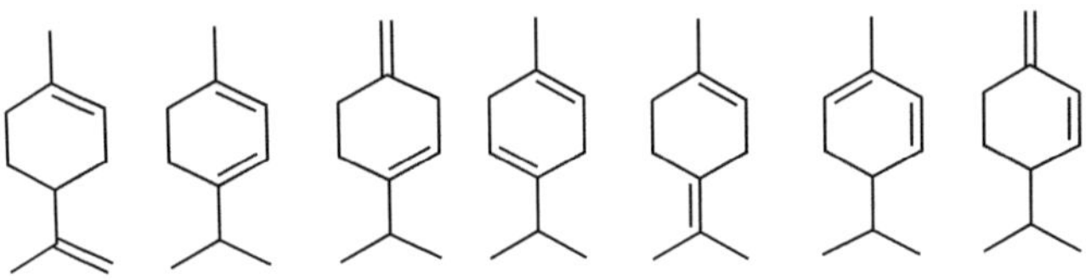

22 23 24 25 26 27 28

Structure 4.7

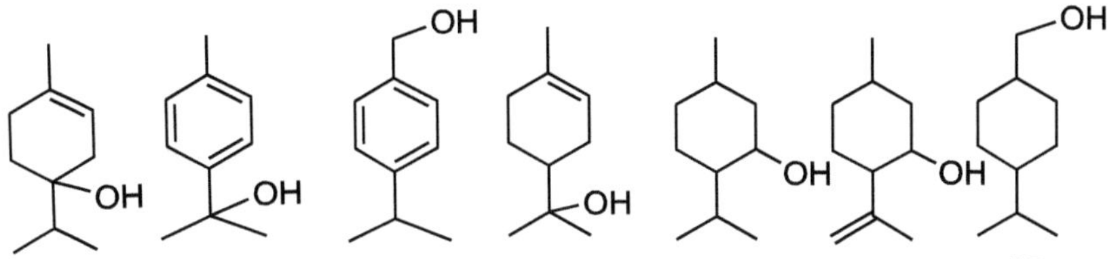

29 30 31 32 33 34 35

Structure 4.8

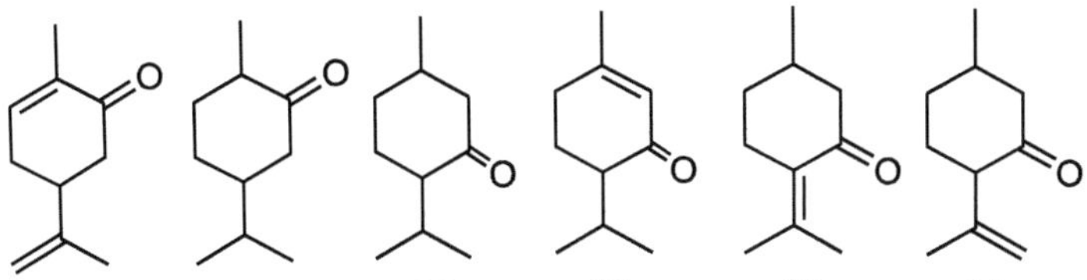

36 37 38 39 40 41

Structure 4.9

Bicyclic Monoterpenes
1,8-Cineole **42** as well as 1,4-cineole **43** are cyclic ethers (Structure 4.10). All including ascaridol **44** are bicyclic oxygenated monoterpenes. Their formation can be seen in Fig. 4.3.

Pinane monoterpenes are bicyclic monoterpenes resulting from intramolecular rearrangement of the α-terpinyl cation yielding the [3.1.1] bicyclic system (Fig. 4.3). α-Pinene **45** and β-pinene **46** (Structure 4.11) are the main constituents of turpentine oil from pines. They occur widely in essential oils.

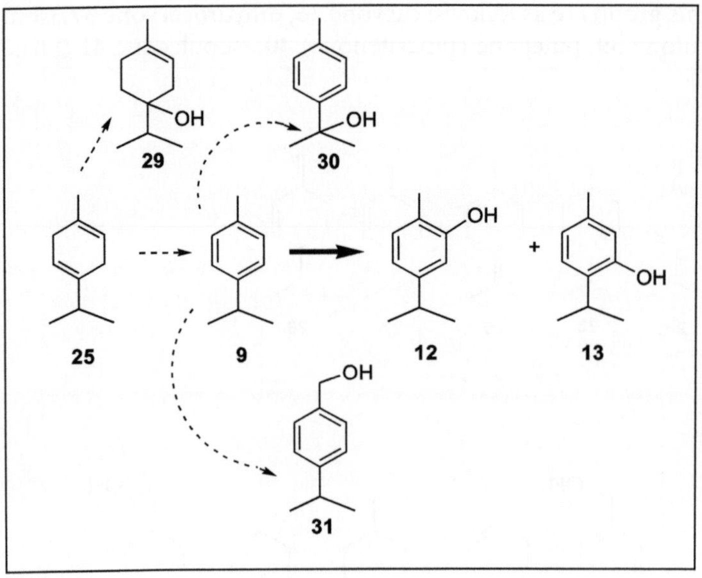

Fig. 4.4 Aromatic monoterpene biosynthesis

The bornane-, camphane- and fenchane-type monoterpenes possess the [2.1.1] bicyclic skeleton formed by different cyclisation of the terpinyl cation. Important members include borneol **47**, isobornyl acetate **48**, camphene **49**, camphor **50**, fenchone **51** (Structure 4.12).

Thujane-type monoterpenes, unusual monoterpenes with a cyclopropane ring in a bicyclo[3.1.0] skeleton, are formed from the terpinen-4-yl cation directly or via the sabinyl cation. Important members include α-thujene **52**, sabinene **53**, the *cis* isomer **54** of sabinene hydrate, sabinol **55**, sabinylacetate **56**, α-thujone **57**, β-thujone **58** and isothujanol **59** (Structure 4.13).

Carane-type monoterpenes possess a cyclopropane ring in a bicyclo[4.1.0] skeleton. δ-3-Carene **10** is a common constituent in various essential oils.

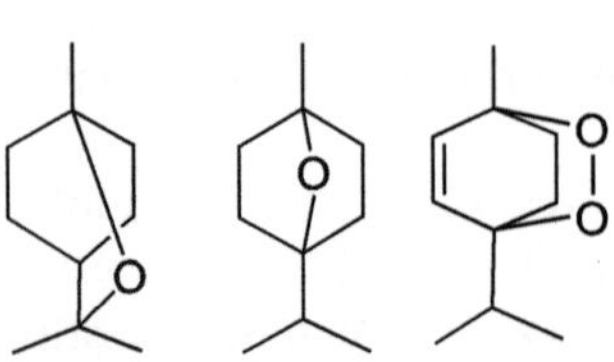

42 **43** **44**

Structure 4.10

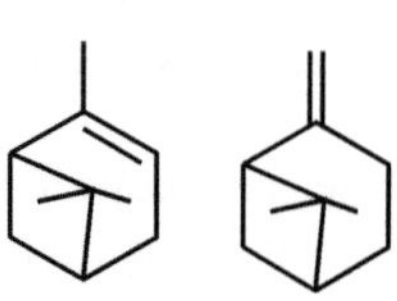

45 **46**

Structure 4.11

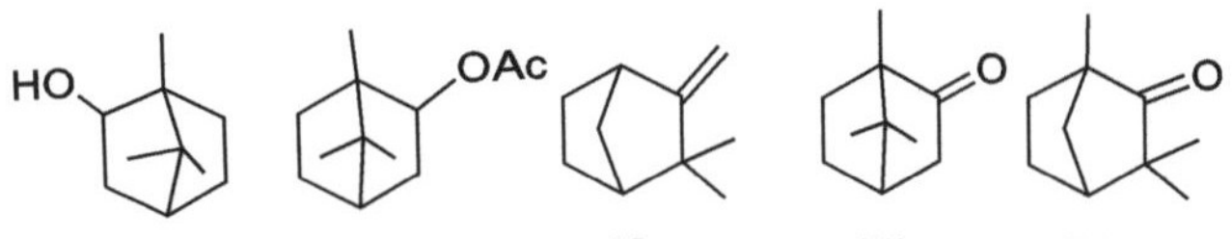

47 48 49 50 51

Structure 4.12

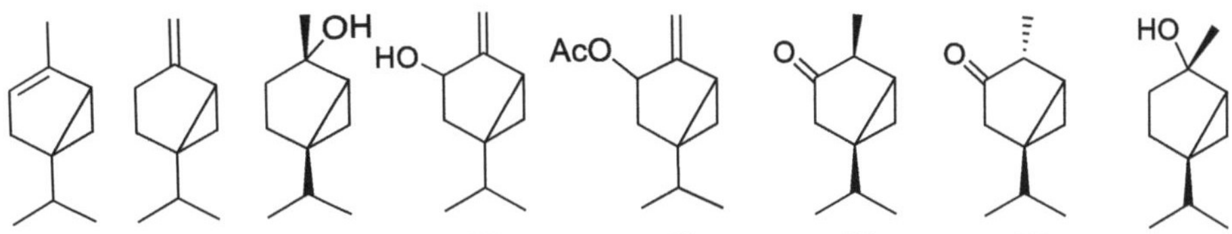

52 53 54 55 56 57 58 59

Structure 4.13

4.1.2.1.1.2.3
Tricyclic Monoterpenes

Tricyclene **11** or 1,7,7-trimethyltricyclo[2.2.1.$0^{2,6}$]heptane, is a good example which frequently occurs in various essential oils.

4.1.2.1.1.3
Irregular Monoterpenes

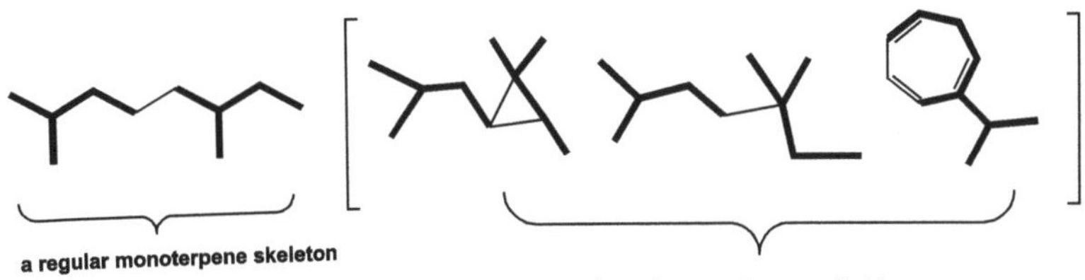

Fig. 4.5 Different classes of monoterpenes

There are two major types of irregular monoterpenes (Fig. 4.5):

1. The substituted cycloheptane monoterpenes, also called tropones. Eucarvone **60**, nezukone (4-isopropyl-2,4,6-cycloheptatrienone) **61** and γ-thujaplicin **62** (Structure 4.14) most probably arise by an unknown ring expansion of the cyclohexane skeleton.

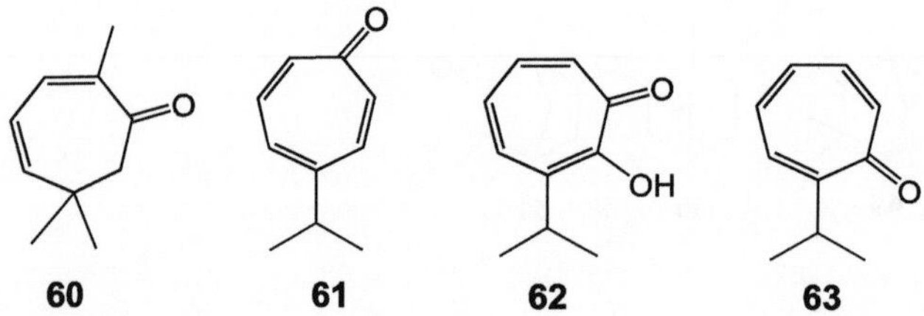

Structure 4.14

2. The other major group of irregular monoterpenes is formed by non-head-to-tail fusion of isoprene units. Important members include artemisia ketone **64**, santolinatriene **65**, chrysanthemol **66**, yomogi alcohol **67** and lavandulol **68** (Structure 4.15). Lavandulane-type compounds occur in the families Lamiaceae (Labiatae) and Apiaceae (Umbelliferae), while chrysanthemane, artemisane and santolinane types occur in the family Asteraceae (Compositae) [47, 48].

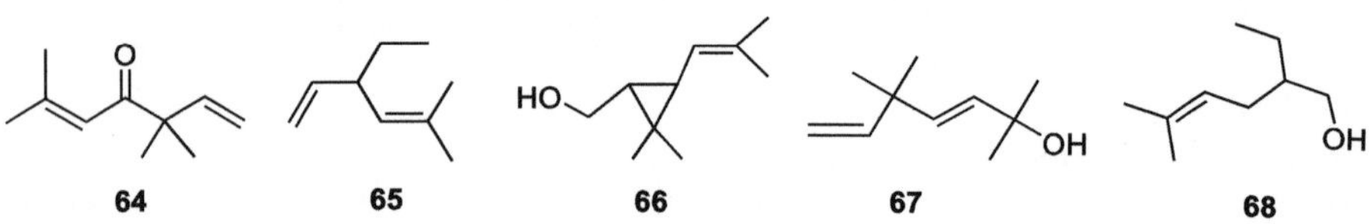

Structure 4.15

4.1.2.1.2
Sesquiterpenes

Sesquiterpenes are formed by the addition of one more isoprene units to a monoterpene molecule, and thus have the molecular formula $C_{15}H_{24}$ (see also Fig. 4.2). There are linear, branched or cyclic sesquiterpenes. Sesquiterpenes are unsaturated compounds. Cyclic sesquiterpenes may be monocyclic, bicyclic or tricyclic. They are the most diverse group among the volatile terpenoids [2, 3, 7–11, 13, 14, 16, 20–24, 37–39, 49]. The DNP treats sesquiterpenoids in 147 different structural types [37]. Various types of sesquiterpenes (**69–109**) can also be seen in Structure 4.16.

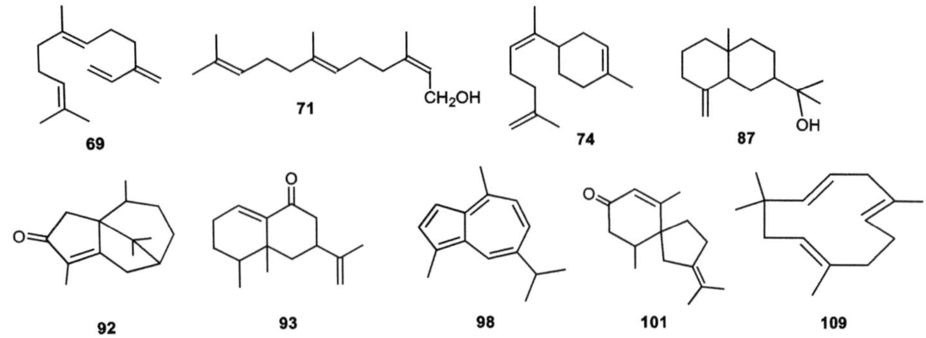

Structure 4.16

4.1.2.1.2.1
Acyclic Sesquiterpenes

β-Farnesene **69** is a constituent of hops oil and many other oils. α-Farnesene **70** is the structural isomer. Structural representations of α-farnesene and β-farnesene are illustrated in Structure 4.17.

Farnesol **71** (Structure 4.18) is widely distributed in flower oils such as rose, acacia and cyclamen.

Nerolidol is isomeric with farnesol and is found in neroli oil and many other oils. Its *E* isomer **72** is more frequently found in nature than its *Z* isomer **73** (Structure 4.19).

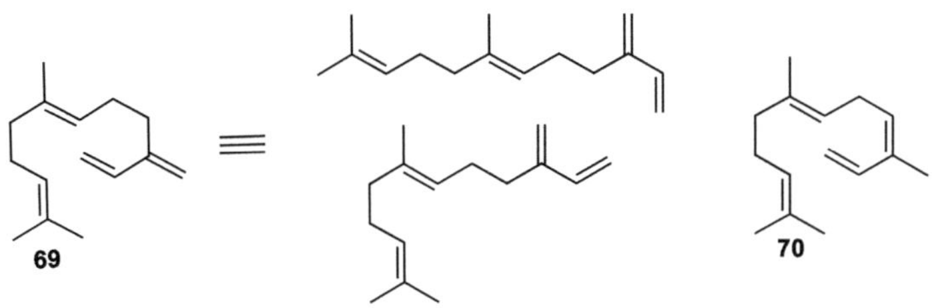

Structure 4.17

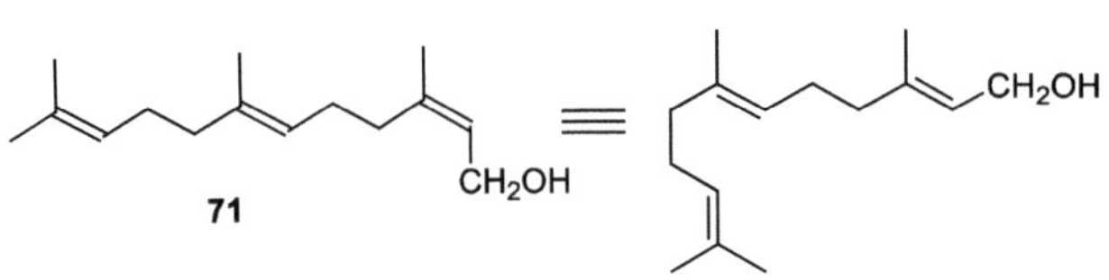

Structure 4.18

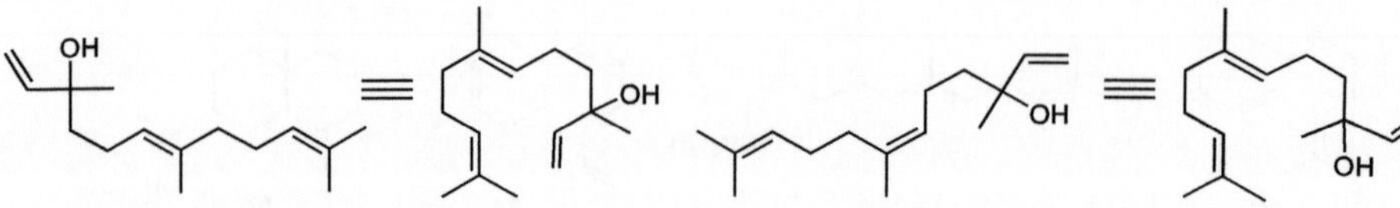

72 73

Structure 4.19

4.1.2.1.2.2
Monocyclic Sesquiterpenes

Bisabolene-type sesquiterpenes, e.g. α-bisabolene **74** (Structure 4.20), are widely distributed in nature. This sesquiterpene hydrocarbon is a constituent of bergamot, myrrh and a wide variety of essential oils. Its oxygenated derivatives α-bisabolol [6-methyl-2-(4-methyl-3-cyclohexen-1-yl)-5-hepten-2-ol] **75** and β-bisabolol [4-methyl-1-(6-methylhept-5-en-2-yl)cyclohex-3-enol] **76** are found abundantly in chamomile.

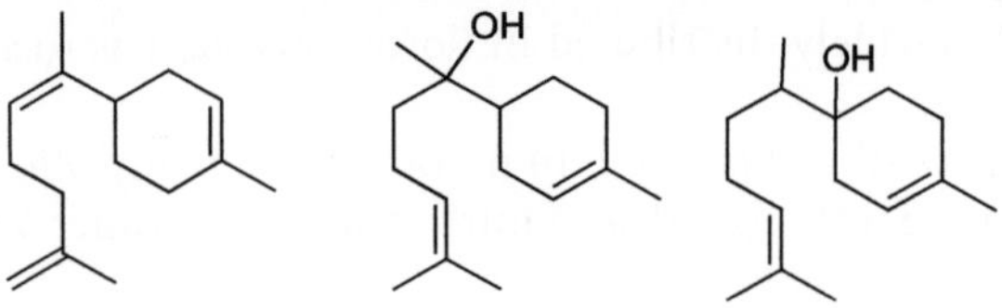

74 75 76

Structure 4.20

Zingiberene [5-(1,5-dimethyl-4-hexenyl)-2-methyl-1,3-cyclohexadien] **77** (Structure 4.21), is a constituent of ginger oil.

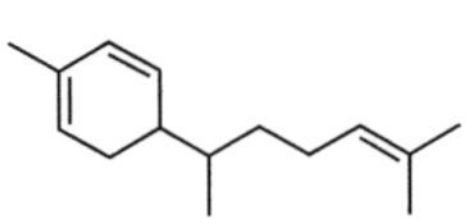

77

Structure 4.21

Lanceol or 2,7(14),10-bisabolatrien-12-ol **78** (Structure 4.22) is a primary alcohol found in the oil of sandalwood (*Santalum lanceolatum*). *Z* and *E* isomers exist.

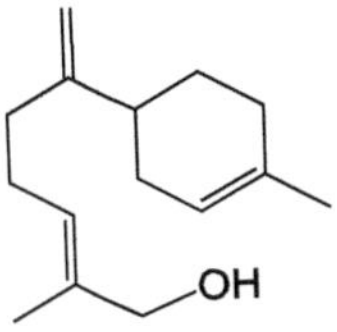

78

Structure 4.22

4.1.2.1.2.3
Bicyclic Sesquiterpenes

Cadinene is a trivial name of a number of isomers which occur in a wide variety of essential oils e.g. cubeb oil. Actually, it is derived from the Cade juniper (*Juniperus oxycedrus* L.). The cadalane (4-isopropyl-1,6-dimethyldecahydronaphthalene) carbon skeleton is the base. Prominent stereochemical isomers are α-cadinene **79**, γ-cadinene **80** and δ-cadinene **81** (Structure 4.23). This group is also known as naphthalene-type sesquiterpenes.

α-Selinene **82**, β-selinene **83**, γ-selinene **84** and δ-selinenes **85** (Structure 4.24) are found in celery oil and many other oils.

α-Eudesmol **86**, β-eudesmol **87** and γ-eudesmol **88** (Structure 4.25) are tertiary alcohols found in many oils. They are practically the oxygenated forms of selinenes.

α-Cyperone **89** (Structure 4.26) is a sesquiterpene ketone found in the essential oil of the tubers of *Cyperus rotundus*. Further hydroxylated derivatives such as α-cyperol **90** and isocyperol **91** can be found along with another ketone with a tricyclic unusual skeleton, namely cyperenone **92**.

The bicyclic ketone eremophilone **93** (Structure 4.27) was first isolated from wood oil of *Eremophila mitchelii*. It is also found in many other oils. *allo*-Eremophilone **94** is also structurally related.

The azulenes are a group of bicyclic sesquiterpenes which are responsible for the blue colour of essential oils. They contain highly conjugated five- and six-membered aromatic carbon rings fused together. Chamazulene **97**, the blue-colouring principle of chamomile oil, is actually formed from matricine **95** during distillation, through the carboxylic acid **96** intermediate, as seen in Fig. 4.6 [1–4], whereas guaiazulene or 1,4-dimethyl-7-isopropylazulene **98** is found in geranium oil.

Vetivanes are sesquiterpene ketones occuring in vetiver oil. Vetivane is basically a spiro[4,5]decane **99** (Structure 4.28). Although structurally different as in the case of α-vetivone **100** and β-vetivone **101**, they are characteristic compounds present in vetiver oil. Analogues such as α-vetispirene **102** and β-vetispirene **103** may occur as well.

The tertiary alcohol guaiol **104** (Structure 4.29), also called 2-(1,2,3,4,5,6,7,8-octahydro-1,4-dimethylazulen-7-yl)propan-2-ol, is found in guaiacum wood oil.

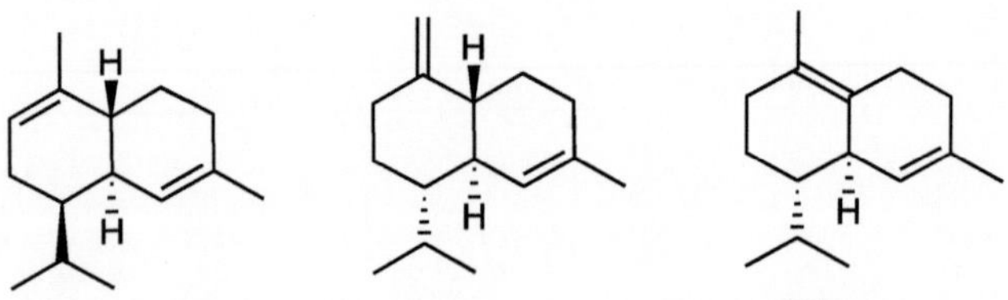

79 80 81

Structure 4.23

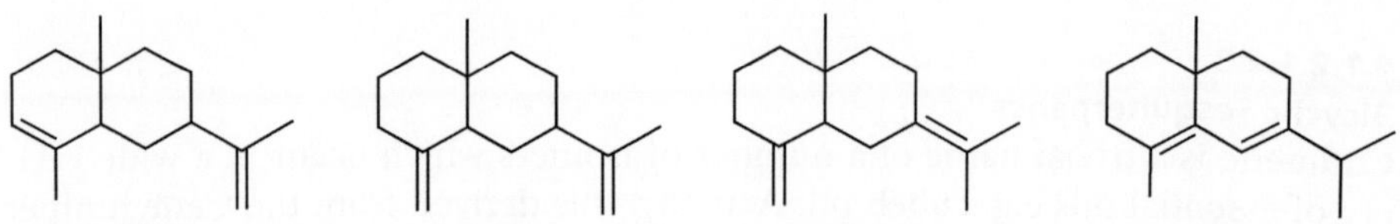

82 83 84 85

Structure 4.24

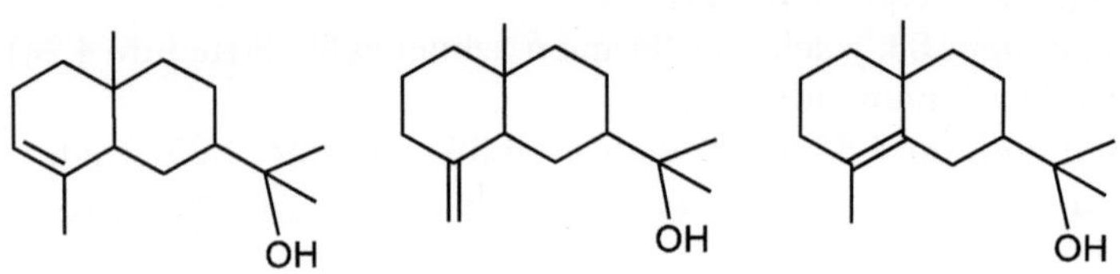

86 87 88

Structure 4.25

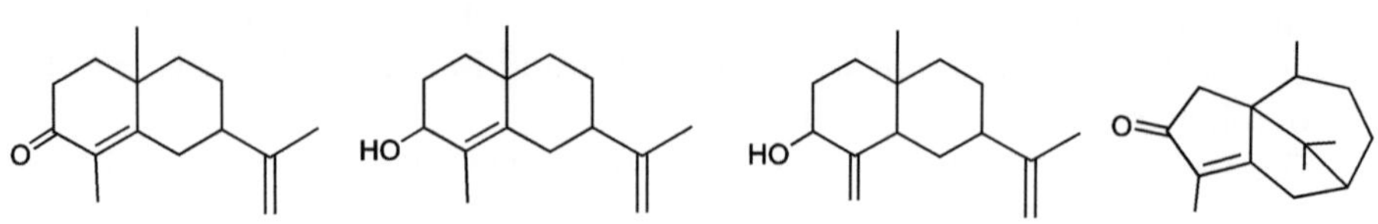

89 90 91 92

Structure 4.26

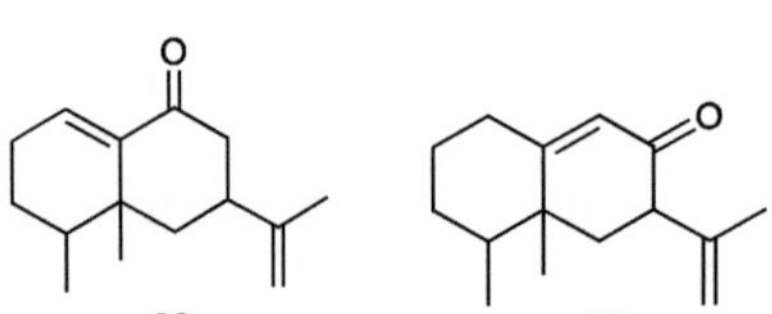

93 94

Structure 4.27

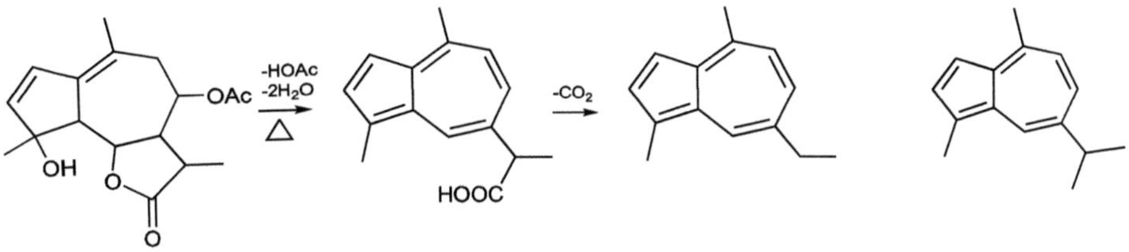

95 96 97 98

Fig. 4.6 Chamazulene chemistry

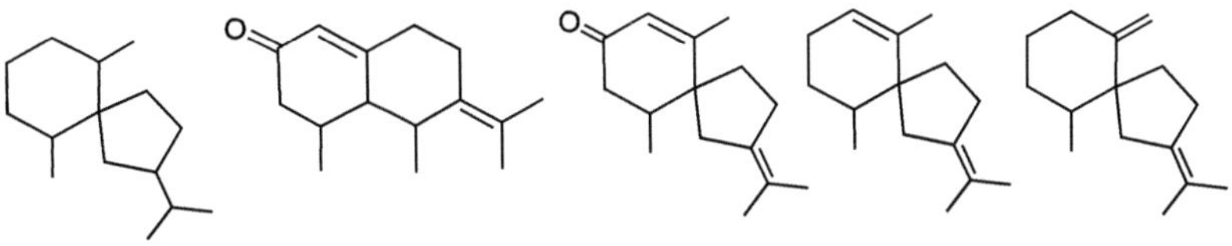

99 100 101 102 103

Structure 4.28

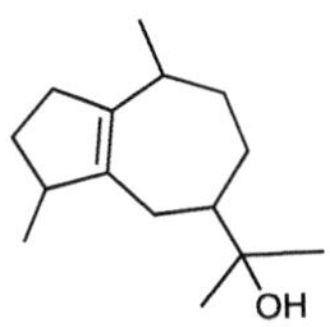

104

Structure 4.29

4.1.2.1.2.4
Miscellaneous Sesquiterpenes

Caryophyllene, a common constituent of essential oils, was first isolated from clove oil. β-Caryophyllene [(*E*)-caryophyllene] **105** (Structure 4.30) is the most widely encountered form of caryophyllenes. Caryophyllene derivatives (**106–108**) are characteristic constituents of most birch oils [49–51].

Humulene **109** (Structure 4.31) is isomeric with caryophyllene. First isolated from hops oil (*Humulus lupulus*), it is a common constituent of essential oils.

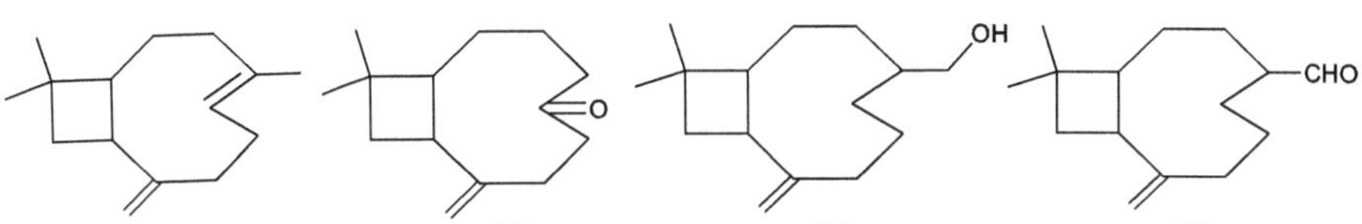

105 106 107 108

Structure 4.30

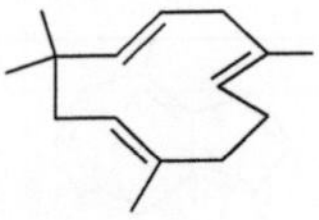

109

Structure 4.31

4.1.2.1.3
Diterpenes

Head-to-tail rearrangement of four isoprene units results in the formation of diterpenes ($C_{20}H_{32}$), as seen also in Fig. 4.2. Diterpenes are generally found in resins, *e.g.* pimaric acid and abietic acid. Some diterpenoids are also constituents of essential oils, e.g. phytol [3, 7–14, 37, 52, 53]. Like sesquiterpenes, diterpenes are heavier than monoterpenes; therefore, they require more energy to go to the vapour phase. For this reason, longer distillation times are necessary for their recovery. The DNP lists 118 different structural types for diterpenoids [37]. Important diterpenes found in essential oils will be detailed. Some representatives of volatile diterpenes are as in Structure 4.32.

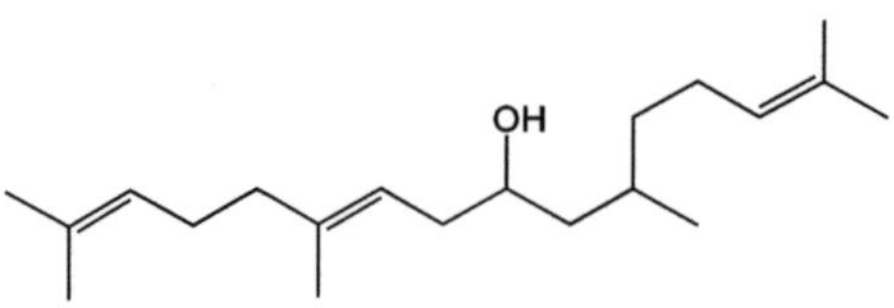

112

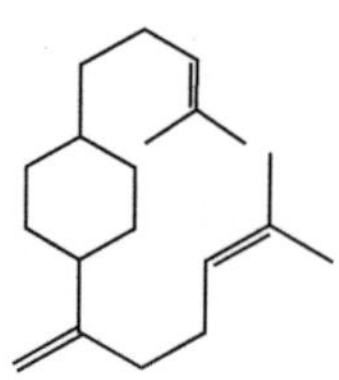

113

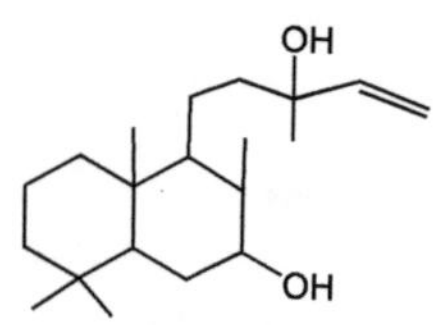

114

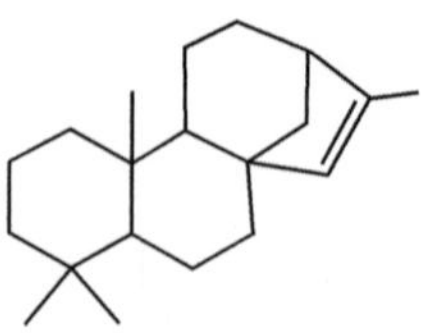

122

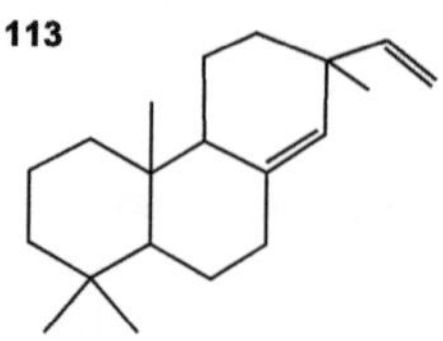

125

Structure 4.32

4.1.2.1.3.1
Acyclic Diterpenes

Phytol, a diterpene alcohol (3,7,11,15-tetramethyl-2-hexadecen-1-ol), occurs in two isomeric forms: *trans*-phytol **110** and *cis*-phytol **111** (Structure 4.33). Phytol was first isolated at the beginning of the nineteenth century during esterification of the chlorophyll molecule. It is a constituent of nettle and many essential oils. Another acyclic diterpene, geranylcitronellol **112**, also occurs in essential oils.

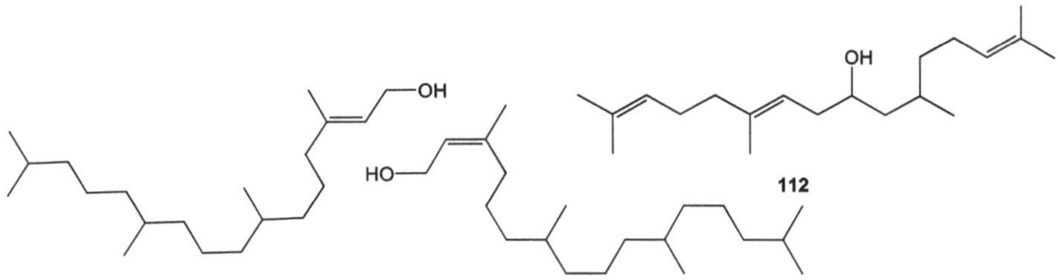

112

110 **111**

Structure 4.33

4.1.2.1.3.2
Cyclic Diterpene

Camphorene **113** (Structure 4.34), which is a constituent of camphor oil, is identical to dimyrcene. Several dimyrcene derivatives are constituents of pistachio oils [54]. The gum resin of *Commiphora mukul* furnishes essential oil (0.4 %) consisting chiefly of myrcene 4 and "dimyrcene" (camphorene 113) [55].

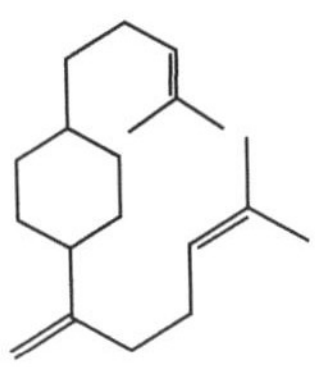

113

Structure 4.34

4.1.2.1.3.3
Bicyclic Diterpenes

Sclareol **114** (Structure 4.35), a ditertiary glycol, is a constituent of clarysage (*Salvia sclarea*) oil [56, 57]. The diterpene ketone sclareolide **115** and the lactone ambrox **116** are important (bio)synthetic derivatives found in clarysage extract.

Manool **117** and manoyl oxide **118** (Structure 4.36) are found in pine oils. They are common diterpenes encountered in many essential oils.

Labdane **119**, abienol **120** and labdanediol **121** (Structure 4.37) are representatives of volatile labdane derivatives.

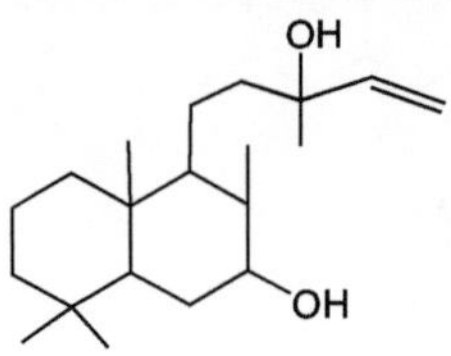

114

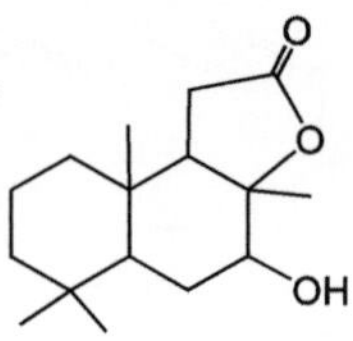

115

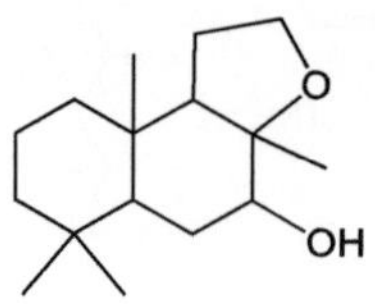

116

Structure 4.35

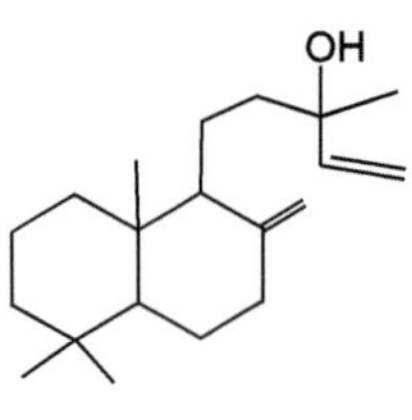

117

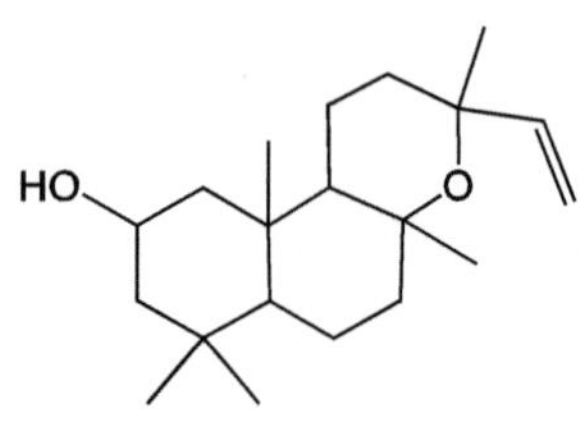

118

Structure 4.36

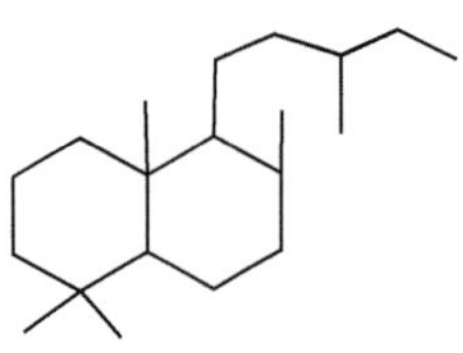

119

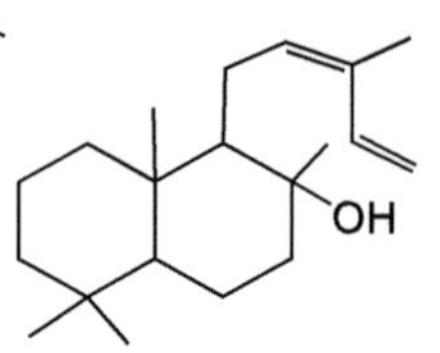

120

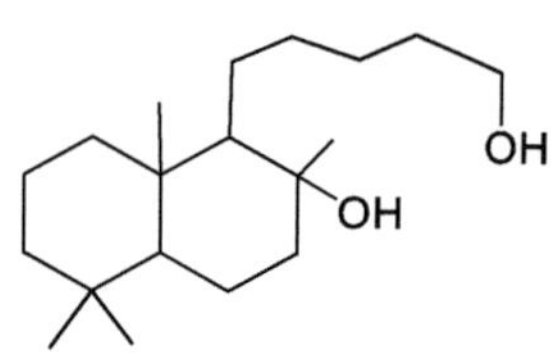

121

Structure 4.37

4.1.2.1.3.4
Tricyclic Diterpenes

Kaur-15-ene **122**, kaur-16-ene **123** and phyllocladene **124** (Structure 4.38) are encountered in essential oils.

The diterpene pimaradiene **125** and sandaracopimaradiene (or isopimaradiene) **126** (Structure 4.39) are found in some essential oils.

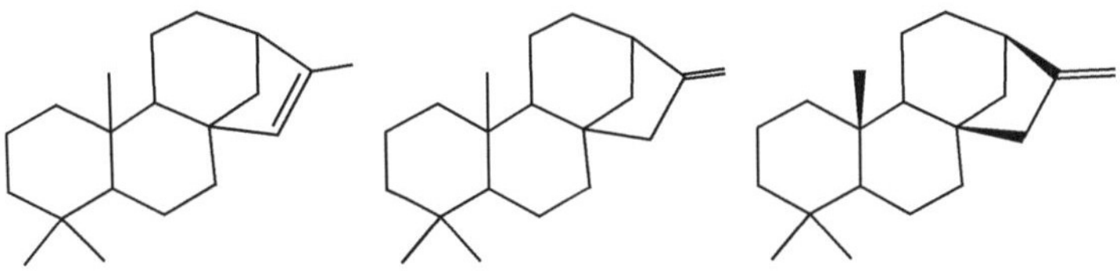

122 **123** **124**

Structure 4.38

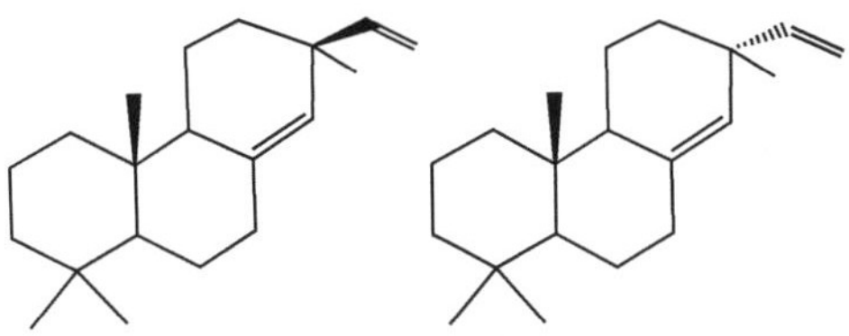

125 **126**

Structure 4.39

4.1.3
C_{13} Norterpenoids

This is a fairly large group of C_{13} compounds generally thought to be degraded carotenoids or catabolites of abscisic acid. *α*-Ionone **127**, *β*-ionone **128**, *γ*-ionone **129**, dihydro-*β*-ionone **130**, (*E*)-geranyl acetone **131** (Structure 4.40), pseudoionones such as *β*-damascenone (3,5,8-megastigmatrien-7-one) **132**, megastigmadienones, megastigmatrienes, edulans such as edulan I **133**, dihydroedulan II **134**, theaspirane **135**, 6-hydroxydihydrotheaspirane **136** (Structure 4.41) and related compounds are found in purple passiflora fruit (*Passiflora edulis*), tea (*Thea sinensis*) and many essential oils [1–4, 8–14, 18–23, 58].

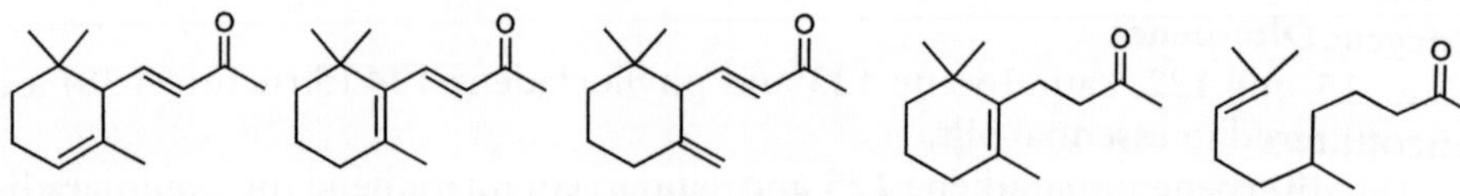

127 128 129 130 131

Structure 4.40

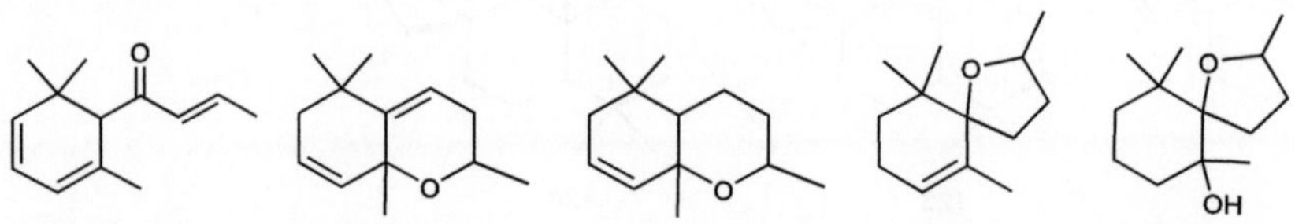

132 133 134 135 136

Structure 4.41

4.1.4
Phenylpropanoids

Phenylpropanoids are biosynthesised by the *shikimic acid pathway* (Fig. 4.7) via the amino acid *l*-phenylalanine by the action of phenylalanine ammonia lyase (PAL), which removes the nitrogen function to generate *trans*-cinnamic acid through which *via* the action of various enzymes, including hydrolases, (ethyl)transferases, oxidoreductases and lygases, a wide range of phenylpropanoids are biosynthesised [12, 41, 59]. Phenylpropanoids contain one or more C_6–C_3 fragments, the C_6 unit being a benzene ring. Simple phenylpropanoids are constituents of essential oils [3, 8, 9, 36]. There is no widely accepted classification method for this class of compounds. Important phenylpropanoids include anethole **137**, methyl chavicol (estragol) **138**, eugenol **139**, cinnamic aldehyde **140** and vanillin **141** (Structure 4.42).

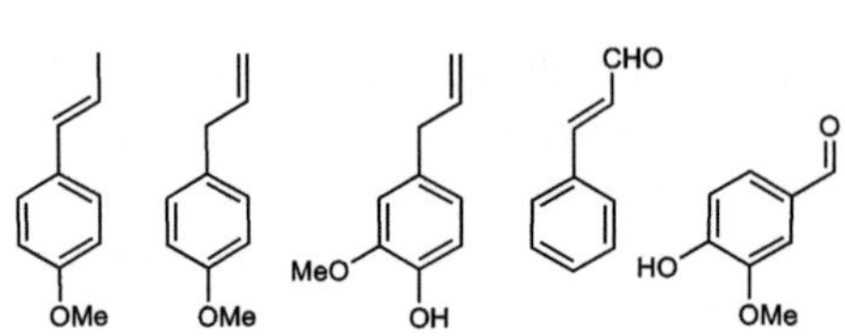

137 138 139 140 141

Structure 4.42

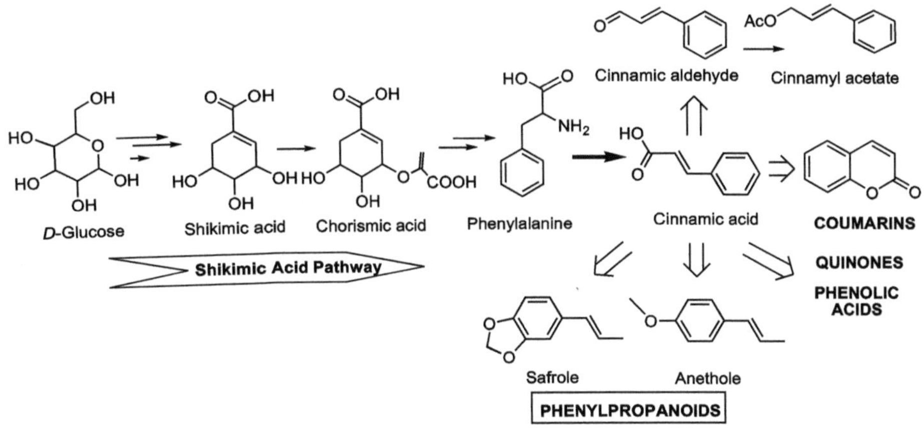

Fig. 4.7 Shikimic acid pathway and volatile phenylpropanoids

4.1.5 Esters

Esters of benzenoid and monoterpenic acids and alcohols as well as unsaturated carboxylic acids such as tiglic acid **142** and angelic acid **143** (Structure 4.43) are found in essential oils [60, 61].

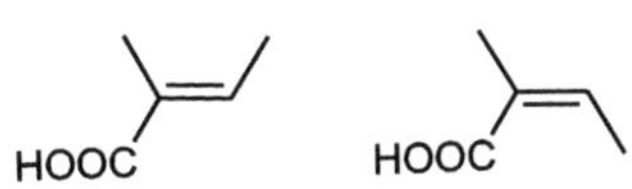

142 **143**

Structure 4.43

Methyl salicylate **144** (Structure 4.44), the main constituent of wintergreen oil, is derived from benzoic acid. Other important esters are linalyl acetate **145**, benzyl benzoate **146** and benzyl isobutyrate **147**.

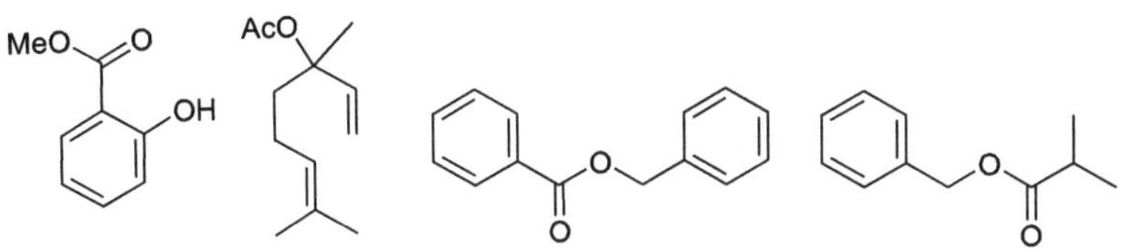

144 **145** **146** **147**

Structure 4.44

4.1.6 Lactones

Lactones are cyclic esters derived from lactic acid ($C_3H_6O_3$). They are constituents of many essential oils and plant volatiles. They contain a heterocyclic oxygen next to a carbonyl function in a five or more membered ring that is saturated or unsaturated. Those with a five-membered ring are called γ-lactones, *e.g.* γ-angelica lactone **148**, whereas compounds containing a six-membered ring are called δ-lactones, *e.g.* δ-valerolactone **149** (Structure 4.45) [1–4, 6, 9, 22, 23, 29, 62].

Some representatives of γ-lactones are γ-valerolactone **150**, γ-decalactone **151** with peach-like flavour, (*Z*)-6-dodecen-4-olide **152**, 3-methyl-4-octanolide (whiskey lactone) **153** and 3-hydroxy-4,5-dimethyl-2(5*H*)-furanone (sotolone) **154** (Structure 4.46), found in fenugreek, coffee and sake [1–4, 21–23, 62].

Additional representatives of six-membered δ-lactones are δ-decalactone **155**, constituent of fruits, cheese and dairy products with creamy-coconut and peachy aroma, jasmolactone **156** as well as δ-2-decenolactone (2-decen-5-olide) **157** (Structure 4.47).

Macrocyclic lactones like ambrettolide (7-hexadecen-1,16-olide) **158**, 15-pentadecanolide (cyclopentadecanolide) **159**, hexadecanolide (cyclohexadecanolide) **160** and cyclohexadec-7-enolide **161** (Structure 4.48) are called musks. They are found in a variety of essential oils, e.g. ambrette seed oil and angelica root oil [1–4, 21–23, 62].

Coumarin **162** (Structure 4.49) is a naturally occurring lactone in crystal form found in hay and tonka beans. It is one of the most used fragrance materials and is responsible for spicy green notes. Dihydrocoumarine **163** is also present in various essential oils with a characteristic sweet herbal odour. Umbelliferone **164**, scopoletin **165**, bergaptene **166** and coumarin are found in Rutaceae, Apiaceae, Lamiaceae and Asteraceae oils. Nepetalactones **167** are confined to the oils of *Nepeta* species [1, 3, 21–23, 63].

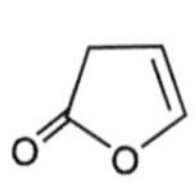
148

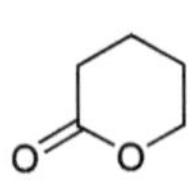
149

Structure 4.45

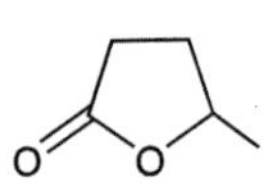
150

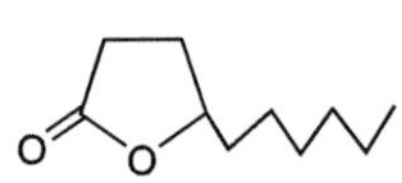
151

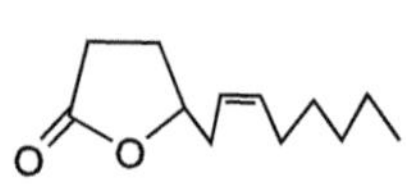
152

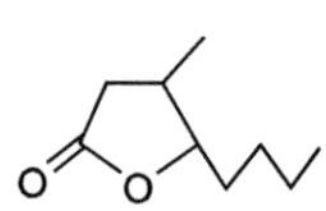
153

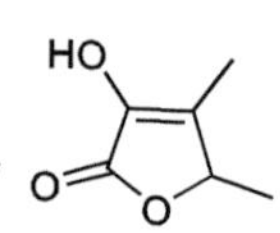

154

Structure 4.46

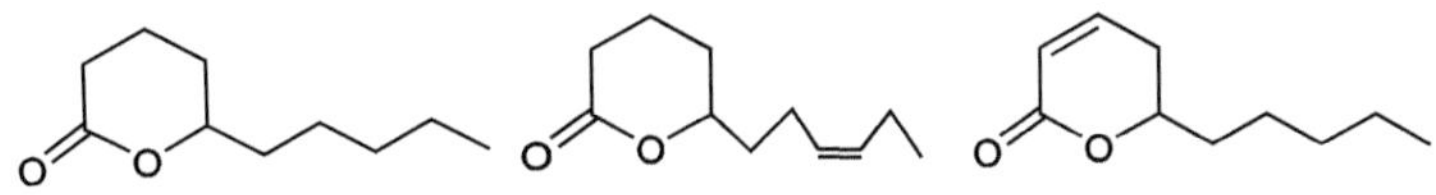

155 156 157

Structure 4.47

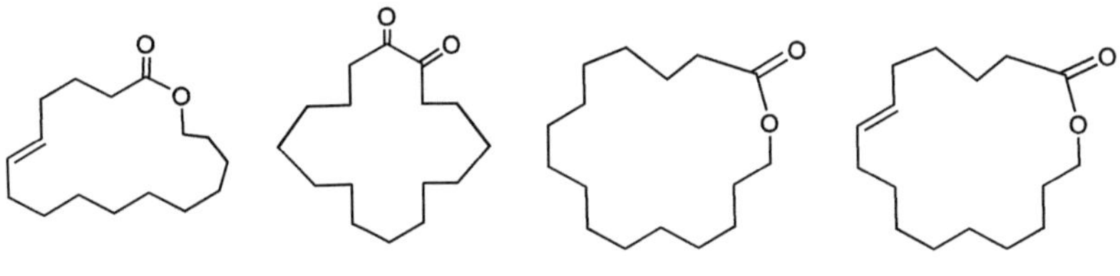

158 159 160 161

Structure 4.48

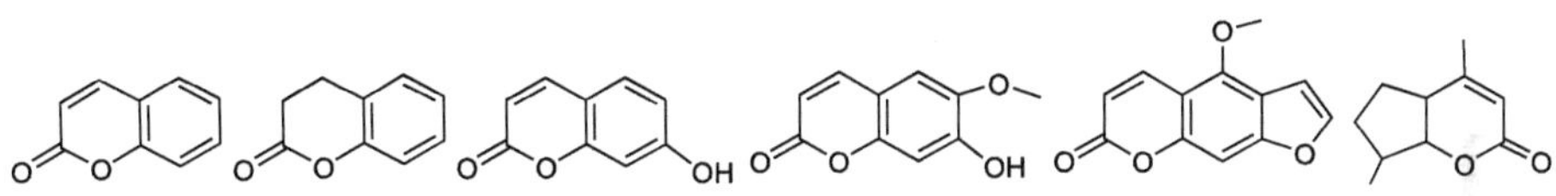

162 163 164 165 166 167

Structure 4.49

4.1.7 Phthalides

Phthalides are lactones of 2-hydroxymethyl benzoic acid. They are also called as benzofuran derivatives. Phthalides are found in some oils of Apiaceae, such

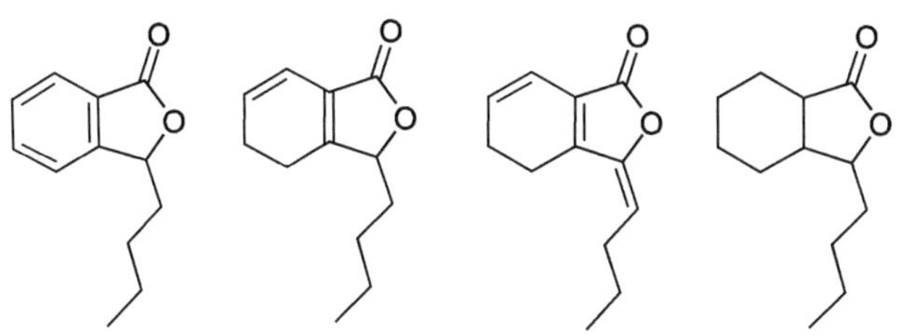

168 169 170 171

Structure 4.50

as celery, lovage and angelica [1–4, 21–23, 25]. Butylphthalides such as 3-butylphthalide **168** (Structure 4.50) are responsible for the celery aroma and odour in leaves, roots, tubers and seeds. The main compound in the oil is sedanolide (3-butyl-4,5-dihydrophthalide) **169**, together with its isomer *cis*-neocnidilide. (*Z*)-Ligustilide also known as 3-butylidene-4,5-dihydro-1(3*H*)-isobenzofuranone and 3-butylidene-4,5-dihydrophthalide **170** is also found along with 3-butylhexahydrophthalide **171** [1–4, 18, 21–23, 25].

4.1.8
Nitrogen-Containing Essential Oil Constituents

Methyl anthranilate **172** (Structure 4.51) is found in the oils of sweet orange, lemon, mandarin, bergamot, neroli and ylang-ylang oils and jasmine and tuberose absolutes. Methyl *N*-methyl anthranilate **173** is the main constituent of mandarin petit grain oil, and occurs also in bitter orange, mandarin and rue oils.

Indole **174** and 3-methyl indole (skatole) **175** (Structure 4.52) are cyclic imines and have a rather unpleasant faecal odour.

2-Methoxy-3-isobutylpyrazine **176** (Structure 4.53) is found in galbanum oil obtained from *Ferula galbaniflua.* 2,4-disubstituted pyridines **177**, N,N-dimethylated amino compounds **178**, alkyl pyrazines **179**, quinoline **180** and methyl quinolines **181** were isolated from fig leaf absolute [64].

Pyridines **177** and pyrazines **179** have been detected in black pepper, sweet orange and vetiver oils [1–4, 21–23, 54, 65].

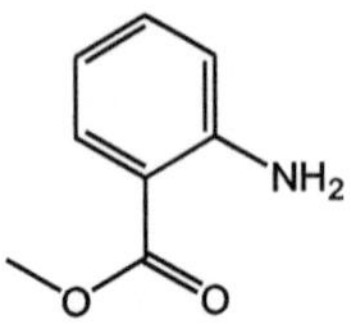

172

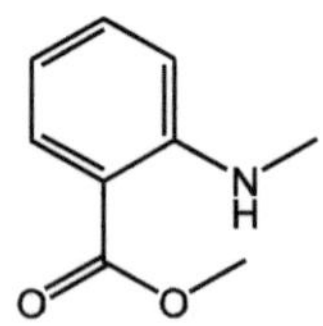

173

Structure 4.51

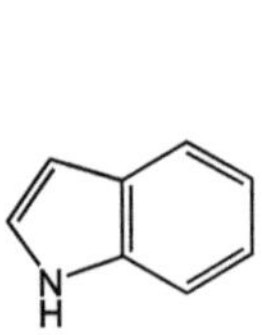

174

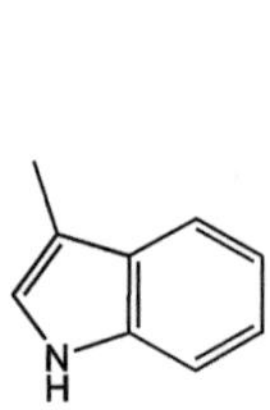

175

Structure 4.52

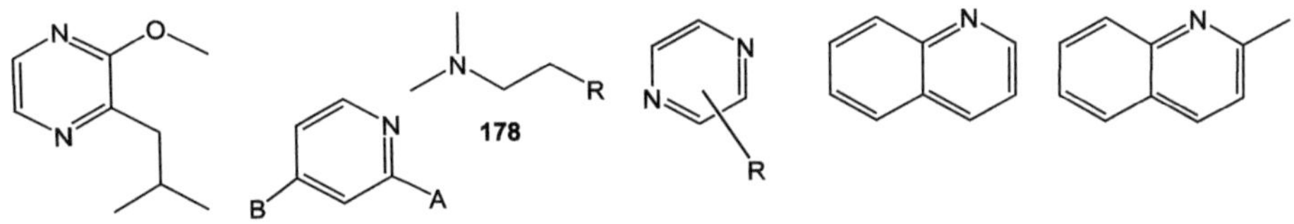

176 177 178 179 180 181

Structure 4.53

4.1.9 Sulphur-Containing Essential Oil Constituents

Several sulphides and thiophenes such as dimethyl sulphide **182**, dimethyl disulphide **183**, diallyl disulphide **184**, and 3,2-dimethylthiophene **185** (Structure 4.54) are volatile constituents of garlic, onion, leek and shallot oils. 4-Mercapto-4-methyl-pentanone **186** is the characteristic component of blackcurrant (*Ribes nigrum*) oil. It has an obnoxious cat-urine smell but in proper dilutions it acquires cassis-like floral and fruity-green aspects [1–4, 21–23, 25, 66]. 8-Mercapto-*p*-menthan-3-one **187** (Structure 4.55), a sulphur derivative of pulegone, is a major constituent of buchu (*Agathosma betulina*) oil together with methylthio and acetylthio derivatives of pulegone and other *p*-menthane molecules [67]. 1-*p*-Menthene-8-thiol **188** is an extremely potent component of grapefruit, orange, yuzu and must oils. 3-Mercaptohexanol **189** derivatives are found in passion fruit flavour. Several *S*-prenylthioesters **190** have been detected in essential oils of Rutaceae genera like *Agathosma* and *Diosma*.

Thiazols **191** (Structure 4.56) were identified in the essential oil of coriander. Sulphur compounds such as dimethyl disulphide, its analogues, rose thiophene (3-methyl-2-prenylthiophene) **192**, the S-analogue of perillene, cyclic disulphides, thiodendrolasin **193**, epithiosesquiterpenes **194**, mint sulphide **195** and isomintsulphide **196** (Structure 4.57) have been detected in rose oil. Mint sulphide occurs in the essential oils of peppermint, spearmint, pepper, ylang ylang, narcissus, geranium, chamomile and davana. Sulphides of humulene **197** and caryophyllene **198** were found in rose and hops oils [1–4, 21–23]. There is a recent review on the comprehensive coverage of sulphur-containing flavour constituents [66].

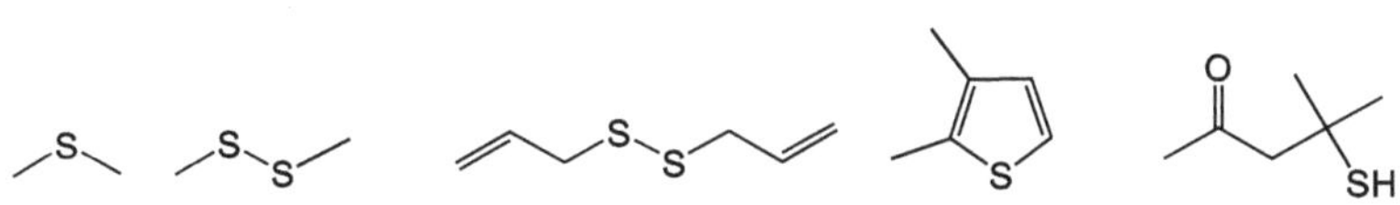

182 183 184 185 186

Structure 4.54

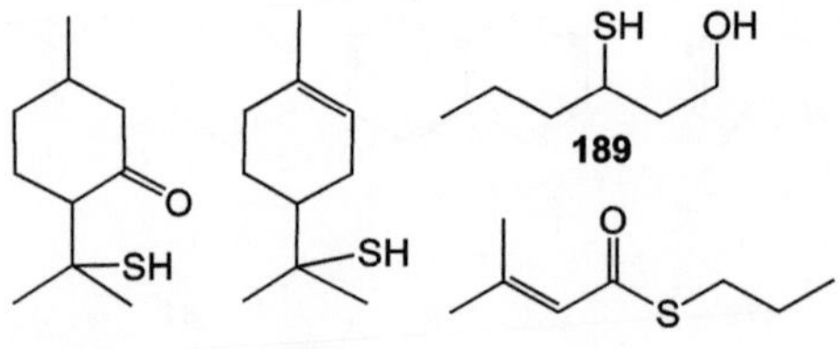

187 188 189 190

Structure 4.55

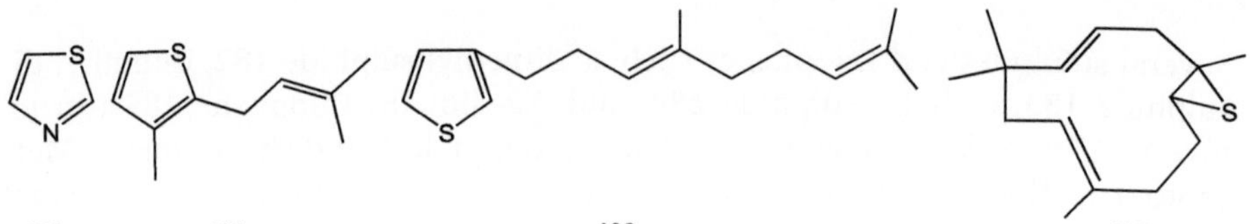

191 192 193 194

Structure 4.56

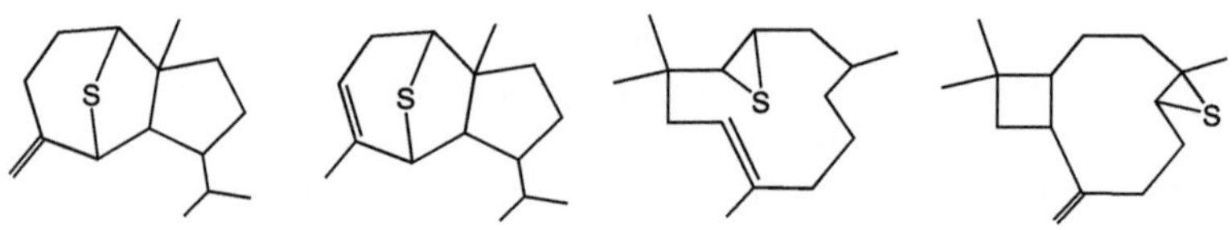

195 196 197 198

Structure 4.57

4.1.10 Isothiocyanates

Isothiocyanates are sulphur- and nitrogen-containing phytochemicals with the general formula R-NSC, *e.g.* phenylethyl isothiocyanate **199**, 3-phenylpropyl isothiocyanate **200** and benzyl isothiocyanate **201** (Structure 4.58). Isothiocyanates occur naturally as glucosinolate conjugates mainly in cruciferous vegetables. Isothiocyanates are also responsible for the typical flavour of these vegetables [1–4, 21–23, 25, 54].

Isothiocyanates can be found in cruciferous vegetables such as mustard, broccoli, cauliflower, kale, turnips, collards, Brussels sprouts, cabbage, radish, turnip and watercress. Glucosinolates are precursors of isothiocyanates along with other metabolites such as thiocyanates, as seen in Fig. 4.8. When the raw vegetables containing glucosinolates are chewed, the plant cells are broken and an enzyme (myrosinase) hydrolyses the glucosinolates into isothiocyanates [1–4, 21–23, 25, 54].

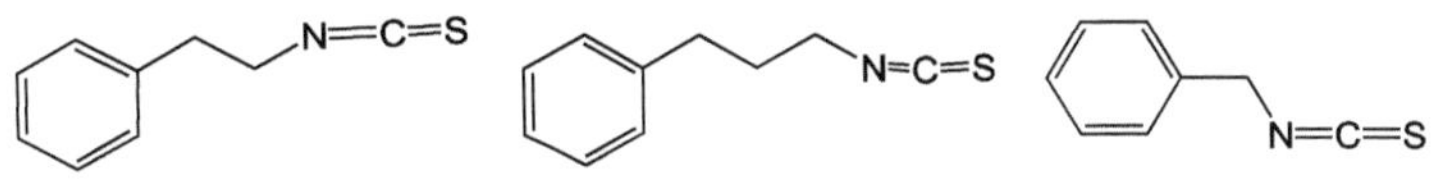

199 200 201

Structure 4.58

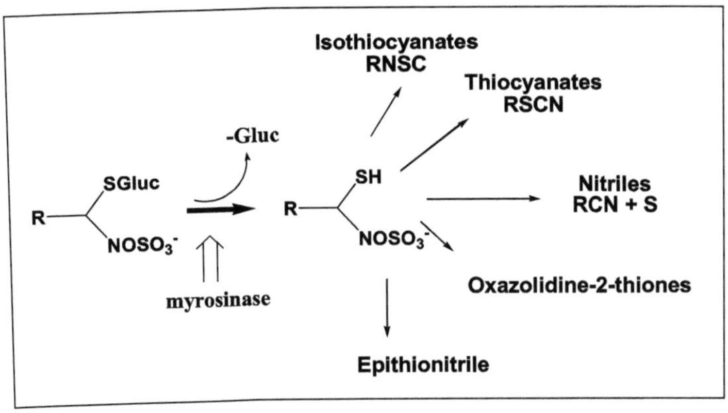

Fig. 4.8 Formation of glucosinolate-derived metabolites

4.2 Impact of Chirality: Enantiomers

Chirality is an important aspect of aroma chemicals since enantiomers of the same compound may possess different organoleptic characters. Chirality means the occurrence of one or more asymmetric carbon atoms in an organic molecule. Such molecules exhibit optical activity and therefore have the ability to rotate plane-polarised light by equal amounts but in opposite directions. In other words, two stereoisomers which are mirror images of each other are said to be enantiomers. If two enantiomers exist in equal proportions, then the compound is called racemic. Enantiomers can be laevorotatory (^{L}, *l*, -, *S*), meaning rotating the plane of the polarised light to the left; or dextrorotatory (^{D}, *d*, +, *R*), that is,

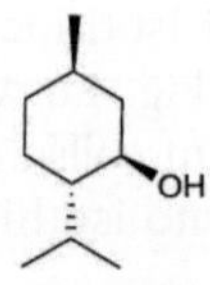

L-menthol **8**
(1*R*,2*S*,5*R*)-2-isopropyl-5-methylcyclohexanol
Minty, fresh, cool

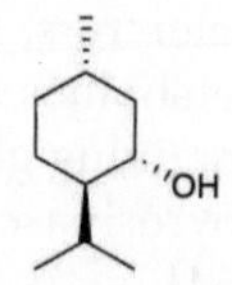

D-menthol **8'**
(1*S*,2*R*,5*S*)-2-isopropyl-5-methylcyclohexanol
Minty, medicinal, camphor-like

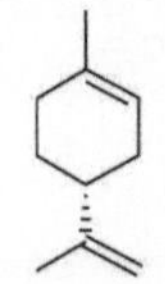

L-limonene **22**
(*R*)-1-methyl-4-(prop-1-en-2-yl)cyclohex-1-ene
Turpentine

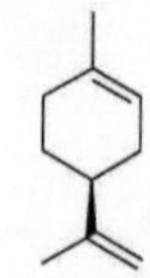

D-limonene **22'**
(*S*)-1-methyl-4-(prop-1-en-2-yl)cyclohex-1-ene
Orange

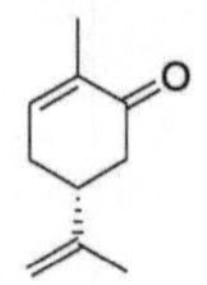

L-carvone **36**
(*R*)-2-methyl-5-(prop-1-en-2-yl)cyclohex-2-enone
Spearmint, costmary

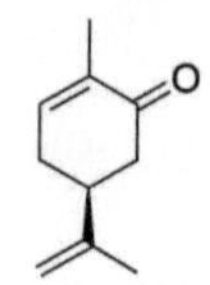

D-carvone **36'**
(*S*)-2-methyl-5-(prop-1-en-2-yl)cyclohex-2-enone
Caraway, dill

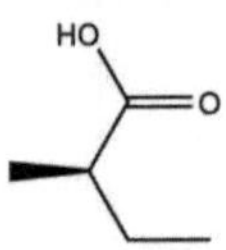

L-2-methylbutanoic acid **201**
(*R*)-2-methylbutanoic acid
Cheesy, sweet

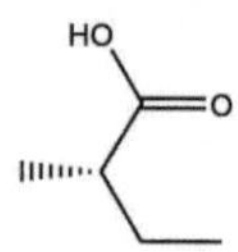

D-2-methylbutanoic acid **201'**
(*S*)-2-methylbutanoic acid
Fruity, sweet

Fig. 4.9 Effect of stereochemistry on flavour and fragrance

rotating the plane of the polarised light to the right. Racemic compounds show zero rotation [1–4, 9, 10, 14, 22, 68–71].

Many natural compounds originating from essential oils which are used in perfumes, flavours and fragrances are optically active. Each enantiomer may display entirely different organoleptic properties. Each enantiomer may be characteristic for a particular essential oil source. Some examples are given in Fig 4.9, illustrating frequently used compounds.

The pattern of distribution of enantiomers may serve as fingerprints to prove the authenticity of a certain essential oil or its adulteration. As high ratio of stereospecificity is achieved in enzyme-catalysed reactions; high enantiomeric purity is expected in chiral natural products. Essential oils generally possess chiral compounds with high enantiomeric purity [70]. "Enantiotaxonomy" can use enantiomeric chemotypes or "enantiotypes" in order to recognise the chemical differences between closely related aromatic plants [72].

Capillary gas chromatography (GC) using modified cyclodextrins as chiral stationary phases is the preferred method for the separation of volatile enantiomers. Fused-silica capillary columns coated with several alkyl or aryl α-cyclodextrin, β-cyclodextrin and γ-cyclodextrin derivatives are suitable to separate most of the volatile chiral compounds. Multidimensional GC (MDGC)–mass spectrometry (MS) allows the separation of essential oil components on an achiral normal phase column and through heart-cutting techniques, the separated components are led to a chiral column for enantiomeric separation. The mass detector ensures the correct identification of the separated components [73]. Preparative chiral GC is suitable for the isolation of enantiomers [5, 73].
The formula for chiral purity is as follows:

$$\frac{A_R}{A_S + A_R} \times 100$$

where A_S is the area of the peak due to the S enantiomer and A_R is the area of the peak due to the R enantiomer.

4.3 Analysis of Essential Oils

Several techniques and criteria are used for the assessment of the quality of essential oils. These are:

1. Sensory evaluations
2. Physical tests
3. Chemical tests
4. Instrumental techniques

Sensory evaluation is carried out by the use of sensory organs and most importantly by the nose. It is considered crucial for the acceptance of an essential oil in perfumery houses. A perfumer or a panel of fragrance experts often have the last word on the acceptance criteria; however, their assessment should be verified and documented by experimental proof [1, 2, 4, 5, 69–73].

Physicochemical tests are required in essential oil monographs published in standards, pharmacopoeias and codices. Chromatospectral techniques are modern methods used to assess the quality of essential oils. The most important technique for the analysis of essential oils is GC. Several detectors may be used in combination with GC. A flame ionisation detector is necessary for quantitative analysis of essential oil constituents. A quadrupole mass detector or an ion-trap detector is indispensable for the characterisation of essential oil constituents. This combination is commonly called GC/MS [1, 2, 4, 5]. This technique is more useful if it is used in conjunction with a reliable computerised library. Several commercial GC/MS libraries exist. Wiley, National Bureau of Standards [74] and National Institute of Standards and Technology libraries [75] contain authentic or keyed-in mass spectra of volatile constituents which may or may not exist in essential oils. The major drawback of such libraries is the lack of retention data; therefore, compounds with identical mass spectra cannot be differentiated. The retention time is the time a compound remains in the column during analysis. The retention index is calculated by a formula and varies with the polarity of a column. Libraries like Adams [76] and MassFinder [77], on the other hand, are specialised libraries for essential oils. They contain the retention index of each compound measured on a non-polar column. Such libraries are more reliable. In case of doubt, coinjection with the suspected compound, checking the retention times in columns with different polarities or isolation and structure elucidation of the compound in question using other spectral techniques may be necessary. The ideal situation is to create a home library if essential oil analysis becomes a major activity. In such a case, mass spectra of known compounds can be entered along with their retention data. It takes several years to create a home library but once created it is more reliable than any other library. We use our own in-house Baser Library of Essential Oil Constituents which contains MS and retention data of over 3,500 genuine compounds found is essential oils.

An atomic emission detector when coupled with GC is capable of separating compounds according to their atoms, such as carbon, hydrogen, oxygen, nitrogen, sulphur and halogens; therefore, it is very useful in detecting compounds containing atoms other than carbon and hydrogen.

MDGC is useful for separating compounds of an essential oil using two columns in line with different polarities. Through column-switching techniques, selected impure compounds in the first column can be diverted to the second column to ensure their complete separation. If the second column is chiral, then enantiomers potentially can be separated. The selected chiral stationary phase affects the resolution and separation drastically [73].

GC/isotope ratio MS and site-specific natural isotope fractionation deuterium NMR spectroscopy are useful more recent tools for detecting sophisticated adulterations [3–5].

Another technique is ^{13}C NMR, which can be successfully utilised in the direct analysis of essential oils without need to separate them by GC [5, 78].

4.4 Conclusions

Essential oils are important natural products used for their flavour and fragrances in food, pharmaceutical and perfumery industries. They are also sources of aroma chemicals, particularly of enantiomers and useful chiral building blocks in syntheses. Biological and pharmacological activities of essential oils and their constituents have been gathering momentum in recent years [79, 80]. Essential oils therefore will continue to be indispensable natural ingredients. The *European Pharmacopeia* contains monographs on 25 essential oils [81]. Many essential oils enjoy generally recognized as safe (GRAS) status. The budding aromatherapy sector is expected to expand the market in coming years.

The compositions of some important essential oils of trade are listed in Table 4.1.

Table 4.1 Composition of important essential oils of trade

Oil	Plant source	Important constituents (%)
Terpene hydrocarbons		
Cade	*Juniperus oxycedrus* L.	Sesquiterpene hydrocarbon (cadinene), guaiacol, cresol
Copaiba	*Copaifera* spp.	Sesquiterpene hydrocarbons: β-caryophyllene (min. 50)
Cypress	*Cupressus sempervirens* L.	Monoterpene hydrocarbons, car-3-ene
Elemi	*Canarium luzonicum* Miq.	Monoterpene hydrocarbons, limonene (40–72), α-phellandrene (10–24) and sesquiterpene alcohol elemol (1–25)
False pepper	*Schinus molle* L.	Fruit oil: Monoterpene hydrocarbons, α-phellandrene (5–26), β-phellandrene (5–7), limonene (4–9) Leaf oil: β-pinene (14), sabinene (13), terpinen-4-ol (11), and sesquiterpene hydrocarbons, bicyclogermacrene (29), germacrene D (12)
Ginger	*Zingiber officinale* Roscoe	Sesquiterpene hydrocarbons, zingiberene (34), β-sesquiphellandrene (12)

CT chemotype

Table 4.1 *(continued)* Composition of important essential oils of trade

Oil	Plant source	Important constituents (%)
Terpene hydrocarbons *(continued)*		
Gurjun balsam	*Dipterocarpus* spp.	Sesquiterpene hydrocarbons, α-gurjunene (min. 60), calarene, α-copaene
Indian curry leaf	*Murraya koenigii* (L.) Spreng.	Sesquiterpene hydrocarbons, β-caryophyllene (29), β-gurjunene (21), α-selinene (13)
Juniper berry	*Juniperus communis* L.	Monoterpene hydrocarbons, pinenes, sabinene, myrcene
Kumquat	*Fortunella japonica* (Thunb.) Swingle	Rind oil: Monoterpene hydrocarbons, limonene (92–95)
Nutmeg	*Myristica fragrans* Houtt.	Monoterpene hydrocarbons, sabinene, pinenes
Opopanax	*Commiphora erythraea* Engl var. *glabrescens* Engl.	Sesquiterpene hydrocarbons, α-santalene, (*E*)-α-bergamotene, (*Z*)-α-bisabolene
Pepper	*Piper nigrum* L.	Monoterpene hydrocarbons (about 80), sabinene (20–25)
Pine silvestris	*Pinus silvestris* L.	Monoterpene hydrocarbons, pinenes, car-3-ene, limonene, myrcene
Sweet orange	*Citrus sinensis* (L.) Osbeck	Monoterpene hydrocarbon, limonene (92–97)
Turpentine	*Pinus* spp.	Monoterpene hydrocarbons, pinenes, camphene
Alcohols		
Ajowan	*Trachyspermum ammi* Spraque	Thymol (4–55)
Amyris	*Amyris balsamifera* L.	Cadinol (50), valerianol (22), cadinene (11), 7-*epi*-γ-eudesmol (11), 10-*epi*-γ-eudesmol (10)
Basil (European type)	*Ocimum basilicum* L.	Linalool (45-62), estragol (trace-30), eugenol (2-15)
Carrot seed	*Daucus carota* L.	Carotol (min. 50)
Cedarwood oil, Chinese	*Cupressus funebris* Endl.	Cedrol (10–16), α-cedrene (13–29), thujopsene (18–31)
Cedarwood oil, Texas	*Juniperus mexicana* Schiede	Cedrol (min. 20), α-cedrene (15–25), thujopsene (25–32)
Cedarwood oil, Virginia	*Juniperus virginiana* L.	Cedrol (5–30), α-cedrene (22–53), thujopsene (10–25)

CT chemotype

Table 4.1 *(continued)* Composition of important essential oils of trade

Oil	Plant source	Important constituents (%)
Alcohols *(continued)*		
Coriander	*Coriandrum sativum* L.	(+)-Linalool (65–78)
Dementholised mint oil	Japanese mint oil	(-)-Menthol (30–50), menthone (17–35), isomenthone (5–13)
Geranium	*Pelargonium* spp.	Citronellol, geraniol
Japanese mint	*Mentha canadensis* L.	(-)-Menthol (about 70)
Matricaria	*Matricaria recutita* L.	(-)-α-Bisabolol (10–65) and bisabolol oxides (29–81) types exist
Neroli	*Citrus aurantium* L. subsp. *aurantium*	(+)-Linalool (28–44), (E)-nerolidol, (E,E)-farnesol, esters
Oregano	*Origanum onites* L., *O. vulgare* L. subsp. *hirtum* (Link) Ietsw. or other *Origanum* spp., *Thymbra spicata* L., *Coridothymus capitatus* Rechb. fil., *Satureja* spp., *Lippia graveolens* Kunth	Carvacrol (min. 60 according to [81])
Palmarosa	*Cymbopogon martini* (Roxb.) W. Wats.	Geraniol (up to 95%)
Patchouli	*Pogostemon cablin* (Blanco) Benth.	(-)-Patchoulol (27–35), nor-patchoulenol (0.4–1)
Peppermint	*Mentha* × *piperita* L.	(-)-Menthol (30–55), menthone(14–32)
Pine, white	*Pinus palustris* Mill.	α-Terpineol (53)
Rose oil	*Rosa* × *damascena* Miller	Citronellol, geraniol, nerol, phenylethyl alcohol
Rosewood	*Aniba rosaeodora* Ducke	(-)-Linalool (up to 86)
Sandalwood, East Indian	*Santalum album* L.	(+)-α-Santalol (45–55), (-)-β-santalol (18–24)
Sweet marjoram	*Origanum majorana* L.	Terpinen-4-ol (min 20), *cis*-sabinene hydrate (3–18)
Tea tree	*Melaleuca alternifolia* (Maiden et Betch) Cheel, *M. linariifolia* Smith, *M. dissitiflora* F. Mueller and other species	Terpinen-4-ol (min. 30), 1,8-cineole (max. 15), γ-terpinene (10–28), α-terpinene (5–13), α-terpineol (1.5–8)
Thyme	*Thymus vulgaris* L., *T. zygis* Loefl. ex L.	Thymol (36–55)

CT chemotype

Table 4.1 *(continued)* Composition of important essential oils of trade

Oil	Plant source	Important constituents (%)
Alcohols *(continued)*		
Vetiver	*Vetiveria zizanoides* (L.) Nash	Sesquiterpene fraction: khusimol (15), α-vetivone and β-vetivone (10)
Esters		
Bergamot	*Citrus aurantium* L. subsp. *bergamia* (Risso et Poit.) Engl.	Linalyl acetate (22–36), linalool (3–15)
Cardamom	*Elettaria cardamomum* Maton	α-Terpinyl acetate (30), 1,8-cineole (30)
Clarysage	*Salvia sclarea* L.	Linalyl acetate (56–78), linalool (6.5–24)
Dwarf pine needle	*Pinus mugo* Turra	Esters calculated as bornyl acetate (1.5–5)
Fir needle, Canadian	*Abies balsamea* Mill.	Esters calculated as bornyl acetate (8–16)
Fir needle, Siberian	*Abies sibirica* Ledeb.	Esters calculated as bornyl acetate (32–44)
Lavandin, abrialis	*Lavandula angustifolia* Mill. × *L. latifolia* Medik.	Linalyl acetate (20–29), linalool (26–38)
Lavandin, grosso	*Lavandula angustifolia* Mill. × *L. latifolia* Medik.	Linalyl acetate (28–38), linalool (24–35)
Lavandin, super	*Lavandula angustifolia* Mill. × *L. latifolia* Medik.	Linalyl acetate (35–47)
Lavender	*Lavandula angustifolia* Miller	Linalyl acetate (25–46), linalool (20–45)
Linaloe	*Bursera* spp.	Linalyl acetate (40–70)
Peru balsam	*Myroxylon pereirae* (Royle) Klotzsch	Benzyl benzoate, benzyl cinnamate
Petitgrain oil, Bigarade	*Citrus aurantium* L. subsp. *Aurantium*	Leaf and twig oil. French: linalyl acetate (51–71), linalool (12–24); Italian: linalyl acetate (51–63), linalool (22–33); Paraguayan: linalyl acetate (40–60), linalool (15–30)
Pine-needle	*Pinus silvestris* L., *P. nigra* Arnold	Esters calculated as bornyl acetate (1.5–5)
Silver fir, European	*Abies alba* Mill.	Esters calculated as bornyl acetate (4–10)
Tolu balsam	*Myroxylon balsamum* (L.) Harms	Benzyl and cinnamyl esters of benzoic and cinnamic acid

CT chemotype

Table 4.1 *(continued)* Composition of important essential oils of trade

Oil	Plant source	Important constituents (%)
Esters *(continued)*		
Valerian	*Valeriana officinalis* L.	(-)-Bornyl acetate
Wintergreen	*Gaultheria procumbens* L.	Methyl salicylate (96-99)
Ylang-ylang	*Cananga odorata* Hook. f. et Thoms.	Benzyl acetate (6–18), geranyl acetate (3–14), *p*-cresyl methyl ether (15–16), methyl benzoate (4–9)
Aldehydes		
Bitter almond	*Prunus amygdalus* Batsch. var. *amara* (DC.) Focke	Benzaldehyde (min. 98)
Cinnamon bark, Ceylon	*Cinnamomum zeylanicum* Nees	Cinnamaldehyde (55–75)
Cinnamon bark, Chinese	*Cinnamomum cassia* Blume	Cinnamaldehyde (70–88), 2-methoxycinnamaldehyde (3–15)
Citronella, Ceylon	*Cymbopogon nardus* (L.) W. Wats.	Citronellal (3–6), geraniol (15–23), citronellol (3–9)
Citronella, Java	*Cymbopogon winterianus* Jowitt.	Citronellal (30–45), geraniol (20–25), citronellol (9–15)
Cumin	*Cuminum cyminum* L.	Cuminaldehyde (20–40), *p*-mentha-1,4-dien-7-al (20–45), *p*-mentha-1,3-dien-7-al (4–12)
Lemon oil	*Citrus limon* (L.) Burman fil.	Geranial (0.5–2), neral (0.2–1.2), limonene (60–80)
Lemongrass, Indian	*Cymbopogon flexuosus* (Nees ex Steud.) W. Wats.	Geranial (35–47), neral (25–35)
Lemongrass, West Indian	*Cymbopogon citratus* (DC.) Stapf	Geranial (40–50), neral (31–40)
Lemon-scented eucalyptus	*Eucalyptus citriodora* Hook.	Citronellal (75)
Litsea cubeba	*Litsea cubeba* C.H. Persoon	Geranial (38–45), neral (25–33)
Ketones		
Armoise	*Artemisia herba-alba* Asso	β-Thujone CT (43–94), camphor CT (40–70); chrysanthnone CT (51), davanone CT (20–70), cis-chrysanthenyl acetate CT (38–71), 1,8-cineole/α-thujone CT (50/27), 1,8-cineole/β-thujone CT (13/12), 1,8-cineole/camphor CT (38/25), *cis*-chrysanthenol CT (25), *cis*-chrysanthenyl acetate CT (25)

CT chemotype

Table 4.1 *(continued)* Composition of important essential oils of trade

Oil	Plant source	Important constituents (%)
Ketones *(continued)*		
Caraway	*Carum carvi* L.	(+)-Carvone (50–65), limonene (30–45)
Common mugwort	*Artemisia vulgaris* L.	α-Thujone (56), 1,8-cineole (27), camphor (20), borneol (19), sabinene (16)
Davana	*Artemisia pallens* Wall.	*cis*-Davanone (38), *trans*-davanone (5)
Dill	*Anethum graveolens* L.	(+)-Carvone (30–40), limonene (30–40), α-phellandrene (10–20), (+)-dill ether (up to 10)
Orris root	*Iris pallida* Lam., *I. germanica* L.	*cis*-γ-Irone (30–40), *cis*-α-irone (20–30) [*I. pallida* oil contains (+) enantiomers; *I. germanica* oil contains (-) enantiomers]
Pennyroyal	*Mentha pulegium* L.	(+)-Pulegone (40–84)
Roman mugwort	*Artemisia pontica* L.	Artemisia ketone (23–46), α-thujone (14–30), 1,8-cineole (12–23)
Sage, Dalmatian	*Salvia officinalis* L.	α-Thujone (18–43), β-thujone (3–9), 18-cineole (6–13), camphor (3–9)
Spearmint	*Mentha spicata* L.	(-)-Carvone (50–80)
Tansy	*Tanacetum vulgare* L.	Thujones (70)
Wormwood	*Artemisia absinthium* L.	There are several chemotypes: (*Z*)-epoxy-ocimene CT (26–54); sabinyl acetate CT (32–85); chrysanthenyl acetate CT (42); β-thujone CT (18–60); β-thujone/(*Z*)-epoxy ocimene CT (21–41/22–29); *cis*-chrysanthenol CT (16–69)
Ethers		
Cajuput	*Melaleuca leucadendron* L.	1,8-Cineole (50–60)
Eucalyptus	*Eucalyptus globulus* Labill.	1,8-Cineole (min. 70); *Eucalyptus globulus* Labill. ssp. *globulus* oil: 1,8-cineole (62–82); *Eucalyptus globulus* Labill. ssp. *maidenii* oil: 1,8-cineole (69–80)
Laurel leaf	*Laurus nobilis* L.	1,8-Cineole (30–70)
Sage, Turkish	*Salvia fruticosa* Mill.	1,8-Cineole (35–51), camphor (7–13)

CT chemotype

Table 4.1 *(continued)* Composition of important essential oils of trade

Oil	Plant source	Important constituents (%)
Phenyl propanoids (phenyl ethers)		
Anis	*Pimpinella anisum* L.	(*E*)-Anethole (87–94)
Basil (Re-union type)	*Ocimum basilicum* L.	Estragol (methyl chavicol) (75–87), linalool (0.5–3)
Bay	*Pimenta racemosa* Moore	Eugenol (44–56), myrcene (20–30), chavicol (8–11)
Bitter fennel	*Foeniculum vulgare* Mill. subsp. *vulgare* var. *vulgare*	(*E*)-Anethole (55–75), fenchone (12–26), limonene (1–5)
Calamus	*Acorus calamus* L.	β-Asarone: diploid variety (0), triploid variety (0–10), tetraploid variety (up to 96%)
Chervil	*Anthriscus cerefolium* (L.) Hoffm.	Estragol (75–80), 1-allyl-2,4-dimethoxy benzene (16–22)
Cinnamon leaf, Ceylon	*Cinnamomum zeylanicum* Nees	Eugenol (70–85)
Clove	*Syzygium aromaticum* (L.) Merill et L.M. Perry	Eugenol (75–88)
India dill	*Anethum sowa* Roxb.	Dill-apiole, limonene, carvone
Parsley seed	*Petroselinum crispum* (Mill.) Nym. ex A.W. Hill	Myristicine (methoxy safrole) (25–50), apiole (dimethoxy safrole) (5–35), 2,3,4,5-tetramethoxy allylbenzene (1–12)
Piper aduncum	*Piper aduncum* L.	Dill-apiole (32–97)
Sassafras, Brazilian	*Ocotea pretiosa* (Nees) Mez.	Safrole (84)
Sassafras, Chinese	*Cinnamomum camphora* Sieb.	High boiling fraction: safrole (80–90)
Star anis	*Illicium verum* Hook fil.	(*E*)-Anethole (86–93)
Sweet fennel	*Foeniculum vulgare* Mill. subsp. *vulgare* var. *dulce*	(*E*)-Anethole (more than 75), fenchone (less than 5)
Tarragon	*Artemisia dracunculus* L.	French tarragon or Italian tarragon oil: β-pinene and sabinene (24–47); Russian tarragon or German tarragon: sabinene(11–47), methyl eugenol (6–36), elemicin (1–60)

CT chemotype

Table 4.1 *(continued)* Composition of important essential oils of trade

Oil	Plant source	Important constituents (%)
Peroxides		
Chenopodium	*Chenopodium ambrosioides* L. var. *anthelminticum* (L.) A. Gray	Ascaridole (60–77)
N- and/or S-containing oils		
Asafoetida	*Ferula foetida* Regel	*R*-2-Butyl-1-propenyl disulphide (a mixture of *E* and *Z* isomers), 1-(1-methylthiopropenyl)-1-propenyl disulphide, 2-butyl-3-methyl-thioallyl disulphide (both as a mixture of diasteromers)
Buchu leaf	*Agathosma betulina* (Bergius) Pillans, *A. crenulata* (L.) Pillans	*trans*-*p*-Menthane-8-thiol-3-one and its *S*-acetate (characteristic minor components), (+)-limonene (10)
Galbanum	*Ferula galbaniflua* Boiss., *F. rubricaulis* Boiss.	2-Methoxy-3-isobutyl pyrazine, 5-*sec*-butyl-3-methyl-2-butenethioate, 1,3,5-undecatriene as minor components, and monoterpene hydrocarbons (75), sesquiterpene hydrocarbons (10), lactones umbellic acid, umbelliferone
Garlic	*Allium sativum* L.	Diallyl disulphide (over 50%)
Mandarin	*Citrus reticulata* Blanco	Methyl *N*-methyl anthranilate (0.3–0.6), limonene (65–75), γ-terpinene (16–22)
Mustard	*Brassica* spp.	Allyl isothiocyanate (over 90)
Onion	*Allium cepa* L.	Methylpropyl disulphide, dipropyl disulphide, propenylpropyl disulphide, 2-hexyl-5-methyl-3(2*H*)-furanone
Lactones		
Ambrette seed	*Hibiscus abelmoschus* L.	(*Z*)-7-Hexadecan-16-olide, ambrettolide (8–9), 5-tetradecen-14-olide, (2*E*,6*E*)-farnesyl acetate (39–59)
Angelica root	*Angelica archangelica* L.	15-Pentadecanolide, 13-tridecanolide as characteristic minor components in addition to terpenoids and sesquiterpenoids (about 90)
Celery seed	*Apium graveolens* L.	3-Butylphthalide and sedanenolide (1.5–11), (+)-limonene (58–79), β-selinene (5–20)

CT chemotype

Table 4.1 *(continued)* Composition of important essential oils of trade

Oil	Plant source	Important constituents (%)
Lactones *(continued)*		
Holy grass	*Hierochloe odorata* L.	Coumarin (10–60)
Lovage root	*Levisticum officinale* Koch	3-Butyl phthalide (32), ligustilide (24), ligusticum lactone
Diterpenes		
Labdanum	*Cistus ladaniferus* L.	Labdane diterpenes and monoterpene hydrocarbons

CT chemotype

References

1. Bauer K, Garbe D, Surburg H (2001) Common Fragrance and Flavor Materials: Preparation, Properties and Uses, 2nd edn. Wiley-VCH, Weinheim
2. Kubeczka KH (1997) Vorkommen und Analytik Ätherischer Öle. Thieme, Stuttgart
3. Dewick PM (2002) Medicinal Natural Products: A Biosynthetic Approach, 2nd edn. Wiley, Chichester
4. Carle R (1993) Ätherische Öle—Anspruch und Wirklichkeit. Wissenschaftliche Verlagsgesellschaft, Stuttgart
5. Baser KHC (1995) In: de Silva KT (ed) A Manual on Essential Oil Industry. UNIDO, Vienna, p 155
6. Breitmaier E (2005) Terpene: Aromen, Dufte, Pharmaka, Pheromone. Wiley-VCH, Weinheim
7. Barton DHR, Meth-Cohn O, Nakanishi K (eds) (1999) Comprehensive natural products chemistry. Isoprenoids Including Carotenoids and Steroids, vol 2. Pergamon, Amsterdam
8. Torssell K (1997) Natural Products Chemistry: A Mechanistic, Biosynthetic and Ecological Approach. Swedish Pharmaceutical Press, Stockholm
9. Evans WC (2002) Trease and Evans' Pharmacognosy, 15th edn. Saunders, London.
10. Mann J, Davidson RS, Hobbs JB, Banthorpe DV, Harborne JB (1994) Natural Products: Their Chemistry and Biological Significance. Longman, London
11. Dev S (1989) In: Rowe JW (ed) Natural Products of Woody Plants, Chemicals Extraneous to the Lignocellulosic Cell Wall, vol 2. Springer, Berlin, Heidelberg New York, pp 691–807
12. Croteau R, Karp F (1991) In: Müller PM, Lamparsky D (eds) Perfumes. Art, Science and Technology. Elsevier, London, pp 101–126
13. Croteau RJ (1986) In: Craker LE, Simon JE (eds) Herbs, Spices, and Medicinal Plants: Recent Advances in Botany, Horticulture, and Pharmacology, vol 1. Food Products, New York, pp 81–133
14. Dey P, Harborne J, Banthorpe D (1991) Methods in Plant Biochemistry. Terpenoids, vol 7. Academic, London
15. Helas G, Slanina J, Steinbrecher R (1997) Biogenic Volatile Organic Compounds in the Atmosphere. SPB, Amsterdam

16. Lawrence BM (2001) Int J Aromather 10:82
17. Craker LE, Simon JE (eds) (1986) Herbs, Spices, and Medicinal Plants: Recent Advances in Botany, Horticulture and Pharmacology, vol 1. Food Products, New York
18. Belitz HD, Grosch, W, Schieberle P (2004) Food Chemistry, 3rd edn. Springer, Berlin Heidelberg New York
19. Hay RKM, Watermann PG (1993) Volatile Oil Crops: Their Biology, Biochemistry and Production. Longman, London
20. McCaskill D, Croteau R (1998) Trends Biotechnol 16:349
21. Pinder AR (1960) The Chemistry of the Terpenes. Wiley, New York
22. Pybus D, Sell C (1999) The Chemistry of Fragrances. Royal Society of Chemistry, London
23. Reineccius G (1999) Source Book of Flavors, 2nd edn. Aspen, Gaithersburg
24. Cheetham PSJ (1997) Adv Biochem Eng Biotechnol 55:1
25. Steward D (2005) The Chemistry of Essential Oils Made Simple. Care Publications, Marble Hill
26. Baser KHC (2005) In: Sacred Incense to Fragrant Elixir: Perfume. YKY, Istanbul, pp 123–125
27. Anonymous (2000) Natural Sources of Flavourings. Report no 1. Council of Europe Publishing, Strasbourg
28. Matheis G (2000) Dragoco Rep 1:23–33
29. Müller PM, Lamparsky D (1991) Perfumes: Art, Science and Technology. Elsevier, London
30. Harlander S (1999) In: Reineccius G (ed) Source Book of Flavors, 2nd edn. Aspen, Gaithersburg, pp 155–175
31. Gabelman A (1994) Bioprocess Production of Flavor, Fragrance and Color Ingredients. Wiley, New York
32. Schreiber WL, Scharpf LG, Katz I (1997) Perfum Flavor 22:11
33. Crabb C (1999) Chem Eng 3:59
34. Tisserand R, Balacs T (1995) Essential Oil Safety: A Guide for Health Care Professionals. Churchill Livingstone, Edinburgh
35. Baytop T (1999) Türkiye'de Bitkiler ile Tedavi, Gecmiste ve Bugün (Therapy with Turkey, Past and Present). Nobel Tıp Basımevi, Istanbul
36. Kurkin VA (2003) Chem Nat Compd 39:123
37. Anonymous (2006) Dictionary of Natural Products on CD-ROM, version. 14.1 1982–2006. Chapman & Hall/CRC, New York
38. Kaufman PB, Csele LJ, Warber S, Duke J.A, Brielmann HL (1998) Natural Products from Plants. CRC, Boca Raton
39. Tetenyi P (1986) In: Craker LE, Simon JE (eds) Herbs, Spices and Medicinal Plants: Recent Advances in Botany, Horticulture and Pharmacology, vol 1. Food Products, New York. pp 11–32
40. Asakawa Y (1995) In: Herz W et al (ed) Progress in the Chemistry of Organic Natural Compounds 65. Springer, Berlin Heidelberg New York
41. Berger RG, Böker, A, Fischer M, Taubert J (1999) In: Teranishi R et al (ed) Flavor Chemistry: 30 Years of Progress. Kluwer/Plenum, New York. pp 229–238
42. Theis N, Lerdau M (2003) Int J Plant Sci 164(3 Suppl):S93
43. Rohdich F, Bacher A, Eisenreich W (2005) Biochem Soc Trans 33:785
44. Steinbacher S, Kaiser J, Eisenreich W, Huber R, Bacher A, Rohdich F (2003) J Biol Chem 278:18401
45. Rohmer M (2003) Pure Appl Chem 75:375

46. Rohmer M (1999) In: Barton D, Nakanishi K (eds) Comprehensive Natural Product Chemistry, vol 2. Isoprenoids Including Carotenoids and Steroids. Elsevier, London, pp 45–67
47. Grayson DH (2000) Nat Prod Rep 17:385
48. Wise ML, Croteau R (1999) In: Barton D, Nakanishi K (eds), Comprehensive Natural Product Chemistry, vol 2. Isoprenoids Including Carotenoids and Steroids. Elsevier, London, pp 97–153
49. Fraga M (2005) Nat Prod Rep 22:465
50. Demirci B, Baser KHC, Demirci F, Hamann MT (2000) J Nat Prod 63:902
51. Klika KD, Demirci B, Salminen J-P, Ovcharenko VV, Vuorela S, Baser KHC, Pihlaja K (2004) Eur J Org Chem 2627
52. Hanson JR (2005) Nat Prod Rep 22:594
53. MacMillan J, Beale MH (1999) In: Barton D, Nakanishi K (eds) Comprehensive Natural Product Chemistry, vol 2. Isoprenoids including Carotenoids and Steroids. Elsevier, London, pp 217–243
54. Boelens MH, van Gemert LJ (1994) Perfum Flavor 19:51
55. Hanus LO, Řezanka T, Dembitsky VM, Moussaieff A (2005) Biomed Pap 149:3
56. Lawrence BM (1989). Perfum Flavor 14: 90
57. Lawrence BM (1994) Proc Remes Rencontres Int Nyons 41
58. Whitfield FB, Last JH (1986) In: Brunke E-J (ed) Progress in Essential Oil Research. de Gruyter, Berlin, pp 3–48
59. Lewinsohn E (1997) In: Baser KHC, Kirimer N (eds) Progress in Essential Oil Research. Anadolu University Press, Eskisehir, pp 219–225
60. Baser KHC (2002) Pure Appl Chem 74:527
61. Baser KHC, Ermin N, Adigüzel N, Aytac Z (1996) J Essent Oil Res 8:297
62. Woda S (2000) In: Green C (ed) Proceedings of the international conference on essential oils and aromas, Florida, 29 October–2 November 2000. IFEAT. London, pp 125–134
63. Baser KHC, Kirimer N., Kurkcuoglu M., Demirci B (2000) Chem Nat Compd 36:356
64. Kaiser R (1986) In: Brunke E-J (ed) Progress in Essential Oil Research. de Gruyter, Berlin, pp 227–239
65. Clery RA, Hammond CJ, Wright AC (2005) J Essent Oil Res 17:591
66. Goeke A (2002) Sulphur Rep 23:243
67. Kaiser R, Lamparsky D, Schudel P (1975) J Agric Food Chem 23:943
68. Linskens HF, Jackson JF (1991) Modern Methods of Plant Analysis, New Series, vol 12. Essential Oils and Waxes. Springer, Berlin Heidelberg New York
69. Teisseire PJ (1994). Chemistry of Fragrant Substances. VCH, New York
70. Werkhoff P, Brennecke S, Bretschneider W, Guntert M, Hopp R, Surburg H (1993) Z Lebensm -Unters Forsch 196:307
71. Woidich H (1992) In: Woidich H, Buchbauer G (eds) Proceedings of the 12th International Congress of Flavours, Fragrances and Essential Oils, Vienna, 4–8 October 1992, vol 2, pp 64–128
72. Ravid U (2006) In: International Symposium The Labiatae: Advances in Production, Biotechnology and Utilization, San Remo, 22–25 February, abstract book, p 23
73. König WA, Hochmuth DH (2004) J Chromatogr Sci 42:423
74. McLafferty FW, Stauffer DB (1989). The Wiley/NBS Registry of Mass Spectral Data. Wiley, New York
75. NIST/EPA/NIH GC-MS Library. http://www.sisweb.com/software/ms/nist.htm

76. Adams RP (2002) Identification of Essential Oil Components by Gas Chromatography/Quadruple Mass Spectroscopy. Allured, Carol Stream
77. Joulain D, König WA, Hochmuth DH (2006) Terpenoids and Related Constituents of Essential Oils. Library of MassFinder 2.1, Hamburg
78. Kubeczka K-H, Formacek V (2002) Essential Oils Analysis by Capillary Gas Chromatography and Carbon-13 NMR Spectroscopy, 2nd edn. Wiley, Chichester
79. Baser KHC (1999) Acta Hortic 503:177
80. Baser KHC (2005) Acta Hortic 676:11
81. Anonymous (2005) European Pharmacopoeia, 5th edn and five supplements (2005–2006). EDQM, Strasbourg

5 Bioactivity of Essential Oils and Their Components

Adolfina R. Koroch, H. Rodolfo Juliani
New Use Agriculture and Natural Plant Products Program,
Cook College, Rutgers University,
59 Dudley Rd, New Brunswick, NJ 08901, USA

Julio A. Zygadlo
Instituto Multidisciplinario de Biologia Vegetal (IMBIV-CONICET),
National University of Córdoba, Argentina

5.1 Introduction

Essential oils (EOs) are secondary metabolites that plants usually synthesized for combating infectious or parasitic agents or generate in response to stress conditions [1]. EOs are aromatic components obtained from different plant parts such as flower, buds, seed, leaves and fruits, and they have been employed for a long time in different industries, mainly in perfumes (fragrances and after-shaves), in food (as flavouring and preservatives) and in pharmaceuticals (therapeutic action) [2].

Ten major EO crops account for 80% of the world market for EOs, the remaining 20% of the world market comprises over 150 crops. The major producers of EOs are developing or emerging countries (Brazil, China, Egypt, India, Mexico, Guatemala and Indonesia), while the major consumers are the industrialized countries (USA, western Europe and Japan). The forecasted annual growth of EO markets of around 4% is thus generating new commercial opportunities for the developing world [3]. The large volumes of EOs produced worldwide and the limited number of species in the world trade show the economic potential of EO plants as new crops [4].

The commercialization of EOs can be targeted around their bioactivity, and in this context the discovery of new uses and applications of EOs will further drive the research and development process [5]. EOs with promising activities are thus reviewed in the present work.

5.2 Antimicrobial Activity

In the last few years, there has been target interest in biologically active compounds, isolated from plant species for the elimination of pathogenic microorganisms, because of the resistance that microorganisms have built against antibiotics [6] or because they are ecologically safe compounds [7].

A wide variety of EOs are known to possess antimicrobial properties and in many cases this activity is due to the presence of active constituents, mainly attributable to isoprenes such as monoterpenes, sesquiterpenes and related alcohols, other hydrocarbons and phenols [8, 9].

The lipophilic character of their hydrocarbon skeleton and the hydrophilic character of their functional groups are of main importance in the antimicrobial action of EO components. Therefore, a rank of activity has been proposed as follows: phenols>aldehydes>ketones>alcohols>esters>hydrocarbons [10].

Some EOs containing phenolic structures, such as carvacrol and thymol, are highly active against a broad spectrum of microorganisms [10-12], including *Shigella* sp. [13]. The importance of the hydroxyl group has been confirmed [9, 14] and the relative position of the hydroxyl group on the phenolic ring does not appear to strongly influence the degree of antibacterial activity [14, 15]; however, it was reported that carvacrol is more active than thymol [9, 16–18]. Furthermore, the significance of the aromatic ring was demonstrated by the lack of activity of menthol [14]. Low activity was observed with components containing only an aromatic ring with alkyl sustituents as in *p*-cymene [9, 13, 19]. However, an aldehyde group with a conjugated double bond and a long hydrocarbon chain link to the aromatic ring might result in a better antibacterial activity [20]. Thus, cinnamaldehyde was highly effective in inhibiting the growth of several strains of bacteria [21] and fungi [22, 23]. Moreover, the strong inhibitory effect against fungi of *Cinnamomum osmophloeum* leaf oil was directly related to cinnamaldehyde content [7, 24].

High antimicrobial and antifungal activities of carvacrol have been reported [17, 25–34] with Gram-positive bacteria being the most sensible germs [35]. Thymol had potential antimicrobial and antifungal properties against plant, animal and human pathogenic fungi [36–38]. When the phenolic group was methylated, components like anethole and estragole still showed antimicrobial activity [8, 39].

EOs rich in 1,8-cineole demonstrated activity against Gram-positive and Gram-negative bacteria [39–43], including *Listeria monocytogenes* [44], against the yeast *Candida albicans* [45, 46] and against phytopathogenic fungi species [47, 48].

The aldehyde citral displayed moderate activity [49–52]. Ketones such as pulegone [53–56], fenchone [39, 57], α-thujone [58] and camphor [48–67] were reported to have antimicrobial activities.

Oxygenated monoterpenes such as menthol and aliphatic alcohols (e.g. linalool) were reported to possess strong to moderate activities against several bacteria [40, 68–73]. The position of the alcohol functional group was found to affect molecular properties of the component, such as a hydrogen-bonding capacity, and hence terpinen-4-ol was active against *Pseudomonas aeruginosa*, while α-terpineol was inactive [8]. The antimicrobial effects of borneol [65, 74, 75] and geraniol [76] were also reported.

Monoterpenes hydrocarbons, such as sabinene [77, 78], terpinenes [12, 31, 32, 79, 80] and limonene [30, 73, 81–83], have also shown antimicrobial properties that appear to have strong to moderate antibacterial activity against Gram-positive bacteria and against pathogenic fungi, but in general weaker activity was observed against Gram-negative bacteria [53, 84].

The bridged bicyclic monoterpenes α-pinene and β-pinene showed considerable antifungal activity [19, 44, 67, 73, 78, 85–90]; however, there is no clear consensus yet as to which pinene isomer is more antimicrobially active [8, 44, 85, 91].

Similarly, EOs that were characterized by high levels of sesquiterpenes, such as 8-cadinene, (*Z*)-β-farnesene, γ-muurolene, spathulenol, hexahydrofarnesyl acetone and α-selinene, exhibited antifungal and antibacterial activity [92, 93]. In addition, caryophyllene oxide has been reported with slight antibacterial activity [55] and was inhibitory to the growth of several agricultural pathogenic fungi [94]. There are reports showing the antimicrobial activity of (*E*)-caryophyllene, [88, 95, 96], cadinanes [79, 97, 98], farnesol [99], α-eudesmol [100], β-eudesmol [101], β-phellandrene [81], biclogermacrene [102] α-cedrene, β-cedrenes and sesquithuriferol [103].

The diterpenes ferruginol and hinokiol [104, 105], geranylgeraniol, teprenone and phytol [106] showed antibacterial activity. β-Hydroxykaurenoic acid produced permeabilisation of the cell membrane of the fungi *Botrytis cinerea* [107, 108].

Antimicrobial activities of garlic and onion oil appeared to be determined by the concentrations of individual constituent sulfides. Sulfides with a single sulfur atom were not active, and sulfides with three or four sulfur atoms were highly inhibitory against the growth of *Candida utilis* and *Staphylococcus aureus* [109, 110].

Usually, major components are mainly responsible for the antibacterial activity in most of the EOs; however, there are some studies where whole EOs have a higher antibacterial activity than the combination of the major isolated components, indicating that minor components are critical to the activity, probably by producing a synergistic effect [111, 112]. The combination of citral with vanillin, thymol, carvacrol or eugenol was demonstrated to have synergistic effects on growth inhibition of *Zygosaccharomyces bailii* [113]. Synergistic activity between carvacrol and thymol [15] and carvacrol and cymene [14, 114] have also been described. Investigation of the two major chemical constituents of *Osmitopsis asteriscoides*, 1,8-cineole and (-)-camphor, both independently and in combination showed that synergistically they have a higher antimicrobial effect on *Candida albicans* than when tested independently [46]. Numerous other examples of synergism have been reported [26, 35, 48, 70, 91, 115, 116].

On the other hand, antagonism was observed in that the activity of different combined components was less than that of the individual components. An antagonistic effect between *p*-cymene, thymol and carvacrol was reported in the oil of *Lippia chevalieri* [38]. It was demonstrated [117] that the physical properties of an aqueous tea tree oil dispersion significantly influenced the actions of the individual components, increasing or reducing antimicrobial efficacy. Thus, non-oxygenated monoterpene hydrocarbons such as γ-terpinene and *p*-cymene appear to create an antagonistic effect with the most active component (terpinen-4-ol) by lowering its aqueous solubility.

It was also reported that there were slight differences in the activity of enantiomers. (*R*)-(+)-Limonene and (*R*)-(+)-carvone were more biologically active than their isomers (*S*)-(-)-limonene and (*S*)-(-)-carvone [115].

The antimicrobial activities and toxicity of terpenes have been documented, but their modes of action are complex and still in some cases unknown. Considering the large number of different groups of chemical compounds present in EOs, it is most likely that their antimicrobial properties are not attributable to one specific mechanism, because of other targets in the cell [118]. Terpenoids are lipophilic agents and consequently disrupt membrane integrity and permeability [14, 119]. Leakage of K^+ ions [99, 119] is usually a sign of damage [120] and is often followed by efflux of cytoplasmic constituents [8, 14, 15, 119]. Terpinen-4-ol inhibited oxidative respiration and induced membrane swelling, increasing its permeability [119]. The antibacterial activity of oregano EO was due to the disruption of membrane integrity, which further affected pH homeostasis and equilibrium of inorganic ions [15]. It has been hypothesized that carvacrol destabilizes the cytoplasmic membrane and, in addition, acts as a proton exchanger, thereby reducing the pH gradient across the cytoplasmic membrane. The resulting collapse of the proton motive force and depletion of the ATP pool eventually leads to cell death [14]. A change in the fatty acid composition of the yeast membrane in *Saccharomyces cerevisiae* with more saturated and fewer unsaturated fatty acids in the membrane was reported after exposure to palmarosa oil [76].

Ergosterol, the predominant sterol in yeast cells, plays an important role in membrane fluidity, permeability and the activity of many membrane-bound enzymes. In terpene-treated cells, ergosterol synthesis was strongly inhibited, and a global upregulation of genes associated with the ergosterol biosynthesis pathway was described in response to terpene toxicity [80, 121].

Different methods to measure the EO activity have been described [10, 122, 123]; however, the diversity of ways of reporting the antibacterial activity of EOs limits comparison between the studies and could lead to duplications [111, 122, 123]. Also, different solvents have been used to facilitate the dispersion of antimicrobial agents in the test media [70, 74, 120], and consequently careful attention should be paid to possible interactive effects of solvents on bactericidal viability [15].

5.3 Antiviral Activity

The development of viral resistance towards antiviral agents enhances the need for new compounds active against viral infections, and therefore natural products may offer a new source of antiviral agents [124].

EO of *Melaleuca alternifolia* and eucalyptus exhibited a high level of antiviral activity against *Herpes simplex virus* type 1 (HSV-1) and *Herpes simplex virus* type 2 (HSV-2) in a viral suspension test [125]. Also, *Santolina insularis* EO had direct antiviral effects on both HSV-1 and HSV-2 and inhibited cell-to-cell transmission of both herpes types [126]. Moreover, it was demonstrated that the incorporation of EOs in multilamellar liposomes greatly improved the antiviral activity against intracellular HSV-1 [127, 128]. EOs from Argentinean aromatic plants exhibited virucidal activity against HSV-1 and Junin virus, and the activity was time- and temperature-dependent [129, 130]. However, the authors were not able to elucidate the nature of the active components of the oils responsible for the inhibitory effect on virions. EOs from *Mentha piperita* [131] and lemon grass [132] had direct virucidal effect against HVS-1. Antiviral activity of EOs against several viruses has been described, such as poliovirus-1 [133], molluscum contagiosum [134], adenoviruses [135] and influenza virus [136].

Isoborneol has been found to be an interesting compound for inhibiting HSV life cycle, on the basis of the specificity of the inhibition of the glycosilation of viral polyptides [137]. Also linalool exhibited the strongest activity against adenoviruses; however, carvone, cineole, β-caryophyllene, farnesol, fenchone, geraniol, β-myrcene and α-thujone did not exhibit activity [135].

A study conducted on EOs from different *Melaleuca* species showed that the EO containing 1,8-cineole and terpinen-4ol exhibited stronger antiviral activity than those with high methyleugenol or 1,8-cineole contents [138].

5.4 Antioxidant Activity

Lipid peroxidation involves the oxidative deterioration of unsaturated fatty acids and the changes resulting from this process. Detrimental events include membrane fragmentation, disruption of membrane-bound enzyme activity, disintegration and swelling of mitochondria and lysosomal lysis. Reactive oxygen species (ROS) may be the causative factor involved in many human degenerative diseases, and antioxidants have been found to have some degree of preventive and therapeutic effects on these disorders. Hydrogen peroxide, one of the main ROS, causes lipid peroxidation and DNA oxidative damage in cells. Vitamins, phenolic compounds and EOs are naturally occurring antioxidants [139, 140]; thus, the commercial development of plants as new sources of antioxidants to enhance health and food preservation is of current interest [141, 142].

The antioxidant activity that some EOs possess is not surprising in view of the presence of phenol groups. It is well known that almost all phenols can function as antioxidants of lipid peroxidation because they trap the chain-carrying lipid peroxyl radicals [143]. Plant phenolics are multifunctional and can act as reducing agents, hydrogen-donating antioxidants and singlet-oxygen quenchers [141, 144]; therefore, dietary antioxidants are needed for diminishing the cumulative effects of oxidative damage [143].

There are numerous antioxidant methods and modifications for the evaluation of antioxidant activity [139, 145–151]. Multiple assays in screening work are highly advisable, considering the chemical complexity of EOs [152].

Many EOs also exhibit antioxidant activity and therefore several studies have been carried out in order to elucidate the activity of the components [139, 153]. For instance, γ-terpinene retarded the peroxidation of linoleic acid [139, 154–156], sabinene showed strong radical-scavenging capacity [139, 157], α-pinene [158] and limonene [146] showed low antioxidant activity in the 2,2-diphenyl-1-picrylhydrazyl (DPPH) test, while terpinene and terpinolene showed high hydrogen-donating capacity against the DPPH radical [146, 150, 155, 158].

The radical-scavenging effect of citronellal showed a strong protective activity in lipid peroxidation processes in a dose-dependent manner [139, 146, 159]. Also, scavenging effects have been described for neral and geranial [146, 152, 160].

Among the oxigenated terpenes, geraniol had a high hydrogen-donating capacity towards the DPPH radical [146] and terpinen-4-ol is a weak antioxidant [146, 149, 158]. Eugenol has been shown to be effective for its scavenging activities against free radicals [160, 162–165], and is more effective than terpinolene [149]. 1,8-Cineole showed scavenger activity [42, 166, 167] and inhibited malonaldehyde formation [168]. However, pro-oxidant activity of linalool and nerolidol has been reported [139].

The monoterpene ketones menthone and isomenthone [159, 166] exhibited OH· radical scavenging activity. Depending on the method employed, different activities for anethole have been reported [153, 169].

At higher concentrations, the antioxidant activities of thymol and carvacrol were close to that of α-tocopherol and were in fact responsible for the antioxidant activity of many EOs which contain them [12, 17, 139, 153, 163, 164, 168, 170–174]. The high potential of phenolic components to scavenge radicals might be explained by their ability to donate a hydrogen atom from their phenolic hydroxyl groups [175].

Germacrene-D, a ten-membered-ring system with three double bonds acting as electron-rich centers, and pinenes and menthadiene of *Xylopia aethiopica* EO showed a significant ability to scavenge superoxide anion radical [176]. EOs with β-caryophyllene as the major compound showed radical-scavenging activity [177].

In many cases, the antioxidant activity of the EOs could not be attributed to the major compounds, and minor compounds might play a significant role in the antioxidant activity, and synergistic effects were reported [158, 171, 176]. For instance, in *Melaleuca* species, EO containing 1,8-cineole (34%) and terpinen-4ol (19%) exhibited stronger antioxidant activity than those with high methyleugenol (97%) or 1,8-cineole (64.30%) contents [138].

The relative effectiveness of antioxidants depends on their antioxidant properties, their concentration, the test system, the emulsion system, the oxidation time and the test method used [155].

5.5 Analgesic Activity

Menthol is a naturally occurring compound of plant origin, and gives plants of the *Mentha* species the typical minty smell and flavour. Menthol is present in the EO of several species of mint plants, such as peppermint and corn mint oil, and it is classified by the US Food and Drug Administration as a topical analgesic [178]. Menthol is a cyclic terpene alcohol with three asymmetric carbon atoms; therefore, it occurs as four pairs of optical isomers named (+)-menthol and (-)-menthol, (+)-neomenthol and (-)-neomenthol, (+)-isomenthol and (-)-isomenthol, and (+)-neoisomenthol and (-)-neoisomenthol. Among the optical isomers, (-)-menthol occurs most widely in nature. It was able to increase the pain threshold, whereas (+)-menthol was completely devoid of any analgesic effect [179]. In contrast to what was observed for the analgesic effect, (+)-menthol and (-)-menthol were able to induce an equiactive anesthetic effect [180]. Applied topically, menthol caused a tingling sensation and a feeling of coolness owing to stimulation of 'cold' receptors by inhibiting Ca^{2+} currents of neuronal membranes [179, 181]. Menthol was able to block voltage-gated Ca^{2+} channels in human neuroblastoma cells [182]. Most research has focused on menthol's effect on cold fibres, where it appeared to accelerate inactivation of L-type Ca^2+ currents [183–185]. The integrity of the central κ-opioid system was fundamental for (-)-menthol antinociception [179]. The ability of a painful stimulus to suppress perception of another one (counterirritation) was assessed for menthol together with other potential analgesics [186, 187]; however, menthol was capable of producing counterirritation when applied in concentrations high enough to cause substantial sensory irritation [188, 189]. Methyl salicylate has been shown to produce significant counterirritation and had a synergic effect with menthol [190]. Menthol, as a topical irritant, may also cause analgesia by reducing the sensitivity of cutaneous apin fibres [191–193]. Earlier psychophysical work on the effects of menthol on thermal perception and heat pain had led to the conclusion that menthol did not desensitize nociceptors [194]. Studies on its supposed antipruritic activity have yielded contradictory results [195–197]. Menthol has shown antitussive activity that might be attributable to its effects on capsaicin-sensitive fibres [198, 199].

Higher analgesic efficacy was exhibited by *Lavandula hybrida* when administration was through the inhalatory route, the noniceptive responses to chemical (writhing test) and thermal (hot plate test) stimuli being significantly reduced [200]. However, linalool and linalyl acetate produced only a scarce or no analgesic effect in the pain models (writhing test and hot-plate test) [201–205]. Although opioidergic neurotransmission seemed to be primarily involved in orally induced analgesia, the cholinergic system appeared to play a significant role in lavender oil analgesia [200]. Another terpene with anticholinesterase activity and an antinocicptive effect was 1,8-cineole [206, 207]. Lavender oil and its principal components, linalool and linalyl acetate [200], and 1,8-cineole

[208] showed antiulcer activities that led to alleviation of pain. The ability of lavender oil to prevent experimental thrombus has been described [209]. The amelioration of gastric microcirculation could be the mechanism underlying the lavender gastroprotection against ethanol injury, which was known to be dependent on microvasculature engulfment in the gastric mucosa [200].

EO of *Lavandula angustifolia* containing 1,8-cineole and borneol as the main components inhibited both phases of the formalin test, reduced the number of abdominal constrictions (writhing test) and suppressed carrageenan-induced paw oedema [211]. EO of *Salvia africana-lutea* and *Dodonaea angustifolia* also showed analgesic activity [211]. The volatile oil of *Cedrus deodara* produced significant inhibition in thewrithing test and the hot-plate reaction in mice [212].

Eucalyptus citriodora, *E. tereticornis*, and *E. globulus* induced analgesic effects in acetic acid induced writhes in mice and hot-plate thermal stimulation in rats [213]. This observation indicated that EOs have both peripheral (writhe reduction) and central (thermal reaction time prolongated) effects. *E. citriodora* contains citronellal as the main component, whereas *E. tereticornis*, and *E. globulus* contain 60–90% of 1,8-cineole; thus, *E. citriodora* EO showed the highest peripheral antinoniceptive effect, whereas *E. tereticornis* EO was the most potent central antinoniceptive substance [214]. Turpentine exudes from *Pinus nigra* subsp. *pallsiana* had a strong analgesic effect when compared with metamizol as a standard analgesic compound [214]. The main components of *Lippia multiflora* EOs (*p*-cymene, thymolacetate and thymol) showed a significant and dose dependent analgesic effect on acetic acid induced writhing in mice [215].

5.6 Digestive Activity

One of the most important uses of many native aromatic plants in popular medicine is for digestive complaints [216]. Some studies suggest that EOs are responsible, at least in part, for the digestive activities of this group of plants, although it is also possible that other components (e.g. caffeic acid esters) also contribute to this activity [217, 218].

Many reports have shown that EOs regulate the digestive process before food reaches the stomach. Lavender and ginger EOs as well as perfumes and strong odours were found to affect gastrointestinal function through activation of the vagus nerve [219, 220] and gastric secretion [221]. The olfactory stimulation generated by lavender oil scent and its main component linalool activated gastric nerves that enhanced food intake and body weight in rodents [222], while grapefruit oil fragrance and its main component limonene showed the opposite effect [223].

Aromatic plants are commonly administered as an infusion or tea, and thus are delivered directly to the site of action, i.e. the gastrointestinal system [216, 224]. Basically, aromatic plants and their EOs exert their digestive action by inhibiting gastric motility (antispasmodics), releasing of bile from the gall bladder (choleretics), inducing the expulsion of gases from the stomach and intestine (carminatives), and more indirectly protecting liver function (hepatoprotectives).

The depressant effect of EOs on smooth muscle in the small intestine is consistent with the therapeutic uses of these aromatic plants as gastrointestinal antispasmodics and carminatives [224]. *In vitro* studies showed that EOs produced the inhibition of gastric motility, and are thus the basis of the treatment of some gastrointestinal disorders [225, 226].

The EOs of *Satureja obovata* (37% camphor, 18% linalool/linlyl acetate) [227], cardamom seed [228], *Acalypha phleoides* (thymol, camphor and γ-terpinene) [229], *Satureja hortensis* (γ-terpinene, carvacrol) [225], *Croton zehnerii* (estragol, anethole) [224], *Croton nepetafolius* (methyleugenol, α-terpineol, 1,8-cineole) [230], *Melissa officinalis* (citral, 60%) [231], *Pelargonium* sp. (citronellol, geraniol, linalool) [232], lavender (linalool/linalyl acetate) [233], *Plectantrus barbatus* (α-pinene, caryophyllene, myrcene) [234], *Pycnocycla spinosa* (14.4% geranyl isopantanoate, 10.6% caryophyllene oxide) [235], *Ferula gummosa* (α-pinene and β-pinene) [226] and peppermint [236] were reported to inhibit gastric motility in isolated segments of rodent intestine.

The EOs reduced the contraction induced by acetylcholine, histamine [226–228, 210, 225, 232, 233], carbachol (muscarinic receptor activator) [237] and 5-hydroxytryptamine [229]. The EOs were found to relax intestinal smooth muscle by reducing the influx of Ca^{2+} [227, 234], K^{+} [210, 224–226, 229, 230] and Ba^{2+} [229, 237]. However, other reports have shown that lavender and geranium EOs were unlikely to act as cationic channel blockers [232]. The activities of the EOs resembled those of dicyclomine and atropine (muscarinic receptor antagonists) and dihydropyridine (calcium antagonist) by producing smooth-muscle relaxation [225, 236].

All these experiments suggested that EOs and their components inhibit muscarinic receptors that block cationic influx and produce smooth-muscle relaxation [238], while *in vivo* studies showed that a commercial peppermint–caraway oil combination had blocking effects on gastroduodenal motility, decreasing the number and amplitude of contractions, thus acting locally to cause smooth-muscle relaxation. All these activities produced symptom-relieving effects in patients suffering from functional dyspepsia [239]. The physiological significance of the inhibition of duodenum mobility was to provide more time to process chyme [240]. The expulsion of gases from stomach and intestine (carminative effect), that was associated with smooth-muscle relaxation, provided additional relief to abdominal complaints (feeling of pressure, heaviness and fullness) [239, 241].

Chemical structure–activity relationships suggested that phenolic monoterpenes (thymol, methyleugenol) seemed to be the most active, followed by alcohols (terpineol) and other oxigenated monoterpenes (1,8-cineole) [225, 229, 230]. Within the monoterpenes, β-pinene was more active than α-pinene [226], and α-pinene was more active than caryophyllene and myrcene [234].

The inhibitory effect of a mixture of α-pinene and β-pinene was reported to be less than the sum of the separate effects [226]. α-Pinene and caryophyllene showed additive effects but did not achieve the maximum effect obtained with the crude oil. The final therapeutic activity was due to the combine effect of several minor constituents of the oil [234].

The choleretic effect induced by EOs that involves the release of bile from the gall bladder is also important for digestion of fats, but this activity of EOs has been less studied in the last decade. The EO of *Salvia desoleana* (1,8-cineole, linalool/linalylacetate and a terpenylacetate), the purified components (linalool and α-terpineol), different chemotypes of the EOs of *Thapsia* sp. (limonene, geranylacetate and methyleugenol), menthol, peppermint oil and a commercial preparation (containing pinenes, camphene, cineole, menthone, menthol and borneol) produced a significant increase in bile secretion [242–244, 252]. *In vitro* studies also showed that *Croton zhenerii* EO increased contractile activity of the bladder in a concentration-dependent manner [224] that could also affect bile release.

Many studies have related the antioxidant activity with liver protection against free radicals [245–247], although other mechanisms also contribute to the hepatoprotective action of EOs and their components [248].

The EO of *Santolina canescens*, its main component santolinediacetylene [249], thymol [250] and *Foeniculum vulgare* (fennel) [251] showed significant hepatoprotective effects against carbon tetrachloride induced hepatotoxicity in rodents. These studies suggested that the protective effect might be mediated through inhibition of lipid peroxidation [249, 250]. Myristicin (the major components of nutmeg EO) exhibited a potent hepatoprotective activity in rats as assessed by marker enzymes of liver injury [248]. The hepatoprotective activity of myristicin might be, at least in part, due to the inhibition of tumour necrosis factor released from macrophages [248]. In *Rosmarinus officinalis*, the hepatoprotective and antimutagenic activities of ethanolic extracts and EO were attributed to the presence of phenolic compounds with high antioxidant activity [253].

Other activities on the gastrointestinal system included antidiarrhoeal and gastroprotective effects. *Satureja hortensis* and *Aloysia triphylla* EOs inhibited castor oil induced diarrhoea in rodents [225, 255]. The EO of lavender and its components (linalool, linalyl acetate) and the EO of *Cryptomeria japonica* (terpin-4-ol and elemol) showed protective activities against acute ethanol/aspirin-induced gastric ulcers in rodents [200, 254].

5.7 Anticarcinogenic Activity

Tumorigenesis is a multistep process that begins with cellular transformation, progresses to hyperproliferation and culminates with the acquisition of invasive potential, angiogenic properties and establishment of metastatic lesions [256].

The major factors for human carcinogenesis are cigarette smoking, industrial emissions, gasoline vapours, infection and inflammation, nutrition and dietary carcinogens. Studies of nutrition and dietary condition will eventually lead to cancer prevention [257–264].

Non-nutrient compounds in the diet have been found to exert inhibitory effects in experimental carcinogenesis [259, 260, 265–269]. Monoterpenes are non-nutritive dietary components found in the EOs of aromatic plants. Several experimental and population-based studies indicate that isoprenoids in the diet play an important role in the ability to avoid cancers [263, 266, 270–276].

Among monoterpenes, perillyl alcohol and *d*-limonene are isoprenoids of great clinical interest. The monocyclic monoterpene limonene, a major component in many citrus EOs, has been used for many years as a flavouring agent, food additive and fragrance [277, 278]. *R*-(+)-limonene exhibited chemopreventive and therapeutic effects against chemically induced mammary tumours in rats [279–281] and metastasis of human gastric cancer [282]. The EO of *Citrus limonum* modulated the apoptosis through the activation of the interleukin-1β-converting enzyme-like caspases [283].

Mechanistic studies revealed that the effects of limonene on cell proliferation and cell cycle progression were preceded by a decrease in cyclin D1 messenger RNA levels [284] and inhibition of posttranslational isoprenylation, rather than through the suppression of cholesterol biosynthesis [271; 279, 285–293]. Limonene and perillyl alcohol and their active serum metabolites inhibit protein isoprenylation [287, 289–291, 294].

Although farnesol did not affect the prenylation of small G-proteins [295], the derivatized forms of farnesol inhibited methyltransferase activity [296–299] and suppressed the prenylation of G-proteins [300].

Limonene was extensively metabolized by a variety of mammalian species [279, 290, 292, 301]. Its principal circulating metabolites identified in the rat were perillic acid and dihydroperillic acid. These components were effective inhibitors of isoprenylation and cellular proliferation *in vitro* [271].

Limonene and perillic acid remarkably reduced the lung metatastatic tumour nodule formation by 65 and 67%, respectively; however, perillyl alcohol was considerably more potent than limonene against breast cancer [284, 302], rat mammary cancer and pancreatic tumours [288]. Phase I studies of *d*-limonene [303, 304] and phase I and phase II [305–311] studies of perillyl alcohol revealed dose-limiting toxicities: nausea, vomiting, anorexia, unpleasant taste and eructation, and thus a maximum tolerated dose for perillyl alcohol was determined [305].

Perillyl alcohol induced apoptosis and was more effective than perillaldehyde at inhibiting the proliferation of human carcinoma cell lines cultured *in vitro* [319]. Perillyl alcohol treatments suppressed cell growth [313–315], reduced cyclin D1 RNA and protein levels and prevented the formation of active cyclin D1 associated with kinase complexes in synchronous cells during the exit of G0 and entry into the cell cycle [284, 316, 317]. In addition, perillyl alcohol treatment induced an increased association of p21 [316–318] with cyclin E-Cdk2 complexes, inhibited the activating phosphorylation of Cdk2 [312, 316, 318–320], initiated apoptosis [321–324] and suppressed small G-protein isoprenylation

[289, 290, 325–328]. All these effects of perillyl alcohol may contribute to the inhibition of the transition out of gap phase (G1) of the cell cycle [271, 284, 317, 329].

Perillyl alcohol represents a novel small molecule that might be effective for treating leukaemia by inducing growth arrest and apoptosis in transformed cells [313]. Blends of isoprenoids suppressed growth of murine melanoma and human leukaemic cells [265, 271].

A phase I clinical trial with limonene indicated its toxic effects in humans; thus, perillyl alcohol is more effective at lower doses [279].

The hydroxylation of limonene affected its chemopreventive potential. The hydroxylated forms carveol, uroterpenol and sobrerol decreased tumour yield, sobrerol being the most potent. These monoterpenes were reported as cancer chemopreventive agents with little or no toxicity [292]. Also, carveol showed chemoprentive activity against carcinogens [293].

The EO of *Syzigium aromaticum* (Myrtaceae), which contains high levels of eugenol, exhibited anticarcinogenic activities and antimutagenic properties [330–336]. Although a single mechanism may not account for chemoprotection exerted by eugenol, it is an effective inducer of detoxifying enzymes [332, 337, 338]. Eugenol is known to inhibit lipid peroxidation by acting as a chain-breaking antioxidant [339, 340], and lipid peroxidation may play a very important role in cell proliferation, especially in tumours [341, 342]; thus, lipid peroxidation control could be a mechanism of action of eugenol as an antitumoral agent. Other reports showed that eugenol is involved in cytotoxic process and can cause apoptotic cell death [343]. Eugenol inhibited the mutagenicity of aflatoxin B1 and *N*-methyl-*N′*-nitro-*N*-nitrosoguanidine [344] and the genotoxicity of cyclophosphamide, procarbazine, *N*-methyl-*N′*-nitro-*N*-nitrosoguanidine and urethane [345].

Carvone prevented chemically induced lung and forestomach carcinoma [346], but had no effect on the lung metatastic tumour growth [347]. Geraniol showed *in vivo* antitumour activity against murine leukaemia, hepatoma and melanoma cells [348, 349]. Geraniol caused 70% inhibition of human colon cancer cell growth, with cells accumulating in the S transition phase of the cell cycle, and concomitant inhibition of DNA synthesis. No signs of cytotoxicity or apoptosis were detected. Geraniol reduced cancer growth by inhibiting polyamine metabolism, which is a process involved in cancer proliferation [350]. Geraniol induced membrane depolarisation with a decrease of membrane resistance owing to local perforation of the cell membrane, caused a 60% reduction of protein kinase C activity and decreased by 50% the amount of active forms of p44/p42 extracellular signal-regulated protein kinases [351]. The combined administration of 5-fluorouracil (20 mg kg^{-1}) and geraniol (150 mg kg^{-1}) caused a 53% reduction of the tumour volume [352]. 3-Hydroxy-3-methylglutarylcoenzyme A (HMG-CoA) reductase catalyses the formation of mevalonate, a precursor of cholesterol that is also required for cell proliferation. Inhibition of mevalonate synthesis could be a useful strategy to impair the growth of malignant cells. Geraniol inhibited HMG-CoA reductase activity in human breast cancer cells, and this effect was closely correlated with the inhibition of cell proliferation [353]. HMG-CoA reductase activity [271] was also inhibited by farnesol and its derivatives [354], as well as by limonene and menthol [355].

The EO of *Matricaria chamomilla* and its main component, the sesquiterpene alcohol α-bisabolol, is considered to be the main component contributing to the mild anti-inflammatory effect of chamomile. Owing to its non-toxic effect in animals, it is widely used in cosmetic preparations [356]. α-Bisabolol was found to have a strong time- and dose-dependent cytotoxic effect on human and rat glioma cells. α-Bisabolol rapidly induced apoptosis through the mitochondrial pathway with no toxic effect on normal glial cells. Glioma is among the most invasive tumours, against which no efficient and non-toxic treatments have so far been reported; thus, α-bisabolol is very promising for the clinical treatment of this highly malignant tumour [357]. The EO of chamomile also inhibited the mutagenic effects induced by daunorubicin and methanesulfonate [358].

Anethole is known to block the nuclear factor kappa B activation process [359] that is linked with cancer proliferation [272, 360]. *trans*-Anethole was also found to inhibit the *in vivo* genotoxicity of xenobionts [345].

Cadalene reduced the incidence of adenomas and inhibited the development of induced lung tumorigenesis in mice [361], while carvacrol inhibited growth of myoblast cells [362]. Menthol exhibited chemopreventive activity against induced rat mammary cancer [363].

Cinnamaldehyde (*Cinnamomum cassia*) is a potent inducer of apoptosis via ROS generation, thereby inducing mitochondrial permeability, depletion of intracellular thiols, activation of caspase-3 and DNA fragmentation [364]. Farnesol was also found to initiate apoptotic cell death [312, 318, 365], while other studies showed that dietary administration of cinnamaldehyde significantly inhibited pulmonary tumorigenesis in mice [366].

The possibility of moderating the response of cells to a particular mutagen by natural substances opens new horizons in cancer control. On this basis, the research for antimutagens could bring about surprises in the discovery of new anticarciongenic substances.

The antimutagenic effect of EOs of *Helichrysum italicum*, *Ledum groenlandicum* and *Ravensara aromatica* could be explained by the interaction of their constituents with cytochrome P450 activation involving in the detoxification system [367].

Linalool showed no toxic or mutagenic effects on erythrocytes and micronucleus [368], or in numerical chromosome aberrations tests [369], indicating that linalool has no potential for carcinogenicity when used as a fragrance ingredient [370]. Linalyl acetate showed neither mutagenic effects in the *Ames* assay nor genotoxic potential [203], nor did it show carcinogenic activity [202, 371]. Coriander oil, dominated by linalool, did not show any significant potential for immunotoxic or neurotoxic effects [370].

Estragole is a natural constituent of a number of plants and their EOs have been widely used in foodstuffs as flavouring agents. Several studies have shown the hepatocarcinogenicity of EOs with estragole and its metabolites [372].

Methyleugenol, a substituted alkenylbenzene found in a variety of food products, caused neoplastic lesions in mice liver. Safrole caused cytotoxicity and genotoxicity in rodents [373]. However, the no-observed-effect level of methyleugenol for rodents was estimated at 10 mg kg^{-1} [374]. The concentrations (1–10 mg kg^{-1}) are approximately 100–1,000 times the anticipated human exposure to these substances. For these reasons it was concluded that the present exposure to methyleugenol and estragole resulting from consumption of food (e.g. spices) does not pose a significant cancer risk. Nevertheless, further studies are needed to define both the nature and the implications of the dose–response curve in rats at low levels of exposure to methyleugenol and estragole [375].

Tumour cells use multiple cell survival pathways to prevail, and thus the terpenes that can suppress multiple pathways have great potential for the treatment of cancer. This review presents evidence that terpenes can be used not only for cancer prevention but also for its treatment.

5.8 Semiochemical Activity

Insect control is becoming difficult because of the development of strains resistant against insecticides, and transgenic varieties [376]. Leaves, flowers, bark and ripe fruits are important for human use and are usually hosts for a wide range of herbivorous insects, and evidence is accumulating that host finding is largely guided by volatile phytochemicals [377–379]. Behaviour-modifying chemicals also have significant potential for commercial application in pest management. In fact, a major impetus for the development of the field of chemical ecology has been generated by the expectation that identified semiochemicals could be used operationally in pest management programmes [380]. Semiochemicals are molecules that carry signals from one organism to another, while pheromones are substances secreted by an individual that induce a specific reaction in another individual of the same species [381].

Gas chromatography linked to electroantennography (EAG) is a technique developed for the identification of a wide range of semiochemicals that could lead to alternative strategies to control economically important insects [376–379].

Male attraction to the female sex pheromone has been studied for the development of environmentally safe control methods. One important drawback of the mating disruption technique is that only male behaviour is affected, so the efficacy of pheromonal methods can be greatly enhanced by compounds that affect also female behaviour [378].

Nine compounds from branches with leaves and green fruit from apple consistently elicited an antennal response in codling female moths (*Cydia pomonella*, Lepidoptera), including methyl salicylate, (*E*)-β-farnesene, β-caryophyllene, 4,8-dimethyl-1,3(*E*)-7-nonatriene, (3*Z*)-hexenol, (*Z*,*E*)-α-farnesene, (*E*,*E*)-α-farnesene, linalool and germacrene D [378].

Straight-chain aliphatic alcohols elicited higher significant EAG responses in *Helicoverpa armigera* (Lepidoptera) female antennae. Hexan-1-ol and hexan-2-ol showed higher responses (hexan-1-ol being dose-dependent) than hexanal, (2*E*)-hexenal and (2*E*)-hexenyl acetate. The responses to ocimene and β-phellandrene were significantly larger than those elicited by the other monoterpenoids. Phenylacetaldehyde and benzaldehyde elicited EAG responses that were significantly larger than those of acetophenone and methyl salicylate, while the corresponding alcohols did not elicit a significant response [376].

Female antennae detected small amounts of (*E*)-β-farnesene, (*Z*,*E*)-α-farnesene, methyl salicylate and germacrene D, while other more abundant compounds, such as (3*Z*)-hexenyl acetate and (*E*)-β-ocimene, gave no significant antennal response [378].

In the weevil *Pissodes notatus* (Coleoptera), single olfactory receptor neurones on the antennae were screened for sensitivity to naturally produced plant volatiles The two most abundant types responded to α-pinene, β-pinene and 3-carene and to isopinocamphone and pinocamphone, respectively. Major as well as minor constituents of plant volatile blends were employed for host and non-host detection, mainly including monoterpenes (bicyclic and monocyclic) [382].

In female *Heliothis* sp. (Lepidoptera) moths, four colocated receptor neurone types were identified, of which three types responded most strongly to the inducible compounds (*E*)-β-ocimene and (*E*,*E*)-α-farnesene. The fourth type responded most strongly to geraniol, which is a common floral volatile [383].

Single receptor neurons on the antennae of tobacco budworm moth responded with high sensitivity and selectivity to germacrene D, suggesting that this component is an important signal for insects in the interaction with plants [384]. Experimental data demonstrated that plants containing germacrene D dispensers had great attractiveness and showed greater ovoposition than plants without them [385].

Single receptor neurons were tuned to a few structurally related components [383–384], while neurons in the antennae of individual insects were more responsive to specific enantiomers, e.g. (+)-linalool [377, 386].

Conifer monoterpenes (mainly α-pinene, β-pinene, myrcene, limonene/phellandrene) elicited antennal responses in tree-killing bark beetles. These components have potential behavioural roles in host location and discrimination [379].

Semiochemicals are being used in commercial products in mass trapping programmes. Only traps baited with ipsenol and/or ipsdienol together with the host volatiles ethanol and α-pinene caught significantly more male and female *Monochamus scutellatus* and *Monochamus clamator* than traps baited with host volatiles alone. Semiochemicals and pheromones thus exhibited synergistic/adding effects [387] and both could be used as the basis of more integrated control strategies [376].

5.9 Other Activities

EOs and their monoterpenes affected bone metabolism when added to the food of rats. It was demonstrated that these lipophilic compounds inhibited bone resorption [388]. It was reported that (2*E*,6*R*)-8-hydroxy-2,6-dimethyl-2-octenoic acid, a novel monoterpene, from *Cistanche salsa* had antiosteoporotic properties [389].

Pine EOs prevented bone loss in an osteoporosis model (ovariectomized rats). The monoterpenes borneol, thymol and camphor directly inhibited osteoclast resorption [388]. It was observed that inactive monoterpenes can be metabolized to their active forms *in vivo*; thus, *cis*-verbenol, a metabolite of α-pinene, inhibited osteoclastic resorption activity, in contrast to the parent compound α-pinene.

Potential activities for the treatment of Alzheimer's disease were demonstrated in a pilot open-label study involving oral administration of the EO of *Salvia lavandulaefolia* Vahl. known as Spanish sage [390].

Chinese angelica (*Angelica sinensis*) is the most important female tonic remedy in Chinese medicine. The effects of angelica EO in three assays in mice (elevated plus maze, light/dark and stress-induced hyperthermia test) suggested that angelica EO exhibited an anxiolytic-like effect [391]. A link to emotion and cognitive performance with the olfactory system was reported [392]. Moreover, the EOs could affect mood, concentration and sleep [393], while other studies had shown that EOs were potentially important to boost the immune system [394, 395].

EOs from different *Lippia alba* chemotypes showed behavioural effects. Greater effects were presented by chemotype 2 (with citral and limonene), while chemotype 1, containing citral, myrcene and limonene, decreased only the number of rearings in the open-field test [396]. The EO of lemon was found to modulate the behavioural and neuronal responses related to nociception, pain and anxiety [397, 398]. Thus, there is widespread and increasing interest in complementary and alternative medicines using EOs [399].

Aloe vera gel enhanced the antiacne properties of *Ocimum gratissimum* L. oil; the oil or its combination with *Aloe vera* gel was more effective than 1% clindamycin in the treatment of *Acne vulgaris* [399]. Linalool-rich EO was potent against promastigotes and amastigotes of *Leishmania amazonensis* [400].

5.10 Conclusions

The present review demonstrates that EOs and their components have many functional properties and exert their action in mammals as well as in other organisms (insects, fungi, bacteria and viruses). The synergistic effect of EO components is a promising field that could lead to the optimisation of a given bioactivity. This phenomenon has been observed in many activities, such as those of antimicrobials, antioxidants, analgesics and semiochemicals. EOs are complex mixtures of components that show higher activities than their isolated components; their final activities are due to the combine effects of several minor components. Thus, EOs contain multifunctional components that exert their activities through different mechanisms. EOs and their components may have new applications against various diseases of different origins (cancer, fungal, bacterial or viral), because some of these complex diseases require multiple components and multifunctional therapies.

The natural product industry is actively seeking natural therapeutics, preservatives, repellents and other agents that can replace synthetic compounds. The scientific literature has identified new applications and uses of both traditional and exotic EOs. These applications can ultimately assist growers and rural communities in the developing world to increase interest in their products.

Acknowledgements

We thank the New Jersey Agricultural Experiment Station and Cook College for their support and the New Use Agriculture and Natural Plant Products Program. We also acknowledge support from the National Council for Scientific and Technical Research from Argentina (CONICET).

References

1. Rauha J-P, Remes S, Heinonen M, Hopia A, Kahkonen M, Kujala T, Pihlaja K, Vuorela H, Vuorela P (2000) Int J Food Microbiol 56:3
2. Zygadlo JA, Juliani HR (2000) Curr Top Phytochem 3:203
3. Anonymous (2003) Alberta Agriculture, Food and Rural Development, Edmonton
4. Lawrence BM (1993) In: Janick J, Simon JE (eds) New Crops. Wiley, New York, p 620
5. Simon JE, Juliani HR, Letchamo W, Smith M, Goliath J, Graven E, Lewinsohn E, Lewinsohn D, Deschamps C, Dudareva N, Morales M, Picerksky E, Gang D, Rotter D, Raskin I, Vieira R, Ranarivelo L, Rasoanaivo P, Acqueye D (2002) New Aromatic Essential Oils. 5th National Symposium. New Crops and New Uses: Strength in Diversity, Atlanta, Georgia, 10–13 November
6. Essawi T, Srour M (2000) J Ethnopharmacol 70:343
7. Lee HC, Cheng SS, Chang ST (2005) J Sci Food Agric 85:2047
8. Griffin SG, Wyllie SG, Markham JL, Leach DN (1999) Flavour Fragrance J 14:322
9. Dorman HJD, Deans SG (2000) J Appl Microbiol 88:308
10. Kalemba D, Kunicka A (2003) Curr Med Chem 10:813
11. Daferera DJ, Ziogas BN, Polissiou MG (2000) J Agric Food Chem 48:2576
12. Gulluce M, Sokmen M, Daferera D, Agar G, Ozkan H, Kartal N, Polissiou M, Sokmen A, Sahin F (2003) J Agric Food Chem 51:3958
13. Bagamboula CF, Uyttendaele M, Debevere J (2004) Food Microbiol 21:33

14. Ultee A, Bennik MHJ, Moezelaar R (2002) Appl Environ Microb 68:1561
15. Lambert RJW, Skandamis PN, Coote PJ, Nychas G-JE (2001) J Appl Microbiol 91:453
16. Salgueiro LR, Cavaleiro C, Goncalves MJ, da Cunha AP (2003) Planta Med 69:80
17. Tepe B, Daferera D, Sokmen M, Polissiou M, Sokmen A (2004) J Agric Food Chem 52:1132
18. Friedman M, Henika PR, Mandrell RE (2002) J Food Protect 65:1545
19. Aligiannis N, Kalpoutzakis E, Mitaku S, Chinou IB (2001) J Agric Food Chem 49:4168
20. Chang ST, Chen PF, Chang SC (2001) J Ethnopharmacol 77:123
21. Ali N, Mohtar M, Shaari K, Rahmanii M, Ali AM, bin Jantan I (2002) J Essent Oil Res 14:135
22. Simic A, Sokovic MD, Ristic M, Grujic-Jovanovic S, Vukojevic J, Marin PD (2004) Phytother Res 18:713
23. Ranasinghe L, Jayawardena B, Abeywickrama K (2002) Lett Appl Microbiol 35:208
24. Wang SY, Chen PF, Chang ST (2005) Bioresource Technol 96:813
25. Sokovic M, Tzakou O, Pitarokili D, Couladis M (2002) Nahrung-Food 46:317
26. Baydar H, Sagdic O, Ozkan G, Karadogan T (2004) Food Control 15:169
27. Aligiannis N, Kalpoutzakis E, Chinou IB, Mitakou S, Gikas E, Tsarbopoulos A (2001) J Agric Food Chem 49:811
28. Karaman S, Digrak M, Ravid U, Ilcim A (2001) J Ethnopharmacol 76:183
29. Hernandez T, Canales M, Avila JG, Garcia AM, Martinez A, Caballero J, de Vivar AR, Lira R (2005) J Ethnopharmacol 96:551
30. Skocibusic M, Bezic N (2004) Phytother Res 18:967
31. Biavati B, Ozcan M, Piccagilia R (2004) Ann Microbiol 54:393
32. dos Santos FJB, Arimateia J, Lopes D, Cito AMGL, de Oliveira EH, de Lima SG, Reis FD (2004) J Essent Oil Res 16:504
33. Chorianopoulos N, Kalpoutzakis E, Aligiannis N, Mitaku S, Nychas GJ, Haroutounian SA (2004) J Agric Food Chem 52:8261
34. Mirjana S, Nada B (2004) J Essent Oil Res 16:387
35. Marino M, Bersani C, Comi G (2001) Int J Food Microbiol 67:187
36. Sonboli A, Fakhari A, Kanani MR, Yousefzadi M (2004) Z Naturforsch C 59:777
37. Couladis M, Tzakou O, Kujundzic S, Sokovic M, Mimica-Dukic N (2004) Phytother Res 18:40
38. Bassole IHN, Ouattara AS, Nebie R, Ouattara CAT, Kabore ZI, Traore SA (2003) Phytochemistry 62:209
39. Dadalioglu I, Evrendilek GA (2004) J Agric Food Chem 52:8255
40. Tzakou O, Pitarokili D, Chinou IB, Harvala C (2001) Planta Med 67:81
41. Cha JD, Jeong MR, Jeong SI, Moon SE, Kim JY, Kil BS, Song YH (2005) Planta Med 71:186
42. Tepe B, Donmez E, Unlu M, Candan F, Daferera D, Vardar-Unlu G, Polissiou M, Sokmen A (2004) Food Chem 84:519
43. Kalpoutzakis E, Aligiannis N, Mentis A, Mitaku S, Charvala C (2001) Planta Med 67:880
44. Mourey A, Canillac N (2002) Food Control 13:289
45. Unlu M, Daferera D, Donmez E, Polissiou M, Tepe B, Sokmen A (2002) J Ethnopharmacol 83:117
46. Viljoen A, van Vuuren S, Ernst E, Klepser M, Demirci B, Baser H, van Wyk BE (2003) J Ethnopharmacol 88:137
47. Kordali S, Cakir A, Mavi A, Kilic H, Yildirim A (2005) J Agric Food Chem 53:1408
48. Pitarokili D, Tzakou O, Loukis A, Harvala C (2003) J Agric Food Chem 51:3294

49. Belletti N, Ndaguimana M, Sisto C, Guerzoni ME, Lanciotti R, Gardini F (2004) J Agric Food Chem 52:6932
50. Araujo C, Sousa MJ, Ferreira MF, Leao C (2003) J Food Protect 66:625
51. Ngassapa O, Runyoro DKB, Harvala E, Chinou IB (2003) Flavour Fragrance J 18:221
52. Martins AP, Salgueiro L, Goncalves MJ, da Cunha AP, Vila R, Canigueral S, Mazzoni V, Tomi F, Casanova J (2001) Planta Med 67:580
53. Tepe B, Sokmen M, Sokmen A, Daferera D, Polissiou M (2005) J Food Eng 69:335
54. Oumzil H, Ghoulami S, Rhajaoui M, Ilidrissi A, Fkih-Tetouani S, Faid M, Benjouad A (2002) Phytother Res 16:727
55. Sibanda S, Chigwada G, Poole M, Gwebu ET, Noletto JA, Schmidt JM, Rea AI, Setzer WN (2004) J Ethnopharmacol 92:107
56. Duru ME, Ozturk M, Ugur A, Ceylan O (2004) J Ethnopharmacol 94:43
57. Bouzouita N, Kachouri F, Hamdi M, Chaabouni MM, Ben Aissa R, Zgoulli S, Thonart R, Carlier A, Marlier M, Lognay GC (2005) J Essent Oil Res 17:584
58. Baser KHC, Demirci B, Demirci F, Kocak S, Akinci C, Malyer H, Guleryuz G (2002) Planta Med 68:941
59. Soylu EM, Yigitbas H, Tok FM, Soylu S, Kurt S, Baysal O, Kaya AD (2005) Z Pflanzenk Pflanzen 112:229
60. Bezic N, Skocibusic M, Dunkic V, Radonic A (2003) Phytother Res 17:1037
61. Zhu SY, Yang Y, Yu HD, Yue Y, Zou GL (2005) J Ethnopharmacol 96:151
62. Alvarez-Castellanos PP, Bishop CD, Pascual-Villalobos MJ (2001) Phytochemistry 57:99
63. Demetzos C, Angelopoulou D, Perdetzoglou D (2002) Biochem Syst Ecol 30:651
64. Setzer WN, Vogler B, Schmidt JM, Leahy JG, Rives R (2004) Fitoterapia 75:192
65. Tabanca N, Kirimer N, Demirci B, Demirci F, Baser KHC (2001) J Agric Food Chem 49:4300
66. Yu HH, Kim YH, Kil BS, Kim KJ, Jeong SI, You YO (2003) Planta Med 69:1159
67. Kim KJ, Kim YH, Yu HH, Jeong SI, Cha JD, Kil BS, You YO (2003) Planta Med 69:274
68. Senatore F, Formisano C, Arnold NA, Piozzi F (2005) J Essent Oil Res 17:419
69. Yu JQ, Lei JC, Yu HD, Cai X, Zou GL (2004) Phytochemistry 65:881
70. Iscan G, Kirimer N, Kurkcuoglu M, Husnu CanBaser K, Demirci F (2002) J Agric Food Chem 50:3943
71. Viljoen AM, Subramoney S, van Vuuren SF, Baser KHC, Demirci B (2005) J Ethnopharmacol 96:271
72. Sonboli A, Eftekhar F, Yousefzadi M, Kanani MR (2005) Z Naturforsch C 60:30
73. Delaquis PJ, Stanich K, Girard B, Mazza G (2002) Int J Food Microbiol 74:101
74. Vardar-Unlu G, Candan F, Sokmen A, Daferera D, Polissiou M, Sokmen M, Donmez E, Tepe B (2003) J Agric Food Chem 51:63
75. bin Jantan I, Yassin MSM, Chin CB, Chen LL, Sim NL (2003) Pharm Biol 41:392
76. Prashar A, Hili P, Veness RG, Evans CS (2003) Phytochemistry 63:569
77. Sokovic MD, Ristic M, Grubisic D (2004) Pharm Biol 42:328
78. Staniszewska M, Kula J, Wieczorkiewicz M, Kusewicz D (2005) J Essent Oil Res 17:579
79. Oyedeji OA, Afolayan A (2005) S Afr J Bot 71:114
80. Parveen M, Hasan MK, Takahashi J, Murata Y, Kitagawa E, Kodama O, Iwahashi H (2004) J Antimicrob Chemother 54:46
81. Al-Burtamani SKS, Fatope MO, Marwah RG, Onifade AK, Al-Saidi SH (2005) J Ethnopharmacol 96:107

82. Jo C, Park BJ, Chung SH, Kim CB, Cha BS, Byun MW (2004) Food Sci Biotechnol 13:384
83. Jirovetz L, Buchbauer G, Stoyanova AS, Georgiev EV, Damianova ST (2003) J Agric Food Chem 51:3854
84. Haznedaroglu MZ, Karabay U, Zeybek U (2001) Fitoterapia 72:829
85. Lis-Balchin M, Ochocka RJ, Deans SG, Asztemborska M, Hart S (1999) J Essent Oil Res 11:393
86. Aligiannis N, Kalpoutzakis E, Kyriakopoulou I, Mitaku S, Chinou IB (2004) Flavour Fragrance J 19:320
87. Rotman A, Ahumada O, Demo MS, Oliva MD, Turina AV, Lopez ML, Zygadlo JA, (2003) Flavour Fragrance J 18:211
88. de Carvalho PM, Rodrigues RFO, Sawaya ACHF, Marques MOM, Shimizu MT (2004) J Ethnopharmacol 95:297
89. Canillac N, Mourey A (2001) Food Microbiol 18:261
90. Martins AP, Salgueiro LR, Goncalves MJ, da Cunha AP, Vila R, Caniguerl SC (2003) Planta Med 69:77
91. Hammer KA, Carson CF, Riley TV (2003) J Appl Microbiol 95:853
92. Cakir A, Kordali S, Kilic H, Kaya E (2005) Biochem Syst Ecol 33:245
93. Cheng SS, Lin HY, Chang ST (2005) J Agric Food Chem 53:614
94. Cakir A, Kordali S, Zengin H, Izumi S, Hirata T (2004) Flavour Fragrance J 19:62
95. Juliani HR, Biurrun F, Koroch AR, Oliva MM, Demo MS, Trippi VS, Zygadlo JA (2002) Planta Med 68:762
96. Martins AP, Salgueiro UR, Goncalves MJ, Vila R, Canigueral S, Tomi F, Casanova J (2005) J Essent Oil Res 17:239
97. Chang ST, Wang SY, Wu CL, Chen PF, Kuo YH (2000) Holzforschung 54:241
98. Cheng SS, Wu CL, Chang HT, Kao YT, Chang ST (2004) J Chem Ecol 30:1957
99. Inoue Y, Shiraishi A, Hada T, Hirose K, Hamashima H, Shimada J (2004) FEMS Microbiol Lett 237:325
100. Wang SY, Wu CL, Chug FH, Chien SC, Kuo YH, Shyur LF, Chang ST (2005) Holzforschung 59:295
101. Yayli N, Yasar A, Gulec C, Usta A, Kolayli S, Coskuncelebi K, Karaoglu S (2005) Phytochemistry 66:1741
102. Lago JHG, de Avila P, de Aquino EM, Moreno PRH, Ohara MT, Limberger RP, Apel MA, Henriques AT (2004) Flavour Fragrance J 19:448
103. Barrero AF, del Moral LFQ, Lara A, Herrador MM (2005) Planta Med 71:67
104. Solis C, Becerra J, Flores C, Robledo J, Silva M (2004) J Chil Chem Soc 49:157
105. Becerra J, Flores C, Mena J, Aqueveque P, Alarcon J, Bittner M, Hernandez V, Hoeneisen M, Ruiz E, Silva M (2002) Bol Soc Chil Quim 47:151
106. Inoue Y, Hada T, Shiraishi A, Hirose K, Hamashima H, Kobayashi S (2005) Antimicrob Agents Chemother 49:1770
107. Cotoras M, Garcia C, Lagos C, Folch C, Mendoza L (2001) Bol Soc Chil Quim 46:433
108. Cotoras M, Folch C, Mendoza L (2004) J Agric Food Chem 52:2821
109. Kim JW, Huh JE, Kyung SH, Kyung KH (2004) Food Sci Biotechnol 13:235
110. Kim JW, Kim YS, Kyung KH (2004) J Food Protect 67:499
111. Burt S (2004) Int J Food Microbiol 94:223
112. Mastelic J, Politeo O, Jerkovic I, Radosevic N (2005) Chem Nat Compd 41:35

113. Rivera-Carriles K, Argaiz A, Palou E, Lopez-Malo A (2005) J Food Protect 68:602
114. Ultee A, Slump RA, Steging G, Smid EJ (2000) J Food Protect 63:620
115. Aggarwal KK, Khanuja SPS, Ahmad A, Kumar TRS, Gupta VK, Kumar S (2002) Flavour Fragrance J 17:59
116. Cimanga K, Kambu K, Tona L, Apers S, De Bruyne T, Hermans N, Totte J, Pieters L, Vlietinck AJ (2002) J Ethnopharmacol 79:213
117. Cox SD, Mann CM, Markham JL (2001) J Appl Microbiol 91:492
118. Carson CF, Mee BJ, Riley TV (2002) Antimicrob Agents Chemother 46:1914
119. Cox SD, Mann CM, Markham JL, Bell HC, Gustafson JE, Warmington JR, Wyllie SG (2000) J Appl Microbiol 88:170
120. Walsh SE, Maillard J-Y, Russell AD, Catrenich CE, Charbonneau DL, Bartolo RG (2003) J Appl Microbiol 94:240
121. Lucini EI (2004) PhD Thesis, Universidad Nacional de Córdoba, Argentina
122. Holley RA, Patel D (2005) Food Microbiol 22:273
123. Rios JL, Recio MC (2005) J Enthopharmacol 100:80
124. Jasim SAA, Naji MA (2003) J Appl Microbiol 95:412
125. Schnitzler P, Schon K, Reichling J (2001) Pharmazie 56:343
126. De Logu A, Loy G, Pellerano ML, Bonsignore L, Schivo ML (2000) Antivir Res 48:177
127. Valenti D, De Logu A, Loy G, Sinico C, Bonsignore L, Cottiglia F, Garau D, Fadda AM (2001) J Liposome Res 11:73
128. Sinico C, De Logu A, Lai F, Valenti D, Manconi M, Loy G, Bonsignore L, Fadda AM (2005) Eur J Pharm Biopharm 59:161
129. Garcia CC, Talarico L, Almeida N, Colombres S, Duschatzky C, Damonte EB (2003) Phytother Res 17:1073
130. Duschatzky CB, Possetto ML, Talarico LB, García CC, Michis F, Almeida NV, De Lampasona MP, Schuff C, Damonte EB (2005) Antivir Chem Chemother 16:247
131. Schuhmacher A, Reichling J, Schnitzler P (2003) Phytomedicine 10:504
132. Minami M, Kita M, Nakaya T, Yamamoto T, Kuriyama H, Imanishi J (2003) Microbiol Immunol 47:681
133. Marongiu B, Porcedda S, Caredda A, De Gioannis B, Vargiu L, La Colla P (2003) Flavour Fragrance J 18:390
134. Burke BE, Baillie JE, Olson RD (2004) Biomed Pharmacother 58:245
135. Chiang LC, Ng LT, Cheng PW, Chiang W, Lin CC (2005) Clin Exp Pharmacol P 32:811
136. Yang ZC, Wang BC, Yang XS, Wang Q (2005) Colloids Surf B 43:198
137. Armaka M, Papanikolaou E, Sivropoulou A, Arsenakis M (1999) Antivir Res 43:79
138. Farag RS, Shalaby AS, El-Baroty GA, Ibrahim NA, Ali MA, Hassan EM (2004) Phytother Res 18:30
139. Ruberto G, MT Baratta (2000) Food Chem 69:167
140. Madsen HL, Bertelsen G (1995) Trends Food Sci Technol 6:271
141. Rice-Evans CA, NJ Miller, G, Paganga (1996) Free Radical Biol Med 20:933
142. Dragland S, Senoo H, Wake K, Holte K, Blomhoff R (2003) J Nutr 133:1286
143. Pietta PG (2000) J Nat Prod 63:1035
144. Kwang-Geun L, Takayuki S (2001) Food Chem 74:443
145. Sanchez-Moreno C (2002) Food Sci Technol Int 8:121
146. Choi HS, Song HS, Ukeda H, Sawamura M (2000) J Agric Food Chem 48:4156

147. Re R, Pellegrini R, Proteggente A, Pannala A, Yang M, Rice-Evans C (1999) Free Radical Biol Med 26:1231
148. Benzie IFF, JJ Strain (1996) Anal Biochem 239:70
149. Dorman HJD, Figueiredo AC, Barroso JG, Deans SG (2000) Flavour Fragrance J 15:12
150. Grassmann J, Hippeli S, Vollmann R, Elstner EF (2003) J Agric Food Chem 51:7576
151. Koleva TA, van Beek JP, Linssen H, de Groot A, Evstatieva LN (2002) Phytochem Anal 13:8
152. Sacchetti G, Maietti S, Muzzoli M, Scaglianti M, Manfredini S, Radice M, Bruni R (2005) Food Chem 91:621
153. Burits M, Bucar F (2000) Phytother Res 14:323
154. Foti MC, Ingold KU (2003) J Agric Food Chem 51:2758
155. El-Ghorab AH, Mansour AF, El-Massry KF (2004) Flavour Fragrance J 19:54
156. Ruberto G, Baratta MT, Deans SG, Dorman HJD (2000) Planta Med 66:687
157. Singh G, Marimuthu R, De Heluani CS, Catalan C (2005) J Food Sci 70:M141
158. Kim HJ, Chen F, Wu CQ, Wang X, Chung HY, Jin ZY (2004) J Agric Food Chem 52:2849
159. Mimica-Dukic N, Bozin B, Sokovic M, Simin N (2004) J Agric Food Chem 52:2485
160. Menut C, Bessiere JM, Samate D, Djibo AK, Buchbauer G, Schopper B (2000) J Essent Oil Res 12:207
161. Teissedre P L, Waterhouse AL (2000) J Agric Food Chem 48:3801
162. Juliani HR, Simon JE, Ramboatiana MMR, Behra O, Garvey AS, Raskin I (2004) Acta Hortic 629:77
163. Lee KG, Shibamoto T (2001) Food Chem 74:443
164. Lee KG, Shibamoto T (2001) Food Chem Toxicol 39:1199
165. Sacchetti G, Medici A, Maietti S, Radice N, Muzzoli M, Manfredini S, Braccioli E, Bruni R (2004) J Agric Food Chem 52:3486
166. Mimica-Dukic N, Bozin B, Sokovic M, Mihajlovic B, Matavulj M (2003) Planta Med 69:413
167. Candan F, Unlu M, Tepe B, Daferera D, Polissiou M, Sokmen A, Akpulat HA (2003) J Ethnopharmacol 87:215
168. Lee KG, Shibamoto T (2001) J Agric Food Chem 81:1573
169. Lado C, Then M, Varga I, Szoke E, Szentmihalyi K (2004) Z Naturforsch C 59:354
170. Puertas-Mejia M, Hillebrand S, Stashenko E, Winterhalter P (2002) Flavour Fragrance J 17:380
171. Agnaniet H, Makani T, Akagah A, Menut C, Bessiere JM (2005) Flavour Fragrance J 20:34
172. Tepe B, Sokmen M, Akpulat HA, Daferera D, Polissiou M, Sokmen A (2005) J Food Eng 66:447
173. Sokmen A, Gulluce M, Akpulat HA, Daferera D, Tepe B, Polissiou M, Sokmen M, Sahin F (2004) Food Control 15:627
174. Sokmen M, Serkedjieva J, Daferera D, Gulluce M, Polissiou M, Tepe B, Akpulat HA, Sahin F, Sokmen A (2004) J Agric Food Chem 52:3309
175. Sawa T, Nakao M, Akaike T, Ono K, Maeda H (1999) J Agric Food Chem 47:397
176. Karioti A, Hadjipavlou-Litina D, Mensah MLK, Fleischer TC, Skaltsa H (2004) J Agric Food Chem 52:8094
177. Yildirim A, Cakir A, Mavi A, Yalcin M, Fauler G, Taskesenligil Y (2004) Flavour Fragrance J 19:367
178. Anonymous (1983) Fed Reg 46:5852
179. Galeotti N, Di Cesare Mannelli L, Mazzanti G, Bartolini A, Ghelardini C (2002) Neuroscience Lett 322:146

180. Galeotti N, Ghelardini C, Di Cesare Mannelli L, Mazzanti G, Baghiroli L, Bartolini A, (2000) Planta Med 66:1
181. Eccles R (1994) J Pharm Pharmacol 46:18
182. Sidell N, Verity MA, Nord EP (1990) J Cell Physiol 142:410
183. Petersen M, Wagner G, Pierau FK (1989) Naunyn Schmiedebergs Arch Pharmacol 339:184
184. Swandulla D, Schafer K, Lux HD (1986) Neurosci Lett 68:23
185. Swandulla D, Carbone E, Shafer K, Lux HD (1987) Pflugers Arch 409:52
186. Hong CZ, Shellock FG (1991) Am J Phys Med Rehabil 70:29
187. White JR (1973) Phys Ther 53:956
188. Green BG (1992) Somatosens Mot Res 9:235
189. Green BG (1996) J Toxicol Cutan Ocul Toxicol 15:277
190. Green BG (1991) Somatosens Mot Res 8:301
191. Green BG, McAuliffe BL (2000) Phys Behav 68:631
192. Cliff MA, Green BG (1994) Physiol Behav 56:1021
193. Cliff MA, Green BG (1996) Physiol Behav 59:487
194. Green BG (1986) Physiol Behav 38:833
195. Bromm B, Scharein E, Darsow U, Ring J (1995) Neurosci Lett 187:157
196. Melton FM, Shelley WB (1950) J Invest Dermatol 15:325
197. Yosipovitch G, Szolar C, Hui XY, Maibach H (1996) Arch Dermatol Res 288:245
198. Laude EA, Morice AH, Grattan TJ (1994) Pulmonary Pharmacol 7:179
199. Morice AH, Marshall AE, Higgins KS, Grattan TJ (1994) Thorax 49:1024
200. Barocelli E, Calcina F, Chiavarini M, Impicciatore M, Bruni R, Bianchi A, Ballabeni V (2004) Life Sci 76:213
201. Peana AT, De Montis MG, Nieddu E, Spano MT, D'Aquila PS, Pippia P (2004) Eur J Pharmacol 485:165
202. Letizia CS, Cocchiara J, Lalko J (2003) Food Chem Toxicol 41:943
203. Letizia CS, Cocchiara J, Lalko J (2003) Food Chem Toxical 41:965
204. Peana AT, D'Aquila PS, Chessa ML, Moretti MD, Serra G, Pippia P (2003) Eur J Pharmacol 460:37
205. Peana AT, D'Aquila PS, Panin F, Serra G, Pippia P, Moretti MD (2002) Phytomedicine 9:721
206. Santos FA, Rao VSN (2000) Phyt Res 14:240
207. Savelev S, Okello E, Perry NS, Wilkins RM, Perry EK (2003) Pharmacol Biochem Behav 75:661
208. Santos FA, Rao VS (2001) Dig Dis Sci 46:331
209. Ballabeni V, Tognolini M, Chiavarini M, Impicciatore M, Bruni R, Bianchi A, Barocelli E (2004) Phytomedicine 11:596
210. Hajhashemi V, Ghannadi A, Sharif B (2003) J Ethnopharmacol 89:67
211. Amabeoku GJ, Eagles P, Scott G, Mayeng I, Springfield E (2001) J Ethnopharmacol 75:117
212. Shinde UA, Phadke AS, Nair AM, Mungantiwar AA, Dikshit VJ, Saraf MN (1999) J Ethnopharmacol 65:21
213. Silva J, Abebe W, Sousa SM, Duarte VG, Machado MIL, Matos FJA (2003) J Ethnopharmacol 89:277
214. Gulcin I, Buyukokuroglu ME, Oktay M, Kufrevioglu OI (2003) J Ethnopharmacol 86:51
215. Abena AA, Diatewa M, Gakosso G, Gbeassor M, Hondi Assah T, Ouamba JM (2003) Fitoterapia 74:231

216. Zygadlo JA, Juliani HR (2003) In: Majundar DK, Govil JN, Singh VK (eds) Phytochemistry and Pharmacology II, vol 8. Stadium, Houston, p 273
217. Meister A, Bernhardt G, Christoffel V, Buschauer A (1999) Planta Med 65:512
218. Juliani H R, Wang M, Moharram H, Asante-Dartey J, Acquaye D, Koroch AR, Simon JE (2005) In: Wang M, Shengmin S, Hwang LS, Ho CT (eds) Challenges in Chemistry and Biology of Herbal Research. ACS Symposium Series 925. American Chemical Society, Washington, p 126
219. Shen J, Niijima A, Tanida M, Horii Y, Maeda K, Nagai K (2005) Neurosci Lett 383:188
220. Triadafilopoulos G, Tsang HP (1996) Am J Gastroenterol 91:2224
221. Van De Graaff KM, Fox SF (1995) Concepts of human anatomy and physiology. Brown, Dubuque
222. Tanida M, Niijima A, Shen J, Nakamura T, Nagai K (2005) Brain Res 1058:44
223. Shen J, Niijima A, Tanida M, Horii Y, Maeda K, Nagai K (2005) Neurosci Lett 380:289
224. Coelho-de-Souza A N, Criddle DN, Leal-Cardoso JH (1998) Phytother Res 12:189
225. Hajhashemi V, Sadraei H, Ghannadi AR, Mohseni M (2000) J Ethnopharmacol 71:187
226. Sadraei H, Asghari GR, Hajhashemi V, Kolagar A, Ebrahimi M (2001) Phytomedicine 8:370
227. Cruz T, Cabo MM, Jiménez J, Zarzuelo A (1990) Fitoterapia 61:247
228. Al-Zuhair H, el-Sayeh B, Ameen HA, al-Shoora H (1996) Pharmacol Res 34:79
229. Astudillo A, Hong E, Bye R, Navarrete A (2004) Phytother Res18:102
230. Magalhaes PJC, Criddle DN, Tavares RA, Melo EM, Mota TL, Leal-Cardoso JH (1998) Phytother Res 12:172
231. Sadraei H, Asghari G, Naddafi A (2003) Phytother Res 17:645
232. Lis-Balchin M, Patel J, Hart S (1998) Phytother Res 12:215
233. Lis-Balchin M, Hart S (1999) Phytother Res 13:540
234. Camara CC, Nascimento NRF, Macedo-Filho CL, Almeida FBS, Fonteles MC (2003) Planta Med 69:1080
235. Sadraei H, Asghari G, Naddafi A (2003) Phytother Res 17:645
236. Hills JM, Aaronson PI (1991) Gastroenterology 101:55
237. Van Den Broucke C, Lemli JA (1982) Planta Med 38:188
238. Uchiyama T, Chess-Williams R (2004) J Smooth Muscle Res 40:237
239. Micklefield GH, Greving I, May B (2000) Phytother Res 14:20
240. Van de Graaf KM, Fox SM (1995) Fundamentals of Anatomy and Physiology, 4th edn. Brown, Dubuque
241. Cavanagh HMA, Wilkinson JM (2002) Phytother Res 16:301
242. Leiss O, von Bergmann K (1985) Gut 26:32
243. Avato P, De Ruvo C, Cellamare S, Carotti A, Mazzoccoli M, Siro-Brigiani G (1998) Pharm Biol 36:335
244. Trabace L, Avato P, Mazzoccoli M, Siro-Brigiani G (1994) Phytother Res 8:305
245. Lin CC, Yen MH, Lo TS, Lin JM (1998) J Ethnopharmacol 60:9
246. Sai-Kato K, Umemura T Takagi A, Hasegawa R, Tanimura A, Kurokawa Y (1995) Food Chem Toxicol 33:877
247. Conti M, Malandrino S, Magistretti MJ (1992) Jpn J Pharmacol 60:315
248. Morita T, Jinno K, Kawagishi H, Arimoto Y, Suganuma H, Inakuma T, Sugiyama K (2003) J Agric Food Chem 51:1560
249. Utrilla MP, Navarro MC, Jimenez J, Montilla MP, Martin A (1995) J Nat Prod 58:1749

250. Alam K, Nagi MN, Baday OA, Al-shabanah OA, Al-Rikabi AC, Al-Bekairi AM (1999) Pharmacol Res 40:159
251. Ozbek H, Ugras S, Dulger H, Bayram I, Tuncer I, Ozturk G, Ozturk A (2003) Fitoterapia 74:317
252. Peana A, Satta M, Moretti MDL, Orecchioni M (1994) Planta Med 60:478-479.
253. Fahim FA, Esmat AY, Fadel HM, Hassan KF (1999) Int J Food Sci Nutr 50:413
254. Matsunaga T, Hasegawa C, Kawasuji T, Suzuki H, Saito H, Sagioka T, Takahashi R, Tsukamoto H, Morikawa T, Akiyama T (2000) Biol Pharm Bull 23:595
255. Perez S, Zavala MA, Vargas R, Perez C, Perez RM, (1998). Phytother Res 12:S45
256. Hahn WC, Weinberg RA (2002) J Med 347:1593
257. Bradlow HL, Sepkovic DW (2002) Ann N Y Acad Sci 963:247
258. Darmon N, Khlar M (2001) Public Health Nutr 4:163
259. Greenwald P, Clifford CK, Milner JA (2001) Eur J Cancer 37:948
260. da Rocha AB, Lopes RM, Schwartsmann G (2001) Curr Opin Pharmacol 1:364
261. Trichopoulou A, Lagiou P, Keper H, Trichopoulo D (2000) Cancer Epidemol Biomarkers Prev 9:869
262. Sugimura T (2000) Carcinogenesis 21:387
263. Crowell PL (1999) J Nutr 129:775S
264. Crosignani P, Russo A, Tagliabue G, Berrino F (1996) Int J Cancer 65:309
265. Tatman D, Mo H (2002) Cancer Lett 175:129
266. Elson CE (1995) J Nutr 125:1666s
267. Wattenberg LW (1983) Cancer Res 43:2488
268. Wattenberg LW (1985) Cancer Res 45:1
269. Hartman PE, Shankel DM (1990) Environ Mol Mutagen 15:145
270. Block KI, Gyllenhaal C, Mead MN (2004) Int Cancer Ther 3:128
271. Mo H, Elson CE (2004) Exp Biol Med 229:567
272. Dorai T, Aggarwal BB (2004) Cancer Lett 215:129
273. Maddocks W (2002) Int J Aromat 12: 30
274. Guba R (2000) Toxicity myths essential oils and their carcinogenic potential. In: Essential Oils and Cancer. Proceedings of the 4th Wholistic Aromatherapy Conference, San Francisco
275. Elson CE, Peffley DM, Hentosh P, Mo H (1999) Proc Soc Exp Biol Med 221:294
276. Elson CE, Yu SG (1994) J Nutr 124:607
277. Matura M, Goossens A, Bordalo O, Garcia-Bravo B, Magnusson K, Wrangsjo K, Karlberg AT (2002) J Am Acad Dermatol 47:709
278. Flavor and Extract Manufacturers' Association (1991) D-Limonene monographa. Flavor and Extract Manufacturers' Association, Washington, p 1
279. Phillips LR, Malspeis L, Supko JG (1995) Drug Metab Dispos 23:676
280. Kaji I, Tatsuta M, Iishi H, Baba M, Inoue A, Kasugai H (2001) Int J Cancer 93:441
281. Maltzman TH, Hurt LM, Elson CE, Tanner MA, Gould MN (1989) Carcinogenesis 10:781
282. Lu XG, Zhan LB, Feng BA, Qu MY, Yu LH, Xie JH (2004) World J Gastroenterol 10:2140
283. Koo HN, Hong SH, Kim CY, Ahn JW, Lee YG, Kim JJ, Lyu YS, Kim HM (2002) Pharm Res 45:469
284. Bardon S, Picard K, Martel P (1998) Nutr Cancer 32:1
285. Izumi S, Takashima O, Hirata T (1999) Biochem Biophys Comm 259:519
286. Hardcastle IR, Rowlands MG, Barber AM, Grimshaw RM, Mohan MK, Nutley BP, Jarman M (1999) Biochem Pharmacol 57:801

287. Kawata S, Hagase T, Yamasaki E, Ishiguro H, Matsuzawa Y (1994) J Cancer 69:1015
288. Crowell PL (1997) Breast Cancer Res Treat 46:191
289. Gelb MH, Tamanoi F, Yokoyama K, Ghomashci F, Esson K, Gould MM (1995) Cancer Lett 91:168.
290. Crowell PL, Elegbede JA, Elson CE, Lin S, Baltey HH, Vedejs E, Gould MN (1994) Cancer Chemother Pharmacol 35:31
291. Crowell PL, Chang RR, Ren Z, Elson CE, Gould MN (1991) J Biol Chem 266:17679
292. Crowell PL, Kennan WS, Haag JD, Ahmad S, Vedejs E, Gould MN (1992) Carcinogenesis 13:1261
293. Crowell PL, Lin S, Vedejes E, Gould MN (1992) Cancer Chemother Pharmacol 31:205
294. Schulz S, Buhiing F, Ansorge S (1994) Eur J Immunol 24:301
295. Miquel K, Pradines A, Terce F, Selmi S, Favre G (1998) J Biol Chem 273:26179
296. Gana-Weisz M, Halaschek Wiener J, Jansen B, Elad G, Haklai R, Kloog Y (2002) Clin Cancer Res 8:555
297. Egozi Y, Weisz B, Gana Weisz M, Ben Baruch G, Kloog Y (1999) Int J Cancer 80:911
298. Haklai R, Weisz MG, Elad G, Paz A, Marciano D, Egozi Y, Ben-Baruch G, Kloog Y (1998) Biochemistry 37:1306
299. Marciano D, Ben-Baruch G, Marom M, Egozi Y, Haklai R, Kloog Y (1995) J Med Chem 38:1267
300. Ura H, Obara T, Shudo R, Itho A, Tanno S, Fujii T, Nishino N, Khogo Y (1998) Mol Carcinog 21:93
301. Haag JD, Gould (1994) Cancer Chemother Pharmacol 34:477
302. Steams V, Coop A, Singh B, Gallagher A, Yamauchi H, Lieberman R, Pennanen M, Trock B, Hayes DF, Ellis MJ (2004) Clin Cancer Res 10:7583
303. Chow HH, Salazar D, Hakin IA (2002) Cancer Epidemiol Biomarkers Prev 11:1472
304. Vigushin DM, Poon GK, Boddy A, English J, Halbert GW, Pagonis C, Jarman M, Coombes RC (1998) Cancer Chemother Pharmacol 42:11
305. Azzoli CG, Miller VA, Nig KK, Krug LM, Spriggs DR, Tong WP, Riedel ER, Kris MG (2003) Cancer Chemother Pharmacol 51:493
306. Liu G, Oettel K, Bailey H, Van Ummersen L, Tutsch K, Staab MJ, Horvath D, Alberti D, Arzoomanian R, Rezazadeh H, McGovern J, Robinson E, De Mets D, Wilding G (2003) Invest New Drugs 21:367
307. Bailey HH, Levy D, Harris LS, Schink JC, Foss F, Beatty P, Wadler S (2002) Gynecol Oncol 85:464
308. Meadows SM, Mulkerin D, Berlin J, Bailey H, Kolesar J, Warren D, Thomas JP (2002) Int J Gastrointest Cancer 32:125
309. Murren JR, Pizzorno G, DiStasio SA, McKeon A, Peccerillo K, Gollerkari A, McMurray W, Burtness BA, Rutherford T, Li X, Ho PT, Sartorelli A (2002) Cancer Biol Ther 1:130
310. Hudes GR, Szarka CE, Adams A, Ranganathan S, McCauley RA, Weiner LM, Langer CJ, Litwin S, Yeslow G, Halberr T, Qian M, Gallo JM (2000) Clin Cancer Res 6:307
311. Ripple GH, Gould MN, Arzoomanian RZ, Alberti D, Feierabend C, Pomplun M, Simon K, Binger K, Tutsch KD, Pomplun M, Wahamaki A, Marnocha R, Wilding G, Bailey HH (2000) Clin Cancer Res 6:390
312. Elegbede JA, Flores R, Wang RC (2003) Life Sci 73:2831

313. Clark SS, Zhong L, Filiault D, Perman S, Ren Z, Gould M, Yang X (2003) Clin Cancer Res 9:4494
314. Sahin MB, Perman SM, Jenkins G, Clark SS (1999) Leukemia 13:1581
315. Stayrook KR, McKinzie JH, Burke YD, Burke YA, Crowell PL (1997) Carcinogenesis 18:1655
316. Bardon S, Foussard V, Fournel S, Loubat A (2002) Cancer Lett 181:187
317. Shi WG,Gould MN (2002) Carcinogenesis 23:131
318. Ariazi EA, Satomi Y, Ellis MJ, Haag JD, Shi W, Sattler CA, Gould MN (1999) Cancer Res 59:1917
319. Elegbede JA, Flores R, Wang RC (2003) Life Sci 73:2831
320. Ferri N, Arnaboldi L, Orlandi A, Yokoyama K, Gree R, Granata A, Hachem A, Paoletti R, Gelb MH, Corsini A (2001) Biochem Pharmacol 62:1637
321. Clark SS, Perman SM, Sahin MB, Jenkins GJ, Elegbede JA (2002) Leukemia 16:213
322. Reddy BS, Wang CX, Samaha H, Lubert R, Steele VE, Kelloff GJ, Rao CV (1997) Cancer Res 57:420
323. Mills JJ, Chari RS, Boyer IJ, Gould MN, Jirtle RL (1995) Cancer Res 55:979
324. Jirte RL, Haag JD, Ariazi EA, Gould MN (1993) Cancer Res 53:3849
325. Lluria-Prevatt M, Morreale J, Gregus J, Alberts DS, Kaper F, Giaccia A, Powell MB (2002) Cancer Epidemiol Biomarkers Prev 11:573
326. Stayrook KR, McKinzie JH, Barbhaiya LH, Crowell PL (1998) Anticancer Res 18:823
327. Ren Z, Elson CE, Gould MN (1997) Biochem Pharmacol 54:113
328. Cates CA, Michael RL, Stayrook KR, Harvey KA, Burke YD, Randall SK, Crowll PL, Crowell DN (1996) Cancer Lett 110:49
329. Berchtold CM, Chen KS, Miyamoto S, Gould MN (2005) Cancer Res 65:8558
330. Zheng GQ, Kenney PM, Lam LKT (1992) J Nat Prod 55:999
331. Miyazawa M, Hisama M (2001) J Agric Food Chem 49:4019
332. Rompelberg CJM, Verhagen H, van Bladeren PJ (1993) Food Chem Toxicol 31:637
333. Rompelberg CJM, Stenhuis WH, de Vogel N, van Osenbruggen WA, Schouten A, Verhagen H (1995) Mutat Res 346:69
334. Rompelberg CJM, Evertz SJCJ, Bruijintjes-Rozier GCDM, van den Heuvel PD, Verhagen H (1996) Food Chem Toxicol 34:33
335. Rompelberg CJM, Steenwinkel MJ, van Asten JG, van Delft JH, Baan RA, Verhagen H (1996) Mutat Res 369:87
336. Rompelberg CJM, Vogels JT, de Vogel N, Bruijntjes-Rozier GC, Stenhuis WH, Bogaards JJ, Verhagen H (1996) Human Exp Toxicol 15:129
337. Yokota H, Hashimoto H, Motoya M, Yuasa A (1988) Biochem Pharmacol 37:799
338. Newbeerne P, Smith RL, Doull J, Goodman JI, Munro IC, Portoghese PS, Wagner BM, Weil CS, Woods LA, Adams TB, Lucas CD, Ford RA (1999) Food Chem Toxicol 37:787
339. Fujisawa S, Atsumi T, Kadoma Y, Sakagami H (2002) Toxicology 177:39
340. Nagababu E, Lakshmaiah N (1992) Biochem Pharmacol 43:2393
341. Udilova N, Jurek D, Marian B, Gille L, Schulte Hermann R, Nohl H (2003) Food Chem Tox 41:1481
342. Gonzalez MJ (1992) Med Hypotheses 38:106
343. Yoo CB, Han KT, Cho KS, Ha J, Park HJ, Nam JH, Kil UH, Lee KT (2005) Cancer Lett 225:41
344. Francis AR, Shetty TK, Bhattacharya RK (1989) Cancer Lett 45:177
345. Abraham SK (2001) Food Chem Toxicol 39:493

346. Wattenberg LW, Spamins VL, Barany G (1989) Cancer Res 48:2688
347. Raphael TJ, Kuttan G (2003) J Exp Clin Cancer Res 22:419
348. Shoff SM, Grummer M, Yatvin MB, Elson CE (1991) Cancer Res 51:37
349. Yu SG, Hildebrandt LA, Elson CE (1995) J Nutr 125:2763
350. Carnesecchi S, Schneider Y, Ceraline J, Duranton B, Gosse F, Seiler N, Raul F (2001) J Pharmacol Exp Ther 298:197
351. Carnesecchi S, Bradaia A, Fischer B, Coelho D, Scholler-Guinard M, Gosse F, Raul F (2002) J Pharmacol Exp Ther 303:711
352. Carnesecchi S, Goncalves RB, Bradaia A, Zeisel M., Gosse F, Poupon MF, Raul F (2004) Cancer Lett 215:53
353. Duncan RE, Lau D, ElSohemy A, Archer MC (2004) Biochem Pharmacol 68:1739
354. Mc Anally JA, Jung M, Mo H (2003) Cancer Lett 202:181
355. Clegg RJ, Middieton B, Bell D, White DA (1982) J Biol Chem 257:2294
356. Budavari S (1996) The Merk Index, 12th edn. Merck Research Laboratories Division of Merck & Co, Whitehouse Station, p 208
357. Cavalieri E, Mariotto S, Fabrizi C, Carcereri de Prati A, Gottardo R, Leone S, Berra LV, Lauro GM, Ciampa AR, Suzuki H (2004) Res Commun 315:589
358. Hernandez Ceruelos A, Madrigal Bujaidar E, de la Cruz C (2002) Toxicol Lett 135:103
359. Surth YJ (2003) Nat Rev Cancer 3:768
360. Bharti AC, Aggarwal BB (2002) Biochem Pharmacol 64:883
361. Kim JH, Lee HJ, Kim GS, Choi DH, Lee SS, Kang JK, Chae C, Paik NW, Cho MH (2004) Cancer Lett 213:139
362. Zeytinoglu H, Incesu Z, Baser KHC (2003) Phytomedicine 10:292
363. Russin WA, Hoesly JD, Elson CE, Tanner MA, Gould MN (1989) Carcinogenesis 10:2161
364. Ka H, Park HJ, Jung HJ, Choi JW, Cho KS, Ha J, Lee KT (2003) Cancer Lett 196:143
365. Wright MM, Henneberry AL, Lagace TA, Ridgway ND, McMaster CR (2001) J Biol Chem 276:25254
366. Imai T, Yasuhara K, Tamura T, Takizawa T, Ueda, Hirose M (2002) Cancer Lett 175:9
367. Idaomar M, El Hamss R, Bakkali F, Mezzoug N, Zhiri A, Baudoux D, Munoz Serrano A, Liemans V, Alonso-Moraga A (2002) Mutat Res 512:61
368. Letizia CS, Cocchiara J, Lalko J, Api AM (2003) Food Chem Toxicol 41:43
369. Sasaki YF, Imanishi H, Phta T, Shirasu Y (1989) Mutat Res 226:103
370. Bickers D, Greim H, Hanifin JH, Rogers AE, Saurat JH, Sipes LG, Smith RL, Tagami H (2003) Food Chem Toxicol 41:919
371. Van Duuren BL, Blazej T, Goldschmidt BM, Katz C, Melchionne S, Sivak A (1971) J Natl Cancer Inst 46:1039
372. De Vincenzi M, Silano M, Maialetti F, Scazzocchio B (2000) Fitoterapia71:725
373. Burkey JL, Sauer JM, McQueen CA, Sipes IG (2000) Mutat Res 453:25
374. Abdo KM, Cunningham ML, Snell ML, Herbert RA, Travlos GS, Eldridge SR, Bucher JR (2001) Food Chem Toxicol 39:303
375. Smith RL, Adams TB, Doull J, Feron VJ, Goodman JI, Marnett LJ, Portoghese PS, Waddell WJ, Wagner BM, Rogers AE, Caldwell J, Sipes IG (2004) Food Chem Toxicol 40:851
376. Burguiere L, Marion-Poll F, Cork A (2001) J Insect Physiol 47:509
377. Borg-Karlson AK, Unelius CR, Valterova I and Nilsson LA (1996) Phytochemistry 41:1477

378. Bengtsson M, Backman AC, Liblikas I, Ramirez MI, Borg-Karlson AK, Ansebo L, Anderson P, Lofqvist J, Witzgall P (2001) J Agric Food Chem 49:3736
379. Pureswaran DS, Gries R, Borden JH (2004) Chemoecology 14:59
380. Allison JD, Borden JH, Seybold SJ (2004) Chemoecology 14:123
381. Gadi V, Reddy P, Guerrero A (2004) Trends Plant Sci 9:253
382. Bichao H, Borg-Karlson AK, Araujo J, Mustaparta H (2003) J Comp Physiol A 189:203
383. Stranden M, Rostelien T, Liblikas I, Almaas TJ, Borg-Karlson AK, Mustaparta H (2003) Chemoecology 13:143
384. Rostelien T, Borg-Karlson AK, Faldt J, Jacobsson U, Mustaparta H (2000) Chem Senses 25:141
385. Mozuraitis R, Stranden M, Ramirez MI, Borg-Karlson AK, Mustaparta H (2002) Chem Senses 27:505
386. Reisenman CE, Christensen TA, Francke W, Hildebrand JG (2004) J Neurosci 24:2602
387. Allison JD, Morewood WD, Borden JH, Hein KE, Wilson IM (2003) Environ Entomol 32:23
388. Mühlbauer RC, Lozano A, Palacio S, Reinli A, Felix R (2003) Bone 32:372
389. Yamaguchi K, Shinohara C, Kojima S, Sodeoka M, Tsuji T (1999) Biosci Biotechnol Biochem 63:731
390. Perry NSL, Bollen C, Perry EK, Ballard C (2003) Pharmacol Biochem Behav 75:651
391. Wei Chen S, Min L, Jing Li W, Xi Kong W, Fang Li J, Jing Zhang Y (2004) Pharmacol Biochem Behav 79:377
392. Broughan C (2002) Int J Aromather 12:92
393. Svoboda KP, Karavia AN, McFarlane V (2002) Int J Aromather 12:67
394. Alexander M (2002) Int J Aromather 12:49
395. Standen MD, Myers SP (2004) Int J Aromather 14:150
396. Vale TG, Matos FJA, de Lima TCM, Viana GSB (1999) J Ethnopharmacol 167:127
397. Aloisi AM, Ceccarelli I, Masi F, Scaramuzzino A (2002) Behav Brain Res 136:127
398. Ceccarelli I, Lariviere WR, Fiorenzani P, Sacerdote P, Aloisi AM (2004) Brain Res 1001:78
399. Spinella M (2001) Epilepsy Behav 2:524
399. Orafidiya LO, Agbani EO, Oyedele AO, Babalola OO, Onayemi O, Aiyedun FF (2004) Int J Aromather 14:15
400. Rosa MSS, Mendonca Filho RR, Bizzo HR, Rodrigues IA, Soares RMA, Souto Padron T, Alviano CS, Lopes AHCS (2003) Antimicrob Agents Chemother 1895

6 Citrus Flavour

Russell Rouseff
Citrus Research and Education Center,
Institute of Food and Agricultural Sciences,
University of Florida, Lake Alfred, FL 33850, USA

Pilar Ruiz Perez-Cacho
CIFA Alameda del Opispo, 14080 Córdoba, Spain

6.1 Introduction

The total world production of citrus fruit grew tremendously during the last four decades of the twentieth century. Oranges constitute the largest single portion of citrus produced and currently contribute over 60% of the total world production. This is a decrease from past decades where oranges constituted as much as 70% of total citrus production. The reason for this diminished portion has not been the result of decreased orange production but rather the increased popularity of mandarin fresh fruit cultivars. Orange production tripled between 1961 and 2001, rising from approximately 18,000,000 t in 1961 to 60,000,000 t in 2001. Drought, disease and hurricanes have diminished total orange production in the last 3 years. Sweet oranges will be the major citrus discussed in this chapter because of their overwhelming predominance. However, other citrus cultivars such as lemon, grapefruit and lime are lesser but still important sources of citrus flavours and will also be discussed.

About two thirds of the citrus produced worldwide is consumed as fresh fruit. Unfortunately, citrus utilised as fresh fruit cannot constitute a source of commercial flavours. However, in certain high-production countries such as the USA (Florida) and Brazil, the majority of the citrus crop is processed. In Florida over 90% of the orange crop is processed and is a major source for citrus flavouring material. Citrus fruits are processed primarily into juice, but oil from the outer layer of the peel, flavedo, and the condensate from making concentrated juice are also major sources of flavour products from citrus fruit.

Citrus has been the source of distinctive flavours that have been esteemed by people throughout the world for centuries. Citrus fruit can be found in a wide range of size, colour and flavour. Sizes range from the 40–45-cm-diameter pummelo (*Citrus grandis*) to the 3-cm Mexican or Key lime (*C. aurantifolia*). Citrus flavours range from the acidic, zesty and distinctive light aroma of limes (*C. aurantifolia*) to the rich sweet, full-bodied taste and aroma of sweet oranges (*C. sinensis*) to the pungent aroma and astringent taste of the citron (*C. medica*). Of these flavours, orange flavour is the most widely recognised and esteemed citrus flavour throughout the world and has been used extensively to flavour a host of foods and beverages. Lemon flavour is the second most popular citrus flavour. Lemon oil has been used extensively to flavour beverages, especially carbonated beverages and to aromatise household products, imparting a clean, light citrus/lemon fragrance. Grapefruit, lime and mandarin oils each possess distinctive aroma profiles but are used to a much lesser extent for fragrance and flavour applications.

Citrus volatiles have been extensively examined over the last several decades and several reviews have summarised composition and concentration data which existed at that time [1–7]. Careful attention should be paid to the analytical technology employed in each study cited. Many of the early studies employed packed-column gas chromatography (GC) which had limited resolving power. Results from these studies should therefore not be accepted uncritically. Studies employing high-resolution capillary GC are less prone to coelution and are probably more reliable.

6.2 Physical Characteristics of Citrus Fruit

Botanically speaking, citrus is a hesperidium, a berry with a leathery aromatic rind and a fleshy interior divided into sections. As shown by the cross section shown in Fig. 6.1, the exocarp or peel consists of an outer layer called the flavedo which contains oil glands and pigments and a white spongy inner layer called the albedo. The fleshy interior or endocarp of the fruit consists of wedge-shaped sections (segments) filled with multiple fluid-filled sacs or vesicles. These juice sacs constitute the edible portion of a citrus fruit. The cytoplasm contents provide the primary source of the citrus juice. The juice consists primarily of water, sugars, pectins, lipids, terpenes, amino acids, phenolics, carotenoids and minerals.

A microscopic section of the flavedo containing a single oil gland is shown in Fig. 6.2. This section of the peel contains the essential oil in circular cavities

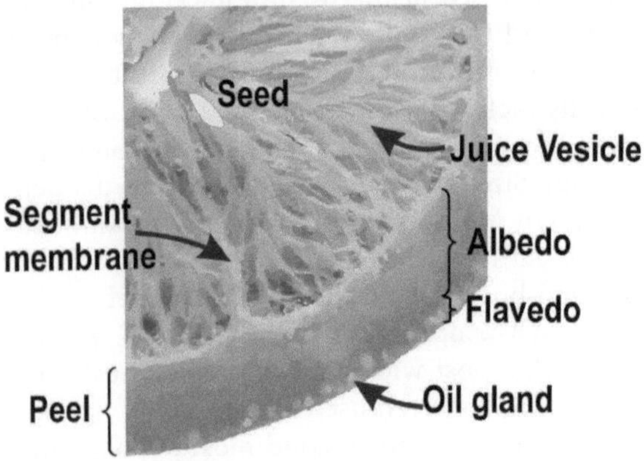

Fig. 6.1 Cross section of a citrus fruit

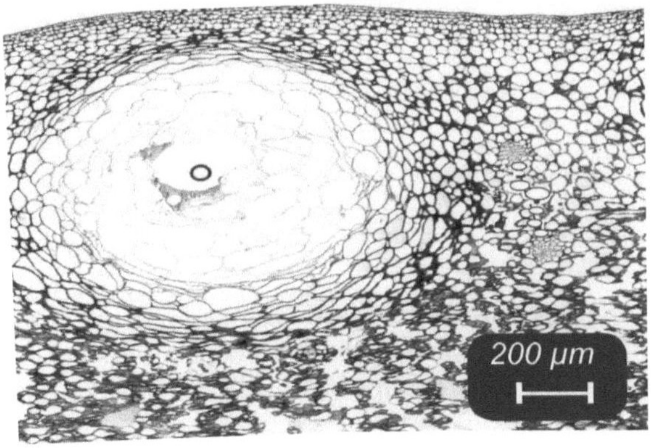

Fig. 6.2 Microscopic section of the flavedo

lined with several layers of specialised epithelial cells which are impervious to the cytotoxic oil. A mature orange will contain between 8,000 and 12,000 small, ductless, oil glands [8]. Essential oils in the oil gland are removed from the peel using a variety of techniques, including maceration and pressing. Most peel oil recovery techniques involve the use of water to physically capture or remove the oil from the fruit.

6.3 Technological Flavour Products

6.3.1 Peel Oil

Fruit oil glands are mechanically ruptured either prior to or during juice extraction and are captured with a steam of water producing an oil–water emulsion containing 0.5–2% oil. Typically, the oil is separated from the water using two centrifuges in series. Polar peel oil components can partition into the aqueous emulsion if allowed sufficient contact time; therefore, the highest-quality oils usually have minimal contact time with water. The first centrifuge (desludging or concentrating centrifuge) concentrates the oil to 70–90%. The final centrifuge (polishing centrifuge) concentrates the oil up to 99% oil. The oil still contains dissolved cuticle wax from the surface of the fruit, along with methoxylated flavones and carotenoids. Concentrations of these nonvolatile components can be reduced by chilling the oil and precipitating the waxes and methoxylated flavones. The resulting oil is often called cold pressed oil or CPO.

Orange peel oil is the major oil produced worldwide and is used extensively in the food industry, primarily as a flavouring in beverages and sweets. It possesses a light, sweet, fresh top note with fruity and aldehydic character. Many household and personal-care products employ orange oil owing to its pleasing character, ability to blend with other aroma components, low cost and availability. Citrus peel and/or essence oils are commonly employed as a top note component in some perfumes and colognes [9].

6.3.2 Essences

Essence oil and aqueous essence (sometimes called aqueous aroma) are both formed from the condensate from steam distillation/evaporation of citrus juices. These products consist of volatile juice compounds and do not contain non-volatile pigments.

6.3.3 Petitgrain Oil

This bitter-sweet, floral, woody smelling oil is a product from the steam distillation of citrus leaves and twigs. Sour orange is the cultivar which produces the highest-quality oil. Oil yields are fairly low, ranging from 0.25 to 0.5%. Even though over 400 components have been reported in this oil [10], the top 25 components comprise 95% of the total oil weight. The combination of linalyl acetate and linalool alone constitutes 80% of the total oil [11]. There are a few monoterpene hydrocarbons in the 1–3% range, including myrcene α-ocimene, β-pinene and β-ocimene. Even though total carbonyl compounds are responsible for only 0.37% of the oil, they are undoubtedly important aroma contributors. The potent β-ionone and β-damascenone are each reported to exist at concentrations 1 million times greater than their odour threshold [10]. Some substituted pyrazines are also present at concentrations 1 million times greater than their odour threshold. Recombination experiments based on quantitative data have not effectively duplicated the aroma of this oil, suggesting the need for the identification of additional trace aroma impact components. Even though analytical concentrations of up to 60 compounds have been reported [12, 13], GC–olfactometry has yet to be employed to determine the aroma-active components in this product.

6.3.4 Oil of Neroli

This highly prized, floral oil is produced from the steam distillation of orange blossoms. Neroli oil is extensively employed in the formulation of perfumes and other high-end fragrances [14]. This oil requires about 850 kg of orange blossoms to produce a single kilogram of neroli oil [15]. Although many of the components found in petitgrain oil are also found in neroli oil [16], their relative compositions differ considerably. Concentrations of α-ocimene and β-ocimene, β-pinene and limonene are considerably higher (6.5, 11 and 17%, respectively) [10]; however, linalyl acetate and linalool are still major components (6 and 36%, respectively). Methyl anthranilate and indole are thought to be the aroma components which primarily differentiate the aroma profile of neroli oil from that of petitgrain oil.

6.4 Botanical Sources of Citrus Flavours

Botanically, *Citrus* is part of the family Rutaceae, subfamily Aurantioideae, containing six closely related genera: *Citrus, Fortunella, Poncitrus, Microcitrus, Eremocitrus* and *Clymenia*. Most flavours of commercial value are found in the *Citrus* genus and subgenus *Eucitrus*. *Citrus* species have been classified using the taxonomic systems of either Swingle [17] or Tanaka [18]. In Swingle's taxonomic system there are 16 species; in Tanaka's system there are 145 citrus species. In this discussion the taxonomy system of Swingle will be used without judging the merit of either system. As citrus has been cultivated and bred for over 2,000 years there are hundreds of named cultivars but only species and cultivars of major commercial interest will be considered in this discussion.

6.4.1 Sweet Orange *(Citrus sinensis)*

This is the major citrus fruit produced worldwide. Since this citrus type has been produced for over 2,000 years, there are a wide range of named cultivars. However, the major cultivars of commercial importance include Valencia, Pera, Navel, Hamlin and Shamouti. The sensory characteristics of juices from a few of these cultivars have been reported [19].

Some of the more thorough studies of orange juice volatile composition were carried out by Schreier et al. [20], Duerr and Schobinger [21] and Nisperos-Carriedo and Shaw [22]. For example, Schreier et al. peeled the oranges before extraction in methanol to inactivate enzymes and prevent contamination from peel oil. Volatiles were separated from the aqueous juice using solvent extraction and were subsequently concentrated. Internal standards were employed to compensate for changes in concentration due to extraction/concentration or variation in sample introduction. Few subsequent studies prepared and analysed juice samples as thoroughly.

Compositional analysis of orange essence oil from Florida was reported by Moshonas and Shaw [23] and more recently by Hognadottir and Rouseff [24].

Sweet orange peel oil composition has been reviewed in [1, 4].

6.4.2
Sour/Bitter Orange *(C. aurantium)*

This species is little used for its juice because it is bitter owing primarily to naringin, a bitter flavanone [25], and it contains high levels of citric acid which produces the sour taste. However, the volatiles of this species are prized by the fragrance industry. The highly appreciated Oil of Neroli is prepared from the flowers of this species grown in the Mediterranean region of Europe and North Africa. More recently, production has shifted to South America, notably Argentina and Paraguay.

Of all citrus cultivars, the compositional information on *C. aurantium* volatiles is the most conflicted. Maekawa et al. [26] reported relative peak area values for 18 components from peel oil of four sour orange cultivars grown in Japan. Terpenes such as limonene (74–86%) and myrcene (1.6–10.9%) comprised the bulk of the oil. A subsequent capillary GC–mass spectrometry (MS) study [27] reported unusually high (24.3% peak area) values for myrcene in sumikan oil, a cultivar of *C. aurantium* grown in Japan. Most recent studies have reported relative limonene concentrations greater than 90% and myrcene in the 1–2% range. Relative terpene composition such as ratios of β-pinene to sabinene has been used as a marker of sour orange authenticity [28]. The relative compositions of oxygenated terpenes have also been used as markers of sour orange oil authenticity [29], especially the absence of citronellal in genuine oil.

Fruit maturity has a major impact on peel oil composition. Terpenes are almost exclusively present in the oil from unripe fruit. As fruit mature, concentrations of aliphatic aldehydes and oxygen-containing terpenes and sesquiterpenes increase [30]. For example, nootkatone and α-selinenone were not detected in the peel oils from fully developed immature fruit, but the oil from ripe fruits contained up to 0.15% of these oxygenated sesquiterpenes.

6.4.3
Lemon *(C. lemon)*

Lemon peel oil is much more valuable than its juice; therefore, extensive research efforts have been expended to determine its natural composition as a way to detect adulteration as well as to determine quality factors [6, 31, 32]. However, a few studies on lemon juice volatiles can be found [33–35]. Lemon oils are notable for possessing relatively low levels of limonene (more than 70%) and relatively high levels of α-pinene (1–2%), β-pinene (6–13%), sabinene (1–2%) and γ-terpinene (8–10%) [32]. The relatively high concentration of β-pinene is thought to instil the green peely odour of lemon oil. The concentrations of aliphatic and monoterpenic aldehydes, (especially citral) as well as those of esters and alcohols are critical components in the perceived quality of the oil. As lemon oil is unstable, quality can deteriorate with improper storage, resulting in the production of quality-degrading components such as *p*-cymene, carvone, *p*-menthadiene-8 ols and *p*-menthen-1,8-diols [36–38].

One of the uncommon compounds observed in lemon oil is methyl jasmonate. This compound, found in two isomeric forms, is thought to contribute to ripe lemon aroma [39].

6.4.4 Grapefruit *(C. paradisi)*

Grapefruit juice volatiles were initially determined from concentrated condensates as aqueous essence or essence oil. Moshonas and Shaw [40] reported finding 32 volatile components in a commercial grapefruit juice aqueous essence after it was further extracted with methylene chloride and concentrated. As might be expected from an aqueous product, the reported components were all relatively polar, consisting of 15 alcohols, six aldehydes, four esters, two ethers, acetal, nootkatone and two other ketones. Limonene was present as a minor component. A direct simultaneous distillation/extraction of a grapefruit juice coupled with GC-MS allowed for the identification of 58 volatiles [41]. Purge-and-trap GC-MS was subsequently used to identify 23 of the most volatile components in fresh grapefruit juice. The advantage of this technique was that it allows the detection and quantification of the most volatile juice components (which are normally obscured by the solvent in solvent extracts). Furthermore, it eliminates distillation and extraction steps required for the other analysis, thus saving time and reducing the possibility of artefact formation. The disadvantage of purge-and-trap techniques is that the relatively important nootkatone [42] is not detected. Although not measured in this study, some of the more volatile sulphur compounds, such as hydrogen sulphide, methyl sulphide and possibly 1-*p*-menthene-8-thiol, which are thought to contribute to the flavour of fresh grapefruit juice [43, 44] could be detected if a suitably sensitive sulphur detector was employed. In a subsequent study employing methylene chloride extraction, 52 volatiles were identified and correlated with flavour preference. Surprisingly, nootkatone was not strongly associated with sensory preference.

Grapefruit peel oil was also included in the earlier mentioned reviews on citrus peel oils [1, 4, 7]. Since grapefruit oils can contain up to 7% non-volatile material in the form of carotenoids, coumarins, furanocoumarins, lipids and waxes, there is some slight disagreement in the literature depending on how this material is taken into account. If just the volatile material is considered, total hydrocarbon (monoterpene) content ranges from 94 to 97%, almost all of which is composed of limonene. The only other terpene present over 1% is myrcene (1–2%). All the other terpenes are generally found to be present at less than 0.5% [45–47]. In the case of grapefruit volatiles, the differences between juice and peel oil composition are quantitative rather than qualitative.

6.4.5
Lime *(C. aurantifolia)*

Lime juice like lemon juice is of less economic value that its peel and essence oils. There are two major cultivars which are responsible for the bulk of lime oil, namely Persian limes and Mexican or Key limes. Mexican or Key lime oils are further separated into two separate classes, type A and type B, depending on how they are prepared. The method of preparation makes a profound difference in their composition. Type A is produced by pricking the peel surface on a needled surface and washing off the oil with water. The water and oil are separated as discussed in Sect. 6.3.1. Type B oil is produced from the distillation of the crushed fruit. Because the oil has come in contact with the hot, acidic juice, acid hydrolysis takes place [48] and this oil contains much higher levels of alcohols than type A juice.

6.4.6
Mandarin *(C. reticulata)*

Mandarin cultivars are among the most popular citrus consumed as fresh fruit because they have brightly coloured peels which are easily removed and possess a balanced sweet–sour taste with a pleasing citrus aroma. The analytical composition of juice volatiles from various mandarin cultivars has been the subject of several studies [49–53]. Most of the volatiles reported were similar to those found in orange juices, but the number of volatiles and the amounts reported varied widely. The wide range in analytical techniques and sample preparation procedures precludes meaningful comparison of results from different reports. For example Moshonas and Shaw [53] reported limonene values from 19–226 μg/mL in a single study involving 15 mandarin and mandarin hybrid juices. Even though the juices were analysed in the same manner (dynamic headspace purge-and-trap GC), the juices were extracted from the fruit using different equipment and treated in different manners; thus, observed differences could not be attributed to cultivar, juice extraction or heating differences alone.

Mandarin peel oil volatiles contain many of the same volatiles as orange peel oil; however, there are a few differences such as elevated levels of dimethyl anthranilate and thymol. It has been reported [54] that the characteristic mandarin peel oil aroma was due to a combination of dimethyl anthranilate, thymol, α-terpinene and β-pinene.

The major volatile components in mandarin peel oil have been separated and quantified using capillary GC with flame ionisation detection/MS detection [7, 55, 56]. The identities and relative composition of 17–85 volatiles were reported.

6.5 Flavour-Impact Compounds

As in most foods of commercial interest, the components of citrus juice and essential oil volatiles found in concentrations greater than 1% have been known for some time. However, it appears that most aroma impact is produced from compounds found at concentrations less than 1%.There is disagreement, however, as to the aroma activity of limonene, the single volatile found in the highest concentrations in citrus juices and oils. Tables 6.1–6.6 contain listings of juice volatiles reported to be aroma-active largely from GC–olfactometry studies. In addition, respective sensory descriptions are listed along with orthonasal and retronasal thresholds and juice concentrations. In each case the original source of the information is cited. Because of space limitations and their relative commercial importance, only orange volatiles have been considered. Orange juice (and essential oil) quality is largely determined from the kinds and relative amounts of aldehydes and esters present. However, until the advent of GC–olfactometry, it was not possible to determine which aldehydes and in what proportions were most responsible for good orange flavour. As seen in Tables 6.1 and 6.2, there are 14 aliphatic and four terpenic aldehydes with reported aroma activity. This is by far the largest group of aroma-active compounds in orange juice and the list does not include all reported aldehydes. Relative amounts are extremely important. Esters are important as they are responsible for the fruity character. The ten esters listed in Table 6.5 are primarily ethyl esters of three-carbon to four-carbon organic acids. Linalool is by far the most important alcohol included in Table 6.3; others are simply alcohol versions of their more potent aldehyde forms. Three of the ten ketones listed in Table 6.4 are off-flavours. They are oxidation products or products of microbial contamination. Their presence above threshold levels severely degrades the quality of the juice/oil and is an indication of microbial contamination, thermal abuse and/or storage abuse. Three of the six aroma-active volatiles listed in Table 6.6 are off-flavours. 4-Vinyl guaiacol is a well-known indicator of thermal abuse and guaiacol is an indicator of microbial contamination most probably from *Alicyclobacillus* bacteria [82].

Table 6.1 Aliphatic aldehydes possessing aroma activity

Compounds	Odour descriptor	Retronasal threshold (μg/L)	Orthonasal threshold (μg/L)	Amount in fresh orange juice (μg/L)	Amount in processed orange juice (μg/L)
Acetaldehyde	Fruity, solvent-like [57–59]	10 [60]	25 [60]	8,305 and 6,400 [58], 3 [61], 3–7 [62], 6,500-15,000 [22]	910–12,000 [22], 5,800–9,700 [63], 1–13,100 [64], 1,300–5,400
Hexanal	Green, grassy fruity, orange, floral [57–59, 65]	3.66 [66], 10.5 [60]	9.18 [66], 10.5 [60]	40–380 [22], 10–290 [63], 44–100 [67], 197 and 65 [58]	Trace to 230 [22], 0–320 [63], 0–330 [64], 0–230 [68]
Octanal	Green, citrus-like fruity, floral, lemon, melon, green grassy [57–59, 65, 69, 70]	0.52 [66], 45 [60]	1.41 [66], 8 [60]	0–40 [22], 4–890 [71], 10–380 [63], 25 and 88 [58]	150–790 [22], 190–830 [63], 30–1,620 [64], 10–1,040 [68]
Nonanal	Soapy, citrus-like, floral [58, 59, 65, 69]	4.25 [66], 3.5 [58],	2.53 [66], 5 [60],	<1-87 [71], 3 and 32 [58],	
Decanal	Green, citrus-like, fatty, soapy [57–59, 65, 69]	3.02 [66], 7 [60]	1.97 [66], 5 [60]	Trace [22], 0–350 [63], 19–500 [71], 45 and 149 [58]	Trace to 1,590 [22], 0–1,730 [63], 20–690 [64], 110–1,700 [68]
Dodecanal	Soapy [57, 58, 65]	1.07 [66]	0.53 [66]		
(*E*)-2-Nonenal	Fatty, tallowy, green [57, 58, 65, 69]	0.08 [60]	0.8 [60]	0.6 and 1.5 [58]	
(*Z*)-2-Nonenal	Green, metallic, fatty [57, 58, 65]				
(*E*)-2-Hexenal	Soapy, fatty, green [57]	49.3 [66]	24.2 [66]	5–58 [71]	
(*Z*)-3-Hexenal	Green, leaf-like, grassy [58, 65, 69]	0.25 [60]	0.03 [60]	187 and 399 [58]	
(*E,E*)-2,4-Decadienal	Fatty, waxy, green [57, 58, 65]	0.05 [60]	0.2 [60]	1.2 [58]	
(*E,E*)-2,4-Nonadienal	Fatty, soapy, green [57, 58, 65]				
(*E,Z*)-2,6-Nonadienal	Cucumber-like, green [57, 58]				
(*E*)-4,5-Epoxi-(*E*)-2-decenal	Metallic, fatty [57, 58]	0.015 [60]	0.12 [60]	4.3 and 5.8 [58]	

Table 6.2 Terpene and sesquiterpene aldehydes with aroma activity

Compound	Odour descriptor	Retronasal threshold (µg/L)	Orthonasal threshold (µg/L)	Amount in fresh orange juice (µg/L)	Amount in processed orange juice (µg/L)
Neral	Lemongrass, lemon-like, citrus, minty [57]			45 [23]	
Geranial	Citrus-like, green, minty [57]	40 [66], 40 [23]		45[23]	270[23]
Citronellal	Citrus-like, minty [57]	35 [66]	66 [66]		
β-Sinensal	Overripe citrus, geranium [5, 57]	3.8 [66]	3.8 [66]		

Table 6.3 Alcohols reported to have aroma activity

Compound	Odour descriptor	Retronasal threshold (μg/L)	Orthonasal threshold (μg/L)	Amount in fresh orange juice (μg/L)	Amount in processed orange juice (μg/L)
Terpinen-4-ol	Metallic, musty, green [57, 69]			<71–200 [71], 80–250 [67]	150–1,000 [64], 100–2,650 [68],
Linalool (3,7-dimethyl-1,6-octadien-3-ol)	Floral, sweet, fruity [57–59, 65, 69, 70]	1.5 [60], 3.8 [66]	6 [60], 5.3 [66]	Trace [22], 13–3,700 [71], 0–1,550 [63], 10–290 [67], 81 and 73 [58]	40–5,300 [22], 0.6 [23], [72], 0–6,060 [63], 90–2,540 [64], 170–1,300 [68],
(*E*)-2-Hexen-1-ol	Green, fruity, leafy [5, 22]			0–100 [22], 0–360 [63], 34–140[67]	0–1,120 [22], 0–140 [63], 0–140 [64], 0–130 [68]
(*Z*)-3-Hexen-1-ol	Woody, green, leafy, fruity [5, 22, 57, 59]	0.070 [73]		60–650 [22], 80–700 [63], 150–840 [71], 9–71 [67]	20–1,900 [22], 0–2,140 [63], 0–850 [64], 30–590 [68]
3-Methyl butanol	Malty [58]		1,000 [60]	0.4–390 [71], 639 and 16 [58]	
1-Octanol	Herbal, green, sweet, floral [57, 59, 70]	54 [66]	190 [66]	73–460 [71], 4–26 [67]	10–470 [64], 0–7,840 [68]

Table 6.4 Orange juice ketones with aroma activity

Compound	Odour descriptor	Retronasal threshold (μg/L)	Orthonasal threshold (μg/L)	Amount in fresh orange juice (μg/L)	Amount in processed orange juice (μg/L)
Carvone[a]	Caraway-like, minty [57, 70]	86 [66]	2.7 [66]	<4–110 [71]	
2,3-Butanedione[a] (diacetyl[a])	Buttery [58, 65, 74]				>1μg/mL to be detected [75]
3-Hydroxy-2-butanone[a] (acetoin[a])	Buttermilk [74]			25–99 [67]	
1-Penten-3-one	Ethereal, pungent [58, 65]	1.2 [66]	0.9 [66]	<8–110 [71]	
1-Octen-3-one	Mushroom [57, 58, 65]	0.01 [60]	1 [60]	4.1 and 5.7 [58]	
(*Z*)-Octa-1,5-dien-3-one	Geranium-like [57, 58, 65]				
2-Propanone	Fruity [59, 76]				
2-Pentanone	Butter, sweet, caramel [65, 70]				
β-Damascenone	Tobacco, floral, apple [57, 77]	0.009 [78], 0.00642 [79]	0.002 [80], 0.0148 [79]	0.122–0.281 [81]	0.145–0.690 [81]
β-Ionone	Floral, raspberry, violet-like, lilac [57–59, 65]	0.461–1,080 [79]	0.0002 [80], 0.521–1,780 [79]		

[a]Off-flavour

Table 6.5 Esters reported to have aroma activity in orange juice

Compounds	Odour descriptor	Retronasal threshold (μg/L)	Orthonasal threshold (μg/L)	Amount in fresh orange juice (μg/L)	Amount in processed orange juice (μg/L)
Methyl butanoate	Fruity, strawberry-like [5, 59]	59 [66]	43 [66]	10–80 [22], 0–110 [63], 0.1–33 [71]	Trace to 40 [22], 0–70 [63], 0–30 [64], 0–120 [68]
Ethyl acetate	Fruity, solvent-like [58, 59, 70]	3.0 [82]		10–580 [22], 60–1810 [63], 77–280 [71]	20–240 [22], 0–0.26 [68], 0–0.13 [68], 0–0.17 [68], 10–320 [63], 0–450 [64]
Ethyl propanoate	Fruity [5, 58]	4.9 [66]	9.9 [66]	3–28 [71]	
Ethyl butanoate	Fruity [5, 57–59, 69, 70]	0.1 [60], 0.13 [66]	1 [60], 0.13 [66]	260–1,020 [22], 230–720 [63], <430–1,530 [71], 1,192 and 50 [58]	20–600 [22], 10–890 [63], 2–4,000 [64], 0–490 [68]
Ethyl-2-methyl propanonate	Fruity [57, 58]	0.03 [60]	0.02 [60]	8.8 and 2.7 [58]	
Ethyl-2-methyl butanoate	Fruity [57, 58]	0.004 [60], 0.0001 [83]	0.006 [60]	48 and 4.2 [58]	
Ethyl hexanoate	Fruity, orange [57–59]	0.5 [60]	5 [60]	63 and 51 [58], 8.7–240 [71]	20–32,200 [64], 0–120 [68]
Ethyl-3-hydroxy hexanoate	-Sweet, fruity [58]	63 [60]	270 [60]	<270–490 [71], 1,136 and 361 [58]	270–6,500 [64], 0–7,500 [68]
Ethyl octanoate	Spicy, floral, fruity [59]			6–63 [71]	
Ethyl decanoate	Roasted meat, cooked, rancid [69]	210 [66]	47 [66]		

Table 6.6 Miscellaneous orange juice volatiles possessing aroma activity

Compounds	Odour descriptor	Retronasal threshold (µg/L)	Orthonasal threshold (µg/L)	Amount in fresh orange juice (µg/L)
2-Isopropyl-3-methoxy-pyrazine	Earthy, beany [58]			
Carvacrol[a]	Fruity, plastic, rubber [59]			
Guaiacol[a]	Medicinal, antiseptic [76]			
4-Vinyl guaiacol[a]	Musty, rancid oil, old fruit, rotten flavour [69, 74]			75 [74]
Furaneol®	Sweet, caramel-like, pineapple [58, 65, 74, 84, 85]			
Wine lactone	Sweet, spicy [58, 65]	0.008 [60]		0.8 and 2.1 [58]

[a]Off-flavour

References

1. Boelens MH (1991) Perfumer and Flavorist 16:17
2. Chamblee TS, Clark BC Jr (1993) In: Teranishi R, Buttery RG, Sugisawa H (eds) Bioactive Volatile Compounds from Plants. ACS Symposium Series, vol 525. American Chemical Society, Washington, p 88
3. Nagy S, Shaw PE (1990) In: Morton ID, Macleod AJ (eds) Food Flavours, Part C: The Flavour of Fruits. Elsevier, New York, p 93
4. Shaw PE (1979) Journal of Agricultural and Food Chemistry 27:246
5. Shaw PE (1991) In: Maarse H (ed) Volatile Compounds in Foods and Beverages. Dekker, New York, p 305
6. Tamura H, Yang RH, Sugisawa H (1993) In: Teranishi R, Buttery R, Sugisawa H (eds) Bioactive Compounds from Plants. ACS Symposium Series, vol 525. American Chemical Society, Washington, p 121
7. Dugo G, Cotroneo A, Verzera A, Bonaccorsi I (2002) In: Dugo G, Di Gaicomo A (eds) Citrus: The Genus Citrus, vol 26. Taylor & Francis, London, p 201
8. Knight TG, Klieber A, Sedgley M (2001) Annals of Botany 88:1039
9. Buccellato F (2000) Perfumer and Flavorist 25: 58
10. Ohloff G (1994) Scent and Fragrances: The Fascination of Odors and Their Chemical Perspectives. Springer, Berlin Heidelberg New York
11. Mondello L, Dugo G, Dugo P, Bartle KD (1996) Journal of Essential Oil Research 8:597
12. Dugo G, Mondello L, Cotroneo A, D'Alcontres IS, Basile A, Previti P, Dugo P, Bartle KD (1996) Perfumer and Flavorist 21:17
13. Mondello L, Basile A, Previti P, Dugo G (1997) Journal of Essential Oil Research 9:255
14. Anonis DP (1985) Perfumer and Flavorist 10:7
15. Peyron L (2002) Medicinal and Aromatic Plants—Industrial Profiles 26:148
16. Prager MJ, Miskiewicz MA (1981) Journal of the Association of Official Analytical Chemists 64:131
17. Swingle WT, Reece PC (1967) In: Reuther W, Webber HJ, Batchelor LD (eds) Citrus Industry, vol 1. University of California Press, Berkley, p 129
18. Tanaka T (1954) Species Problems in *Citrus* (Revisio Aurantiacerum, IX). Japanese Society for the Promotion of Science, Tokyo
19. Fellers P (1985) In: Pattee H (ed) Evaluation of Quality in Fruits and Vegetables. AVI, Westport, p 83
20. Schreier P, Drawert F, Heindze I (1979) Chemie Mikrobiologie Technologie der Lebensmittel 6:71
21. Duerr P, Schobinger UAIAWFRS (1981) In: Schreier P (ed) Flavour '81. Springer, Berlin Heidelberg New York
22. Nisperos-Carriedo MO, Shaw PE (1990) Journal of Agricultural and Food Chemistry 38:1048
23. Moshonas MG, Shaw PE (1986) Food Technology 40:100
24. Hognadottir A, Rouseff R (2003) Journal of Chromatography A 998:201
25. Rouseff RL (1988) Journal of the Association of Official Analytical Chemists 71:798
26. Maekawa K, Kodama M, Kushii M, Mitamura M (1967) Agricultural and Biological Chemistry 31:373

27. Kusunose H, Sawamura M (1980) Journal of the Japanese Society of Food Science and Technology 27:517
28. Calvarano I (1966) Essenze, Derivati Agrumari 36:5
29. Di Giacomo A, Rispoli G, Tracuzzi ML (1964) Essenze, Derivati Agrumari 34:3
30. Boelens MH, Jimenez R (1989) Flavour and Fragrance Journal 4:139
31. Staroscik JA, Wilson AA (1982) Journal of Agricultural and Food Chemistry 30:507
32. Boelens MH, Jimenez R (1989) Journal of Essential Oil Research 1:151
33. Allegrone G, Belliardo F, Cabella P (2006) Journal of Agricultural and Food Chemistry 54:1844
34. Wan X, Tang J, Yuan S, Wang C, Liu Y (1991) Shipin Yu Fajiao Gongye 31
35. Moshonas MG, Shaw PE (1972) Journal of Agricultural and Food Chemistry 20:1029
36. Clark BC, Powell CC, Radford T (1977) Tetrahedron 33:2187
37. Kimura K, Nishimura H, Iwata I, Mizutani J (1983) Journal of Agricultural and Food Chemistry 31:801
38. Schieberle P, Grosch W (1988) Journal of Agricultural and Food Chemistry 36:797
39. Nishida R, Acree TE (1984) Journal of Agricultural and Food Chemistry 32:1001
40. Moshonas MG, Shaw PE (1971) Journal of Agricultural and Food Chemistry 19:119
41. Nunez AJ, Maarse H, Bemelmans JMH (1985) Journal of the Science of Food and Agriculture 36:757
42. Shaw PE, Wilson CW (1981) Journal of Agricultural and Food Chemistry 29:677
43. Shaw PE, Ammons JM, Braman RS (1980) Journal of Agricultural and Food Chemistry 28:778
44. Demole E, Enggist P, Ohloff G (1982) Helvetica Chimica Acta 65:1785
45. Wilson CW III, Shaw PE (1980) Journal of Agricultural and Food Chemistry 28:919
46. Correra M, Tapanes R, Pino J (1985) Acta Alimentaria 14:303
47. Pino JA, Acevedo A, Rabelo J, Gonzalez C, Escandon J (1999) Journal of Essential Oil Research 11:75
48. Clark BC, Chamblee TS (1992) In: Charalambous G (ed) Off-Flavors in Foods and Beverages. Developments in Food Science, vol 28. Elsevier , Amsterdam, p 229
49. Kaneko K, Katayama O (1980) Shokuryo Kenkyusho Kenkyu Hokoku 36:57
50. Shimoda M, Osajima Y (1981) Journal of the Agricultural Chemical Society of Japan 55:319
51. Ohta H, Yoshida K, Hyakudome K, Aoyagi H, Okabe M, Susukida W (1983) Nippon Shokuhin Kogyo Gakkaishi 30:200
52. Araki C, Sakakibara H (1991) Agricultural and Biological Chemistry 55:1421
53. Moshonas MG, Shaw PE (1997) Journal of Agricultural and Food Chemistry 45:3968
54. Wilson CW III, Shaw PE (1981) Journal of Agricultural and Food Chemistry 29:494
55. Cotroneo A, Dugo G, Favretto L, Favaretto LG (1990) Journal of Chemometrics 4:379
56. Choi H-S, Sawamura M (2002) Journal of Food Science and Nutrition 7:5
57. Mahattanatawee K, Rouseff R, Valim MF, Naim M (2005) Journal of Agricultural and Food Chemistry 53:393
58. Buettner A, Schieberle P (2001) Journal of Agricultural and Food Chemistry 49:2387
59. Rega B, Fournier N, Guichard E (2003) Journal of Agricultural and Food Chemistry 51:7092
60. Rychlik M, Schieberle P, Grosch W (1998) Compilation of Odor Thresholds, Odor Qualities and Retention Indices of Key Food Odorants. Deutsche Forschungsanstalt fuer Lebensmittelchemie, Garching

61. van Straten S, de Beauveser JC, Visscher CA (1982). TNO-CIVO Food Analysis Institute, Zeist, The Netherlands
62. Byrne B, Sherman G (1984) Food Technology 38:57
63. Shaw PE, Buslig BS, Moshonas MG (1993) Journal of Agricultural and Food Chemistry 41:809
64. Shaw PE, Moshonas MG, Buslig BS, Barros S, Widmer W (1999) Journal of the Science of Food and Agriculture 79:1949
65. Hinterholzer A, Schieberle P (1998) Flavour and Fragrance Journal 13:49
66. Ahmed EM, Dennison RA, Dougherty RH, Shaw PE (1978) Journal of Agricultural and Food Chemistry 26:187
67. Maccarone E, Campisi S, Fallico B, Rapisarda P, Sgarlata R (1998) Journal of Agricultural and Food Chemistry 46:2293
68. Shaw PE, Roussef RL, Goodner KL, Bazemore R, Nordby HD, Widmer WW (2000) Lebensmittel-Wissenschaft und -Technologie 33:331
69. Bazemore R, Goodner K, Rouseff R (1999) Journal of Food Science 64:800
70. Tonder D, Peterson MA, Poll L, Olsen CE (1998) Food Chemistry 61:223
71. Moshonas MG, Shaw PE (1994) Journal of Agricultural and Food Chemistry 42:1525
72. Swift LJ (1961) Journal of Agricultural and Food Chemistry 9:298
73. Buttery RG, Seifert RM, Guadagni DG, Ling LC (1971) Journal of Agricultural and Food Chemistry 19:524
74. Rouseff R, Nagy S, Naim M, Zahavi U (1992) In: Charalambous G (ed) Off-Flavors in Foods and Beverages. Developments in Food Science, vol 28. Elsevier, Amsterdam, p 211
75. Hill EC, Wenzel FW (1957) Food Technology 11:240
76. Gocmen D, Elston A, Williams T, Parish M, Rouseff RL (2005) Letters in Applied Microbiology 40:172
77. Bezman Y, Rouseff R, Naim M (2001) Journal of Agricultural and Food Chemistry 49:5425
78. Ohloff G (1978) Perfumer and Flavorist 3:1
79. Plotto A, Barnes KW, Goodner KL (2006) Journal of Food Science 71:S401
80. Buttery RG, Teranishi R, Ling LC, Turnbaugh JG (1990) Journal of Agricultural and Food Chemistry 38:336
81. Mahattanatawee K, Rouseff RL, Goodner KL, Valim FM (2004) Abstracts of Papers of the American Chemical Society 227:U31
82. Keith ES, Powers JJ (1968) Journal of Food Science 33:213
83. Shaw PE (1986) In: Morten ID, MacLeod AJ (eds) Developments in Food Science, vol 3B. Elsevier, Amsterdam, chap 7
84. Tatum JH, Nagy S, Berry RE (1975) Journal of Food Science 40:707
85. Walsh M, Rouseff R, Naim M (1997) Journal of Agricultural and Food Chemistry 45:1320

7 Fruits and Vegetables of Moderate Climate

Lars P. Christensen, Merete Edelenbos, Stine Kreutzmann
Department of Food Science,
Danish Institute of Agricultural Sciences,
Research Centre Aarslev,
Kirstinebjergvej 10, 5792 Aarslev, Denmark

7.1 Introduction

The flavour of fruits and vegetables is determined by taste and odour-active compounds. Taste is perceived on the tongue and odour in the olfactory system. The olfactory system is extremely sensitive; it can detect odours in amounts of parts per trillion, whereas receptors on the tongue can detect flavour compounds in amounts of parts per hundred. Sugars, acids, salts and compounds that contribute to bitterness, e.g. isocoumarins and polyacetylenes in carrots and related vegetables [1, 2] and sesquiterpene lactones in chicory and lettuce [3, 4], and to astringency such as phenolic acids, flavonoids, alkaloids, tannins [5, 6], are important for the taste of fruits and vegetables. The perception of sweetness, which is mainly due to fructose, glucose and sucrose, is one of the most important flavours of fruit and vegetables. Sweetness may be modified by sourness or acid levels from, e.g., citric, malic, oxalic and tartaric acids and odour-active compounds. The contribution of odour-active compounds to the flavours of fresh and processed fruits and vegetables has gained increasing attention because these compounds are important for the characteristic flavours of fruits and vegetables. The present chapter contains information on odour-active volatiles of fruits and vegetables of moderate climate.

Many factors affect the volatile composition of fruit and vegetables, e.g. genetics, maturity, growing conditions and postharvest handling. Furthermore, preparation of the fruits and vegetables for consumption and the method for isolation of volatile compounds may change the volatile profile and key aroma compounds compared to non-processed fruits and vegetables.

The most difficult problem in flavour research is to interpret the results of the volatile analysis, which gives information on the identity and the quantity of the volatile compounds collected from a given product. Many volatile compounds are not flavour-active, i.e. they cannot be detected in the olfactory system, while others may even in trace amounts have significant effects on flavour owing to their low odour-threshold values that is defined as the minimum concentration needed to produce an olfactory response. Consequently, the most abundant volatiles are not necessarily the most important contributors to flavour. Much attention has been given to identify the odour-active or character-impact compounds in fruits and vegetables by various techniques based on gas chromatography–olfactometry (GC-O). In the classic GC-O procedure, the effluent of the GC column is split, with one portion of the eluted volatiles flowing to the instrument detector and the rest to a sniff port where the odour-active compounds are identified and described [7]. In recent years, the GC-O technique has been combined with methods that determine the intensity of the odour-active compounds by dilution techniques and determination of odour-detection threshold values [7–11] as in CharmAnalysis and aromatic extract dilution analysis (AEDA). More recently, the Osme method, which determines the quality, intensity and duration of odour-active compounds, was introduced. Although all these techniques ignore synergism and antagonism between compounds, they seem to be the best methods to identify odour-active compounds in fruits and vegetables at present. The information on key odour compounds given in this chapter was mainly obtained by the use of these techniques.

7.2 Formation of Flavours in Fruits and Vegetables

A large number of volatile compounds are formed in fruits and vegetables during maturation and preparation such as cutting, chewing and mild heat treatment. The typical flavour of most fruits is not present during early fruit growth and development but develops after a ripening process. During this period, metabolism changes to catabolism and volatile compounds are formed from major plant constituents through various biochemical pathways [12, 13]. Many climacteric fruits, e.g. apples, pears, peaches, nectarines, apricots and plums, have a green note when unripe [14]. This note disappears during ripening and the characteristic aroma for the intact fruit becomes prominent [14, 15]. However, this profile may change again during preparation. In stone fruits, for example, glycoside-bound monoterpene alcohols and lactones are released upon maceration [16, 17].

The release of volatile compounds owing to cutting, chewing and mild heat treatment is an uncontrolled effect, where enzymes are mixed with primary and secondary metabolites that are separated in the intact tissue. Cooking for a long time or at high temperature can result in the formation of a whole new group of volatile flavour compounds that are usually a result of the breakdown of carbohydrates, proteins, lipids and carotenoids. Volatile compounds produced by severe cooking may completely overshadow key flavour compounds of fruits and vegetables, but they are not included in this chapter.

Volatile compounds formed by anabolic or catabolic pathways include fatty acid derivatives, terpenes and phenolics. In contrast, volatile compounds formed during tissue damage are typically formed through enzymatic degradation and/or autoxidation reactions of primary and/or secondary metabolites and includes lipids, amino acids, glucosinolates, terpenoids and phenolics.

7.2.1
Compounds Formed by Degradation of Fatty Acids

Fatty acids originate from triglycerides, phospholipids or glycolipids that are important parts of the cell membranes. Fatty acids are precursors for a large number of volatile compounds of which many are important character-impact aroma compounds responsible for the fresh, green and fruity notes of fruits and vegetables. Degradation of fatty acids occurs mainly by three different oxidative routes: (1) β-oxidation, (2) oxidation by the lipoxygenase (LOX) pathway and (3) autoxidation. However, fatty acids do not accumulate in healthy plant tissue and therefore the initial phase in the oxidative degradation process of fatty acids is their liberation by acyl hydrolases before an oxidative degradation [18].

β-oxidation is the classical biochemical pathway involved in fatty acid degradation [19, 20] that typically occurs in intact tissue during ripening of fruits and vegetables. β-oxidation acts on acylcoenzyme A (acetyl-CoA) and consists of a four-step reaction sequence, yielding an acyl-CoA, which has two carbons less and an acetyl-CoA. This sequence is repeated several times until the complete breakdown of the compound (Scheme 7.1). Depending on many factors, the breakdown can be stopped, resulting in the liberation of medium-chain-length or short-chain-length volatile compounds. These metabolites can exit the pathway between β-oxidation cycles or inside the sequence. This can lead to a variety of volatile compounds such as saturated and unsaturated lactones, esters, alcohols, ketones and acids (Scheme 7.1).

The volatiles produced by the LOX pathway and autoxidation are typically volatile aldehydes and alcohols responsible for fresh and green sensorial notes. In the LOX pathway these volatile compounds are produced in response to stress, during ripening or after damage of the plant tissue. The pathway is illustrated in Scheme 7.2. Precursors of the LOX (EC 1.13.11.12) catalysed reactions are C_{18}-polyunsaturated fatty acids with a (*Z*,*Z*)-1,4-pentadiene moiety such as linoleic and α-linolenic acids that are typically oxidised into 9-, 10- or 13-hydroperoxides depending on the specificity of the LOX catalyst. These compounds are then cleaved by hydroperoxide lyase (HPL) into mainly C_6, C_9 and C_{10} aldehydes, which can then be reduced into the corresponding alcohols by alcohol dehydrogenase (ADH; EC 1.1.1.1) (Scheme 7.2) [21, 22]. The production of volatile compounds by the LOX pathway depends, however, on the plants as they have different sets of enzymes, pH in the cells, fatty acid composition of cell walls, etc.

Many of the compounds derived from enzyme-catalysed oxidative breakdown of unsaturated fatty acids may also be produced by autoxidation [23]. While the enzymatically produced hydroperoxides in most cases yield one hydroperoxide as the dominant product, non-enzymatic oxidation of unsaturated fatty acids yields a mixture of hydroperoxides which differ in the position of the peroxide group and in the geometrical isomerism of the double bonds [24]. As the number of double bonds increases, the number of oxidation and oxygen-addition sites increases proportionally and thus the number of possible volatile degradation products increases [24]. Autoxidation of linoleic acid produces the 9- and 13-hydroperoxides, whereas linolenic acid in addition also produces 12- and 16-hydroperoxides [25]. Hexanal and 2,4-decadienal are the primary oxidation products of linoleic acid, whereas autoxidation of linolenic acid produces 2,4-heptadienal as the major product. Further autoxidation of these aldehydes leads to the formation of other volatile products [23].

Unsaturated fatty acids also seem to undergo oxidative breakdown during cooking. The volatile compounds found in cooked products are generally the same as in the raw product. Frequently there are, however, quantitative differences between the cooked and the raw product. However, not much is known about the thermal fatty acid breakdown, but possibly it involves decomposition of already formed hydroperoxides in the raw product and/or oxidation of already formed volatile compounds. For example, 1-octen-3-ol occurs in raw cut mushroom, whereas 1-octen-3-one cannot be detected. On the other hand, 1-octen-3-one is found in relatively large amounts in cooked mushroom [26].

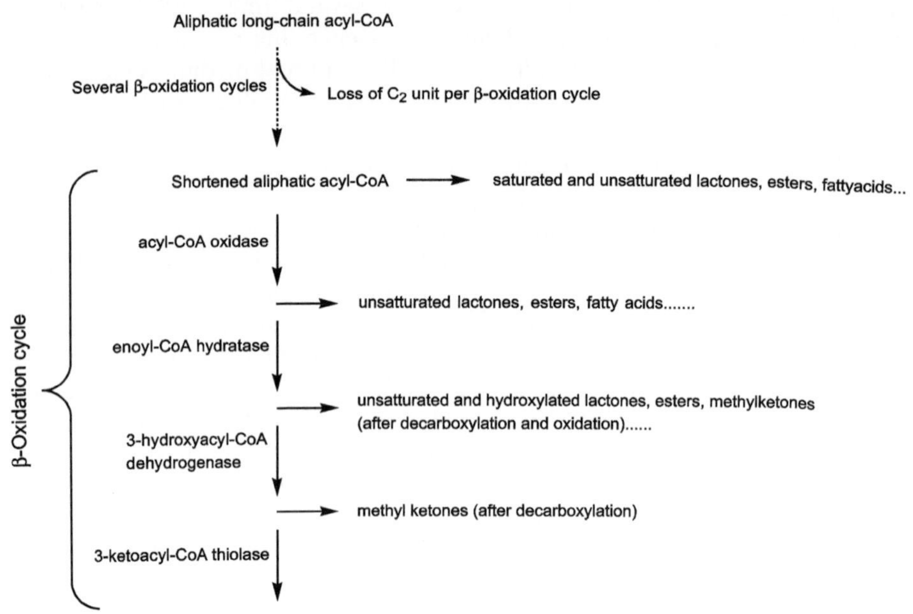

Scheme 7.1 Enzymatic degradation of fatty acids by the β-oxidation cycle and formation of various types of aroma compounds in fruits and vegetables

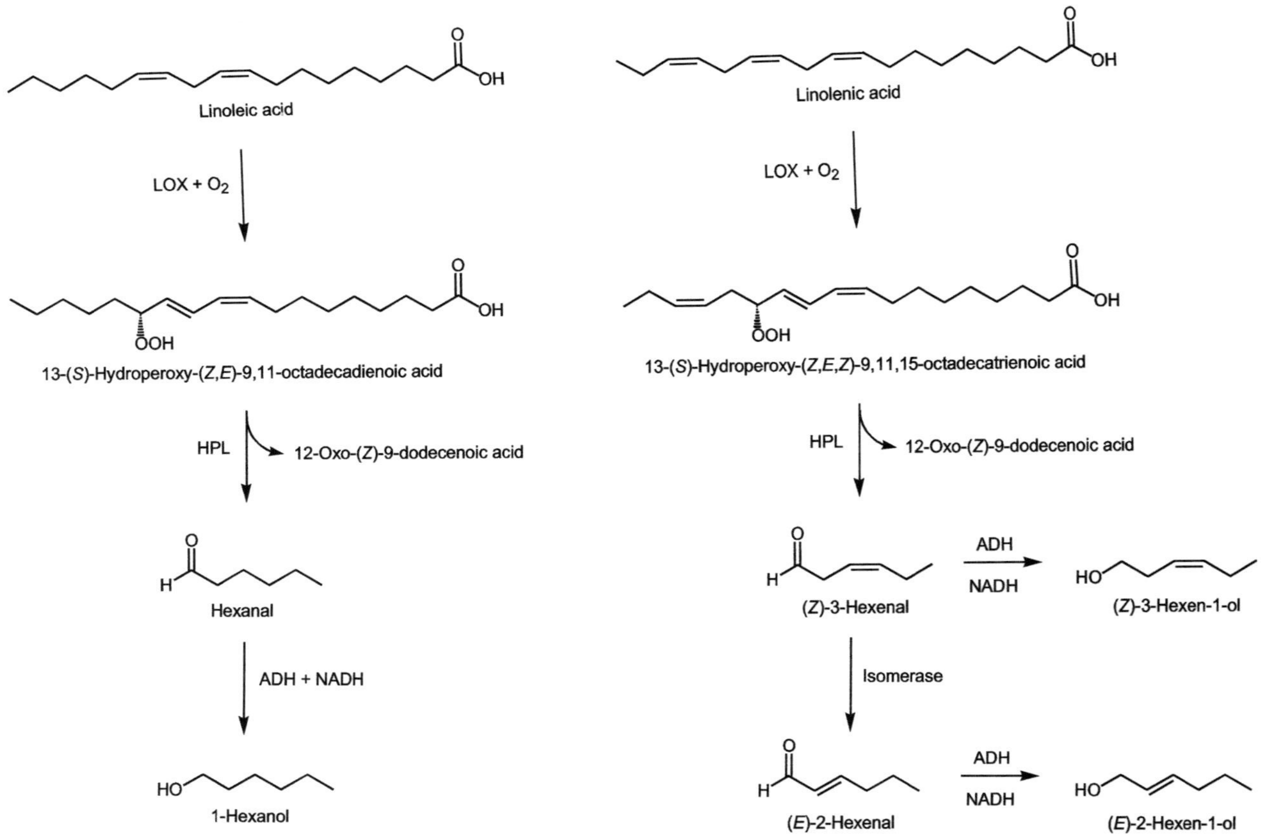

Scheme 7.2 Pathway for the enzymatic degradation of linoleic acid and linolenic acid via the lipoxygenase (*LOX*) pathway to C_6 key aroma compounds in fruits and vegetables responsible for green notes. *HPL* hydroperoxide lyase, *ADH* alcohol dehydrogenase

7.2.2
Compounds Formed from Amino Acids

Some volatile compounds are produced by the action of enzyme systems on amino acids when the tissue of the vegetable is damaged. This seems to be particular true for sulfur-containing amino acids in vegetables of the Alliaceae and Brassicaceae families. The distinct aroma of freshly cut *Allium* species (Alliaceae) is dominated by numerous sulfur-containing volatile compounds originating from the decomposition of the odourless non-volatile precursors (+)-*S*-alk(en)yl cysteine sulfoxides by the action of the enzyme alliinase (EC 4.4.1.4) as shown in Scheme 7.3 [27–29]. Owing to the compartmentation of alliinase in the vacuole and the cysteine sulfoxides in the cytoplasm, volatile compounds are not produced until cell rupture, e.g., by cutting. The products of this are pyruvate, ammonia and various sulfenic acids depending on the (+)-*S*-alk(en)yl cysteine sulfoxides present in the tissue. At least five different cysteine sulfoxides occur commonly in *Allium* species, which gives rise to different sulfenic acids and hence different volatile sulfur compounds (Scheme 7.3) [30, 31]. The sulfenic acids are highly reactive and will quickly combine to form thiosulfinates, which are responsible for the odour of freshly cut *Allium* species. The thiosulfinates are also unstable and will rearrange to form disulfides and thiosulfonates. The thiosulfonates expel sulfur dioxide to give the corresponding monosulfides, and the disulfides can rearrange to form monosulfides and trisulfides, so the final products of the reaction will end up being a combination of monosulfides and polysulfides (Scheme 7.3a). Further, the amino acid (*E*)-*S*-1-propenyl cysteine sulfoxide (isoalliin) can apart from taking part in the formation of polysulfides as described earlier result in the formation of thiopropanal-*S*-oxide (the lachrymatory factor) (Scheme 7.3b). Thiopropanal-*S*-oxide is also unstable and rearranges spontaneously to form propanal and sulfur. Propanal may undergo an aldol condensation with a further propanal molecule and give rise to 2-methyl-2-pentenal and other volatile aldehydes [31, 32].

If the disulfides are methylpropenyl disulfide or propylpropenyl disulfide (Scheme 7.3a) this may lead to thiophene compounds [31, 32]. This is not a very common process in freshly cut *Allium* species, but heating seems to promote this process [33, 34].

Amino acids may also undergo thermal degradation, which is almost always coupled with some other food components, particular sugars. The major types of volatile compounds formed from amino–sugar interactions include Strecker degradation aldehydes, alkyl pyrazines, alkyl thiazolines and thiazoles and other heterocycles [35, 36]. As the subject has mainly relevance for baked and roasted vegetable food products, this subject will not be discussed in further detail.

Finally amino acids are precursors for some branched aliphatic compounds such as 2-methyl-1-butanol and 3-methyl-1-butanol that are formed during the amino acid catabolism [20].

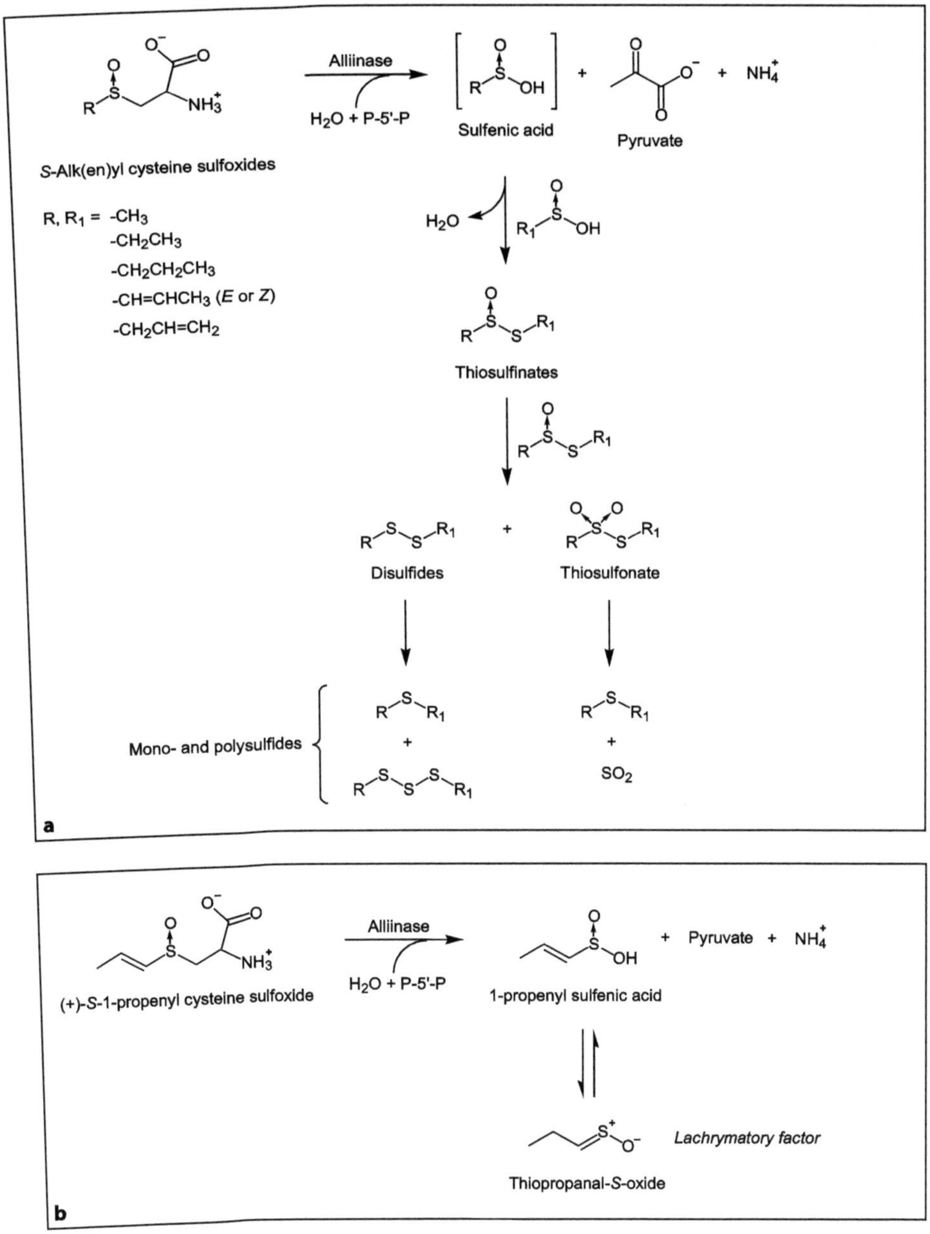

Scheme 7.3 Enzymatic production of sulfur-containing flavour compounds in *Allium* species from amino acid flavour precursors. **a** *S*-Alk(en)yl cysteine sulfoxides and **b** (+)-*S*-1-propenyl cysteine sulfoxide (isoalliin) *P-5′-P* pyridoxal-5′-phosphate

7.2.3 Compounds Formed from Glucosinolates

In a number of vegetables, in particular those of the cabbage family (Brassicaceae), glucosinolates are present. Glucosinolates are thioglucosides that consist of a common basic skeleton containing a β-thioglucose grouping, a side chain and a sulfonated oxime moiety (Scheme 7.4). When the plant tissue is damaged, e.g. by cutting or chewing, glucosinolates are hydrolysed enzymatically by the enzyme myrosinase (EC 3.2.3.1), which is physically separated from the glucosinolates in intact plant tissue. The products of this reaction are initially isothiocyanates, nitriles, glucose and a sulfate (Scheme 7.4). Some glucosinolates also give rise to the formation of thiocyanates. The nature of the hydrolysis products depends primarily on the side chain of the glucosinolate, the conditions of the hydrolysis, such as pH, and the presence of cofactors [37, 38]. In the cabbage family, the major breakdown products from the glucosinolates are 2-propenyl isothiocyanate, 3-butenyl isothiocyanate and the corresponding nitriles. The shredding of cabbage tissue in the preparation of coleslaw is particular effective in bringing about the enzymatic conversion of the glucosinolates. The nitriles can also be produced by the thermal decomposition of the glucosinolates.

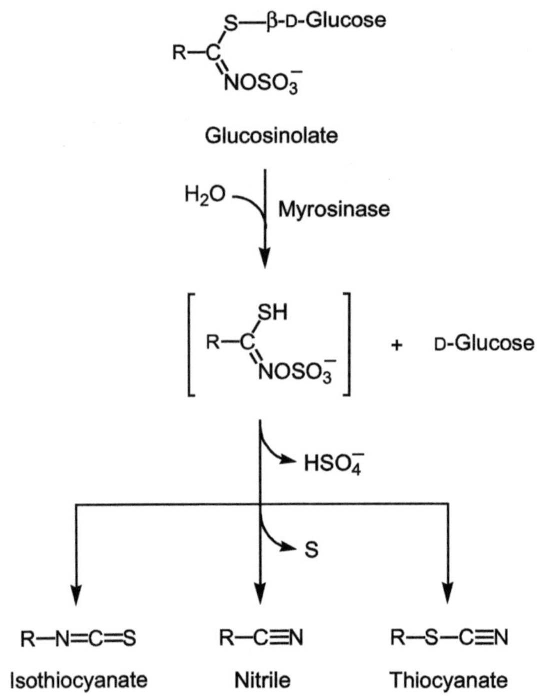

Scheme 7.4 Products of thioglucosidase (myrosinase) hydrolysis of glucosinolates. Volatile isothiocyantes and their corresponding nitriles are important flavour compounds, in particular in vegetables of the cabbage family. At low pH the formation of nitrile is favoured, whereas neutral or high pH favours the formation of the isothiocyanate

7.2.4 Compounds of Terpenoid Origin

Terpenoids are widely distributed among vegetables and fruits, and in some vegetables, e.g. carrots, they are the major contributor to the flavour of this vegetable. There are two main types of terpenoids that may contribute significantly to the flavour of vegetables and fruits and these are (1) monoterpenes and sesquiterpenes and (2) irregular terpenes mainly produced by catabolistic pathways and/or autoxidation. The monoterpenes and sesquiterpenes are mainly formed by anabolic processes and are therefore present in intact plant tissue [39]. Tissue disruption therefore does not normally alter the profile of monoterpenes and sesquiterpenes significantly in the raw product, although changes in the concentration of some monoterpenes and sesquiterpenes may occur owing to oxidation and release of glycoside-bound oxygenated terpenoids.

α-Terpineol and terpinen-4-ol might result from oxidation of terpinolene and further it cannot be ruled out that some monoterpenes and sesquiterpenes, such as geraniol and geranial, may arise from oxidative cleavage of carotenoids. Finally, glycoside-bound oxygenated terpenoids that are released enzymatically may be a source of volatile oxygenated terpenoids, especially in fruits during ripening or cell disruption [40].

The formation of some irregular terpenes cannot be explained by anabolic pathways in plants. These terpenoids are primarily oxidative degradation products of the carotenoids. The oxidative breakdown of carotenoids seem somewhat related to the oxidative breakdown of unsaturated fatty acids discussed earlier in Sect. 7.2.1. As with fatty acids, carotenoid oxidation occurs whenever the plant tissue is damaged and/or during senescence (ripening or bleaching) and the volatile degradation products generated obviously depend on the carotenoids present in the different vegetables and fruits [19, 41, 42]. For example, the tomato volatiles 6-methyl-5-hepten-2-one, geranyl acetone and farnesyl acetone may result from the oxidative cleavage of acyclic carotenoids (Scheme 7.5a). Similarly, α-ionone, β-ionone and β-damascenone probably result from the oxidative breakdown of cyclic carotenoids (Scheme 7.5b) and as for other terpenoids may exist in intact plant tissue bound as glycosides [40].

Heating (cooking) seems to produce certain terpenoids. In some vegetables, such as tomatoes and potatoes, there is a considerable increase in the formation of some terpene alcohols, including linalool, α-terpineol and terpinen-4-ol during heat treatments.

7.2.5 Phenols and Related Compounds

A large number of volatile phenols and related compounds occur in vegetables and fruits, and some of them are potent aroma compounds. The majority of volatile phenols and related compounds in plants are formed mainly through the shikimic acid pathway, and are present in intact plant tissue either as free

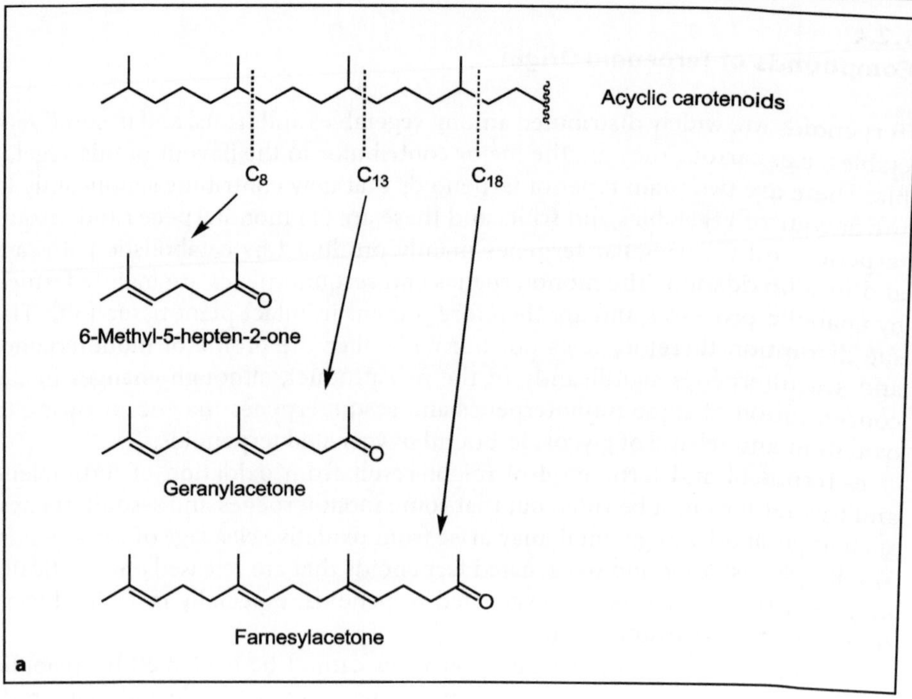

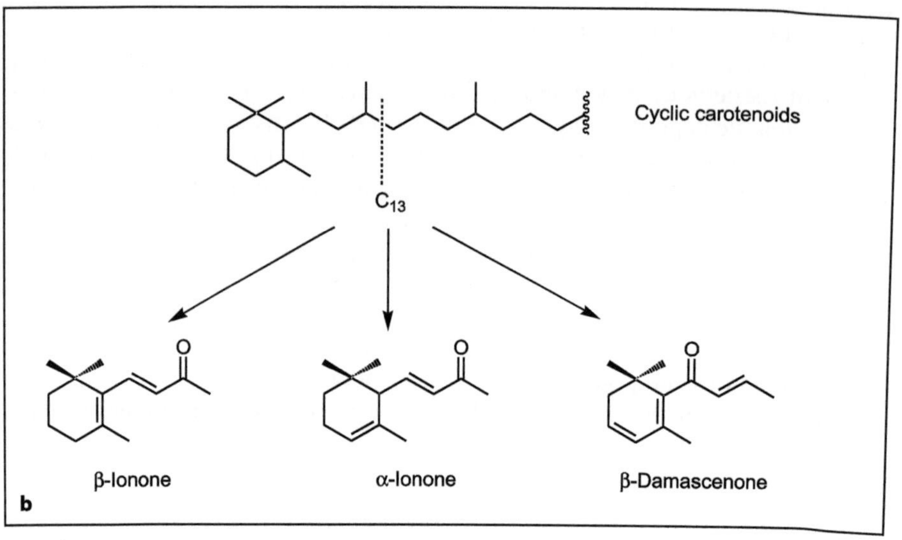

Scheme 7.5 Formation of some aroma compounds after oxidative cleavage of **a** acyclic carotenoids (e.g., lycopene, phytofluene and phytoene) and **b** cyclic carotenoids (e.g. α-carotene and β-carotene)

aglycones or bound as glycosides that can be liberated by enzymatic hydrolysis [40]. Although many of the phenols and related compounds, in particular the phenylpropanoids, originate from some of the "building blocks" of lignin such as ferulic acid and *p*-coumaric acid, these compounds are not breakdown products of lignin. Generally the volatile phenols and related compounds are substituted benzene derivatives with methoxy and phenolic groups with often an allyl, a vinyl or an aldehyde group. Common flavour compounds of this group are eugenol, vanillin, myristicin, apiole, elemicin and benzaldehyde.

7.3 Fruits

Volatile compounds in fruits are diverse, consisting of hundreds of different chemical compounds comprising only 0.001–0.01% of the fruit's fresh weight [36, 43]. This diversity is partially responsible for the unique flavours found in different species of fruit as well as differences among individual cultivars.

7.3.1 Pome Fruits

7.3.1.1 Apple

More than 350 volatile compounds have been identified in apples [44]. Only a few of these volatiles have been identified as being responsible for apple aroma [45]. The most abundant volatile components in apples are esters (78–92% of total volatiles), alcohols (6–16% of total volatiles), aldehydes, ketones and ethers [35, 45], which are present in various amounts in different cultivars [46]. Esters are the principal compounds responsible for apple odour (Table 7.1, Fig 7.1) [47]. The ultimate levels of esters in fresh and stored apples are determined by the amount of precursors for ester formation, e.g. lipids, which are influenced by cultivar, growing conditions, harvest maturity and storage conditions [47]. In Fuji apples, acetate ester concentrations increase during maturation, 2-methylbutyl acetate being the major ester component in the volatile compound profile [48]. Ethyl 2-methylbutanoate, 2-methylbutyl acetate and hexyl acetate contribute most to the characteristic aroma of Fuji apples [49]. In Red Delicious apples, ethyl butanoate, ethyl 2-methylbutanoate, propyl 2-methylbutanoate and hexyl acetate contribute to the characteristic aroma as determined by Charm-Analysis and/or AEDA [50, 51]. In a comparative study of 40 apple cultivars, the highest odour potency or Charm value was found for β-damascenone [52]. This compound usually occurs in a glycosidically bound form and is present primarily in processed products owing to hydrolysis of the glycoside bond after crushing fruit cells [53]. β-Damascenone has a very low odour threshold with a sweet, fruity, perfumery odour and is not typical of apple aroma in general [54]. Sensory evaluation of Gala apples revealed that 2-methylbutyl acetate and hexyl acetate contribute to the flavour of this cultivar [55, 56]. In a study of Gala apple aroma, the Osme method revealed that butyl acetate, hexyl acetate, butyl 2-methylbutanoate, hexyl 2-methylbutanoate and hexyl propanoate contributed to apple-like, fruity aroma and methyl 2-methylbutanoate, ethyl 2-methylbutanoate and propyl 2-methylbutanoate to sweet and berry-like odours [54]. Fuhrmann and Grosch [44] showed that the character impact odours of Elstar and Cox Orange apples depend on sample preparation. Ethyl butanoate and ethyl 2-methylbutanoate were the odour-active compounds in intact Elstar apples and ethyl butanoate, acetaldehyde, 2-methyl-1-butanol and ethyl methylpropanoate in that of Cox Orange. Ethyl 2-methylbutanoate had also a direct impact on Granny Smith apple flavour [57].

7.3.1.2
Pear

Pears are divided into European pears, which combine a buttery juicy texture with rich flavour and aroma, and Asian pears, which are characterised by a crisp texture and sweet but subacid flavour [58]. European pears are considered to be

Table 7.1 Key flavour compounds in pome fruits

Key flavour compounds	Apple (*Malus domestica*)	Pear (*Pyrus communis*)
Esters		
Butyl acetate	[44, 48, 54, 231]	[58–60, 62]
Pentyl acetate	[57]	[58]
Hexyl acetate	[44, 48, 49, 54, 231]	[58–60, 62]
2-Methylbutyl acetate	[44, 48, 49, 54, 57, 231]	
Hexyl propanoate	[54]	
Ethyl butanoate	[44]	
Butyl butanoate		[58]
Hexyl butanoate	[231]	
Methyl 2-methylbutanoate	[54]	
Ethyl 2-methylbutanoate	[44, 48, 49, 54, 57]	
Propyl 2-methylbutanoate	[54]	
Butyl 2-methylbutanoate	[54]	
Hexyl 2-methylbutanoate	[54]	
Ethyl hexanoate		[58]
Ethyl octanoate		[58]
Ethyl (*E*)-2-octenoate		[58]
Methyl (*E*,*Z*)-2,4-decadienoate		[58–62]
Ethyl (*E*,*Z*)-2,4-decadienoate		[59–62]
Terpenoids		
β-Damascenone	[44, 52, 54]	

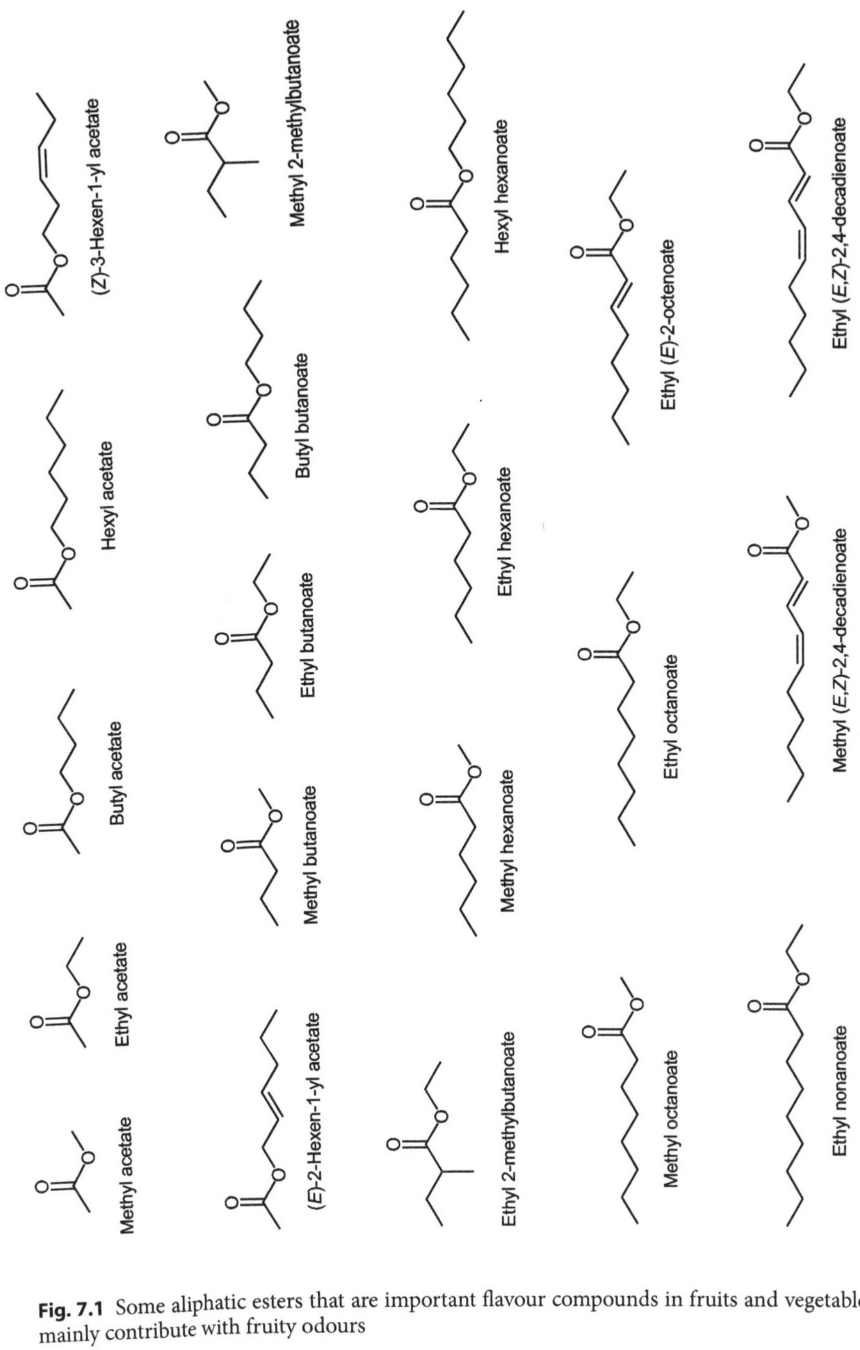

Fig. 7.1 Some aliphatic esters that are important flavour compounds in fruits and vegetables that mainly contribute with fruity odours

cultivars of *Pyrus communis*, whereas Asian cultivars are derived from *Pyrus pyrifolia*. More than 300 volatile compounds have been identified in pears, including hydrocarbons, aldehydes, alcohols, esters, ketones and sulfur compounds [58]. Some of the most important character-impact compounds of pears are summarised in Table 7.1 and Fig. 7.1. Methyl to hexyl esters of decadienoate are the character-impact compounds of the European pear [58–62]. Other volatile esters, e.g. hexyl acetate, 2-methylpropyl acetate, butyl acetate, butyl butanoate, pentyl acetate, and ethyl hexanoate also possess strong pear-like aromas (Table 7.1, Fig. 7.1) [58]. Ethyl octanoate and ethyl (*E*)-2-octenoate contribute with floral, sweet or fruity odours in pears [58]. Pears with a high concentration of 2,4-decadienoates in the fruit flesh are more accepted by consumers than those with a low content [59]. The acetate ester concentrations increase in La France pears during maturation, butyl acetate and hexyl acetate being the major ester components in the volatile compound profile [60]. However, the metabolism of the volatile compounds can be reactivated in pears after cold storage, controlled-atmosphere storage or treatment with 1-methylcyclopropene (1-MCP) [58]. In Passa Crassana pears, the concentration of butyl acetate, hexyl acetate and decadienoate esters increased during maturation following storage for 25 weeks at 5 °C [59]. D'Anjou pears treated with 1-MCP developed a volatile profile similar to that of untreated fruits during ripening, while lower amounts of volatile compounds were produced in Packham's Triumph pears during ripening [61, 63].

7.3.2 Stone Fruits

γ-Lactones and δ-lactones (Fig. 7.2) from chain length C_6 to C_{12} are important for the typical flavour of stone fruit [64]. These compounds are actively formed in the final period of fruit maturation only [13, 14, 65–67]. Stone fruits that are picked early for easy handling and shipping may lack γ-lactones and δ-lactones and their characteristic aroma [13, 64]. The most important character-impact compounds of stone fruits are summarised in Table 7.2.

7.3.2.1 Peach and Nectarine

Peaches and nectarines are members of the same species (*Prunus persica*). There is controversy over whether nectarine is a separate and distinct fruit or merely a variety of peach [68]. Nectarines lack skin fuzz or pubescence. Approximately 100 volatile compounds have been identified in peaches and nectarines, including alcohols, aldehydes, alkanes, esters, ketones, lactones and terpenes [14, 15, 17, 64, 65, 68–71]. Among them, lactones, particularly γ-decalactone and δ-decalactone, have been reported as character-impact compounds in peaches and nectarines where they process a strong peach-like aroma [66]. Lactones act in association with C_6 aldehydes, aliphatic alcohols and terpenes (Table 7.2,

Table 7.2 Key flavour compounds in stone fruits (*Prunus* spp.)

Key flavour compounds	Peach (*P. persica*)	Nectarine (*P. persica* var. *nucipersica*)	Apricot (*P. armeniaca*)	Plum (*P. domestica*)	Sweet cherry (*P. avium*)	Sour cherry (*P. cerasus*)
Esters						
Ethyl acetate		[66]	[74]			
Butyl acetate		[66]	[76]			
Propyl acetate		[66]				
Hexyl acetate		[65, 66]	[74, 76, 77]			
(*E*)-2-Hexen-1-yl acetate		[65]	[74]			
(*Z*)-3-Hexen-1-yl acetate	[17]	[66]	[74]			
2-Methylpropyl acetate		[66]				
Ethyl butanoate	[66]		[16]			
Butyl butanoate			[16]			
Ethyl hexanoate			[16]			
Ethyl nonanoate				[35]		
Methyl cinnamate				[35]		
Alcohols						
(*E*)-2-Hexen-1-ol	[66]	[66]			[35]	
Benzyl alcohol						[35, 83]
Aldehydes						
(*E*)-2-Hexenal						
(*Z*)-3-Hexenal	[17, 69]	[15, 64]	[16, 74, 76]	[78]	[80]	
Hexanal	[17]	[64]				
(*E,E*)-2,4-Decadienal	[66, 69]	[15, 64, 66]	[16, 74, 76]	[78]	[35, 80]	
Benzaldehyde	[68]		[16]	[35, 78]	[35, 80]	[82–84]

Table 7.2 *(continued)* Key flavour compounds in stone fruits (*Prunus* spp.)

Key flavour compounds	Peach (*P. persica*)	Nectarine (*P. persica* var. *nucipersica*)	Apricot (*P. armeniaca*)	Plum (*P. domestica*)	Sweet cherry (*P. avium*)	Sour cherry (*P. cerasus*)
Phenols						
Eugenol					[35]	[35, 83, 84]
Vanillin						[83, 84]
Lactones						
γ-Octalactone	[17, 65, 66]	[64, 65]	[16, 77]			
δ-Octalactone	[66]	[66]		[35, 78]		
γ-Decalactone	[17, 65, 66, 68, 69]	[15, 64, 66]	[13, 16, 73, 75, 77]	[35, 78]		
δ-Decalactone	[17, 65, 68]	[15, 64]	[13, 73]	[35, 78]		
γ-Dodecalactone	[17, 65]	[64]	[13, 16]			
δ-Dodecalactone	[65]					
γ-Jasmolactone	[17]					
Terpenoids						
Terpinolene	[65]	[65]				
Geraniol			[76]			[35]
Linalool	[64, 65, 68, 69]	[15, 64–66]	[16, 76]	[35, 78]	[35]	[35]
α-Terpineol			[76]			
β-Damascenone	[17]					
β-Ionone			[13, 16, 73]			

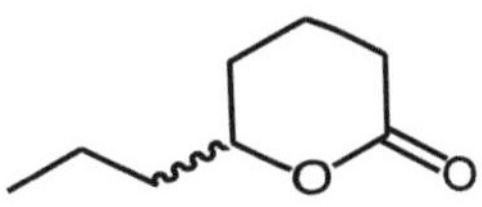

d-Octalactone (5-octanolide)

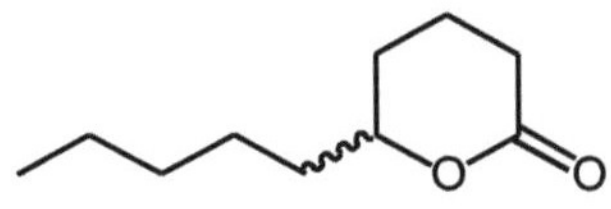

d-Decalactone (5-decanolide)

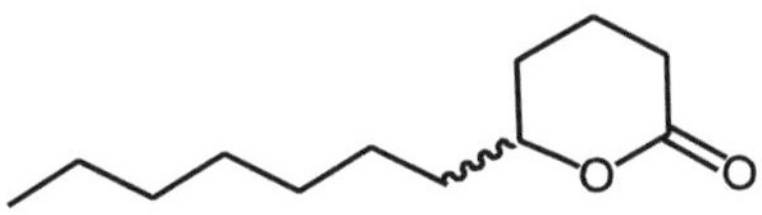

d-Dodecalactone (5-dodecanolide)

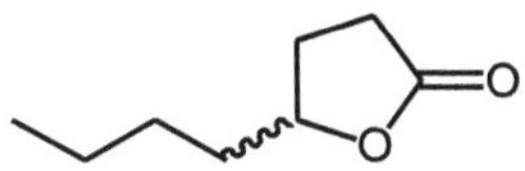

g-Octalactone (4-octanolide)

g-Decalactone (4-decanolide)

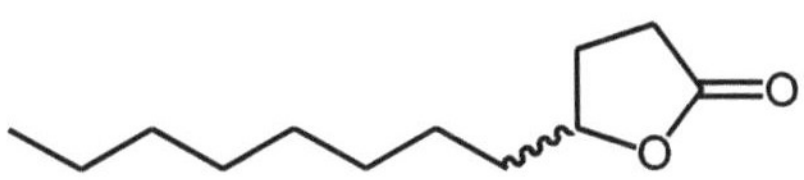

g-Dodecalactone (4-dodecanolide)

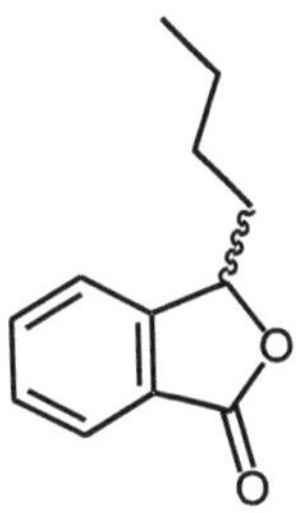

3-Butylphthalide

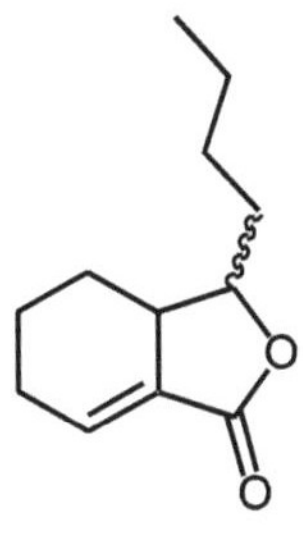

Sedanolide

Fig. 7.2 Examples of volatile lactones important for the flavour of fruits and/or vegetables

Figs. 7.1–7.4), which are responsible for a spicy, floral and fruity characteristic of stone fruits [17, 64, 71, 72]. C_6 compounds are the major volatiles in immature, green fruits but the levels of these compounds decrease drastically during maturation, and lactones (lactonic note), aldehydes (benzaldehyde with an almond, nutty and stone fruit note), terpenes (linalool with a floral note) and esters become prominent [14, 15, 65, 66, 68, 70].

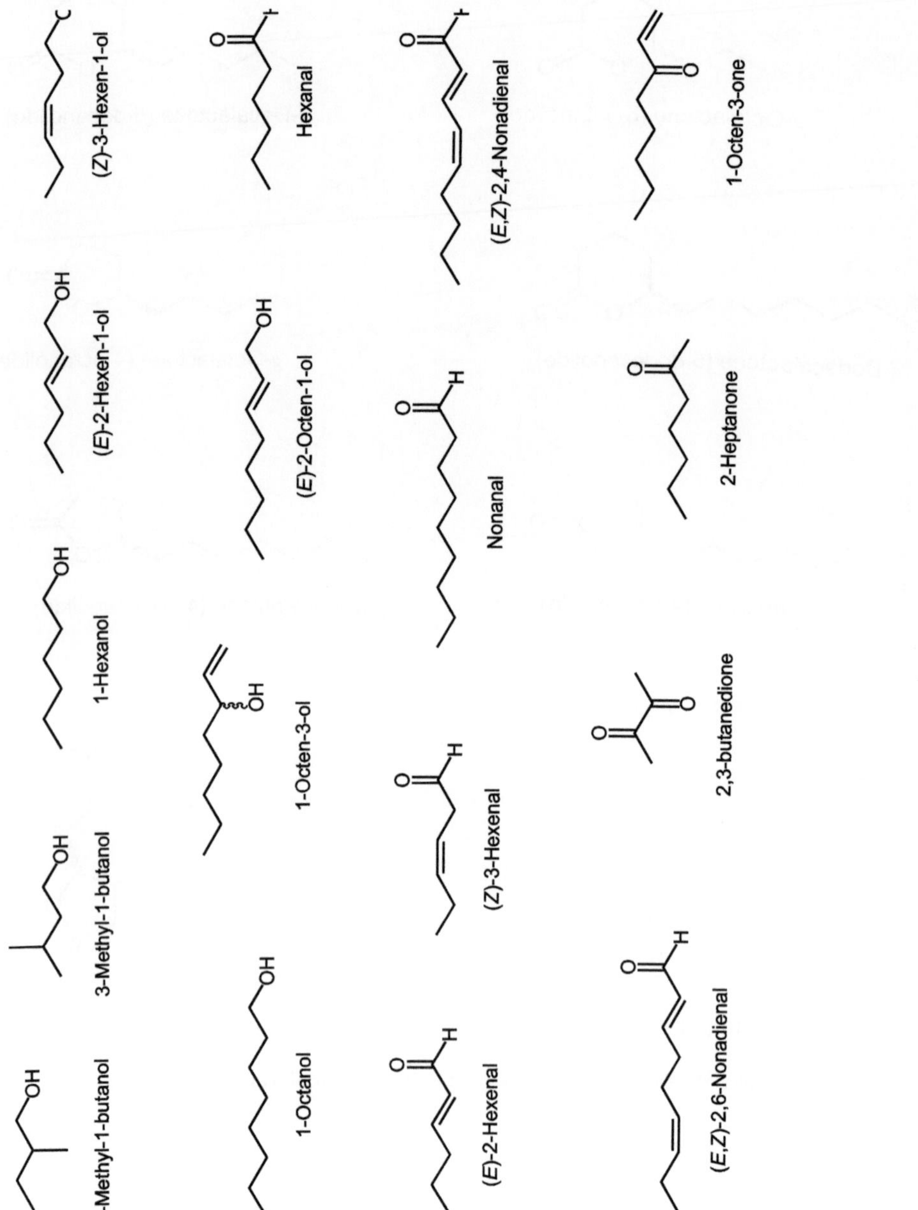

Fig. 7.3 Some aliphatic alcohols, aldehydes and ketones which are important flavour compounds in fruits and vegetables that mainly contribute with green and/or sweet notes

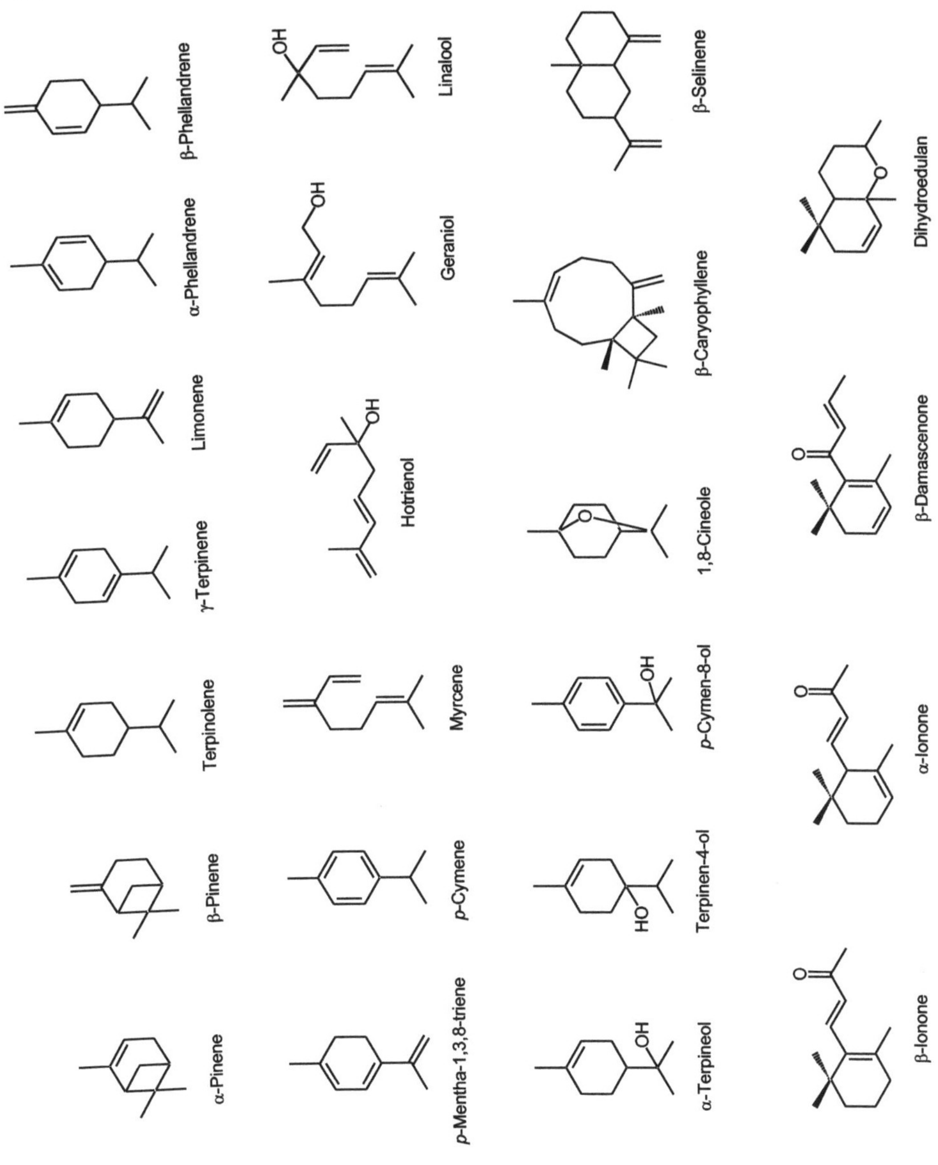

Fig. 7.4 Examples of some terpenes that contribute to the flavour of fruits and vegetables

7.3.2.2 Apricot

Approximately 80 volatile compounds have been identified in apricot [73], including alcohols, aldehydes, alkanes, esters, ketones, lactones and terpenols [13, 16, 67, 73–76]. Toth-Markus et al. [76] identified γ-decalactone (apricot, apricot jam-like odour), linalool, nerol and geraniol (floral, rose-like odour), α-terpineol (spicy, turpentine-like odour) and 2-methylbutyric acid (spicy odour) as important for the flavour of macerated apricots. Takeoka et al. [16] listed linalool and β-ionone as responsible for the floral character of apricots, while γ-octalactone, γ-decalactone and γ-dodecalactone provided a fruity, peach and coconut-like background odour. Ethyl butanoate, ethyl 2-methylbutanoate, butyl butanoate, ethyl hexanoate, butyl 2-methylbutanoate and hexyl 2-methylbutanoate seemed to play a role in the fruity odour of fresh, intact apricot fruits [16]. Similarly, Guichard et al. [77] identified hexyl acetate, γ-octalactone and γ-decalactone as the key flavours of apricots by combining sensory and instrumental data. The *R* form of γ-octalactone, which predominates in apricots, has a spicy-green, coconut and almond note and that of γ-decalactone has a strong, fatty-sweet fruity note somewhat reminiscent of coconut and caramel [75]. Guichard and Souty [73] reported that apricots with a high concentration of C_6 volatiles have a herbaceous note, while apricots that possess irregular terpenes, e.g. β-ionone, have a flowery aroma. Apricots that contain a broad range of C_6 volatiles, terpenes and lactones have the most pleasant aroma.

7.3.2.3 Plum

Approximately 75 volatile compounds have been identified in juices prepared from plums (*Prunus domestica*) [35]. Lactones from C_6 to C_{12} are the major class of compound in plums [78]. The distribution of plum lactones differs from that found in peaches in that the C_{12} γ-lactones are found in higher concentrations than the corresponding C_{10} γ-lactones and δ-decalactone (Fig. 7.2) [78]. GC sniffing has uncovered benzaldehyde, linalool, ethyl nonanoate, methyl cinnamate, γ-decalactone and δ-decalactone as volatile compounds contributing to plum juice aroma (Table 7.2, Figs. 7.1, 7.2, 7.4, 7.5) [35].

7.3.2.4 Cherry

Cherries are divided into sweet cherries (*Prunus avium*) and sour cherries (*Prunus cerasus*). The majority of sweet cherry volatile compounds are alcohols, aldehydes, esters and acetic acid. Sweet cherry fruits contain many volatile

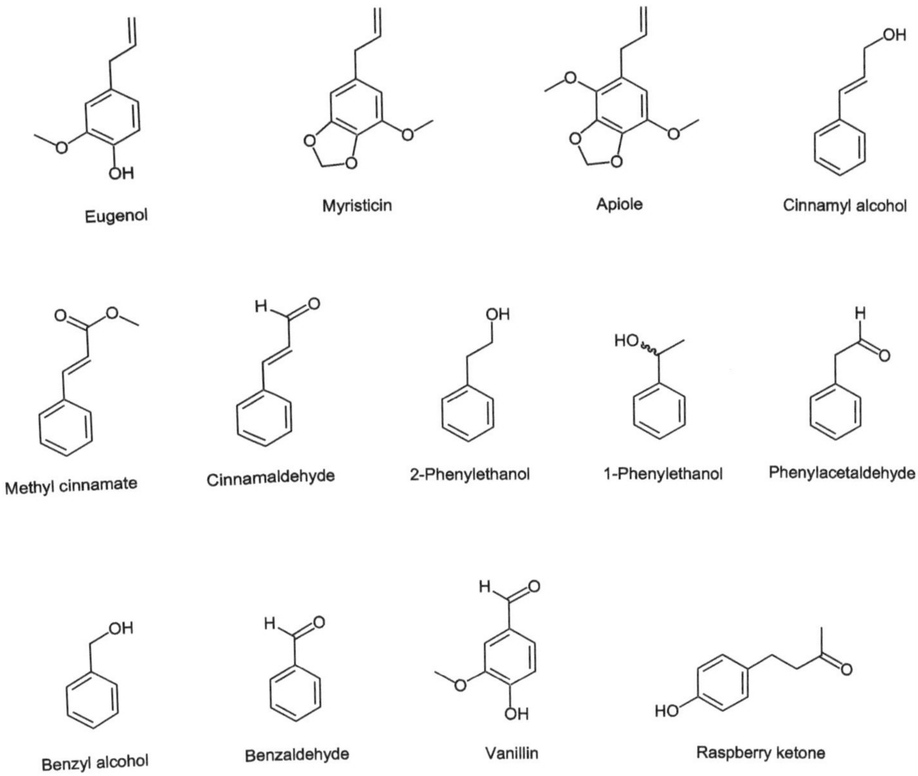

Fig. 7.5 Some phenols and related compounds that are important for the flavour of fruits and vegetables

compounds [79], and a number of these compounds, including benzaldehyde, (*E*)-2-hexenal and hexanal, contribute to fruit flavour and the aroma of macerated sweet cherry fruits, juice and jam [80]. Quantitative and qualitative changes occur in the volatile production during fruit development and ripening and during controlled-atmosphere storage [80, 81].

The typical flavour of sour cherries is produced during processing into wine, liqueur, juice, jam or fruit sauce. Benzaldehyde has been determined to be the most important aroma compound in sour cherries [82], but benzyl alcohol, eugenol and vanillin are also important flavour compounds (Table 7.2, Fig. 7.5) [83]. Growing and storage conditions affect the concentration of benzaldehyde, benzyl alcohol, eugenol and vanillin [83, 84], and cold and rainy weather produces sour cherries with a less delicate sour cherry aroma [83].

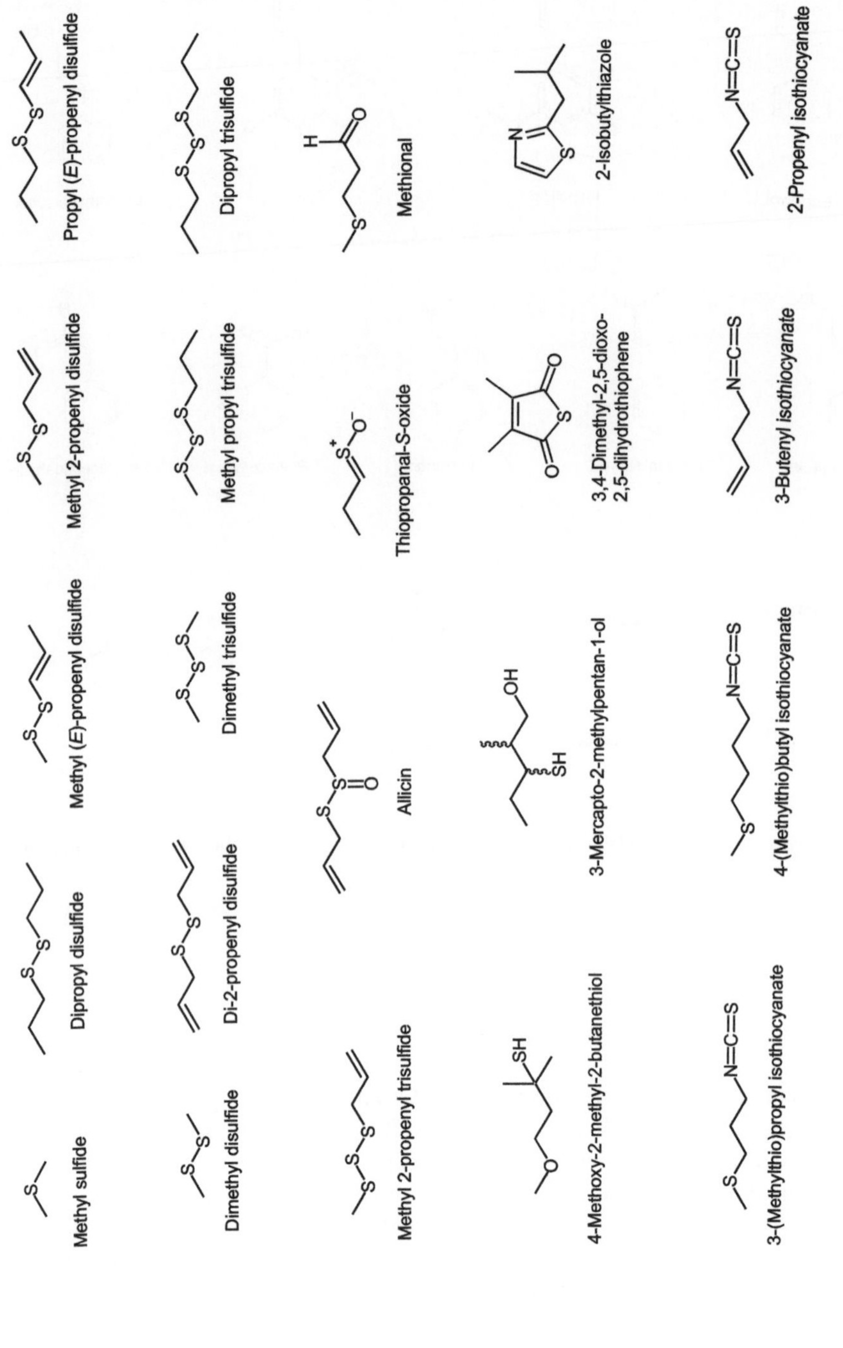

Fig. 7.6 Some sulfur-containing compounds that are important for the flavour cof fruits and vegetables

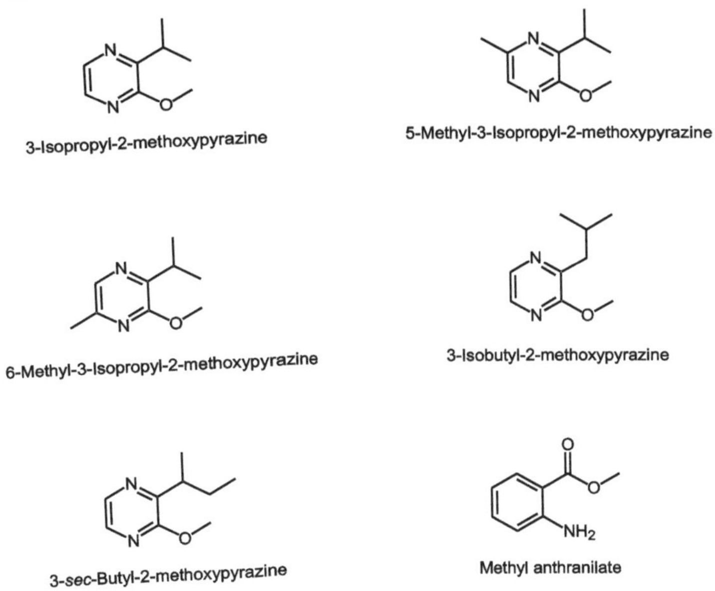

Fig. 7.7 Some pyrazines that are important contributors to the flavour of fruits and vegetables

7.3.3 Berry Fruits

The berry or the small fruits consist of strawberry, raspberry, blackberry, black currant, blueberry, cranberry and elderberry. The volatiles responsible for the flavour of small fruits are esters, alcohols, ketones, aldehydes, terpenoids, furanones and sulfur compounds (Table 7.3, Figs. 7.1–7.7). As fruit ripen, the concentration of aroma volatiles rapidly increases, closely following pigment formation [43].

7.3.3.1 Strawberry

Sugars, acids and aroma compounds contribute to the characteristic strawberry flavour [85]. Over 360 different volatile compounds have been identified in strawberry fruit [35]. Strawberry aroma is composed predominately of esters (25–90% of the total volatile mass in ripe strawberry fruit) with alcohols, ketones, lactones and aldehydes being present in smaller quantities [85]. Esters provide a fruity and floral characteristic to the aroma [35, 86], but aldehydes and furanones also contribute to the strawberry aroma [85, 87]. Terpenoids and sulfur compounds may also have a significant impact on the characteristic strawberry fruit aroma although they normally only make up a small portion of the strawberry volatile compounds [88, 89]. Sulfur compounds, e.g. methanethiol,

Table 7.3 Key flavour compounds in berry fruits

Key flavour compounds	Strawberry (*Fragaria* spp.)	Raspberry (*Rubus idaeus*)	Blackberry (*Rubus fruticosus*)	Black currant (*Ribes nigrum*)	Blueberry (*Vaccinium corymbosum*)	Cranberry (*Vaccinium macrocarpon*)	Elderberry (*Sambucus nigra*)
Esters							
Methyl acetate				[7]			
Butyl acetate	[95]						
Hexyl acetate	[91, 94]						
2-Methylpropyl acetate				[7]			
2-Methylbutyl acetate				[7]			
(*E*)-2-Hexen-1-yl acetate	[91, 94]						
Ethyl propanoate	[85]						
Ethyl 2-methylpropanoate			[109]				
Methyl butanoate	[85, 87, 91–94, 96, 98–100]			[7, 113, 115, 116]	[122]		
Ethyl butanoate	[85, 87, 92–96, 99, 100]		[111]	[7, 113, 115, 116]			
Methyl 2-methylbutanoate	[91, 94, 98]		[109–111]		[122]		[127, 129]
Methyl 3-methylbutanoate					[122]		
Ethyl 2-methylbutanoate					[122]		
Ethyl 3-methylbutanoate	[43, 85, 93]				[122]		[127, 129]
Methyl hexanoate	[43, 91, 93, 95, 96, 99, 100]						
Ethyl hexanoate	[86, 87, 92–96, 98, 99]		[66]	[7, 113, 115, 116]			
Methyl octanoate							[127, 129]
Methyl nonanoate							[127, 129]
Ethyl 9-decenoate							[127]
Alcohols							
2-Methyl-1-butanol							[127–129]
3-Methyl-1-butanol							[127–129]

Table 7.3 *(continued)* Key flavour compounds in berry fruits

Key flavour compounds	Strawberry (*Fragaria* spp.)	Raspberry (*Rubus idaeus*)	Blackberry (*Rubus fruticosus*)	Black currant (*Ribes nigrum*)	Blueberry (*Vaccinium corymbosum*)	Cranberry (*Vaccinium macrocarpon*)	Elderberry (*Sambucus nigra*)
Alcohols *(continued)*							
1-Hexanol							[127–129]
1-Octanol							[127–129]
(*E*)-2-Hexen-1-ol	[95]				[121, 123]		[127–129]
(*Z*)-3-Hexen-1-ol		[101, 103, 105]			[120, 122]		[127–129]
Benzyl alcohol		[101]					
(*E*)-Cinnamyl alcohol						[35]	
1-Phenylethanol						[35]	
2-Phenylethanol						[35]	[127, 128]
Aldehydes							
Hexanal	[93–95]		[109, 110]		[123]		[127–129]
Nonanal				[7, 113]			[127–129]
(*E*)-2-Hexenal	[93, 94]		[110]		[121, 123]		[127–129]
(*E,Z*)-2,6-Nonadienal			[111]				
Benzaldehyde					[43]	[35]	
(*E*)-Cinnamaldehyde						[35]	
Phenylacetaldehyde							[127, 128, 130]
Ketones							
4-(*p*-Hydroxyphenyl)-2-butanone		[101]					
Raspberry ketone		[101]					

Table 7.3 *(continued)* Key flavour compounds in berry fruits

Key flavour compounds	Strawberry (*Fragaria* spp.)	Raspberry (*Rubus idaeus*)	Blackberry (*Rubus fruticosus*)	Black currant (*Ribes nigrum*)	Blueberry (*Vaccinium corymbosum*)	Cranberry (*Vaccinium macrocarpon*)	Elderberry (*Sambucus nigra*)
Ketones *(continued)*							
3-Hydroxy-2-butanone							[127, 129]
2,3-Butanedione				[7, 119]			
2-Heptanone	[99]		[110]				
1-Octen-3-one	[94]			[7, 113, 115]			
(*Z*)-1,5-Octadien-3-one			[111]				
Acids							
Acetic acid		[101, 103]					
Benzoic acid						[35]	
Lactones							
γ-Butyrolactone					[43]		
γ-Decalactone	[91, 94, 99]						
γ-Dodecalactone	[91, 94]						
Terpenoids							
α-Pinene		[103–105]	[110]	[115]			
β-Pinene		[103, 104]					
α-Phellandrene		[103–105]					
1,8-Cineole				[7, 113, 115, 116]			
Limonene					[123]		
Hotrienol							[127–130]
Nerol					[123]		

Table 7.3 *(continued)* Key flavour compounds in berry fruits

Key flavour compounds	Strawberry (*Fragaria* spp.)	Raspberry (*Rubus idaeus*)	Blackberry (*Rubus fruticosus*)	Black currant (*Ribes nigrum*)	Blueberry (*Vaccinium corymbosum*)	Cranberry (*Vaccinium macro-carpon*)	Elderberry (*Sambucus nigra*)
Terpenoids *(continued)*							
Linalool	[94, 99]	[101, 103, 104]	[110, 111]	[7, 116, 119]	[43, 120–123]		
Geraniol	[94, 99]	[101, 103]	[111]		[121, 123]		
Citral		[104]					
Myrtenol			[110]				
α-Terpineol				[7, 115, 116, 119]	[43, 123]		
Terpinen-4-ol				[7, 115, 119]			
β-Caryophyllene		[104, 105]					
Farnesyl acetate	[94]						
β-Damascenone			[111]	[7, 115, 116, 119]			[127–130]
α-Ionone		[101, 104, 105]	[110]				
β-Ionone		[101, 103–105]	[110]				
Dihydroedulan							[128, 129]
Miscellaneous							
Furaneol	[85, 91–95, 97–99]		[109, 111]				
Mesifurane	[43, 85, 91, 95, 97, 99]						
Methional			[109, 111]				
4-Methoxy-2-met 7-yl-butanethiol				[119]			
Methyl anthranilate	[100]						

dimethyl sulfide and dimethyl disulfide (Fig. 7.6), are also considered to be important compounds, particularly in some "older" cultivars [88]. The most important aroma compounds in strawberry include those with a sweet, fruity and green note, e.g. ethyl butanoate, methyl butanoate, methyl hexanoate, ethyl hexanoate, ethyl 3-methylbutanoate, hexyl acetate, (*E*)-2-hexen-1-yl acetate, and those with a caramel-sweet note, e.g. furaneol, mesifurane and linalool (Table 7.3) [43, 85, 90–94]. The concentration of these key flavour compounds depends on the maturation of the fruits and the level of light at harvest as a low light level reduces the concentration of glucose and sucrose in the fruit [91, 95–98]. Larsen et al. [99] proposed that furaneol, linalool and ethyl hexanoate were important for general strawberry aroma and that ethyl butanoate, methyl butanoate, γ-decalactone and 2-heptanone were important for cultivar-specific aroma. Ulrich et al. [100] divided strawberries into three aroma groups: a methyl anthranilate type, which contain methyl anthranilate (spicy-aromatic and flowery note) (Fig. 7.7) as in wood strawberries (*Fragaria vesca*); an ester type, which has a high content of fresh and fruity ester aroma; and a furaneol type, which has a high content of furaneol and mesifurane, but has a medium to poor strawberry flavour.

7.3.3.2 Raspberry

Approximately 230 volatile compounds have been identified in raspberry fruit [35]. The aroma of raspberries is composed of a mixture of ketones and aldehydes (27%) and terpenoids (30%), alcohols (23%), esters (13%) and furanones (5%). The raspberry ketone (Fig. 7.5) along with α-ionone and β-ionone have been found to be the primary character-impact compounds in raspberries. Other compounds such as benzyl alcohol, (*Z*)-3-hexen-1-ol, acetic acid, linalool, geraniol, α-pinene, β-pinene, α-phellandrene, β-phellandrene and β-caryophyllene contribute to the overall aroma of mature red raspberries [101–105]. The most important character-impact compounds of raspberries are summarised in Table 7.3.

7.3.3.3 Blackberry

Wild and cultivated blackberries have been used as food and medicine for hundreds of years [106]. Approximately 150 volatiles have been reported from blackberries [107]. The aroma profile is complex, as no single volatile is described as characteristic for blackberry [108, 109]. Several compounds have been suggested as prominent volatiles in blackberries using AEDA, e.g. ethyl hexanoate, ethyl 2-methylbutanoate, ethyl 2-methylpropanoate, 2-heptanone, 2-undecanone, 2-heptanol, 2-methylbutanal, 3-methylbutanal, hexanal, (*E*)-2-hexenal, furaneol, thiophene, dimethyl sulfide, dimethyl disulfide, dimethyl trisulfide, 2-methylthiophene, methional, α-pinene, limonene, linalool, sabinene,

α-ionone and β-ionone [109, 110]; however, these volatiles may vary between growing regions [109–111]. Recently, Wang et al. [111] demonstrated that the same cultivar grown in different regions in the USA had similar aroma compositions; however, in one region ethyl butanoate (fruity, apple-like), linalool (floral, perfume), methional (cooked potato), (*E*,*Z*)-2,6-nonadienal (green cucumber), (*Z*)-1,5-octadien-3-one (green grass) and furaneol (sweet, strawberry-like) were prominent, while ethyl butanoate, linalool, methional, methyl 2-methylbutanoate (fruity), β-damascenone (rose-like, berry) and geraniol (sweet, rose-like) were prominent volatiles in another region. The most important character-impact compounds of blackberries are summarised in Table 7.3.

7.3.3.4 Black Currant

The aroma of intact black currant fruit is mostly produced by anabolic pathways of the plant, and production of fruit volatiles occurs mainly during a short ripening period [112]. The aroma profile of black currant shares similarities with that of other berry fruits, although terpenes are more abundantly present in black currant [107]. Black currant is mainly used for the production of juice. Over 150 volatile compounds have been reported from either black currant berries and/or juice, of which the major groups are monoterpenes, sesquiterpenes, esters and alcohols [107]. Processing of berries to juice has been shown to lead to major changes in the aroma composition [113–118].

Important aroma compounds of black currant berries have been identified mainly by GC-O techniques by Latrasse et al. [119], Mikkelsen and Poll [115] and Varming et al. [7] and those of black currant nectar and juice by Iversen et al. [113]. The most important volatile compounds for black currant berry and juice aroma include esters such as 2-methylbutyl acetate, methyl butanoate, ethyl butanoate and ethyl hexanoate with fruity and sweet notes, nonanal, β-damascenone and several monoterpenes (α-pinene, 1,8-cineole, linalool, terpinen-4-ol and α-terpineol) as well as aliphatic ketones (e.g. 1-octen-3-one) and sulfur compounds such as 4-methoxy-2-methyl-butanethiol (Table 7.3, Figs. 7.3, 7.4, 7.6). 4-Methoxy-2-methylbutanethiol has a characteristic "catty note" and is very important to black currant flavour [119].

7.3.3.5 Blueberry

Blueberry consists of cultivated highbush blueberries (*Vaccinium corymbosum*) and wild lowbush blueberries (*Vaccinium angustifolium*). The aroma of cultivated and wild blueberries is dominated by long-chain alcohols, esters and terpenoids. Forney [43] reported that γ-butyrolactone, α-terpineol, 6-ethyl 2,6-decadiene-4,5-diol, linalool, benzaldehyde and 2-ethyl-2-hexenal contribute to the aroma of fresh, whole highbush blueberries using GC-O analysis. In another study, Parliament and Scarpellino [120] determined that a combination of linalool and (*Z*)-3-hexen-1-ol produced a blueberry-like flavour, while Horvat and Senter [121] reported that a mixture of (*Z*)-3-hexen-1-ol, (*E*)-2-hexen-1-ol, (*E*)-2-hexenal, linalool and geraniol gave an aroma similar to the aroma isolated from blueberries. The odour-active volatiles of intact lowbush blueberries (*Vaccinium angustifolium*) include methyl 2-methylbutanoate, ethyl 3-methylbutanoate, ethyl 2-methylbutanoate, ethyl 3-methylbutanoate, methyl butanoate and linalool [122]. The major contributors to the aroma profile of blueberry juice are hexanal, (*E*)-2-hexenal and (*E*)-2-hexen-1-ol, limonene, linalool, α-terpineol, geraniol and nerol (Table 7.3) [123].

7.3.3.6 Cranberry

The American cranberry (*Vaccinium macrocarpon*) is larger than the European cranberry (*Vaccinium oxycoccous*) but poorer in aroma. The European cranberry is a valuable raw material in the production of alcoholic drinks, liqueurs and jams in Scandinavia [35]. A few older studies report approximately 70 volatile compounds in cranberry [124, 125]. Cranberry aroma is characterised by several aromatic compounds, such as 1-phenylethanol, 2-phenylethanol, 3-phenylpropanol, (*E*)-cinnamyl alcohol, 2-(4-hydroxyphenyl)ethanol, 2-(4-methoxyphenyl)ethanol, salicylaldehyde and 4-methoxybenzaldehyde. A tart flavour has been attributed to the levels of benzoic acid, although benzaldehyde, 4-methoxybenzaldehyde, benzoate and benzyl esters might significantly contribute to the overall cranberry aroma [35]. The most important character-impact compounds of cranberry are summarised in the Table 7.3.

7.3.3.7 Elderberry

Elderberry (*Sambucus nigra*) is cultivated on small scale in Europe. The fruits have a high concentration of red and purple anthocyanins and a relatively low concentration of sugars, organic acids and aroma compounds, which make this juice attractive as a natural colour ingredient in other red fruit products [126–129]. The fresh green odour of elderberry juice is associated with volatile compounds with typical green notes such as 1-hexanol, 1-octanol, (*Z*)-3-hexen-1-ol, (*E*)-2-hexen-1-ol, hexanal and (*E*)-2-hexenal, whereas the floral aroma is mainly due to the presence of hotrienol and nonanal [127–130].

The characteristic elderberry odour has been shown to be correlated to β-damascenone, dihydroedulan and ethyl 9-decenoate with elderberry-like notes, and to some extent also to 2-phenylethanol, phenylacetaldehyde and nonanal with elderflower-like notes [127, 128, 130, 131]; however, only nonanal, dihydroedulan and β-damascenone have repeatedly been identified in various investigations as character-impact compounds for elderberry odour.

The fruity-sweet flavours in elderberry juice and products have primarily been associated with aliphatic esters such as ethyl 2-methylbutanoate, ethyl 3-methylbutanoate, methyl heptanoate, methyl octanoate, methyl nonanoate, alcohols (2-methyl-1-propanol, 2-methyl-1-butanol and 3-methyl-1-butanol) and the aldehydes pentanal, heptanal and octanal [127, 129, 130, 132].

The most important character-impact compounds of elderberries are summarised in Table 7.3.

7.3.4 Soft Fruits

7.3.4.1 Grapes

The flavour of grapes is made up of volatile alcohols, esters, acids, terpenes and carbonyl compounds. Grapes (genus *Vitis*) are used for winemaking or as table grapes. Grape varieties may be divided into aromatic and non-aromatic varieties. Most wine-producing varieties belong to the non-aromatic type [133] which mainly produce C_6 alcohols and aldehydes such as hexanal, (*E*)-2-hexenal, 1-hexanol, (*Z*)-3-hexen-1-ol and (*E*)-2-hexen-1-ol, formed after crushing of the skin [133, 134]. Octanoic acid and alcohols, particularly 2-phenylethanol, are also recognised after crushing [133]. Free terpenols, e.g. linalool and geraniol, have been identified as major aroma compounds in red grapes and in white Muscat grapes (Table 7.4) [133, 135]. Fruity flavour, sweetness and skin friability are highly correlated with consumer likings of table grapes [136]; however, the key flavour compounds of table grapes still need to be identified.

7.3.4.2 Kiwi

The kiwi fruit is a cultivar group of the species *Actinidia deliciosa*. More than 80 compounds have been identified in fresh and processed kiwi [137]. Methyl acetate, methyl butanoate, ethyl butanoate, methyl hexanoate and (*E*)-2-hexenal have the most prominent effect on consumer acceptability of kiwi fruit flavour [137–140]. The volatile composition of kiwi fruit is very sensitive to ripeness, maturity and storage period [138, 139]. Bartley and Schwede [140] found that (*E*)-2-hexenal was the major aroma compound in mature kiwi fruits, but on further ripening ethyl butanoate began to dominate. Ripe fruits had sweet and fruity flavours, which were attributed to butanoate esters, while unripe fruits had a green grassy note due to (*E*)-2-hexenal [140]. The most important character-impact compounds of kiwi fruits are summarised in Table 7.4.

Table 7.4 Key flavour compounds in soft fruits

Key flavour compounds	Kiwi (*Actinidia deliciosa*)	Grapes (*Vitis vinifera*)
Esters		
Methyl acetate	[139]	
Methyl butanoate	[138–140]	
Ethyl butanoate	[137–140]	
Methyl hexanoate	[139]	
Alcohols		
1-Hexanol		[133, 134]
(*E*)-2-Hexen-1-ol		[134]
(*Z*)-3-Hexen-1-ol		[134, 232]
Aldehydes		
Hexanal		[134, 232]
(*E*)-2-Hexenal	[137–140]	[134, 232]
Terpenoids		
Geraniol		[133]
Linalool		[133, 232]

7.4 Vegetables

The modern distinction between vegetable and fruit has been applied and therefore those plants or plant parts that are usually consumed with the main course of a meal will be regarded as vegetables; thus, cucumber, tomato and pumpkin that botanically are classified as fruits are included in this section. The flavour compounds found in vegetables are diverse and include fatty acid derivatives, terpenes, sulfur compounds as well as alkaloids. This diversity is partially responsible for the unique flavours found in different species of vegetables.

7.4.1 Alliaceae

7.4.1.1 Onion and Shallot

The bulb of the onion (*Allium cepa* L.) can be eaten raw or cooked after boiling, roasting or frying. More than 140 volatile compounds have been identified in onions. The characteristic onion flavour develops when the cells are disrupted, allowing the enzyme alliinase to act upon the aroma precursors (+)-*S*-alk(en)yl cysteine sulfoxides (Sect. 7.2.2), yielding a large number of volatile sulfur compounds that contribute significantly to the aroma of raw onion [35, 36, 141, 142]. The chemistry of onion volatiles is, however, quite complex, in particular, because significant changes occur during storage and/or processing in the volatile spectrum owing to disruption of the cell walls [143, 144].

The most important flavour compound in raw onions is thiopropanal-*S*-oxide, the lachrymatory factor [145, 146]. Other important flavour compounds are 3,4-dimethyl-2,5-dioxo-2,5-dihydrothiophene and alkyl alkane thiosulfonates such as propyl methanethiosulfonate and propyl propanethiosulfonate with a distinct odour of freshly cut onions [35, 36, 147]. Various thiosulfinates that have a sharp and pungent odour may also contribute to the flavour of onions. These compounds, however, are rapidly decomposed to a mixture of alkyl and alkenyl monosulfides, disulfides and trisulfides (Scheme 7.3) of which dipropyl disulfide, methyl (*E*)-propenyl disulfide, propyl (*E*)-propenyl disulfide, dipropyl trisulfide and methyl propyl trisulfide are the most important contributors to the aroma of raw and cooked onions (Table 7.5, Fig. 7.6) [148–150]. Recently, 3-mercapto-2-methylpentan-1-ol was identified in raw and cooked onions eliciting intense meat broth, sweaty, onion and leek-like odours [142, 151].

Shallots (*Allium ascalonicum*) are an allium wherein the bulb laterals separate into individual bulbs. Apparently, shallots do not develop a lachrymatory factor, such as thiopropanal-*S*-oxide upon maceration [35]. The major aroma constituents in shallots are similar to those found in *A.cepa*. In raw shallots, the most important aroma compounds appear to be dipropyl disulfide, propyl (*E*)-propenyl disulfide, methyl propyl trisulfide, dimethyl trisulfide and dipropyl trisulfide (Table 7.5, Fig. 7.6) [35, 152, 153].

7.4.1.2 Garlic

The bulb-like root of garlic (*Allium sativum*) consists of several cloves. Garlic is used principally as a flavouring agent, fresh, dried or as an oil obtained by steam distillation. More than 30 volatiles have been identified in garlic [35, 154–157]. The characteristic flavour of crushed raw garlic is due to formation of dialkyl thiosulfinates by the action of alliinase upon *S*-alk(en)yl cysteine sulfoxides. Allicin, which is formed from alliin (*S*-allyl cysteine sulfoxides) is the most abundant and important dialkyl thiosulfinate formed in garlic [35, 141, 146, 157]. However, allicin is very unstable and will undergo non-enzymatic disproportionation and form symmetrical and mixed monosulfides, disulfides and trisulfides, many of which contribute to garlic flavour [35, 141, 146, 157]. Volatile sulfur compounds with a characteristic *Allium* flavour found in garlic include allicin, di-2-propenyl disulfide, methyl 2-propenyl disulfide, dimethyl trisulfide, methyl 2-propenyl trisulfide and di-2-propenyl trisulfide (Table 7.5, Fig. 7.6) [154–160].

Table 7.5 Key flavour compounds in *Allium* species (Alliaceae)

Key flavour compounds	Onion (*A. cepa*)	Garlic (*A. sativum*)	Leek (*A. ampeloprasum*)	Shallot (*A. ascalonicum*)
Sulfur compounds				
1-Propanethiol			[163]	
Dipropyl disulfide	[148–150]		[148, 163, 164]	[35, 152, 153]
Methyl (*E*)-propenyl disulfide	[148–150]		[148, 150, 164]	
Methyl 2-propenyl disulfide		[155, 156]		
Propyl (*E*)-propenyl disulfide	[148–150]		[148, 163, 164]	[35, 152, 153]
Di-2-propenyl disulfide		[155, 156]		
Methyl propyl trisulfide	[148–150]			[35, 152, 153]
Dimethyl trisulfide		[155, 156]		[35, 152, 153]
Dipropyl trisulfide	[148–150]		[148, 163, 164]	[35, 152, 153]
Di-2-propenyl trisulfide		[155, 156]		
Methyl 2-propenyl trisulfide		[35]		
Allicin				
Propyl methanethiosulfonate	[35, 36, 147]			
Propyl propanethiosulfonate	[35, 36, 147]			
3,4-Dimethyl-2,5-dioxo-2,5-dihydrothiophene	[35, 36, 147]			
3-Mercapto-2-methylpentan-1-ol	[142, 151]			
Thiopropanal-*S*-oxide	[145, 146]			
Alcohols				
2-Propen-1-ol		[156]		
1-Octen-3-ol			[163]	
Aldehydes				
Pentanal			[148, 163, 164]	
Hexanal			[148, 163, 164]	
Decanal			[148, 163, 164]	

7.4.1.3
Leek

The edible portion of leek (*Allium ampeloprasum* var. *porrum*) is a false stem or elongated bulb. More than 90 volatile compounds have been reported from leek, including numerous sulfur-containing volatile compounds. It is the thiosulfinates that originate from alliinase-catalysed decomposition of (+)-*S*-alk(en)yl cysteine sulfoxides [161, 162] that are responsible for the odour of freshly cut leeks [144, 145, 163]. The thiosulfinates readily rearrange to thiosulfonates, which then transform to various monosulfides, disulfides and trisulfides (Scheme 7.3). 1-Propanethiol, dipropyl disulfide, dipropyl trisulfide, methyl (*E*)-propenyl disulfide and propyl (*E*)-propenyl disulfide are the most important sulfur-containing aroma compounds possessing leek aroma notes in fresh and blanched leek (Table 7.5, Fig. 7.6) [31, 35, 148, 163, 164].

Products of the LOX pathway or compounds formed by autoxidation of fatty acids (Scheme 7.2) are also important for leek aroma [31, 163]. Volatile compounds of the LOX pathway are not pronounced in the aroma profile of freshly cut leeks owing to a high content of thiosulfinates and thiopropanal-*S*-oxide [30]. In processed leeks that have been stored for a long time (frozen storage), however, these aliphatic aldehydes and alcohols have a greater impact on the aroma profile owing to volatilisation and transformations of sulfur compounds [31, 165]. The most important volatiles produced from fatty acids and perceived by GC-O of raw or cooked leeks are pentanal, hexanal, decanal and 1-octen-3-ol (Table 7.5) [31, 35, 148, 163, 164].

7.4.2
Brassicaceae (Formerly Cruciferae)

7.4.2.1
Broccoli

The edible portion of broccoli (*Brassica oleracea* var. *italica*) is the inflorescence, and it is normally eaten cooked, with the main meal. Over 40 volatile compounds have been identified from raw or cooked broccoli. The most influential aroma compounds found in broccoli are sulfides, isothiocyanates, aliphatic aldehydes, alcohols and aromatic compounds [35, 166–169]. Broccoli is mainly characterised by sulfurous aroma compounds, which are formed from glucosinolates and amino acid precursors (Sects. 7.2.2, 7.2.3) [170–173]. The strong off-odours produced by broccoli have mainly been associated with volatile sulfur compounds, such as methanethiol, hydrogen sulfide, dimethyl disulfide and trimethyl disulfide [169, 171, 174, 175]. Other volatile compounds that also have been reported as important to broccoli aroma and odour are dimethyl sulfide, hexanal, (*Z*)-3-hexen-1-ol, nonanal, ethanol, methyl thiocyanate, butyl isothiocyanate, 2-methylbutyl isothiocyanate and 3-isopropyl-2-methoxypyrazine

(Table 7.6) [166, 168, 169, 174, 175]. Some of the odour sensations characteristic of these volatile compounds are "cabbage", "boiled potato", "cut grass", "floral", "citrus", "sour", "laundry" and "vegetation" [166, 167].

Table 7.6 Key flavour compounds in *Brassica* species (Brassicaceae)

Key flavour compounds	Broccoli (*B. oleraceae* var. *italica*)	Brussel sprout (*B. oleraceae* var. *gemmifera*)	Cabbage (*B. oleraceae* var. *capita*)	Cauliflower (*B. oleraceae* var. *botrytis*)
Sulfur compounds				
Methanethiol			[35, 177]	[183]
Dimethyl sulfide (methyl sulfide)	[166]	[35]		[177, 183]
Dimethyl disulfide	[166, 169, 174]	[176]		
Dimethyl trisulfide	[166, 174]	[35]	[178]	[177, 183]
3-(Methylthio)propyl isothiocyanate				[35, 177]
4-(Methylthio)butyl isothiocyanate	[35]			
2-Propenyl isothiocyanate		[35, 176]	[35, 178]	[183]
Butyl isothiocyanate	[168, 169, 174]			
2-Methylbutyl isothiocyanate	[168, 169, 174]			
3-Butenyl isothiocyanate			[35]	
Esters				
Methyl acetate			[35]	
Ethyl acetate			[35]	
Alcohols				
Ethanol	[166, 168, 169]		[35]	
(*Z*)-3-Hexen-1-ol	[166]		[35]	
Aldehydes				
Hexanal	[166, 167, 169]		[35]	
Nonanal	[166]			[175, 177]
(*E*)-2-Hexenal			[35]	
Pyrazines				
3-Isopropyl-2-methoxypyrazine	[166, 168, 169]			

7.4.2.2
Brussels Sprout

The buds and the leaves (less often) of the Brussels sprout plant (*Brassica oleracea* var. *gemmifera*) are eaten cooked with the main meal. In Brussels sprouts, breakdown products from glucosinolates are dominant and represent about 80–90% of the volatiles in headspace samples [176]. The residual volatiles are mostly sulfur compounds [176]. Compounds likely to be associated with the aroma of Brussels sprouts are 2-propenyl isothiocyanate, dimethyl sulfide, dimethyl disulfide and dimethyl trisulfide (Table 7.6) [35, 176].

7.4.2.3
Cabbage

The leaves of cabbage (*Brassica oleracea* var. *capitata*) can be eaten cooked as part of the main meal, raw as coleslaw or as a fermented product. They can be cooked after processing such as by dehydration. A total of approximately 160 volatile compounds have been identified in the raw, cooked, and dehydrated material, and includes aliphatic alcohols, aldehydes and esters as well as isothiocyanates and other sulfur containing compounds [35, 177–180]. 2-Propenyl isothiocyanate is generally considered one of the desirable flavour compounds in cabbage where it provides characteristic fresh cabbage notes and hotness. This component appears to be important in very fresh cabbage, since it is found to be the major flavour-bearing sulfur compound detected soon after blending [177]. Other major compounds identified in raw cabbage include methanethiol, dimethyl trisulfide, ethanol, methyl acetate, ethyl acetate, hexanal, (*E*)-2-hexenal and (*Z*)-3-hexen-1-ol [35, 177, 178]. The most important flavour compounds of cabbage leaves are summarised in the Table 7.6.

7.4.2.4
Cauliflower

The roundish flower head, the curd, of the cauliflower plant (*Brassica oleracea* var. *botrytis*) is the edible portion of this vegetable. It can be eaten raw in salads or as a pickled condiment in vinegar. More often it is boiled and eaten with the main meal or is converted into sauces and soups. Over 80 volatile compounds have been identified in raw and cooked cauliflower. Among the compounds potentially active in cooked cauliflower, certain sulfides such as methanethiol, dimethyl sulfide and dimethyl trisulfide have often been incriminated in objectionable sulfurous aromas and overcooked off-flavours [169, 177, 178, 181–183]. Additional aldehydes have been found to be the most abundant cauliflower volatiles, with nonanal as a major component [175, 177]. A recent study showed that volatiles such as 2-propenyl isothiocyanate, dimethyl trisulfide, dimethyl sulfide and methanethiol were the key odorants of cooked cauliflower "sulfur" odours, whereas different glucosinolates were correlated with bitterness intensity [183]. Some of the most important character-impact compounds of raw and/or cooked cauliflower are summarised in Table 7.6.

7.4.3 Cucurbitaceae

7.4.3.1 Cucumber

The fruit of the cucumber plant (*Cucumis sativus*) is mainly eaten raw or as pickle. Approximately 30 volatile compounds have been detected in the volatile fraction of cucumber, with aliphatic alcohols and carbonyl compounds being most abundant [35]. Fresh cucumber flavour develops as a result of enzymatic degradation of linoleic and linolenic acid rapidly after the tissue is disrupted (Scheme 7.2), by which (*E*,*Z*)-2,6-nonadienal and (*E*)-2-nonenal mainly are formed [184]. (*E*,*Z*)-2,6-Nonadienal is the main flavour volatile of cucumber fruit, with (*E*)-2-nonenal as the second most important compound (Table 7.7) [185, 186].

Table 7.7 Key flavour compounds in Cucurbitaceae fruits

Key flavour compounds	Cucumber (*Cucumis sativus*)	Pumpkin (*Cucurbita pepo*)
Alcohols		
(*Z*)-3-Hexen-1-ol		[35]
Aldehydes		
Hexanal		[35]
(*E*)-2-Hexenal		[35]
(*E*,*Z*)-2,6-Nonadienal	[184–186]	
(*E*)-2-Nonenal	[184–186]	
Ketones		
2,3-Butanedione		[35]
Pyrazines		
3-Isopropyl-2-methoxypyrazine	[35]	

7.4.3.2
Pumpkin

The fruit of pumpkin (*Cucurbita pepo*) is eaten boiled or baked. About 30 compounds have been identified in the volatile extracts of raw pumpkin, with the major classes of compounds being aliphatic alcohols and carbonyl compounds, furan derivatives and sulfur-containing compounds. Hexanal, (*E*)-2-hexenal, (*Z*)-3-hexen-1-ol and 2,3-butanedione have been identified as important for the flavour of freshly cooked pumpkins (Table 7.7) [35]; however, studies using GC-O techniques are needed to get a better understanding of the character-impact compounds of pumpkins.

7.4.4
Fabaceae (Formerly Leguminosae) and Solanaceae

7.4.4.1
Potato

The tuber of potato (*Solanum tuberosum*) is eaten boiled, baked or fried, and after rehydration or reheating of dried, frozen or canned products.

Raw potato possesses little aroma. Approximately 50 compounds have been reported to contribute to raw potato aroma. Raw potatoes have a high content of LOX, which catalyses the oxidation of unsaturated fatty acids into volatile degradation products (Scheme 7.2) [187]. These reactions occur as the cells are disrupted, e.g. during peeling or cutting. Freshly cut, raw potatoes contain (*E*,*Z*)-2,4-decadienal, (*E*,*Z*)-2,6-nonadienal, (*E*)-2-octenal and hexanal, which are all products of LOX-initiated reactions of unsaturated fatty acids [188, 189]. It is reported that two compounds represent typical potato aroma in raw potato: methional and (*E*,*Z*)-2,6-nonadienal [189]. Other important volatiles in raw potatoes produced via the LOX pathway are 1-penten-3-one, heptanal, 2-pentyl furan, 1-pentanol and (*E*,*E*)-2,4-heptadienal [189]. Pyrazines such as 3-isopropyl-2-methoxypyrazine could be responsible for the earthy aroma of potato [35]. Some of the most important character-impact compounds of raw potatoes are summarised in Table 7.8. Aroma compounds from cooked, fried and baked potatoes have previously been reviewed [35].

7.4.4.2
Tomato

The fruit of the tomato plant (*Lycopersicon esculentum*) is eaten raw, boiled, baked or fried. Tomato is also canned whole or pureed. More than 400 volatile compounds have been identified in tomato [190, 191], of which 16 or so have odour-

Table 7.8 Key flavour compounds in Fabaceae (pea) and Solanaceae (potato, tomato) vegetables

Key flavour compounds	Pea (*Pisum sativum*)	Potato (*Solanum tuberosum*)	Tomato (*Lycopersicon esculentum*)
Alcohols			
3-Methyl-1-butanol			[234]
1-Pentanol		[189]	
1-Hexanol	[206]		[192]
1-Octen-3-ol		[35]	
(*Z*)-3-Hexen-1-ol	[206]		[35, 192]
(*E*)-2-Octenol		[35]	
2-Phenylethanol			[192]
Aldehydes			
3-Methylbutanal			[192]
Hexanal	[206]	[188]	[196, 198, 234, 235]
Heptanal		[189]	
(*E*)-2-Hexenal			[192]
(*Z*)-3-Hexenal			[190, 198, 235, 236]
(*E*)-2-Octenal	[206]	[188]	
(*E*,*E*)-2,4-Heptadienal		[189]	
(*E*,*Z*)-2,6-Nonadienal		[189]	
Ketones			
1-Penten-3-one		[189]	[196, 234]
1-Octen-3-one			[196]

Table 7.8 *(continued)* Key flavour compounds in Fabaceae (pea) and Solanaceae (potato, tomato) vegetables

Key flavour compounds	Pea (*Pisum sativum*)	Potato (*Solanum tuberosum*)	Tomato (*Lycopersicon esculentum*)
Sulfur compounds			
Methional		[189]	[192]
2-Isobutylthiazole			
Pyrazines			
3-Isopropyl-2-methoxypyrazine	[203, 204, 206]	[35]	
3-*sec*-Butyl-2-methoxypyrazine	[203, 204, 206]		
3-Isobutyl-2-methoxypyrazine	[206]		
5-Methyl-3-isopropyl-2-methoxypyrazine	[206]		
6-Methyl-3-isopropyl-2-methoxypyrazine	[206]		
Terpenoids			
6-Methyl-5-hepten-2-one			[192]
β-Ionone			[192]
β-Damascenone			[35]
Miscellaneous			
Furaneol			[202]

threshold values that indicate that they contribute to tomato flavour. The nature and relative amount of volatiles in tomato seem to depend on species, maturity and preparation of the product more than in any other vegetable. No character-impact compound has been identified in tomatoes, although 2-isobutylthiazole is unique to tomato flavour [192]. The most important compounds in tomatoes are 3-methylbutanal, hexanal, (*Z*)-3-hexenal, (*E*)-2-hexenal, 3-methyl-1-butanol, 1-hexanol, (*Z*)-3-hexen-1-ol, 1-penten-3-one, 6-methyl-5-hepten-2-one, β-ionone, β-damascenone, 2-phenylethanol, methyl salicylate, furaneol and 2-isobutylthiazole, and of these, (*Z*)-3-hexenal and β-ionone have the highest odour units [190–202].

7.4.4.3 Pea

The seed and immature pod of the pea plant (*Pisum sativum*) are traditionally eaten raw or cooked or fried. Approximately, 120 volatile compounds have been identified in peas, with 1-hexanol, 1-propanol, 2-methylpropanol, 1-pentanol, 2-methyl-1-butanol, 3-methyl-1-butanol and (*Z*)-3-hexen-1-ol found in the highest concentrations [35, 203–206]. Compounds contributing to the aroma profile of peas seem to be grouped in two main categories: (1) the fatty acid breakdown products, which contribute to the pea aroma with "strong, green", "perfume, sweet", "orange, sweet" and "mushroom" odours and (2) the methoxypyrazines, responsible for the characteristic pea aroma also associated with bell pepper [206]. The most important volatile compounds of the first category include hexanal, (*E*)-2-heptenal, (*E*)-2-octenal, 1-hexanol and (*Z*)-3-hexen-1-ol, and those of the second category include 3-alkyl-2-methoxypyrazines, such as 3-isopropyl-2-methoxypyrazine, 3-*sec*-butyl-2-methoxypyrazine, 3-isobutyl-2-methoxypyrazine, 5-methyl-3-isopropyl-2-methoxypyrazine and 6-methyl-3-isopropyl-2-methoxypyrazine (Table 7.8) [35, 206].

7.4.5 Apiaceae (Formerly Umbelliferae)

7.4.5.1 Carrots

The root of carrot (*Daucus carota*) is eaten raw or cooked. The characteristic aroma and flavour of carrots are mainly due to volatile compounds, although non-volatile polyacetylenes and isocoumarins contribute significantly to the bitterness of carrots [1, 2]. More than 90 volatile compounds have been identified from carrots (Table 7.9) [207–215]. The carrot volatiles consist mainly of terpenoids in terms of numbers and amounts and include monoterpenes, sesquiterpenes and irregular terpenes. Monoterpenes and sesquiterpenes account

Table 7.9 Key flavour compounds in Apiaceae vegetables

Key flavour compounds	Carrot (*Daucus carota*)	Celery and celeriac (*Apium graveolens*)	Parsnip (*Pastinaca sativa*)	Parsley (*Petroselinum crispum*)
Terpenoids				
p-Cymene	[209, 216, 219]			
Limonene	[207, 216, 219]	[220–222]		
p-Mentha-1,3,8-triene				[227, 237–241]
Myrcene	[207, 209, 216, 219]	[222]		[240]
β-Ocimene	[216]			
β-Phellandrene				[229, 237, 241]
α-Pinene	[207, 216, 219]			
Sabinene	[207, 209, 212, 216]			
γ-Terpinene	[207, 212, 216]	[221]		
Terpinolene	[207, 209, 212, 217]		[35]	[229, 237]
Linalool				[227]
β-Caryophyllene	[207, 209, 212, 217]			
β-Selinene		[35]		
β-Ionone	[209]			
(*E*)-γ-Bisabolene	[207, 216, 217]			
Aldehydes				
(*Z*)-3-Hexenal				[227]
(*Z*)-6-Decenal				[227]

Table 7.9 *(continued)* Key flavour compounds in Apiaceae vegetables

Key flavour compounds	Carrot (*Daucus carota*)	Celery and celeriac (*Apium graveolens*)	Parsnip (*Pastinaca sativa*)	Parsley (*Petroselinum crispum*)
Phthalides				
3-Butylphthalide		[35, 220]		
Sedanolide		[35, 220, 226]		
Pyrazines				
3-*sec*-Butyl-2-methoxypyrazine			[35]	[227]
Miscellaneous				
Apiole				[237]
Myristicin			[35]	[227]
4-Isopropenyl-1-methylbenzene				[229, 237, 241]

for about 98% of the total volatile mass in carrots [208, 213]. The characteristic flavour of carrots depends on the composition of different volatiles. α-Pinene, sabinene, myrcene, limonene, β-ocimene, γ-terpinene, *p*-cymene, terpinolene, β-caryophyllene, α-humulene, (*E*)-γ-bisabolene and β-ionone are found to be the key flavour compounds of raw carrots [207, 209, 210, 212, 216, 217]. Some of the odour sensations characteristic for the volatiles are "carrot top", "terpene-like", "green", "earthy", "fruity", "citrus-like", "spicy", "woody" and "sweet". Monoterpenes like sabinene, myrcene and *p*-cymene seems to be important contributors to "green", "earthy" or "carrot top" flavour with relatively high odour activity values. Sesquiterpenes like β-caryophyllene and α-humulene contribute to "spicy" and "woody" notes, whereas a "sweet" note is mainly due to β-ionone [209].

7.4.5.2 Celery and Celeriac

Cultivated celery (*Apium graveolens* var. *dulce*) and celeriac (*Apium graveolens* var. *rapaceum*) are closely related members of the Apiaceae. The leaf stem is the edible part of celery and the swollen base of the stem the edible part of celeriac. Both vegetables are eaten raw in salads or cooked. Terpenes and phthalides are the volatiles responsible for the aroma of celery and celeriac. The phthalides are represented in smaller amounts than the terpenes, but their contribution to celery aroma is dominant. Over 165 volatile components have been characterised in celery and celeriac [218–225]. Major aroma components of celery are 3-butylphthalide and 3-butyltetrahydrophthalide (sedanolide) (Fig. 7.2) with strong characteristic celery aroma [220, 221, 224, 226]. Other main volatile compounds found in celery and celeriac include (*Z*)-3-hexen-1-ol, myrcene, limonene, α-pinene, γ-terpinene, 1,4-cyclohexadiene, 1,5,5-trimethyl-6-methylene-cyclohexene, 3,7,11,15-tetramethyl-2-hexadecen-1-ol and α-humulene [218–223]. The most important character-impact compounds of cranberry are summarised in Table 7.9.

7.4.5.3 Parsley

Parsley (*Petroselinum crispum*) is a member of the Apiaceae family. The fresh leaves of parsley and the dried herb are widely used as flavouring. More than 80 compounds have been identified in the volatile fraction, and the aromatic volatiles of parsley are mainly monoterpenes and the aromatics myristicin and apiole. It is suggested that the characteristic odour of parsley is due to the presence of *p*-mentha-1,3,8-triene, myrcene, 3-*sec*-butyl-2-methoxypyrazine, myristicin, linalool, (*Z*)-6-decenal and (*Z*)-3-hexenal [227, 228]. Furthermore, β-phellandrene, 4-isopropenyl-1-methylbenzene and terpinolene contribute significantly to parsley flavour [229]. Studies have shown that a decrease in the intensities of parsley-like and green notes in the odour profile during storage is particularly due to losses of *p*-mentha-1,3,8-triene, myrcene and (*Z*)-6-decenal [227, 230].

7.4.5.4 Parsnip

The root of parsnip (*Pastinaca sativa*) is eaten boiled or baked. The major classes of compounds identified in raw and cooked parsnip are monoterpenoids, aliphatic sulfur compounds, and 3-alkyl-2-methoxypyrazines [35]. To the best of our knowledge, no investigations have been performed to elucidate the character-impact compounds in parsnip by modern GC-O techniques; however, it has been suggested that volatile compounds such as terpinolene, myristicin and 3-*sec*-butyl-2-methoxypyrazine may be important contributors to the flavour of parsnip owing to either their high concentrations or their low threshold values, or both [35].

7.5 Conclusions

The flavour of fruits and vegetables is a very important aspect of quality. This review has focused on the most important aroma compounds in fruits and vegetables of moderate climate and demonstrated that a wide variety of volatile compounds are formed naturally in the products or after processing that influence the aroma and flavour of fresh and processed fruits and vegetables.

It is characteristic for many of the compounds responsible for the aroma of fruits and vegetables that they have strong penetration odours with low threshold values. Recent advances in isolation techniques combined with more sensitive and advance chromatographic and spectroscopic techniques for identifying and quantifying volatile compounds in various types of extracts have increased our knowledge about volatile compounds of fruits and vegetables. Recent advances in olfactometric techniques on how to interpret the results of the volatile analysis have also increased our knowledge; however, we still know too little about the synergistic or antagonistic interactions between aroma compounds and nonvolatile flavour compounds such as sugars, acids and bitter compounds in fruits and vegetables. Recent advances in methods for measuring flavour release in complex matrices and sensory techniques combined with advanced chemometric methods may give some answers in the future to this central aspect of flavour science.

A complete understanding of the flavour chemistry and biochemistry of volatile components of fruits and vegetables is important in order to improve the flavour quality of fresh and processed produce that complies with the consumer needs for better quality vegetable and fruit products.

References

1. Czepa A, Hofmann T (2003) J Agric Food Chem 51:3865
2. Czepa A, Hofmann T (2004) J Agric Food Chem 52:4508
3. Peters AM, Van Amerongen A (1998) J Am Soc Hortic Sci 123:326
4. Sessa RA, Bennett MH, Lewis MJ, Mansfield JW, Beale MH (2000) J Biol Chem 275:26877
5. Waterhouse AL (2002) N Y Acad Sci. 957:21
6. Mail S, Borges RM (2003) Biochem Syst Ecol 31:1221
7. Varming C, Petersen, MA, Poll L (2004) J Agric Food Chem 52:1647
8. Blank I (2002) In: Marsili R (ed) Flavor, fragrance, and odor analysis. Dekker, New York, p 297
9. Acree TE (1993) In: Acree TE, Teranishi R (eds) Flavor science. Sensible principles and techniques. American Chemical Society, Washington, p 120
10. Mayer F, Takeoka G, Buttery R, Nam Y, Naim M, Bezman Y, Rabinowitch H (2003) ACS Symp Ser 836:144
11. Deibler KD, Acree TE, Lavin EH (1999) J Agric Food Chem 47:1616
12. Peréz A, Rios j, Sanz C, Olias J (1992) J Agric Food Chem 40:2232
13. Gomez E, Ledbetter CA (1997) J Sci Food Agric 74:541
14. Engel KH, Ramming DW, Flath RA, Teranishi R (1988) J Agric Food Chem 36:1003
15. Aubert C, Gunata Z, Ambid C, Baumes R (2003) J Agric Food Chem 51:3083
16. Takeoka GR, Flath RA, Mon TR, Teranishi R, Guentert M (1990) J Agric Food Chem 38:471
17. Derail C, Hofmann T, Schieberle T (1999) J Agric Food Chem 47:4742
18. Tomás-Barberán FA, Robins RJ (1997) (eds) Phytochemistry of fruits and vegetables. Clarendon, Oxford
19. Aguedo M, Ly MH, Belo I, Teixeira JA, Belin J-M, Waché Y (2004) Food Technol Biotechnol 42:327
20. Graham IA, Eastmond PJ (2002) Prog Lipid Res 41:156
21. Haslbeck F, Grosch W (1985) J Food Biochem 9:1
22. Yilmaz E, Baldwin EA, Shewfelt RL (2002) J Food Sci 67:2122
23. Chan HWS (ed) (1987) Autoxidation of unsaturated lipids. Academic, London
24. German JB, Zhang HJ, Berger R (1992) ACS Symp Ser 500:74
25. HO CT, Chen QY (1994) Lipids Food Flavours 558:2
26. Picardi SM, Issenberg P (1973) J Agric Food Chem 21:959
27. Krest I, Glodek J, Keusgen M (2000) J Agric Food Chem 48:3753
28. Lancaster JE, Shaw ML, Joyce MDP, McCallum JA, McManus MT (2000) Plant Physiol 122:1269
29. Won T, Mazelis M (1989) Physiol Plant 77:87
30. Fenwick GR, Hanley AB (1985) Crit Rev Food Sci Nutr 22:273
31. Nielsen GS, Larsen LM, Poll L (2003) J Agric Food Chem 51:1970
32. Boelens M, de Valois PJ, Wobben HJ, van der Gen A (1971) J Agric Food Chem 19:984
33. Wu JL, Chou CC, Chen MH, Wu CM (1982) J Food Sci 47:606
34. Whitaker JR (1976) Adv Food Res 22:73
35. Maarse H (ed) (1991) Volatile compounds in foods and beverages. Dekker, New York
36. Buttery RG (1981) In: Teranishi R, Flath RA, Sugisawa H (eds) Flavor research. Dekker, New York, p 175

37. Bones AM, Rossiter JT (1996) Physiol Plant 97:194
38. Verkerk R, Dekker M, Jongen WMF (2001) J Sci Food Agric 81:953
39. Loomis WD, Croteau R (1980) In: Stumpf PK (ed) The biochemistry of plants, vol. 4. Lipids: structure and function. Academic, London, p 363
40. Stahl-Biskup E, Inert F, Holthuijzen J, Stengele M, Schulz G (1993) Flavour Frag J 8:61
41. Bonnie TYP, Choo YM (1999) J Oil Palm Res 2:62
42. Wintherhalter P, Rouseff R (eds) (2001) Carotenoid-derived aroma compounds. American Chemical Society, Washington
43. Forney CF (2001) HortTech 11:529
44. Fuhrmann E, Grosch W (2002) Nahrung-Food 46:187
45. Dixon J, Hewett EW (2000) N Z J Crop Hortic Sci 28:155
46. Schulz I, Ulrich D, Fischer C (2003) Nahrung-Food 2:136
47. Fellman JK, Miller TW, Mattinson DS, Mattheis JP (2000) HortSci 35:1026
48. Echeverria G, Graell J, Lopez ML, Lara I (2004) Posharvest Biol Technol 31:217
49. Echeverria G, Fuentes MT, Graell J, Lopez ML (2004) J Sci Food Agric 84:5
50. Acree TE, Barnard J, Cunningham DG (1984) Food Chem 14:273
51. Kollmannsberger H, Berger RG (1992) Chem Mikrobiol Technol Lebensm 14:81
52. Cunningham, DG, Acree, TE, Barnard, J, Butts RM, Braell PA (1986) Food Chem 19:137
53. Roberts DD, Acree TE (1995) In: Rouseff RL, Leahy MM (eds) Fruit flavours: biogenesis, characterization, and authentication. ACS symposium series 596. American Chemical Society, Washington, p 190
54. Plotto A, McDaniel MR, Mattheis JP (2000) J Am Soc Hortic Sci 125:714
55. Young H Gilbert JM, Murray SH, Ball RD (1996) J Sci Food Agric 71:329
56. Plotto A, Mattheis JP, Lundahl DS, McDaniel MR (1998) In: Mussinan CJ, Morello MJ (eds) Flavor analysis. Developments in isolation and characterization. American Chemical Society, Washington, p 290
57. Lavilla T, Puy J, Lopez ML, Recasens I, Vendrell P (1999) J Agric Food Chem 47:3791
58. Rapparini F, Predieri S (2003) In: Janick J (ed) Horticultural reviews, vol 28. Wiley, New York, pp. 237–324
59. Rizzolo A, Sodi C, Polesello A (1991) Food Chem 42:275
60. Shiota H (1990) J Sci Food Agric 52:421
61. Argenta LC, Fan XT, Mattheis JP (2003) J Agric Food Chem 51:3858
62. Kahle K, Preston C, Richling E, Heckel F, Schreier P (2005) Food Chem 91:449
63. Chervin C, Speirs J, Loveys B, Patterson BD (2000) Postharvest Biol Technol 19:279
64. Engel KH, Flath RA, Buttery RG Mon TR Ramming DW, Teranishi R (1988) J Agric Food Chem 36:549
65. Visai C, Vanoli M (1997) Sci Hortic 70:15
66. Lavilla T, Recasens I, Lopez ML, Puy J (2002) J Sci Food Agric 82:1842
67. Toth-Markus M, Boross F, Blazso M, Kerek M (1989) Nahrung-Food 33:423
68. Takeoka GR, Flath RA, Guntert M, Jennings W (1988) J Agric Food Chem 36:553
69. Horvat RJ, Chapman GWJ, Robertson JA, Meredith FI, Scorza R, Callahan AM, Morgens P (1990) J Agric Food Chem 38:234
70. Horvat RJ, Chapman GW (1990) J Agric Food Chem 38:1442
71. Narain N, Hsieh TCY, Johnson CE (1990) J Food Sci 55:1303
72. Rizzolo A, Lombardi P, Nanoli M Polesello S (1995) J High Resolut Chromatogr 18:309

73. Guichard E, Souty M (1988) Z Lebensm-Unters Forsch A 186:301
74. Botondi R, DeSantis D, Bellincontro A, Vizovitis K, Mencarelli F (2003) J Agric Food Chem 51:1189
75. Guichard E, Kustermann A, Mosandl A (1990) J Chromatogr 498:396
76. Toth-Markus M, Boross F, Blazso M, Kerek M (1989) Nahrung-Food 33:433
77. Guichard E, Schlich P, Issanchou S (1990) J Food Sci 55:735
78. Horvat RJ, Chapman GW Jr, Senter SD, Robertson JA, Okie WR, Norton JD (1992) J Sci Food Agric 60:21
79. Mattheis JP, Buchanan DA, Fellman, JK (1992) Phytochemistry 31:775
80. Mattheis JP, Buchanan DA, Fellman JK (1992) J Agric Food Chem 40:471
81. Mattheis JP, Buchanan DA, Fellman JK (1997) J Agric Food Chem 45:212
82. Schwab W, Schreier P (1990) Z Lebensm-Unters Forsch A 190:228
83. Petersen MB, Poll L (1999) Eur Food Res Technol 209:251
84. Poll L, Petersen MB, Nielsen GS (2003) Eur Food Res Technol 216:212
85. Forney CF, Kalt W, Jordan MA (2000) HortSci 35: 1022
86. Gomes da Silav MDR, Chaves das Neves, HJ (1999) J Agric Food Chem 47:4568;
87. Bood KG, Zabetakis I (2002) J Food Sci 67:2
88. Gomes da Silav MDR, Chaves das Neves, HJ (1999) J Agric Food Chem 47:4568
89. Dirinck PJ, De Pooter HL, Willaert GA, Schamp NM (1981) J Agric Food Chem 29:316
90. Schreier P (1980) J Sci Food Agric 31:487
91. Menager I, Jost M, Aubert C (2004) J Agric Food Chem 52:1248
92. Larsen M, Poll L (1992) Z Lebensm-Unters-Forsch A 195:120
93. Gomes da Silav MDR, Chaves das Neves, HJ (1999) J Agric Food Chem 47:4568
94. Schulbach KF, Rouseff RL, Sims CA (2004) J Food Sci 69:S273
95. Azodanlou R, Darbellay C, Luisier JL, Villettaz JC, Amado R (2004) Eur Food Res Technol 218:167
96. Hakala MA, Lapvetelainen AT, Kallio HP (2002) J Agric Food Chem 50:1133
97. Lavid N, Schwab W, Kafkas E, Koch-Dean M, Bar E, Larkov O, Ravid U, Lewinsohn E (2002) J Agric Food Chem 50:4025
98. Watson R, Wright CJ, McBurney T, Taylor AJ, Linforth RST (2002) J Exp Bot 53:2121
99. Larsen M, Poll L, Olsen CE (1992) Z Lebensm-Unters Forsch A 195:536
100. Ulrich D, Hoberg E, Rapp A, Kecke S (1997) Z Lebensm-Unters-Forsch A 205:218
101. Larsen M, Poll L (1990) Z Lebensm-Unters-Forsch A 191:129
102. Larsen M, Poll L, Callesen O, Lewi M (1991) Acta Agric Scand 41:447
103. Klesk K, Qian M, Martin RR (2004) J Agric Food Chem 52:5155
104. de Ancos B, Ibanez E, Reglero G, Cano MP (2000) J Agric Food Chem 48:873
105. Robertson GW, Griffiths DW, Woodford JAT, Birch ANE (1995) Phytochemistry 38:1175
106. Mazza G, Miniati E (1993) (eds) Anthocyanins in fruits, vegetables, and grains. CRC, Boca Raton
107. Nijssen LM, Visscher CA, Maarse H, Willemsens LC, Boelsens MH (1996) (eds) Volatile compounds in food: qualitative and quantitative data. 7th edn. TNO Nutrition and Food Research Institute, Zeist
108. Klesk K, Qian M (2003) J Food Sci 68:697
109. Klesk K, Qian M (2003) J Agric Food Chem 51:3436
110. Qian MC, Wang YY (2005) J Food Sci 70:C13

111. Wang YY, Finn C, Qian MC (2005) J Agric Food Chem 53:3563
112. Schreier P (1984) In: Bertsch W, Jennings WG, Kaiser RE (eds) Chromatographic studies of biogenesis of plant volatiles. Hüthig, Heidelberg, p 52
113. Iversen CK, Jakobsen HB, Olsen CE (1998) J Agric Food Chem 46:1132
114. Nijssen LM, Visscher CA, Maarse H, Willemsens LC, Boelsens MH (1996) (eds) Volatile compounds in food: qualitative and quantitative data, 7th edn. TNO Nutrition and Food Research Institute, Zeist
115. Mikkelsen BB, Poll L (2002) J Food Sci 67:3447
116. Varming C, Poll L (2003) In: Le Quéré JL, Eriévant PX (eds) Flavour research at the dawn of the twenty-first century. Proceedings of the 10th Weurman flavour research symposium. Dijon. Lavoisier, Cachan, p 741
117. Varming C, Andersen ML, Poll L (2004) J Agric Food Chem 52:7628
118. Bagger-Sørensen R, Meyer AS, Varming C, Jonsson G (2004) J Food Eng 64:23
119. Latrasse A, Rigaud J, Sarris J (1982) Sci Aliment 2:145
120. Parliament TH, Scarpellino, R (1977) J Agric Food Chem 25:97
121. Horvat RJ, Senter SD (1985) J Food Sci 50:429
122. Lugemwa FN, Lwe W, Bentley MD, Mendel MJ, Alford AR (1989) J Agric Food Chem 37:232
123. Di Cesare LF, Nani R, Proietti M, Giombelli R (1999) Alimentari 38:277
124. Anjou K, von Sydow E (1967) Acta Chem Scand 21:2076
125. Hirvi T, Houkanen E, Pyysalo T (1981) Z Lebensm-Unters-Forsch A 172:365
126. Kaack K (1996) Fruit Var J 51:28
127. Poll L, Lewis MJ (1986) Lebens-Wiss Technol 19:258
128. Jensen K, Christensen LP, Hansen M, Jørgensen U, Kaack K (2000) J Sci Food Agric 81:237
129. Kaack K, Christensen LP, Hughes M, Eder R (2005) Eur Food Res Technol 221:244
130. Eberhardt R, Pfannhauser W (1985) Microchim Acta 1:55
131. Miková K, Havlíková L, Velíšek J, Víden I, Pudil F (1984) Lebensm-Wiss Technol 17:311
132. Davidek J, Pudil F, Velíšek J, Kubelka V (1982) Lebensm-Wiss Technol 15:181
133. Rosillo L, Salinas MR, Garijo J, Alonso GL (1999) J Chromatogr A 847:155
134. Gomez E, Martinez A, Laencina J (1995) J Sci Food Agric 67:229
135. Garcia-Moruno E (1999) Sci Aliments 19:207
136. Cliff MA, Dever MC, Reynolds AG (1996) J Enol Vitic 47:301
137. Jordan MJ, Margaria CA, Shaw PE, Goodner KL (2002) J Agric Food Chem 50:5386
138. Wan XM, Stevenson RJ, Chen XD, Melton LD (1999) Food Res Int 32:175
139. Paterson VJ, Macrae EA, Young H (1991) J Sci Food Agric 57:235
140. Bartley JP, Schwede AM (1989) J Agric Food Chem 37:1023
141. Lancaster JE, Reynolds PHS, Shaw ML, Dommisse EM, Munro J (1989) Phytochemistry 28:461
142. Granvogl M, Christlbauer M, Schieberle P (2004) J Agric Food Chem 52:2797
143. Block E, Naganathan S, Putman D, Zhao SH (1992) J Agric Food Chem 40:2418
144. Block E, Putman D, Zhao SH (1992) J Agric Food Chem 40:2431
145. Ferary S, Auger J (1996) J Chromatogr A 750:63
146. Ferary S, Thibout E, Auger J (1996) Rapid Commun Mass Spectrom 10:1327
147. Boelens M, de Valois PJ, Wobben HJ, van der Gen A (1971) J Agric Food Chem 19:984
148. Schulz H, Kruger H, Liebmann J, Peterka H (1998) J Agric Food Chem 46:5220
149. Jarvenpaa EP, Zhang ZY, Huopalahti R, King JW (1998) Eur Food Res Technol 207:39

150. Tokitomo Y, Kobayashi A (1992) Biosci Biotechnol Biochem 56:1865
151. Widder S, Lüntzel CS, Dittner T, Pickenhagen W (2000) J Agric Food Chem 48:418
152. D'Antuono LF, Moretti A, Neri R (2002) Genet Resour Crop Evol 49:175
153. Wu JL, Chou CC, Chen MH, Wu CM (1982) J Food Sci 47:606
154. Mondy N, Duplat D, Christides JP, Arnault I, Auger J (2002) J Chromatogr A 963: 89
155. Pino J, Rosado A, Gonzalez A (1991) Acta Aliment 20:163
156. Laakso I, Seppanenlaakso T, Hiltunen R, Muller B, Jansen H, Knobloch K (1989) Planta Med 257
157. Yu TH, Wu CM, Liou YC (1989) J Agric Food Chem 37:725
158. Faheid SMM (1998) Dtsch Lebensm-Rundsch 94:187
159. Lee SN, Kim NS, Lee DS (2003) Anal Bioanal Chem 377:749
160. Edris AE, Fadel HM (2002) Eur Food Res Technol 214:105
161. Lancaster JE, Shaw ML, Joyce MDP, McCallum JA, McManus MT (2000) Plant Physiol 122:1269
162. Krest I, Glodek J, Keusgen M (2000) J Agric Food Chem 48:3753
163. Nielsen GS, Poll L (2004) J Agric Food Chem 52:1642
164. Hanum T, Sinha NK, Guyer DE, Cash JN (1995) Food Chem 54:183
165. Petersen MA, Poll L, Larsen L M (1999) Lebensm-Wiss Technol 32:32
166. Jacobsson A, Nielsen T, Sjoholm I (2004) J Agric Food Chem 52:1607
167. Ulrich D, Krumbein A, Schonhof I, Hoberg E (1998) Nahrung-Food 42:392
168. Buttery RG, Gaudagni DG, Ling LC, Seifert RM, Lipton W (1976) J Agric Food Chem 24:829
169. Hansen M, Buttery RG, Stern DJ, Cantwell MI, Ling LC (1992) J Agric Food Chem 40:850
170. Dan K, Nagata M, Yamashita I (1997) J Jpn Soc Hortic Sci 66:621
171. Dan K, Todoriki S, Nagata M, Yamashita I (1997) J Jpn Soc Hortic Sci 65:867
172. Chin HW, Lindsay RC (1994) Sulfur Compd Foods 564:90
173. Kubec R, Drhova V, Velisek J (1992) J Agric Food Chem 46:4334
174. Forney CF, Jordan MA (1999) HortSci 34:696
175. Derbali E, Makhlouf J (1998) Postharvest Biol Technol 13: 191
176. Van Langenhove HJ, Cornelis CP, Schamp NM (1991) J Sci Food Agric 55:483
177. Chin HW, Lindsay RC (1993) J Food Sci 58:835
178. Buttery RG, Guadagni DG, Ling LC, Seifert RM, Lipton W (1976) J Agric Food Chem 24:829
179. Bailey SD (1961) J Food Sci 26:163
180. Kushad MM, Brown AF, Kurilich AC, Juvik JA, Klein BP, Wallig MA, Jeffery EH (1999) J Agric Food Chem 47:1541
181. Forney CF, Mattheis JP, Austin RK (1991) J Agric Food Chem 39:2257
182. Maruyama FT (1970) J Food Sci 35:540
183. Engel E, Baty C, le Corre D, Souchon I, Martin N (2002) J Agric Food Chem 50:6459
184. Grosch W, Schwarz JM (1971) Lipids 6:351
185. Buescher RH, Buescher RW (2001) J Food Sci 66:357
186. Schieberle P, Ofner S, Grosch W (1990) J Food Sci 55:193
187. Galliard T, Phillips DR (1971) Biochem J 124:431
188. Josephson DB, Lindsay RC (1987) J Food Sci 52:328
189. Petersen MA, Poll L, Larsen LM (1998) Food Chem 61:461
190. Maneerat C, Hayata Y, Kozuka H, Sakamoto K, Osajima Y (2002) J Agric Food Chem 50:3401

191. Hayata Y, Maneerat C, Kozuka H, Sakamoto K, Ozajima Y (2002) J Jpn Soc Hortic Sci 71:473
192. Buttery RG, Ling LC (1993) In: Teranishi R, Buttery RG, Sugisawa H (eds) American Chemical Society, Washington, p 23
193. Brauss MS, Linforth RST, Taylor AJ (1998) J Agric Food Chem 46:2287
194. Baldwin EA, Scott JW, Einstein MA, Malundo TMM, Carr BT, Shewfelt RL, Tandon KS (1998) J Am Soc Hortic Sci 123:906
195. Buttery RG, Takeoka G, Teranishi R, Ling LC (1990) J Agric Food Chem 38:2050
196. Krumbein A, Auerswald H (1998) Nahrung-Food 42:395
197. Buttery RG, Takeoka GR (2004) J Agric Food Chem 52:6264
198. Buttery RG, Teranishi R, Ling LC (1987) J Agric Food Chem 35:540
199. Mayer F, Takeoka G, Buttery R, Nam Y, Naim M, Bezman Y, Rabinowitch H (2003) Freshness Shelf Life Foods 836:144
200. Bezman Y, Mayer F, Takeoka GR, Buttery RG, Ben Oliel G, Rabinowitch HD, Naim M (2003) J Agric Food Chem 51:722
201. Maul F, Sargent SA, Sims CA, Baldwin EA, Balaban MO, Huber DJ (2000) J Food Sci 65:1228
202. Buttery RG, Takeoka GR, Ling LC (1995) J Agric Food Chem 43:1638
203. Murray KE, Shipton J, Whitfield FB, Last JH (1976) J Sci Food Agric 27:1093
204. Shipton J, Whitfiel FB, Last JH (1969) J Agric Food Chem 17:1113
205. Murray KE, Shipton J, Whitfiel FB, Kennett BH, Stanley G (1968) J Food Sci 33:290
206. Jakobsen HB, Hansen M, Christensen MR, Brockhoff PB, Olsen CE (1998) J Agric Food Chem 46:3727
207. Alasalvar C, Grigor JM, Quantick PC (1999) Food Chem 65:391
208. Kjeldsen F, Christensen LP, Edelenbos M (2001) J Agric Food Chem 49:4342
209. Kjeldsen F, Christensen LP, Edelenbos M (2003) J Agric Food Chem 51:5400
210. Shamaila M, Durance T, Girard B (1996) J Food Sci 61:1191
211. Alasalvar C, Grigor JM, Zhang DL, Quantick PC, Shahidi F (2001) J Agric Food Chem 49:1410
212. Buttery RG, Seifert RM, Guadagni DG, Black DR, Ling LC (1968) J Agric Food Chem.16:1009
213. Howard LR, Braswell D, Heymann H, Lee Y, Pike LM, Aselage J (1995) J Food Sci 60:145
214. Simon PW, Peterson CE, Lindsay RC (1980) J Agric Food Chem 28:559
215. Yoo KS, Pike LM, Hamilton BK (1997) HortSci 32:714
216. Schnitzler WH, Broda S, Schaller RG (2003) J Appl Bot 77:53
217. Tóth-Markus M, Takács-Hájos M (2001) Acta Aliment 30:219
218. Deng CH, Song GX, Zheng XH, Hu YM, Zhang XM (2003) Chromatographia 57:805
219. Macku C, Shibamoto T (1991) Food Chem 42:121
220. Macleod G, Ames JM (1989) Phytochemistry 28:1817
221. Van Wassenhove F, Dirinck P, Vulsteke G, Schamp N (1990) HortSci 25:556
222. Tirillini B, Pellegrino R, Pagiotti R, Pocceschi N, Menghini L (2004) J Food Sci 16:477
223. Rao LJM, Nagalakshmi S, Naik JP, Shankaracharya NB (2000) J Food Sci Technol-Mysore 37:631
224. Thappa RK, Dhar AK, Balyan SS, Khan S, Raina P, Dhar PL, Choudhary DK (2003) J Food Sci Technol-Mysore 40:426
225. Habegger R, Schnitzler WH (2000) Obst-, Gemüse-, Kartoffelverarbeitung 85:162
226. Macleod AJ, Macleod G, Subramanian G (1988) Phytochemistry 27:373

227. Masanetz C, Grosch W (1998) Flavour Fragrance J 13:115
228. Jung HP, Sen A, Grosch W (1992) LWT Food Sci Technol 25:55
229. Macleod AJ, Snyder CH, Subramanian G (1985) Phytochemistry 24:2623
230. Masanetz C, Grosch W (1998) Z Lebensm-Unters Forsch A 206:114
231. Hermann K (1999) Gemüseverwertung 84:106
232. Lopez-Tamames E, Carro-Marino N, Gunata YZ, Sapis C, Baumes R, Bayonove C (1997) J Agric Food Chem 45:1729
233. Palma-Harris C, McFeeters RF, Fleming HP (2001) J Agric Food Chem 49:4203
234. Linforth RST, Savary I, Pattenden B, Taylor AJ (1994) J Sci Food Agric 65:241
235. Ruiz JJ, Alonso A, Garcia-Martinez S, Valero M, Blasco P, Ruiz-Bevia F (2005) J Sci Food Agric 85:54
236. Tandon KS, Baldwin EA, Shewfelt RL (2000) Postharvest Biol Technol 20: 261
237. Diaz-Maroto MC, Vinas MAG, Cabezudo MD (2003) Eur Food Res Technol 216:227
238. Diaz-Maroto MC, Perez-Coello MS, Cabezudo MD (2002) Eur Food Res Technol 215:227
239. Hashem FA, Sahab AF (1999) Food Chem 65:29
240. Masanetz C, Grosch W (1998) Flavour Fragrance J 13:115
241. Broda S, Habegger R, Hanke A, Schnitzler WH (2001) J Appl Bot 75:201

[illegible]
[illegible]
[illegible]
[illegible]
[illegible]
[illegible]
[illegible]
[illegible]
[illegible]
[illegible]
[illegible]
[illegible]
[illegible]
[illegible]
[illegible]

8 Tropical Fruit Flavour

Mário Roberto Maróstica Jr, Gláucia Maria Pastore
Department of Food Science,
State University of Campinas,
Monteiro Lobato Street 80,
13083-862 Campinas, São Paulo, Brazil

8.1 Introduction

The characteristic exotic flavour of fruits from the tropics is one of the most attractive attributes to consumers. Nowadays, food industries are looking at how to use these volatiles to produce amazing new products that can accommodate this new demand. The following sections report some of the relevant research data on volatiles of some important tropical fruits.

8.2 Guava (Genus *Psidium*)

Guava is native to Central America. It was distributed into other parts of tropical and subtropical areas such as Asia, South Africa, Egypt, and Brazil by the early seventeenth century [49]. Some examples of impact-flavour compounds have already been identified in guava: β-ionone [58], terpene hydrocarbons [63], and esters [43] could be mentioned.

Essences of pink and white fresh guava obtained by direct extraction of flesh juices with dichloromethane revealed that the total amount of C_6 aldehydes, alcohols, and acids comprised 20 and 44% of the essence of fresh white and pink guavas, respectively [49]. The flavour of the Costa Rican guava has been described as sweet with strong fruity, woody-spicy, and floral notes [53]. One hundred and seventy-three volatile compounds were isolated by simultaneous steam distillation–solvent extraction. The terpenes and terpenic derivatives were found in this fruit in major concentrations and were strong contributors to tropical fruit notes (Fig. 8.1). The aliphatic esters contributed much to its typical flavour.

Characterisation of the aromatic profile in commercial guava essence and fresh fruit puree extracted with solvent yielded a total of 51 components [29]. Commercial essence was shown to be rich in components with low molecular weight, especially alcohols, esters, and aldehydes, whereas in the fresh fruit puree terpenic hydrocarbons and 3-hydroxy-2-butanone were the most abundant components.

Volatile compounds isolated from strawberry guava fruit by simultaneous steam distillation–solvent extraction were identified by capillary gas chromatography–mass spectrometry (GC-MS) and were characterised sensorially by sniffing GC [52]. Terpenes and terpenic derivatives were identified and were shown to contribute much to the typical strawberry guava flavour. The presence of many aliphatic esters and terpenic compounds is thought to contribute to the unique flavour of the strawberry guava fruit.

Some plagues that jeopardise guava cultivars are caused by *Timocrata albella*, which attacks the stalk, and *Conotrachelus psidii* (a beetle that attacks the fruits). Diseases in guava are also caused by *Puccinia psidii*, a fungus that attacks leaves, flowers, and fruits (http://www.seagri.ba.gov.br).

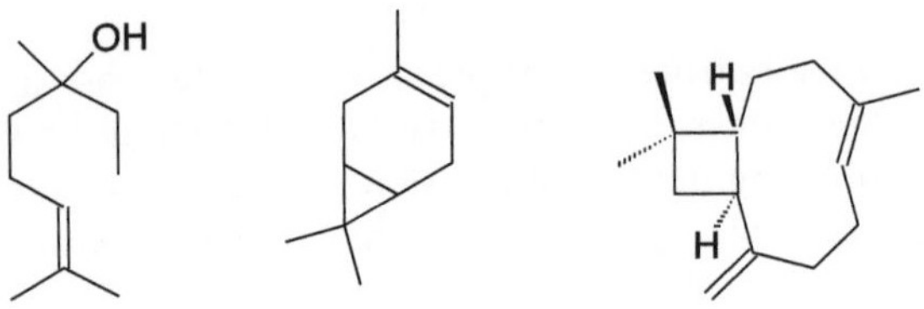

1 2 3

Fig. 8.1 Acyclic, monocyclic, and bicyclic terpenes contribute to tropical fruit flavours; **1** linalool to papaya; **2** Δ^3-carene to mango; **3** *ß*-caryophyllene to guava fruit

8.3 Banana (Genus *Musa*)

Banana (*Musa sapientum* L.) is one of the most common tropical fruits, and one of Central America's most important crops. It is grown in all tropical regions and is one of the oldest known fruits [45]. From a consumer perspective, bananas are nutritious with a pleasant flavour and are widely consumed throughout the world [57]. Esters predominate in the volatile fraction of banana (Fig. 8.2). Acetates are present in high concentrations in the fruit and generally possess a low threshold. Isopentyl acetate and isobutyl acetate are known as the two most important impact compounds of banana aroma. Alcohols are the second most important group of volatiles in banana extracts. 3-Methyl-1-butanol, 2-pentanol, 2-methyl-1-propanol, hexanol, and linalool are the alcohols present in higher concentrations in the fresh fruit [45].

The concentrations of acetates and butanoates seemed to increase during ripening of Valery bananas [40]. This was confirmed by an investigation in which bananas were treated with the ethylene antagonist 1-methylcyclopropene (1-MCP) [22]. The volatiles were recovered by a Tenax TA trap. The 1-MCP treatments caused quantitative changes in the amounts and the composition of the aroma volatile compounds, resulting in a substantial increase in the concentration of alcohols and a decrease in the concentration of their related esters.

Later, another research group suggested that not all of the volatile components found in large concentration in the commercial banana essence contributed to the aroma, such as 2-pentanone, 2-pentanol, butanol, and isobutyl acetate. However, isoamyl acetate, 2-pentanol acetate, 2-methyl-1-propanol, 3-methyl-1-butanol, 3-methylbutanal, acetal, isobutyl acetate, hexanal, ethyl butanoate, 2-heptanol, and butyl butanoate contributed and defined the aroma in the commercial fruit essence [30]. Isoamyl alcohol was the most abundant compound found in headspace flavour compounds of Taiwanese banana recovered by solid phase microextraction [39].

Aroma compounds of fresh banana from different countries (Martinique, Canary Islands, and Côte d'Ivoire) were examined using the same extraction technique. As expected, differences in aroma composition were detected in the fruits of different origins. Isoamyl alcohol, isoamyl acetate, butyl acetate, and elemicine were detected by olfactometric analyses as characteristics of banana odour [7].

Among the diseases of banana cultivars, yellow sigatoka (caused by *Mycosphaerella musicola*), black sigatoka (caused by the ascomycete *Mycosphaerella fijiensis*) and mal-do-Panamá (caused by *Fusarium oxysporum* strains) are the most important. Damage caused by yellow sigatoka can reach a loss of 50–100%. But the black sigatoka is a more severe disease, causing destruction of the leaves leading to a loss of 100%. The damages caused by mal-do-Panamá vary with soil and cultivar (http://www.embrapa.br).

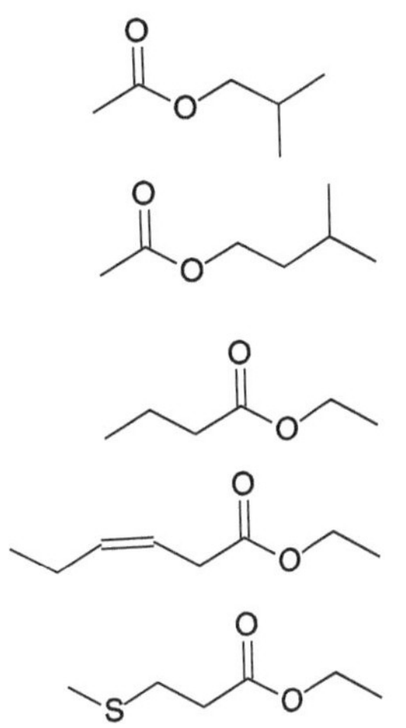

4
5
6
7
8

Fig. 8.2 Esters as character-impact compounds: **4** isobutyl acetate in passion fruit; **5** isopentyl acetate in banana; **6** ethyl butanoate in cupuacu; **7** ethyl (3*Z*)-hexenoate and **8** ethyl-3-(methylthio)propanoate in pineapple

8.4 Mango (*Mangifera indica*)

Mango is one of the most popular and best known tropical fruits [44] and possesses a very attractive and characteristic flavour. Some authors reported great differences in flavour compounds (including esters, lactones, monoterpenes, and sesquiterpenes) [14].

A wide range of volatile compounds from Indian mango were identified by pioneer group research [20, 21]. Esters, lactones, monoterpenes, sesquiterpenes, and furanones were among the volatiles. It has been suggested that the ratio of palmitic to palmitoleic acids determines the flavour quality of the ripe fruit, a ratio of less than 1 resulting in strong aroma and flavour [44].

Terpenes were identified as the most abundant compounds (over 54% of the solvent-extracted volatiles) from Venezuelan mangoes [44]. Figure 8.1 shows structures of some monoterpenes found in tropical fruits. Car-3-ene, described as having the characteristic aroma of mango leaves, was the major contributor with 26% of volatiles in the sample. This result was confirmed for African mangoes [55]. The acids, esters, and lactones found were considered to be produced by the lipid metabolism in the development of the aroma and flavour of mango fruit during ripening.

Volatiles of three cultivars of mango (Jaffna, Willard, and Parrot) from Sri Lanka were analysed, and among the 76 components identified, monoterpenes and sesquiterpenes hydrocarbons were described as the major contributors [42]. Variations in the amounts of esters, ketones, and alcohols were also related.

The importance of glycosidically bound volatile compounds (GBVC) and their contribution to fruit aroma were evidenced in African mango [54]. Some terpenes (like α-terpineol and linalool oxide isomers), phenolic compounds, carotenoid aroma derivatives like 9-hydroxymegastigma-4,7-dien-3-one, and acids (like palmitic and stearic) were reported. In the same way, another investigation reported that the composition and concentrations of GBVC in pulp and skin of the Kensington Pride variety were strongly influenced by the fruit part and maturity stages [36]. Most of the GBVC increased in the pulp as maturity progressed.

Fifteen Brazilian varieties of Mango fruit were divided in three groups according to the component present in the greatest concentration [3]. The first group comprised the varieties rich in α-terpinolene: Cheiro (66.1% of α-terpinolene), Chana (62.4%), Bacuri (57.0%), Cametá (56.3%), Gojoba (54.8%), Carlota (52.0%), Coquinho (51.4%), and Comum (45.0%). The second group, with the varieties rich in car-3-ene, was represented by Haden (71.4%), Tommy (64.5%), and Keith (57.4%) and the third group, rich in myrcene, was composed of the varieties Cavalo (57.1%), Rosa (52.4%), Espada (37.2%), and Paulista (30.3%).

The changes in the production of volatile aroma compounds during fruit ripening seemed to be mediated by ethylene. The production of most terpenes during ripening of the Kensington Pride mango has been reported to occur parallel to the production of ethylene; however, the exact role of ethylene in biosynthesis of volatiles was not well established. Experiments carried out with ethylene inhibitors clearly suggested that biosynthesis of monoterpenes, esters, and aldehydes in the mango fruit were strongly dependent on ethylene production and action [36–38].

In mango cultivation, the tree may be attacked by several plagues (mosquitoes and mites). Two particular species of mosquitoes (*Anastrepha* spp. and *Ceratitis capitata*) can severely damage the tree, causing a great decrease in the production. Other plagues that result in minor damage are caused by *Eriophyes mangiferae, Selenothrips rubrocinctus*, and *Aphis gossypii* (hpp://www.embrapa.br).

8.5 Melon (*Cucumis melo*)

The species *Cucumis melo* comprises a great number of varieties that exhibit considerable diversity in their biological characteristics [65]. The dessert melons of commercial importance exhibit a wide variation in flavour and aroma profiles [66].

Charentais cantaloupe melon (*Cucumismelo* L. var. *cantalupensis* Naud.) was characterised by abundant sweetness and a good aromatic flavour [68]. The aroma volatiles of Charentais-type cantaloupe melons, as with other cantaloupes, comprise a complex mixture of compounds including esters, saturated and unsaturated aldehydes and alcohols, as well as sulfur compounds [26, 65]. Among these compounds, volatile esters were quantitatively the most important and therefore represent key contributors to the aroma [68]. The linear saturated and unsaturated aldehydes seem to originate from the degradation of linolenic and linoleic acids [26, 32, 33, 67].

The aroma volatiles of some melon species consist of a complex mixture of esters together with other components, including C9 unsaturated aldehydes, alcohols, and acetates whose sensory properties have been described as "melon-like" [10, 31–33, 35]. Several esters and alcohols were described among the volatiles of muskmelons [33, 34].

Melons stored at low temperatures showed different relative amounts of volatiles recovered by solvent extraction [34]. Some of the C9 unsaturated esters and alcohols presumably originated as a result of lipoxygenase activity. The previous results together with some investigations of the C9 unsaturated esters and alcohols suggested that the activity of lipoxygenase on melons seems to be dependent on cultivar, age, storage conditions, and sample location [65].

More recently, static headspace GC analysis of eight cultivars of melons detected esters as the major volatile components. Differences among the compositions of the volatiles of the cultivars studied were also reported and are probably due to different efficiencies of biosynthetic pathways of each variety [56].

Sulfur compounds are also likely to be of considerable sensory importance in melon aroma. Dimethyl sulfide and ethyl (methylthio)acetate were found to be sulfur volatiles from Golden Crispy melons [65]. 2-(Methylthio)ethyl acetate was present in greater amounts in the majority of the melon cultivars analysed. The structures of the sulfur volatile compounds suggested that they may have been derived from methionine [66].

The extraction technique can play an important role in the recovery of volatiles, resulting in different profiles of volatiles for the same variety [67]. The direct extraction of *Cucumis melo* L. var. *cantaloupensis* with Freon 11 under low temperature was capable of recovering compounds never found before in melons [26]. The authors attributed this to the non-destructive extraction at low temperatures and the very efficient capillary chromatographic system used for the analysis.

It has been shown that suppression of ethylene production results in a strong inhibition of aroma volatiles in Charentais-type melons [4, 68].

Cultivars of *Cucumis melo* L. can be attacked by *Didymella bryoniae* Auersew, which can cause considerable losses, because fruits attacked by this organism do not have commercial value anymore. Mosquitoes, mainly *Bemisia tabaci*, also attack melon cultivars. Reductions in weight, size, and sugar content are evident consequences of mosquitoes attack. *Diaphania nitidalis* attacks flowers and fruits, whereas *Diaphania hyalinata* generally attacks the leaves of the melon tree.

8.6 Papaya (*Carica papaya*)

Papaya is a native fruit from America and is widely planted throughout the tropics [41], and is a crop of economic importance to tropical countries [11]. It has become a commercially important fresh fruit crop, particularly in the USA and Europe [51]. Papaya possesses a characteristic aroma, which is due to several volatile components, such as alcohols, esters, aldehydes, and sulfur compounds [11].

Several volatile components of papaya (Solo variety) were recovered by four methods: vacuum trapping train (distillation under low temperatures with liquid nitrogen traps), codistillation–extraction (vacuum), vacuum distillation, and codistillation–extraction (1 atm) [17]. In spite of great variations due to the recovery method, the results showed that linalool was always the major compound detected for all the methods.

In another investigation, linalool (Fig. 8.1) was detected in relatively low concentration in the solvent-extracted volatiles of fresh papaya pulp from Sri Lanka [41]. The authors attributed the characteristic sweaty note of this papaya fruit mainly to methyl butanoate. Phenylacetonitrile was also found in high amounts (17.7%), which, according to the authors, combined with lesser concentrations of benzyl isothiocyanate (1.5%) can play a role in the aroma of papaya.

The concentrations of linalool and benzyl isothiocyanate in papaya are clearly affected by the addition of Hg^{2+}. It is suggested that the mercurous ion could block the activity of different enzymes taking part in monoterpene formation [25]. Oxygenated terpenoids derived from linalool can play an important role in Brazilian papaya aroma [64]. Several oxygenated derivatives of linalool were identified in the solvent-extracted samples, such as the two diasteroisomers of 6,7-epoxy-linalool: 2,6-dimethyl-octa-1,7-diene-3,6-diol and 2,6-dimethyl-octa-3,7-diene-2,6-diol. The terpenes linalool and terpinen-4-ol showed an increased production ratio in a Cuban papaya variety (*Carica papaya* L. var. *maradol roja*) [1].

Fifty-one volatile components from intact Hawaiian papayas in different ripeness stages were recovered by trapping with Tenax [18]. As expected, the greatest number of components were found in the fully ripe fruits. Linalool, followed by linalool oxide A, linalool oxide B, and ethyl acetate were the major components in the fully ripe fruits. Several compounds were also present in the four ripeness stages: linalool and all aldehydes can be mentioned.

Another investigation reported the esters as the predominant volatile components of the Maradol variety (about 41% w/w of the total volatiles) [51]. The major representative compounds in the simultaneous steam distillation–solvent extraction were methyl butanoate and ethyl butanoate. Previous work described the esters as the predominant compounds among the volatiles; papayas, for example, from Sri Lanka and Colombia had 52 and 63% of esters in the total volatiles respectively [25, 41].

Plagues in papaya cultivation are generally mites (*Polyphagotarsonemuis latus*, *Tetranychus urticae*). *Polyphagotarsonemuis latus* is known as "tropical mite" and attacks the leaves causing death of the tree. Other diseases are caused by *Phytophthora palmivora* and *Colletotrichum gloesosporioides*. The fruits do not possess commercial value after the attack (http://www.seagri.ba.gov.br).

8.7 Passion Fruit (*Passiflora edulis*)

Owing to their unique and delicate flavour, species of the genus *Passiflora* have been the subject of intensive research on their volatile constituents [13]. The purple passion fruit (*Passiflora edulis* Sims) is a tropical fruit native to Brazil but is now grown in most tropical and subtropical countries [50]. Purple passion fruit is cultivated in Australia, India, Sri Lanka, New Zealand, and South Africa [48]. Yellow passion fruit (*Passiflora edulis* f. *flavicarpa*) is one of the most popular and best known tropical fruits, having a floral, estery aroma with an exotic tropical sulfury note [62]. Yellow passion fruit is cultivated in Brazil, Hawaii, Fiji, and Taiwan [48]. Because of its more desirable flavour, the purple passion fruit is preferred for consumption as fresh fruit, whereas the yellow passion fruit is considered more suitable for processing [28].

The first report about volatile constituents in purple passion fruit (*Passiflora edulis* Sims) described the identification of 20 volatiles in the solvent extract of passion fruit juice from New Guinea [50]. The author attributed the unique aroma of the purple passion fruit to the several esters (Fig. 8.2) identified (about 80% of neutral essence). Similarly, the volatile fraction of passion fruit juice (*Passiflora edulis* Sims) was reported as a complex mixture of components [47], but the majority was esters derived largely from various combinations of alkanols with acids. Sulfur compounds can play an important role in the overall flavour characteristics of passion fruit [13]. The attractive tropical flavour note of ripe yellow passion fruits has been shown to be associated with trace levels of sulfur volatiles [62].

The analyses of the flavour composition of yellow passion fruits were performed by four different isolation techniques, namely vacuum headspace sampling (VHS), the dynamic headspace method, simultaneous distillation and extraction at atmospheric pressure, and simultaneous distillation and extraction under reduced pressure [62]. Significant differences were found not only in the chemical composition of the resultant extracts but also in their sensory properties. The most representative and typical extract was obtained by VHS.

Later, the chemical characterisation of the volatiles from yellow passion fruit essence and from the juice of the fruit was done by GC-MS and GC–olfactometry (GC-O) [27]. Esters were the components found in the largest concentrations in passion fruit juice and essence extracted with methylene chloride. Analysis by GC-O yielded a total of 66 components which appeared to contribute to the aroma of passion fruit juice and its aqueous essence. Forty-eight compounds were identified in the pulp of Brazilian yellow passion fruits (*Passiflora edulis* f. *flavicarpa*) [48]. The predominant volatile compounds belonged to the classes of esters (59%), aldehydes (15%), ketones (11%), and alcohols (6%).

Plagues in passion fruit are mainly caterpillars (*Dione juno juno* and *Agraulis vanillae vanillae*), which attack mainly the leaves, decreasing the growth and the production of fruits. Passion fruit rot and withering can be caused by *Colletotrichum, Rhizopus, Cladosporium, Fusarium, Lasiodiplodia, Phomopsis, Alternaria alternata, A. passiflorae, Septoria passiflorae,* and *Sclerotinia sclerotium.* (http://www.seagri.ba.gov.br)

8.8 Pineapple (*Ananas comosus*)

Pineapple, one of the most popular tropical fruits in the world, has been cultivated in South America since the fifteenth century [61]. It has been very popular throughout the world for many years [16]. Native to Central America and South America, pineapples grow in several tropical countries, such as Hawaii, India, Malaysia, the Philippines, and Thailand [12]. Owing to its attractive sweet flavour, pineapple is widely consumed as fresh fruit, processed juice, canned fruit, and as an ingredient in exotic foods. The volatile constituents of pineapple have been studied for over 60 years by many researchers. More than 280 compounds have been found among volatiles of pineapples so far [60].

The earliest investigations concerning pineapple volatiles date from 1945 [23, 24]. The great majority of pineapple components are contributed by ethyl and methyl esters (Fig. 8.2) In 1970, a North American group reported that aliphatic esters were the predominant compounds among the solvent extract of Smooth Cayenne pineapple. Monoterpene alcohols (linalool, α-terpineol, and terpinen-4-ol) were also identified [16].

Sesquiterpenes recovered by solvent extraction were identified in pineapple fruit (*Ananas comosus* Merr.) from Côte d'Ivoire. The authors suggested that some of the sesquiterpenoids found were derived from germacrene precursors [6]. The same authors studied the identification of trace compounds with impact character in pineapple fruit (*Ananas comosus* Merr.). Some potent compounds were an undecatriene, an undecatraene, and ethyl esters [5].

The sulfur components ethyl *S*-(+)-2-methylbutanoate and dimethyl trisulfide (with 0.006 and 0.01 µg/L odour thresholds in water, respectively) were reported as impact-flavour compounds in fresh Hawaiian pineapple essence prepared by solvent extraction. The major volatile components were methyl and ethyl esters [59].

The volatile compounds of juices made from freshly cut pineapple fruits from different cultivars from Costa Rica, Ghana, Honduras, Côte d'Ivoire, the Philippines, Réunion, South Africa, and Thailand were studied in comparison to that of commercial water phases/recovery aromas, juice concentrates as well as commercially available juices [12]. The qualitative pineapple fruit flavour profile showed several methyl esters, some characteristic sulfur-containing esters, and various hydroxy esters were responsible for the typical pineapple flavour profile.

Twenty-nine odour-active compounds were detected by using aroma extract dilution analysis (AEDA) [60]. The results of AEDA together with GC-MS analysis showed ethyl 2-methylbutanoate (described as 'fruity' flavour), followed by methyl 2-methylbutanoate and 3-methylbutanoate (fruity, apple-like), 4-hydroxy-2,5-dimethyl-3(2*H*)-furanone (sweet, pineapple-like, caramel-like), δ-decalactone (sweet, coconut-like), 1-(*E*,*Z*)-3,5-undecatriene (fresh, pineapple-like), and a unknown compound (fruity, pineapple-like) as the most odour-active compounds.

The most common plagues in pineapple cultivation are caused by *Thecla basalides* and *Dysmicoccus brevipes*. *Fusarium subglutinans* also causes an important disease, leading to the most significant losses. (http://www.seagri.ba.gov.br).

8.9 Cupuacu (*Theobroma grandiflorum*)

Cupuacu is an Amazonian forest tree from Para state, Brazil. The fruits are 15–25 cm in length, 10–12 cm in diameter, and weigh between 0.8 and 2 kg. They are oblong fruits with a hard skin. The seeds contain caffeine and theobromine, alkaloids with stimulant properties. The seeds contain about 48% of a white fat similar to cocoa butter. The creamy-white pulp has an attractive and characteristic aroma and flavour. The fruits are consumed mainly as juice.

Volatile constituents of cupuacu were isolated by steam distillation–extraction of pulp or juice [2].The identification of volatile constituents was based on mass spectral analysis. The pleasant aroma compounds were mainly esters (Fig. 8.2). Large amounts of ethyl butanoate and small amounts of ethyl acetate, butyl acetate, and butyl isobutanoate were described.

More recently, several aroma compounds were isolated from cupuacu pulp by vacuum distillation, solid-phase extraction, and simultaneous steam distillation–extarction and were analysed by GC, GC-MS, and GC-O [8]. The olfaction of the extracts obtained by solid-phase extraction indicated linalool, α-terpineol, 2-phenylethanol, myrcene, and limonene as contributors of the pleasant floral flavour. In this study, the esters ethyl 2-methylbutanoate, ethyl hexanoate, and butyl butanoate were involved in the typical fruity characteristics.

In another investigation, the volatile compounds were isolated [19] using a Porapack Q trap by vacuum for 2 h and were then eluted with hexane. The esters were the chemical class of compounds that predominated in the samples among 21 volatile compounds detected. Ethyl butanoate, ethyl 2-methylbutanoate, 1-butanol, ethyl hexanoate, 3-hydroxy-2-butanone, ethyl octanoate, acetic acid, linalool, palmitic acid, and oleic acid were identified in cupuacu pulp by solid-phase extracton [15].

Plagues in cupuacu cultivation are mainly caused by beetles (*Costalimaita* sp.) which attack the leaves, grasshoppers and ants. Vassoura de bruxa disease is caused by the fungus *Crinipellis perniciosa* (http://www.seagri.ba.gov.br)

8.10 Bacuri (*Platonia insignis* M. or *Platonia sculenta*)

The bacuri tree grows in the south of the Amazonian forest in Para state, Brazil. The fruits are ovoid to subglobose, are 7–15 cm in diameter and weigh 200–1,000 g. They have a creamy white mucilaginous, fibrous, juicy pulp with a very attractive exotic flavour. The fruit is consumed as such, as a juice, or in ice cream or jellies.

The first study on the volatile composition of bacuri revealed linalool, 2-heptanone, and *cis*-3-hexenyl acetate as the most important flavour compounds [2]. Volatiles were isolated by a steam distillation–extraction of pulp.

The main volatile components isolated from bacuri shells using various isolation methods, such as steam distillation and supercritical CO_2 extractions were (*Z*)-linalool oxide, (*E*)-linalool oxide, 2-pentanone, 2-nonanone, *cis*-hexenyl acetate, methyl dodecacanoate, and several hydrocarbons, including bisabolene [46].

The free and bound flavour components of bacuri fruits were analysed by GC and GC-MS using XAD-amberlite separation. Seventy-five components were identified in the free volatile fraction, and the most abundant components were terpene alcohols. Among the saturated and unsaturated alcohols present in the volatile fraction of bacuri, hexanol and (*Z*)-hex-3-en-1-ol seemed to contribute to the herbaceous odour detected in intact fruits [8]. Bacuri glycosides identified in this work, like benzyl, 2-phenylethyl, and (*E*)-linalool furanooxides, linalool glucosides, and benzyl, 2-phenylethyl, and linalool rutinosides, were considered by the authors as the most important glycosides. The formation of volatile compounds using heat treatment of the bacuri pulp at its natural pH and at pH 7 was verified during the simultaneous distillation and extraction technique [9]. Increased amounts of oxygenated and hydrocarbon terpenes and of aldehydes were observed after simultaneous distillation and extraction at pH 3. More particularly, linalool, linalaool furanoxides, α-terpineol, hotrienol, nerol oxide, nerol, and geraniol were described as the main components. These results were partially explained by the hydrolysis of the glycosidically bound compounds.

8.11 Sustainability of Tropical Cultivation

Tropical soils may seem fertile when covered with luxuriant vegetation, but they are sometimes surprisingly poor in nutrients. The low fertility of tropical areas may result from natural and anthropogenic causes. The soil composition and the pluviometer index are the main factors determining fertility; the flushing out of nutrients is higher at higher temperatures and at higher incidence of rain. The amount of organic matter and nutrient elements accumulated in vegetation relative to that in soil is generally larger in tropical forests than in temperate ones; therefore, exhaustion of nutrients by removing forest vegetation is more serious in the tropics than in temperate regions.

Desertification is caused by overcultivation, overgrazing, and deforestation. This may result in soil exhaustion and erosion. This will in turn decrease the soil productivity, reduce food production, deprive the land of its vegetative cover, and negatively impact areas not directly affected by its symptoms, by causing floods, soil salinisation, deterioration of water quality, and silting of rivers, streams, and reservoirs (http://www.fao.org).

The aroma compounds from the tropical fruits described in this chapter can be very important for consumers and industry as they are exotic and extremely pleasant; however, the production of these compounds by biotechnological processes should be emphasised since the extraction from the fruits is a hard task. Many tropical soils contain less nitrogen and phosphorus, have lower capacity to absorb fertilisers, and therefore have lower conventional productive capacity, but some tropical soils have been very intensively farmed and further intensification is possible in other areas. Thus, the evaluation of a sustainable agriculture in tropical regions requires a sophisticated approach including the estimation of the risk of microbial or insect infestations. As many fruits go directly to fresh markets or to immediate processing, a continuing supply of the flavour manufacturers in the future is not completely assured.

References

1. Almora K, Pino JA, Hernández M, Duarte C, González J, Roncal E (2004) Food Chem 86:127
2. Alves S, Jennings WC (1978) Food Chem 4:149
3. Andrade EHA, Maia JGS, Zoghbi MGB (2000) J Food Comp Anal 13:27
4. Bauchot AD, Mottram DS, Dodson AT, John P (1998) J Agric Food Chem 46:4787
5. Berger RG, Drawert F, Kollmannsberger H, Nitz S, Schraufstetter B (1985) J Agric Food Chem 33:232
6. Berger RG, Drawert F, Nitz S (1983) J Agric Food Chem 31:1237
7. Boudhrioua N, Giampaoli P, Bonazzi C (2003) Lebensm Wiss Technol 36:633
8. Boulanger R, Chassagne D, Crouzet J (1999) Flavour Fragrance J 14:303
9. Boulanger R, Crouzet J (2001) J Agric Food Chem 49:5911
10. Buttery RG, Seifert RM, Ling LC, Soderstom EL, Ogawa JM, Turnbaugh JG (1982) J Agric Food Chem 30:1208
11. Chan HT Jr, Flath RA, Forrey RR, Cavaletto CG, Nakayama TOM, Brekke JE (1973) J Agric Food Chem 21:566
12. Elss S, Preston C, Hertzig C, Heckel F, Richling E, Schreier P (2005) Lebensm Wiss Technol 38:263
13. Engel K-H, Tressl R (1991) J Agric Food Chem 39:2249
14. Engel K-H, Tressl R (1983) J Agric Food Chem 31:798
15. Fischer N, Hammerschmidt F-J, Brunke E-J (1995) Fruit Process 3:61
16. Flath RA, Forrey RR (1970) J Agric Food Chem 18:306
17. Flath RA, Forrey RR (1977) J Agric Food Chem 25:103
18. Flath RA, Light DM, Jang EB, Mon TR, John JO (1990) J Agric Food Chem 38:1060
19. Franco MRB, Shibamoto T (2000) J Food Chem 48:1263
20. Gholap AS, Bandyopadhyay C (1975) J Food Sci Technol 12:262
21. Gholap AS, Bandyopadhyay C (1977) J Sci Food Agric 28:885
22. Golding JB, Shearer D, McGlasson WB, Wyllie SG (1999) J Agric Food Chem 47:1646
23. Haagen-Smit AJ, Kirchner JG, Deasy CL, Prater AN (1945) J Am Chem Soc 67:1646
24. Haagen-Smit AJ, Kirchner JG, Deasy CL, Prater AN (1945) J Am Chem Soc 67:1651
25. Heidlas J, Lehr M, Idstein H, Schreier P (1984) J Agric Food Chem 32:1020
26. Homatidou VI, Karvouni SS, Dourtoglou VG, Poulos C (1992) J Agric Food Chem 40:1385
27. Jordán MJ, Goodner KL, Shaw PE (2000) Proc Fla State Hortic Soc 113:284
28. Jordán MJ, Goodner KL, Shaw PE (2002) J Agric Food Chem 50:1523
29. Jordán MJ, Margaria CA, Shaw PE, Goodner KL (2003) J Agric Food Chem 51:1421
30. Jordán MJ, Tandon K, Shaw PE, Goodner KL (2001) J Agric Food Chem 49:4813
31. Kemp TR, Knavel DE, Stoltz LP (1971) Phytochemistry 10:1925
32. Kemp TR, Knavel DE, Stoltz LP (1972) J Agric Food Chem 20:196
33. Kemp TR, Knavel DE, Stoltz LP (1972) Phytochemistry 11:3321
34. Kemp TR, Knavel DE, Stoltz LP (1973) Phytochemistry 12:2921
35. Kemp TR, Knavel DE, Stoltz LP, Lundin RE (1974) Phytochemistry 13:1167
36. Lalel HJD, Singh Z, Tan SC (2003) J Hortic Sci Biotechnol 78:485
37. Lalel HJD, Singh Z, Tan SC (2003) Postharv Biol Technol 27:323
38. Lalel HJD, Singh Z, Tan SC (2003) Postharv Biol Technol 29:205

39. Liu T-T, Yang T-S (2002) J Agric Food Chem 50:653
40. Macku C, Jennings WG (1987) J Agric Food Chem 35:845
41. MacLeod AJ, Pieris NM (1983) J Agric Food Chem 31:1005
42. MacLeod AJ, Pieris NM (1984) Phytochemistry 23:361
43. MacLeod AJ, Troconis NG (1982) Phytochemistry 21:1339
44. MacLeod AJ, Troconis NG (1982) Phytochemistry 21:2523
45. Mayra D, Märka T, Lindinger W, Brevard H, Yeretzian C (2003) Int J Mass Spectrom 223–224:743
46. Monteiro AR, Meireles MA, Marques MOM, Petenate AJ (1997) J Supercrit Fluids 11:91
47. Murray KE, Shipton J, Whitfield FB (1972) Aust J Chem 25:1921
48. Narain N, Almeida JN, Galvão MS, Madruga MS, Brito ES (2004) Ciênc. Tecnol Aliment 24:212
49. Nishimura O, Yamaguchi K, Mihara S, Shibamoto T (1989) J Agric Food Chem 37:139
50. Parliament TH (1972) J Agric Food Chem 20:1043
51. Pino JA, Almora K, Marbot R (2003) Flavour Fragrance J 18:492
52. Pino JA, Marbot R, Vázquez (2001) J Agric Food Chem 49:5883
53. Pino JA, Marbot R, Vásquez C (2002) J Agric Food Chem 50:6023
54. Sakho M, Chassagne D, Crouzet J (1997) J Agric Food Chem 45:883
55. Sakho M, Crouzet J, Seck S (1985) J Food Sci 50:548
56. Senesi E, Lo Scalzo R, Prinzivalli C, Testoni A (2002) J Sci Food Agric 82:655
57. Shiota H (1993) J Agric Food Chem 41:2056
58. Stevens K, Brekke JE, Stern DJ (1970) J Agric Food Chem 18:598
59. Takeoka GR, Buttery RG, Teranishi R, Flath RA, Güntert M (1991) J Agric Food Chem 39:1848
60. Tokitomo Y, Steinhaus M, Büttner A, Schieberle P (2005) Biosci Biotechnol Biochem 69:1323
61. Umano K, Hagi Y, Nakahara K, Shoji A, Shibamoto T (1992) J Agric Food Chem 40:599
62. Werkhoff P, Güntert M, Krammer G, Sommer H, Kaulen J (1998) J Agric Food Chem 46:1076
63. Wilson CW, Shaw P E (1978) Phytochemistry 17:1435
64. Winterhalter P, Katzenberger D, Schreier P (1986) Phytochemistry 25:1347
65. Wyllie SG, Leach DN (1990) J Agric Food Chem 38:2042
66. Wyllie SG, Leach DN (1992) J Agric Food Chem 40:253
67. Yabumoto K, Jennings WG (1977) J Food Sci. 42:32
68. Yahyaoui FEL, Wongs-Aree C, Latche A, Hackett R (2002) Eur J Biochem 269:2359

9 Vanilla

H. Korthou
Plant Science, Fytagoras BV,
Zernikedreef 9, 2333 CK Leiden, The Netherlands

R. Verpoorte
Section Metabolomics, Department of Pharmacognosy, IBL,
Leiden University, PO Box 9502, 2300 RA Leiden, The Netherlands

9.1 Introduction

Vanilla is widely used in food, beverages and cosmetics. It is produced from the beans of *Vanilla planifolia* Andrews, a member of the orchid family (Orchidaceae). The plant originates from Mexico where it was already used when the Spaniards arrived. Now it is cultured in various tropical countries, such as Madagascar, Indonesia, Uganda, Comoro, Tahiti, Papua Guinea, India and Mexico. Each of these growth sites yields vanilla with different flavour characteristics.

The total world consumption of vanilla beans is decreasing, mainly owing to the very high price ($300–500 per kilogram in 2004); in 2004 the demand was about half of the 2,200 t used in 1999 [5]. Madagascar produces more than half of the world production (1,000–1,200 t). Indonesia is the second largest producer, with some 350 t. Owing to various diseases of the plant and strong international competition, including from new production regions, Indonesian production has considerably decreased in the past few years, but the quality has increased [3].

The plant requires special growth conditions and the formation of beans requires pollination by specialised insects, which means that in most places pollination is done by hand. After 9 months of maturation, the vanilla beans undergo an elaborate processing known as curing, a process which takes about 6 months. Basically curing is a sort of fermentation process at elevated temperature, in which the beans are dried and the flavour develops, among others through the hydrolysis of the vanillin glucoside, resulting in free vanillin (Fig. 9.1), the most important flavour compound in the beans. The curing process is a highly traditional process; it is still not well understood and differs in the various producing regions. Despite various studies concerning the biosynthesis of vanillin and its role for the plant, many questions still remain. But to increase and ensure reproducible quality and to improve the efficiency of the curing process, further insight into the biochemistry of the vanillin production in the plant is required, as well as of other characteristic flavour compounds occurring in the beans.

Vanilla and vanillin are very versatile flavours, at any concentration they are acceptable, and most people enjoy the flavour, making it the world's most popular flavour. It is used in food (e.g. ice cream, various other diary products, chocolates and cakes), beverages (cola-type drinks), cosmetics and tobacco. There is a distinct difference between vanilla extract and vanillin, and most people prefer the extract-based products. Hoffman et al. [21] reviewed the analysis of vanilla constituents and flavours of vanilla. Improved or altered methods of curing and new cultivars may lead to a further diversification of the vanilla flavour.

Because of the large consumption of vanilla-flavoured products, vanillin is also made by other routes, such as (bio)conversion of related natural products or via synthesis. Only 0.2% of the approximately 6,000 t of vanillin used in the flavour market is derived from plants, for which vanilla is the major source [60, 61]. Most vanillin is synthetic; some several tons comes from microbial processes [38, 52]. About 60% of the vanillin goes into food and beverages, 33% into perfumes and cosmetics and 7% into pharmaceuticals [44]. The price of natural vanillin extracted from vanilla is estimated to be between $1,200 and $4,000 per kilogram [60, 61]. Natural vanillin derived from microbial production has a price of about $1,000 per kilogram [52]. Synthetic vanillin costs about $11–15 per kilogram [52, 53]. Consequently counterfeiting occurs because of the large price differences between natural vanillin and synthetic vanillin. Special analytical tools such as NMR spectrometry are applied to analyse the source from which the vanillin is derived.

Here we will review the current knowledge about the vanillin curing process, the biosynthesis of vanillin and alternative biotechnological production methods.

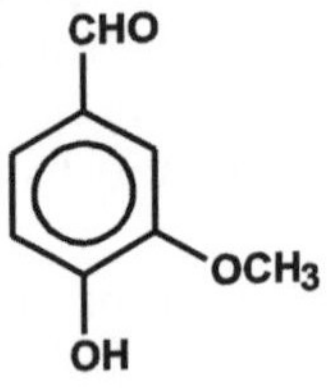

Fig. 9.1 Vanillin

9.2 The Plant

The genus *Vanilla* belongs to the family Orchidaceae, one of the largest plant families, with more than 18,500 species. The *Vanilla* Swartz genus has more than 100 species, amongst which 15 are aromatic. Three species have economic value, one of which is *Vanilla planifolia* Andrews (previously known as *V. fragrans* (Salisb.) Ames). Two other species, *V. pompona* Schiede and *V. tahitensis* J.W. Moore, are cultivated on a small scale for vanillin production. The former is more resistant against diseases, but gives pods of an inferior quality. The latter species is grown in Tahiti; it has a distinctly different flavour. It might be a man-made hybrid. Because of its unique flavour, the beans are more expensive than those of *V. planifolia* [46]

The diseases include fungal infestations caused by, among others, *Calospora vanillae* (anthracnose, whole plant), *Fusarium* sp. (root rot, fruit rot), *Phytophthora* sp. (fruit rot), *Colletotrichum* sp. and *Gloeralla vanilliae* (root rot). Besides these fungal diseases, viral diseases also pose a serious problem, e.g. Cymbidium mosaic virus and the cucumber mosaic virus. Suboptimal growing conditions and excessive rain or drought are the major reasons for diseases [46].Vanilla is a fleshy perennial vine, and requires a tree or artificial support for its growth. Adventitious aerial roots adhere to the supporting tree. The plant can grow as high as 10–15 m, but for cultivation the plants are kept low to facilitate the hand pollination and the harvest of the beans. The plant, propagated by cuttings, needs about 2–4 years before it flowers, and can produce for a period of 5–6 years. Each plant has about ten to 20 flower clusters of 15–20 flowers each. After hand pollination eight to 12 of these flowers will develop into pods.

The plant grows best in humid, tropical conditions; drought can easily kill the plant. Direct sun should also be avoided for growth sites, but full shade is also detrimental for the plant; therefore, vanilla is often planted between small shade-giving trees such as bananas and coffee, which should reduce the full sunshine to about one third to half of its intensity. Also artificial nets are used to create the right growth conditions. The plant grows well from sea level to altitudes of more than 760 m at a temperature ranging from 20 to 30 °C [46].

9.3 Vanillin

Vanillin is the most widely appreciated flavour compound in the world. Its odour threshold for humans is 11.8×10^{-14} M [6]. It has the unique characteristic that even at very high dose the flavour is still pleasant. Vanillin has various activities. The antimicrobial effects on the fungi *Aspargillus flavus*, *A. niger*, *A. ochraeus* and *A. parasiticus* and the bacteria *Escherichia coli*, *Bacillus subtilus* and *Staphylococcus aureus* were reviewed by Tipparaju et al. [57]. This makes vanillin a potential food preservative for a wide variety of products like diary products, soft drinks and fruit juices [13, 61]. In several studies the antioxidant, antimutagenic, anticlastogenic and anticarcinogenic activities of vanillin were demonstrated [12, 27, 54].

9.4 Biosynthesis

The biosynthesis of vanillin has been extensively studied both in the plant and in plant cell cultures (Scheme 9.1). There are some contradictions between the results of these studies, and consequently several questions about the biosynthetic pathway remain unanswered. At least all studies support the involvement of the shikimate pathway and the phenylalanine (phenylpropanoid) pathway. The first question is at what stage the C_3 side chain is oxidised to yield the aldehyde function. This could be before or after the formation of the typical 3-methoxy, 4-hydroxy substitution pattern in the aromatic ring. It is also unknown whether vanillin is derived from the lignan precursors having an alcohol function in the C_3 part of the phenylpropanoid, or from the cinnamic acid type (acid group). Zenk [63] showed that labelled ferulic acid was incorporated into vanillin. However, Kanisawa et al. [26] proposed that the major pathway would go via 4-coumaric acid glucoside, which is the precursor for *p*-hydroxybenzaldehyde glucoside, the central intermediate for the biosynthesis of the glucosides A and B as well as vanillin (Scheme 9.1). But their hypothesis left the possibility that the more oxidised compounds such as ferulic acid glucoside also can be converted into the corresponding aldehyde.

Various experiments in *Vanilla* plant cell cultures, however, gave different results [14–18, 30, 51]. This might be due to the fact that different biosynthetic pathways operate in the beans and in the cell culture. In fact most of the work in cell cultures showed only conversion of non-glucosylated products. Ferulic acid feeding resulted in increased vanillin levels. The fact that the *V. planifolia* cell cultures do not produce vanillin in any significant amount means that the results from studies using vanilla cell cultures for elucidation of the pathway should be considered with caution. Finally, it cannot be excluded that different pathways may contribute to the vanillin production in the beans. Scheme 9.1 shows that vanillin can be formed through different ways in a complex network of compounds.

For a general review of the biosynthesis of C_6C_1 compounds, see Mustafa and Verpoorte [39].

9.5 Enzymes

Only a few steps of the biosynthesis of vanillin are known down to the level of the enzymes and the genes. Particularly the glucosidases involved in the formation of vanillin from its glucoside have been studied extensively. As vanillin is almost completely stored in the form of a glucoside, the role of the glucosidase is crucial for the quality of the final product, as a high level of vanillin is required for good quality. Concerning the glucosidases, different results have been reported.

Kanisawa et al. [26] reported that two glucosidases are present in vanilla pods. A non-specific enzyme occurs in both leaves and beans, whereas a specific vanillin glucosidase was detected only in the beans. Using the *p*-nitrophenylglucoside (NPG) assay for detecting activity, Odoux et al. [42] and Ranjoanisafy [47] (cited in Odoux and Havkin-Frenkel [41]) purified and characterised a glucosidase from the beans. The enzyme with a molecular mass of 201 kDa con-

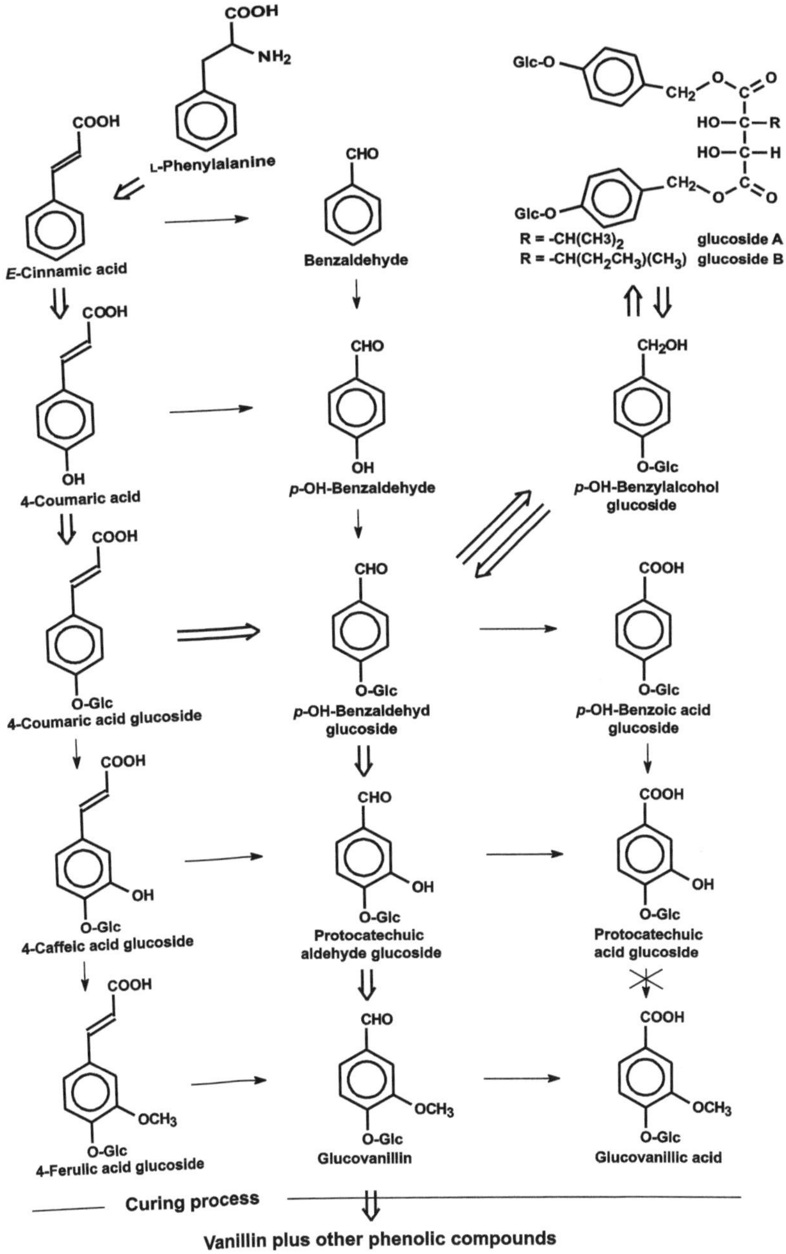

Scheme 9.1 Proposed pathway for vanillin biosynthesis in *Vanilla planifolia* beans according to Kanisawa et al. [26]. The *thick arrows* represent the most likely pathway

sisted of four subunits (50 kDa each). Because no specific enzyme assay was used, the occurrence of a highly specific vanillin glucoside hydrolysing enzyme, as reported by Kanisawa et al. [26], cannot be excluded.

Havkin-Frenkel (cited in Odoux and Havkin-Frenkel [41]) found a series of glucosyl hydrolases in green vanilla beans. The enzymes included α-glucosidase, β-glucosidase, α-galactosidase, β-galactosidase, α-mannosidase and β-mannosidase. The β-glucosidase showed maximum activity at 50 °C; the α-galactosidase and the β-galactosidase had optima at 55 and 60 °C, respectively, temperatures which are similar as those during the curing process.

Dignum et al.[9] followed glucosidase activity during the curing process using the NPG assay. They could not detect glucosidase activity anymore after the autoclaving step, though vanillin glucoside was still hydrolysed in the beans. The presence of a non-NPG-assay-detectable glucosidase can thus not be excluded. The glucosidase activity measured in green beans was also strongly dependent on the type and pH of the incubation buffer used. The highest activity was obtained at pH 8 with a [bis(2-hydroxyethyl)amino]tris(hydroxymethyl)methane propane buffer. Freezing of the green beans caused a dramatic loss of glucosidase activity, and the enzyme extract lost much of its activity when stored at -20 °C; storage at -80 °C particularly in the presence of glycerol gave better results. Dignum et al. [10] measured the kinetic properties of the glucosidase from the green beans for several glucosides occurring in vanilla. For vanillin glucoside the V_{max} was 9.4 IU mg^{-1} protein and K_m was 5.1 mM; values in the same range were found for the glucosides of vanillic acid, *p*-hydroxybenzaldehyde and ferulic acid, Also for the synthetic substrate NPG, a similar activity was found; however, for creosol, and guaiacol glucosides a much higher K_m was found, whereas for 2-phenylethanol and *p*-cresol glucosides no activity was detected. Hanum [20] reported a much higher V_{max} and a lower K_m for vanilla glucosidase.

The enzyme is localised in the placental tissue of the beans, i.e. where the vanillin glucoside is also found [41, 42]. The enzyme seems to be localised in the cytoplasm. Though not proven, it is hypothesised that the enzyme and the glucoside are in different cellular compartments, similar to the case for many other secondary metabolite glucosides in other plants (e.g. see Geerlings et al. [19]). The compartmentation is part of the plant defence. Once a cell has been destroyed by, e.g., an insect, the glucosidase will come together with the glucoside, resulting in the formation of a toxic aglycone. A common defence response in plants, a phytoanticipin present in the plant cell is converted immediately into a highly toxic and reactive compound after attack by a microorganism or an insect. During the curing process when the cell integrity is destroyed, the vanillin glucoside will diffuse through the bean and in contact with the enzyme it will be converted into vanillin.

Roeling et al. [50] studied the possibility that microorganisms may be involved in the hydrolysis of the glucoside during the curing process; however, they could not find any evidence for the presence of specific microorganisms growing on the beans after the killing step.

For the biosynthesis of vanillin, several other enzymes are of interest. First of all, phenylalanine ammonia lyase (PAL); this enzyme converts phenylalanine into the cinnamic acid type of compounds, the first intermediates in the vanillin biosynthesis after the primary metabolism. PAL activity could be detected in green beans, but after scalding this activity is lost. The chain shortening enzyme (CSE), responsible for the conversion of a C_6C_3 compound into a C_6C_1 compound, was found to be localised in the cytosol of cells of the placental trichomes in the green beans [23].

Peroxidases might play a role in decomposition of vanillin and in flavour generation during the curing process. Their activity is high in green beans and the enzymes also remain active during the curing. The same applies for proteases, which might be a reason for not recovering glucosidase activity from the beans in the curing process.

9.6 Curing

Dignum et al. [10] studied the curing process in Bali (Indonesia). The curing process starts with the so called killing or scalding. After harvesting, the green beans are thrown into hot water (60–70 °C) for 1–2 min. After that they are stored in boxes for 2–5 days; this phase of the process is called autoclaving, a confusing name as the temperature never goes over 50 °C. After this step the beans are spread in the sun on blankets and wrapped up again during the night in boxes; this phase (sunning and sweating) goes on for about 2 weeks. As the beans are everyday in the hands of the labourers, they sort the beans during this process into different quality classes. In the next phase (2–4 weeks) the beans are dried further; at the end of this step they have reached a water content of 25–40%. The final stage consists of storing the beans for several months in small bundles in a sealed box, or in plastic vacuum bags. After this stage the beans are ready for use. The total curing process may last as long as 6 months.

Dignum et al. [8, 10] followed this process in detail on a production site in Bali to measure the various parameters of the processing in order to mimic these in a laboratory model curing system. The parameters are summarised in Table 9.1. On the basis of the observations, a model curing system was set up to study different parameters under controlled conditions and the effect on some enzymes and the vanillin production.

The curing process is an essential step for the production as the flavour develops gradually during the process, in part due to enzymatic conversion of the vanillin glucoside, in part due to other enzymatic and chemical reactions, including oxidations. Further knowledge of these chemical and biochemical processes is thus of great importance for optimising the production of high-quality vanilla beans.

Table 9.1 Parameters of laboratory curing processes under traditional Indonesian conditions [10]

Stage	Temperature (°C)	Relative humidity (%)	Time
Scalding (killing)	70		1.5 min
Autoclaving	60	95	3 h
	55	95	3 h
	50	95	3 h
	45	95	3 h
Sunning/sweating	40	70	1 h
	47.5	62.5	3 h
	55	55	2 h
	50	95	6 h
	42.5	95	12 h
Slow drying	30	80	3 weeks

9.7 Chemistry

More than 250 compounds have already been identified in vanilla beans, representing a broad variety of classes of natural products such as monoterpenoids, fatty acids and various esters thereof, benzoic acid derivatives, hydrocarbon ketones and alcohols, phenylpropanoids and other phenolics [8, 21]. Some of these phenolics occur also as glucoside [8, 25, 26]. Major components in cured beans, besides vanillin (0.3–2%), are *p*-hydroxybenzaldehyde (0.2%), *p*-hydroxybenzylmethyl ether (0.02%) and acetic acid (0.02%). In green beans glucovanillin, bis[4-(*β*-D-glucopyranosyloxy)benzyl-2-isopropyltartrate] (glucoside A) and bis[4-(*β*-D-glucopyranosyloxy)benzyl-2-(2-butyl)tartrate] (glucoside B) are the major compounds [8, 25, 26] .

More than 95% of the volatile components are present at very low level (below 10 ppm) [21]. For a review on the various compounds identified in vanilla, see Dignum et al. [8]. For studies on the specificity of the glucosidase(s) in vanilla a series of glucosides occurring in green beans have been synthesised [11, 31, 40]. Glucovanilline can also be produced by feeding vanillin to plant cell cultures [28, 55]—an almost 50% yield of glucosylation was obtained from *V. planifolia* cell cultures [62].

Synthetic vanillin is a major intermediate in the production of various chemicals, including medicines and herbicides. Because of the large difference in price between vanillin from a natural origin and synthetic vanillin, counterfeiting is not uncommon. As natural and synthetic vanillin are chemically identical, isotope ratios of hydrogen (D/H) and carbon ($^{13}C/^{12}C$) isotopes are used to determine the source of vanillin. Such analyses can be done by means of mass spectrometry or NMR. The latter has the advantage that the position-specific incorporation is determined (site-specific natural isotope fractionation NMR spectrometry). This highly specific method enables the differentiation of vanillin of all known sources [49].

To cater for the large demand for vanillin, besides different synthetic methods also biotechnological processes have been developed. Synthetic vanillin has a major drawback that products containing this compound cannot be labelled as containing a natural flavour. On the other hand, biotechnological products can be labelled as natural.

For the synthesis several processes have been described using different natural starting materials, such as coniferin, eugenol, guaiacol and lignin (for reviews, see [22, 44, 48, 60, 61].

9.8 Biotechnological Production of Vanillin

To produce natural vanillin at a lower price, various biotechnological approaches have been explored, such as plant cell cultures and bioconversion of natural compounds by means of microorganisms or isolated enzymes.

The production of fine chemicals by means of large-scale plant cell cultures is feasible [59]. But although *V. planifolia* cell cultures have been studied extensively, no economically feasible vanillin production has resulted from this (for reviews, see [22, 44, 48, 61]. *Capsicum frutescens* cell cultures were able to produce some vanilla flavour compounds upon being fed various precursors [48]. The amount of vanilla flavour compounds could be enhanced by treating the cultures with methyl jasmonate [56]. Cell suspension cultures of *Capsicum annuum* were able to produce vanillin after being fed with exogenous ferulic acid [24]. Next to plant cell cultures, it was shown that crude enzyme extracts from plants could be used for bioconversions of readily available precursors. Enzymes from soybean are able to convert isoeugenol into vanillin after addition of powdered activated carbon and peroxide [34]. A soybean lipoxygenase can produce vanillin from esters of coniferyl alcohol [35]. A vanillyl alcohol oxidase with broad specificity was obtained from *Penicillium*; this enzyme can convert vanillylamine (e.g. obtainable from the hydrolysis of capsaicin) or creosol [58].

Various microorganisms (e.g. *Bacillus fusiformis, Pseudomonas fluorescens, Pseudomonas acidovorans, Penicillium simplicissimum, E. coli, Corynebacterium glutamicum, Saccharomyces cerevisiae, Pycnoporus cinnabarinus, A. niger*) are able to convert fed natural phenylpropanoids precursors, such as ferulic acid, eugenol, isoeugenol, coniferyl alcohol, vanillyl alcohol and vanillylamineisorhapotin (a stilbene), into vanillin [1, 2, 29, 38, 44, 48, 52, 60, 61, 64]. All these precursors have the same aromatic substitution pattern as vanillin; thus, they only require a chemical modification in the aliphatic carbon side chain. Priefert et al. [44] distinguished four different mechanisms for the shortening of the side chain of ferulic acid: non-oxidative decarboxylation, side chain reduction, and coenzyme A (CoA) dependent as well as independent deacetylation. In all cases the toxic and highly reactive aldehyde formed is rapidly converted to the corresponding alcohol or acid. Rabenhorst and Hopp [45] using an *Amycolatopsis* species and Mueller et al. [37] using *Streptomyces setonii* were able to obtain high yields of vanillin (more than10 g l^{-1}) in the conversion of ferulic acid. The molar yields were about 75%. In both cases the bacterial species used had a high tolerance for vanillin. Muheim and Lerch [38] reported that the *Streptomyces setonii* strain mentioned could be the basis of an economical microbial production of vanillin from ferulic acid with a production of more than 6.4 g l^{-1}. The major bottleneck for these processes is the high price of ferulic acid. Eugenol is a much cheaper alternative ($9 per kilogram) [44], but so far the reported vanillin yields are lower than for ferulic acid conversion. When ferulic acid can be obtained from agricultural by-products for a low cost, it might be an interesting alternative for the production of vanillin. Ferulic acid is the most abundant hydroxycinnamic acid in the plant world, since it is an important structural component of the plant cell wall. Feruloyl esterases from a wide range of microorganisms can be used to release ferulic acid from the plant [36]. Cheap agricultural by-products like sugar beet pulp and maize bran are a good source for ferulic acid that could be released by the filamentous fungus *Pycnoporus cinnabarinus* [4, 32]. Interestingly, addition of a culture filtrate of the fungus *A. niger* resulted in direct biotransformation of autoclaved maize bran into vanillin [32]. A novel strain of *Bacillus fusiformis* was described that produced high amounts of vanillin from isoeugenol [64]. The cost of vanillin from a microbial production was estimated to be $1,000 per kilogram [52].

High-rate bioconversion of eugenol to ferulic acid was reported for an *E. coli* XL1-blue strain expressing the *vaoA* gene from *Penicillium simplicissimum* encoding vanillyl alcohol oxidase, which converts eugenol to coniferyl alcohol, together with the genes *calA* and *calB* encoding coniferyl dehydrogenase and coniferyl aldehyde dehydrogenase of *Pseudomonas*. This transgenic bacterial strain was able to convert eugenol to ferulic acid (14.7 g l^{-1}) with a molar yield of 93.3% [43]. The enzyme 4-hydroxycinnamoyl-CoA hydratase/lyase (HCHL) from *Pseudomonas fluorescens* converted ferulic acid CoA into vanillin. This gene in combination with 4-hydroxycinnamoyl-CoA ligase was overexpressed in *E. coli*; this strain is capable of converting ferulic acid into vanillin.

E. coli has been genetically engineered to convert shikimate into vanillin by introducing the genes encoding a shikimate dehydrogenase yielding 3-dehydroshikimic acid, a dehydratase converting this into protocatechuic acid and a catechol-*O*-methyltransferase converting this acid into vanillic acid. Finally a reductase yielded vanillin [33, 44]. The various patents for biotechnological production of vanillin were reviewed by Priefert et al. [44] and Daugsch and Pastore [7].

The HCHL encoding gene (see above) has been overexpressed in various plant cells and plants by Walton and co-workers (tobacco, *Datura stramonium*). In none of these systems could vanillin be detected; however, various benzoic acid derivatives were found, including vanillic acid glucoside [61]. This might be due to a lack of ferulic acid as a substrate in the plant cells and/or due to the toxic aldehydes being immediately converted into the corresponding acids or alcohols, similar to what is found in experiments on feeding vanillin to various cell cultures [62]. Genetic engineering of vanilla plants to overexpress this enzyme is not likely to be very successful, as the plants already contain a very high level of vanillin (2–6%) in the producing tissues.

Whether genetically engineered organisms will be successful for the production of vanillin not only depends on the economic feasibility of the process, but also on the acceptance by the public of GMO-produced vanillin.

9.9 Conclusions

Vanillin is the most important flavour compound in vanilla, and is often used to replace the extract. Vanillin can be obtained from vanilla beans, but because of the high costs of the beans, various other production methods have been developed. By far the cheapest production method is chemical synthesis, but vanillin made in this way cannot be labelled as natural. Of course this provokes counterfeiting and necessitates advanced quality control methods (NMR). It also initiates many studies in alternative natural production methods. These include microbial production and genetic engineering of microorganisms and plants. Microbial production of vanillin has been achieved, but the price is high ($1,000 per kilogram). Genetic engineering might be possible to either increase vanillin production in vanilla or introduce the pathway into other plants. However, public opinion against GMOs will be a major hurdle for this approach, besides the fact that the vanillin biosynthetic pathway is not known and no transformation–regeneration system for vanilla has been developed yet. In any case, all these methods only focus on vanillin, whereas the flavour of vanilla is more than only vanillin. Therefore, improvement of the agricultural practices and the curing system might be a more important strategy, also as it would offer farmers in the developing countries higher yields of better quality and thus higher incomes. To improve agricultural practice and yields, more knowledge about the pest and disease resistance of the plant and about the regulation of flowering, fruit ripening and vanillin biosynthesis is required. Further studies on vanilla are thus of great interest.

References

1. Achterholt S, Rabenhorst J, Steinbuechel A and Preifert H (2003) Process for the preparation of vanillin and suitable microorganisms. US Patent 2,003,092,143
2. Achterholt S, Rabenhorst J, Steinbuechel A and Preifert H (2004) Process for the preparation of vanillin and suitable microorganisms. US Patent 2,004,203,123

3. Bernard F (2005) Vanilla in Indonesia. In: Vanilla: The first international congress. Allured, Carol Stream, pp 33–40
4. Bonnina E, Brunel M, Gouy Y, Lesage-Meessen L, Asther M, Thibault J (2001) *Aspergillis niger* I-1472 and *Pycnoporus cinnabarinus* MUCL39533, selected for the biotransformation of ferulic acid to vanillin, are able to produce cell wall polysaccharide-degrading enzymes and feruloyl esterases. Enzyme Microb Technol 28:70–80
5. Brownell R (2005) The commercial survival of natural vanilla. In: Vanilla: The first international congress. Allured, Carol Stream, pp 1–3
6. Buccellato (2005) The various uses of vanilla in perfumery. In: Vanilla: The first international congress. Allured, Carol Stream, pp 67–70
7. Daugsch A, Pastore G (2005) Obtencao de vanilina: Oportunidade biotecnologica. Quim Nova 28:642–645
8. Dignum M, Kerler J, Verpoorte R (2001a) Vanilla production: technological, chemical and biosynthetic aspects. Food Res Int 17:199–219
9. Dignum M, Kerler J, Verpoorte R (2001b) Alpha-glucosidase and peroxidase stability in crude enzyme extracts from green beans of *Vanilla planifilolia* Andrews. Phytochem Anal 12:174–179
10. Dignum M, Kerler J, Verpoorte R (2002) Vanilla curing under laboratory conditions. Food Chem 79:165–171
11. Dignum M, van der Heijden R, Kerler J, Winkel C, Verpoorte R (2004) Identification of glucosides in green beans of *Vanilla planifolia* Andrews and kinetics of vanilla beta-glucosidase. Food Chem 85:199–205
12. Durant S, Karran P (2003) Vanillins–a novel family of DNA-PK inhibitors. Nucleic Acids Res 31:5501–5512
13. Fitzgerald DJ, Stratford M, Gasson MJ, Narbad A (2004) The potential application of vanillin in preventing yeast spoilage of soft drinks and fruit juices. J Food Prot 67:391–395
14. Funk C, Brodelius P (1990a) Influence of growth regulators and an elicitor on phenylpropanoid metabolism in suspension cultures of *Vanilla planifolia* Andr. Phytochemistry 29:845–848
15. Funk C, Brodelius P (1990b) Phenylpropanoid metabolism in suspension cultures of *Vanilla planifolia* Andr. II Effects of precursor feeding and metabolic inhibitors. Plant Physiol 94:95–101
16. Funk C, Brodelius P (1990c) Phenylpropanoid metabolism in suspension cultures of *Vanilla planifolia* Andr. III Conversion of 4-methoxycimnnamic acids into 4-hydroxybenzoic acids. Plant Physiol 94:102–108
17. Funk C, Brodelius P (1992) Phenylpropanoid metabolism in suspension cultures of *Vanilla planifolia* Andr. IV Induction of vanillinic acid formation. Plant Physiol 99:256–262
18. Funk C, Brodelius P (1994) *Vanilla planifolia* Andrews: in vitro biosynthesis of vanillin and other phenylpropanoids derivatives. In: Bajaj YPS (ed) Biotechnology in agriculture and forestry. Medicinal and aromatic plants VI, vol 26. Springer, Berlin Heidelberg New York, pp 377–402
19. Geerlings A, Martinez-Lozano Ibanez M, Memelink J, van der Heijden R, Verpoorte R (2000) The strictosidine ß-D-glucosidase gene from *Catharanthus roseus* is regulated coordinately with other terpenoid-indole alkaloid biosynthetic genes and the encoded enzyme is located in the endoplasmic reticulum. J Biol Chem 275:3051–3056
20. Hanum T (1997) Changes in vanillin and activity of β-glucosidase and oxidases during post harvest processing of vanilla beans (*Vanilla planifolia*). Bull Teknol Ind Pangan 8:435–443

21. Hoffman P, Harmon A, Ford P, Zapf M, Weber A, King S, Grypa R, Philander E, Gonzalez L, Lentz K (2005) Analytical approaches to vanilla quality and authentication. In: Vanilla: The first international congress. Allured, Carol Stream, pp 41–49
22. Hrazdina G (2006) Aroma production by tissue cultures Agric Food Chem 54:1116–1123
23. Joel DM, French JC, Graf N, Kourtiva G, Dixon AD, Havkin-Frenkel D (2003) A hairy tissue produces vanillin. Isr J Plant Sci 51:157–159
24. Kang SM, Jung HY, Kang YM, Min JY, Karigar CS, Yang JK, Kim SW, Ha YR, Lee SH and Choi MS (2005) Biotransformation and impact of ferulic acid on phenylpropanoid and capsaicin levels in *Capsicum annuum* L.cv.P1482 cell suspension cultures. J Agric Food Chem 53:3449–3453
25. Kanisawa T (1993) Flavor development in vanilla beans. Kouryou 180:113–123
26. Kanisawa T, Tokoro K, Kawahara S (1994) Flavor development in the beans of Vanilla planifolia. In: Kurihara K, Suzuki N, Ogawa H (eds) Olfaction taste XI, Proceedings of the international symposium. Springer, Berlin Heidelberg New York, pp 268–270
27. Keshava C, Keshava N, Whong WZ, Nath J, Ong TM (1998) Inhibition of methotrexate-induced chromosomal damage by vanillin and chlorophyllin in V79 cell. Teratogen Carcinogen Mutagen 17:313–326
28. Kometani T, Tanimoto H, Nishimura T, Osaka S (1993) Glucosylation of vanillin by cultured plant cells. Biosci Biotechnol Biochem 57:1157–1161
29. Krings U: Berger RG (1998) Biotechnological production of flavours and fragrances. Appl Microbiol Biotechnol 49:1–8
30. Labuda IM, Goers KA, Keon KA (1993) Microbial bioconversion process for the production of vanillin. In: Schreier P, Winterthaler P (eds) Progress in flavour precursor studies: analysis, generation, biotechnology. Proceedings of the international conference, Wuerzburg. Allured, Carol Stream, pp 477–482
31. Leong G, Uzio R, Derbesy M (1989) Synthesis, identification and determination of glucosides presnt in green vanilla beans. Flavour Fragrance J 4:163–167
32. Lesage-Meessen L, Lomascolo A, Bonnin E, Thibault JF, Buleon A, Roller M, Asther M, Record E, Ceccaldi BC and Asther M (2002) A biotechnology process involving filamentous fungi to produce natural crystalline vanillin from maize bran. Appl Biochem Biotechnol 102–103:141–153
33. Li K and Frost JW (1998) Synthesis of vanillin from glucose. J Am Chem Soc 120:10545–10546
34. Li YH, Sun ZH, Zao LQ, Xu Y (2005) Bioconversion of isoeugenol into vanillin by crude enzyme extracted from soybean. Appl Biochem Biotechnol 125:1–10
35. Markus PH, Peters ALJ, Roos R (1992) Process for the preparation of phenylaldehydes. Eur Patent Appl EP 0 542 348 A2
36. Mathew S, Abraham TE (2004) Ferulic acid: an antioxidant found naturally in plant cell walls and feruloyl esterases involved in its release and their applications. Crit Rev Biotechnol 24:59–83
37. Mueller B, Muench T, Muheim A, Wetli M (1998) Process for the production of vanillin. Patent Appl EPO885968
38. Muheim A, Lerch K (1999) Towards a high-yield bioconversion of ferulic acid to vanillin. Appl Biochem Biotechnol 51:456–461
39. Mustafa NK, Verpoorte R (2005) Chorismate derived C6C1 compounds in plants. Planta 222:1–5

40. Negishi I, Ozawa T (1996) Determination of hydroxycinnamic acids, hydroxybenzoic acids, hydroxybenzaldehydes, hydroxybenzyl alcohols and their glucsides y high-performance liquid chromatography. J Chromatogr A 756:129–136
41. Odoux E, Havkin-Frenkel D (2005) Hydrolysis of glucovanillin by β-glucosidase during curing of vanilla been (*Vanilla planifolia* Andrews). In: Vanilla: The first international congress. Allured, Carol Stream, pp 95–100
42. Odoux E, Chauwin A, Brillouet JM (2003) Purification and characterization of vanilla bean (*Vanilla planifolia* Andrews) β-glucosidase. J Agric Food Chem 51:3168–3173
43. Overhage J, Steinbuechel A and Priefert H (2003) Highly efficient biotransformation of eugenol to ferulic acid and further conversion to vanillin in recombinant strains of *Escherichia coli.* Appl Environ Microbiol 69:6569–6576
44. Priefert H, Rabenhorst J, Steinbuechel (2001) Biotechnological production of vanillin. Appl Microbiol Biotechnol 56:296–314
45. Rabenhorst J, Hopp R (2000) Process for the preparation of vanillin and suitable microorganisms. DE Patent 19,532,317
46. Ranadive AS (2005) Vanilla cultivation. In: Vanilla: The first international congress. Allured, Carol Stream, pp 25–32
47. Ranjoanisafy X (1998) Purification et étude de quelques propriétés de la β-glucosidase de la vanilla. Implications dans la préparation de la vanilla. Master's thesis, Ensia-Siarc, Montpellier
48. Rao SR and Ravishankar GA (2000) Vanilla flavour: Production by conventional and biotechnological routes. J Sci Food Agric 80:289–304
49. Remaud GS, Martin YK, Giles G, Martin GJ (1997) Detection of natural vanilla flavors and extracts: Application of the SNIF-NMR to vanillin and p-hydroxybenzaldehyde. J Agric Food Chem 45:859–866
50. Roeling WFM, Kerler J, Braster M, Apriyantono A, Stam H, Van Verseveld HW (2001) Microorganisms with a taste for vanilla: Microbial ecology of traditional Indonesian vanilla curing. Appl Environ Microbiol 67:1995–2003
51. Romagnoli LG, Knorr D (1988) Effects of ferulic acid treatment on growth and flavor development of cultured *Vanilla planifolia* cells. Food Biotechnol 2:93–104
52. Schrader J, Etschmann MMW, Sell D, Hilmer J-M, Rabenhorst J (2004) Applied biocatalysis for the synthesis of natural flavour compounds—current industrial processes and future prospects. Biotechnol Lett 26:463–472
53. Serra S, Fuganti C, Brenna E (2005) Biocatalytic preparation of natural flavours and fragrances. Trends Biotechnol 23:193–199
54. Sinigaglia M, Reguly ML, de Andrade HH (2004) Effect of vanillin on toxicant-induced mutation and mitotic recombination in proliferating somatic cells of *Drosophila melanogaster.* Environ Mol Mutagen 44:394–400
55. Sommer J, Schroeder C, Stoeckigt J (1997) In vivo formation of vanillin glucoside. Plant Cell Tissue Organ Cult 50:119–123
56. Suresh B, Ravishankar GA (2005) Methyl jasmonate modulated biotransformation of phenylpropanoids to vanillin related metabolites using *Capsicum frutescens* root cultures. Plant Physiol Biochem 43:125–131
57. Tipparaju S, Ravishankar S, Slade PJ. (2004) Survival of *Listeria monocytogenes* in vanilla-flavored soy and dairy products stored at 8 degrees C. J Food Prot 67:378–382

58. Van den Heuvel RRH, Fraaije MW, Laane C, van Berkel WJH (2001) Enzymatic synthesis of vanillin. J Agric Food Chem 49:2954–2958
59. Verpoorte R, Contin A, Memelink J (2002) Biotechnology for the production of plant secondary metabolites. Phytochem Rev 1:13–25
60. Walton NJ, Narbad A, Faulds CB, Williamson G (2000) Novel approaches to the biosynthesis of vanillin. Curr Opin Biotechnol 11:490–496
61. Walton NJ, Mayer MJ, Narbad A (2003) Vanillin. Phytochemistry 63:505–515
62. Yuana, Dignum MJW, Verpoorte R (2002) Glucosylation of exogenous vanillin by plant cell cultures. Plant Cell Tissue Organ Cult 69:177–182
63. Zenk MH (1965) Biosynthesis von Vanillin in *Vanilla planifolia* Andr. Z Pflanzenphysiol 53:404–414
64. Zhao LQ, Sun ZH, Zheng P, Zhu LL (2005) Biotransformation of isoeugenol to vanillin by novel strain of *Bacillus fusiformis*. Biotechnol Lett 27:1505–1509

10 Flavour of Spirit Drinks: Raw Materials, Fermentation, Distillation, and Ageing

Norbert Christoph, Claudia Bauer-Christoph
Bavarian Health and Food Safety Authority,
Luitpoldstr. 1, 97082 Würzburg, Germany

10.1 Introduction

Spirit drinks are food products and represent a major outlet for the agricultural industry all over the world. This outlet is largely the result of the flavour quality and reputation these products have acquired on the world market over hundreds of years; various national and international legal decrees, standards, and specifications lay down rules on the definition, description, and presentation of the different categories of spirit drinks [1–4] which can be separated in two main categories, distilled spirits and liqueurs. Distilled spirits have alcoholic strengths between 30 and 50% v/v and are produced by distillation from fermented agricultural products containing carbohydrates; their flavour is not only characterised by aroma compounds originating from the raw material and the alcoholic fermentation, but also from distillation, storage, and ageing. Liqueurs are spirits with a minimum ethanol content of 15% v/v and a sugar content of 100 g L^{-1}; they are produced by flavouring ethanol of agricultural origin, distillates of agricultural origin, or one or more spirit drinks with natural plant materials such as herbs, fruits, fruit juice, cream, chocolate, steam-distilled essential oils, distilled spirit drinks, or natural or artificial flavouring extracts.

Aroma compounds in distilled spirits and liqueurs, their levels, odour attributes, and thresholds are most important for quality and authenticity. Using gas chromatography and mass spectrometry, especially the composition of volatile aroma compounds in distilled spirits has been widely investigated [4–8]. By direct injection of an alcoholic distillate it is possible to determine more than 50 components within levels between 0.1 and 1,000 mg L^{-1}; special methods of extraction can be used to increase this number up to more than 1,000 volatile substances [6]. However, sensory analysis is still indispensable to describe and evaluate spirit drinks.

The following review focuses on the composition of flavour compounds in spirit drinks, their origin, and their sensory attributes like odour quality and threshold value. Important information on flavour-related aspects of technology like distillation and ageing, as well as the main categories and brands of spirits to be found on the national and international markets are summarised. Finally, aspects of sustainability in the production of distilled spirits are discussed.

10.2
Flavour Compounds in Distilled Spirits

Flavour compounds of distilled spirits originate from the raw materials used for fermentation and from alcoholic fermentation by yeasts (*Saccharomyces cerevisiae)* and other microorganisms which metabolise carbohydrates, amino acids, fatty acids, and other organic compounds. Figure 10.1 shows a basic scheme of precursors, intermediates, and metabolites of the main groups of flavour compounds which are produced during alcoholic fermentation in yeast cells [4–9]. Fermentation of carbohydrates not only leads to the main products ethanol, glycerol and carbon dioxide, but also to a typical fingerprint of volatile metabolites at relatively low levels, like aldehydes, ketones, higher alcohols, organic acids, and esters, which are called 'fermentation by-products' or 'congeners'.

Table 10.1 gives a summary of the main by-products of fermentation by yeasts and other microbiological activities which can be identified in distilled spirits from different raw materials, like fruits, wine, grain, sugar cane, or other carbohydrate-containing plants. Since the sensory relevance of a flavour compound is related to its odour thresholds and odour quality, Table 10.1 presents also odour qualities and a review of threshold values of the fermentation by-products in ethanol solutions [9–10] and/or water [11–14] (Christoph and Bauer-Christoph 2006, unpublished results).

The concentration range of the flavour compounds is given in milligram per litre for distilled spirits adjusted to about 40% v/v ethanol. In order to compare distillates with different ethanol content, it is also common to calculate the relative concentrations of volatile compounds in milligram 0.1 L^{-1} pure ethanol (p.e.); thus a propanol concentration of 400 mg L^{-1} in a distilled spirit with 40% v/v ethanol would correspond to a concentration of 100 mg 0.1 L^{-1} p.e. The reason for the variations of the absolute concentrations of the volatile compounds in commercial products is mainly a result of the different conditions of fermentation and the distillation technique. The threshold data to be found in the literature [9–14] are rather different and threshold values in water are significant lower than in ethanol solutions owing to the masking effect of the high ethanol concentration present in spirits.

Some of the volatile substances which are produced during fermentation, like acrolein, diacetyl, 2-butanol, allyl alcohol, or acetic acid, are a result of enhanced microbiological activities and may cause an unpleasant flavour (off-flavour) at certain levels; thus, elevated concentrations of such compounds are markers for spoilage of the raw material, negative microbiological influences during or after the fermentation process, or a poor distillation technique.

10.2.1
Carbonyl Compounds

Acetaldehyde is the major important carbonyl compound of alcoholic fermentation and is formed as an intermediate compound by degradation of pyruvate;

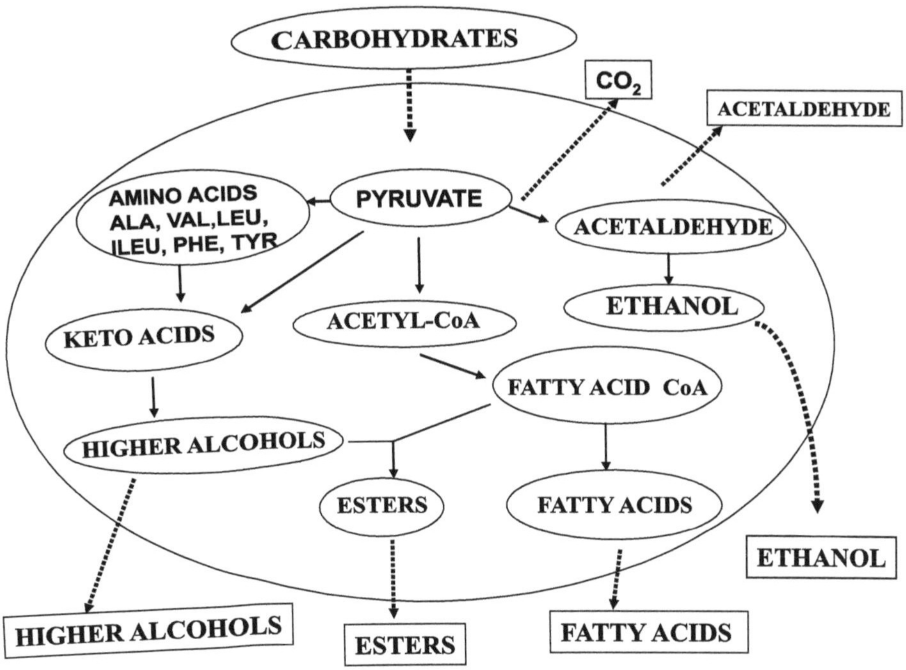

Fig. 10.1 Precursors, intermediates, and metabolites of the main groups of flavour compounds produced during alcoholic fermentation of carbohydrates by *Saccharomyces cerevisiae* yeast

its production by yeasts depends on the pyruvate decarboxylase activity of the yeast. Since acetaldehyde is one of the most volatile compounds, the highest levels are in the 'head cut' of the distillation and thus can be separated from the 'heart cut'. In vodka-type spirits 10 mg L^{-1} may cause off-flavours owing to the pungent odour, whereas in fruit or wine distillates higher concentrations do not affect the quality owing to an odour threshold of about 100 mg L^{-1}. Acrolein (2-propenal) has a peppery, horseradish-like smell and is either formed by dehydration of glycerol during distillation in the presence of acids on hot metallic surfaces or especially by bacteria during fermentation of spoiled raw materials. The biochemical pathway of the formation of acrolein is initiated by a bacterial dehydratase enzyme which converts glycerol to 3-hydroxypropionaldehyde. The compound undergoes slow, spontaneous dehydration to acrolein in an acidic medium [15, 16]. Acetaldehyde as well as acrolein reacts with ethanol to form the acetals 1,1-diethoxyethane and 1,1,3-triethoxypropane, respectively, with the consequence of a reduction of the pungent odour of the aldehydes; the equilibrium concentration of the 1,1-diethoxyethane formed is close to 10% in relation to the amount of acetaldehyde present [17]. Higher aldehydes and their acetals can be found at concentrations less than 0.1 mg L^{-1}. Diacetyl (2,3-butandione) is responsible for buttery flavour notes that sometimes arise in beer or wine distillates. It is formed by lactic acid bacteria; thus, distillates from wine

Table 10.1 Odour quality, odour threshold value in water and/or ethanol solution, and concentration range of single volatile compounds in distilled spirits produced during alcoholic fermentation from carbohydrates by yeasts and other microorganisms

Compound	Flavour quality	Threshold ($mg\ L^{-1}$ water)[a]	Threshold ($mg\ L^{-1}$ ethanol solution)[a]	Typical concentration [$mg\ L^{-1}$ (40% v/v)][a]
Carbonyl compounds				
Acetaldehyde	Pungent, sweet,	0.025	10	<2–160
Diethoxyethane	Fruity, sherry-like	0.005	1	<3–72
Triethoxyethane	Pungent			<0.5–6
Acrolein (2-propenal)	Horse radish, peppery	0.04		<0.1–1.2
Diacetyl (2,3-butandione)	Buttery	0.1; 2.5		<0.1–12
Alcohols				
Ethanol	Alcoholic	24.9		
Methanol	Alcoholic		668	20–1,000
1-Propanol	Stupefying	500	830	40–800
1-Butanol	Alcoholic	0.5; 1.3	820	1–80
2-Methyl-1-propanol	Alcoholic		40; 75	40–400
2-Butanol	Alcoholic		1,000	0.4–320
2-Methyl-1-butanol	Malty	0.32	7; 30	8–720
3-Methyl-1-butanol	Malty	1	7; 30	4–1,200
Allyl alcohol	Unpleasant	19		4–52
Phenethyl alcohol	Rose-like	1	7.5; 10	4–32
Esters				
Ethyl acetate	Solvent-like, nail polish	17.6	7.5	4–800
Ethyl butanoate	Fruity, floral	0.001	0.02	<0.1–3.2
Methylbutyl acetate	Fruity, banana, pear	0.3	0.03	1.2–12
2-Phenethyl acetate	Rose, honey, fruity	0.02	0.25	4–12
Ethyl hexanoate	Apple, banana, violet	0.005	0.005	0.4–3.2
Ethyl octanoate	Pineapple, pear	0.07	0.26; 0.002	4–20
Ethyl decanoate	Floral, fatty	0.5		4–36
Ethyl dodecanoate	Floral			1.6–32
Diethyl succinate			100	2–12
Ethyl lactate			100	<10–400
Acids				
Acetic acid	Vinegar-like, pungent	100; 1,000		1–50
Butyric acid	Buttery	1	4; 10	<0.1
Hexanoic acid	Rancid, fatty		8; 3	1–19
Octanoic acid	Oily, fatty , soapy		15; 8.8	1–4
Decanoic acid	Fatty, citrus		10; 15	0.3–5

[a]Minimum and maximum threshold values cited from the literature

or cider which have undergone a malolactic fermentation may have rather high levels of diacetyl [5–9].

10.2.2 Aliphatic and Aromatic Alcohols

Methanol, 1-butanol, and 2-butanol are not compounds of alcoholic fermentation but characteristic substances for the type and authenticity of specific raw materials in distilled spirits; their threshold values are rather high and therefore they do not contribute significantly to the flavour. High methanol concentrations are typical for fruit spirits as a result of the enzymatic degradation of the pectin in the fruits and grapes, respectively. 1-Butanol levels below 3 mg 0.1 L^{-1} p.e. are typical for cherry distillates, whereas in other fruit distillates the level may rise to 100 mg 0.1 L^{-1} p.e. [5–8]. 2-Butanol levels higher than 50 mg 0.1 L^{-1} p.e. indicate a bacterial spoilage of raw materials or mash; also 1-propanol levels higher than 500 mg 0.1 L^{-1} p.e are an indicator for the spoilage of fruit mashes [5–8].

Higher alcohols, also called 'fusel alcohols', are quantitatively the largest group of volatile flavour compounds produced as metabolites from the degradation of amino acids via keto acids (2-oxo acids). The most important alcohols are 1-propanol, 2-methyl-1-propanol, 2-methylbutanol, 3-methylbutanol, and the aromatic alcohol 2-phenylethanol. The term 'fusel alcohols' refers to their malty and burnt flavour, with the exception of 2-phenethyl alcohol, which has a rose-like odour. The concentrations of aliphatic alcohols in distilled spirits vary over a large range and depend mainly on the type of distillation, separation, and fractionation, respectively. Excessive concentrations of higher alcohols can result in a strong pungent and 'fusel-like' smell and taste, whereas optimal levels impart fruity character [5–9]. 2-Methylbutanol and 3-methylbutanol, the mixture of which is also called isoamyl alcohol, are the most abundant minor components of distilled spirits synthesised by yeasts; depending on the nature of the raw material, these alcohols comprise 40–70% of the total fusel alcohol fraction [9].

10.2.3 Fatty Acids

The biosynthesis of fatty acids produced during alcoholic fermentation is initiated in the yeast cell by the formation of acetylcoenzyme A, which reacts with malonylcoenzyme A to form mainly saturated straight-chained fatty acids with an even number of four to 18 carbon atoms; the appearance of relatively low levels of fatty acids with odd numbers of carbon atoms as well as unsaturated fatty acids depends on the fermentation conditions [6]. The volatile fatty acids contribute to the flavour of fermented beverages like wine or beer and their concentration usually lies between 100 and 250 mg 0.1 L^{-1} p.e. In distilled spirits the concentration of free fatty acids is significantly lower owing to the esterification and separation by distillation; thus, the concentration in wine distillates like Cognac is in the range of 50 mg 0.1 L^{-1} p.e. Acetic acid can be produced during and/or after fermentation by oxidation of ethanol under aerobic conditions by the acetic acid bacteria *Acetobacter*; acetic acid levels should not be higher than 100 mg 0.1 L^{-1} p.e in distilled spirits, since higher levels may contribute to a typical vinegar-like off-flavour.

10.2.4
Esters

Esters are the largest group of flavour compounds [5–7] with mostly pleasant flavour properties. Their quantities and mutual proportions are of great importance for the perceived flavour of a spirit drink since their concentrations are generally above the sensory threshold values. Especially the low-boiling ethyl esters like ethyl 2-methylbutanoate, ethyl hexanoate, and ethyl octanoate, and the acetates like ethyl acetate, isoamyl acetate, isobutyl acetate, hexyl acetate, and 2-phenethyl acetate are of great importance for the flavour of distilled spirits. Ethyl acetate, mainly produced as a result of esterification of acetic acid, is the main ester which occurs in fermented products and their distillates; it contributes significantly to a solvent-like nail polish off-flavour at levels higher than 400 mg 0.1 L^{-1} p.e. The flavour fraction with the lowest volatility is composed of C_{14}–C_{18} fatty acid esters; these esters as well as the long-chain fatty alcohols may contribute to the stearine-like smell that is characteristic of Scotch malt whisky in particular. Malolactic fermentation also has an influence on the concentration of these compounds; distillates show a loss of fruitiness and aroma intensity with decreasing levels of ethyl hexanoate, hexyl acetate, 2-phenethyl acetate, and with increasing levels of ethyl lactate, acetic acid, and diethyl succinate [18].

10.3
Important Flavour Compounds from Raw Materials

Table 10.2 presents a summary of odour qualities, odour thresholds in water, and concentrations of some selected volatile compounds, which are characteristic flavour impact compounds, owing to their typical flavour quality and their rather low odour thresholds. These compounds are not formed during fermentation but originate from the raw material and contribute significantly to the typical flavour of a fruit. The components summarised in Table 10.2 are important compounds in wine and different fruits and are discussed later.

Table 10.2 Odour quality and minimum and maximum odour thresholds in water [11] (Christoph and Bauer-Christoph 2006, unpublished results) of selected volatile impact compounds of raw materials for distilled spirits

Compound	Odour quality	Threshold (μg L^{-1})[a]	Raw material
Hexanol	Green, flowery	500; 2,500	Wine, fruits
(E)-2-Hexenol	Green, apple	20; 500	Apple
Hexyl acetate	Fruity	20; 50	Wine, fruits
Ethyl (*S*)-2-methylbutanoate	Apple	0,006	Apple
Ethyl (2*E*,4*Z*) decadienoate	Pear	300	Bartlett, Williams pear
Benzaldehyde	Almond	35	Stone fruits
Hydrocyanic acid	Bitter almond	2,000	Stone fruits
1-(*p*-Hydroxyphenyl)-3-butanone	Raspberry	5; 100	Raspberry
α-Ionone	Flowery, violet	0.05; 5	Raspberry
β-Ionone	Flowery, violet	0.007; 0.5	Raspberry
(*R*)-Linalool	Flowery, citrus	0.00014	Wine, fruits
Geraniol	Rose	5; 75	Wine, fruits
Citronellol	Citrus fruits	10; 40	Wine, fruits
γ-Decalactone	Fruity, peach	5; 10	Fruits
γ-Dodecalactone	Fruity, peach	7	Fruits
Ethyl cinnamate	Fruity	15; 0.06	Stone fruits

[a]Minimum and maximum threshold values cited from the literature

10.4 Distillation—Separation and Fractionation of Flavour

Distillation of wines, fermented juices, and mashes with an alcoholic strength between 5 and 15% v/v is the main technological step in the production of distilled spirits by which ethanol and flavour compounds are separated and transferred into the distillate; ethanol is distilled as a water–ethanol azeotropic mixture at 78.15 °C, together with other more or less volatile compounds. Two types of distillation apparatus (stills) are used [4, 8]. The most basic type of still is the batch distillation with a pot still for the production of heavily flavoured distillates, which is an enclosed copper vessel (the 'kettle' or 'pot') that narrows into an overhead-vapour pipe at the top to collect alcohol vapour. The pipe bends downwards off the top of the pot to a water-cooled condenser which causes the alcohol vapour to condense back into liquid. The first distillate ('head cut') with an alcoholic strength of less than 30% v/v has to be distilled a second time in order to increase the alcoholic strength to more than 60% v/v. The traditional pot stills mostly are used for production of brandies, whiskies, and fruit distillates. Modern distillation columns are equipped with up to three column plates. The vapour is fractionated at these plates via reflux from the descending liquid and is carried over into a second and third plate where it is once more circulated and concentrated to the desired percentage of alcohol. Finally a so-called dephlegmator, a device placed in the upper part of the column, also increases the ethanol content in the vapour phase by partial condensation of water. Column stills are more efficient than pot stills in extracting a higher concentration of alcohol. With both stills it is possible to separate different fractions ('cuts'). The 'head cut' contains rather volatile compounds like acetaldehyde whereas the 'tail cut' (below 40% v/v) is characterised by high-boiling compounds such as ethyl esters of long-chained fatty acids; since the flavours of both fractions contain undesirable aroma compounds, these substances can be separated off from the 'heart cut' which is rich in aroma compounds important for sensory quality. However, since regulations on distillation limits and minimum amounts of volatiles exist for many types of distilled spirits [1–3], it is important to take the distillation behaviour of the different volatile components into consideration [19]. Using distillation plants with high fractionation capacity as well as further purification techniques (activated charcoal), 'ethyl alcohol of agricultural origin' [1] with a minimum alcoholic strength of 96.0% v/v and a maximum level of other aroma compounds lower than 3.8 mg 0.1 L^{-1} p.e. can be produced.

10.5 Flavour Compounds Originating from Ageing

Ageing of distilled spirits is an important technological step to improve the flavour since fresh distillates are often characterised by a raw, pungent odour and taste. Different components of a fresh distillate may react during the maturation period, which is favoured by a high ethanol content of the distillates to be stored. Thus, the concentrations of ethyl esters of fatty acids increase during ageing, but the concentrations of esters of other alcohols, such as 3-methylbutyl acetate, decrease by transesterification. Further reactions during ageing are the evaporation of aldehydes or their reaction to form acetals [15, 20]. By storing distillates in wooden casks, volatile aroma compounds like *cis*-β-methyl-γ-octalactone and *trans*-β-methyl-γ-octalactone ('whisky lactone'), vanillin, guajacol, eugenol, cresols, and other phenolic compounds migrate from the toasted wood into the distillate. These compounds are responsible for the characteristic oak wood and vanilla-like flavour [21]. Table 10.3 gives a summary of typical aroma compounds of oak wood and their odour thresholds.

Semivolatile and non-volatile compounds of wood change the colour of the distillate and contribute to an up-rounded flavour. The wooden barrels which are permeable allow air to pass in and cause ethanol to evaporate; thus, the ethanol content decreases and the aroma gets more intense, complex, and concentrated. Also harsher aroma constituents are removed and the spirit changes to mellow. The period of maturation depends on the size of the casks used, the alcoholic strength, as well as the temperature and humidity in the warehouse which leads to a smoother flavour. For production of neutral highly rectified distilled spirits like vodka, grain spirit, or white rum, the quality of water is of utmost importance to the flavour. In vodka production different treatments of water like de-

Table 10.3 Odour qualities and threshold values of aroma impact compounds from toasted oak wood

Compound	Odour quality	Threshold (mg L^{-1} water)
Furfural	Smoky, almond	8
Guajacol	Smoky	0.005
cis-Methyl-γ-octalactone/ *trans*-Methyl-γ-octalactone	Oak, wood	0.02
Vinylguajacol	Phenolic, clove	0.03
4-Methylguajacol	Smoky, burnt wood	0.01
4-Ethylguajacol	Smoky, phenolic	0.02
4-Ethylphenol	Stable-like, horse	0.13
Eugenol	Spicy, clove	0.007
Vanillin	Vanilla, spicy	0.1
o-Cresol	Medicinal, tar	0.04
m-Cresol	Medicinal, smoky	0.2

ionisation, alkalinisation, or neutralisation are used. Thus, anions like chloride, nitrate, or sulphate can be used as marker compounds in ion chromatography for such special technologies and for authentication of specific brands of colourless neutral spirits [22].

10.6 Flavour and Flavour-Related Aspects of Distilled Spirits

Distilled spirits are produced from different fermented plant materials grown all over the world. The most important criteria of quality and authenticity of each type of distilled spirits are the typical flavour composition originating from the raw material and/or the special techniques of fermentation, distillation, and ageing. In the following sections, specific flavour compounds and flavour-related technologies as well as peculiarities of the most important categories of distilled spirits are summarised.

10.6.1 Wine and Wine-Pomace Brandies

The term 'brandy' traces back to the Dutch word brandewijn ('burnt wine') which was introduced by Dutch traders in the sixteenth century to describe wine that had been 'burnt' or boiled, in order to distil it. Actually only spirits distilled from wine are called brandy. These spirits are normally aged in wooden casks (usually oak), a process by which colour, mouthfeel and flavour are significantly changed. They are also frequently blended with a so-called typage which may contain wine, grape juice, caramel sugar, and flavourings like extracts of dried plums, green walnuts, and almond peels. Each wine-growing country produces typical brandies mostly labelled with a specific geographic designation. Brandies from different countries like France (Cognac, Armagnac), Germany (Weinbrand), Spain (Brandy de Jerez), Italy, Mexico, and Chile (Pisco from muscat grapes) are usually distilled in column stills and aged in oak casks for varying periods of time. The most famous region-specific brandies Cognac and Armagnac must be produced from wines exclusively from those specific geographical regions; the flavour of the final products is not characterised by single specific flavour compounds, but to a greater degree the quality is determined by the grape varieties of the wines used for distillation, the distillation itself, the ageing technology, as well as the blending [23, 24]. By gaschromatography–mass spectrometry analysis, 34 volatile compounds were found to be responsible for the typical sensory descriptors attributed to freshly distilled Cognac not matured in oak barrels [25]; the most relevant sensory descriptors were diacetyl (butter), nerolidol (hay), (*Z*)-3-hexen-1-ol (grass), 2-phenylethyl acetate (rose), 2-methylbutyl acetate (pear-like), and 3-methylbutyl acetate (banana-like).

Brandies like Italian Grappa and French Marc are the best-known examples of distillates which are made from pomace, i.e. the pressed grape pulp, skins, and stems that remain after the grapes have been crushed and pressed to extract most of the juice for wine. Pomace brandies usually are minimally aged in wood but have an acquired flavour; they often tend to be rather raw, although they can offer a fresh, fruity aroma of the type of grape used, a characteristic that is lost in regular oak-aged brandy. The products have comparatively high concentrations of methanol, higher alcohols, acetaldehyde, ethyl acetate, higher esters, and distinctive aroma compounds of the wines' grape cultivars which were used for distillation [26].

10.6.2 Fruit Spirits

According to the EEC Regulations distilled spirits from fruits shall be called 'spirit' proceeded by the name of the fruit, such as cherry spirit. They may be also called Wasser (Germany) or eau de vie (France) with the name of the fruit [1]. These distilled spirits are made by fermentation and distillation from fresh and fleshy fruits and their musts, respectively. The aroma of these spirits is significantly influenced by the specific flavour compounds (Table 10.2) of the fruits. Stone fruits like cherry, plum, apricot, and yellow plum as well as pomaceous fruits like apples, pears, and single varieties like Bartlett (Williams Christ) pear, and Golden Delicious or Cox Orange apples are mostly used; however, also rare fruit varieties or even fruits with a relatively low sugar content like quince, strawberry, and elderberry are fermented.

The flavour of distillates from apple and pear is characterised by typical aroma compounds from these fruits formed by enzymatic degradation of fatty acids to C_6-fragments like hexanol, *trans*-2-hexenol, as well as ethyl esters and acetates of hexanoic acid. In distillates of pears, especially of the variety Bartlett pear, the characteristic pear flavour is mainly dominated by the ethyl and methyl esters of *trans*-2-*cis*-4-decadienoic acid and *trans*-2-*trans*-4-decadienoic acid [27–29]. The biogenesis of these monounsaturated, diunsaturated, and triunsaturated esters may be explained by β-oxidation of unsaturated linoleic and linolenic acid in the fruits. The sesquiterpene compound α-farnesene, which is formed during postharvest ripening and storage of Bartlett pears [28], shows that quality and intensity of distilled pear spirits is mainly influenced by the quality and degree of ripeness of the fruits.

A special fruit spirit called Calvados is distilled in the French region of the Normandy from cider (cidre). It is necessary to mix different apple varieties to obtain a well-balanced flavour of the cider; apple size, ripeness, and storage are important to achieve the highest aromatic level before pressing. The fermentation of the must (mout) lasts approximately 2 months at low temperatures (10–15 °C) in order to reduce the loss of volatile esters which occur at high fermentation temperatures [30]. After ageing for about 1 year, the cider is distilled with a pot still; the resulting *petite eau*, is distilled a second time. The aroma of the resulting Calvados is gained through the ageing process in young oak barrels, which should give the Calvados its vanilla taste; this storage should not be too long to keep the apple flavour. In a next stage of maturation, the distillate is stored in older barrels. A total of 169 volatile compounds were identified in dichloromethane extracts obtained by liquid-liquid extraction in studies on the chemical and sensorial aroma characterisation of freshly distilled Calvados [31, 32]. Esters have a probable maximum level around 500 mg 0.1 L^{-1} p.e. Acetaldehyde and acetals are typical compounds that contribute to the "green" note, whereas higher alcohols do not have a direct impact on quality. Especially unsaturated alcohols and aldehydes as well as phenolic derivatives are specific for flavour.

Distilled spirits are produced from stone fruits like cherry (Kirschwasser, Cherry, Kirsch), plum (Zwetschgenwasser, Slivovitz), yellow plum, and apricots not only in many regions of Europe but also in many other parts of the world. The flavour of stone fruit spirits is mostly affected by the aroma compound benzaldehyde, which originates from the enzymatic degradation of amygdalin in the stones of the fruits, passing into the mash during fermentation and later into the distillate at rather high levels. Since the aroma-active compound hydrocyanide which is also a product of enzymatic degradation of amygdalin is known as the precursor of the genotoxic compound ethylcarbamate, several technologies (addition of copper salt before distillation, use of copper inlets in stills) have been developed since 1989 to reduce and eliminate hydrocyanic acid during production [33, 34]. These distillates are characterised by more fruity and flowery notes owing to less benzaldehyde and hydrocyanic acid and the enhanced perception of aroma compounds like ethyl hexanoate, γ-decalactone, γ-dodecalactone, geraniol, and eugenol [27, 35, 36]. Fruits like peach, apricot, raspberry, blackberry, bilberry, sloe, or rowanberry have very intensive and pleasant flavour notes, but have generally not enough sugar to yield ethanol. Especially in Austria, France, Germany, Hungary, and northern Italy, distilled spirits from these fruits are produced by macerating the fresh fruits in high-proof ethyl alcohol of agricultural origin in order to extract their flavour and aroma; the macerate is distilled once at a low proof. The flavour of these spirits, which are called Geist, like for raspberry Himbeergeist or Framboise, are characterised by the typical flavour compounds of the fruits only. The flavour impact compound of raspberry 1-(4-hydroxyphenyl)-3-butanone is detected only at trace levels (below 10 μg 0.1 L^{-1} p.e.) in authentic distillates, since it is a very high boiling component; other aroma compounds like linalool, α-terpineol, α-ionone, β-ionone, benzyl alcohol, γ-lactones, and *cis*-3-hexenol seem to be important for the flavour of Himbeergeist [27, 37].

10.6.3 Grain Spirits

Grain spirits are produced by distillation of a fermented mash of cereals, malted or not malted, and they have the organoleptic characteristics derived from the raw material used. German grain spirit may be labelled as Korn or Kornbrand, provided that it is traditionally and exclusively produced by the distillation of a fermented mash of whole grains of wheat, barley, oats, rye, or buckwheat with all component parts or by redistillation of such distillates [1]. Whisky in the sense of the EEC decree 1576/89 [1] is also a spirit drink produced by the distillation of a mash of cereals, which is saccharified by diastase of the malt contained therein, with or without natural enzymes. It must be distilled at less than 94.8% v/v to keep the aroma and taste derived from the raw material. Whisky must be matured for at least 3 years in wooden oak barrels not exceeding 700-L capacity. This barrel-ageing smoothes the rough palate of the raw spirit and adds aromatic and flavouring nuances (Table 10.3), all of which set whiskies apart from colourless grain spirits which are distilled closer to neutrality in taste and generally are not aged in wood. Scotch whisky can be differentiated into grain whisky, produced by continuous distillation, and malt whisky, which is distilled twice in large copper pot stills [38]. Blending is a process of mixing different whiskies in order to reach uniformity in a product with definite standard colour and flavour. The typical flavour of Scotch whisky is mainly related to the presence of volatile phenolic compounds due to the burning of peat during the barley kilning. Among these aroma compounds, cresols, guajacol, ethylphenol, and vinylguajacol are responsible for the phenolic, medicinal, smoky, and burnt flavour [38]. Whiskies from North America, bourbon whiskey, and Canadian whiskey are grain spirits that have been produced from a mash bill that usually mixes together corn, rye, wheat, barley, and other grains in different proportions, and then the mixture is generally aged for an extended period of time in wooden barrels. Whisky is also produced in other countries; in Asia the most important product is Thai whisky.

10.6.4 Vodka

Vodka is the dominant spirit in Russia, Finland, Poland, Sweden, and other eastern European countries. It is made by rectifying ethyl alcohol of agricultural origin mainly produced from grain, potatoes, molasses of beets, and other plants. Rye and wheat are the main raw materials of vodka, with most of the best Russian vodkas being produced from wheat, while in Poland they are mostly distilled from a rye mash; Swedish and Baltic distillers only partially use wheat mashes. Molasses, a sticky, sweet residue from sugar production, is widely used for inexpensive, mass-produced brands of vodka [39]. The aim to produce high-quality vodka with its typical mellow and soft flavour is reached by filtration through activated charcoal, further fractionation, and special treatments of the water used for dilution [22]. These special technologies reduce the levels of volatile compounds down to traces of some few congeners. Except for a few minor styles, vodka is not put in wooden casks or aged for an extensive period of time. It can, however, be flavoured or coloured with a wide variety of fruits, herbs, and spices. In Russia and Poland different flavours are used, like dried lemon and orange peels, ginger, cloves, coffee, anise and other herbs and spices, fruit tree leaves, port, Malaga wine, or buffalo (bison) grass, an aromatic grass favoured by the herds of the rare European bison. In recent years numerous other flavoured vodkas have been launched on the world market e.g. with fruit flavours such as currant or orange.

10.6.5 Rum, Cachaça

Rum is a product obtained from distillation of fermented sugar-cane juice, molasses, or mixtures of both. The molasses contain over 50% sugar, but also significant amounts of other components, which may contribute to the final typical rum flavour. Rums made from cane juice, primarily on Haiti and Martinique, have a naturally smooth palate. Depending on the recipe, the 'wash' (the cane juice, or molasses and water) is fermented, using either cultured yeast or airborne wild yeasts, for a period ranging from 24 h for light rums up to several weeks for heavy, full varieties. The choice of stills has a profound effect on the final flavour of rum. White rums are primarily used as mixers and blends particularly well with fruit flavours. Golden rums, also known as amber rums, are generally medium-bodied. Most have spent several years ageing in oak casks, which give them smooth, mellow palates. Dark rums are traditionally full-bodied, and caramel-dominated rums. The best are produced mostly from pot stills and are frequently aged in oak casks for extended periods. The composition of aroma compounds is related to these different categories. Heavy, full varieties are characterised by high concentrations of fusel alcohols and ethyl esters. Ethyl esters of acetic, propionic, butyric, and valeric acids and higher homologues contribute to the distinct aroma of rum [40]. Also other flavour-active compounds like heterocyclic nitrogen compounds originating from the Maillard reaction as well as phenolic compounds are important for rum flavour.

Cachaça and aguardente de cana are the most consumed distilled spirits in Brazil exclusively made from cane-sugar juice. Sugar and caramel may be added for colour adjustment. The total content of congeners is between 200 and 650 mg 0.1 L^{-1} p.e. Like other spirits, the flavour of cachaça is mainly characterised by the presence of fermentation by-products such as higher alcohols, esters, carboxylic acids, and carbonyl compounds [41–43].

10.6.6 Juniper-, Caraway-, and Aniseed-Flavoured Spirits

Spirit drinks called gin (genever) are white spirits flavoured with the highly aromatic berries of juniper, a low-slung evergreen bush (genus *Juniperus*). Additional botanical flavourings can include anise, angelica root, cinnamon, orange peel, coriander, and cassia bark. All gin and genever producers have their own secret combination of botanicals, the number of which can range from as few as four to as many as 15. Unlike liqueurs, where flavourings are added to the distilled spirits, gin is made by redistilling the spirit with the flavourings, either with the flavouring ingredients in the still, or by passing the vapour through the flavouring agents during distillation. The spirit base of gin is primarily grain (usually wheat or rye), which results in a light-bodied spirit. Top-quality gins and genevers are flavoured in a unique manner. After one or more distillations the base spirit is redistilled one last time. During this final distillation the alcohol vapour wafts through a chamber in which the dried juniper berries and botanicals are suspended. The vapour gently extracts aromatic and flavouring oils and compounds from the berries and spices as it travels through the chamber on its way to the condenser. The resulting flavoured spirit has a noticeable degree of complexity. The main components detected in gin are the monoterpenes α-pinene, β-myrcene, limonene, γ-terpinene, and p-cymene, reflecting the typical composition of character-impact compounds of juniper berries; further oxygenated monoterpenes like linalool, α-terpineol, 4-terpineol, and bornyl acetate as well as sesquiterpenes like γ-cadinene, δ-cadinene, caryophyllene, and β-elemene were detected [44].

Spirit drinks which are produced by flavouring ethyl alcohol of agricultural origin with distillates of caraway or dill are called akvavit or aquavit and mainly come from Denmark and Scandinavia; these spirits are flavoured using neutral alcohol distillates of caraway (*Carvum carvi*) and/or dill (*Anethum graveolens*); the use of essential oils is prohibited. The impact compounds of these spirits are (+)-carvone and anethol.

Aniseed-flavoured spirit drinks are produced in Greece (ouzo), Turkey (raki), or France (pastis); they are produced by flavouring ethyl alcohol of agricultural origin with natural extracts of star anise, anise, fennel, or any other plant containing *trans*-anethol, the principal aromatic constituent of aniseed and further aroma compounds like *cis*-anethol, estragol, anisaldehyde, and anise alcohol. For flavouring, different technologies like maceration and distillation, redistillation in the presence of the plant materials, or addition of natural distilled extracts may be used. Pastis must also contain natural extracts of liquorice root (*Glycyrrhiza glabra*), which implies the presence of the colorants known as chalcones, as well as glycyrrhetinic acid between a minimum of 0.05 and a maximum of 0.5 g L^{-1}; the anethol level of pastis must be between 1.5 and 2 g L^{-1} [1]. The concentration of anethol in raki is between 1 and 1.7 g L^{-1}, whereas ouzo contains less than 1 g L^{-1} [45].

10.6.7 Tequila, Mezcal

The typical Mexican distilled spirits tequila and mezcal are made by distilling the fermented juice of the agave plant, a spiky-leafed member of the lily family. By Mexican law the agave spirit called tequila can be made only from one particular type of agave, the blue agave (*Agave tequiliana Weber*), and can be produced only in specifically designated geographic areas, primarily the state of Jalisco in west-central Mexico [41]. Mezcal is made from the fermented juice of other species of agave. Both tequila and mezcal are prepared for distillation in similar ways. When the plant reaches sexual maturity, it starts to grow a flower stalk, which is cut off just as it is starting to grow; in consequence, the central stalk swells into a large bulbous shape that contains a sweet juicy pulp. The so-called piña, which resembles a giant green and white pineapple, is cut into quarters, and is slowly baked in steam ovens or autoclaves until all of the starch has been converted to sugars. For mezcal it is baked in underground ovens heated with wood charcoal, which gives mezcal its distinctive smoky flavour. In consequence, Maillard compounds like 5-hydroxymethylfurfural, 2-furanmethanol, or 2-furancarboxyaldheyde result from these thermal processings [45]. The piña is then crushed and shredded to extract the sweet juice, called aguamiel (honey water). The fermentation stage determines whether the final product will be 100% agave; this highest-quality tequila is made from agave juice only mixed with some water. 'Mixto' is made by fermenting and then distilling a mix of agave juice and other sugars, usually cane sugar with water. Traditionally tequila and mezcal are distilled in pot stills; the resulting spirit is clear, but contains a significant amount of congeners and other flavour compounds like different esters, terpenes, phenoles, and thiazoles; the most important flavour compounds are isovaleraldehyde, isoamyl alcohol, β-damascenone, 2-phenethyl alcohol, phenethyl acetate, and eugenol [46, 47]. Colour in tequila and mezcal comes mostly from the addition of caramel, although barrel ageing is a factor in some high-quality brands. Additionally, some distillers add small amounts of natural flavourings such as sherry, prune concentrate, and coconut to smooth out the often hard-edged palate of agave spirits. Beyond the two basic designations of '100% tequila' and 'mixto', there are four further categories: silver or blanco tequilas, which are clear, with little or no ageing, gold tequila, which is an unaged silver tequila that has been coloured and flavoured with caramel, reposado tequila, which is aged in wooden tanks or casks for a legal minimum period of at least 2 months and añejo tequila, which is aged in wooden barrels (usually old bourbon barrels) for a minimum of 12 months. Ageing tequila for more than 4 years is a matter of controversy; most tequila producers oppose doing so because they feel that excessive oak ageing will overwhelm the distinctive earthy and vegetal agave flavour notes.

10.6.8
Shochu, Soju, Awamori

Shochu is Japan's other indigenous alcoholic beverage, but unlike sake, which is the wine-like rice brew, shochu is distilled. The Korean counterpart is called soju. Shochu and soju are made from one of several raw materials like rice, soba (buckwheat), or barley, but even from sweet potato (imo-shochu), brown sugar, chestnuts, and other grains. Each of these raw materials gives a very distinct flavour and aroma profile to the final sake, which ranges from smooth and light (rice) to peaty, earthy, and strong (sweet potato). For distillation of the sake, two different methods are used: the first is the traditional single-round (batch) distillation of individual raw material (otsu-rui or honkaku shochu); using the second method, kou shochu produced from different raw materials goes through continuous distillation (kou-rui shochu). The alcoholic content usually is 25% v/v although sometimes it can be as high as 42% v/v or more. Awamori is made from long-grain indica rice imported from Thailand [48, 49].

10.6.9
Absinth

Spirit drinks with a predominantly bitter taste are produced by flavouring ethyl alcohol of agricultural origin with natural and/or nature-identical flavouring substances. Absinth is the most famous representative of this category, a high-alcoholic sometimes anise-flavoured spirit drink derived from herbs including the flowers and leaves of the medicinal plant *Artemisia absinthum*, also called wormwood. The main aroma compounds of wormwood essential oil are α-thujone and β-thujone (40–90%), absinthin, and artabsin, whereas the sesquiterpene absinthin is the most bitter compound [48, 49]. Thujone is restricted for bitter spirits to a level of 35 mg kg^{-1} in the EU; in commercial products the thujone level investigated was lower than 2 ppm in 51% of cases, between 2 and

10 ppm in 26% of cases, between 10 and 35 ppm in 14% of cases, and more than 35 ppm in 9% of cases [50].

10.6.10 Liqueurs and Speciality Products

Liqueurs and speciality products are a very important group of spirit drinks on the world market with an enormous global consumption, representing an extremely wide range of traditional brands and products of special composition. Liqueurs are, by definition, coloured or colourless sweet spirits which are produced by adding products of agricultural origin or flavourings to ethyl alcohol or distillates of agricultural origin. According to the European Council Regulation 1576/1989 [1], liqueurs have a minimum ethanol content of 15% v/v and a minimum sugar content of 100 g L^{-1}. The flavour of liqueurs can originate from plant materials such as herbs, fruits or fruit juice, from different food products like wine, cream or chocolate, from steam distilled essential oils, distilled spirit drinks and/or from natural and artificial flavouring extracts or flavour compounds. The natural extracts can be obtained by infusion (digestion), percolation, distillation, or any combination of these processes [51]. Fruit liqueurs of cherry, blackcurrant, raspberry, bilberry, pineapple, and citrus fruits are produced by adding juices of the named fruits and natural aroma; the use of nature-identical aroma is not allowed for these fruit juices, whereas fruit liqueurs from peach, apricot, plums, banana, apple, pear, and strawberry can be produced with nature-identical aroma compounds too. Bitter liqueurs have a bittersweet flavour and are produced with spices, herbs, and bitter-tasting drugs like quinine or calmus. Egg liqueurs contain a minimum sugar or honey content of 150 g L^{-1} and a minimum content of egg yolk of 140 g L^{-1} (70 g L^{-1} for liqueurs with eggs). The additional descriptor 'crème' with the name of a specific fruit or raw material used, excluding milk products, is reserved for liqueurs with a minimum sugar content of 250 g L^{-1}. In the USA, the manufacture and definition of liqueurs and so-called cordials is controlled by federal regulations; boosted natural flavours are allowed, which means that flavours may contain up to 0.1% of artificial (synthetic) flavour components and still be classified as natural. Regulations in other countries may also differ.

Table 10.4 summarises some of the most famous international brands of liqueurs and their composition; a summary of more than 400 liqueurs and speciality products and their composition is given by Clutton [51].

The world's top brands of liqueurs and speciality products are products from companies like De Kuyper, Berentzen, Bols, and Marie Brizard. Famous brands are Kahlua (coffee liqueur), Bailey's Original Irish Cream, Grand Marnier, Cointreau, Amaretto, and Sambuca as well as the bitter liqueurs or aperitifs like Campari, Jägermeister, Fernet Branca, Ramazzotti, Averna, Unicum, and Suze.

Table 10.4 Composition of some selected liqueurs and brands of liqueurs

Liqueur brands	Raw material
Fruit liqueurs	
Crème de Cassis	Black currant juice
Grand Marnier, Cointreau	Orange, orange peels
Curacao	Citrus fruits, bitter orange
Maraschino	Marasca cherries
Apricot brandy	Apricot flavour, brandy
Apfelkorn	Apple juice, Korn, apple flavour
Bitter liqueurs, Amaro:	
Campari, Picon,	Wormwood, quinine
Fernet Branca, Ramazzotti, Averna, Jägermeister, Unicum	Different herbs, wormwood
Cynar	Artichokes
Angostura	Extracts of bark, roots, spices, vegetables, gentian
Boonekamp	Herbs, spices
Miscellaneous	
Crème de Cacao, chocolat, Bailey's	Cacao, chocolate, cream
Kahlua	Coffee, 490 g L^{-1} sugar
Irish Cream	Whisky, cream, chocolate
Coconut Liqueur	Rum, coconut
Advocaat	Egg yolk, vacilla
Crème de menthe	Peppermint
Crème de vanille	Vanilla
Rose Liqueur	Rose flowers, rose oil
Allasch	Caraway seeds, bitter almonds, aniseed
Amaretto	Almonds

10.7 Sustainability in Production of Flavour of Spirits

Ethanol and distilled spirits are produced from various renewable raw materials such as fruits, grapes, grain, sugar beet, and sugar cane, and even from waste materials like pomace and other fermentation residues; thus, production of spirit drinks, especially distilled spirits, is based on further essential issues and components of sustainability:

- Spirit drinks with their manifold and unique composition of ethanol and flavour compounds are not only used for direct human consumption but also as flavourings in many other food products, like bakery products and sweets; their consumption and the regeneration of their raw materials are well balanced because all raw materials are regenerative plant materials.
- Even the waste which arises during production of ethanol can be recycled, for example for fertilisation of agricultural areas or as cattle fodder.
- Fruit brandies produced in Europe are a very good example of a sustainable and ecological production; the fruits which are used are not cropped from plantations but mainly are fruits from trees grown in special meadows; these areas are important ecological systems within monoagricultural areas. Since these fruits cannot be put on the market as table fruits, the conservation of these ecological areas is only possible by growing fruits which are used as raw materials for distilled spirits by small-scale distillers.
- The social component which is also very important for sustainability of a production of spirits is fulfilled too; the production of fruit brandy, cachaça, tequila, rum, etc. is a very important basis of existence or additional earning for small-scale agricultural producers. These producers either directly sell their distillates as spirit drinks or they offer them to bigger distilleries for mass marketing.

10.8 Conclusions

Although researchers have been successful in identifying a great number of flavour compounds in spirit drinks and their raw materials, knowledge on factors which contribute to the quality and the typical, unique flavour of these food products is still fragmentary. The flavour of spirit drinks is mainly influenced by the quality and flavour of the raw materials, their varieties and their geographical origin. Flavour quality is also influenced by the various special, mostly traditional technologies of fermentation, distillation, and maturation. Thus, the composition of the flavour of spirit drinks will remain unique and even in future it will not be possible to replicate or displace it by synthetic flavourings [52].

References

1. European Commission Council Regulation (EEC) no 1576/89 (1989) Laying Down General Rules on the Definition, Description and Presentation of Spirit Drinks. Official Journal of the European Community no L160/1. http://europa.eu.int/eur-lex/en/consleg/pdf/1989/en_1989R1576_do_001.pdf
2. Definitions, Descriptions and Presentation of Spirit Drinks. http://europa.eu.int/scadplus/leg/en/lvb/l21093.htm
3. Specifications, Definitions, and Regulations for the Composition of Industrial, Beverage, and Fuel Alcohol Products in Various Countries. http://www.distill.com/specs/
4. Lea AGH, Pigott JR (2003) (eds) Fermented Beverage Production. Kluwer/Plenum, New York
5. Nykänen L, Suomalainen H (1983) Aroma of Beer, Wine and Distilled Alcoholic Beverages. Kluwer, Dordrecht
6. Nykänen L (1986) Am J Enol Vitic 37:84
7. Nykänen L, Nykänen I, (1991) In: Maarse H (ed) Volatile Compounds in Foods and Beverages. Dekker, New York, pp 547–580
8. Kolb E, Fauth R, Frank W, Simson I, Strömer G (2002) (eds) Spirituosentechnologie. Behr, Hamburg
9. Lambrechts MG, Pretorius IS (2000) S Afr J Enol Vitic 21:97
10. Moreno JA, Zea L, Moyano L, Medina M (2005) Food Control 16:333
11. Fritsch HT, Schieberle P (2005) J Agric Food Chem 53:7544
12. Fazzalari FA (ed) (1978) Compilation of Odor and Taste Threshold Data. ASTM Data Series DS 48A. American Society for Testing and Materials, Philadelphia
13. http:// www.wittenberg.edu/academics/chem/LabSafety/odor-in-water.htm
14. Rychlik M, Schieberle P, Grosch W (1998) Compilation of Odour Thresholds, Odour Qualities and Retention Indices of Key Food Odorants. Deutsche Forschungsanstalt für Lebensmittelchemie and Institut für Lebensmittelchemie der TU München, Garching
15. Butzke, CE, Boßmeyer M, Scheide K, Misselhorn K (1990) Branntweinwirtschaft 130:286
16. Butzke, CE, Scheide K, Misselhorn K (1992) Branntweinwirtschaft 132:27
17. Brandes W, Karner M, Eder R (2005) Mitt Klosterneuburg 55:76
18. Plessis HW, Steger CLC, du Toit M, Lambrechts MG (2002) J Appl Microbiol 92:1005
19. Adam L, Postel W (1992) Chem Mikrobiol Technol Lebensm 14:95
20. Rodriguez Madrera R, Blanco Gomis D, Mangas Alonso JJ (2003) J Agric Food Chem 51:5709
21. Boidron JN, Chatonnet P, Pons M (1988) Connaiss Vigne Vin 22:275
22. Lachenmeier DW, Attig R, Frank W, Athanasakis C (2003) Eur Food Res Technol 218:105
23. Cantagrel R, Galy B (2003) In: Lea AGH, Pigott JR (eds) Fermented Beverage Production. Kluwer/Plenum, New York, p 195
24. Bertrand A (2003) In: Lea AGH, Pigott JR (eds) Fermented Beverage Production. Kluwer/Plenum, New York, p 213
25. Ferrari G, Lablanquie O, Contagrel R, Ledauphin J, Payot T, Fournier N, Guichard E (2004) J Agric Food Chem 52:5670

26. Silva LM, Malcata FX (1999) Z Lebensm Unters Forsch A 208:134
27. Postel W, Adam L (1989) In: Piggott JR, Paterson A (eds) Distilled Beverage Flavour. Harwood, Chichester, p 133
28. Jennings WG, Tressl R (1974) Chem Mikrobiol Technol Lebensm 2:52
29. Brandes W, Karner M, Eder R (2003) Mitt Klosterneuburg 53:103
30. Lea AGH, Drilleau JF (2003) In: Lea AGH, Pigott JR (eds) Fermented Beverage Production. Kluwer/Plenum, New York, p 59
31. Ledauphin J, Saint-Clair JF, Lablanqui O, Guichard H, Founier N, Guichard E, Barillier D (2004) J Agric Food Chem 52:5124
32. Ledauphin J, Guichard H, Saint-Clair JF, Picoche B, Barillier D (2003) J Agric Food Chem 51:433
33. Christoph N (1989) Process for Purifying Fruit Brandies and Spirits. German Patent DE3734400A1. http://publikationen.dpma.de
34. Christoph N, Schmitt A, Hildenbrand K (1988) Alkohol-Ind 102:342
35. Tuttas R, Beye F (1977) Branntweinwirtschaft 117:349
36. Adam L, Postel W (1992) Branntweinwirtschaft 132:110
37. Postel W, Adam L (1983) Dtsch Lebensm-Rundsch 79:117
38. Pigott JR, Conner JM (2003) In: Lea AGH, Pigott JR (eds) Fermented Beverage Production. Kluwer/Plenum, New York, p 239
39. Aylott RI (2003) In: Lea AGH, Pigott JR (eds) Fermented Beverage Production. Kluwer/Plenum, New York, p 289
40. Nicol DA (2003) In: Lea AGH, Pigott JR (eds) Fermented Beverage Production. Kluwer/Plenum, New York, p 263
41. Boscolo M, Bezerra CWB, Cardoso DR, Lima Neto BS, Franco DW (2000) J Braz Chem Soc 11:86
42. Nonato EA, Carazza F, Silva FC, Carvalho CR, de L Cardeal Z (2001) J Agric Food Chem 49:3533
43. Faria JB, Loyola E, Lopez MG, Dufour JP (2003) In: Lea AGH, Pigott JR (eds) Fermented Beverage Production. Kluwer/Plenum, New York, p 335
44. Vichi S, Riu-Aumatell M, Mora-Pons M, Buxaderas S, Lopez-Tamames E (2005) J Agric Food Chem 53:10154
45. Yvas I, Rapp A (1991) Dtsch Lebensm-Rundsch 87:41
46. Lopez MG, Mancilla-Margalli NA (2000) In Schieberle P, Engel KH (eds) Frontiers of Flavour Science. Deutsche Forschungsanstalt für Lebensmittelchemie, Garching
47. Benn SM, Peppard TL (1996) J Agric Food Chem 44:557
48. Nose A, Hamasaki T, Hojo M, Kato R, Uehara K, Ueda T (2005) J Agric Food Chem 53:7074
49. http://www.sake-world.com
50. Lachenmeier DW, Frank W, Athanasakis C, Padosch SA, Madea B, Rothschold MA, Kröner LU (2004) Dtsch Lebensm-Rundsch 100:117
51. Clutton DW (2003) In: Lea AGH, Pigott JR (eds) Fermented Beverage Production. Kluwer/Plenum, New York p 309
52. Cole VC, Noble AC (2003) In: Lea AGH, Pigott JR (eds) Fermented Beverage Production. Kluwer/Plenum, New York, p 393

11 Wine Aroma

Ulrich Fischer
Dienstleistungszentrum Ländlicher Raum (DLR) Rheinpfalz,
Department of Viticulture & Enology,
Breitenweg 71, 67435 Neustadt a.d. Weinstraße, Germany

11.1 Introduction

Wine is not only considered as one of the oldest beverages of the world, it may be also the beverage with the most sophisticated diversity, which in turn attracts enormous attention of consumers worldwide across many cultures. This phenomenon, as illustrated by the great number of wine competitions and the abundance of wine magazines, is at least partially explained by the enormous sensory variation of wine. This complexity originates from three major sources: the raw material, which originates from thousands of grape varieties growing on a wide array of geological formations in different climates and altitudes, the fermentation process accomplished by a multitude of yeast and malolactic bacteria species and strains, and the ageing process, which varies owing to different storage methods, container size and material, such as oak barrels of varying origin, but also owing to stocking time, which may range from a few weeks to more than several decades. Finally, wine is not only produced by industrial wineries, applying standardised protocols as it is often common for many other food items, but also by a myriad of artisan wine producers employing traditional processes for their local vine varieties.

Although wine is principally considered as a traditional beverage, production of wine is subject to stylistic changes initiated by changing consumer sentiments and expectations, and is also subject to the planting of new varieties and the implementation of novel viticultural practices and enological techniques. Most recently, global warming has resulted in dramatic changes, including replacement of cool-climate varieties for hot-climate vine varieties, a tendency to switch from white to red grapes, and new viticulture techniques to limit the stress due to water shortage or enhanced UV irradiation. Consequently, research has increased to find chemical markers for stress-induced sensory aberrations and a growing trend is towards use of ethanol-reducing techniques.

According to the scope of this book, the discussion of key factors contributing to the wine sensory variation will be limited to aroma compounds. In general, research on wine aroma follows four major goals: determination of the key components explaining the sensory properties of varieties and geographical origin, comprehension of the role of microorganisms during winemaking, examination of the modifications encountered by viticultural and enological measures as well as ageing, and understanding the biochemical and chemical pathways leading to those results. Emphasis will be placed on the impact of environmental factors related to climatic changes and aromas generated and modified by the application of modern viticulture and enology.

11.2 Logic behind Varietal Aroma

Most wines are identified by variety. Although traditional wine producing countries such as France, Italy and Spain usually label their wines by region of origin and not variety, often one leading grape variety such as Cabernet Sauvignon or Merlot for Bordeaux, Pinot noir or Chardonnay for Burgundy, Tempranillo for Rioja or Sangiovese for Chianti dominates their sensory properties. To date, over 700 aroma compounds have been identified [1–5], which is strong evidence for the complexity of wine. However, with the exception of potent character-impact compounds such as linalool or *cis*-rose oxide for Muscat varieties or methoxypyrazine derivates for Sauvignon blanc and Cabernet Sauvignon, the aroma of varietal wines arises from specific combination of odour-active aroma compounds. Only recently, the application of sensory-based analytical strategies involving aroma extracted dilution analysis (AEDA) or multivariate statistics relating aroma compounds with descriptive sensory analysis made it possible to reconstitute the aroma of some neutral vine varieties [6–8] and to build some models to explain important single sensory characters such as tropical fruit [9, 10]. Examining the contribution of volatile compounds to characteristic varietal aromas, Ferreira [11] suggested three patterns. The most obvious is to produce a huge amount of distinctive volatiles, which are absent or not detectable in other varieties, as is the case for monoterpenes in Muscat varieties. The second, somewhat overlapping, mechanism is based on non-odorous precursors, such as glycoside or cysteine conjugates which are specific for the particular wine variety. In order to make these aroma compounds accessible for sensory perception, acid-catalysed hydrolysis or enzymatic release by microorganisms or technical enzymes has to take place during winemaking or ageing. In the study of the aroma of neutral wine varieties, such as Airen or Chenin blanc, which lack impact odorants, the focus shifts towards identification of the by-products of alcoholic fermentation, protein metabolism and utilisation of unsaturated fatty acids of the grape. As outlined in more detail in Chap. 10 by Christoph and Bauer-Christoph, amino acids are important aroma precursors for the yeast. Thus, the amino acid profile in a grape must, which varies significantly among wine varieties and during ripening, has a strong influence on the wine aroma of neutral varieties. This has been directly demonstrated by supplementing a synthetic grape juice with amino acids, resembling those natural profiles found in different grape varieties [12], yielding an aroma composition close to the expected varietal specific profiles. Among 28 aroma compounds analysed, 17 varied significantly with the amino acid supplementation, including fusel alcohols and their acetate esters and iso-acids and their ethyl esters. However, aroma formation is not only governed by the amino acid composition of the grape juice alone, but also by complex interaction with yeast strains and their specific fermentation behaviour and nitrogen requirements.

11.3 Chemical Basis of Varietal Aroma

Focusing more closely on the aroma compounds, four distinct classes of varietal aromas are well defined and comprise the groups of monoterpenes, C_{13} norisoprenoids, substituted methoxypyrazines and sulphur compounds with a thiol function. Typically, grape juice has very little flavour and is not varietally distinct. Only a few impact compounds, such as the monoterpene linalool or the methoxypyrazines, are present in their free form in the grape and in the juice after pressing. In contrast, the majority of varietal aroma compounds are present in a bound form, making them non-volatile and hence they have no odour. Examples of non-volatile precursors are monoterpenes or norisoprenoids bound to monosaccharides or disaccharides, thiols as cysteine conjugates as well as fatty acids, carotenoids and phenolic acids which are enzymatically cleaved to powerful odorants such as 3-*cis*-hexenol, *β*-damascenone or 4-vinylguaiacol, respectively. A third source for varietal aroma is acid-catalysed rearrangements of odourless or barely volatile compounds yielding highly active odorants, such *cis*-rose oxide/*trans*-rose oxide of Gewürztraminer or 1,1,6-trimethyl-1,2-dihydronaphthaline of aged Riesling wines.

11.3.1 Monoterpenes

During the last decade, the parents of most grape varieties were identified by application of molecular biology techniques. For many traditional European white *Vitis vinifera* varieties such as Riesling, Sauvignon blanc, Pinot gris or Silvaner, Gewürztraminer has been identified as one parent [13]. Thus, monoterpenes are found in a great number of white wine varieties, although they are classified as character-impact compounds only for Muscat, Gewürztraminer and Morio Muskat. Besides the acyclic alcohols linalool, geraniol and nerol, cyclic ethers such as *cis*-rose oxide/*trans*-rose oxide or wine lactone [(3*S*,3*aS*,7*aR*)-3*a*,4,5,7*a*-tetrahydro-3,6-dimethylbenzofuran-2(3*H*)-one] are potent odorants with thresholds in the low nanogram per litre range (Table 11.1). A more complete list is provided in the excellent review of Francis and Newton [14].

The odour thresholds given in Table 11.1 should be interpreted with caution. They differ according to the matrix in which they were determined (air, water, water–ethanol model, real wine), the sensitivity of the judges, the methodology applied and the rigour of sensory testing. Furthermore, odour activity of orthonasal and retronasal perception is different for the same compound as well as the psychophysical function, explaining the relation between volatile concentration and perceived sensory intensity. Still, odour thresholds give an orientation to which order of magnitude the compounds are sensorialy active.

Synthesis of monoterpenes has been located in the grape berries on the basis of an intuitive experiment where the inflorescence of a Muscat variety was grafted on a Shiraz shoot and vice versa [26]. Indeed, the Muscat grape grafted on a Shiraz vine produced high amounts of monoterpenes but no anthocyanins, while the red Shiraz grapes provided no monoterpenes, although they were grown on a Muscat vine. In order to elucidate the biosynthesis of monoterpenes in grapes, precursors which were labelled by stable radioactive isotopes were injected directly into the grape berry. Metabolites were extracted by a stir-bar–sorptive extraction coupled with multidimensional gas chromatography–mass spectrometry [27]. As a result, two independent biosynthesis pathways were revealed which take place in different cell compartments of the berry. While the classical mevalonate pathway is located in the cytoplasm of the grape, the newly described 1-desoxy-d-xylulose-5-phosphate (DOXP) pathway takes place in the plastids [27]. The biosynthesis of *cis*-rose oxide/*trans*-rose oxide could be explained by a stereoselective reduction of geraniol to (*S*)-citronellol, which is rearranged under acidic conditions to the odour-active rose oxide [28], which is considered as one impact aroma compound of Gewürztraminer [4, 29]. The wine lactone (3*S*,3*aS*,7*aR*)-3*a*,4,5,7*a*-tetrahydro-3,6-dimethylbenzofuran-2(3*H*)-one may be formed in the grape by an acid-catalysed rearrangement of (*S*)-linalool through an intermediate, which has been identified as a glucose ester in a Riesling wine [30].

Since the initial research suggesting the presence of non-volatile precursors of wine aroma compounds in grape [31], extensive work has established a good understanding of the chemical nature of the glycosidic precursors [32–35]. The majority of the aglycones are not linked to a single β-d-glucopyranose, but to disaccharides combing the β-d-glucopyranose with a second sugar molecule such as α-l-rhamnopyranose, α-l-arabinofuranose or β-d-apiofuranose. The release from these precursors can be achieved by acidic hydrolysis at low pH or by enzymatic hydrolysis [36]. Enzymatic hydrolysis using a pectinase with β-glucosidase side activities releases only a small portion of the total precursor potential, since the lack of rhamnosidase, arabinosidase and apiosidase activities prevents the cleavage of the complete disaccharides moiety. However, in modern winemaking pectinase or yeast strains exhibiting a specific β-glucosidase side activity are used as so-called aroma enzymes or aroma yeast to enhance the floral odour of Muscat varieties, as well as Gewürztraminer or Riesling wines [37]. Acid-catalysed hydrolysis not only releases the aglycones, but also induces rearrangements of odour-active monoterpene alcohols to less volatile diols [2].

Table 11.1 Odour-active compounds involved in varietal aroma of wine

Compound	Odour description	White wine (μg/L)	Red wine (μg/L)	Odour threshold (μg/L)
Monoterpenes				
Linalool	Lily of the valley, lychee, floral with citrus notes	4.7 [6], 307 [8]	1.7–220 [15]	15 [16], 25 [17]
Geraniol	Sweet, rose blossom, geranium	221 [18, 19]	0.91–44.4 [18, 19]	30 [8]
cis-Rose oxide	Green, grassy, lycee, rose	3–21 [8]		0.2 [8]
Wine lactone	Sweet, coconut like, spice	0.1 [8]	<0.01–0.09 [15]	0.01 [8]
C_{13} norisoprenoids				
β-Damescenone	Apple, rose, honey, lemon balm	0.089–9.4 [8]	0.29 [15], 6.2 [18, 19]	0.05 [8]
β-Ionone	Violet, flower, raspberry,	0.059–0.11 [6]	0.032–1.95 [18, 19]	0.09 [17], 0.8 [20]
Vitispiran (*E*)-6-methylene-2,10,10-trimethyl-1-oxaspiro[4.5]dec-7-en	Balsamic, resinous		20–320, >800 in port wine [5]	800 [5]
1,1,6-Trimethyl-1,2-dihydronaphthalene	Petroleum-like, kerosene-like		1–59 [5]	20 [5]
Methoxypyrazines				
3-Isobutyl-2-methoxypyrazine	Green bell pepper	<0.006 [5], 0.042 [18, 19]		0.002 in water [5]
3-Isopropyl-2-methoxypyrazine	Green bell pepper, earthy, raw potato, musty	0.035 [5]		0.002 in water [5]
3-*sec*-Butyl-2-methoxypyrazine	Green bell pepper	0.0005 [5]		0.001 in water [5]
Thiols				
4-Mercapto-4-methylpentan-2-one	Box tree, passion fruit, cat urine	<0.01–30 ng/L [21]		0.0008 [22], 0.030 [21]
3-Mercaptohexan-1-ol	Passion fruit, grapefruit	<0.05–5 [21]	0.07–4 [23]	0.06 [22]
3-Mercapto-hexyl acetate	Grapefruit, passion fruit, black currant	0.12-1.3 [24]	ND–0.02 [23]	0.004 [22]
Miscellaneous				
2-Aminoaceto-phenone	Acacia blossom	0.1–5		1.29 in wine [25]

ND not detected

11.3.2
C_{13} Norisoprenoids

This very diverse group of natural compounds is presumably generated by an oxidative cleavage of the carotenoidal molecule between the C_9 and C_{10} positions, yielding norisoprenoids with 13 carbon atoms. Although other norisoprenoids of nine to 20 carbon atoms are present in nature, for wine only the C_{13} norisoprenoids are of importance. Comparable to the monoterpenes, the majority of the C_{13} norisoprenoids are present as glycosides; however, they exist only as monoglucosides. Acid-catalysed rearrangements in the wine yield very potent aroma compounds, such as (*E*)-*β*-damescenone, which is formed via the intermediate "grasshopper" ketone from the breakdown of neoxanthin [38]. (*E*)-*β*-Damescenone not only has a very low odour threshold of 50 ng/L in a model wine [6], but also exhibits different odours. While low concentrations at the odour threshold levels are described as lemon balm, 100 times higher concentrations are likely to exhibit apple, rose and honey notes [39]. (*E*)-*β*-Damascenone was identified for the first time in Chardonnay and Riesling wines [40, 41], and has been reported lately in many gas chromatography–olfactometry studies utilising aroma-extract-dilution analysis owing to its omnipresent precursor carotene [42]. Similar to *β*-damascenone, *β*-ionone with its odour reminiscent of violets has been identified in a wide range of varieties, but occurs at higher concentrations up to 2.45 μg/L in red wines only [20]. However, owing to a recognition threshold of 1.5 g/L in a red wine, its sensory contribution to white and red wine is rather limited.

This is not the case for vitispiran, Riesling acetal or 1,1,6-trimethyl-1,2-dihydronaphthalene (TDN), which arise from the breakdown of the carotenoids antheraxanthin, violaxanthin and neoxantin and subsequent enzyme- and acid-catalysed rearrangements [16]. TDN is linked to the famous ageing flavour of Riesling, which is described as petroleum, kerosene, diesel, Band-Aid® or the German expression *Firne* [43]. Especially in wines made from Riesling grapes grown in warm climate areas such as Australia or South Africa, evolution of this ageing flavour is accelerated and may impart the wine quality as soon as after 6 months after harvest [44–46]. Other varieties such as Chardonnay or Silvaner and even grape varieties descending from a crossing involving Riesling, such as Müller-Thurgau, do not exhibit the TDN flavour to such an extent as is the case in Riesling. While TDN is generally absent in grapes or young wines, it may develop up to 200 μg/L in aged wines, exceeding the odour threshold of 20 μg/L by a factor of 10. Tasting Riesling wines from the same vineyard in a vertical tasting of more than 30 years, formation of TDN cannot be exclusively linked to hot and dry climatic conditions alone. Other factors, such as infection with *Botrytis cineria*, have an impact as well and have not been completely revealed yet. In a comparison of Riesling grapes grown in northern and southern Italy, a second precursor for TDN was identified [47] from which TDN is released by a different mechanism from that previously published [48], which may explain why Riesling grapes grown in very hot years at higher latitude have lower TDN levels than Riesling grapes grown in regions of lower latitude at the same temperature.

11.3.3 Methoxypyrazines

These extremely potent odorants with very low odour thresholds in water and wine are N-heterocycles and are formed during the reaction of glycine with leucine, isoleucine and valine, explaining the different moieties at the C_3-position [49]. Methoxypyrazines have been identified in a wide range of varieties, but their aroma impact is restricted to Cabernet Sauvignon, Sauvignon blanc and to some extent Cabernet franc, Merlot and recent crossings involving the latter varieties. Although it has been reported that methoxypyrazines are predominantly located in the berry skin [50], the so-called saigner juice of Cabernet Sauvignon or Merlot, which is removed directly after crushing of the grapes in order to increase the skin-to-juice ratio for improved colour and tannin extraction shows extremely high levels of methoxypyrazines. These saigner juices may even be used as a methoxypyrazine "reserve" to enhance the green bell pepper aroma of an overripe Sauvignon blanc. The appreciation of the green bell pepper aroma may change strongly from region to region. In Sauvignon blanc wines from New Zealand, a strong, green aroma is highly appreciated, while in the Bordeaux region the same flavour is regarded as a marker for unripe grapes, especially for Cabernet and Merlot. Most recently, low levels of methoxypyrazines in South African Sauvignon blanc wines led winemakers to add illegally green bell pepper extracts in order to enhance the varietal flavour.

11.3.4 Sulphur Compounds with a Thiol Function

Sulphur compounds are generally viewed as being responsible for a range of off-flavours caused by the smell of rotten eggs exhibited by H_2S and the odour of onions, green asparagus, burnt rubber or even garlic due to methyl and ethyl sulphides, disulphides and thiols [51, 52]. While specific thiols were identified as impact aromas in several fruits such as blackcurrant, grapefruit, passion fruit or guava in the 1980s, their strong impact for the aroma of Sauvignon blanc was reported for the first time in 1993 in Sauvignon blanc [53]. The first compound found to exhibit a typical Sauvginon blanc aroma was 4-mercapto-4-methylpentan-2-one (4-MMP), whose odour is reminiscent of black currant, boxwood and broom, exhibiting an extremely low odour threshold of 0.8 ng/L in a wine model solution [20]. The tropical fruit of Sauvginon blanc could be linked to 3-mercapto-hexan-1-ol (3-MH) and its acetate ester (3-MHA), the latter exhibiting an odour threshold of 4.2 ng/L, close that one of 4-MMP [20]. Higher odour thresholds have been reported for 4-mercapto-4-methylpentan-2-ol (4-MMPOH) exhibiting a smell of citrus, while 3-mercapto-3-methylbutan-1-ol (3-MMB) yields an odour reminiscent of cooked leeks [20]. Several of these thiols have been analysed in an array of varietal wines ranging from Cabernet Sauvignon and Merlot to the white varieties Gewürztraminer, Scheurebe, Riesling, Muscat, Pinot gris, Pinot blanc, Semillion, Colombard and even Silvaner

[8, 20, 54]. Reviewing the odour thresholds, only 4-MMP, 3-MH and 3-MHA should be considered of sensory importance.

In contrast to many tropical fruits, the thiols are not present in their free and odorous form in the grape berries, but as their odourless cystein conjugate [55], which of course hindered and delayed the identification of these powerful odorants in grapes. On the other hand, this research led to the detection of a novel type of precursors in grapes, which was described in plants for the first time. The recent identification of 3-MH glutathione in a Sauvginon blanc juice [56] supports the role of a glutathione transferase in the biosynthesis of the cysteine conjugates, which reacts with an unsaturated α,ß-unsaturated carbonyl compound such as 4-methyl-3-penten-2-one acting as an electrophile towards the mercapto group of glutathione. Further cleavage of the glutathione moiety by a peptidase in the vacuole results in the specific cysteine conjugate [16]. While cysteine conjugates of 4-MMP and 4-MMPOH are equally distributed between berry skin and pulp of Sauvignon blanc grapes, the precursors of 3-MH were found in concentrations 8 times higher in the berry skin [57], indicating an aroma-enhancing effect of skin maceration in the case of 3-MH.

According to Dubourdieu [20], the French enologist Emile Peynaud showed remarkable intuition when he described on tasting a Sauvignon blanc grape, "the initial flavour is quite discreet. 20 to 30 seconds later, after you have swallowed it, an intense, aromatic Sauvignon blanc aftertaste suddenly appears in the rear nasal cavity". He concluded that "fermentation brings out the primary aroma hidden in the fruit". Indeed, the enzymatic activity of a specific β-lyase exhibited by yeasts during fermentation is responsible for the release of the odorous thiols, which will be addressed in more detail in Sect. 11.5.2. Extensive screening of different *Saccharomyces cerevisiae* strains has led to a range of commercially available dry yeast cultures, which provide the desired β-lyase activity. The sensory effect can be demonstrated by a descriptive analysis made from two Scheurebe wines made from the same juice, but fermented by two different yeast strains on an industrial scale in Fig. 11.1 [58]. Besides the H_2S odour, the only significant difference was observed for the cassis aroma, presumably owing to the release of 4-MMP by the β-lyase activity of the Maurivin 350 yeast strain.

Most concentration data regarding the thiols are based on winemaking which was not aware of the special role of the yeast strain and hence neglected the use of dry cultured yeasts exhibiting enhanced β-lyase activity. Thus, it can be speculated that in the future we will see much higher amounts of these potent odorants in the wines in general and more grape varieties exhibiting these varietal aromas than to date.

Recently, methods for recovering and identifying thiols were developed, including a preservation of the highly reactive mercapto group with *p*-hydroxymercuribenzoate [22]. As a consequence, other thiols have been identified which have an impact on wine aroma: 2-furanmethanthiol evolves in wines fermented in oak barrels [59]; levels of benzenemethanethiol, 2-furanmethanethiol and ethyl 3-mercaptopropionate increase during ageing of champagne [60].

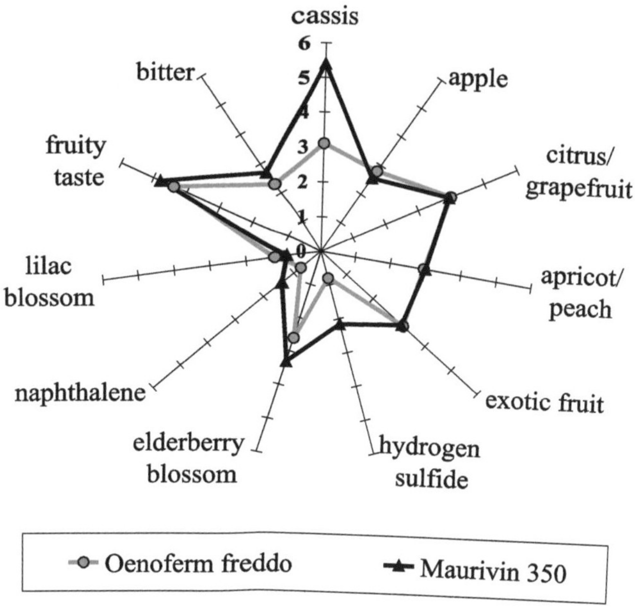

Fig. 11.1 Descriptive analysis of two Scheurebe wines made from the same juice but fermented by two different commercial yeast strains (ten judges × two replications) [58]

11.4 Impact of Viticulture and Growing Conditions

Viticultural practices in the vineyard have been increasingly used to modify the flavour of grapes and wines rather than solely focusing on controlling crop yield. Increasing the exposure of grape clusters to sunlight by removing basal leaves affects the formation and concentration of several important flavour compounds.

11.4.1 Sun Exposure

In cool viticultural climates increased sun exposure enhanced the glycosidic aroma precursors [61, 62], including monoterpene and C_{13} norisoprenoid aglycones. However, increased sunlight exposure may have detrimental effects as well: berry temperatures may rise up to 50 °C, leading to cell disruption and the socalled sunburn may cause crop losses up to 30%. Especially in cool climates, ripening could be hindered by a low leaf-to-fruit ratio, due to severe leaf removal. Delayed picking of very mature, but still sound grapes increases overall wine aroma in the wines, as has been demonstrated for monoterpenes in Muscat varieties [63, 64].

Monoterpene accumulation proceeds in three phases [64]: high concentrations in young berries are diluted by water incorporation during berry growth until *véraison*, succeeded by a strong increase during ripening. Similar patterns have been observed for glycosidically bound 2-phenylethanol and benzyl alcohol [65]. In respect to global warming, the scientific dissent about the situation in overripe fruit is noteworthy. While some authors report further increase even beyond the point where the maximum sugar level is reached [64, 66], others found at least the free monoterpenes to decrease before the sugar maximum [65].

Exposure of grapes to sunlight during ripening generally accelerates carotenoid breakdown [67]. Increase of glycosylated C_{13} norisoprenoids has been reported in Riesling and Syrah grapes [68]. Grape enzymes are involved in oxidative carotenoid breakdown as well as the following glycosylation mechanism [68].

Increased levels of methoxypyrazines are found in rather unripe grapes which are grown in a cool climate [69] or on heavy limestone or clay soils [20], which enhance vegetative growth. Concentrations of methoxypyrazines gradually decrease during ripening, which is at least partially explained by the light sensitivity of methoxypyrazines [70]. Thus, limiting vegetative growth by planting vines in well-drained, gravely soils, using less vigorous root stocks, establishing trellising systems and a canopy management with reduced shading of the grapes and active leaf removal during ripening are successful measures to reduce methoxypyrazines. Vice versa, increasing shading in hot climates may preserve an important residual of methoxypyrazines, contributing to the varietal aroma of Sauvignon varieties, making canopy management a reliable and powerful tool to determine the final expression of methoxypyrazines in relevant varietal wines [71].

It is a common reaction of grapes against sun exposure to increase the concentration of polyphenols and carotenoids in the berry skin. Additional anthocyanins and flavonols may enhance colour and tannins in red wine, but in white wines higher levels of polyphenols may enhance a bitter taste and a undesired astringency. Ferulic and coumaric acid and their tartrate esters fertaric and coutaric acid may act as precursors for the volatile phenols 4-vinylguaiacol and 4-vinylphenol, respectively [72]. Tartrate esters are cleaved by a cinnamyl esterase activity, which belongs to the spectrum of less purified pectinases made from *Aspergillus niger* cultures [20]. Free ferulic and coumaric acids and those liberated from their tartrate esters are decarboxylated by a highly specific cinnamate decarboxylase (CD) expressed by *Saccharomyces cerevisiae*, which is only active during alcoholic fermentation. Other cinnamic acids are not decarboxylated. Lower amounts of 4-vinylguaiacol may contribute to the varietal odour of Gewürztraminer and Pinot gris [73], while in most cases the medicinal and smoky smell masks other more desirable flavours. In red wines, increased amounts of polyphenols inhibit the CD activity of *Saccharomyces cerevisiae*, leading to low levels of grape-related volatile phenols in red wines. In measurements of cinnamate tartrates in Riesling grapes grown in a sun-exposed steep slope vineyard in the hot vintage 2003, the highest levels were observed in the water-stressed control, owing to the senescence of basal leaves, and followed by the irrigated trial [74]. The lowest amounts, however, were found in a non-irrigated, but sun-protected trial, where partially transparent gauze was covering the fruit zone. In Germany, Riesling wines from the extremely hot and dry vintage of 2003 and in some regions 2005 as well showed masked varietal aroma, due to 4-vinylguaiacol concentrations above their sensory threshold. Presumably, these unusually high levels were due to enhanced generation of cinnamate tartrates. Strategies to circumvent this stress-related off-flavour include more shading of the fruit zone, application of pectinases free of cinnamyl esterase activity during grape and juice processing, removal of hydroxyl cinnamic acids and their tartrate esters by polyvinyl polypyrrolidone fining or by hyperoxidation in the grape juice and fermentation with yeast strains with low or no CD activity [20, 75].

Comparing sun-exposed with shaded grape bunches in South African Riesling during the ripening period [45], sun-exposed grapes had up to 3 times more acid-releasable TDN than shaded bunches in the same vineyard. In the wines made from sun-exposed and shaded grapes, sun exposure induced a 50% increase in TDN.

Research on the impact of viticulture, soil composition and climate on the cysteine conjugates is still very limited. Only recently, it was demonstrated that severe water stress reduced the levels of the cysteine conjugates of 4-MMP and 3-MH, respectively [76]; however, moderate water stress enhanced the content of the cysteine conjugates. Low nitrogen supply of the grapes also limited the precursor formation, as well as excessive nitrogen levels [76]. Optimising the nitrogen nutrition with respect to precursor formation, nitrogen deficiency can be induced not only by reduced fertilisation but also by limited water supply. On the other hand, excess of nitrogen favours infestation by *Botrytis cineria*, which seems to be able to metabolise the cysteine conjugates [76] and hence reduce the flavour potential.

11.4.2
Stress-Induced Aroma Compounds

It is extremely difficult to study the impact of the climatic changes observed during the last decades for wine, because several highly correlated climatic variables such as temperature, water supply and UV radiation have an impact on vine physiology and grape constituents. Furthermore, it is not the grape which really matters to consumers, it is the wine. Wine is produced by a multistage winemaking process, which by itself can be manipulated by the grape composition. For example, nutrition of yeast and malolactic bacteria through the native grape constituents has a strong impact on flavour formation. At the same time, concurrent analysis of several important aroma compounds is not trivial at all.

Thus, much research regarding environmental stress had been done with red grapes, limiting the impact of stress to changes in colour formation, which is easily measured in grapes and wines by straightforward photometric analysis.

A global-warming trend is suggested by the fact that the seven warmest years in global records have all occurred since 1990. Longer vegetation periods, enhanced evaporation and reduced water availability will definitely change the portfolio of grape varieties planted in moderate latitudes, proliferation of irrigation and altered soil management. In general, a continuing rise in CO_2 concentration will stimulate photosynthesis in grapes as shown for Riesling in Montpellier, France, where doubling the CO_2 concentration enhanced photosynthesis by 35% [77]. Increased leaf area and vegetative dry weight as consequences of raised photosynthesis may lead to more fruit shading and could translate into higher levels of methoxypyrazines in some varieties. Plant responses to increased UV-B radiation during the last few decades vary from species to species, but it seems a general adaptation to enhance the accumulation of UV-absorbing compounds, such as red anthocyanins or the antioxidants ascorbic acid and glutathione. Concurrently, carotenoid pigment formation and incorporation of nitrogen into amino acids can be inhibited [77]. In order to study the effects of these direct aroma precursors or at least flavour-modulating constituents, the fruit zone of Riesling grapevines were shielded by a UV-B absorbing polyester film and by a UV-A and UV-B absorbing diacetate film. Protection from UV radiation leads to a nearly complete absence of visible pigmentation of the berries, leading to a 15% decrease in reflectance over the entire visible range. From analysis of the berry skin samples, it was found that current ambient levels of UV-B radiation reduced significantly both amino acid and carotenoid concentrations at harvest [77]. Degradation of carotenoids was more pronounced in berries under natural UV-B exposed conditions than in UV-B protected berries [77]. With respect to the impact of carotenoids on the evolution of C_{13} norisoprenoids and the impact of amino acids on fermentation aroma [11], but also with respect to the impact on stress-induced off-flavour such as 2-aminoacetophenone [78], more data should be gathered in order to study the complex interaction of UV-B radiation with fruit composition and subsequent aroma development during winemaking.

In the late 1980s, the hot and dry seasons in Germany led to the appearance of an off-flavour which has been described by acacia blossom, naphthalene, fusel alcohol, furniture polish and wet wool, combined with a loss of varietal aromas and increased bitterness in a study using sensory descriptive analysis [79]. This off-flavour was named "untypical ageing flavour" (UTA), owing to the premature loss of fermentation and varietal odours occurring 4–6 months after harvest. After its precise sensory description, wines exhibiting UTA off-flavours were also reported in northern Italy, Oregon, southern France and eastern Europe. 2-Aminoacetophenone (2-AAP), an odorant reminiscent of acacia blossom and an integral part of the labrusca grape flavour, has been identified as a chemical marker for this off-flavour [78].

Formation of 2-AAP could be traced back to the plant hormone indole-3-acetic acid (IAA) [80], which is formed in the grape berry. The oxidative degradation of IAA is started by superoxide radicals, which are formed in wine by the co-oxidation of sulphite to sulphate, following the addition of the antioxidant and antimicrobial preservative SO_2. After decarboxylation, pyrrole oxidation and ring cleavage, 2-formylacetophenone (2-FAP) was the main volatile compound of the non-enzymatic degradation of IAA induced by sulphite addition. In a last step, 2-FAP is completely hydrolysed to 2-AAP. The formation of 2-AAP and 2-FAP was significantly lower in white wines than in ethanolic solutions spiked with IAA. Owing to a low odour threshold of 1.29 μg/L in white wines [25], low formation rates of less than 5% are sufficient to cause the UTA off-flavour. Formation of 2-AAP continues during ageing. At the same time, acid-catalysed ester hydrolysis and oxidation of monoterpenes during wine ageing will decrease fermentation and even partially varietal aroma; thus, the sensory significance of UTA will increase with time. Even though IAA is accepted as the major precursor of 2-AAP, no correlation could be established between the IAA content in grape juice and the 2-AAP formation in the subsequent wines [81]. Bound precursors of IAA, their enzymatic cleavage by yeasts as well as the nitrogen content of the grape juice seem to govern 2-AAP formation as well [82]. Although sensory UTA ratings are highly correlated with 2-AAP concentrations, enrichment of wines with 2-AAP alone failed to describe the whole sensory spectrum of UTA. Especially the occurrence of notes reminiscent of naphthalene, fusel alcohol and the long-lasting bitter phenolic sensation has not been explained yet on a molecular level. Thus, further research is essential to fully explore and predict the UTA potential in grape juice.

In red wine, polyphenolic compounds act as scavengers towards the superoxide radicals, preventing the appearance of UTA. Besides the legal additive ascorbic acid [80], an increase of gallic acid, catechine and grape seed extracts of 30 mg/L proved to be sufficient to limit the formation of 2-AAP below its sensory threshold in a wine spiked with 1 mg/L IAA [25]. Thus, skin contact of 4–12 h of white grapes after crushing as well as the addition of ascorbic acid after fermentation and prior to SO_2 addition is part of a UTA prevention regime. However, the major action has to be devoted to the vineyard, where a combination of crop reduction, delayed harvesting, avoidance of water stress and sufficient nitrogen fertilisation has been proven to be a successful strategy to prevent the occurrence of UTA. Unfortunately, irrigation of stressed vines, which is the most efficient measure to prevent UTA, is not easy to establish in all vineyards. However, even without much irrigation, an intelligent vineyard management and enological provisions successfully avoided UTA formation in the extremely hot and dry 2003 vintage in Germany.

11.5
Impact of Enology

Governed by worldwide consumer attitude and purchasing habits, wine production tends to separate more and more into two different wine segments. The first is composed of high-volume/low-price wines with rather standardised sensory properties, restricted to a few varieties that are planted worldwide such as Chardonnay, Sauvignon blanc, Cabernet Sauvignon, Merlot and Syrah. The second segment comprises low-volume/high-price wines and a strong focus on geographic origin: autochthonous grape varieties such as Sangiovese, Tempranillio and Riesling and winemaking employing traditional methods in small-scale operations. Membrane processes such as reverse osmosis or nanofiltration, vacuum distillation and even upscaled countercurrent chromatography and the addition of toasted oak particles (oak chips) are widely implemented in the winemaking process of the high-volume wines to achieve a low-cost wine whose sensory properties precisely match those which market research has identified as drivers of consumer preferences. Triggered by bilateral negotiations between the EU and the USA or Australia, most recently (2005/2006) consumers have been confronted with the application of these techniques, predominantly outside the EU, which started an intense public discussion about modern versus traditional winemaking among consumers as well as in the wine industry.

Winemaking can be divided into three important phases. During grape and juice processing, it is the objective to transfer as much of the desired grape constituents such as flavour precursors or anthocyanins as possible in the grape juice. Owing to the breakdown of cell compartments during grape crushing, many precursors are subject to acidic hydrolysis, which will continue during the whole shelf life of wine. Substances leading to detrimental sensory properties or increased instabilities in wines may be removed by fining and clarification of the juice or wine. In the second phase, alcoholic and malolactic fermentation not only convert sugar to ethanol and malic acid to lactic acid, respectively, but also generate de novo a wide range of flavour compounds, such as esters or fusel alcohols. Selected yeast strains even enzymatically release important varietal odours such as monoterpenes or carbonyl thiols. For red wines, maceration on the skins will provide extensive extraction of anthocyanins and polyphenols from grape skins and seeds. Prolonged yeast contact (*sur lie*) extends the reductive environment and may protect highly reactive aroma compounds such as the carbonyl thiols from oxidation owing to an overall reductive environment.

The third phase is dominated by stabilisation efforts in order to prevent formation of tartrates or protein hazes, and also in red wines to enhance polymerisation of anthocyanins and polyphenols to achieve a stable colour and smooth-tasting tannins. In the course of stabilisation, some wines are stored and aged in small oak barrels, which additionally act as a source of oak-derived volatiles such as vanillin or oak lactones. In conclusion, winemaking has the objective to gain the maximum from the aroma potential generated in the grapes, then to convert this potential to the maximum of free odorants accessible to human sensory perception and finally to maintain this pleasant composition of sensory stimuli as long as possible during the shelf life of wine.

11.5.1 Grape Processing

During grape processing, prolonged maceration regimes increase the transfer of water-soluble free aroma compounds, glycosidic precursors and cysteine conjugates into the grape juice. To monitor aroma precursors in day-to-day operations during harvest and to facilitate a more complete grape assessment, the glycosidic glucose assay (GG assay) was developed [34, 83]. During the GG assay, glycosidically bound juice constituents are separated via solid-phase extraction, followed by elution of these bound compounds with methanol, subsequent release of the bound glucose at elevated temperature by acid hydrolysis and a final enzymatic determination of the released glucose moiety as a measure for the glyosidically bound aroma precursors. According to the GG assay, the maceration time of 6–24 h increased significantly the transfer of aroma precursors into the juice in Riesling, Gewürztraminer and Müller-Thurgau, which can be further enhanced by the application of a pectinase [84]. Sensory analysis revealed increased floral, peach, passion fruit and citrus aroma (Fig. 11.2), showing good correlation of glycosidic glucose concentration and floral aroma (R^2 = 0.66) and body (R^2 = 0.56), but only weak correlation for peach and passion fruit (R^2 = 0.40) [84], the aroma of which is presumably determined more by thiols such as 3-mercaptohexan-1-ol than by monoterpenes. The limited application of the GG assay outside Australia may not only be explained by the sophisticated analysis, but also by the fact that monoterpenes make up only a small portion of the precursors determined by glycosidic glucose: in a Gewürztraminer, for example, three different pectinases released 40–50 μmol of glycosidic glucose, but only 2.19–2.38 μmol of total monoterpenes [84]. The length of the macera-

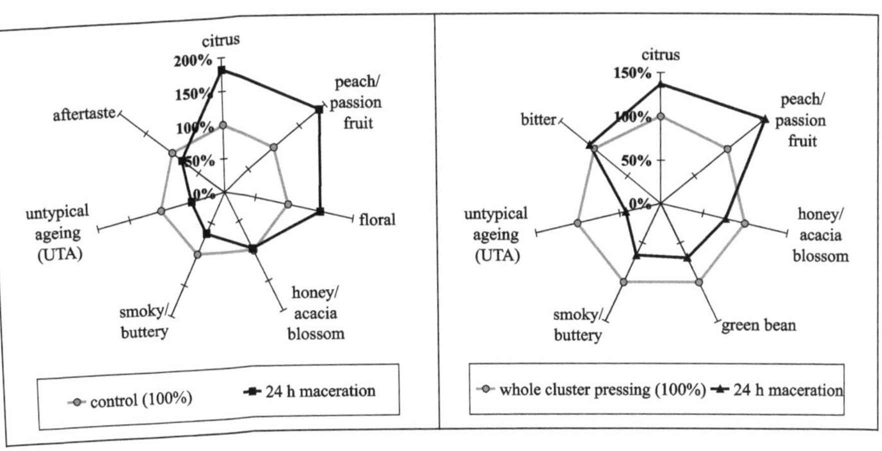

Fig. 11.2 Sensory impact of skin maceration during white wine making in Muscat (*left*) and Riesling (*right*) [37]

tion time in white wines is limited by excessive extraction of bitter polyphenols and the loss in acidity due to increased potassium extraction and subsequent precipitation of potassium bitartrate.

11.5.2
Impact of Yeast

The generation of general fermentation flavours such as fusel alcohols and esters was covered in Chap. 10 by Christoph and Christoph-Bauer dealing with flavours in spirits, and an excellent review on microbial modulation of wine flavour is provided by Swiegers et al. [21]. According to the main focus of this book, only the impact of microorganisms on flavour precursors generated by grapes in vineyards will be addressed.

Many varietal aromas such as monoterpenes or C_{13} norisoprenoids are present as odourless glucose conjugates in the juice. During and after fermentation they are liberated by a rather slow acid-catalysed hydrolysis and a much faster enzymatic hydrolysis, due to side activities of added technical pectinases or native enzymes expressed by the grape itself or microorganisms. While most C_{13} norisoprenoids are only bound to glucose, the majority of monoterpenes are linked to disaccharides. Before a β-glucosidase may release the aglycon, it is first necessary to remove the terminal sugar moiety by a specific β-glycosidase (arabinosidase, rhamnosidase, xylosidase or apiosidase). With use of a chemically defined grape juice medium supplemented by a precursor extract of Muscat Frontignac, the enzymatic aroma release of three yeast strains was investigated, by excluding any grape enzymes. As expected, the liberation of monoterpenes by acid hydrolysis alone was very low in the control and was restricted to the release of linalool, α-terpineol and geraniol [85]. The yeast strains varied not only in their release of aglycons, but also in their further transformation to diols or oxidised monoterpenes. Furthermore, four commercial strains differed in the time course of their liberation of aglycons during fermentation [85]. In view of the complex enzymatic requirements for aglycon liberation, it was noteworthy that fermentation released α-l-rhamopyranosyl glucopyranoside and α-l-arabinofuranosyl glucopyranoside glycons at the same rate of 30–40% as did β-d-glucopyranoside alone; only the β-d-apiofranosyl glucopyranoside remained stable [85]. Since no significant differences occurred among the three yeast strains, these β-glycosidase activities seem to be quite common in *Saccharomyces* strains.

The role of microorganisms for the release of thiols from odourless cysteine conjugates were first proposed after a cell-free enzyme extract of the gastrointestinal bacterium *Eubacterium limosum* was found to be able to release 4-MMP from cysteine–4-MMP [55]. Owing to the modulation of 4-MMP with regard to different yeast strains, a yeast carbon–sulphur lyase was suggested [86]. Indeed, deletion of genes encoding putative carbon–sulphur lyases in laboratory strains of *Saccharomyces cerevisiae* led to reduced 4-MMP levels. It was also shown that ester-forming alcohol-acetyl transferase of yeast is responsible for the conversion of 3-MH to its subsequent acetate ester 3-MHA [87]. At the same time, chemically synthesised cysteine–4-MMP and cysteine–3-MH decreased during fermentation, while free 4-MMP and 3-MH increased [88]. However, only 3.2% of the cysteine–3-MH was liberated and the overall amounts of free 3-MH correlated with the initial concentration of its precursor cysteine–3-MH present in the grape juice [89]. In conclusion, an enormous potential of non-released cysteine conjugates remains in the wine after fermentation, and could be utilised more efficiently, if yeast with enhanced carbon–sulphur lyase activity could be selected. In analogy to the highly restricted release rate of thiols from cysteine conjugates, the release of grape-derived glycosidic precursors by yeast is limited as well, presumably because the majority of sugar moieties are disaccharides, which cannot be liberated by yeast β-glucosidase activities alone. Comparing yeast strains in general, *Sacchromyces bayanus* strains release more 4-MMP than *Saccharomyces cerevisiae* and even hybrids obtained by crossing of both strains release more thiols than *Saccharomyces cervisiae* [23]. Anecdotally, the typical flavour reminiscent of passion fruit of the German Scheurebe variety can be enhanced owing to a spontaneous fermentation, including wild yeast such as *Kloeckera apiculata* in the initial phase of fermentation. Current sensory studies with German Riesling, comparing spontaneous fermentations with those conducted with a commercial *Saccharomyces cerevisiae* strain, revealed higher intensities of passion fruit and elderberry blossom, suggesting an enhanced release of 3-MH, 3-MHA and 4-MMP, respectively, by the spontaneous yeast flora [90]. Screening commercially available yeast strains, the VIN7 strain could triple the released amount of 4-MMP compared with the industry standard "Sauvignon" strain VL3 [91]. Enhancement of 3-MH and 3-MHA ranged between 20 and 120%. According to this screening, yeast strains vary regarding the release of 4-MMP and 3-MH and even more with respect to the ester formation leading to 3-MHA, as well as altering the 3-MHA to 3-MH ratio [91]. Yeast strains showing a high "thiol release" activity do not exhibit concurrently the strongest 3-MH to 3-MHA conversion and vice versa; thus, carbon–sulphur lyase and acetate acetyl transferase activities do not seem to be coupled in natural yeast [91]. Since 3-MHA has a much lower odour threshold than 3-MH, a strategy to maximise flavour generation would combine a yeast strain with a high "thiol release" with a second strain exhibiting a maximal "thiol-conversion" rate.

Besides yeast selection, the choice of a proper temperature regime during fermentation is a major factor in practical winemaking. While some authors report higher release rates for 4-MMP or 3-MH at higher temperatures [92], others found no significant differences [91]. The benefits of higher temperature, ranging between 20 and 28 °C in the experiments, seem to be limited to the start of fermentation, where in general the bulk of sensory-relevant flavour generation during fermentation takes place [91] and which may be related to active yeast growth. For the rest of fermentation, lower temperatures will be beneficial by limiting volatility of the aroma compounds formed and preventing them from being removed by CO_2 percolation. Temperature seems to have an impact as well during skin maceration, where higher temperatures enhanced the extraction of cysteine-3-MH from its major source, the exocarp [57]. Because cysteine-4-MMP is mostly present in the pulp, skin maceration and temperature do not have a profound impact on 4-MMP levels in the wines [57].

Over the last few years, concurrent application of skin contact for flavour enhancement and use of yeast strains with a cysteine lyase activity at moderate to cool fermentation temperatures led to an unusual increase of "Sauvignon blanc" aroma even in varieties classified as "neutral" regarding their impact aromas. In Germany, varieties such as Silvaner, Müller-Thurgau, Pinot blanc, Pinot gris and even Riesling exhibit enhanced aroma characters such as passion fruit, elderberry blossom and black currant, which are likely to be induced by thiols. These observations highlight the critical impact of yeast not only on fermentation aromas, but also varietal aroma and suggest a wider distribution of cysteine conjugate precursors in white wine varieties than so far anticipated. In addition to the aroma compounds involved in the varietal aroma of grapes in Table 11.1, the Table 11.2 summarises the impact of grapes and microorganisms on the release of the most important aroma compounds in wine, excluding off-flavors [11].

11.5.3 Impact of Modern Wine Technology

New enological technologies aim to lower volatile acidity, enhance sugar content in must in cool climates and vice versa reduce the alcohol content of wines from hot climates, modify pH, cations, anions and acidity to achieve tartrate stability, complement traditional ageing in oak barrels with the use of small oak wood particles and most recently, extract phenolic compounds by a countercurrent chromatography process from wine to diminish or enrich tannins in red wines.

To meet growing public concern about chemical treatments and additives to wine, a general trend towards the application of physical processes can be observed. This trend goes hand in hand with the objective of modern enology, to be as gentle as possible to grape must and wine. In order to reach this goal, in some cases physical methods are employed first to separate those compounds which should be removed or modified from the general wine matrix with its precious constituents. In a second step, only the fraction obtained will be treated. For example, to lower volatile acidity, a water–alcohol–acetic acid fraction is obtained through reverse osmosis and only this fraction, which is free of valuable aroma, colour or phenolic compounds, will undergo ion exchange, removing specifically acetic acid, while alcohol and water are redirected into the wine.

Global warming, improved viticultural techniques, irrigation and continuing selection of highly productive clones of traditional *Vitis vinifera* varieties give rise to the grape's sugar content. This worldwide trend is indirectly supported by the highly acclaimed benefits of a long hang time, utilising an extended vegetation period in order to achieve a maximum of flavour, colour and tannin generation in the grapes. Although grapes of *Vitis vinifera* are not classified as cli-

Table 11.2 Summary of most important aroma compounds in wine, excluding off-flavours, and the impact of grapes and microorganisms (modified according to [11])

Compound	Impact of grape variety	Source in grape				Impact of yeast	Role of yeast/bacteria	
		Free form	Glycosidic precursors	Cysteine conjugates	Other precursors		De novo synthesis	Release / modification
Linalool	Very strong	X	X			Weak		X
cis-Rose-oxide	Very strong	X	X			Weak		X
Wine lactone	Very strong	X				No		
4-Methyl-4-mercap-topentan-2-one	Very strong			X		Strong		X
3-Mercaptohexan-1-ol	Very Strong			X		Strong		X
3-Mercaptohexyl acetate	Very strong			X		Strong		X
Isoamyl acetate	Strong				X	Strong	X	
Methoxypyrazines: 3-isobutyl-2-methoxy-pyrazine, 3-isopropyl-2-methoxypyrazine	Strong	X				No		
Volatile phenols: 4-ethylphe-nol, 4-vinylphenol, 4-ethyl-guaiacol, 4-vinylguaiacol	Strong		X		X	Strong		X
Vanillin and re-lated compounds	Strong		X		X	Weak		X
β-Damescenone	Average		X		X	Weak		X

Table 11.2 *(continued)* Summary of most important aroma compounds in wine, excluding off-flavours, and the impact of grapes and microorganisms (modified according to [11])

Compound	Impact of grape variety	Source in grape				Impact of yeast	Role of yeast/bacteria	
		Free form	Glycosidic precursors	Cysteine conjugates	Other precursors		De novo synthesis	Release / modification
2-Phenylethanol	Average	X				Weak	X	
2-Phenylacetaldehyde	Unknown					Strong	X	
Fusel alcohols	Average					Strong	X	
Iso-acids	Average					Strong	X	
Iso-acid ethyl esters	Average					Strong	X	
γ-Lactones	Average					Unclear	X	
Fatty acid ethyl esters	Average					Strong	X	
Methional	Unknown					Unknown	X	
Diacetyl	Average				X	Strong	X	
Sotolon	Unknown				Botrytis	Unknown		
Flor sherry aldehydes	Unknown					Unknown	Flor yeast	
(*E*)-Oak lactone	No				Oak wood	No		
Burnt sugar compounds	Probably strong				oak wood	Unknown		

macteric fruits, the term "physiological ripeness" is prevalently discussed with respect to the determination of an optimal picking time. In both cool-climate and hot-climate viticulture, "physiological ripeness" refers to sufficient colour, tannin and flavour development and in case of cooler growing regions moderate acidity as well. Without much scientific basis for the benefits of stable colour, soft tannins and improved varietal aroma, a late picking time is recommended, accepting a high sugar content and subsequently enhanced ethanol levels in the fermented wines.

It is a major objective of modern enology to make flavour development independent of sugar accumulation. In cool climates, reverse osmosis and vacuum distillation are utilised to enhance sugar content by the removal of water at the juice stage and outside the EU also at the wine stage. In hot climates, and more and more former cool climates are becoming hot owing to global warming, the same methods are used to remove excessive ethanol, which is detrimental from a sensory and health point of view. As ethanol is a better solvent for aroma compounds than water, an increase in alcohol content in wines will reduce the volatility of odorants and will subsequently diminish their sensory perception. In a simple experiment the odour threshold for hexyl acetate decreased by 30% when the ethanol concentration in a white wine was raised from 11 to 14% vol [93]. For less-volatile aroma compounds, the effect could be even more pronounced.

To facilitate must concentration by the removal of water, reverse osmosis is widely utilised (Fig. 11.3). This technique applies high pressure to cause water to move through a membrane against the osmotic pressure, while valuable odour, colour and phenolic compounds cannot pass the membrane owing to their high molecular weight. Alternatively, vacuum distillation can be applied for must concentration, where differences in volatility govern separation. Application of reverse osmosis to concentrating 30 must samples from different German grape varieties to the legally defined maximum increase of 2% vol potential alcohol yielded an average increase of 5–15% of esters, monoterpenes, fusel alcohols and C_6-alcohols in the final wines [94]. Vacuum distillation showed similar results for aroma compounds such as esters or monoterpenes, which are generated or released during fermentation. However, volatiles either formed or already present in their free form in the grape juice such as *Z*-3-hexen-1-ol and the *Botrytis cineria* marker 1-octen-3-ol were strongly diminished by vacuum distillation. This was even true for hexyl acetate, which was explained by the loss of the precursor hexanol in the juice stage [94].

In hot-climate viticulture it is a common practice to lower the high ethanol content of wines made from overripe fruit by partial dealcoholisation. This objective can be achieved by vacuum distillation, where the spinning cone column technique allows even more viscous liquids to be processed. Alternatively, a water–ethanol fraction can be separated from wine by reverse osmosis, followed by distillation of the water–ethanol permeate to yield high-grade ethanol and pure water. The latter will be added back to the treated wine.

Applying vacuum distillation in an one-stage process removes nearly 75% of the wine volatiles, predominantly owing to the transfer of esters and fusel alcohols into the distillate. More polar aroma compounds, however, such as monoterpenes, lactones, volatile phenols and short chain fatty acids, are retained [39]. To counteract these flavour losses, the spinning cone column process first produces a highly volatile flavour fraction at lower temperatures, before the alcohol is removed using higher temperatures. Applying reverse osmosis, more than 60% of the total volatiles remained in the dealcoholised wine, 25% were transferred into the permeate and a small portion was absorbed by the membrane itself. In contrast to vacuum distillation, where volatility and polarity mainly determined the degree of separation, permeation through the reverse osmosis membrane occurred across all different chemical classes, only limited by molecular weight [39]. Although some flavour loss still occurs during dealcoholisation, this shortcoming does not pose a critical question, since only a small portion of the high-alcohol wine is dealcoholised, while the major fraction is not treated at all.

However, from a legal point of view, new questions arise from these fractioning techniques in general. Should fractions be considered as wine? Should a recombination of vitivinicultural fractions, containing alcohol, water, flavour, tannins or pigments, still be accepted as wine production? If the volatile flavour fraction, which has been removed from the wine prior to dealcoholisation, is not added back completely to the original wine, but is added to another, more valuable wine for sensory improvement, we may call this practice flavorisation, which is currently illegal in all wine-producing countries of the world.

In general, new enological treatments not only offer more possibilities for modification, but owing to their complex technology they are highly versatile While a traditional enological treatment such as the removal of H_2S off-flavour by copper sulphate is limited to a defined effect, the reverse osmosis unit can be used for concentration of must as well as wine, dealcoholisation, and removal of acetic acid or even volatile phenols derived from *Brettanomyces* yeast. In conclusion, new enological technologies are not just an advancement of traditional enology, which relies primarily on grape quality and reacts to specific shortcomings of a must or wine. In fact, their application introduces a new concept of enology, where winemaking is a steered process to produce well-defined wine styles, according to results obtained by consumer research and demand expressed by worldwide markets.

11.6 The Mystery of Wine Ageing

As outlined in Sect. 11.1, part of the fascination of wine is due to its nearly unlimited ageing potential. If we had access to the hidden treasures of the top wine collectors, we may still be able to drink wines which were produced decades and even centuries ago. Apart from this more intellectual fascination, more and more wine is consumed relatively young and even top-class red wine producers were deprived of the privilege to market a wine not earlier than when it has reached its sensory peak.

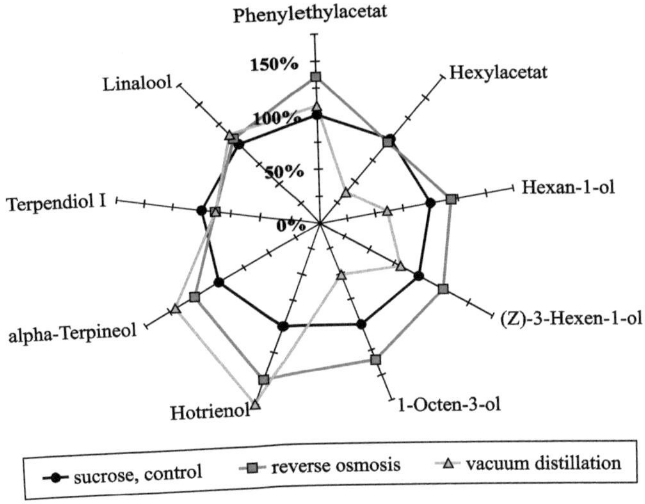

Fig. 11.3 Impact of must concentration technologies on aroma compounds in Riesling wines (n=3) [94]

During the first few years of ageing, esters generated during fermentation undergo cleavage by acid-catalysed ester hydrolysis [46]. Predominantly acetate esters are diminished, while a range of ethyl esters even increase, and the ratio of acetate to ethyl esters has been suggested as an ageing marker as well as the slow build-up of ethyl tartrate [95]. Wines of neutral varieties and of low ripeness will lose a lot of their sensory edge during this phase, while wines with a strong varietal character may even benefit from the loss of fermentation aroma and are able to reveal their true values. While free monoterpenes slowly undergo oxidation and sensory extinction [2], acid-catalysed hydrolysis is able to replenish the lost free monoterpenes from the reservoir of still-bound monoterpenes [95]. For Riesling, in some years the build-up of TDN will give the wines their so-called kerosene or diesel ageing flavour [96], while in many red wines methional seems to be a strong ageing marker [20]. While slow oxidation is a threat to fruity and reductive white wines such as Riesling or Sauvignon blanc, wines aged *sur lie* for more than 1 year in oak barrels as well as most red wines are not threatened at all, because most oxidative changes already happened during winemaking before bottling. Consequently, further slow oxidation during bottle storage will not make a strong difference during the next 5–10 years.

11.7 Conclusion

The impact of global warming can be well documented in worldwide viticulture by a invariably earlier initiation of blossoming, *véraison* and grape maturation with increasing sugar accumulation. At the same time, grapes have to adapt to increased stress exerted by water deficiency and enhanced UV-B radiation. While cool climates may alter their portfolio from early-ripening varieties to late-ripening varieties, hot climates have to counteract the impact of global warming by technological measures to reduce sugar or more likely remove excessive ethanol after fermentation.

With respect to flavour formation in grapes, knowledge has tremendously increased during the last two decades and the chemical basis of numerous varietal aromas as well as stress-induced off-flavours has been elucidated. This enables viticulturists to develop new techniques to gain the maximum varietal aroma, by either protecting the grapes against excessive sun exposure in hot climates or enhancing the benefits of sun exposure in cool climates.

At the same time, improved grape processing and a better understanding of the role of yeast in the release of varietal aromas during fermentation facilitates enologists to use the aroma potential present in the grapes to a greater extent and to produce wines of high extinction.

Application of modern technologies adapted from other food-processing areas, such as that of milk, introduces the possibility to freely recombine fractions obtained from wine. This may be beneficial for large-scale winemaking in order to produce mass-market wine styles according to consumer demands, but it also threatens the common public perception of wine as being an authentic image of unique growing conditions defined by geologic and climatic diversity, as well as regional wine varieties and traditional winemaking techniques. In this respect it seems to be of utmost importance to maintain the worldwide ban on adding any flavours to wine or grape juice. Only by sustaining this ban, the enormous sensory variation perceived in wines will reflect exclusively the natural flavour formation in grapes and during grape processing, fermentation and bottle maturation.

Acknowledgements

The critical review of the manuscript by Ann C. Noble (University of California Davis) and Peter Winterhalter (TU Braunschweig) is highly appreciated.

References

1. Schreier P (1979) Crit Rev. Food Sci. Nutr. 12:59
2. Rapp A, Mandery H (1986) Experientia 42:873
3. Maarse H, Vissher CA (1994) Volatile Compounds in food, qualitative and quantitative data. TNO, Zeist, p 593
4. Guth H (1997) J Agric. Food Chem. 45:3022
5. Etiévant PX (1991) In: Maarse H (ed) Volatile compounds in foods and beverages. Dekker, New York p 483

6. Escudero A, Gogorza MA, Melús A, Ortín A, Cacho J, Ferreira V (2004) J. Agric. Food Chem. 52:3516
7. Ferreira V, Ortín A, Escudero A, López R, Cacho J (2002) J. Agric. Food Chem. 50:4048
8. Guth H (1997) J. Agric. Food Chem. 45:3027
9. Aznar M, López R, Cacho J, Ferreira V (2003) J. Agric. Food Chem. 51:2700
10. Campo E, Cacho J, Ferreira V (2005) J. Agric. Food Chem. 53:5682
11. Ferreira V (2005) Yeast's contribution to the sensory profile of wine. Lallemand, La Rioja, p 19
12. Hernández-Orte P, Cacho J, Ferreira V (2002) J. Agric. Food Chem. 50:2891
13. Regner F, Stadlbauer A, Eisenheld C (1999) Vitic. Enol. Sci. 53:114
14. Francis IL, Newton JL (2005) Aust. J. Wine Res. 11:114
15. Guth H, Sies A (2002) 11th Australian Wine Industry Technical Conference. Australian Wine Industry Technical Conference, Adelaide, p 128
16. Wüst M (2003) Chem. Unserer Zeit 37:8
17. Ferreira V, Lopez R, Cacho JF (2000) J. Sci. Food Agric. 80:1659
18. Kotseridis Y, Baumes RL, Bertrand A, Skouroumounis GK (1999) J. Chromatogr. A 841:229
19. Kotseridis Y, Baumes RL, Bertrand A, Skouroumounis GK (1999) J. Chromatogr. A 841:317
20. Ribéreau-Gayon P, Glories Y, Maujean A, Dubourdieu D (1998) Traité d'oenologie 2. Chimie du vin, stabilization et traitements. Dunod, Paris
21. Swiegers JH, Bartowsky EJ, Henschke PA, Pretorius IS (2005) Aust. J. Wine Res. 11:139
22. Tominaga T, Murat M-L, Dubourdieu D (1998) J. Agric. Food Chem. 46:1044
23. Murat ML, Tominaga T, Dubourdieu D (2001) J. Agric. Food Chem. 49:5412
24. López R, Ferreira V, Hernández P, Cacho JF (1999) J. Sci. Food Agric. 79:1461
25. Fischer U, Simat TJ, Hoenicke K, Rauhut D, Rapp A, Scholten G, Sponholz WR (2001) Deutscher Lebensmittelchemiker Tag, Braunschweig, p 65
26. Gholami M, Hayasaka Y, Coombe BG, Jackson JF, Robinson SP, Williams PJ (1995) Aust. J. Wine Res. 1:29
27. Luan F, Wüst M (2002) Phytochemistry 60:451
28. Wüst M, Mosandl A (1999) Z. Lebensm. -Unters. -Forsch. 209:3
29. Ong PKC, Acree TE (1999) J. Agric. Food Chem. 47:665
30. Winterhalter P, Bonnlander B (2001) ACS Symp. Ser. 794:21
31. Cordonnier RE, Bayonove CL (1974) C. R. Acad. Sci. Ser. D 278:3387
32. Günata YZ, Bayonove CL, Baumes RL, Cordonnier RE (1985) J Chromatogr. 331:83
33. Günata YZ, Bayonove CL, Baumes RL, Cordonnier RE (1985) J. Sci. Food Agric. 36:857
34. Williams PJ, Cynkar W, Francis IL, Gray JD, Iland PG, Coombe BG (1995) J. Agric. Food Chem. 43:121
35. Williams PJ, Sefton MA, Marinos VA (1993) 3rd international Haarmann & Reimer symposium on recent developments in flavor and fragrance chemistry, Tokyo, p 283
36. Winterhalter P, Skouroumounis GK (1997) Adv. Biochem. Eng. 55:73
37. Fischer U, Trautmann S, Binder G, Wilke A, Göritz S (2000) Intensivierung des Weinaromas. KTBL, Darmstadt, p 121
38. Skouroumounis GK, Sefton MA (2002) ACS Symp. Ser. 802:241
39. Fischer U (1995) Mass balance of aroma compounds during the dealcoholization of wine: correlation of chemical and sensory data. Universität Hannover, Hannover
40. Simpson RF, Miller GC (1984) Vitis 23:143

41. Schreier P, Drawert F, Junker A (1976) J. Agric. Food Chem. 24:331
42. Kotseridis Y, Baumes RL, Skouroumounis GK (1999) J. Chromatogr. A 849:245
43. Simpson RF, Miller GC (1983) Vitis 22:51
44. Marais J, Van Wyk CJ, Rapp A (1992) S. Afr. J. Enol. Vitic. 13:33
45. Marais J, Van Wyk CJ, Rapp A (1992) S. Afr. J. Enol. Vitic. 13:23
46. Rapp A, Guentert M, Ullemeyer H (1985) Z. Lebensm. -Unters. -Forsch. 180: 109
47. Versini G, Rapp A, Marais J, Mattivi F, Spraul M (1996) Vitis 35:15
48. Winterhalter P (1991) J. Agric. Food Chem. 39:1825
49. Bungert M, Jahns T, Becker H (2001) Flavour Fragrance J. 329
50. Roujou de Boubée AM, Cumsille P, Dubourdieu D (2002) Am. J. Enol. Vitic. 53:1
51. Goniak OJ, Noble AC (1987) Am. J. Enol. Vitic. 38:223
52. Rauhut D, Kuerbel H, Dittrich HH (1993) Vitic. Enol. Sci. 48:214
53. Darriet P, Tominaga T, Demole E, Dubourdieu D (1993) C. R. Acad. Sci. Ser. III 316:1332
54. Tominaga T, Baltenweck-Guyot R, Des Gachons CP, Dubourdieu D (2000) Am. J. Enol. Vitic. 51:178
55. Tominaga T, Masneuf I, Dubourdieu D (1995) J. Int. Sci. Vigne Vin 29:227
56. Peyrot des Gachons C, Tominaga T, Dubourdieu D (2002) J. Agric. Food Chem. 50:4076
57. Peyrot des Gachons C, Tominaga T, Dubourdieu D (2002) Am. J. Enol. Vitic. 53:144
58. Fischer U (2000) In: Rantz JM (ed) ASEV 50th anniversary annual meeting. ASEV, Seattle, p 3
59. Tominaga T, Blanchard L, Darriet P, Dubourdieu D (2000) J. Agric. Food Chem. 48:1799
60. Tominaga T, Guimbertau G, Dubourdieu D (2003) J. Agric. Food Chem. 51:1016
61. Reynolds AG, Wardle DA (1997) S. Afr. J. Enol. Vitic. 18:3
62. Zoecklein BW, Wolf TK, Marcy JE, Jasinski Y (1998) Am. J. Enol. Vitic. 49:35
63. Dimitradis E, Williams PJ (1984) Am. J. Enol. Vitic. 35:291
64. Wilson B, Strauss CR, Williams PJ (1984) J. Agric. Food Chem. 33:919
65. Günata YZ, Bayonove CL, Baumes RL, Cordonnier RE (1985) J. Sci. Food Agric. 36: 857
66. Park SK, Morrison JM, Adams DC, Noble AC (1991) J. Agric. Food Chem. 39:514
67. Razungles A, Bayonove CL, Cordonnier RE, Sapis JC (1988) Am. J. Enol. Vitic. 39:44
68. Razungles A, Bayonove CL (1996) In: Bouard J, Guimberteau G (eds) La viticulture à l'aube du 3rd millénaire. J. Int. Sci. Vigne Vin 85 l'aube
69. Lacey MJ, Allen MS, Harris RLN, Brown WV (1991) Am. J. Enol. Vitic. 42:103
70. Heymann H, Noble AC, Boulton RB (1986) J. Agric. Food Chem. 34:268
71. Marais J, Hunter JJ, Haasbroek PD (1999) S. Afr. J. Enol. Vitic. 20:19
72. Chatonnet P, Dubourdieu D, Boidron JN, Lavigne V (1993) J. Sci. Food Agric. 62:191
73. Versini G (1985) Vignevini 12:57
74. Piccoli M (2004) Untersuchungen zur Entwicklung phenolischer und aromatischer Substanzen in Beeren der Sorte Riesling (Vitis vinifera L.) bei Wassermangel. University of Udine
75. Wyk CJ, Rogers IM (2000) S. Afr. J. Enol. Vitic. 21:52
76. Peyrot des Gachons C, van Leeuwin C, Tominaga T, Soyer J-P, Gaudillére J-P, Dubourdieu D (2005) J. Sci. Food Agric. 85:73
77. Schultz HR (2000) Aust. J. Wine Res. 6:2
78. Rapp A, Versini G, Ullemeyer H (1993) Vitis 32:61
79. Fischer U, Sponholz W (2000) Dtsch Weinbau 55:16

80. Christoph N, Bauer-Christoph C, Geßner M, Köhler HJ, Simat TJ, Hoenicke K (1998) Vitic. Enol. Sci. 53:82
81. Hoenicke K, Simat TJ, Steinhart H, Christoph N, Kohler HJ, Schwab A (1999) Adv. Exp. Med Biol 467:671
82. Hoenicke K, Christoph N, Schwab A, Simat TJ, Steinhart H (2000) Czech J Food Sci. 18:52
83. Iland PG, Cynakar W, Francis IL, Coombe BG (1996) Aust. J. Wine Res. 2:171
84. Fischer U, Trautmann S, Binder G (2001) In: Schmitt A (ed) 6th international enology symposium. Forschungsring Deutscher Weinbau, Stuttgart, p 180
85. Ulgiano M, Rinaldi A, Gambuti A, Moio L, Bartowsky EJ, Pretorius, Henschke PA (2005) 17th Lallemand scientific conference: Yeast's contribution to the sensory profile of wine, La Rioja, Spain, p 47
86. Tominaga T, Peyrot des Gachons C, Dubourdieu D (1998) J. Agric. Food Chem. 46:5215
87. Swiegers JH, Willmott R, Hill-Ling A, Capone DL, Pardon KH, Esey GM, Howell KS, de Barros Lopes MA, Sefton MA, Lilly M, Pretorius IS (2005) Weurman Flavour Research Symposium, Roskilde
88. Dubourdieu D, Tominaga T, Maneuf I, Peyrot des Gachons C, Murat M-L (2000) ASEV 50th anniversary annual meeting. ASEV, Seattle, p 196
89. Murat M-L, Masneuf I, Darriet P, Lavigne V, Tominaga T, Dubourdieu D (2001) Am. J. Enol. Vitic. 52:136
90. Fischer U (2005) Dtsch Weinmag. 12:14
91. Swiegers JH, Francis IL, Herderich MJ, Pretorius IS (2006) Aust. Wine Ind. J. 21:34
92. Murat ML, Dumeau F (2005) Aust. N. Z. Grapegrower Winemaker 2005:49
93. Fischer C, Fischer U, Jakob L (1996) In: Henick-Kling T, Wolf TE, Harkness EM (eds) 4th international symposium on cool climate enology & viticulture. ASEV Eastern Section, Rochester, p VII42
94. Clos D (2003) Chemical and sensory impact of physical concentration techniques in must and wine. Technical University Kaiserslautern, Kaiserslautern
95. Rapp A, Marais J (1993) Dev. Food Sci. 33:891
96. Rapp A, Guentert M (1986) Dev. Food Sci. 12:141

81. [illegible] D, [illegible] M, [illegible] H, [illegible] (1999) [illegible]
[illegible]
82. [illegible] (1997) [illegible]
[illegible]
83. [illegible] (2000) [illegible] 2002 [illegible]
[illegible] Oxford [illegible] 1990 [illegible]
84. [illegible] (2003) [illegible] International [illegible]
[illegible]
85. [illegible] Flavor [illegible]
[illegible] (eds) [illegible]
[illegible]
86. [illegible] Food Chem [illegible]
87. [illegible]
[illegible] Flavor [illegible]
[illegible]
88. [illegible]
[illegible]
89. [illegible] (2002) [illegible]
[illegible]
[illegible]
[illegible]

12 The Maillard Reaction: Source of Flavour in Thermally Processed Foods

Donald S. Mottram
Department of Food Biosciences,
University of Reading,
Whiteknights, Reading RG6 6AP, UK

12.1 Introduction

Since man first discovered fire, thermal treatment of foods has been one of the most common ways to prepare food. The use of heating improved the eating quality of food in terms of flavour and digestibility and it also became apparent that cooked food could be stored for longer time than the raw material. Cooked foods develop characteristic flavour and colour and the main reactions which take place are the breakdown of lipid, sugars, amino acids, carotenes, thiamine and other trace food components.

The Maillard reaction, which occurs between amino compounds and reducing sugars, has been recognised for over 60 years as one of the most important routes to flavour and browning in cooked foods [1]. This extremely complex reaction has been the subject of much research by food scientists seeking to identify compounds that provide the flavour and colour characteristics of heated foods (see reviews by Hodge [2], Hurrell [3], Mauron [4], Mottram [5] and Nursten [6, 7]). The reaction has implications in other areas of the food industry, including the deterioration of food during processing and storage (owing to the loss of essential amino acids and other nutrients) and the protective effect of the antioxidant properties of some Maillard reaction products [7]. In recent years the physiological significance of the reaction has been recognised in relation to in vivo glycation of proteins and the link to diabetic complications and cardiovascular and other diseases [7, 8]. The possibility of mutagenic compounds being formed in the Maillard reaction has also been recognised for many years and this was given particular attention in the 1980s when carcinogenic heterocyclic aromatic amines were isolated from well-grilled or charred steaks and were shown to derive from Maillard reactions involving amino acids, reducing sugars and creatinine [9]. In 2002 the Maillard reaction between the amino acid asparagine and reducing sugars was shown to be responsible for the formation of the suspect carcinogen acrylamide (2-propenamide) in fried and oven-cooked potato and cereal products at concentrations as high as 5 mg/kg [10, 11]. This illustrates the complexity of the reaction and its important place in food science.

The Maillard reaction is inextricably linked to the desirable flavour and colour characteristics of cooked foods and this review provides an insight into some of the chemistry associated with flavour generation in the reaction and the different aromas which are involved. The chemical pathways associated with the initial and intermediate stages of the Maillard reaction are presented and routes by which the important classes of aroma compounds may be formed from Maillard intermediates are discussed.

12.2 The Chemistry of the Maillard Reaction

Thermal reactions between amino acids and carbonyl compounds were first observed by Strecker [12] in 1862, who described the formation of aldehydes through oxidative degradation of amino acids. Soon after this Schiff [13] started investigating the addition reactions between amino and carbonyl groups. However, it was a French scientist, Louis-Camille Maillard, who in 1912 first reported the formation of colour through the interaction of amino acids with glucose [14]. The chemical interpretation of the reaction had to wait another 40 years until Hodge in 1953 drew up a scheme to explain the essential steps in the complex reaction [15]. It is noteworthy that some 50 years later the Hodge scheme still provides the basis for our understanding of the reaction.

12.2.1 Stages in the Maillard Reaction

The chemical mechanisms involved in initial stages of the Maillard reaction have been studied in some detail and involve the condensation of the carbonyl group of the reducing sugar with the amino compound to give a glycosylamine. During thermal processing this breaks down to various sugar dehydration and degradation products. These compounds interact with other reactive components such as amines, amino acids, aldehydes, hydrogen sulphide and ammonia, and it is these interactions which provide the basis for the colours and aromas which characterise cooked foods.

The scheme devised by Hodge divides the Maillard reaction into three stages. The reaction is initiated by the condensation of the carbonyl group of a reducing sugar with an amino compound (Scheme 12.1), producing a Schiff base. If the sugar is an aldose, this cyclises to an N-substituted aldosylamine. Acid-catalysed rearrangement gives a 1,2-enaminol, which is in equilibrium with its keto tautomer, an N-substituted 1-amino-2-deoxyketose, known as an Amadori rearrangement product. Ketosugars, such as fructose, give Heyns rearrangement products by related pathways. It is also considered that the N-substituted aldosylamine can degrade to fission products via free radicals without forming the Amadori or Heyns rearrangement products [16].

The Amadori and Heyns rearrangement products are unstable above ambient temperature. They have various keto-enol tautomers, which undergo enolisation, deamination, dehydration and fragmentation steps giving rise to a collection of sugar dehydration and fragmentation products, containing one or more carbonyl groups, as well as furfurals, furanones and pyranones (Scheme 12.2). In this intermediate stage of the Maillard reaction the amino acid also undergoes deamination and decarboxylation through Strecker degradation (Sect. 12.2.2). The aldehydes, furfurals, furanones and other carbonyls produced at this stage may contribute to flavour characteristics associated with the Maillard reaction.

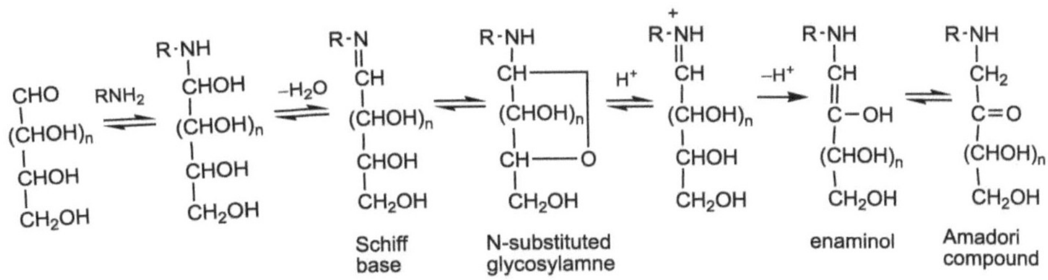

Scheme 12.1 Initial steps in the Maillard reaction showing the formation of an Amadori compound

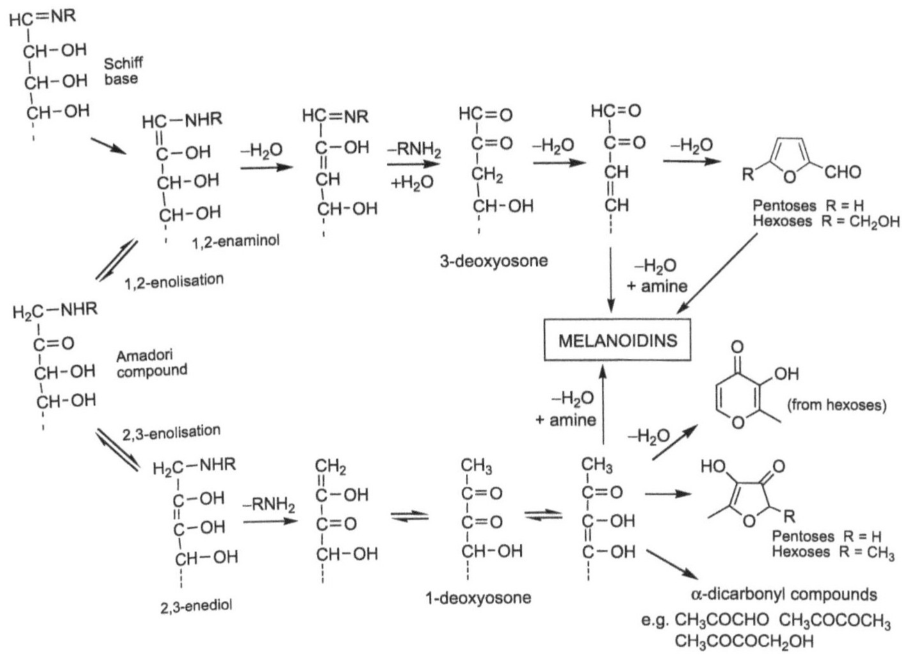

Scheme 12.2 Intermediate stages of the Maillard reaction showing the formation of carbonyl compounds

The products of the initial and intermediate stages of the Maillard reaction are colourless or pale yellow and Hodge attributed colour formation to the final stage of the reaction, where condensation between carbonyls (especially aldehydes) and amines occurs to give high molecular mass, coloured products known as melanoidins. These have been shown to contain heterocyclic ring systems, such as pyrroles, pyridines and imidazoles, but their detailed structures are unknown. The final stage of the reaction is of great importance for flavour formation when carbonyl compounds react with each other, as well as with amino compounds and amino acid degradation products, such as hydrogen sulphide and ammonia. It is these interactions that lead to the formation of flavour compounds, including important heterocyclics, such as pyrazines, pyrroles, furans, oxazoles, thiazoles and thiophenes.

12.2.2 Strecker Degradation

An important reaction associated with the Maillard reaction is the Strecker degradation of amino acids [12, 17]. In the initial and intermediate stages of the Maillard reaction the mechanisms focus on the degradation of sugar, initiated or catalysed by amino compounds. Strecker degradation, on the other hand, can be seen as the degradation of α-amino acids initiated by carbonyl compounds. It is usually considered as the reaction between an amino acid and an α-dicarbonyl compound in which the amino acid is decarboxylated and deaminated, yielding an aldehyde, containing one less carbon atom than the original acid (termed a Strecker aldehyde), and an α-aminoketone (Scheme 12.3). However, the reaction need not be restricted to dicarbonyls. Any active carbonyl group which can form a Schiff base with the amino group of an amino acid should, under appropriate conditions, promote the decarboxylation and deamination of an amino acid. Thus, α-hydroxycarbonyls and deoxyosones, formed as Maillard intermediates, as well as dicarbonyls, can act as Strecker reagents and produce Strecker aldehydes. Other carbonyl compounds found in foods which could act as Strecker reagents include 2-enals, 2,4-decadienals and dehydroascorbic acid.

Strecker degradation is very important in flavour generation, as it provides routes by which nitrogen and sulphur can be introduced into heterocyclic compounds in the final stage of the Maillard reaction. The α-aminoketones are key precursors for heterocyclic compounds, such as pyrazines, oxazoles and thiazoles. In the case of alkylpyrazines, the most direct and important route for their formation is thought to be via self-condensation of α-aminoketones, or condensation with other aminoketones [18]. If the amino acid is cysteine, Strecker degradation can lead to the production of hydrogen sulphide, ammonia and acetaldehyde, while methionine will yield methanethiol (Scheme 12.4). These compounds, together with carbonyl compounds produced in the Maillard reaction, provide intermediates for reactions giving rise to important aroma com-

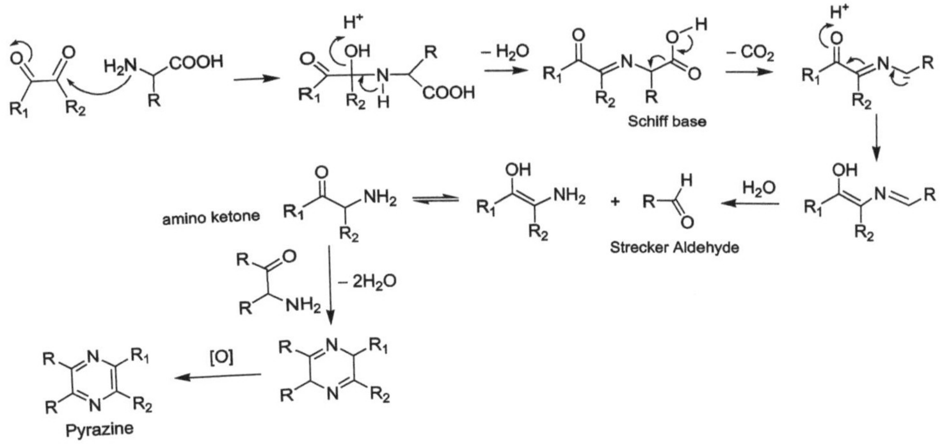

Scheme 12.3 Strecker degradation

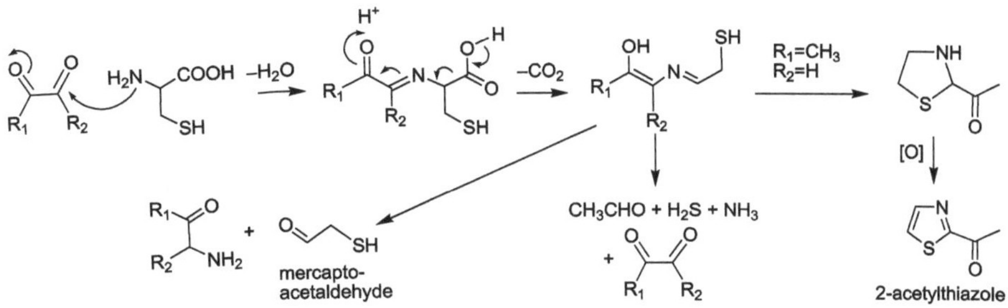

Scheme 12.4 Strecker degradation of cysteine

pounds, including sulphur-containing compounds such as thiophenes, thiazoles, trithiolanes, thianes, thienothiophenes and furanthiols and disulphides.

Proline and hydroxyproline differ from the other amino acids in that they contain a secondary amino group in a pyrrolidine ring; therefore, they do not produce aminoketones and Strecker aldehydes in the reaction with dicarbonyls. However, nitrogen heterocyclics are produced, including 1-pyrroline, pyrrolidine, 2-acetyl-1-pyrroline and 2-acetyltetrahydropyridine (Scheme 12.5) [19].

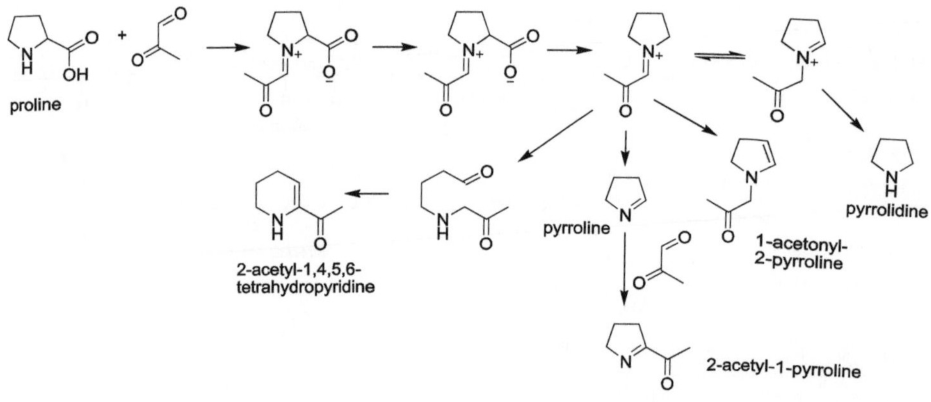

Scheme 12.5 Routes to the formation of compounds with bread-like aromas from the reaction of proline with 2-oxopropanal

12.3 Classes of Aroma Compounds Formed in the Maillard Reaction

The aroma volatiles produced in the Maillard reaction have been classified into three groups by Nursten [6], and this provides a convenient way of viewing the origin of the complex mixture of volatile compounds derived from the Maillard reaction in foods:

1. "Simple" sugar dehydration/fragmentation products: furans, pyrones, cyclopentenes, carbonyl compounds, acids
2. "Simple" amino acid degradation products: aldehydes, sulphur compounds (e.g. hydrogen sulphide, methanethiol), nitrogen compounds (e.g. ammonia, amines)
3. Volatiles produced by further interactions: pyrroles, pyridines, pyrazines, imidazoles, oxazoles, thiazoles, thiophenes, dithiolanes, trithiolanes, dithianes, trithianes, furanthiols

The first group contains compounds produced in the early stages of the reaction by the breakdown of the Amadori or Heynes intermediates, and includes similar compounds to those found in the caramelisation of sugars. Many of these compounds possess aromas that could contribute to food flavour, but they are also important intermediates for other compounds. The second group comprises simple aldehydes, hydrogen sulphide or amino compounds that result from the Strecker degradation occurring between amino acids and dicarbonyl compounds.

The products in these two groups are capable of further reaction, and the subsequent stages of the Maillard reaction involve the interaction of furfurals, furanones and dicarbonyls with other reactive compounds such as amines, amino acids, hydrogen sulphide, thiols, ammonia, acetaldehyde and other aldehydes.

These reactions lead to many important classes of flavour compounds that comprise the third group of compounds in the classification.

12.3.1 Oxygen-Containing Compounds

Furans and pyrans with oxygenated substituents (furfurals, furanones, pyranones) occur in the volatiles of all heated foods, and are among the most abundant products of the Maillard reaction. Compounds such as furfural, 5-hyroxymethylfurfural, 2-acetylfuran, maltol and isomaltol generally impart caramel-like, sweet, fruity characteristics to foods. 2,5-Dimethyl-4-hydroxy-3(2*H*)-furanone and its 5-methyl homologue, which have been found in many heated and non-heated foods, have aromas described as caramel-like, burnt pineapple-like, although at low concentrations the dimethyl derivative attains a strawberry-like note. These furanones are believed to be important contributors to the aroma of cooked meat in their own right and as precursors of other aroma compounds [20, 21]. The odour threshold values of furfurals and furanones are generally at the parts per million level [22]. Oxygenated furans may contribute to caramel-like, sweet aromas in heated foods; however, they are important intermediates to other flavour compounds, including thiophenes, furanthiols and other sulphur-containing compounds.

Aliphatic carbonyl compounds, such as diacetyl, which has a butter-like odour, also may contribute to the aromas derived from the Maillard reaction, and many of the Strecker aldehydes also have characteristic aromas (Table 12.1).

Table 12.1 Aldehydes and some other related intermediates formed in by Strecker degradation

Amino acid	Strecker aldehyde	Odour description
Valine	2-Methylpropanal	Green, overripe fruit
Leucine	3-Methylbutanal	Malty, fruity, toasted bread
Isoleucine	2-Methylbutanal	Fruity, sweet, roasted
Phenylalanine	Phenylacetaldehyde	Green, floral, hyacinths
Methionine	Methional, methanethiol, 2-propenal	Vegetable-like aromas
Proline	Pyrrolidine, 1-pyrroline. No *Strecker* aldehyde	Important intermediates for bread-like aromas
Cysteine	Mercaptoacetaldehyde, acetaldehyde, hydrogen sulphide, ammonia	Important intermediates for meat-like aromas

12.3.2 Nitrogen-Containing Compounds

12.3.2.1 Pyrazines

These important aroma compounds are believed to contribute to the pleasant and desirable flavour of many different foods. Although tetramethylpyrazine was first isolated from the molasses of sugar beet in 1879 and several alkyl pyrazines were found in coffee in 1928, it was not until the mid-1960s that their occurrence in foods was widely reported, and since then this class of aroma compound has received considerable attention [23]. The alkylpyrazines generally have nutty, roast aromas with some eliciting earthy or potato-like comments [22]. The odour threshold values of the monomethylpyrazines, dimethylpyrazines, trimethylpyrazines and tetramethylpyrazines are all relatively high (above 1 mg/kg), and these pyrazines probably only play minor roles in food aromas. However, replacing one or more of the methyl groups with ethyl can give a marked decrease in the threshold value [24], and some ethyl-substituted pyrazines have sufficiently low threshold values for them to be important in the roast aroma of cooked foods.

Several mechanisms have been proposed for the formation of pyrazines in food flavours [18, 23, 25], but the major route is from α-aminoketones, which are products of the condensation of a dicarbonyl with an amino compound via Strecker degradation (Scheme 12.3). Self-condensation of the aminoketones, or condensation with other aminoketones, affords a dihydropyrazine that is oxidised to the pyrazine.

12.3.2.2 Oxazoles and Oxazolines

Oxazoles have been found in relatively few cooked foods, although over 30 have been reported in coffee and cocoa, and 9 in cooked meat. Oxazolines have been found in cooked meat and roast peanuts, but not to any extent in other foods. 2,4,5-Trimethyl-3-oxazoline has been regularly detected in cooked meat [26], and when it was first identified in boiled beef [27] it was thought that the compound possessed the characteristic meat aroma; however, on synthesis it was shown to have a woody, musty, green flavour with a threshold value of 1 mg/kg [28]. Other 3-oxazolines have nutty, sweet or vegetable-like aromas and the oxazoles also appear to be green and vegetable-like [28]. The contribution of these compounds to the overall aroma of heated foods is probably not as important as the closely related thiazoles and thiazolines.

12.3.2.3
Pyrroles, Pyrrolines and Related Compounds

Pyrroles are found in the volatiles of most heated foods [29], although they have received less attention than some other classes of aroma volatiles. Some pyrroles may contribute desirable aromas, e.g. 2-acetylpyrrole has a caramel-like aroma, and pyrrole-2-carboxaldehyde is sweet and corn-like, but alkylpyrroles and acylpyrroles have been reported to have unfavourable odours [22]. Many more volatile pyrroles have been found in coffee than in other foods [30], and they are common products of amino acid–sugar model systems. Pyrroles are closely related in structure to the furans, and they are probably formed in a related manner from the reaction of a 3-deoxyketose with ammonia or an amino compound followed by dehydration and ring closure (cf. Scheme 12.2).

The characteristic aroma of wheat bread crust has been attributed to 2-acetyl-1-pyrroline, and its formation depends on the presence of bakers' yeast [31]. In model systems it was demonstrated that the acetylpyrroline is formed from the reaction of proline with pyruvaldehyde or dihydroxyacetone. Other compounds with bread-like aromas formed in the reaction of proline with pyruvaldehyde include 1-acetonyl-2-pyrroline and 2-acetyltetrahydropyridine (Scheme 12.5). These compounds are unstable, which explains why the characteristic aroma of freshly baked bread disappears quickly during storage.

Since proline already contains a pyrrolidine ring it provides a potential source of nitrogen heterocyclics in the Maillard reaction, and a number of proline-containing model systems have been examined. Tressl et al. [32] identified more than 120 proline-specific compounds in the reaction of proline or hydroxyproline with various sugars. These include pyrrolines, pyrroles, pyridines, indolines, pyrrolizines and azepines, but relatively few of the compounds have been identified among food volatiles.

The roasting of foods such as malt or coffee can result in bitter-tasting compounds; however, until recently little was known about the chemistry of any compounds formed in the Maillard reaction that could be responsible for such tastes. Frank et al. [33] identified a new class of compound, 1-oxo-2,3-dihydro-1*H*-indolizinium-6-oxalates, from reaction mixtures containing xylose, rhamnose and alanine (Fig. 12.1). A number of such compounds have been reported and they appear to have low taste thresholds (below 1×10^{-3} mmol/L).

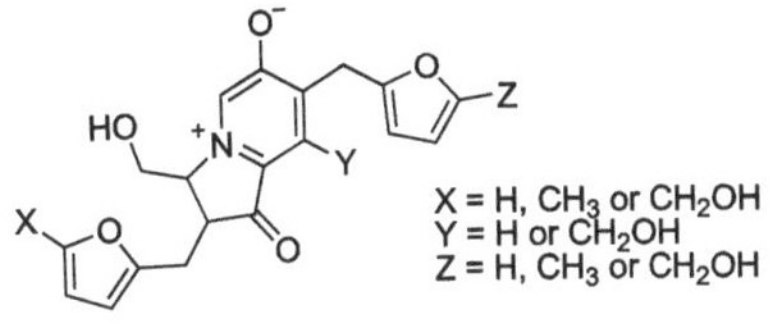

Fig. 12.1 Structures of some bitter tasting 1-oxo-2,3-dihydro-1*H*-indolizinium-6-oxalates found in Maillard reaction systems

12.3.3
Sulphur-Containing Compounds

Aliphatic thiols, sulphides and disulphides are found in the volatiles of heated foods; however, the majority of the sulphur compounds produced as a result of thermal treatment of food contain heterocyclic sulphur. These include thiophenes, thiopehnones, thiazoles, dithiazines, trithiolanes and trithaines. Over 250 different sulphur-containing volatiles have been reported in heated foods, with the largest numbers in coffee and meat [34]. It is interesting to note that foods from cereals and other plant sources appear to have many more nitrogen-containing than sulphur-containing volatiles, whilst in meat the opposite trend is observed. This may reflect the higher protein content of meat and, therefore, the greater availability of sources of sulphur in the form of the sulphur amino acids.

Hydrogen sulphide is a key intermediate in the formation of many heterocyclic sulphur compounds. It is produced from cysteine by hydrolysis or by Strecker degradation; ammonia, acetaldehyde and mercaptoacetaldehyde are also formed (Scheme 12.4). All of these are reactive compounds, providing an important source of reactants for a wide range of flavour compounds. Scheme 12.6 summarises the reactions between hydrogen sulphide and other simple intermediates formed in other parts of the Maillard reaction.

12.3.3.1
Thiazoles and Thiazolines

Most cooked foods contain thiazoles. Simple alkyl-substituted thiazoles generally have odour threshold values in the range 1–1,000 μg/kg. Odour descriptions include green, vegetable-like, cocoa, nutty, and some are claimed to have meaty characteristics [22]. Although most alkylthiazoles result from thermal

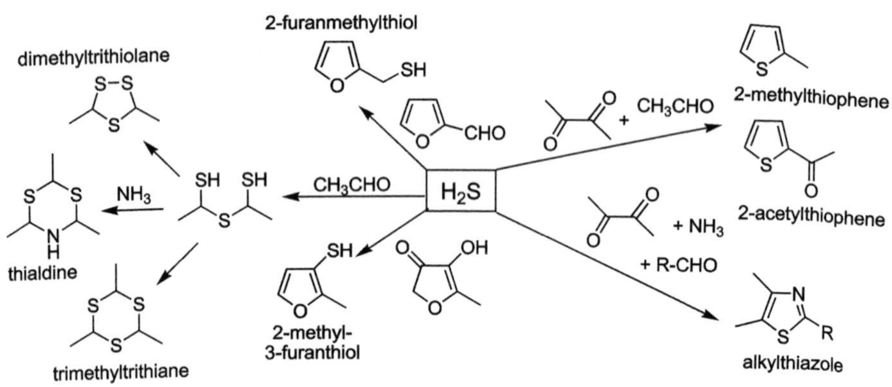

Scheme 12.6 The formation of heterocyclic aroma compounds from the reaction of hydrogen sulphide with intermediates of the Maillard reaction

reactions, some, such as 2-isobutylthizole, are biosynthesised. This compound makes a very important contribution to the aroma of fresh tomatoes [35]. 2-Acetylthiazole has been reported in a number of cooked foods, including meats, shellfish, coffee, nuts, cereals and some heated vegetables, and it probably makes important contributions to roast, nutty aromas in cooked foods. Mulders [36] proposed a pathway for its formation from the mercaptoiminenol intermediate in the Strecker degradation of cysteine and pyruvaldeyhde (Scheme 12.4). The route to alkylthiazoles probably involves the reaction of α-dicarbonyls, such as 2,3-butanedione or 2-oxopropanal (pyruvaldehyde), with ammonia and hydrogen sulphide (Scheme 12.7). This mechanism requires the participation of an aliphatic aldehyde, whose alkyl chain becomes substituted in the 2-position of the thiazole. This aldehyde may be acetaldehyde or a simple Strecker aldehyde, resulting from Strecker degradation of an amino acid. Alternatively, it may be a lipid oxidation product, such as hexanal or nonanal. Several thiazoles with C4–C8 *n*-alkyl substituents have been found in the volatiles of cooked meat [37–39] and, recently, 48 2-alkyl-3-thiazolines were reported in the headspace volatiles of boiled beef from animals in which the meat contained raised levels of polyunsaturated fatty acids [40]. However, these compounds with long alkyl chains were not found to be potent odorants.

12.3.3.2
Dithiazines

Thialdine (2,4,6-trimethyldihydro-1,3,5-dithiazine) is a six-membered heterocyclic compound containing sulphur and nitrogen in the ring. It was first reported in a food product in 1972 by Brinkman et al. [41], who identified it in heated pork. Subsequently it has been found in other meat species, as well as in peanuts, dry red beans, soybeans, boiled shrimp and several other seafoods [42]. Thialdine was reported to be the major volatile product obtained from a sample of boiled mutton [43]. Thialdine was first reported over 150 years ago by Wöhler and von Liebig [44], who showed that it was formed by the reaction of acetaldehyde, hydrogen sulphide and ammonia. The reaction occurs very readily without heating and, therefore, it is possible that it is formed during flavour-extraction procedures. Nevertheless, there is evidence that dihydrodithiazines do occur in food products and contribute to aroma [45].

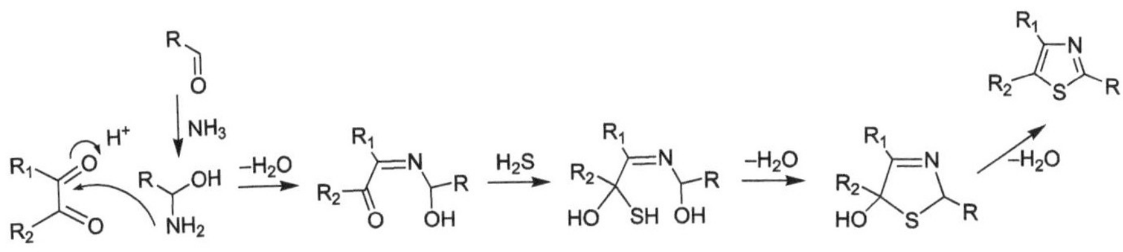

Scheme 12.7 Route for the formation of thiazoles

In the 1980s, several other dithiazines were identified in Antarctic krill [46] and later in shrimp [47] and dried squid [48]. They were considered to make important contributions to the aroma of these seafoods. Over 40 different dithiazine derivatives have now been identified in other foods, including beef, pork, chicken, grilled liver, roast peanuts, peanut butter and cocoa [42, 49]. The occurrence and sensory properties of these compounds have been discussed in an excellent review by Werkhoff et al. [42]. They also discuss the formation of these compounds in model systems comprising aldehydes, ammonia and hydrogen sulphide. The odour thresholds are reported to be in the range 5–500 μg/kg and the odour properties of 42 synthesised dithiazines are given in the review. Typical odour descriptors are roasted, onion, garlic-like, meaty, roast peanut, egg-like and sulphury.

12.3.3.3
Furanthiols and Sulphides

A number of furans with thiol, sulphide or disulphide substitution have been reported as aroma volatiles, and these are particularly important in meat and coffee. In the early 1970s, it was shown that furans and thiophenes with a thiol group in the 3-position possess strong meat-like aromas and exceptionally low odour threshold values [50]; however, it was over 15 years before such compounds were reported in meat itself. In 1986, 2-methyl-3-(methylthio)furan was identified in cooked beef and it was reported to have a low odour threshold value (0.05 μg/kg) and a meaty aroma at levels below 1 μg/kg [51]. Gasser and Grosch [52] identified 2-methyl-3-furanthiol and the corresponding disulphide, bis(2-methyl-3-furanyl) disulphide, as major contributors to the meaty aroma of cooked beef. The odour threshold value of this disulphide has been reported as 0.02 ng/kg, one of the lowest known threshold values [53]. Other thiols which may contribute to meaty aromas include mercaptoketones, such as 2-mercapto-pentan-3-one. 2-Furylmethanethiol (2-furfurylmercaptan) has also been found in meat, but is more likely to contribute to roasted rather than meaty aromas. Disulphides have also been found, either as symmetrical disulphides derived from two molecules of the same thiol or as mixed disulphides from two different thiols [54].

Disulphides and thiols containing a furan ring have also been found among the volatiles of coffee; however, those containing the 2-furylmethyl moiety are more abundant than compounds with the 2-methyl-3-furyl moiety. 2-Furylmethanethiol was first described as an important constituent of coffee in a patent published in 1926 [55]. Since then its 5-methyl homologue and various other thiols and disulphides have also been found [30]. These thiols have coffee-like characteristics at low concentrations, but are sulphurous and unpleasant at higher concentrations. An interesting bicyclic compound 2-methyl-3-oxa-8-thiabicyclo[3.3.0]-1,4-octadiene (kahweofuran), which is closely related to the 2-methyl-3-furanthio compounds, has also been identified in coffee.

The routes involved in the formation of the various furan sulphides and disulphides involve the interaction of hydrogen sulphide with dicarbonyls, furanones and furfurals. Possible pathways are shown in Scheme 12.8. Furanthiols have been found in heated model systems containing hydrogen sulphide or cysteine with pentoses [56–58]. 2-Methyl-3-furanthiol has also been found as a major product in the reaction of 4-hydroxy-5-methyl-3(2*H*)-furanone with hydrogen sulphide or cysteine [21, 59]. This furanone is formed in the Maillard reaction of pentoses; alternatively it has been suggested that it may be produced by the dephosphorylation and dehydration of ribose phosphate, and that this may be a route to its formation in cooked meat [21, 60].

12.4 Conclusion

The Maillard reaction is a major source of flavour in cooked foods. The reaction is complex and, because different foods have different profiles of amino acids and sugars, a wide range of flavours are produced when foods are heated. Research over the past 50 years has provided some understanding of the chemical pathways that are involved in the reaction. The identification of a large number of volatile compounds, including many heterocyclic structures, in heated foods has helped flavour scientists understand some of the relationships between the structure of flavour compounds and the perceived flavour. An understanding of the Maillard reaction also provides the potential for improving the sensory quality of heated foods through better control of processing conditions and through the enhancement of the important precursors in the raw materials during the production of both plant and animal foods.

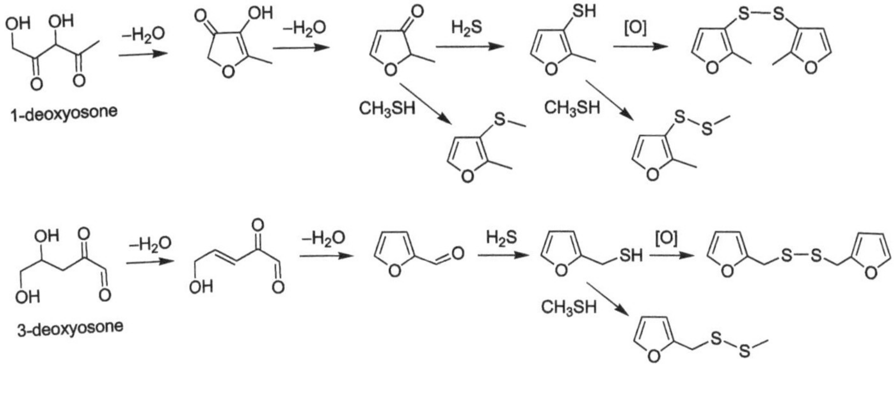

Scheme 12.8 Routes for the formation of furanthiols, sulphides and disulphides in the Maillard reaction

References

1. Kawamura S (1983) In: Waller GR, Feather MS (eds) The Maillard Reaction in Foods and Nutrition. ACS Symposium Series 215. American Chemical Society, Washington, p 3
2. Hodge JE (1967) In: Schultz HW, Day EA, Libbey LM (eds) Chemistry and Physiology of Flavors. AVI, Westport, p 465
3. Hurrell RF (1982) In: Morton ID, MacLeod AJ (eds) Food Flavors. Elsevier, Amsterdam, p 399
4. Mauron J (1981) In: Eriksson C (ed) Maillard Reactions in Food. Pergamon, Oxford, p 3
5. Mottram DS (1994) In: Parliment TH, Morello MJ, McGorrin RJ (eds) Thermally Generated Flavors: Maillard, Microwave, and Extrusion Processes. ACS Symposium Series 543. American Chemical Society, Washington, p 104
6. Nursten HE (1980) Food Chem. 6:263
7. Nursten HE (2005) The Maillard Reaction. Royal Society of Chemistry, Cambridge
8. Ledl F, Schleicher E (1990) Angew. Chem. Int. Ed. Engl. 29:565
9. Negishi C, Wakabayashi M, Tsuda M, Sato S, Sigimura T, Saito H, Maeda M, Jagerstad M (1984) Mutat. Res. Lett. 140:55
10. Tareke E, Rydberg P, Karlsson P, Eriksson S, Törnqvist M (2002) J. Agric. Food Chem. 50:4998
11. Mottram DS, Wedzicha BL, Dodson AT (2002) Nature 419:448
12. Strecker A (1862) Liebigs Ann. Chem. 123:362
13. Schiff H (1866) Ann. Chem. Pharm. 140:92
14. Maillard LC (1912) C. R. 154:66
15. Hodge JE (1953) J. Agric. Food Chem. 1:928
16. Namiki N (1988) Adv. Food Res. 32:115
17. Schonberg A, Moubacher R, Mostafa A (1948) J. Chem. Soc. 176
18. Vernin G, Parkanyi C (1982) In: Vernin G (ed) Chemistry of Heterocyclic Compounds in Flavors and Aromas. Harwood, Chichester, p 151
19. Hodge JE, Mills FD, Fisher BE (1972) Cereal Sci. Today 17:34
20. Guth H, Grosch W (1994) J. Agric. Food Chem. 42:2862
21. Whitfield FB, Mottram DS (1999) J. Agric. Food Chem. 47:1626
22. Fors S (1983) In: Waller GR, Feather MS (eds) The Maillard Reaction in Foods and Nutrition. ACS Symposium Series 215. American Chemical Society, Washington, p 185
23. Maga JA (1982) In: Morton ID, MacLeod AJ (eds) Food Flavors. Elsevier, Amsterdam, p 283
24. Guadagni DG, Buttery RG, Turnbaugh JG (1972) J. Sci. Food Agric. 23:1435
25. Amrani-Hemaimi M, Cerny C, Fay LB (1995) J. Agric. Food Chem. 43:2818
26. Mottram DS (1991) In: Maarse H (ed) Volatile Compounds in Foods and Beverages. Dekker, New York, p 107
27. Chang SS, Hirai C, Reddy BR, Herz KO, Kato A (1968) Chem. Ind. 1639
28. Mussinan CJ, Wilson RA, Katz I, Hruza A, Vock MH (1976) In: Charalambous G, Katz I (eds) Phenolic, Sulphur and Nitrogen Compounds in Food Flavors. ACS Symposium Series 26. American Chemical Society, Washington, p 133
29. Maga JA (1981) J. Agric. Food Chem. 29: 691
30. Flament I (1991) In: Maarse H (ed) Volatile Compounds in Foods and Beverages. Dekker, New York, p 617

31. Grosch W, Schieberle P (1991) In: Maarse H (ed) Volatile Compounds in Foods and Beverages. Dekker, New York, p 41
32. Tressl R, Grunewald KG, Kersten E, Rewicki D (1985) J. Agric. Food Chem. 33:1137
33. Frank O, Jezussek M, Hofmann T (2003) J. Agric. Food Chem. 51:2693
34. Nijssen LM, Visscher CA, Maarse H, Willemsen LC (1996) Volatile Compounds in Food—Qualitative and Quantitative Data. TNO-CIVO Food Analysis Institute, Zeist
35. Buttery RG, Teranishi R, Ling LC (1987) J. Agric. Food Chem. 35:540
36. Mulders EJ (1973) Z. Lebensm. -Unters. -Forsch. 152:193
37. Elmore JS, Mottram DS (2000) J. Agric. Food Chem. 48:2420
38. Hartman GJ, Jin QZ, Collins GJ, Lee KN, Ho C-T, Chang SS (1983) J. Agric. Food Chem. 31:1030
39. Tang J, Jin QZ, Shen GH, Ho C-T, Chang SS (1983) J. Agric. Food Chem. 31:1287
40. Elmore JS, Mottram DS, Enser M, Wood JD (1997) J. Agric. Food Chem. 45:3603
41. Brinkman HW, Copier H, de Leuw JJM, Tjan SB (1972) J. Agric. Food Chem. 20:177
42. Werkhoff P, Güntert M, Hopp R (1992) Food Rev. Int. 8:391
43. Nixon LN, Wong E, Johnson CB, Birch EJ (1979) J. Agric. Food Chem. 27:355
44. Wöhler F, von Liebig J (1847) Ann. Chem. Pharm. 61:1
45. Ohloff G, Flament I, Pickenhagen W (1985) Food Rev. Int. 1:99
46. Kubota K, Kobayashi A, Yamanishi T (1982) Agric. Biol. Chem. 46:2835
47. Kubota K, Shijimaya H, Kobayashi A (1986) Agric. Biol. Chem. 50:2867
48. Kawai T, Ishida Y, Kakiuchi H, Ikeda N, Higashida T, Nakamura S (1991) J. Agric. Food Chem. 39:770
49. Velluz A, Brönner H, Näf R, Wüest H, Büchi G, Pickenhagen W (1994) Flavour Fragr. J. 9:81
50. Evers WJ, Heinsohn HH, Mayers BJ, Sanderson A (1976) In: Charalambous G, Katz I (eds) Phenolic, Sulfur and Nitrogen Compounds in Food Flavors. ACS Symposium Series 26. American Chemical Society, Washington, p 184
51. MacLeod G, Ames JM (1986) Chem. Ind. 175
52. Gasser U, Grosch W (1988) Z. Lebensm. -Unters. -Forsch. 186:489
53. Buttery RG, Haddon WF, Seifert RM, Turnbaugh JG (1984) J. Agric. Food Chem. 32:674
54. Mottram DS (1998) Food Chem. 62:415
55. Reichstein T, Staudinger H (1955) Perfum. Essent. Oil Rec. 46:86
56. Farmer LJ, Mottram DS, Whitfield FB (1989) J. Sci. Food Agric. 49:347
57. Hofmann T, Schieberle P (1995) J. Agric. Food Chem. 43:2187
58. Mottram DS, Whitfield FB (1995) J. Agric. Food Chem. 43:984
59. van den Ouweland GAM, Peer HG (1975) J. Agric. Food Chem. 23:501
60. Mottram DS, Nobrega ICC (1998) In: Contis ET, Ho C-T, Mussinan CJ, Parliment TH, Shahidi F, Spanier AM (eds) Food Flavours: Formation, Origin, Analysis and Packaging Influences. Developments in Food Science 40. Elsevier, Amsterdam, p 483

13 Chemical Conversions of Natural Precursors

Peter H. van der Schaft
I.F.F. (Nederland) B.V.,
P.O. Box 5021, 5004 EA Tilburg, The Netherlands

13.1 Introduction

Over many decades the flavour industry has built up the ability to prepare flavours mimicking nature by using flavour ingredients that are isolated from natural sources like plants. Isolation of flavour ingredients from natural sources results in the natural status of such an ingredient, which is seen as an advantage for the labelling of the food or beverage in which it is applied. Natural sources are from vegetable or animal origin; also isolation of a flavouring from a fermentation broth using natural-source raw materials is regarded as natural; fermentation products are also regarded as renewable resources. In North America flavour substances prepared by chemical synthesis using natural raw materials and synthetic catalysts are also regarded as natural. In Europe the use of a synthetic catalyst results in a nature-identical chemical, whether the raw materials are natural or not. The European nature-identical status of a chemical means that the chemical has been identified in an edible food or beverage, but it has been prepared from materials from natural or petrochemical origin using organic chemistry, including the use of synthetic catalysts. In North America a flavour ingredient can also be synthesised from petrochemical origin leading to the artificial status of the ingredient, but it has to be generally regarded as safe (GRAS) as declared by the Flavors and Extracts Manufacturing Association (FEMA).

The reasons to use raw materials from renewable resources can be various. When a natural flavour ingredient has to be prepared, a natural raw material is essential, and natural raw materials are renewable, because they come from plants, animals or fermentation. For nature-identical flavour ingredients, a renewable raw material can be a good choice from a chemical point of view and quite often also from a cost point of view; if turpentine is readily available in a country with limited or no petrochemical resources, β-pinene from the renewable source is cheaper than chemically synthesised β-pinene. A manufacturer chooses only for sustainable production if it is remunerative and at least as attractive as other options.

Estimation of the volume of renewable resources involved in the flavour industry is very difficult. It is assumed that the total flavour and fragrance market is worth $12 billion and that flavours constitute 50% of this. When the average global market price of a flavour is assumed to be $10 per kilogram, then the volume involved would be 600,000 t of flavour. This flavour volume includes a high variety of products, such as flavour chemicals, essential oils, liquid flavour compounds, including their carriers like ethanol and propylene glycol and dry flavour compounds, including their carriers like maltodextrin, salt and gums. Many of the solvents and carriers used are obtained from renewable resources themselves. It is even more difficult to estimate the volatile flavour part of this very heterogeneous group of flavour products representing this 600,000 t of flavour. A rough estimate may be 10%, which would represent 60,000 t of flavour volatiles. If a hypothetical 50% of this volume is obtained from renewable resources, the corresponding volume is 30,000 t. Although this is a very rough figure it does not seem unrealistic when compared with estimates of the total world production of essential oils. In 1993 the total global production of the top 20 essential oils was estimated at 56,000 t [1]. The top three included orange oil (26,000 t), corn mint oil (4,000 t) and eucalyptus oil (4,000 t), but did not include turpentine, which was estimated at 250,000 t. The majority of these oils and turpentine are used in fragrance applications and for the preparation of other chemicals not used as flavours and fragrances. In addition, the estimated 30,000 t of flavour volatiles obtained from renewable resources is not only derived from essential oils and turpentine, but also includes materials like vanillin from vanilla or wood pulp and process flavour volatiles. Very recently the world production of essential oils for flavours was estimated at 21,670 t [2]. Compared with this figure, the estimated 30,000 t of flavour volatiles obtained from renewable resources seems to be the right order of magnitude.

In this chapter chemical conversions of natural precursors resulting in flavour chemicals are discussed. The main groups of natural precursors are terpenes for all kinds of terpene derivatives, vanillin precursors like lignin and eugenol, sugars for Maillard-associated flavour chemicals, amino acids and molecules obtained by fermentation or available as residual streams of renewable resources.

13.2 Terpenes as Renewable Resources for Terpene Flavour Molecules

13.2.1 Pinenes from Turpentine

Many terpenes are derived from renewable plant oil resources like essential oils. α-Pinene and β-pinene from turpentine may be the best known examples, because they represent a very large volume. Turpentine was originally obtained from pine trees by tapping gum oleoresin from the stem of the living trees followed by steam distillation of the crude oleoresin and subsequently separation into rosin and turpentine by distillation. The ratio of α-pinene to β-pinene in this turpentine varies considerably and depends a lot on the pine species from which the turpentine is derived, but in general the oil is much more abundant in α-pinene than β-pinene. Another more recent form of available turpentine is called crude sulfate turpentine which is obtained as a by-product of paper manufacturing from softwood (pine, fir, spruce). Sulfate turpentine contains around 20–25% β-pinene and 60–70% α-pinene. The rest of the turpentine consists of light fractions (1–2%), dipentene (limonene) (3–10%), pine oil (3–7%) and other volatiles like estragole, anethole and caryophyllenes (1–2%).

Scheme 13.1 shows the main commercial chemical pathways based on α-pinene and β-pinene from turpentine. These pathways lead to four important target molecules or groups of molecules. These are the terpineols, menthol, cam-

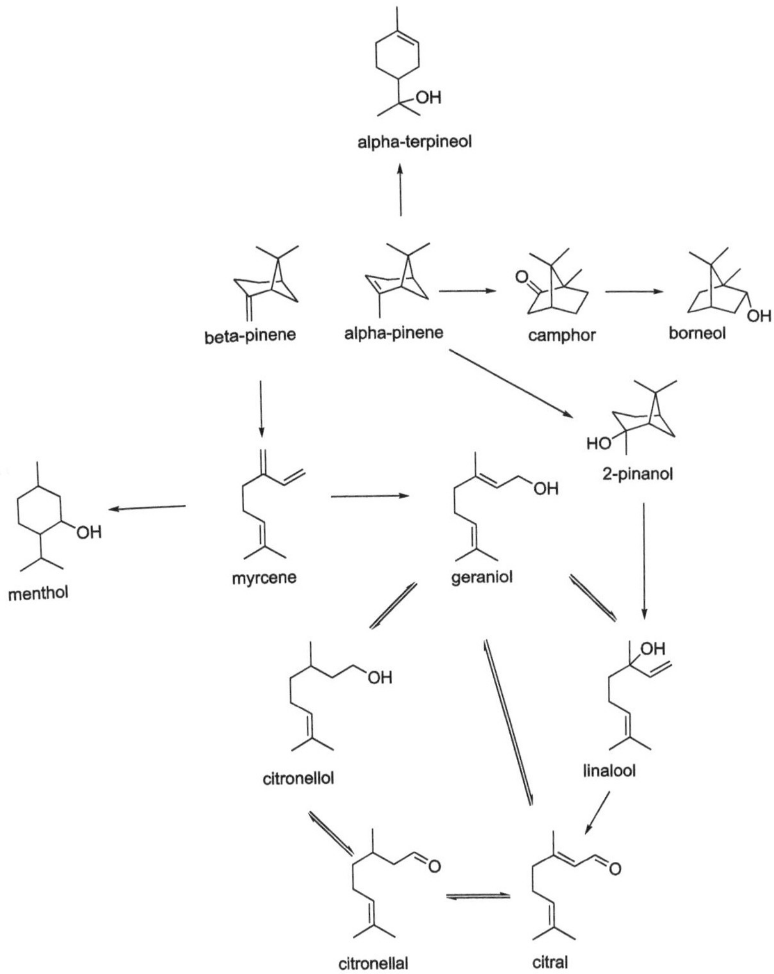

Scheme 13.1 Overview of the most important commercial chemical pathways based on α-pinene and β-pinene in turpentine

phor and the related borneol and the group of rose alcohols and citral-related materials (geraniol, linalool, citral and citronellal/citronellol). These chemicals are used in the manufacturing of fragrances, which represents a big volume, and also for the production of flavours.

By hydration to terpin and subsequent dehydration, pinenes can be converted into terpineols; the main representative is α-terpineol, which is used in lime, among many other flavours.

By isomerisation, α-pinene can be converted into camphene, and this can then be esterified to obtain an ester of isoborneate, which can be saponified to isoborneol. Isoborneol can be dehydrogenated to camphor, which can be reduced again to borneol, which is used in many fruit flavours.

α-Pinene can also be reduced to pinane, which can be oxidised to 2-pinanol. Pyrolysis of this alcohol results in the formation of linalool, from which the other rose alcohols and citral-related chemicals can be formed.

Pyrolysis of β-pinene results in the triene myrcene, which leads to menthol and its derivatives, on one hand, and the rose alcohols and citral-related chemicals, on the other hand.

As indicated before, Scheme 13.1 shows only a summary of the most important commercial chemical pathways based on the pinenes in turpentine. Terpenoid chemistry is very well developed and a lot more monoterpenes can be synthesised using one of the chemicals in Scheme 13.1 as the starting material.

13.2.2 Citral

13.2.2.1 Sources of Citral

Citral is another important starting material for the chemical synthesis of many linear monoterpenes, sesquiterpenes and diterpenes. Citral is the main ingredient (60–80%) of *Litsea cubeba* oil, which is obtained by distillation of the fruits from this tree at a yield of 3–5%; China produced 1,500 t of this oil annually in the 1990s [3]. Also lemon grass oil is an important source of citral. On the other hand, a high volume of citral is manufactured by the petrochemical industry starting from isobutylene, to which formaldehyde is added to form isoprenol, which is a good starting material for citral synthesis. Citral can also be prepared from isoprene, which is primarily produced from petrochemicals, but it can also be obtained by pyrolysis of limonene, which is available from sustainable resources such as sulfate turpentine or it can be synthesised from the pinenes present in turpentine (Scheme 13.1).

The choice by a company for a sustainable or a petrochemical source of an ingredient will depend on various aspects, but cost will be the primary driving force. Availability of a low-cost feedstock is crucial, but also the available and appropriate technology in the company will have a large influence on this decision.

Terpenes important for both fragrances and flavours can be prepared from citral, such as citronellol, linalool, nerolidol, geraniol, farnesol and bisabolol. Citral is also an important starting material for the synthesis of vitamins A and E, carotenoids and other flavour and fragrance compounds like ionones. Most of the β-ionone synthesised is probably used for vitamin A synthesis.

13.2.2.2
Ionones from Citral

α-Ionone is used in large quantities in the fragrance industry. β-Ionone is a costly specialty chemical that is used in the manufacture of vitamin A, which is widely used in the animal feed industry, and carotenoids such as β-carotene. Also large quantities β-ionone are used as an additive in fragrances and flavourings. α-Ionone and β-ionone are important flavour substances for all kinds of fruit flavours, especially berry flavours such as raspberry, but also for the violet note in many fragrances. These substances can be prepared from citral, which can be obtained from lemon grass oil or from petrochemical sources.

For the preparation of α-ionone, citral and acetone are reacted in an aldol condensation catalysed by a base to form so-called pseudo-ionone (Scheme 13.2). The pseudo-ionone can be cyclised to form α-ionone catalysed by an acid.

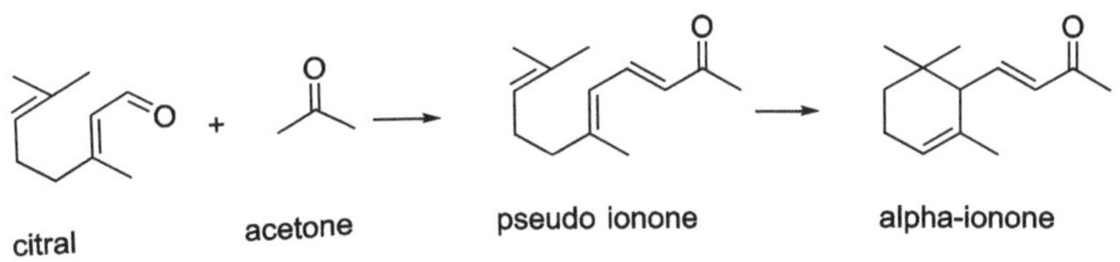

Scheme 13.2 Preparation of α-ionone from citral and acetone

13.2.3
The Mint Components L-Menthol and L-Carvone

Most menthol is isolated from peppermint oils, especially from crude oil from *Mentha arvensis* from India. But menthol can also be prepared by chemical synthesis. There are two important commercial processes for the synthesis of menthol. One is based on a renewable resource, β-pinene from turpentine, and the other on *m*-cresol from petrochemical origin (Scheme 13.3).

Alkylation of *m*-cresol with propene in the presence of an aluminium catalyst results in the formation of thymol, which upon hydrogenation gives a mixture of all eight isomers of menthol, D-menthol, L-menthol, neomenthol, isomenthol and neoisomenthol (Scheme 13.3). The preferred isomer is L-menthol, because of its ability to induce physiologically the sense of cold which is desired in many products such as chewing gum and toothpaste; L-menthol is about

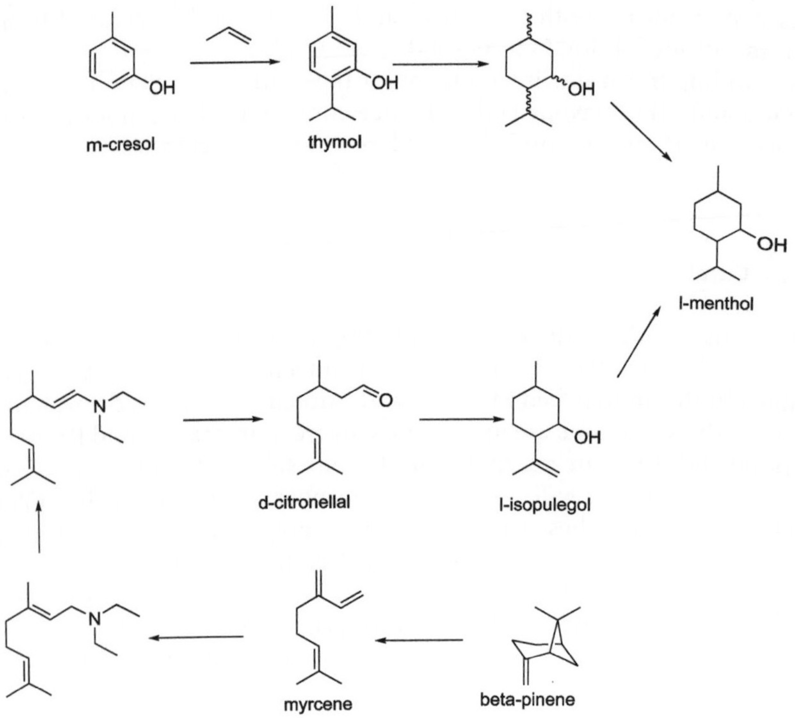

Scheme 13.3 Chemical synthesis of menthol from *m*-cresol and from *β*-pinene

50 times more cooling than D-menthol. This synthesis from *m*-cresol is cheap from a raw material and processing point of view, but nearly 70% of the product is not L-menthol and an extensive separation process is required, which in this case is a combination of fractional distillation of the isomers resulting in a D,L-menthol fraction and crystallisation of the benzoate esters of D,L-menthol to finally obtain pure L-menthol [4].

The other synthesis of L-menthol (Scheme 13.3) is based on the pyrolysis of *β*-pinene to produce myrcene. Diethylamide addition to myrcene results in the formation of *N,N*-diethylgeranylamine, which is subsequently isomerised to the *N,N*-diethylenamine of citronellal. Citronellal is obtained by hydrolysis of the enamine, which can be cyclised to L-isopulegol catalysed by zinc chloride. L-menthol is finally obtained by hydrogenation of isopulegol using a nickel catalyst. In this synthesis enantiomerically pure product is prepared, because of the control of the stereochemistry during the reactions where new chiral centres are obtained. This eliminates the need for a laborious working-up procedure as required in the synthesis starting from *m*-cresol.

Also syntheses based on 3-carene from Indian turpentine and on D-pulegone from pennyroyal oil have been practised on a commercial scale in the past, but these approaches have been abandoned, because they were considered uneconomical compared with the current processes. When these syntheses were used in India and southern Europe, respectively, local factors such as the availability of a chiral feedstock affected the economics of the process in a positive way.

The main renewable resource for L-carvone is spearmint oil (*Mentha spicata*), which contains up to 75% of this flavour chemical. There also exists a synthetic process for the manufacturing of L-carvone, which is based on (+)-limonene, which is available as a by-product of the citrus juice industry as a major component of orange peel oil (Scheme 13.4). The synthesis was developed in the nineteenth century and starts with the reaction of (+)-limonene and nitrosyl chloride, which ensures the asymmetry of the ring. Treatment with base of the nitrosyl chloride adduct results in elimination of hydrogen chloride and rearrangement of the nitrosyl function to an oxime. Acid treatment of the oxime finally results in l-carvone.

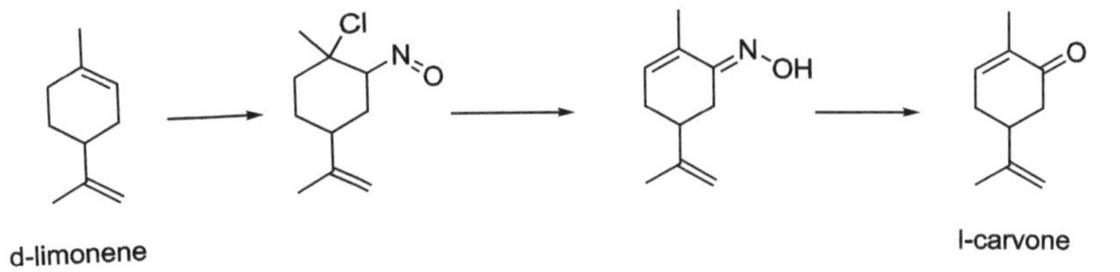

Scheme 13.4 Chemical synthesis of l-carvone from (+)-limonene

13.2.4 Terpene Sulfur Compounds

13.2.4.1 *p*-1-Menthen-8-thiol from *β*-Pinene

p-1-Menthene-8-thiol is a character-impact constituent of grapefruit flavour which has been found in grapefruit juice at very low levels around 0.02 ppb [5]. The chemical can be synthesised by hydrogen sulfide addition to several monoterpenes like *α*-terpineol, limonene and *β*-pinene which are all derived from renewable resources (Scheme 13.5).

The chemical is used both in flavours and fragrances, but owing to its very strong smell and low odour threshold value the volume is limited.

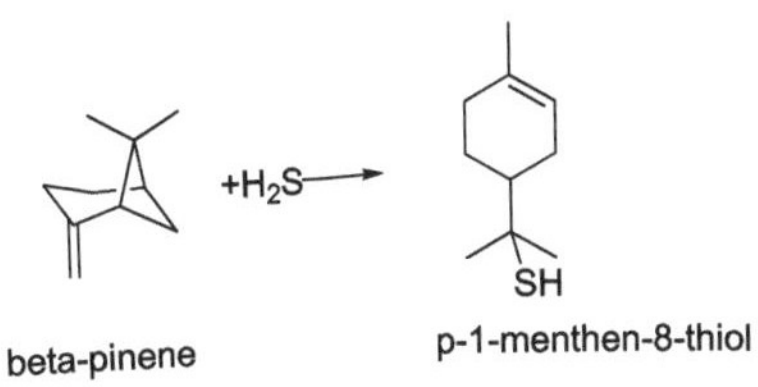

Scheme 13.5 Chemical synthesis of *p*-1-menthen-8-thiol from *β*-pinene

13.2.4.2
8-Mercapto-*p*-menthan-3-one from Pulegone

8-Mercapto-p-menthan-3-one has been identified as a constituent of *Buchu* leaf oil [6] and is a very useful substance in black currant and tropical flavours. The reaction of pulegone with hydrogen sulfide in an alkaline medium results in 8-mercapto-*p*-menthan-3-one formation (Scheme 13.6).

Pulegone is the major constituent of pennyroyal oil from *Mentha pulegium* L., which is available in large quantities.

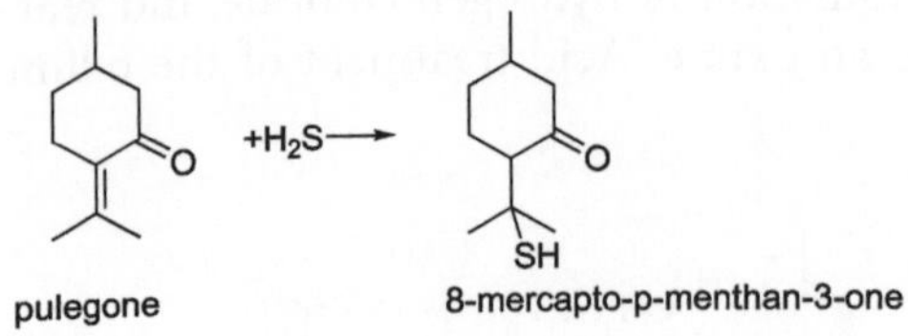

Scheme 13.6 Chemical synthesis of 8-mercapto-*p*-menthan-3-one from pulegone

13.2.5
Other Terpene Derivatives

13.2.5.1
Anisaldehyde from Anethole

Another example is the synthesis of anisaldehyde from anethole obtained from star anise oil from the fruit and leaves of *Illicium verum*. Anethole can be oxidised, for instance, by chromic acid (a mixture of sodium dichromate and sulfuric acid) to anisaldehyde (Scheme 13.7). Star anise oil is produced in China in annual quantities of 500–800 t [3].

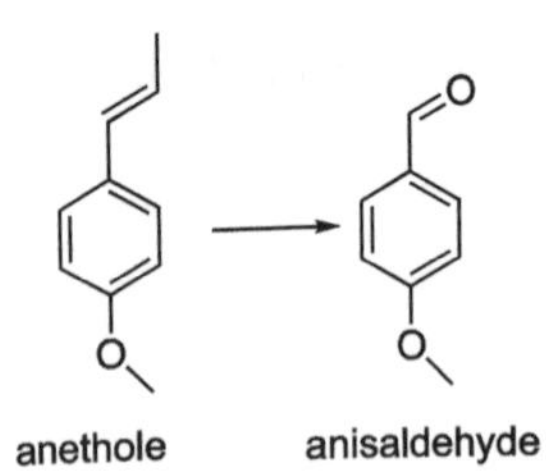

Scheme 13.7 Conversion of anethole into anisaldehyde

13.2.5.2
Hotrienol from Linalool

Hotrienol was found for the first time in Ho leaf oil as the *S* enantiomer [7], but has been found since then in many natural sources; for instance, the *R* enantiomer was found in black tea and in green tea. The product can be used in many flavours, such as elderflower, grape, berry and honey flavours. It can be prepared from linalool obtained from citrus oils or Chinese Ho oils, but most linalool is obtained by synthesis from isoprene from petrochemical sources.

Recently a practical and convenient synthesis was described starting from linalool via linalyl acetate [8]. It involves the ene-type chlorination of linalyl acetate prepared from linalool which results in the formation of γ-chloro-α-linalyl acetate (Scheme 13.8). Dehydrochloronation with lithium bromide and lithium carbonate in dimethylformadide followed by hydrolysis of dehydro-α-linalyl acylate results in hotrienol.

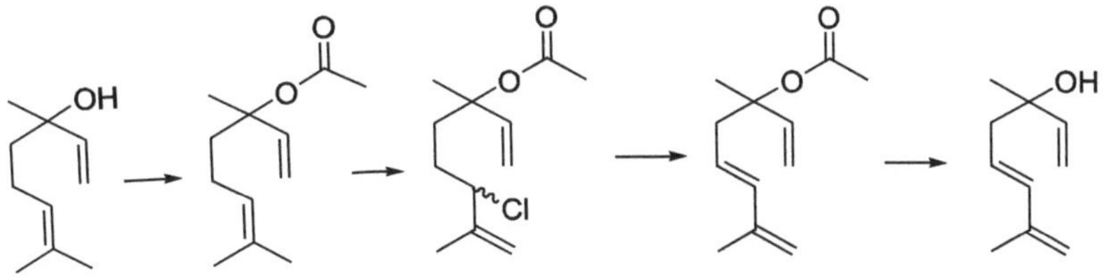

Scheme 13.8 Chemical synthesis of hotrienol from linalool

13.2.5.3
Nootkatone from Valencene

Nootkatone is an important constituent from grapefruit flavour. It is synthesised by oxidation of valencene, which is obtained and isolated from orange peel oil where it occurs at a maximum level of 0.4% [9].

For the oxidation of valencene to nootkatone, strong oxidation agents like oxygen in the presence of metal salts, peroxides or chromate compounds are required. There are also several patents on the bioconversion of valencene to nootkatone which results in natural nootkatone (Scheme 13.9) [10, 11].

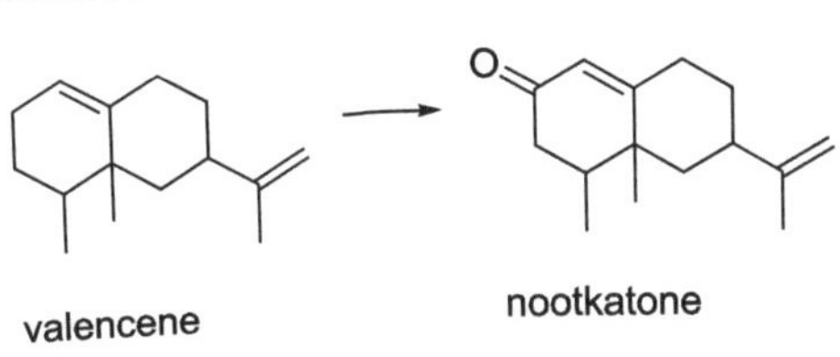

Scheme 13.9 Conversion of valencene into nootkatone

13.2.5.4 Terpene Esters

The most important and frequently used terpene esters in flavours are the acetates of nerol, geraniol, citronellol, linalool and isoborneol [12]. As discussed before, all these terpene alcohols are available both from renewable resources and from petrochemical origin. Acetic acid can be obtained from renewable resources by pyrolysis of wood as wood vinegar, and also by synthesis from petrochemical origin.

13.3 Vanillin

13.3.1 Vanillin Synthesis

Eugenol obtained from clove oil is an important precursor for the preparation of vanillin (Scheme 13.10). The reaction consists of two steps. First, eugenol needs to be converted into isoeugenol, which requires alkaline treatment or ruthenium or rhodium catalysis. Second, the isoeugenol is oxidised to vanillin using, for instance, chromic acid. This method results in nature-identical vanillin.

Lignin waste from the wood pulp industry also serves as a renewable source of precursors for the preparation of vanillin. Lignin can be degraded by treatment with alkali and oxidising agents to coniferyl alcohol like structures which can be oxidised to vanillin. Curcumin, a yellow colouring material from the roots of *Curcuma longa*, can also serve as a precursor for vanillin; from each molecule of curcumin two molecules of vanillin can be derived. Also a lot of synthetic vanillin is prepared from petrochemical sources such as phenol, catechol and guaiacol.

13.3.2 Vanillin Derivatives

Vanilla flavour is not only determined and characterised by the vanillin molecule, but also by many more phenolic compounds and vanillin derivatives. Two examples of molecules that recently obtained FEMA-GRAS status are vanillyl ethyl ether and vanillin 2,3-butanediol acetal (Scheme 13.11). Vanillin can be hydrogenated to form vanillyl alcohol, which is also used in vanilla flavours. Vanillyl alcohol can be reacted with ethanol to form vanillyl ethyl ether. Vanillin can also form an acetal with 2,3-butanediol (obtained by fermentation of sugars) catalysed by *p*-toluene sulfonic acid in toluene.

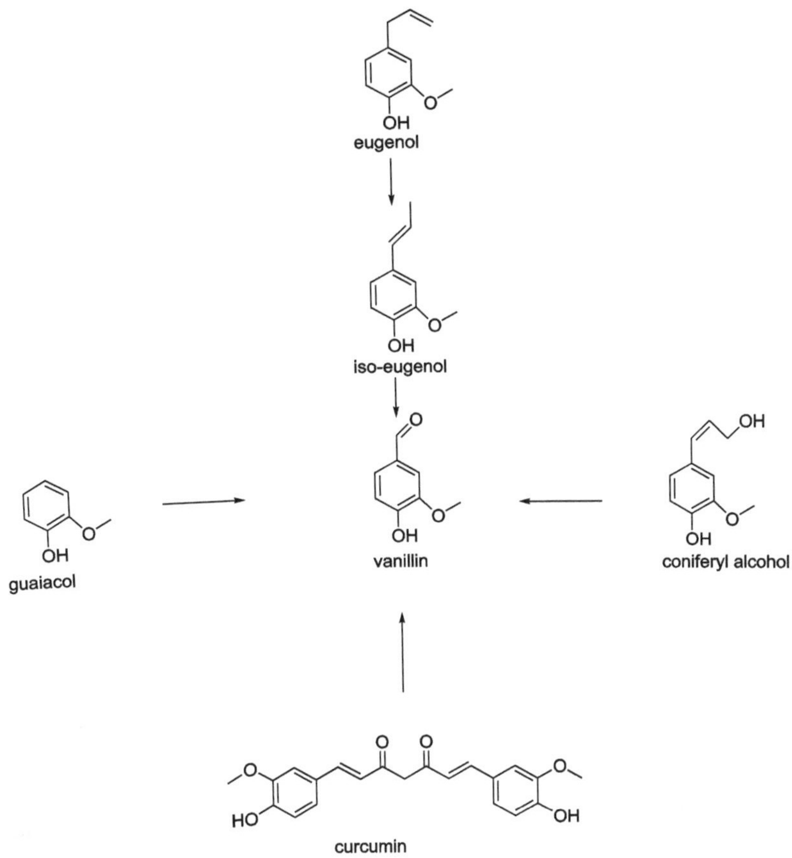

Scheme 13.10 Chemical synthesis of vanillin from various sources

13.3.3
Heliotropine from Safrole

Safrole is used for the preparation of heliotropine, which is mainly used in vanilla flavours. Safrole has to be converted into isosafrole by alkali treatment or ruthenium or rhodium catalysis analogous to the eugenol to isoeugenol conversion before it can be oxidised to heliotropine using chromic acid as a catalyst [13] (Scheme 13.12).

Safrole, which is both toxic and carcinogenic, occurs in sassafras oil up to 90%. There are two commercially important sassafras oils: the Brazilian oil is obtained from the trunk wood of *Ocotea pretiosa* and the Chinese oil is obtained from *Cinnamomum camphora* by steam distillation of wood chips. Safrole from

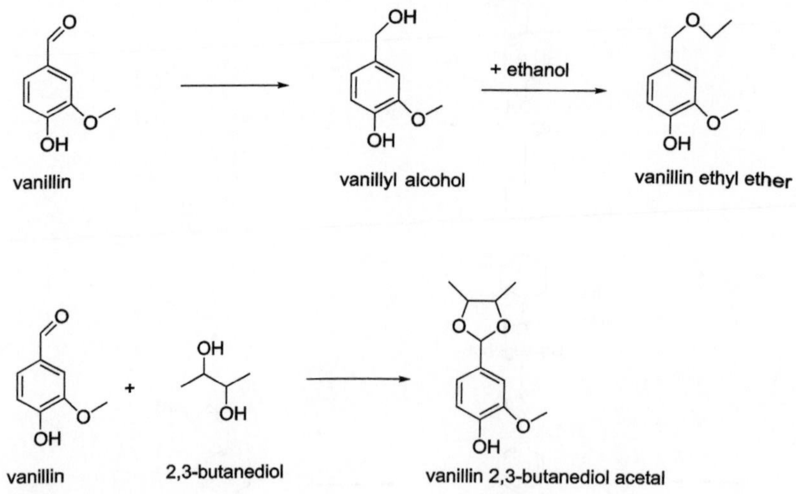

Scheme 13.11 Formation of vanillin ethyl ether and vanillin 2,3-butanediol acetal from vanillin

sassafras oil is not only used to prepare heliotropine, but also for the synthesis of piperonal butoxide, which is a vital ingredient for pyrethroid insecticides. The demand for sassafras oil is 2,000 t. Brazil is confronted with a depletion of its natural resource for sassafras oil, because more trees are felled than are replanted and mature [14]. This has resulted in measures involving restrictions on felling trees, resulting in a production decline of sassafras oil. This illustrates the fact that also renewable resources can be depleted and result in a shortage, which can be a long-term problem, especially when the wood from full-grown trees is required. Currently *Piper* species, which are weeds or shrubs indigenous in Central and South America, are being investigated as an alternative source of safrole. Propagation studies and growing trials have shown that these plants may act as renewable resources when cultivated on plantations. Biomass yields and oil productivity per hectare per year together with projected oil prices will determine whether economic returns to the farmer can be guaranteed.

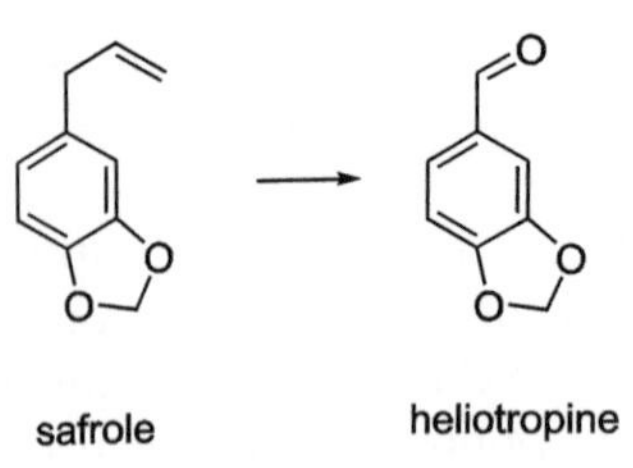

Scheme 13.12 Conversion of safrole into heliotropine

13.4 Sugars as Precursors

13.4.1 Sources of Xylose and Rhamnose

Sugars are important as renewable resources for the preparation of process flavours. On the other hand, some rare sugars like xylose and rhamnose are also important for the preparation of specific flavour chemicals. The C5 sugar xylose can be obtained from plant materials like wood, straw and hulls; acid hydrolysis of its precursor xylan, which is a polysaccharide built up from D-xylose, results in xylose. The C6 deoxy sugar rhamnose also occurs in many plants. Commercially available rhamnose is obtained by chemical hydrolysis of arabic and karaya gums, or from rutin (the rhamnoglucoside of quercetin), which is present in many plants. Also naringin, which is the main bitter compound in grapefruit, occurs in the rind of citrus fruits and is a source of rhamnose.

13.4.2 Examples of Flavour Chemicals Derived from Sugars

13.4.2.1 2-Methylfuran-3-thiol from Xylose and Hydrogen Sulfide

2-Methylfuran-3-thiol is responsible for a boiled note in milk, chicken and beef flavours and this chemical as well as its disulfide are widely used in flavours. Both molecules can be prepared by heating xylose and hydrogen sulfide in an autoclave (Scheme 13.13). Alternatively, these sulfur compounds can also be synthesised from 2-methylfuran made from furfural derived from oat hulls.

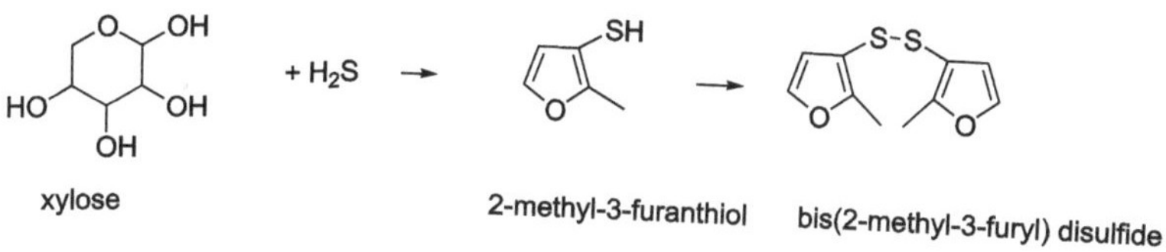

Scheme 13.13 Formation of 2-methylfuran-3-thiol and its disulfide from xylose

13.4.2.2 4-Hydroxy-2,5-dimethyl-3(2*H*)-furanone from Rhamnose

4-Hydroxy-2,5-dimethyl-3(2*H*)-furanone is a widely used flavour molecule for fruit and brown flavours which is prepared by heating of rhamnose. A nitrogen-containing base like proline or piperidine serves as a catalyst for its formation (Scheme 13.14).

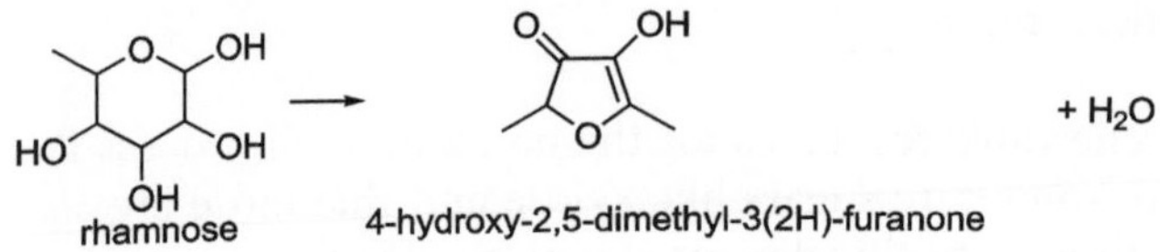

Scheme 13.14 Formation of 4-hydroxy-2,5-dimethyl-(2*H*)-furanone from rhamnose

13.4.2.3
Pyrazines from Rhamnose and Ammonia

Pyrazines are mainly used in roasted, peanut and chocolate flavours. Alkyl pyrazines can be obtained by the reaction in an autoclave at high temperature (120–160 °C) of reducing sugars like rhamnose and ammonia.

13.4.2.4
Furfural As a Precursor for Furfuryl Mercaptan

Furfural comes from pentose sugars in cereal straws and brans. Furfural is the precursor of furfuryl mercaptan and its disulfide, difurfuryl disulfide, which are both important chemicals for coffee, meat and roasted flavours. They are prepared by the reaction of furfural and hydrogen sulfide (Scheme 13.15).

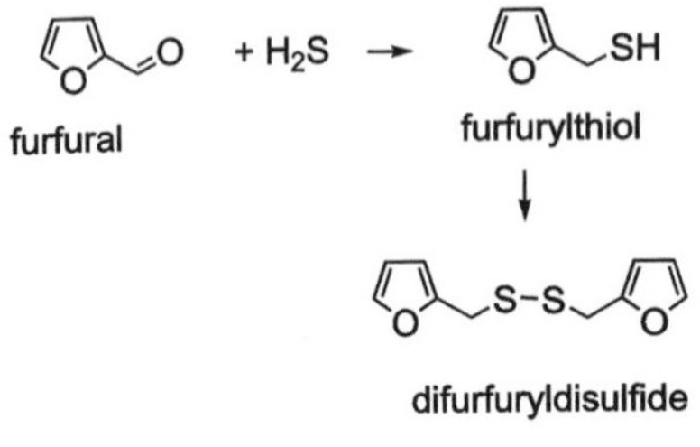

Scheme 13.15 Formation of furfurylthiol and its disulfide from furfural

13.5 L-Cysteine and L-Methionine as Sources of Hydrogen Sulfide and Methanethiol

13.5.1 Cysteine

Cysteine can be obtained by hydrolysis from cysteine-rich proteins in hair or feathers or from petrochemical sources. Cysteine is an important raw material in Maillard reactions for the preparation of process flavours, but it can also serve as a source of ammonia and hydrogen sulfide for the preparation of flavour chemicals, such as the terpene sulfur compounds mentioned in Sect. 13.2.4 and furfuryl mercaptan mentioned in Sect. 13.4.2.4.

13.5.2 Methionine

Methionine can be obtained from enzymatic protein hydrolysates or from petrochemical sources. To a lesser extent than cysteine, it is a raw material in Maillard reactions for the preparation of process flavours and it can also be utilised as a precursor for the chemical preparation of the sulfide methional, which is an important flavour constituent for potato, malt, seafood and many other flavours. Methional can be reduced to methionol, which can be esterified with organic acids to, for instance, methionyl acetate and methionyl butyrate, which are useful compounds for pineapple and other fruit flavours (Scheme 13.16).

Scheme 13.16 Methionine as a source of flavour chemicals

13.6 Chemical Conversions of Natural Precursors Obtained by Fermentation or from Residual Streams

13.6.1 Aliphatic and Aromatic Esters

A very important group of flavour molecules is the esters. Many flavour esters can be prepared from organic acids and alcohols from renewable resources. Important alcohols readily available from renewable resources are methanol from wood pyrolysis, ethanol and butanol by fermentation of sugars, propanol and (iso)amyl alcohol from fusel oil, and also benzyl and phenyl ethyl alcohol from essential oils. Many of these alcohols are also prepared in large quantities from petrochemical sources. The same applies to organic acids like acetic acid from wood pyrolysis, lactic and butyric acids from fermentation of sugars, hexanoic, octanoic, decanoic, myristic and oleic acids from vegetable oils and levulinic acid from cellulose; most of these acids also have economical petrochemical sources.

13.6.2 Heterocyclic Flavour Molecules

Also heterocyclic flavour molecules can be formed from renewable resources. 3,5-Diethyl-1,2,4-trithiolane is an important molecule for onion flavours and can easily be prepared from propanal obtained by biotransformation and hydrogen sulfide (Scheme 13.17). A meat flavour molecule like thialdine [dihydro-2,4,6-trimethyl-1,3,5(4*H*)-dithiazine] can be prepared from acetaldehyde isolated from molasses and ammonium sulfide (Scheme 13.18). The bacon flavour substance 2,4,6-triisobutyl-5,6-dihydro-4*H*-1,3,5-dithiazine can be prepared from isovaleraldehyde prepared from essential oils and ammonium sulfide (Scheme 13.19).

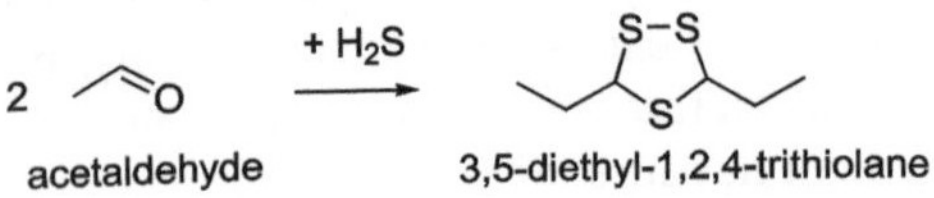

Scheme 13.17 Formation of 3,5-diethyl-1,2,4-trithiolane from acetaldehyde

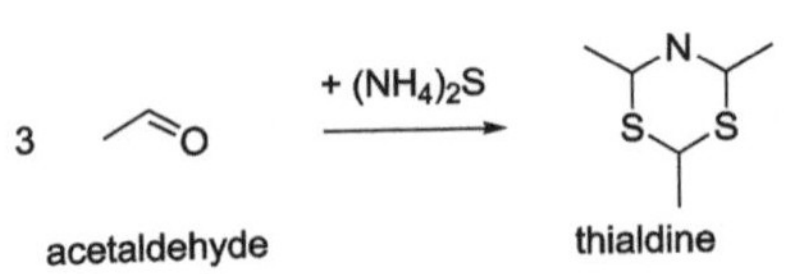

Scheme 13.18 Formation of thialdine from acetaldehyde

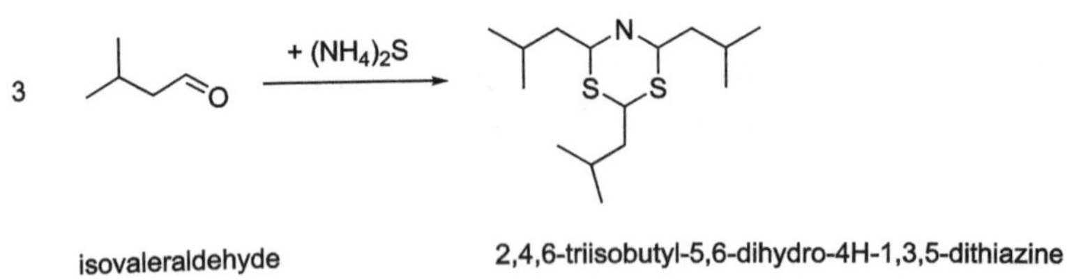

Scheme 13.19 Formation of 2,4,6-triisobutyl-5,6-dihydro-4*H*-1,3,5-dithiazine from isovaleraldehyde

13.7 Conclusions

Many products from the flavour industry are primary products from renewable resources or secondary products obtained by chemical conversions of the primary products. In general these secondary products are key flavour chemicals with a high added value. The cost difference between a precursor, the primary product and the flavour chemical can easily amount to a factor 20–1,000, especially when it concerns a natural flavour chemical. A large part of this cost reflects, of course, the efficiency of the reaction, the labour involved and the cost of the other reagents.

Although quite often these flavour chemicals can be prepared from petrochemical sources, renewable resources are preferred by the flavour industry, because access to these renewable resources is very good and already existed when these companies were started. In addition, chemicals from renewable resources are natural, so they can be used in natural flavours and offer the possibility to be used for the production of natural secondary products.

References

1. Lawrence BM (1993) In: Janick J, Simon JE (eds) New crops. Wiley, New York, p 620
2. Sanganeria S (2005) Perfum Flavor 30(7):24
3. Zaobang S (1995) CIFOR occasional paper 6. Center for International Forestry Research, Bogor
4. Sell CS (2003) A fragrant introduction to terpenoid chemistry. Royal Society of Chemistry, Cambridge
5. Demole E, Enggist P, Ohloff G (1982) Helv Chim Acta 65:1785
6. Lamparsky D, Schudel P (1971) Tetrahedron Lett 36:3323
7. Yoshida T, Muraki S, Kawamura H, Komatsu A (1969) Agric Biol Chem 33:343
8. Yuasa Y, Kato Y (2003) J Agric Food Chem 51:4036
9. MacLeod AJ, MacLeod G, Subramanian G (1988) Phytochemistry 27:2185
10. Muller B, Dean C, Schmidt C, Kuhn JC (1997) WO 9722575
11. Huang R, Christensen PA, Labuda IM (2001) US Patent 6,200,786
12. Wright J (2004) Flavor creation. Allured, Carol Stream
13. Dorsky J (1991) In: Mueller PM, Lamparsky D (eds) Perfumes: art, science and technology. Elsevier, Amsterdam, chap 14
14. Coppen J (1995) Flavors and fragrances of plant origin. FAO, Rome

14 Industrial Quality Control

Herbert J. Buckenhueskes
LWB—Lebensmittelwissenschaftliche Beratung,
Hirschstrasse 25, 71282 Hemmingen, Germany

14.1 Introduction

During the last two decades the term "quality" has become one of the most stressed words in the field of food and food production. The facts behind this are, on the one hand, the traditionally different meanings of the word "quality" and, on the other hand, the advanced importance of quality and quality management systems as tools for an economical and safe production of food. "Quality" originates from the Latin language meaning as much as "property" or "characteristic". In relation to food it originally was used as a synonym for "freshness" and "unspoilt".

From antiquity up to now, many philosophers, scientists and economists have tackled with the sense and the meaning of the term "quality", leading to numerous quality models, for example metaphysical, product management, economical, ecological and cognitive approaches. To get an overview of the major quality models, see [1].

In the book *Flavor Science—Sensible Principles and Techniques*, Acree and Teranishi [2] distinguished two different meanings of the word "quality". One meaning is that of an attribute, for example sweet, bitter or floral. The second meaning depends on whether someone likes these attributes, by which quality relates to acceptance and the question how people interact with it.

14.2 Quality and Quality Management Systems

According to the International Organisation for Standardisation (ISO) "quality" is the entirety of attributes and characteristics of a product or a service which are necessary to fulfil its defined or assumed requirements. As already indicated, quality can be seen to be more or less comprehensive, for which reason many organisations define or describe their commitment to quality in a so-called *quality policy* statement. A quality policy typically is based on three fundamental principles:

1. Ensuring that the customer's needs are identified and that these are conformed.
2. Examination of all production and service processes in order to identify the potential for errors and to take necessary actions to eliminate them.
3. Ensuring that each employee understands how to do his/her job and is doing it right.

In order to implement the quality policy in the daily work, quality management systems are installed, covering quality planning, quality control, quality assurance and quality improvement. To ensure that the quality assurance system is in place and effective, external standards are used, for example the DIN EN ISO 9000 ff. standard system, commonly shortened to ISO 9000 (DIN is an acronym for *Deutsches Institut für Normung*, meaning "German Industry Standard").

The best known international quality management standard seems to be the so-called DIN EN ISO 9001 ff. standard. ISO standards are agreements developed by technical committees. Since the members of these committees come from many countries, ISO standards tend to have very broad support. Conformance to that is said to guarantee that a company provides quality services and products. This standard was amended in 2000, so the current standard is called DIN EN ISO 9000:2000.

The most important steps to follow the ISO 9000 standard are:

- Deciding quality assurance policies and objectives
- Formally writing down the company's policies and requirements and how the staff can implement the quality assurance system
- Implementation of the quality assurance system
- Examination of the quality assurance system by an outside assessor to see whether it complies with the ISO standard
- Describing the parts of the standard the company is missing and correction of any problem
- Certification that the company is in conformance with the standard

Beside the ISO standards there are some other standards which are set up by different organisations and sometimes it is really a problem to fulfil the requirements of the different partners in the food market. Not least because of this situation it is not possible here to go in more details. For further information on quality systems and quality management systems see the specific standards as well as the respective literature.

Very special quality demands and quality systems are those for the production and certification of kosher and halal products, which at the time are gaining in importance all over the world. Kashrut is the body of Jewish law dealing with what foods Jews can and cannot eat and how those foods must be prepared. "Kashrut" originates from the Hebrew and means "fit", "proper" or "correct". The more commonly known word "kosher" comes from the same roots and refers to foodstuffs that meet these dietary requirements of Jewish law. "Halal" is an

Arabic word meaning “lawful” or “permitted”, and eating halal is obligatory for every Muslim. The opposite of “halal” is *haram*, which means “prohibited”. Whether a company fulfils the requirements for a kosher or halal production or not can be examined and certificated by specially qualified people or organisations. For further information, see the recently published books concerning kosher [3] and halal [4] production.

14.3 Quality Control

In the frame of a quality management system, quality control is defined as a set of activities or techniques whose purpose is to ensure that all quality requirements are being met. Every raw material used, all intermediate products as well as all flavours and flavouring products which are delivered to the customers have to be controlled by appropriate physicochemical, biotechnological (e.g. enzymic or immunologic procedures), sensory or if necessary microbiological methods. The quality control of flavourings as well as their raw materials is a highly complex field and quality control laboratories in the flavour industry may have more than 500 defined analytical procedures [5].

A prerequisite for any quality control is the definition of how the characteristics of a specific raw material, an intermediate product or a final product of a manufacturing process should be described. This means that all characteristics for every single product have to be defined in adequate standards and specifications so that the results obtained can be compared with these data. Numerous standards and specifications have been established in more or less official specification collections, for example pharmacopoeias, the aforementioned ISO or DIN standards, standards of the Essential Oil Association or the American Spice Trade Organization (ASTA).

According to [6], the main objectives of the quality control in the flavour industry concern the following items:

- *Identity*: Does the raw material delivered or the manufactured product correspond with the order?
- *Purity*: Are the raw materials or the manufactured product free from unacceptable impurities, e.g. filth?
- *Contamination*: Is there contamination, e.g. heavy metals, pesticides, aflatoxins, microorganisms?
- *Adulterations*: Are the raw materials free from adulterations?
- *Quantitatively limited substances*: Are legal regulations concerning limited amounts of specific substances observed?
- *Spoilage*: Ageing and unsuitable storage conditions can lead to quality changes of raw materials or products up to the complete spoilage of the product.
- *Authenticity*: Conformity of the declared and real origin of a raw material. Are materials which are declared as “natural” really natural and not synthetic?

The extensive quality control tests of raw materials, intermediate and final products represent a flood of data which have to be evaluated and documented according to the different aims of the quality control system. Considering the fact, that quality control often has to work under deadline pressure this work can only be done by using powerful electronic labour information and management systems (LIMS).

14.4 Physicochemical Methods

Supported by the overall development in all fields of analysis during the past few decades, a precise analytical methodology has been developed for the different aspects of quality control, comprising physicochemical, biotechnological, sensory and microbiological methods. In order to meet the sense of the quality control system and by that the customers requirements, all methods applied have to be validated by adequate quality assurance tools.

In the frame of this short review it is not even possible to discuss only the major methods and techniques used in industrial quality control in detail, so they will only be summarised here [5, 7].

For sample preparation, isolation and separation traditional methods like distillation (e.g. essential oil content of raw materials) or Soxhlet extraction are still in use. Beyond that, more recent methods are employed, for example supercritical fluid extraction with liquid carbon dioxide.

Even in modern quality control laboratories you will find a number of traditional methods for the identification of single flavour compounds, for example the estimation of optical rotation, refractive index, density and melting point, since these methods are generally accepted, effective and less time-consuming. Especially for the purpose of fast identification checks of more complex systems, spectroscopic methods, above all infrared (IR) and near-IR spectroscopy, are gaining more and more importance.

Numerous analyses in the quality control of most kinds of samples occurring in the flavour industry are done by different chromatographic procedures, for example gas chromatography (GC), high-pressure liquid chromatography (HPLC) and capillary electrophoresis (CE). Besides the different IR methods mentioned already, further spectroscopic techniques are used, for example nuclear magnetic resonance, ultraviolet spectroscopy, mass spectroscopy (MS) and atomic absorption spectroscopy. In addition, also in quality control modern coupled techniques like GC-MS, GC–Fourier transform IR spectroscopy, HPLC-MS and CE-MS are gaining more and more importance.

14.5
Sensory Evaluation

Over the last few decades scientist have developed sensory testing from the earliest individual examinations into a formalised, structured and codified methodology. Subsequently, sensory tests have become valuable, important and precise tools in quality control, which are equivalent to the physical and chemical methods used. However, sensory testing is not only a tool in quality assurance, but also in grading, product development and marketing, as well as for the correlation between specific chemical/physical properties of a food and the effect on the human sensorial perception.

Besides smell and taste, the sensorial evaluation of raw materials and final products covers trigeminal impressions (e.g. hot) and visual impressions like colour, opacity and particle size.

In order to obtain reproducible results, special care must be given to panel selection and panel education, testing facilities, sample presentation and the design of each test. Modern sensory facilities display a kitchen/laboratory for sample preparation as well as separate sensory booths with controlled air and lighting. The evaluation of the sensory results can be supported by specialised computer software packages.

The most frequently used tests in quality control in the flavour industry are paired-sample comparison tests, and triangle tests, which are often combined with the description of deviation from a reference item. For the selection and training of panellists, further test methods are used, for example ranking tests for colour, taste and odour, threshold detections (taste, off-flavour), colour blindness tests and odour identification tests [6].

14.6
Specific Safety Aspects

Over the last few decades, safety has become one of the most important topics related to food. From this view, quality control of vegetable raw materials has at first to cover the following issues: natural and anthropogenic contaminants (e.g. heavy metals, pollution from industrial and private combustions, not professionally deposited waste products, radionuclides), residues of fertilisers (e.g. nitrate), plant-conditioning and plant-protective agents, filth, pests, the microbial status and the occurrence of microbial toxins. It is not possible to discuss all these aspects in detail; however, with a focus on herbs and spices, two of them should be stressed more thoroughly. For further information, see [8].

As background, it has to be remembered that a large number of our spices are imported from less-developed tropical and subtropical countries which often lack the necessary consistency in the application of a quality-oriented cultivation and adequate processing. The hot, humid climate prevailing in many cultivation countries, the mostly simple, unpretentious production conditions, and the often inadequate instruction of farmers and farm hands give rise to fundamental problems [8].

14.7 Microbial Aspects and Microbiological Methods

The fact that food-borne infections and intoxications increased during the last decade of the twentieth century in several European countries underlines the necessity of microbial investigations of food as well as their additives and ingredients [9]. However, compared with the physicochemical methods, microbiological tests play a less important role in the quality control of the flavour industry. Most of the liquid flavours contain solvents like ethanol, propylene glycol or edible oils in concentrations between 70 and 90% so that they possess bacteriostatic or bactericidal properties. Moreover some of the aroma compounds possess the same properties, so routine microbial investigations are not necessary for many of the raw materials and final products [6]. Microbiologically critical products of the flavour industry are above all emulsified, pasty or dry products as well as the agricultural raw materials of animal and plant origin.

According to [10], for microorganisms each particular type of vegetable provides a unique environment in terms of type, availability and concentration of substrate, buffering capacity, competing microorganisms and perhaps plant antagonists. So it is not surprising that every natural plant material harbours numerous and varied types of microorganisms, among which pathogenic or food-spoiling species like salmonella, staphylococci, bacilli, clostridia, enterohaemorrhagic *Escherichia coli* or moulds may occur. In the case of herbs and spices—one of the important sources for the flavour industry—the number and composition of the microflora is above all influenced by the part of the plant used for spice (leaves, seeds, flowers, etc.) and beyond that by harvesting, post-harvest treatment, the drying process as well as storage and transport conditions. In order to give an idea about the microbial load of herbs and spices Table 14.1 surveys the microbial numbers which can be found on untreated ground spices.

Among the pathogenic microorganisms which might be available, special attention has to be paid to the *Enterobacteriaceae*. A salmonella outbreak in 1993 caused by paprika-powdered potato chips showed the importance of these microorganisms even with spices [12]. Burow and Pudich [13] reported that 5.4% of 317 samples of paprika powder and 10.1% of 139 snack products investigated in 1993 and 1994 were salmonella-positive. The differentiation of the isolated salmonellae revealed a number of seldom-found serotypes, which led to the conclusion that the contamination had its origin in the producing countries.

The evaluation of the aforementioned outbreak seems to be of practical consequence for quality control throughout the food industry. On the basis of the results of this evaluation and on the basis of the knowledge that there are strains of salmonella which owing to the existence of plasmids or prophages are able

Table 14.1 Microbial counts (cfu/g) of untreated ground spices [11]

Spice	Total germ count	Coliforms	Moulds
Allspice	1×10^5–5×10^6	$<10^2$–1×10^3	5×10^2–5×10^4
Anise	2×10^5–2×10^6	$<10^2$–1×10^4	$<10^2$–1×10^3
Basil	2×10^4–4×10^5	$<10^2$	$<10^2$
Caraway	1×10^5–5×10^6	$<10^2$–1×10^3	$<10^2$–1×10^4
Cinnamon	2×10^3–5×10^4	$<10^2$	2×10^2–5×10^3
Cloves	5×10^4–1×10^7	$<10^2$	$<10^2$
Coriander	1×10^3–1×10^7	$<10^2$	$<10^2$–5×10^4
Fennel	5×10^4–1×10^5	$<10^2$	1×10^2
Garlic	5×10^4	$<10^2$–5×10^2	$<10^2$–5×10^2
Ginger	1×10^4–1×10^7	$<10^2$	$<10^2$
Laurel (bay)	1×10^3–5×10^4	$<10^2$	2×10^2–2×10^4
Mace	1×10^4–5×10^4	$<10^2$–1×10^3	1×10^2–5×10^4
Marjoram	2×10^5–1×10^6	5×10^2	5×10^3–5×10^4
Oregano	5×10^3–5×10^4	$<10^2$	5×10^2–5×10^3
Paprika	1×10^5–5×10^5	$<10^2$	2×10^2–5×10^2
Pepper black	5×10^5–1×10^7	1×10^2–5×10^4	5×10^2–2×10^4
Pepper white	1×10^4–5×10^5	$<10^2$–1×10^3	$<10^2$–1×10^5
Tarragon	5×10^4	$<10^2$–1×10^3	$<10^2$
Thyme	5×10^5–1×10^7	$<10^2$	5×10^2–1×10^4
Turmeric	1×10^4–2×10^7	$<10^2$–1×10^3	$<10^2$–3×10^3

to change their virulence and thereby their minimal infectious doses, it is assumed today that under special circumstances one single salmonella could be infectious. In this case, any finding of salmonellae in a food has to be regarded as a health risk. Consequently it was decided by the European authorities that salmonella should not be detectable in 25 g of herbs or spices.

As a consequence, the importing and manufacturing companies have established comprehensive and efficient examination procedures for raw material acceptance. The major problem in this relation is the fact that microorganisms as well as other contaminants normally are not homogenously distributed within the product. For that reason detailed sampling plans have been developed; the best known one is the Foster plan [14], which is specialised for salmonella detection.

Industrial microbiological quality control normally covers the total count of germs, the counts of yeasts and moulds, coliform bacteria and *E. coli*. In special cases these investigations are complemented by the detection of *Staphylococcus aureus*, salmonella and listeria.

The German Association for Hygiene and Microbiology has published microbial approximate and warning values for spices which should be given to the consumer or which should be used in the production of foods which do not

Table 14.2 Microbial approximate and warning values for spices which should be given to the consumer or which should be used in foods without any further germ reduction treatment [15]

Organism	Approximate value (cfu/g)	Warning value (cfu/g)
Salmonella	–	Not detectable in 25 g
Staphylococcus aureus	1.0×10^2	1.0×10^3
Bacillus cereus	1.0×10^4	1.0×10^5
Escherichia coli	1.0×10^4	–
Sulphite-reducing clostridia	1.0×10^4	1.0×10^5
Moulds	1.0×10^5	1.0×10^6

undergo further heat treatment. The data are summarised in Table 14.2 and can be used as a guideline for the microbiological assessment of spices. Once again, it can be seen that salmonella should not be detectable in 25 g of the spice.

The traditional microbiological methods are very time consuming and sometimes limited concerning their interpretation. For that reason fast analysis methods as well as automated methods have been developed; the latter are often used in specialised microbiological laboratories. During the last few years more and more modern biotechnological methods have been implemented into quality control, for example the enzyme-linked immunosorbent assay or more recently the polymerase chain reaction, which allows the detection of very specific microorganisms.

The occurrence of moulds on or within vegetable raw materials represents a serious problem, since the topic is of increasing interest and is doubly problematic: for one thing, in many countries mouldy foodstuffs are considered disgusting, regardless of the sanitary risk they present; secondly, the presence of moulds always implies the risk of mycotoxin formation. Besides their acute toxicity, some of the mycotoxins have been found to be teratogenic, mutagenic and/or carcinogenic. The most critical spices concerning the possible occurrence of mycotoxins are coriander, paprika, chillies and nutmeg, a special problem being that moulds growing within capsicum fruits as well as within nutmeg nuts cannot be detected from the outside.

14.8 Residues of Plant-Conditioning and Plant-Protective Agents

For several reasons, residues of plant-conditioning and plant-protective agents represent a highly complex problem. So is it often not quite clear what to look for since for different reasons sometimes substances are used which are not intended for the particular crop. This can occur when the intended preparations are not available or by crossing-over effects from intensively used plantations which may be located close to the often small spice parcels. Otherwise up to now there are no global methods for the detection of residues of plant-conditioning and plant-protective agents available, which is above all owing to the high number of active substances as well as to their multifarious chemical structures.

Another serious problem is the fact that the maximal accepted residues for the different plant-conditioning and plant-protective agents are not harmonised and that, for example, in Germany for some substances the maximum value for residues in herbs and spices is generally set at 0.01 ppm. This value was solely set by a political decision and has no proven toxicological background. Moreover in a number of cases this value is near the lowest detection value of the particular substance.

14.9 Biologically Active Substances

One critical subject concerning the quality of herbs and spices as well as of flavours is the discussion of so-called biologically active substances. Among the estimated 60,000–100,000 different secondary plant metabolites numerous substances can be found which have considerable effects on the human organism, i.e. they are biologically active. This is not astonishing, since most of the vegetables used as herbs and spices today were originally used for their pharmaceutical properties. Since Paracelsus (1493–1541) we know that it is only a question of dose whether something is poisonous or not, and so it is not surprising that among the secondary plant metabolites substances can be found which have to be seen as critical from a toxicological point of view. Often we have to ask the question how to reconcile the obvious contradiction that animal experiments with isolated substances indicate their carcinogenic potential, whereas daily experience obviously does not indicate a serious health risk for humans. Moreover, a recently undertaken comparison of the genotoxicity of estragol with that of tarragon (the plant which contains the highest percentage of estragol) showed that untreated tarragon has a genotoxic potential under the conditions of the test carried out. However, this activity is clearly lower than that observed with methylchavicol at the same concentrations. Dried tarragon, on the other hand, showed no genotoxic potential under the same experimental conditions (study conducted for the European Spice Association, 2005, unpublished).

Since the idea of a zero risk of food in principle cannot be realised, it is not least an ethical request to set the assessment of naturally occurring biological active substances on a new basis corresponding to the rules of a scientifically substantiated risk assessment.

References

1. Dürrschmid K, Zenz H (2000) Ernährung/Nutrition 24:119
2. Acree TE, Teranishi R (1993) Flavor Science—Sensible Principles and Techniques. ACS Professional Reference Book. American Chemical Society, Washington
3. Blech ZY (2006) Kosher Food Production. Blackwell, Oxford
4. Riaz MN, Chaudry MM (2004) Halal Food Production. CRC, Boca Raton
5. Matheis G (1999) Quality Control of Flavourings and Their Raw Materials. In: Ashurst R (ed) Food Flavorings, 3rd edn. Aspen, Gaithersburg, p 153
6. Lösing G, Matheis G, Romberg H, Schmidt V (1997) Dragoco Rep. 42:93
7. Bauer K, Garbe D, Surburg H (1997) Common Fragrance and Flavor Materials. Wiley-VCH, Weinheim
8. Buckenhueskes HJ (1998) Z. Arznei- Gewürzpflanz. 3:21
9. Schmidt K, Kolb H (1996) Verbraucherdienst 41:4
10. Daeschel MA, Andersson RE, Fleming HP (1987) FEMS Microbiol. Rev. 46:357
11. Buckenhueskes HJ, Rendlen M (2004) Food Sci. Biotechnol. 13:262
12. Lehmacher A, Bockemuehl J, Aleksic S (1995) Epidemiol. Infect. 115:501
13. Burow H, Pudich U (1996) Fleischwirtschaft 76:640
14. Foster EM (1971) J. AOAC 54:259
15. DGHM—Deutsche Gesellschaft für Hygiene und Mikrobiologie (1988) Dtsch. Lebensm.-Rundsch. 88:78

15 Advanced Instrumental Analysis and Electronic Noses

Hubert Kollmannsberger, Siegfried Nitz
Lehrstuhl für Chemisch Technische Analyse, TU München,
Weihenstephaner Steig 23, 85350 Freising, Germany

Imre Blank
Science Department, Nestlé Product Technology Center,
1350 Orbe, Switzerland

15.1 Introduction

Since the development of suitable gas chromatographic methods in the 1960s, researchers have been able to identify thousands of volatile compounds in foods, essential oils and fragrances. High-resolution gas chromatography (GC) combined with mass spectrometry (MS) became the key technique used for quantification and identification of flavour compounds. Individual flavours were found to be complex mixtures containing hundreds of compounds in concentrations ranging from percent to trace (nanograms per kilogram) levels. Additional information is obtained by smelling the GC effluent after its separation. This technique is called GC-sniffing or GC-olfactometry (GC-O). To determine the relative sensorial importance of a volatile compound to the overall flavour, odour activity values are calculated by relating the measured concentration of a compound to its odour threshold. Since this does not work with unidentified and trace compounds, other methods using GC-O for the determination of the specific contribution of gas chromatographically separated compounds have emerged: aroma extract dilution analysis (AEDA), combined hedonic aroma response measurements (CHARM) and OSME (a time-intensity rating method) [1, 2]. These sniffing techniques clearly demonstrated that of all these volatile components only a few contribute to the characteristic odour and only in some cases "character-impact compounds" could be found. Therefore, the comprehensive identification of all substances is no longer the main goal. The aim of modern flavour research is now focused on those volatile constituents which either independently or in combination produce a characteristic aroma response. This forced flavour chemists to develop new advanced techniques, both for isolation and concentration as well as for separation and identification.

Besides classical headspace analysis, simultaneous distillation–extraction and solvent extraction, new sampling and enrichment developments include solvent-assisted flavour evaporation (SAFE) [3] and sorptive techniques like SPME solid-phase microextraction (SPME) [4,5] and stir-bar sorptive extraction (SBSE) [6], which are treated in a dedicated chapter in this book. This contribution will deal with advanced developments of GC techniques for improvement of separation and identification (classical multidimensional GC, or

MDGC, and comprehensive two-dimensional GC, or GC×GC), faster separation techniques (fast GC), fast methods for quality assessment or process control in the flavour area ("electronic noses" and fingerprinting MS) and on-line time-resolved methods for analysis of volatile organic compounds (VOCs) such as proton-transfer reaction MS (PTR-MS) and resonance-enhanced multi-photon ionisation coupled with time-of-flight MS (REMPI-TOFMS). The scope of this contribution does not allow for lengthy discussions on all available techniques; therefore, only a selection of developments will be described.

15.2 Multidimensional Gas Chromatography

Single-column gas chromatographic analysis has become the standard approach for separation of volatile and semivolatile constituents in numerous applications; however, this does not necessarily provide the best analytical result in terms of unique separation and identification. There is considerable opportunity for peak overlaps, both on a statistical basis and also on the basis of observed separations achieved for real samples [7–9].

15.2.1 Classical Multidimensional Gas Chromatography

In order to expand the analytical separation, chromatographers have developed a range of solutions based on more than one separation space. Termed MDGC, it consists of an arrangement of two or more columns where distinctive segments of effluent from the first column are fed into one or more secondary columns (Fig. 15.1).

The entire procedure is enabled by the presence of a specific transfer device between the two columns. Cortes [10] and Bertsch [11, 12] have presented a comprehensive discussion of conventional MDGC technology and their contributions are recommended to readers who wish to have a more detailed outline of MDGC and its applications.

In conventional two-dimensional GC (Fig. 15.1a), discrete fractions of effluent (heart cut) are diverted into a secondary column, generally with a different polarity. This arrangement has the disadvantage that components from different cuts may intermingle in the second column and thus can no longer be correlated. Application of parallel traps is one possibility (Fig. 15.1b) to solve this problem. Of course, this is achieved at the expense of increasing the total analysis time, since each fraction delivered to the trap must be individually analysed in a conventional GC procedure. This disadvantage can be circumvented by using the highly complex arrangement shown in Fig. 15.1c, where each cut is directed into a separate column.

The mechanism by which effluent is switched from the first to the second column is critical. Column flows are diverted basically by using valves or valve-

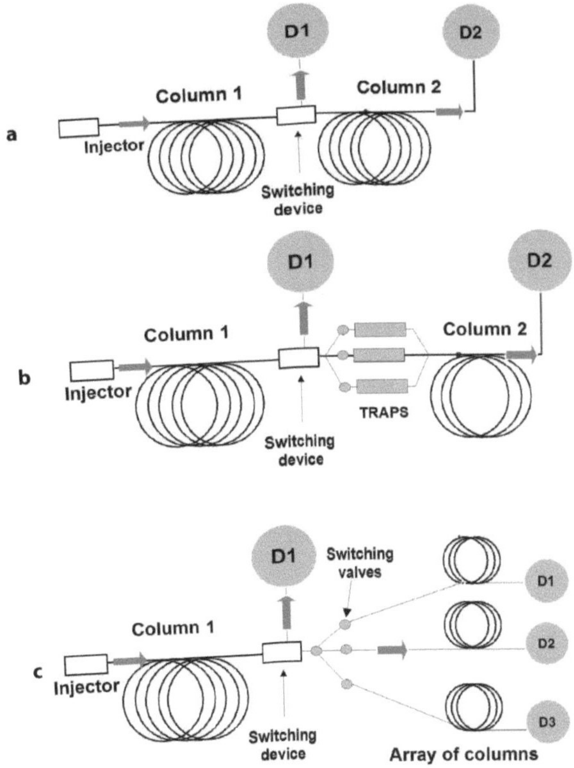

Fig. 15.1 Basic arrangements in multidimensional gas chromatography (*MDGC*): **a** conventional, **b** multitrap, **c** multicolumn

less Deans-type switching devices. Accurate descriptions of the most important MDGC interfaces have been reported in the literature [12]. A truly versatile arrangement is presented in Fig. 15.2 by way of example.

It consists of a dual oven with a valveless Deans-type interface (Live-T switching device) between the columns, an intermediate effluent monitor detector and facilities for flow reversal. Additionally, a device (total transfer) to focus and reinject the desired effluent fraction after the first column allows the combination of a high-capacity precolumn (packed or thick-film wide bore) with a high-resolution capillary or chiral column. Trapping a heart cut before introduction into the second dimension makes the second separation independent of the first. Depending on the application, the instrument incorporates inlets for liquid injection, dynamic headspace enrichment, thermal desorption and facilities for SPME, SBSE and headspace sorptive extraction (HSSE). After suitable enrichment and initial separation on a high-capacity column, appropriate heart cuts are transferred to the main column. After final separation, the substances of interest can be directed via a second Live switching device to an MS, isotope

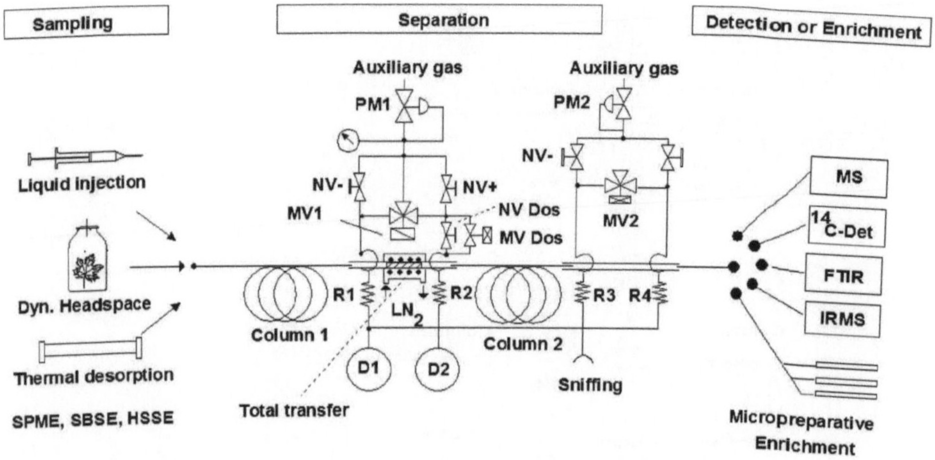

Fig. 15.2 A two-dimensional GC (*GC×GC*) system. *D1* intermediate effluent monitor detector, *column 1* high-capacity precolumn, *column 2* high-resolution capillary or chiral column

ratio MS (IRMS), ^{14}C or Fourier transform IR detector or into individual traps. Accumulation of sufficient material (preparative scale isolation) for further characterisation by off-line instruments such as an NMR spectrometer can be achieved by collection of selected heart cuts from the second column in separate traps after a certain number of GC enrichment cycles [13,14].

Apart from petrochemical and environmental applications [12], classical MDGC was/is used in the flavour field mainly for enrichment and identification of odorous trace compounds in complex mixtures, or for authenticity evaluation by chiral separation or isotopic ratio determination. In Table 15.1 some typical applications are given.

In order to illustrate the potential of MDGC when dealing with complex mixtures, an application to determine off-flavour compounds in defective coffee beans is given by way of example (Fig. 15.3)

As can bee seen, MDGC is a targeted analysis applicable to a specific application. Heart cutting can be effectively applied to a small number of regions of interest in the chromatogram. Transferring them to a second column gives enhanced resolution of the heart-cut zones owing to the different column selectivity. Care should be taken that compounds from one cut do not interfere with the separation of another cut. With respect to the maximum numbers of cuts achievable in a single MDGC run, this is dependent on the sample type and on the analytical conditions. A practical implementation of a large number of heart cuts in order to completely separate a complex mixture in a single run is just not possible. Nevertheless, heart-cutting MDGC may be considered for target applications the most appropriate analytical choice.

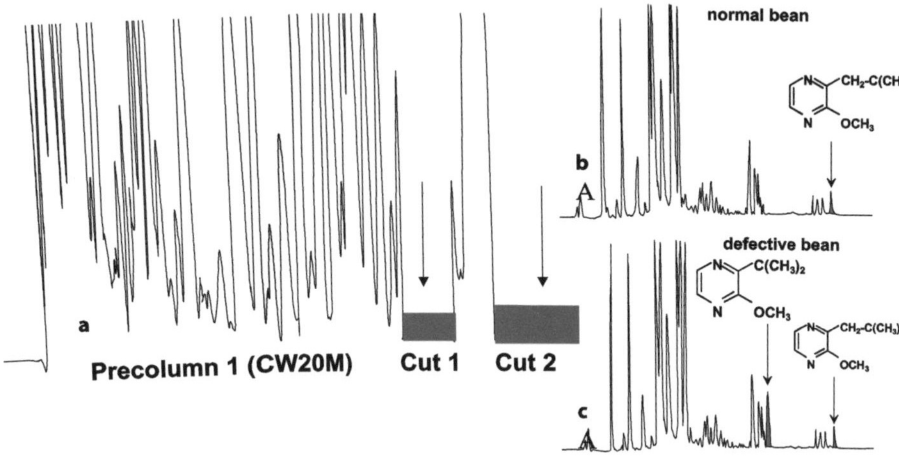

Fig. 15.3 Dynamic headspace MDGC analysis of three coffee beans. **a** Chromatogram of precolumn showing heart cuts at retention of peasy-like odours detected by sniffing the eluent from the precolumn. **b** Heart cut from normal beans on the main column. **c** Heart cut from defective beans on the main column [14]

If the speed of the secondary separation is high enough to separate a cut from the first separation while the next cut is being collected, the complete two-dimensional chromatogram could be constructed. A new type of instrumentation was developed to accomplish this goal. The technique, called comprehensive GC×GC, was introduced in 1991 by Liu and Phillips [30]. It expands the MDGC method into a generally usable format that does not rely on targeting specific zones of a first-dimension analysis.

15.2.2 Comprehensive Two-Dimensional Gas Chromatography

Comprehensive GC×GC uses the whole two-dimensional separation space to generate resolution, provided that the individual dimensions are orthogonal. GC×GC consists of two chromatography columns, serially coupled, with a modulation mechanism at their junction. A typical column set is composed of a standard low-polarity column (first dimension), with typical dimensions of 25 m × 0.25-mm internal column diameter × 0.25-μm film thickness, coupled to a much shorter and more polar second column (second dimension) (or a column providing a separation mechanism capable of further differentiating target sample components), with dimensions of 1 m × 0.1-mm internal column

Table 15.1 Selected applications of multidimensional gas chromatography in flavour research

	Application	References
Enrichment of trace compounds from interfering complex mixtures	Identification of odour-active undecaenes in fruits and vegetables	[15, 16]
	Identification of a peasy off-flavour in coffee beans	[14, 17, 18]
	Identification of musty/earthy off-flavours in wheat grains	[19]
	Identification of a rotten off-flavour in car mats	[14]
Selective transfer of compounds of interest into a second column	Enantioresolution of lactones for authenticity control of fruit products	[20–24]
	Enantioresolution of terpenes for authenticity control in essential oils	[24–26]
	Enantioresolution of nor-carotenoids	[27]
	$^{13}C/^{12}C$ isotope ratio determination for authenticity control of flavours	[23, 24, 28]
Preparative-scale enrichment	Isolation of terpenes from essential oils	[13, 29]

diameter × 0.1-μm film thickness. The modulator interfaces the two coupled columns, and is responsible for the quantitative transfer and compression of all solutes, or a representative fraction thereof, from column 1 to column 2. To accumulate the analyte in narrow bands in the modulator, either elevated temperature (the thermal sweeper) to accelerate the solute into a narrow band or cryogenic means to retard the analyte and cause on-column trapping is used [31]. The operation of a dual-jet cryogenic system is explained in Fig. 15.4.

The resulting very sharp peaks are then released onto the short fast column 2. The modulator actually collects eluent from column 1 every few seconds (generally 2–9 s), and so an individual chromatographic peak is actually sliced into many fragments. Figure 15.5 demonstrates how two overlapping peaks are effectively deconvoluted into two interleaved series of pulses.

Each fragment is focussed and pulsed to column 2 for fast analysis. Because modulation is a mass-conservative process, the peak height increases to accommodate the reduction in peak width; thus, greater analytical sensitivity is obtained. Provided that column 2 can resolve the substances focussed by the

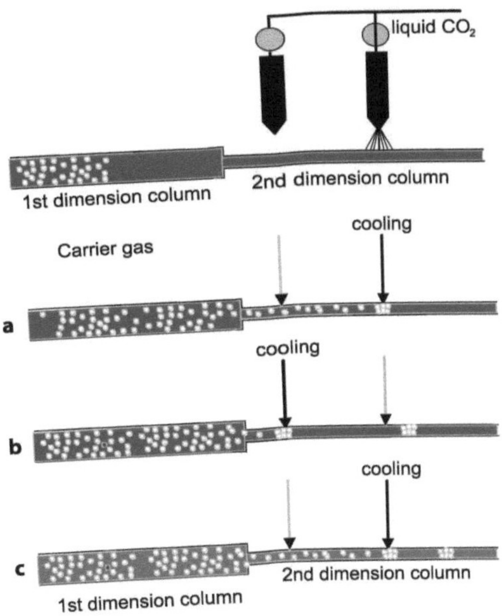

Fig. 15.4 A dual-jet cryogenic modulator. **a** Right-hand-side jet traps analytes eluted from the first column; **b** right-hand-side jet switched off, cold spot heats up rapidly and analyte pulse is released into the second column; simultaneously, left-hand-side jet switched on to prevent leakage of first-column material; **c** next modulation cycle is started (adapted from [32])

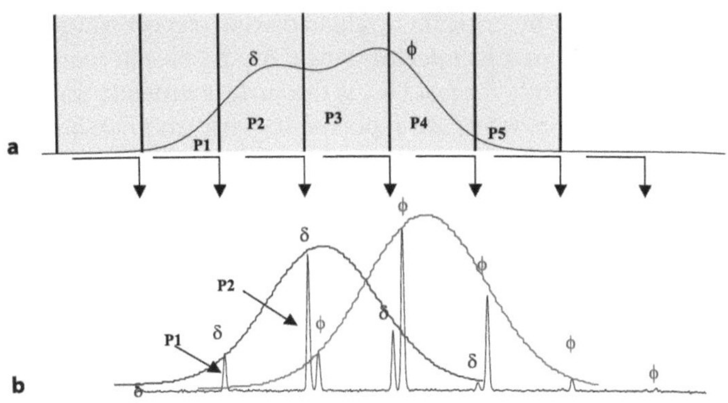

Fig. 15.5 Illustration of how two overlapping peaks δ and φ emerging from the first column (**a**) are resolved in GC×GC after passage to the second column (**b**) [31]. Reprinted from Marriott, P., Shellie, R., Principles and applications of comprehensive two-dimensional gas chromatography. Trends Anal. Chem. (2002), 21:573–583 with permission from Elsevier

modulator, compounds which would coelute under conventional single-column GC (Fig. 15.5a) will be separated because of the modulation process, and different pulses can be assigned to different compounds depending upon their retention times (Fig. 15.5b).

The data are generally presented in a two-dimensional plane, which plots the retention time on column 1 (minutes) against the retention time on column 2 (seconds), or a three-dimensional plot where the detector response is also included. In performing a carefully tuned GC×GC experiment, the peak capacity of the overall separation is approximately equal to the product of the peak capacities of the individual separation steps [33]. Thus, the opportunity to characterise mixtures fully is far greater using GC×GC than for both single-column GC and MDGC.

Regardless of the type of analysis, many variables need to be adjusted for optimal performance. It is generally desirable that chromatography in the second column is complete before another aliquot is transferred from the primary column. The need for rapid elution from the second column sets practical limits. In order to get maximum separation performance, each individual first-dimension peak should be modulated into several fractions (in general some five to ten). For that purpose, the ratio of separation speeds between the second and first dimensions must be at least on the order of 50 [30]. Generally, complete elution within a time frame of 2–8 s is required for column 2. With a half peak width of 0.2 s in the second dimension, and the acceptance that at least ten points per peak half width are required to be suitably measured by a chromatographic detector, fast electronics for detection and data collection are needed. In GC×GC to date, detection techniques employed include flame ionisation detection (FID), sulphur chemiluminescence detection (SCD), atomic emission detection (AED), electron capture detection (ECD), nitrogen chemiluminescence detection (NCD) and MS detection (both time-of-flight MS, or TOFMS, and quadrupole MS, or qMS) [34]. In order to conduct GC×GC, the scan speed of the detector is critical: each detector used must be critically evaluated with respect to operational considerations that may limit or affect performance. A possible alternative detection technology, described recently for fast GC, is the surface acoustic wave (SAW) sensor [35].

Several approaches are reported to perform peak quantification in GC×GC. The most common one integrates all individual second-dimension peaks by means of conventional integration algorithms, and then sums all peak areas belonging to one compound. For another method, firstly a so-called base plane is subtracted, and subsequently three-dimensional peak volumes are calculated by means of imaging procedures. Although the peak capacity of GC×GC is high, peak overlapping in two-dimensional separation is very possible, especially for highly complex samples. Chemometric methods, like the generalised rank annihilation method (GRAM), have been used to resolve and quantify severely overlapped GC×GC peaks. Some other methods have also been used, like curve fitting, wavelet analysis [34] and orthogonal projection resolution [36].

The practicability and potential of comprehensive GC×GC coupled to TOFMS (GC×GC-TOFMS) for the analysis of complex mixtures is illustrated in the following application [37].

Figure 15.6 shows the separation achieved for the essential oil of *Coriandrum sativum* using GC×GC. The identity of the compound was elucidated and confirmed primarily from the MS library matches as well as by comparing the first-dimension retention index with reference libraries. GC×GC-TOFMS allowed the identification of 81 compounds, compared with only 41 compounds identified by conventional GC-qMS.

A great advantage of GC×GC is that homologous series of compounds form linear relationships in the two-dimensional separation plane. Figure 15.7 shows the homologous series of compounds identified in *C. sativum* essential oil. This provides another method to confirm compound identity and allows easy discrimination between series of isomers. For example, (*E*)-2-alkenals and (*Z*)-2-alkenals exhibit very similar mass spectra but are easily distinguished by GC×GC (Fig. 15.7). In addition, many of the heavier compounds were not present in the mass spectral libraries (and so were often misidentified as lower-mass homologues) but were easily identified using their homologous series.

It is important to appreciate that whilst the GC×GC analysis might not be any faster overall than normal capillary GC, within a similar analysis time, higher sensitivity, greater peak resolution (and hence one could expect greater precision of analysis) and a fingerprint pattern that may contain much subtle information on the chemical class composition of samples, which cannot be achieved in any other way, is obtained.

Selected examples of application areas of GC×GC are presented in Table 15.2. For further details about instrumentation and applications of comprehensive GC×GC, the contributions in [32, 38, 39] should be consulted.

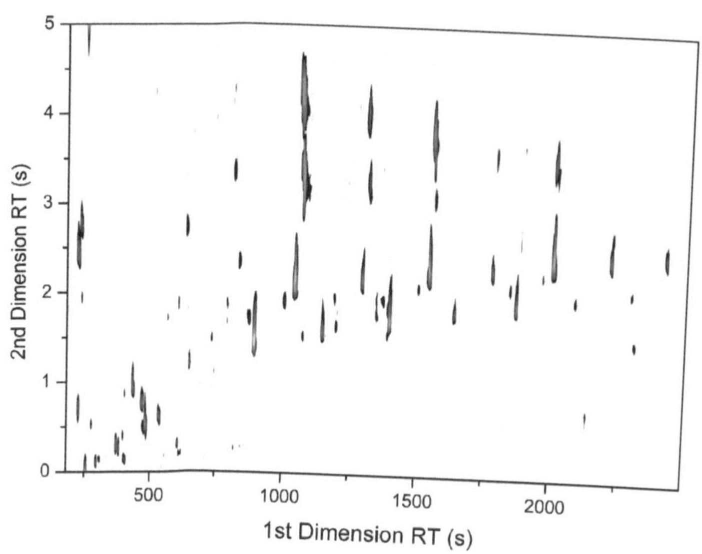

Fig. 15.6 GC×GC time-of-flight mass spectrometry (*TOFMS*) total ion chromatogram of coriander essential oil as a contour plot. Reprinted with permission from [37]. Copyright (2005)

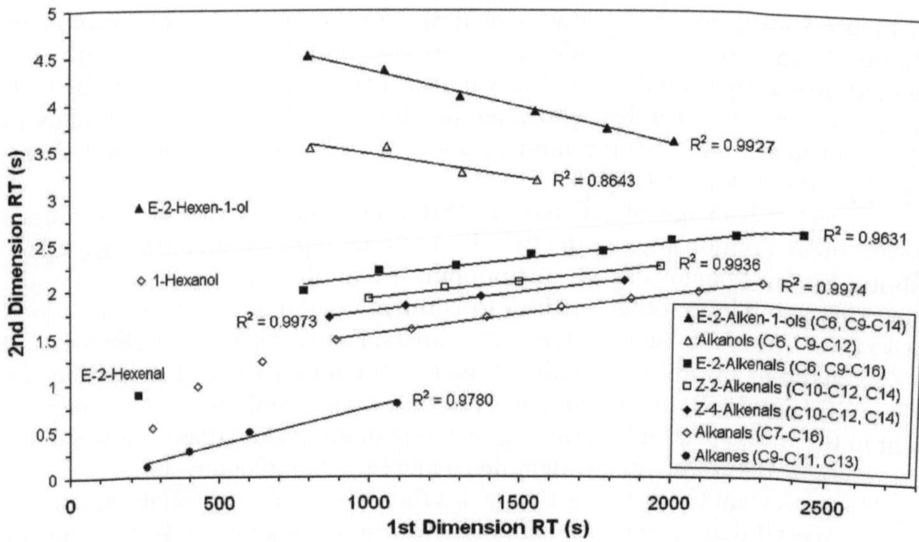

Fig. 15.7 Apex plot showing the homologous series of compounds present in coriander essential oil. The apex of each peak is plotted as the second -dimension retention time against the first-dimension retention time. Reprinted with permission from [37]. Copyright (2005) Wiley

15.3 Fast Gas Chromatography

Conventional GC allows effective separations of complex natural mixtures, but this is frequently achieved at a high cost in time. This becomes a limiting factor, especially for laboratories with a high sample throughput and/or where the need for quick results for the determination of quality and authenticity are required. In recent years there has been increasing interest, within the chromatographic community, towards the development of faster separation methods without considerable loss of resolving power. Various approaches have been theorised and developed with various proposals: shorter column lengths [66], reduced internal column diameter and stationary phase thickness (narrow-bore column) [67, 68], microparticle-packed capillary columns [69, 70], multicapillary columns [71], vacuum-outlet conditions [72], turbulent flow [73] and helically coiled columns [74]. Reviews describing the most important existing high-speed GC methods, also trade-offs and compromises in terms of sensitivity and/or selectivity in combination with MS, have been published [75, 76].

The narrow-bore column approach is a very effective and is the most popular way of increasing analysis speed. Substantial reductions in analysis times are achieved by exploiting two factors: a shorter column length and the application of higher than optimum average linear velocities. Operating under optimum

Table 15.2 Selected applications of comprehensive two-dimensional gas chromatography

Plant constituents	Enantiomeric alkaloids [40]
	Volatiles from germander [41]
	Tobacco essential oil [44]
	Lipids in lanolin [50]
	Coriander leaves essential oil [37]
	Pistacia vera essential oil [58]
	Sandalwood oil [55]
	Tea tree and lavender essential oils [60]
	Origanum micranthum essential oil [61]
	Hop essential oil [65]
Food	Flavour compounds in butter [48]
	Volatiles in strawberry cultivars [54]
	Roasted coffee bean volatiles [56]
	Methoxypyrazines in wine [51]
	PCBs in milk and cheese [53]
	Trace odorants in sour cream [63]
	Fatty acid composition in foods [62]
	Citrus essential oil [64]
	Yeast cell metabolites [45]
Fragrances	Perfume analysis [52]
	Allergens in fragrances [59]
Human breath	Volatile organic compounds [42]
Petrochemicals	Oil spill [43]
	Composition [46]
	Diesel fuel hydrocarbons [47]
	Sulphur compounds in crude oil [57]
Environmental	Polychlorinated alkanes in dust [49]

experimental conditions, a 10 m × 0.1-mm internal diameter, 0.1-μm film thickness column is characterised approximately by the same resolving power as a 25 m × 0.25-mm internal diameter, 0.25-μm film thickness column [67]. Figure 15.8 shows the chromatograms of conventional GC and fast GC analysis of a lemon oil with the aforementioned columns [77].

As can bee seen, 57 components were separated with both methods. The fast GC technique performs the same separation within 9 min, a speed gain of a factor of 5 compared with the conventional method. A lime oil sample, in a application aimed at quality control, was separated satisfactorily in only 90 s on a 5 m × 0.5-mm internal diameter, 0.05-μm film thickness column [78]. Other applications include essential oil analyses [79,80], flavour volatiles in fruits [81], fatty acid composition [82] and pesticides [83].

Fast GC requires instrumentation provided with high split ratio injection systems because of low sample column capacities, increased inlet pressures, rapid oven heating and fast electronics for detection and data collection. Hydrogen is generally used as a carrier gas because of the flatness of its Van Deemter height equivalent to the theoretical plate (HETP)–μ curve, which allows higher linear gas velocities to be applied than the optimum without substantial loss of resolution. Shorter columns, thinner films and smaller internal diameter columns used in fast GC require smaller amounts of sample to be injected to prevent overloading of the column. This in turn causes the detection limits to be higher in fast GC. This is a problem when working with trace levels of analytes.

It has been shown that using fast temperature programming is a better way than using faster flow rates to decrease the analysis times [84]. This parameter has been ignored in many studies, but it offers valuable time savings with some added benefits. Shorter columns with typical internal diameters (e.g. 0.25 mm) and film thicknesses can be used, without much loss in sample capacity.

It is important to ensure that the data collection rate is fast enough for peaks with low retention times in order to ensure good reproducibility of all peak parameters. For modern instrumentation, this is generally not a problem; for example, FID detectors are typically able to achieve a data acquisition rate of 50–250 Hz using the standard instrument configuration.

The data sampling rate of a typical quadrupole benchtop mass spectrometer is not always fast enough for very fast GC analyses, especially for quantitation. Typical spectral acquisition rates of scanning mass spectrometers, such as the ion trap, the quadrupole and the sector instruments, are limited to a maximum of ten to 20 spectra per second in the full-scan mode. This is just on the edge of applicability for fast GC. If faster detection is required, non-scanning TOF analysers are an alternative. TOFMS can provide up to 500 full spectra per second and allow accurate detection of peaks with peak widths in the millisecond range (very fast GC), while still providing high-quality spectra [85]. In terms of ultimate potential instrument performance, sensitivity is sacrificed for gains in speed in TOF [76]. In the literature, much of the discussion about fast GC-MS originates from the chromatographer's point of view, and a chromatographer tends to prefer baseline resolution between peaks. Although more selectivity in the separation can be beneficial in some respects, in other respects the time spent to resolve coeluting compounds by GC is wasted if the compounds can be adequately resolved by the MS detector. Mass-spectral deconvolution software is an effective and efficient tool to resolve coeluting peaks in GC-MS and thus is very important for fast GC-MS. Humans simply cannot conduct adequate background subtraction in a complex chromatogram, and a highly trained person could spend hours trying to do what an adequate deconvolution program can do in seconds. The future usefulness of fast GC-MS depends to some extent upon improvement of existing approaches and commercialisation of interesting

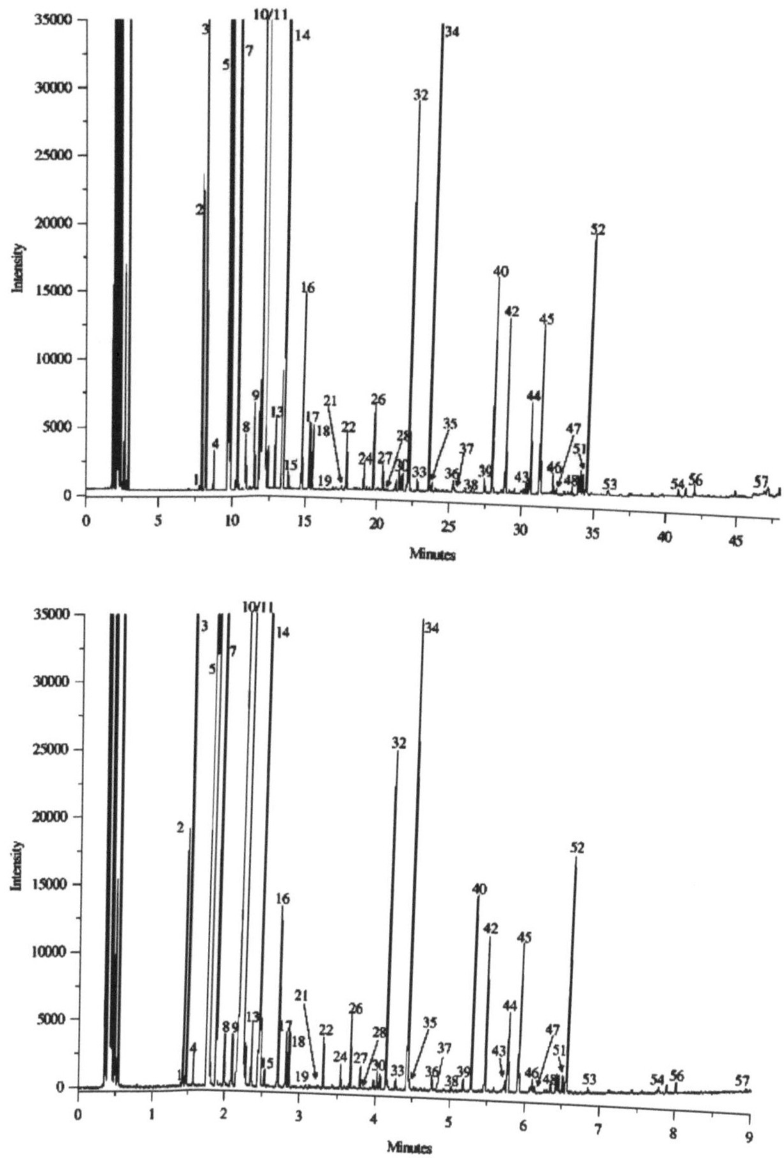

Fig. 15.8 Conventional GC (**a**) and fast GC (**b**) chromatograms of a lemon essential oil [77]. Copyright (2003) American Chemical Society

new techniques. Moreover, a greater emphasis is needed to rationalise overall laboratory operations and sample preparation procedures if fast GC-MS is to become implemented in routine applications [76].

15.4 Electronic Noses

In the food industry there exists an increased demand for fast, simple and sensitive methods of assessing volatiles, for identification, authentication, process control, and product blending or formulation. Since the quality of raw materials and processed products is determined frequently by the volatiles characteristic for a particular odour, objective methods for aroma and flavour evaluation are needed. By this means contaminations or off-flavours of such products could also be detected. Actually for this purpose human sensory panels of trained experts are used in the food and aroma industry, complemented by more objective, but time-consuming analytic methods, such as GC. In order to be able to interpret the laboratory analysis, those GC methods often have to be combined with MS or simultaneously used "sniffing" lines. However, even the most sophisticated analytic methods cannot fully replace the human nose, as only our olfactory sense can determine whether a compound is relevant to a specific odour. Therefore, it is not surprising that repeated efforts have been made over the years to introduce instruments operating on a similar principle to the human nose. The instruments comprise an array of electronic chemical sensors with partial specificity and an appropriate pattern-recognition system, capable of recognising simple or complex volatile mixtures. These systems would in most cases not replace but would complement conventional analyses of volatile compounds by sensory methods and by classic analytical techniques. The arrangements realised are often called "electronic noses" or "artificial noses" although the only aspect in common with our odour-sensing organ is the function. The operating principle, the number of sensors as well as the sensitivity and selectivity are, however, very different. This is why they should better be called "multisensor array technology".

Within the last few years, gas sensors together with associated pattern-recognition techniques have been used to differentiate and identify complex mixtures of volatile compounds. Potential areas of application include the food industry [86], perfumery, the chemical industry, the pharmaceutical industry, the tobacco industry [87], cosmetics [88], health and environmental control [89], where they are intended to be used for quality control of raw and manufactured products, monitoring of freshness and maturity, evaluating process effectiveness, prediction of shelf life, microbial pathogen detection [90, 91] and authenticity profiling. The major categories involved in the development of electronic noses and the principles of sensor design and technology will be described in the following section.

15.4.1 Catalytic or Metal Oxide Sensor

A catalytic or metal oxide sensor (MOS) consists of an electrically heated (250–450 °C) ceramic pellet upon which a thin film of tin(II) oxide doped with precious metals is deposited [92]. Tin(II) oxide is an n-type semiconductor and when oxygen adsorbs on the surface, one of the negatively charged oxygen species is generated, depending on the temperature. This results in the surface potential becoming increasingly negative and the electron donors within the material become positively charged. When an oxidisable material comes into contact with the sensor surfaces, the adsorbed oxygen is consumed in the resulting chemical reaction. This reduces the surface potential and increases the conductivity of the film. As a result the electrical resistance of the sensors will change. Disadvantageous are the relatively poor selectivity, which can be to some extent improved by dopants and temperature adjustment during the measurement, sensor drift over time, and finally the high power-consuming operation temperature.

15.4.2 Metal Oxide Semiconductor Field-Effect Transistor

The metal oxide semiconductor field-effect transistor (MOSFET) sensor device is based on a field-effect transistor with a catalytic metal as the gate contact. The gate voltage controls the current through the MOSFET device. The gas molecules will affect the voltage to the gate contact and thus change the current through the transistor. In a field-effect sensor, the interaction of gases with the catalytic gate metal induces dipoles or charges, which give an additional voltage to the gate contact. The choice of operation temperature, type of catalytic metal and structure of the metal influence the chemical reactions on the gate of the sensor, and thus the selectivity and sensitivity of the sensor.

15.4.3 Conducting Polymer Sensors

Polymer materials like polypyrrole and polyaniline are conducting (or semiconducting) and show a variation in conductivity with sorption of different gases and vapours. The sensor response is not necessarily a linear relationship between the analyte concentration and conductivity. Owing to their molecular structure, they show good sensitivities to polar compounds. The sensors display rapid adsorption and desorption at room temperature and specificity can be achieved by incorporating different metal ions in the structure of the polymer. Disadvantages include the reproducibility of fabrication (poor batch-to-batch reproducibility), strong humidity interference and the baseline drift over time owing to oxidation processes or changes in the conformation owing to exposure to inappropriate compounds.

15.4.4 Acoustic Wave Sensors

One of the first sensors to be introduced was the thickness-shear mode (TSM) sensor, which, if the substrate is quartz, may commonly be termed the quartz crystal microbalance (QMB) or bulk acoustic wave (BAW) sensor. The sensor consists of overlapping metal electrodes at the top and bottom. This type can be used with up to 10-MHz fundamental resonance frequency with a standing resonant wave being generated where the wavelengths are related to the thickness. As the thickness increases (e.g. owing to added mass by deposition on the surface), the wavelength increases and the frequency decreases. Mass-sensitive devices, such as QMB or SAW oscillators, can detect a change of mass accurately via resonance frequency shifts. In contrast to other sensor technologies, these kinds of transducers generate a fully digital electrical output signal with all its advantages to further signal processing (e.g. no analogue to digital conversion, fewer electromagnetic compatibility problems). In a SAW device, an acoustic wave propagates along the surface, whereas in a QMB crystal the acoustic wave propagates in the bulk. Compared with the SAW devices, the resonance frequency of QMB sensors is an order of magnitude lower, allowing a wider tolerance for temperature control of the oscillator electronics. When coated with gas-sensitive layers, both devices can detect gases or vapours. In contrast to SAW sensors, which use ultrathin layers to avoid damping of the oscillation, QMB sensors can also be coated with bulky layers. These coatings should have a high vapour permeability. SAW sensors can also be operated in the liquid phase and are theoretically more sensitive owing to their higher resonance frequency. In practice this advantage can often be compensated by using thicker coatings on the QMB devices. A large variety of chemical-sensitive materials can be deposited onto mass-sensitive devices. The long-term characteristics of these devices are mainly dependent on the ageing and/or bleeding of coating materials. Stationary phases used in GC columns (e.g. silicone, carbowax) have been optimised in respect to these features over the last 20 years. A QMB sensor, coated with these polymers, shows reproducible and stable behaviour for a wide range of chemical substances. The abilities and limitations of this type of sensor have been described in detail recently [93].

15.4.5 Mass Spectrometry Based Systems

While most of the commercially available gas sensors are based on one of the aforementioned four major sensor technologies, MS sensors are based on a measuring technique well known and widely used for almost 30 years: MS.

Volatile sample components are introduced into the mass spectrometer without separation, thus creating a mass spectrometric pattern of fragment ions that describes the mixture of volatiles in the headspace [94–96]. Each fragment ion represents a potential sensing element and the intensity of the fragment ion is equivalent to the sensor signal. Theoretically, when performing a full-scan measurement from, for example, m/z 50 to m/z 300, one can choose up to 251 sensors to form a sensor array. However, it is not useful to work with such a great number of sensors, as in most cases only a very small number of fragment ions is needed for setting up a sensor array. The choice of fragment ions that should be selected to build up an array is based on knowledge and has to be correlated with the sample properties to be determined.

Ion mobility spectrometry (IMS), which has the ability to separate ionic species at atmospheric pressure, is another technique that is useful for detect and characterising organic vapours in air [97]. This involves the ionisation of molecules and their subsequent drift through an electric field. Analysis is based on analyte separations resulting from ionic mobilities rather than ionic masses. A major advantage of operation at atmospheric pressure is that it is possible to have smaller analytical units, lower power requirements, lighter weight and easier use.

Other MS-fingerprinting techniques that are in commercial development are based on atmospheric pressure ionisation (API), resonance-enhanced multiphoton ionisation (REMPI) TOF and proton-transfer reaction (PTR). They are rapid, sensitive and specific and allow measurements in real time and may play an increasingly important role in the future development of electronic noses and tongues.

15.4.6 Other Sensor Technologies

Apart from the aforementioned most frequently used sensor technologies, also selective electrochemical sensor combinations have been commercialised for use in dedicated applications. The combination of electrochemical CO, H_2S, SO_2 and NH_3 sensors was used for quality and freshness control of foods like fish [98] and meat [99]. Combinations of MOSs and MOSFETs supplemented with a selective IR absorption sensor for carbon dioxide and a humidity sensor for measuring relative humidity were also described [100].

One of the technologies which may become relevant for gas sensing is bundled fibre optics, through which fluorescence is measured from photodeposited polymer-sensing elements. On one end of a fibre optic bundle, as many as 30 small regions of polymer fluorescent dye mixtures are photodeposited. A flash of light at an excitation wavelength is applied to the other end of the fibre optic, and fluorescence intensity at selected wavelengths from the polymer/dye mix is subsequently measured back through the fibre optic. Different polymer/dye combinations interact with gases differently, such that upon exposure to a given sample, the different regions or sensors provide unique information [101].

Advances in the production, immobilisation and characterisation of mammalian olfactory receptors led to the development of biosensors where isolated olfactory binding proteins were deposited on the surface of QMBs [102, 103] or were connected to nanoelectrodes [104]. Although still at the development stage, such an array-type device coated with different olfactory receptors will be a powerful and useful tool for detecting and discriminating odorants in the future.

15.4.7 Data Processing

A generalised structure of an electronic nose is shown in Fig. 15.9. The sensor array may be QMB, conducting polymer, MOS or MS-based sensors. The data generated by each sensor are processed by a pattern-recognition algorithm and the results are then analysed. The ability to characterise complex mixtures without the need to identify and quantify individual components is one of the main advantages of such an approach. The pattern-recognition methods may be divided into non-supervised (e.g. principal component analysis, PCA) and supervised (artificial neural network, ANN) methods; also a combination of both can be used.

PCA reduces multidimensional, partly correlated data, to two or three dimensions. Projections are chosen so that the maximum amount of information is retained in the smallest number of dimensions. This technique allows the similarities and differences between objects and samples to be better assessed [105].

A neural network is a program that processes data like (a part of) the nervous system. Neural networks are especially useful for classification problems and for function approximation problems which are tolerant of some imprecision, which have lots of training data available, but to which hard and fast rules (such as laws of nature) cannot easily be applied.

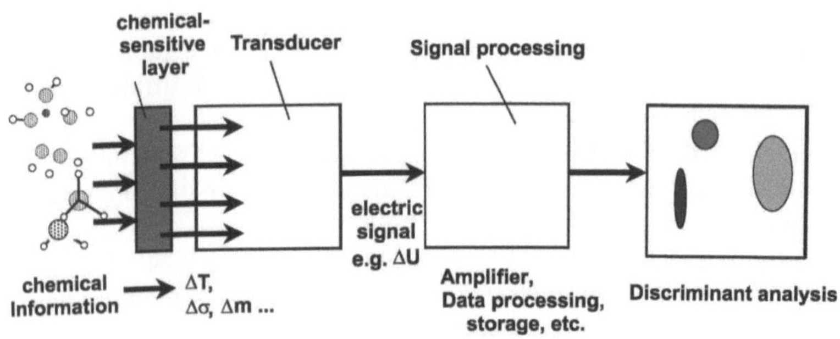

Fig. 15.9 Generalised structure of an "electronic nose"

Neural networks are trained with complete data sets consisting of input and output data. Typically one starts with a random configuration and calculates output data from the given input data. One compares the calculated output data with the output data of the complete data set and tries to minimise the error of the output data by varying the parameters of the network When the network has learned the complete data sets, one takes an independent collection of complete data sets to test the generalisation capability of the network. Both collections of complete data sets must be large enough and correctly distributed within the range of possible data. Perhaps the greatest advantage of ANNs is their ability to be used as an arbitrary function approximation mechanism which learns from observed data. However, using them is not so straightforward and a relatively good understanding of the underlying theory is essential. The danger is that the network overfits the training data and fails to capture the true statistical process generating the data, resulting in worse predicting ability [106].

15.4.8 Applications, Potential and Limitations

Most publications deal with the application to foods (Table 15.3), but published studies are also available covering other products, such as tobacco, cosmetics, health diagnostics and the environment [86].

The feasibility and limitations of using multisensor array systems in food and aroma applications will be discussed with an application intended to discriminate hop varieties [149] by way of example

The sensor responses generated in a measurement result from physical and/or chemical interactions between the sensors and the volatile compounds present in the headspace above the measured sample. By using a QMB sensor system with an array of six sensors, good discrimination between three hop varieties can be observed (Fig. 15.10a). In this example only 12 measurements per sample were analysed. The distance between clusters is reduced if the data set is increased to 50 measurements per sample (Fig. 15.10b).

The reason for this effect has to be attributed to a better and adequate ratio between sample size and array dimensionality. For a significant clustering of the patterns, with an array of six sensors a sample size of at least 18 is required [149, 184]. As a consequence, the discrimination based on only 12 measurements has poor statistical relevance. Most of the applications with sensor arrays found in the literature do not consider this fact; frequently discriminations with 12–32 sensors in an array and with a sample size of three to four are described. All of them are of limited feasibility with concurrent poor validation, especially in terms of reproducibility and predictive ability. In other words, if there are not enough calibration measurements one can separate data in a predetermined way, but will fail to verify the result using independent test samples.

A great disadvantage of MOS, MOSFET, conducting polymer and QMB sensor arrays is that system-to-system matching is not possible in practice, as can be seen in Fig. 15.11. A discrimination of hop varieties with QMB and MOS arrays is partly obtained, but the results are not comparable. The hop variety Tradition, for example, is well separated from other varieties with the MOS sensor, whereas with the QMB sensor overlapping with other varieties is observed. The different responses and sensitivities of the sensors for chemical compounds is the reason for non-comparable discriminations when different sensor systems are used; therefore, a standardisation of these electronic noses is not possible.

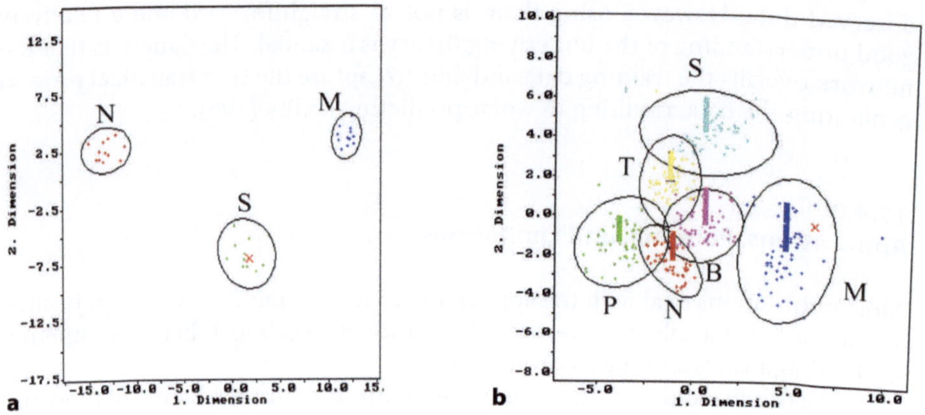

Fig. 15.10 Discrimination of hop varieties with six quartz crystal microbalance (*QMB*) sensors with 12 (**a**) and 50 (**b**) measurements per sample. *N* Nugget, *S* Select, *M* Magnum, *P* Perle, *T* Tradition, *B* Northern Brewer

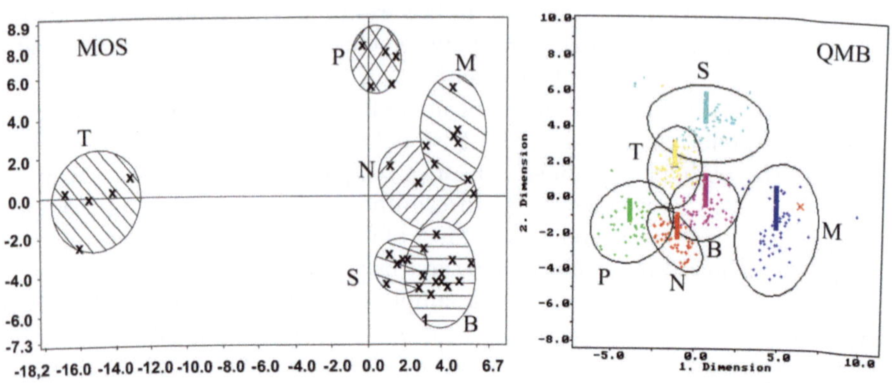

Fig. 15.11 Discrimination of six hop varieties (see Fig. 15.10) by means of metal oxide sensor (*MOS*) and QMB sensor arrays

The interesting question now is whether the sensors classified the different hop samples on the basis of interaction with substances in the headspace which are highly correlated with the varieties. The GC analyses clearly show that myrcene is the main component in the headspace and is therefore responsible for the signals generated by the sensors. But the content of myrcene in hops is highly dependent on climatic, soil, growth and processing conditions, and cannot be regarded as a specific indicator for a variety assessment. Therefore, not the different hop varieties, but the different content of myrcene in the samples was discriminated. Frequently, authors tend to correlate an observed discrimination with the property they wanted to measure, without any additional chemical information. Electronic noses as commonly employed do not allow for chemical differentiation. Owing to the unspecific nature of the sensors, the reasons for a successful discrimination of samples are usually unknown.

In order to correlate a discrimination to the different varieties, a sensor system that selectively interacts with variety-specific compounds in the headspace is needed. The GC analysis of the essential oil reveals that there are some minor volatile compounds, which can be used for a differentiation of different hop varieties (e.g. Nugget and Tettnanger), as shown in Fig. 15.12.

By choosing appropriate fragment ions a virtual sensor array based on a mass spectrometer in single ion monitoring mode can be implemented to discriminate the hop varieties (Fig. 15.13) without appreciable interference of the main component myrcene. In contrast to the analyses performed with QMB and MOS sensors, the discrimination obtained with the MS sensor is based on chemical knowledge and not on assumptions.

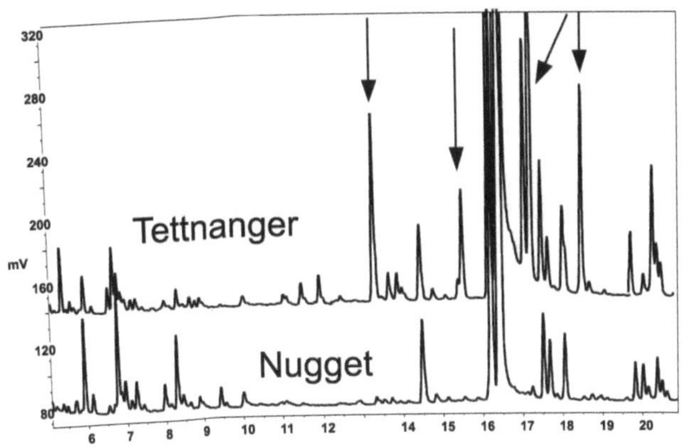

Fig. 15.12 Expanded region of a GC chromatogram of the hop varieties Nugget and Tettnanger

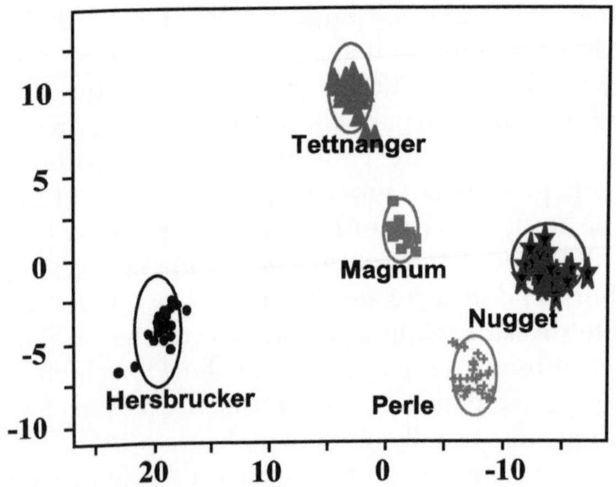

Fig. 15.13 Discrimination of hop varieties with a mass spectrometry based sensor array

15.4.9 Conclusions

The electronic nose technology applied to food must be regarded as being in its early stage. There is rapidly advancing research and development going on both for sensors and instrument hardware and software in order to enhance selectivity, sensitivity and reproducibility of the gas sensors. Much effort is also being put into solving the drift problem of the sensors by increasing their stability and lifetime, and into developing improved mathematical algorithms for drift counteraction, automatic calibration and standardisation, as well as transferability between gas sensor array instruments.

The performance of common multisensor arrays is ultimately determined by the properties of their constituent parts. Key parameters such as number, type and specificity of the sensors determine whether a specific instrument is suitable for a given application. The selection of an appropriate set of chemical sensors is of utmost importance if electronic nose classifications are to be utilised to solve an analytical problem. As this requires time and effort, the applicability of solid-state sensor technology is often limited. The time saved compared with classic analytical methods is questionable, since analysis times of electronic nose systems are generally influenced more by the sampling method utilised than the sensor response time [185].

Common electronic noses are so called as they are often aimed at detection of odorous compounds; it is generally not clear that discriminations are based on odorous rather than non-odorous, and possibly incidental, components of the headspace. In the headspace of a food sample, odorants contributing to the flavour may be present in low concentrations, whereas non-odorous molecules can be present in much larger numbers and higher concentration. In such cases,

Table 15.3 Selected applications of electronic nose systems to different food products

Product	Type of application	Sensor technology	References
Grains	Off-odour caused by microbial infection	MOS	[107]
		MOSFET, MOS, IR	[108, 109]
		CP	[110]
Meat	Quality assessment, lipid oxidation, fermentation, storage spoilage/shelf life	MOSFET, MOS	[111, 112]
		MOSFET, MOS, IR	[110, 103]
		CP	[114, 115]
		CO, H_2S, SO_2, NH_3	[99]
Fish	Quality assessment, lipid oxidation, spoilage, freshness, storage	MOSFET	[116]
		MOSFET, MOS	[117]
		MOS	[118–120]
		CP	[121, 122]
		CO, H_2S, SO_2, NH_3	[98, 123]
		QMB	[123, 124]
		IMS	[97]
Dairy products	Flavour quality, cheese characterisation, heat treatment, flavour differences, off flavours, microbial contaminants	BAW	[125]
		MOS	[126, 131–133, 139]
		CP	[127, 129, 134, 135], [137, 138]
		MOS, CP, QMB	
		MOS, CP	[128, 130, 136, 140–144]
		MS	
Fruits	Flavour quality, harvest dates, storage, maturity, processing	QMB	[145]
		MOS, MS	[146]
		MOS	[147–149]
		CP	[150]
Alcoholic beverages	Wines, spirits, origin, variety, barrel ageing, cork taint	MOS	[151]
		MS	[152–155]
		MOS, CP, QMB	[156]
Beer and hops	Characterisation of aroma, ageing, raw materials, hop varieties	MS	[157]
		MOS	[158]
		CP	[159, 160]
		QMB, MOS, MS	[149]
Spices	Characterisation, differentiation, composition of mixtures, microencapsulation, γ-irradiation	MOSFET, MOS	[161]
		CP	[162]
		MOS	[163, 167, 168]
		QMB	[93, 164, 165]
		MS	[94, 166]
Olive oil	Oxidation, rancidity, vinegary defects, distinguish different qualities, shelf life, geographical origin?	MOSFET, MOS	[169]
		SAW	[170]
		MOS	[171, 172]
		CP	[173]
		CP, MOS, MOSFET	[174]
		MS	[175]
Coffee	Discrimination, roasting	CP	[176]
		MOS	[177, 178, 180]
		REMPI-TOFMS	[179]
Packaging	Retained solvents, printing inks, colouring agents, foil adhesives	MOS, CP, QMB	[181]
		QMB	[182]
		MOSFET, MOS, IR	[183]
		QMB, MOS, MS	[149]

MOS metal oxide sensor, *MOSFET* metal oxide semiconductor field-effect transistor, *IR* infrared, *CP* conducting polymer, *QMB* quartz crystal microbalance, *IMS* ion mobility spectrometry, *BAW* bulk acoustic wave, *MS* mass spectrometry, *SAW* surface acoustic wave, *REMPI-TOFMS* resonance-enhanced multiphoton ionisation time-of-flight mass spectrometry

MS-based systems have considerable advantages over the commonly used gas-sensor arrays, particularly in terms of selectivity, adaptability, sensitivity and standardisation. Array selection and deselection can be done rapidly by changing the scanning method and/or simply by changing the fragment ions used for pattern recognition. Furthermore, taking into account that fragment ions contain chemical information about the sample, the information that can be obtained with an electronic nose improves substantially with the MS-based devices. An optimal instrument configuration also allows the same instrument to be used as a rapid screening tool (electronic nose) and also as a research tool for revealing further chemical information about doubtful samples. A great advantage is obtained with soft ionisation methods like PTR-MS or REMPI because molecular information is easier to deconvolute in the case of overlapping fragments or parent ions.

Although common electronic noses are generally not suitable for odour assessment, they can be successfully used in applications where the main components in the headspace are directly correlated with the property to be determined (e.g. quality of spice mixtures) or in cases where substances are formed and released into the headspace, for example owing to oxidation processes, fermentation, microbial contamination, thermal treatment, etc.

15.5 Time-Resolved Analysis of Volatile Organic Compounds

Over the last decade, interest in release and delivery of VOCs has been steadily growing, with a particular focus on food, environmental and medical applications [186–190]. Consequently, considerable effort was invested to develop analytical methods capable of capturing such dynamic VOC release processes (Fig. 15.14) [179, 191]. This led to improvements in electronic sensor methods (often termed "electronic noses") [192].

One other approach is direct-inlet MS. A prerequisite for mass analysis is ionisation, a process that critically influences the sensitivity and selectivity of the experiment. Electron impact ionisation (EI) causes considerable fragmentation. Because of overlapping fragment and parent ions, the molecular information is difficult to deconvolute, and little chemical information can be extracted.

Therefore, application of direct-inlet MS for monitoring complex mixtures of VOCs requires using ionisation techniques which produce little or no fragmentation (soft ionisation). Chemical ionisation in combination with a quadrupole mass filter, either in atmospheric pressure chemical ionisation MS (APCI-MS) [188, 189] or in PTR-MS [193–195], have been successfully applied to VOC analyses. The advantages and limitations of direct-inlet MS with soft-ionisation approaches have been discussed [196].

One particularly well-performing technique is PTR-MS [193–195]. On-line trace-gas analysis by proton transfer [197] has become a powerful approach, mainly owing to the higher sensitivity and lower ionisation-induced fragmen-

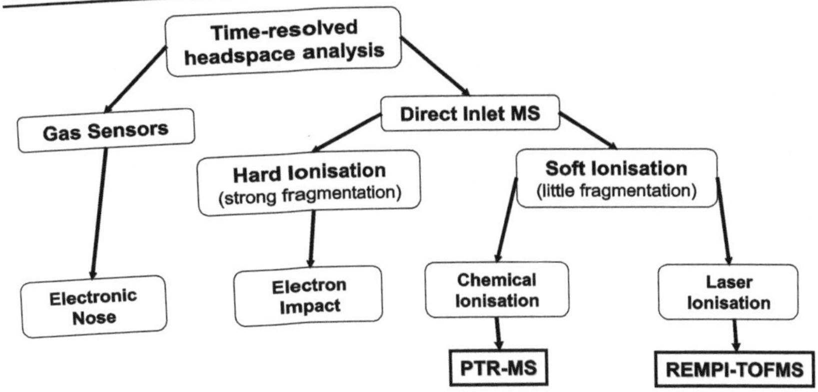

Fig. 15.14 Analytical techniques for time-resolved headspace analysis. An electronic nose can be used as a low-cost process-monitoring device, where chemical information is not mandatory. Electron impact ionisation mass spectrometry (*EI-MS*) adds sensitivity, speed and some chemical information. Yet, owing to the hard ionisation mode, most chemical information is lost. Proton-transfer-reaction MS (*PTR-MS*) is a sensitive one-dimensional method, which provides characteristic headspace profiles (detailed fingerprints) and chemical information. Finally, resonance-enhanced multiphoton ionisation (*REMPI*) TOFMS combines selective ionisation and mass separation and hence represents a two-dimensional method. (Adapted from [190])

tation relative to EI, the latter preventing efficient on-line trace-gas analysis of volatile mixtures via direct MS. In contrast, fragmentation by EI-MS is advantageous if used as a detector for GC to unequivocally identify pure compounds.

An alternative to chemical ionisation is resonant (and non-resonant) laser ionisation methods [179], i.e. selective and soft laser photoionisation, such as REMPI. A particularly interesting setup is the combination of REMPI with TOFMS for monitoring coffee brew headspace. This chapter deals with technical features and applications of time-resolved analytical methods with particular focus on PTR-MS and resonant and laser ionisation methods (REMPI-TOFMS).

15.5.1 Proton-Transfer-Reaction Mass Spectrometry

PTR-MS combines a soft, sensitive and efficient mode of chemical ionisation, adapted to the analysis of trace VOCs. Briefly, headspace gas is continuously introduced into the chemical ionisation cell, which contains besides buffer-gas a controlled ion density of H_3O^+. VOCs that have proton affinities larger than water (proton affinity of H_2O is 166.5 kcal/mol) are ionised by proton transfer from H_3O^+, and the protonated VOCs are mass-analysed. The chemical ionisation source was specifically designed to achieve versatility, high sensitivity and little fragmentation, and to allow for absolute quantification of VOCs. To achieve these targeted specifications, the generation of the primary H_3O^+ ions and the chemical ionisation process—proton transfer from H_3O^+ to VOCs—are spatially and temporally separated and individually controlled. This allows (1) maximising signal intensity by increasing the generation of primary reactant ions, H_3O^+, in the ion source, (2) reducing fragmentation and clustering by optimising the conditions for proton transfer in the drift tube and (3) quantifying VOCs from measured count rates.

The four key features of PTR-MS can be summarised as follows. First, it is fast. Time dependent variations of headspace profiles can be monitored with a time resolution of better than 1 s. Second, the volatile compounds do not experience any work-up or thermal stress, and very little fragmentation is induced by the ionisation step; hence, measured mass spectral profiles closely reflect genuine headspace distributions. Third, measured mass spectral intensities can be directly related to absolute headspace concentrations, without calibration or use of standards. Finally, it is not invasive and the process under investigation is not affected by the measurements. All these features make PTR-MS a particularly suitable method to investigate fast dynamic process.

15.5.1.1
Technical Features

PTR-MS was introduced in 1993 by Lindinger and co-workers at the university of Innsbruck. A schematic drawing of the apparatus is given in Fig. 15.15. Here, only a brief description will be given. A more detailed discussion of the technical aspects of PTR-MS has been published in a series of review papers [193–195].

Primary (reactant) ions A^+, generated in a hollow cathode ion source, travel through a buffer gas within the drift tube, to which the reactant gas (VOC) is added in small amounts, so that the density of the buffer gas is much larger than the density of the VOC. On their way through the reaction region, ions perform many non-reactive collisions with buffer gas atoms or molecules; however, once they collide with a reactant gas particle, they may undergo a reaction:

$$A^+ + \text{VOC} \rightarrow \text{products}. \tag{1}$$

When H_3O^+ is used as the proton donor, most of the organic trace components R in air are ionised by proton-transfer processes:

$$H_3O^+ + \text{VOC} \xrightarrow{k} \text{VOC.H}^+ + H_2O. \tag{2}$$

These reactions are invariably fast, whenever they are exoergic, with rate coefficients, k, close to the collisional limiting values, $k_0 \approx 10^{-9}$ cm^3/s$^-$ [197]. Water has a proton affinity of 7.22 eV (166.5 kcal/mol), and common organic molecules have proton affinities in the range from 7 to 9 eV (161–208 kcal/mol), as shown in Table 15.4.

Hence, most of the relevant proton-transfer reactions involving H_3O^+ are slightly exoergic, and H_3O^+ will perform proton-transfer reactions with nearly any kind of VOC in the headspace of food products. However, H_3O^+ does not react with the natural components of air such as O_2, N_2, CO_2, CO or others (see Table 15.4). The exoergicity of the proton-transfer reaction with most VOCs, however, is low enough that breakup seldom occurs. On the basis of this ionisation principle, a PTR-MS setup was developed applicable to trace-gas analysis, and aimed at speed, sensitivity, versatility and simple handling.

The example shown in Fig. 15.16 was obtained by reconstituting a powdered beverage with hot water while measuring the headspace VOCs on-line by PTR-MS. It shows the relative ratio of the compounds released into the headspace and their dynamic behaviour. However, it is hardly possible to assign the mass traces to individual VOCs. For that, coupling of PTR-MS with GC-MS is required, which will be discussed in the next section.

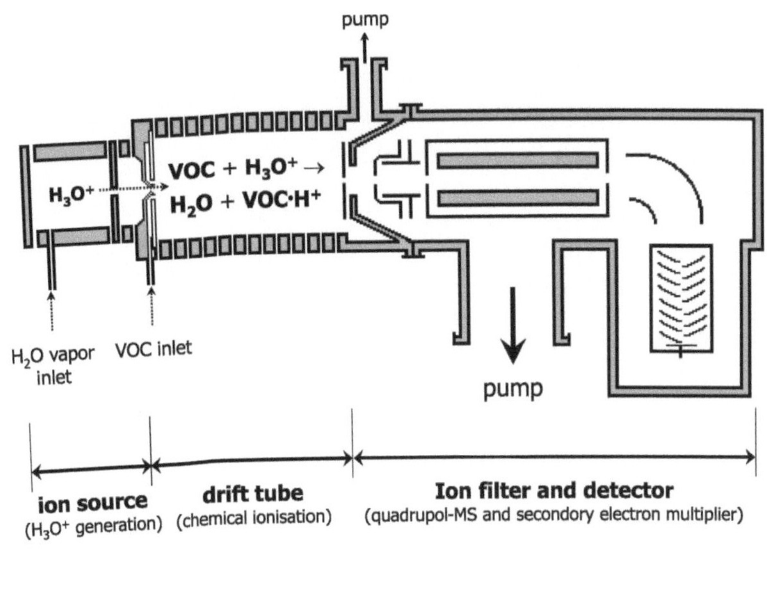

Fig. 15.15 The PTR-MS apparatus. It consists of a series of three main chambers. In the first chamber, H_2O is introduced and protonated in an electrical discharge to form H_3O^+. These ions are then driven by a small field through an orifice into the drift tube (chemical ionisation chamber). Coaxial to this orifice, neutral volatile organic compounds (*VOCs*) are introduced into the drift tube and collide at thermal energies with H_3O^+. VOCs with proton affinities exceeding 166.5 kcal/mol are ionised by proton transfer from H_3O^+ and are accelerated out of the drift tube into the quadrupole mass filter and onto the detector. (Adapted from [190])

Table 15.4 Proton affinities of the constituents of clean air and of various volatile organic compounds. All volatile organic compounds with a higher proton affinity than H_2O (166.5 kcal/mol) will be protonated with a very high efficiency when colliding with H_3O^+. This is the case for most of the volatile organic compounds in the headspace of coffee, with the exception of the natural constituents of clean air. In contrast, if NH_4^+ is used as a chemical ionisation agent, only compounds with a proton affinity exceeding 204.0/kcal · mol are ionised (*below dotted line*). (Adapted from [190])

Compounds		Proton affinities
Name	Formula	(kcal/mol)
	He	42.5
	Ne	48.1
	Ar	88.6
	O_2	100.9
	N_2	118.2
	CO_2	130.9
	CH_4	132.0
	N_2O	136.5
	CO	141.9
Water	**H_2O**	**166.5**
Butane	C_4H_{10}	163.3
Hydrogen sulphide	H_2S	170.2
Hydrogen cyanide	HCN	171.4
Formic acid	HCOOH	178.8
Propane	C_3H_6	179.8
Benzene	C_6H_6	181.9
Methanol	CH_3OH	181.9
Acetaldehyde	CH_3COH	186.6
Acetonitrile	CH_3CN	188.0
Ethanol	C_2H_5OH	188.3
Furane	C_4H_4O	192.2
2,3-Butanedione	$C_4H_6O_2$	194.8
Acetone	CH_3COCH_3	196.7
2,3-Methylbutanal	$C_5H_{10}O$	~195
Ammonia	**NH_3**	**204.0**
Pyrrole	C_4H_5N	207.6

Table 15.4 *(continued)* Proton affinities of the constituents of clean air and of various volatile organic compounds.

Compounds		Proton affinities
Name	Formula	(kcal/mol)
Oxazole	C_3H_3NO	208.2
Pyrazine	$C_4H_4N_2$	209.0
Pyrazole	$C_3H_3N_2$	212.8
Dimethylamine	C_2H_7N	217.0
Pyridine	C_5H_5N	220.8
Trimethylamine	C_3H_9N	225.1

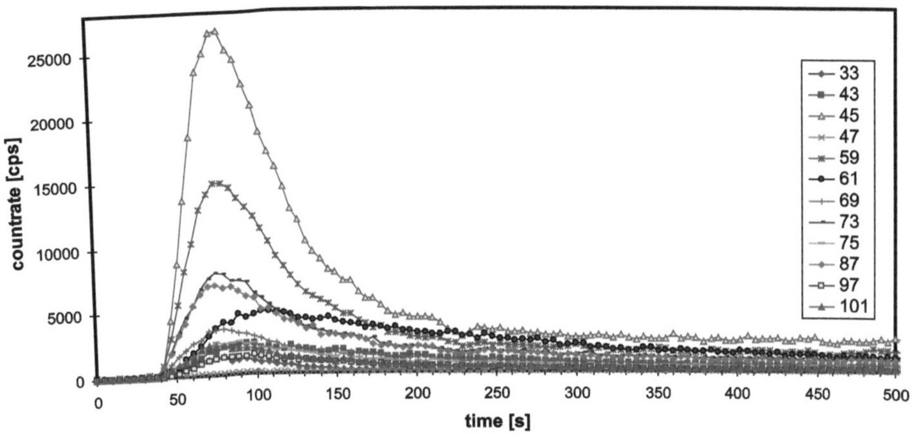

Fig. 15.16 Ion traces (in *m/z*) of VOCs released upon reconstitution of an instant beverage and analysed on-line by PTR-MS

15.5.1.2 Coupling of Proton-Transfer-Reaction Mass Spectrometry with Gas Chromatography–Mass Spectrometry

The success of PTR-MS triggered interest in further improving its performance. Indeed, PTR-MS is a one-dimensional technique, and ions from a complex headspace, e.g. coffee, can often only be tentatively assigned. Ions from different compounds (parent and fragment ions) can overlap in PTR-MS and prevent an unambiguous identification of VOCs in a complex mixture [198]. Therefore, the attempt has been made to address this problem and propose an extension of PTR-MS which allows for an unambiguous identification of headspace compounds. This is achieved by coupling GC with simultaneous PTR-MS and EI-MS detection. In this chapter, we can only introduce the basic features of the new setup: a technical and analytical extension of PTR-MS which removes this shortcoming, while preserving its salient and unique features. Combining separation of VOCs by GC with simultaneous and parallel detection of the GC effluent by PTR-MS and EI-MS, an unambiguous interpretation of complex PTR-MS spectra becomes feasible. A more detailed description of characteristic performance parameters, such as resolution, linear range and detection limit, has been published in a recent paper [199].

As an example, the novel setup was applied to the characterisation of coffee headspace as a complex food system Basically, an aliquot of the headspace is trapped in defined time periods on several Tenax® adsorbents for characterisation by GC-MS. Figure 15.17 shows the simultaneously recorded total ion counts of the EI-MS (top frame) and PTR-MS (bottom frame) for VOCs trapped on the first Tenax® cartridge. The GC-separated pure compounds are identified

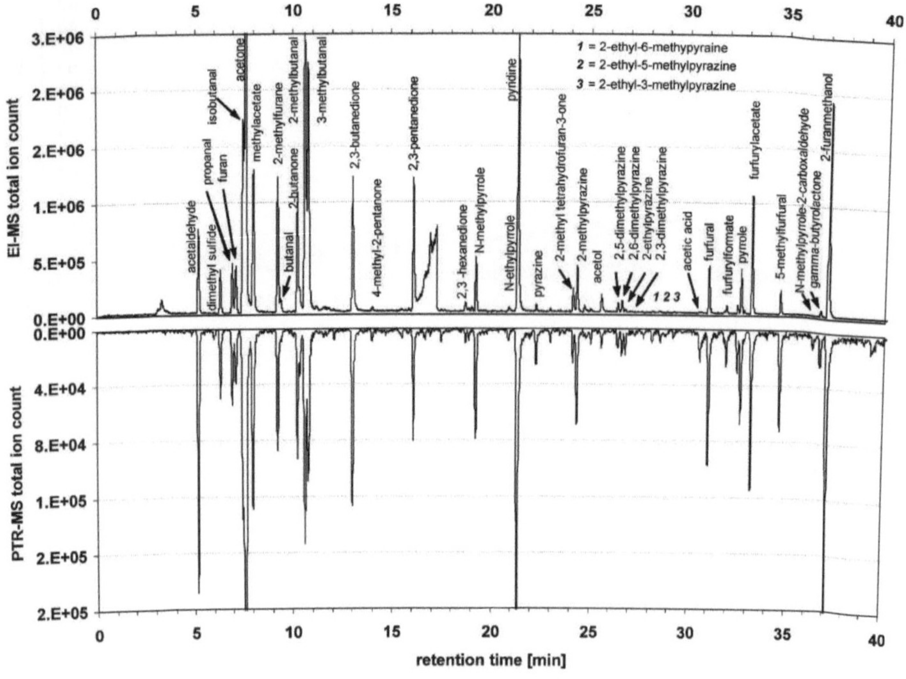

Fig. 15.17 Simultaneous EI-MS (*top trace*) and PTR-MS (*bottom trace*) total ion count analysis of coffee headspace. Identification was based on MS spectra obtained at 70 eV and the retention index of the reference compounds. (Adapted from [199])

by comparison of their EI-MS fragmentation patterns with the Wiley database (Wiley 7th edition) as well as their retention indices obtained with reference compounds. The PTR-MS spectrum allows the PTR-MS fragmentation pattern of the GC-separated pure compounds to be identified.

GC traces over the entire 40 min of the GC run are shown in Figs. 15.18 and 15.19, for the compounds desorbed from Tenax® cartridge no. 1 (trapping time window between 1 and 3 min). The data reveal that the PTR-MS ion signal at *m*/*z* 111 is a superposition of ions originating from two different compounds, i.e. 2-acetylfuran and 5-methylfurfural, contributing with 29 and 71%, respectively, to the total ion peak intensity at *m*/*z* 111. Similarly, the PTR-MS ion signal at *m*/*z* 87 is a superposition of 57% 2-methyl-1-propanal and 39% 2-butanone, with traces from 4-methyl-2-pentanone and 2-methyl tetrahydrofuran-3-one (2% each). While the single PTR-MS traces shown in Fig. 15.18 represent a superposition of several compounds, a series of PTR-MS ion traces are shown in Fig. 15.19 that are nearly pure (more than 89%), indicating that essentially only one single compound contributes to the ion signal (with only traces from other VOCs). Hence, in an on-line PTR-MS measurement of coffee headspace, the ion masses at *m*/*z* 68, 75, 80 and 95 can be assigned to pyrrole, acetol, pyridine and

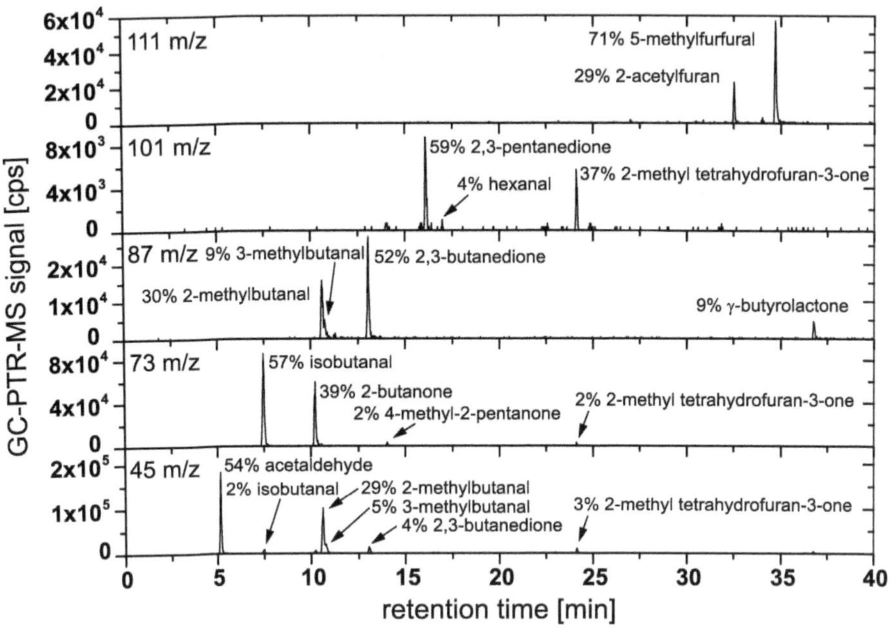

Fig. 15.18 Unambiguous identification of the molecules assigned to the trace ions. This identification is only valid for the first 120-s period of Tenax® trapping. (Adapted from [199])

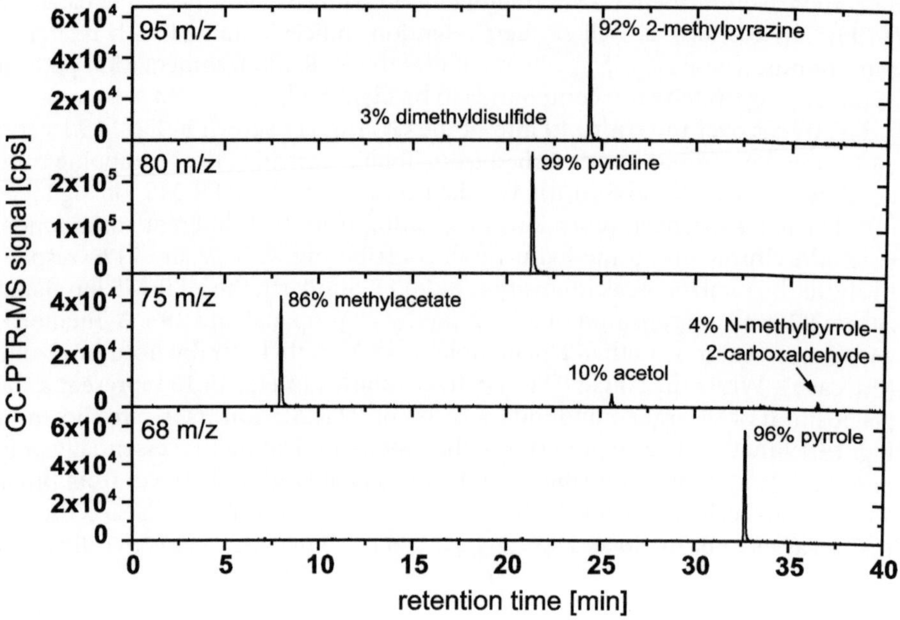

Fig. 15.19 Unambiguous identification of the molecules assigned to trace ions. This identification is only valid for the first 120-s period of Tenax® trapping. (Adapted from [199]

2-methylpyrazine, respectively. The coupling of PTR-MS with GC-MS, as introduced here, allows identification and quantification of the VOCs that contribute to a single PTR-MS ion signal.

15.5.2 Resonance-Enhanced Multiphoton Ionisation Time-of-Flight Mass Spectrometry

Selective and time-resolved monitoring can be achieved by REMPI at 266 nm coupled to a direct-inlet TOFMS device. Selectivity was introduced into the ionisation step by resonant ionisation at a fixed UV laser wavelength. The photoexcitation energy scheme for REMPI is illustrated in Fig. 15.20.

Depending on molecular resonances, VOCs with an optical (electronic) absorption at 266 nm absorb a laser photon, while those transparent at 266 nm remain in the ground state. The width of optical absorptions is given by the ground-state population, and broadens with the molecule's temperature, which itself depends on the expansion conditions at the inlet system.

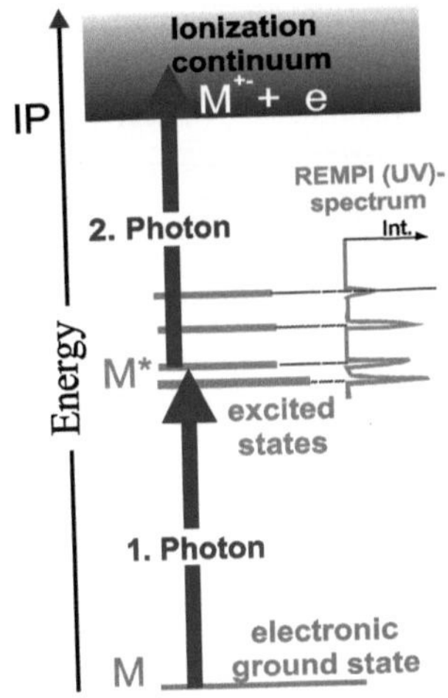

Fig. 15.20 The REMPI process

Since an effusive molecular beam was used (no cooling), a range of rotational and vibrational states was populated, resulting in broad absorption bands. Consequently, a range of compounds may be ionised simultaneously, owing to overlapping absorption bands [200].Technical reviews on REMPI can be found in the literature [200–202].

In a typical REMPI scheme, molecules absorb a first photon and are excited into a UV electronic state. These excited molecules are subsequently ionised by absorbing a second photon. For effective and selective REMPI detection, the following conditions have to be fulfilled:

1. *Resonance condition:* the molecule has a UV-active excited state, whose energy corresponds to the energy of the laser photon.
2. *Lifetime condition:* the excited state has a lifetime which is long enough for it to absorb a second photon for ionisation.
3. *Ionisation condition:* the energy of two photons is equal to or higher than the ionisation energy of the molecule.

The on-line VOC sampling depicted in Fig. 15.21 gives a schematic overview of the experimental setup, to illustrate the sampling of the roaster gas and the introduction of the volatiles into the TOF mass spectrometer [203]. A quartz tube with a passivated inner surface of 10-mm inner diameter was used to sample gas from the roaster. The tube reached about 2 cm into the rotating drum. A constant off-gas sampling stream of 1.5 l/min was pumped through the sampling system. A quartz wool paper filter was integrated into the tube to prevent solid contamination such as dust or silver skins reaching the capillary inlet system. All sampling lines were heated to 250 °C, to minimise condensation of low-volatile compounds.

A typical REMPI at 266 nm mass spectrum is shown Fig. 15.22, obtained by roasting 80 g of *Arabica* coffee at 225 °C. The laser power density was adjusted to 10^6–10^7 W/cm^2 in order to avoid non-resonant ionisation processes. The spectrum contains predominantly molecular ions. Chemical assignment of the ion peaks was based on three distinct pieces of information: the literature on coffee flavour compounds [204], the mass as observed in TOFMS and optical absorption properties. With this information, many volatiles observed in Fig. 15.22 were unambiguously identified.

A full three-dimensional representation—mass, time, intensity—of a typical roasting process at 200 °C, recorded at 10 Hz by REMPI at 248 nm is shown in Fig. 15.23, panel a [179]. Characteristic cross-sections through the three-dimensional surface are given in Fig. 15.23, panels b and c. Figure 15.23, panel b gives a cross-section of the roast gas composition at a fixed time (approximately 12min). In Fig. 15.23, panel c two cross-sections at fixed masses *m*/*z* 94 and *m*/*z* 150 are shown, corresponding to *t*–*I* profiles of phenol and 4-vinylguaiacol.

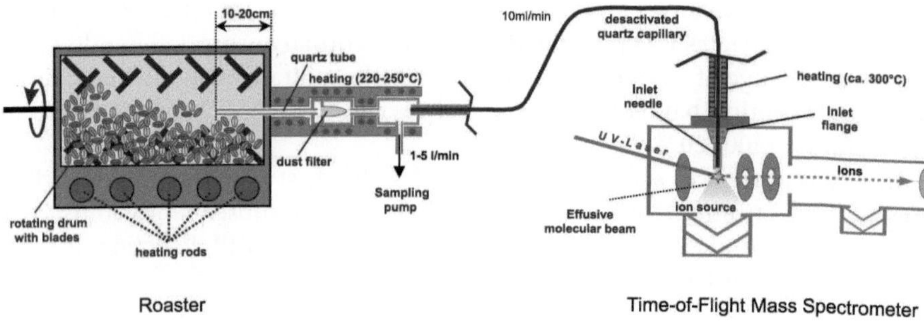

Fig. 15.21 The experimental setup including the laboratory-scale coffee roaster with a sampling unit and a laser mass spectrometer. The homebuilt mobile device consisted of a Reflectron TOFMS analyser, an effusive beam inlet system and a built-in laser operated at 266 nm (Continuum Nd: YAG laser SURELIGHT™, 266 nm). (Adapted from [203])

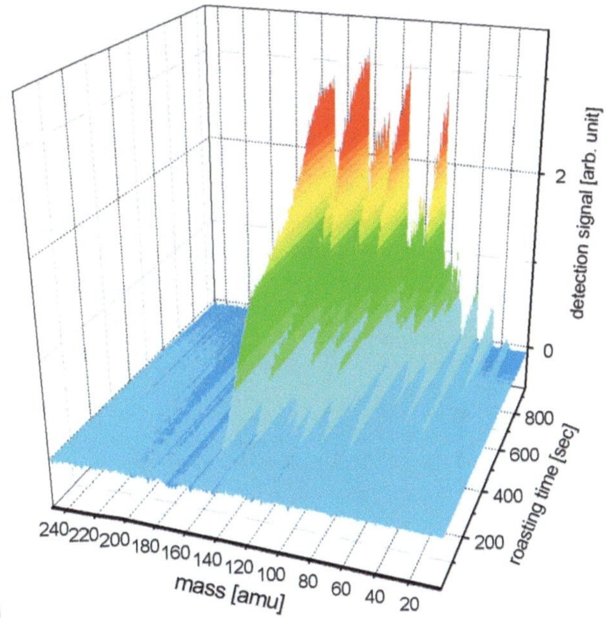

a

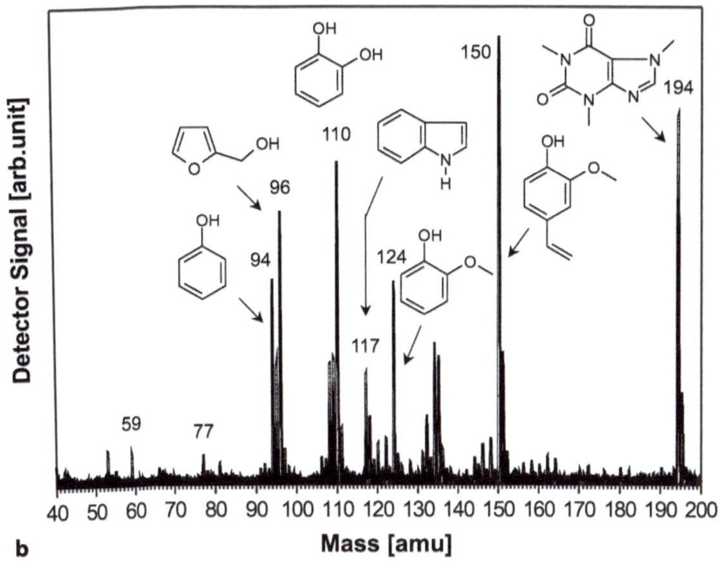

b

Fig. 15.22 On-line REMPI-TOFMS (at 266 nm) analysis of roast gas while roasting 80 g *Arabica* coffee. **a** The full-time–mass–intensity three-dimensional plot as recorded during roasting. **b** A time–intensity cross-section from **a** at a fixed time (medium roast level). The three phenolic VOCs, phenol (*m*/*z* 94), guaiacol (*m*/*z* 124) and 4-vinylguaiacol (150 *m*/*z*), are efficiently ionised at 266 nm. In addition, furfurylacohol (*m*/*z* 96), dihydroxybenzene (*m*/*z* 110), indol (*m*/*z* 117) and caffeine (*m*/*z* 194) were also detected. (Adapted from [203])

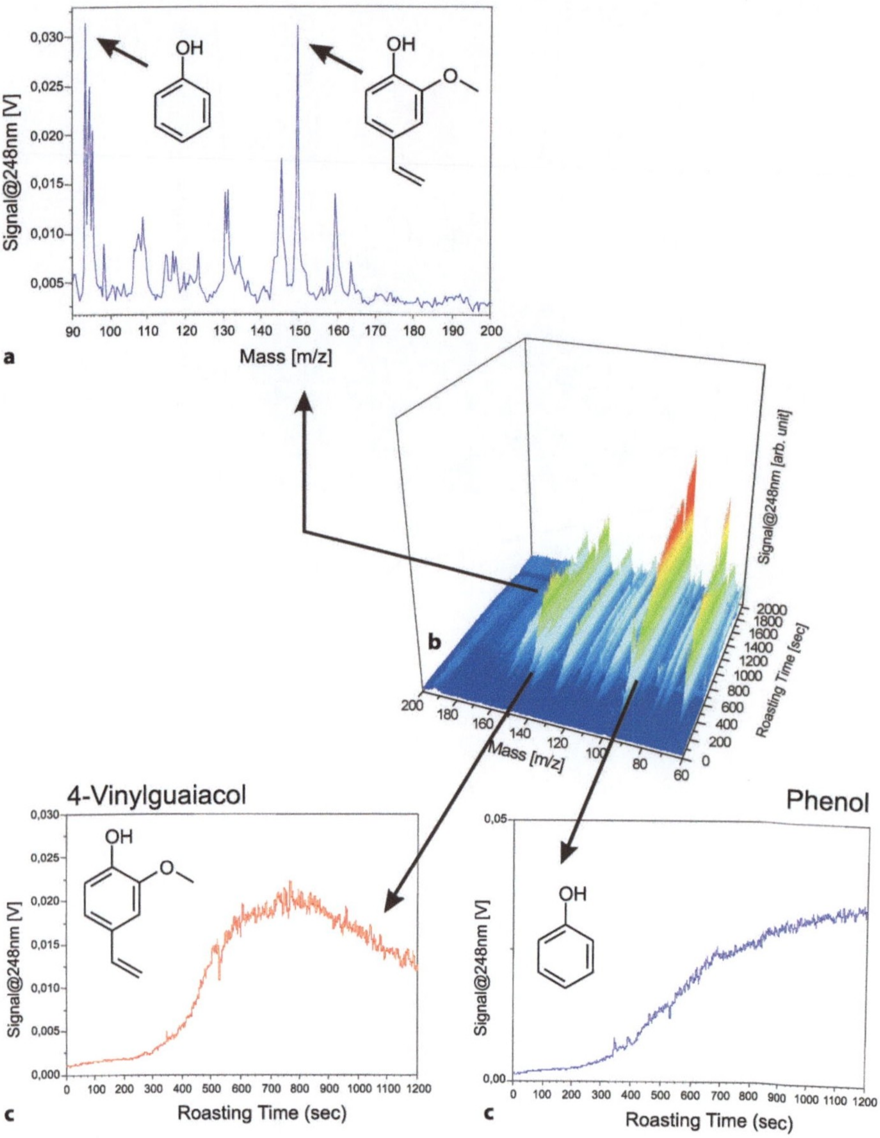

Fig. 15.23 **a** Three-dimensional REMPI at 248 nm TOFMS mass spectrum of coffee roasting off-gas while roasting in a steel cylinder at 200 °C. The three dimensions are mass, time and intensity. **b** Cross section of **a** at a fixed time. **c** Time–intensity REMPI at 248 nm TOFMS profiles of phenol (*m*/*z* 94) and 4-vinylguaiacol (*m*/*z* 150), corresponding to two cross-sections from **a** at fixed masses. (Adapted from [179])

Acknowledgements

We thank Chahan Yeretzian (Nespresso SA, Paudex, Switzerland) and Christian Lindinger (Nestlé Research Center, Lausanne, Switzerland) for fruitful discussions. We would like to thank P.J. Marriott and L. Mondello for providing some figures.

References

1. Mistry, B.S., Reineccius, T., Olson, L.K. (1997) Gas chromatography olfactometry for the determination of key odorants in foods. In: Marsili R (ed) Techniques for Analysing Food Aroma. Dekker, New York, pp 265–292.
2. Grosch, W. (1990) Analyse von Aromastoffen. Chem. Unserer Zeit 24:82–89.
3. Engel, W., Bahr, W., Schieberle, P. (1999) Solvent assisted flavour evaporation—a new and versatile technique for the careful and direct isolation of aroma compounds from complex food matrices. Eur. Food Res. Technol. 209:237–241.
4. Arthur, C.L., Pawliszyn, J. (1990) Solid phase microextraction with thermal desorption using fused silica optical fibers. Anal. Chem. 62:2145–2148.
5. Zhang, Z., Yang, M.J., Pawliszyn, J. (1994) Solid phase microextraction: A new solvent-free alternative for sample preparation. Anal. Chem. 66:844A–853A.
6. Baltussen, E., Sandra, P., David, F., Cramers, C. (1999) Stir bar sorptive extraction (SBSE), a novel extraction technique for aqueous samples: theory and principles. J. Microcolumn Sep. 11:737–747.
7. Giddings, J.C. (1995) Sample dimensionality: predictor of order-disorder in component peak distribution in multidimensional separation. J. Chromatogr. A 703:3–15.
8. Davis, J.M., Giddings, J.C. (1985) Statistical method for estimation of number of components from single complex chromatograms: theory, computer-based testing, and analysis of errors. Anal. Chem. 57:2168–2177.
9. Davis, J.M., Giddings J.C. (1985) Statistical method for estimation of number of components from single complex chromatograms: application to experimental chromatograms Anal. Chem. 57:2178–2182.
10. Cortes, H. (ed) (1990) Multidimensional Chromatography: Techniques and Applications. Dekker, New York.
11. Bertsch, W. (1978) Methods in high resolution gas chromatography: Two-dimensional techniques. J. High Resolut. Chromatogr. Chromatogr. Commun. 1:85–90, 187–194, 289–297.
12. Bertsch, W. (1999) Two-dimensional gas chromatography. Concepts, instrumentation, and applications–part 1: Fundamentals, conventional two-dimensional gas chromatography, selected applications. J. High Resolut. Chromatogr. 22:647–665.
13. Nitz, S., Drawert, F., Albrecht, M., Gellert, U. (1988) A micropreparative system for enrichment of capillary GC-effluents. J. High Resolut. Chromatogr. 11:322–327.
14. Nitz, S., Kollmannsberger, H., Drawert, F. (1989) Determination of sensorial active trace compounds by multidimensional gas chromatography combined with different enrichment techniques. J. Chromatogr. 471:173–185.

15. Nitz, S. (1985) Multidimensional gas-chromatography in aroma research. In: Berger, R.G., Nitz, S., Schreier, P. (eds) Topics in Flavour Research. Eichhorn, Hangenham, pp 43–57.
16. Berger, R.G., Drawert, F., Kollmannsberger, H., Nitz, S. (1985) Natural occurrence of undecaenes in some fruits and vegetables. J. Food Sci. 50:1655–1656, 1667.
17. Becker, R., Döhla, B., Nitz, S., Vitzthum, O.G. (1987) Identification of the "peasy" off-flavour note in Central African coffees. In: Proceedings of the XII International Conference on Coffee Science, Montreaux, pp 203–215.
18. Nitz, S., Kollmannsberger, H., Drawert, F. (1988) Analysis of flavours by means of combined cryogenic headspace enrichment and multidimensional GC. In: Schreier, P. (ed) Bioflavour '87. de Gruyter, Berlin, pp 123–135.
19. Wasowicz, E., Kaminski, E., Kollmannsberger, H., Nitz, S., Berger, R.G., Drawert, F. (1988) Volatile components of sound and musty wheat grains. Chem. Mikrobiol. Technol. Lebensm. 11:161–168.
20. Nitz, S., Kollmannsberger, H., Drawert, F. (1989) Determination of non natural flavours in sparkling fruit wines. I: Rapid method for the resolution of enantiomeric gamma lactones by multidimensional GC. Chem. Mikrobiol. Technol. Lebensm. 12:75–80.
21. Nitz, S., Kollmannsberger, H., Drawert, F. (1989) Über den Nachweis von nicht natürlichen Aromen in Fruchtschaumweinen. II: Enantiomere gamma-Lactone in Passionsfrüchten und Passionsfruchtprodukten. Chem. Mikrobiol. Technol. Lebensm. 12:105–110.
22. Kollmannsberger, H., Nitz, S., Drawert, F. (1991) Über den Nachweis von nicht natürlichen Aromen in Fruchtschaumweinen. III: Enantiomere γ-Lactone in Ananasfrüchten und Fruchtprodukten. Chem. Mikrobiol. Technol. Lebensm. 13:58–63.
23. Nitz, S., Kollmannsberger, H., Weinreich, B., Drawert, F. (1991) Enantiomeric distribution and 13C/12C isotope ratio determination of γ-lactones—appropriate methods for the differentiation between natural and non-natural Flavours? J. Chromatogr. 557:187–197.
24. Mosandl, A. (1995) Enantioselective capillary gas chromatography and stable isotope ratio mass spectrometry in the authenticity control of flavours and essential oils. Food Rev. Int. 11:597–664.
25. Weinreich, B., Nitz, S. (1992) Influences of processing on the enantiomeric distribution of chiral flavour compounds partA: Linalyl acetate and terpene alcohols. Chem. Mikrobiol. Technol. Lebensm. 14:117–124.
26. Kreck, M., Scharrer, A., Bilke, S., Mosandl, A. (2002) Enantioselective analysis of monoterpene compounds in essential oils by stir bar sorptive extraction (SBSE)-enantio-MDGC-MS. Flavour Fragrance J. 17:32–40.
27. Full, G., Winterhalter, P., Schmidt, G., Herion, P., Schreier, P. (1993) MDGC-MS—a powerful tool for enantioselective falvor analysis. J. High Resolut. Chromatogr. 16:642–644.
28. Nitz, S., Weinreich, B., Drawert, F. (1992) Multidimensional gas chromatography–isotope ratio mass spectrometry (MDGC-IRMS) J. High Resolut. Chromatogr. 15:387–391.
29. Schroll, W., Nitz, S. (1992) Präparative Multidimensional Gaschromatography with packed columns. Chem. Mikrobiol. Technol. Lebensm. 14:104–107.
30. Liu, Z., Phillips, J.B. (1991) Comprehensive two-dimensional gas chromatography using an on-column thermal modulator interface. J. Chromatogr. Sci. 29:227–231.
31. Marriott, P., Shellie, R. (2002) Principles and applications of comprehensive two-dimensional gas chromatography. Trends Anal. Chem. 21:573–583.
32. Beens, J., Brinkman, U.A.T. (2005) Comprehensive two-dimensional gas chromatography–powerful and versatile technique. Analyst 130:123–127.

33. Lee, A.L., Bartle, K.D., Lewis, A.C. (2001) A model of peak amplitude enhancement in orthogonal two-dimensional gas chromatography. Anal. Chem. 73:1330–1335.
34. Ryan, D., Marriott, P. (2006) Studies on thermionic ionisation detection in comprehensive two-dimensional gas chromatography. J. Sep. Sci. 29:2375–2382.D
35. Lammertyn, J., Veraverbeke, E.A., Irudayaraj, J. (2004) zNoseTM technology for the classification of honey based on rapid aroma profiling. Sens. Actuators B 98:54–62.
36. Kong, H., Ye, F., Lu, X., Guo, L., Tian, J., Xu, G., (2005) Deconvolution of overlapped peaks based on the exponentially modified Gaussian model in comprehensive two-dimensional gas chromatography. J. Chromatogr. A 1086:160–164.
37. Eyres, G., Dufour, J.P., Hallifax, G., Sotheeswaran, S., Marriott, P.J. (2005) Identification of character-impact odorants in Coriander and wild coriander leaves using gas chromatography-olfactometry (GCO) and comprehensive two-dimensional gas chromatography-time-of-flight mass spectrometry (GCxGC-TOFMS). J. Sep. Sci. 28:1061–1074.
38. Adahchour, M., Beens, J., Vreuls, R.J.J., Brinkman, U.A.T. (2006) Recent developments in comprehensive two-dimensional gas chromatography (GC×GC). Trends Anal. Chem. 25:438–454, 540–553.
39. Gorecki, T., Panic, O., Oldridge, N. (2006) Recent advances in comprehensive two-dimensional gas chromatography (GC×GC). J. Liquid Chromatogr. Relat. Technol. 29:1077–1104.
40. Wang, M., Marriott, P.J., Chan, W.H., Lee, A.W.M., Huie, C.W. (2006) Enantiomeric separation and quantification of ephedrine-type alkaloids in herbal materials by comprehensive two-dimensional gas chromatography. J. Chromatogr. A 1112:361–368.
41. Oezel, M.Z., Goegues, F., Lewis, A.C., (2006) Determination of Teucrium chamaedrys volatiles by using direct thermal desorption–comprehensive two-dimensional gas chromatography-time-of-flight mass spectrometry. J. Chromatogr A 1114:164–169.
42. Sanchez, J.M., Sacks, R.D. (2006) Development of a multibed sorption trap, comprehensive two-dimensional gas chromatography, and time-of-flight mass spectrometry system for the analysis of volatile organic compounds in human breath. Anal. Chem. 78:3046–3054.
43. Nelson, R.K., Kile, B.M., Brian, M., Plata, D.L., Sylva, S.P., Xu, L., Reddy, C.M., Gains, R.B., Frysinger, G.S., Reichenbach, S.E. (2006) Tracking the weathering of an oil spill with comprehensive two-dimensional gas chromatography. Environ. Forensics 7:33–44.
44. Zhu, S., Lu, X., Xing, J., Kong, H., Xu , G., Wu, C. (2006) Determination of volatile compounds in tobacco essential oil by comprehensive two-dimensional gas chromatography/time-of-flight mass spectrometer. Fenxi Huaxue 34:191–195.
45. Mohler, R.E., Dombek, K.M., Hoggard, J.C., Young, E.T., Synovec, R.E. (2006) Comprehensive two-dimensional gas chromatography time-of-flight mass spectrometry analysis of metabolites in fermenting and respiring yeast cells. Anal. Chem. 78:2700–2709.
46. Von Mühlen, C., Zini, C.A., Caramao, E.B., Marriott, P.J. (2006) Applications of comprehensive two-dimensional gas chromatography to the characterization of petrochemical and related samples. J. Chromatogr. A 1105:39–50.
47. Arey, J.S., Nelson, R.K., Xu, L., Reddy, C.M. (2005) Using comprehensive two-dimensional gas chromatography retention indices to estimate environmental partitioning properties for a compete set of diesel fuel hydrocarbons. Anal. Chem. 77:7172–7182.
48. Adahchour, M., Wiewel, J., Verdel, R., Vreuls, R.J.J., Brinkman, U.A.T. (2005) Improved determination of flavour compounds in butter by solid-phase (micro) extraction and comprehensive two-dimensional gas chromatography. J. Chromatogr. A 1086:99–106.

49. Korytar, P., Parera, J., Leonards, P.E.G., Santos, F. J., de Boer, J., Brinkman, U.A.T. (2005) Characterization of polychlorinated n-alkanes using comprehensive two-dimensional gas chromatography-electron-capture negative ionisation time-of-flight mass spectrometry. J. Chromatrogr. A 1086:71–82.
50. Jover, E., Adahchour, M., Bayona, J.M., Vreuls, R.J.J., Brinkman, U.A.T. (2005) Characterization of lipids in complex samples using comprehensive two-dimensional gas chromatography with time-of-flight mass spectrometry. J. Chromatogr. A 1086:2–11.
51. Ryan, D., Watkins, P., Smith, J., Allen, M., Marriott, P. (2005) Analysis of methoxypyrazines in wine using headspace solid phase microextraction with isotope dilution and comprehensive two-dimensional gas chromatography. J. Sep. Sci. 28:1075–1082.
52. Mondello, L., Casilli, A., Tranchida, P.Q., Dugo, G., Dugo, P. (2005) Comprehensive two-dimensional gas chromatography in combination with rapid scanning quadrupole mass spectrometry in perfume analysis. J. Chromatogr. A 1067:235–243.
53. Bordajandi, L.R., Korytar, P., De Boer, J. Gonzalez, M.J. (2005) Enatiomeric separation of chiral polyclorinated biphenyls on ß-cyclodextrin capillary columns by means of heart cut multidimensional gas chromatography and comprehensive two-dimensional gas chromatography. Application to food samples. J. Sep. Sci. 28:163–171.
54. Williams, A., Ryan, D., Olarte Guasca, A., Marriot, P., Pang, E. (2005) Analysis of strawberry volatiles using comprehensive two-dimensional gas chromatography with headspace solid-phase microextraction. J. Chromatogr. B 817:97–107.
55. Shellie, R., Marriott, P., Morrison, P. (2004) Comprehensive two-dimensional gas chromatography with flame ionization and time-of-flight mass spectrometry detection: Qualitative and quantitative analysis of west Australian sandalwood oil. J. Chromatogr. Sci. 42:417–422.
56. Ryan, D., Shellie, R., Tranchida, P., Casilli, A., Mondello, L., Marriott, P. (2004) Analysis of roasted coffee bean volatiles by using comprehensive two-dimensional gas chromatography-time-of-flight mass spectrometry. J. Chromatogr. A 1054:57–65.
57. Hua. R., Wang, J., Kong, H., Liu, J., Lu, X., Xu, G. (2004) Analysis of sulfur-containing compounds in crude oils by comprehensive two-dimensional gas chromatography with sulfur chemiluminescense detection. J. Sep. Sci. 27:691–698.
58. Ozel, M.Z., Gogus, F., Hamilton, J.F., Lewis, A.C. (2004) The essential oil of Pistacia vera L. at various temperatures of direct thermal desorption using comprehensive gas chromatography with time-of-flight mass spectrometry. Chromatographia 60:79–83.
59. Debonneville, C., Chaintreau, A. (2004) Quantitation of suspected allergens in fragrances. Part II. Evaluation of comprehensive chromatography-conventional mass spectrometry. J. Chromatogr. A 1027:109–115.
60. Shellie, R., Marriott, P., Cornwell, C. (2000) Characterization and comparison of tea tree and lavender oils by using comprehensive gas chromatography. J. High Resolut. Chromatogr. 23:554–560.
61. Gogus, F., Ozel, M.Z., Lewis, A.C. (2005) Superheated water extraction of essential oils of Origanum micranthum. J. Chromatogr. Sci. 43:87–91.
62. Mondello, L., Casilli, A., Tranchida, P.Q., Dugo, P., Dugo, G. (2003) Detailed analysis and group-type separation of naturalfats and oils using comprehensive two-dimensional gas chromatography. J. Chromatogr. A 1019:187–196.

63. Adahchour, M., van Stee, L.L.P., Beens, J., Vreuls, R.J.J., Batenburg, M.A., Brinkman, U.A.T. (2003) Comprehensive two-dimensional gas chromatography with time-of-flight mass spectrometric detection for the trace analysis of flavour compounds in food. J. Chromatogr. A 1019:157–172.
64. Mondello, L., Casilli, A., Tranchida, P.Q., Dugo, P., Dugo, G. (2005) Comprehensive two-dimensional GC for the analysis of citrus essential oils. Flavour Fragrance J. 20:136-140.
65. Roberts, M.T., Dufour, J.P., Lewis, A.C. (2004) Application of comprehensive multidimensional gas chromatography combined with time of flight mass spectrometry (GC×GC-TOFMS) for high resolution analysis of hop essential oil. J. Sep. Sci. 27:473–478.
66. Bicchi, C., Brunelli, C., Galli, M., Sironi, A. (2001) Conventional inner diameter short capillary columns: an approach to speeding up gas chromatographic analysis of medium complexity samples. J. Chromatogr. A 931:129–140.
67. Sandra, P., Proot, M., Diricks, G., David, F. (1987) In: Sandra, P., Bicchi, C. (eds) Capillary Gas Chromatography in Essential Oil Analysis. Hüthig, Heidelberg, pp 34–42.
68. van Es, A. (1992) High Speed Narrow Bore Capillary Gas Chromatography. Hüthig, Heidelberg.
69. Shen, Y., Lee, M.L. (1997) High-speed gas chromatography using packed capillary columns. J. Microcolumn Sep. 9:21–27.
70. Shen, Y., Yang, Y.J., Lee, M.L. (1997) Fundamental considerations of packed-capillary GC, SFC, and LC using nonporous silica particles. Anal. Chem. 69:628–635.
71. van Lieshout, M., van Deursen, M., Derks, R., Janssen, H.G., Cramers, C.A. (1999) A practical comparison of two recent strategies for fast gas chromatography: Packed capillary columns and multicapillary columns. J. Microcolumn Sep. 11:155–162.
72. Mondello, L., Dugo, P., Basile, A., Dugo, G., Bartle, K.D. (1995) Interactive use of linear retention indices, on polar and apolar columns, with a ms-library for reliable identification of complex mixtures J. Microcolumn Sep. 7:581–591.
73. van Es, A., Rijk, J., Cramers, C.A. (1989) Turbulent flow in capillary gas chromatography. J. Chromatogr. A 477:39–47.
74. Tijssen, R., van den Hoed, N., van Kreveld, M.E. (1987) Theoretical aspects and practical potentials of rapid gas analysis in capillary gas chromatography. Anal. Chem. 59:1007–1015.
75. Korytár, P., Janssen, H.G., Matisová, E., Brinkman, U.A.T. (2002) Practical fast gas chromatography: methods, instrumentation and applications. Trends Anal. Chem. 21:558–572.
76. Mastovska, K., Lehotay, S.J. (2003) Practical approaches to fast gas chromatography–mass spectrometry. J. Chromatogr. A 1000:153–180.
77. Mondello, L., Casilli, A., Tranchida, P.Q., Cicero, L., Dugo, P., Dugo, G. (2003) Comparison of fast and conventional GC analysis for citrus essential oils. J. Agric. Food Chem. 51:5602–5606.
78. Mondello, L., Shellie, R., Casilli, A., Tranchida, P.Q., Marriott, P., Dugo, G. (2004) Ultra-fast essential oil characterization by capillary GC on a 50 μm ID column. J. Sep. Sci. 27:699–702.
79. Dugo, G., Tranchida, P.Q., Cotroneo, A., Dugo, P., Bonaccorsi, I., Marriott, P., Shellie, R., Mondello, L. (2005) Advanced and innovative chromatographic techniques for the study of citrus essential oils. Flavour Fragrance J. 20:249–264.
80. Veriotti, T., Sacks, R. (2001) High-speed GC and GC/time-of-flight MS of lemon and lime oil samples. Anal. Chem. 73:4395–4402.

81. Song, J., Fan, L:, Beaundry, R.M. (1998) Application of solid phase microextraction and gas chromatography/time-of-flight mass spectrometry for rapid analysis of flavor volatiles in tomato and strawberry fruits. J. Agric. Food Chem. 46:3721–3726.
82. Mondello, L., Tranchida, P.Q., Costa, R., Casilli, A., Dugo, P., Cotroneo, A., Dugo, G. (2003) Fast GC for the analysis of fats and oils. J. Sep. Sci. 26:1467–1473.
83. Kirchner, M., Matisova, E., Otrekal, R., Hercegova, A., de Zeeuw, J (2005) Search on ruggedness of fast gas chromatography-mass spectrometry in pesticide residues analysis. J. Chromatogr. A 1084:63–70.
84. Reed, GL (1999) Fast GC: Applications and theoretical studies. Dissertation, Faculty of Virginia Polytechnic Institute and State University Blacksburg, Virginia.
85. van Deursen, M.M., Beens, J., Janssen, H.G., Leclercq, P.A., Cramers, C.A. (2000) Evaluation of time-of-flight mass spectrometric detection for fast gas chromatography. J. Chromatogr. A 878:205–213.
86. Schaller, E., Bosset, J.O., Escher, F. (1998) "Electronic noses" and their application to food. Lebensm.-Wiss. Technol. 31:305–316.
87. Shurmer, H.V., Gardner, J.W., Chan, H.T. (1989) The application of discrimination technique to alcohols and tobaccos using tin-oxide sensors. Sens. Actuators 18:361–371.
88. Buet, D., Burgaud, H., Rossi, P. (1996) Electronic nose: a real interface between sensory panels and fine analytical procedures in cosmetics. In: Olfaction and Electronic Nose, 3rd International Symposium, Toulouse.
89. Hodgins, D. (1995) The development of an electronic 'nose' for industrial and environmental applications. Sens. Actuators B 27:255–258.
90. Gardner, J.W., Craven, M., Dow, C., Hines, E.L. (1998) The prediction of bacteria type and culture growth phase by an electronic nose with a multi-layer perceptron network. Meas. Sci. Technol. 9:120–127.
91. Gibson, T.D., Prosser, O., Hulbert, J.N., Marshall, R.W., Corcoran, P., Lowery, P., Ruckkeene, E.A., Heron, S. (1997) Detection and simultaneous identification of microorganisms from headspace samples using an electronic nose. Sens. Actuators B 44:413–422.
92. Persaud, K.C., Travers, P.J. (1997) Arrays of broad specificity films for sensing volatile chemicals. In: Kress-Rogers, E. (ed) Handbook of Biosensors and Electronic Noses. CRC, Frankfurt, pp 563–592.
93. Nitz, S., Kollmannsberger, H., Lachermeier, C., Horner, G. (1999) Odour assessment with piezoelectric quartz crystal sensor array, a suitable tool for quality control in food technology? Adv. Food. Sci. 21:136–150.
94. Dittmann, B., Nitz, S. (2000) Strategies for the development of reliable QA/QC methods when working with mass spectrometry-based chemosensory systems. Sens. Actuators B 69:253–257.
95. Dittmann, B., Horner, G., Nitz, S., Parlar, H. (1999) Verfahren und Anordnung zum Erkennen komplexer Gas-, Geruchs- und Aromamuster auf der Basis der Massensepktroskopie. Patent 197 13 194, 01.04.1999.
96. Dittmann, B., Nitz, S., Horner, G. (1998) A new chemical sensor on a mass spectrometric basis. Adv. Food Sci. 20:122–131.
97. Raatikainen, O., Reinikainen, V., Minkkinen, P., Ritvanen, T., Muje, P., Pursiainen, J., Hiltunen, T., Hyvönen, P., von Wright, A., Reinikainen, S.P. (2005) Multivariate modelling of fish freshness index based on ion mobility spectrometry measurements. Anal. Chim. Acta 544:128–134.

98. Olafsdottir, G., Jonsdottir, R., Lauzon, H.L., Luten, J., Kristbergsson, K. (2005) Characterization of volatile compounds in chilled cod (Gadus morhua) fillets by gas chromatography and detection of quality indicators by an electronic nose. J. Agric. Food Chem. 53:10140–10147.

99. Hansen, T., Agerlin-Petersen, M., Byrne, D.V. (2005) Sensory based quality control utilising an electronic nose and GC-MS analyses to predict end-product quality from raw materials. Meat Sci. 69: 621–634.

100. Rajamaki, T., Alakomi, H.L., Ritvanen, T., Skytta,E., Smolander, M., Ahvenainen, R. (2005) Application of an electronic nose for quality assessment of modified atmosphere packaged poultry meat. Food Control 17:5–13.

101. Sarig, Y. (2000) Potential applications of artificial olfactory sensing for quality evaluation of fresh produce. J. Agric. Eng. Res. 77:239–258.

102. Wu, T.Z. (1999) A piezoelectric biosensor as an olfactory receptor for odour detection: electronic nose. Biosens. Bioelectron. 14:9–18.

103. Ko, H.J., Park, T.H. (2005) Piezoelectric olfactory biosensor: ligand specificity and dose-dependence of an olfactory receptor expressed in a heterologous cell system. Biosens. Bioelectron. 20:1327–1332.

104. Gomila, G., Casuso, I., Errachid, A., Ruiz, O., Pajot, E., Minic, J., Gorojankina, T., Persuy, M.A., Aioun, J., Salesse, R., Bausells, J., Villanueva, G., Rius, G., Hou, Y., Jaffrezic, N., Penneta, C., Alfinito, E., Akimov, V., Reggiani, L., Ferrari, G., Fumagalli, L., Sampietro, M., Samitier, J. (2006) Advances in the production, immobilization, and electrical characterization of olfactory receptors for olfactoric nanobiosensor development. Sens. Actuators B 116:66–71.

105. Deisingh, A.K., Stone, D.C., Thompson, M. (2004) Review: Applications of electronic noses and tongues in food analysis. Int. J. Food Sci. Technol. 39:587–604.

106. Tetko, I.V., Livingstone, D.J., Luik, A.I. (1995) Neural network studies. 1. Comparison of overfitting and overtraining. J. Chem. Inf. Comput. Sci. 35:826–833.

107. Jelen, H.H., Majcher, M., Zawirska-Wojtasiak, R., Wiewiorowska,M., Wasowicz, E. (2003) Determination of geosmin, 2-methylisoborneol, and a musty-earthy odor in wheat grain by SPME-GC-MS, profiling volatiles, and sensory analysis. J. Agric. Food Chem. 51:7079–7085.

108. Olsson, J., Borjesson, T., Lundstedt, T., Schnurer, J. (2000) Volatiles for mycological quality grading of barley grains: determinations using gas chromatography-mass spectrometry and electronic nose, Int. J. Food Microbiol. 59:167–178.

109. Borjesson, T., Eklov, T., Jonsson, A., Sundgren, H., Schnurer, J. (1996) Electronic nose for odor classification of grains. Cereal Chem. 73:457–461.

110. Magan, N., Evans, P. (2000) Volatiles as an indicator of fungal activity and differentiation between species, and the potential use of electronic nose technology for early detection of grain spoilage. J. Stored Prod. Res. 36:319–340.

111. Olsen, E., Vogt, G., Veberg, A., Ekeberg, D., Nilsson, A. (2005), Analysis of early lipid oxidation in smoked, comminuted pork or poultry sausages with spices. J. Agric. Food Chem. 53:7448–7457.

112. Eklöv, T., Johansson, G., Winquist, F., Lundström, I. (1998) Monitoring sausage fermentation using an electronic nose. J. Sci. Food Agric. 76:525–532.

113. Blixt, Y., Borch, E. (1999) Using an electronic nose for determining the spoilage of vacuum-packed beef. Int. J. Food Microbiol. 46:123–134.

114. Panigrahi, S., Balasubramanian, S., Gu, H., Logue, C., Marchello, M. (2006) Neural-network-integrated electronic nose system for identification of spoiled beef. Food Sci. Technol. 39:135–145.

115. Arnold, J.W., Senter, S.D., (1998) Use of digital aroma technology and SPME, GC-MS to compare volatile compounds produced by bacteria isolated from processed poultry. J. Sci. Food Agric. 78:343–348.
116. Haugen, E., Undeland, I. (2003) Lipid oxidation in herring fillets (Clupea harengus) during ice storage measured by a commercial hybrid gas sensor array system. J. Agric. Food Chem. 51:752–759.
117. Kent, M., Oehlenschlager, J., Mierke-Klemeyer, S., Manthey-Karl, M., Knoechel, R., Daschner, F., Schimmer, O. (2004) New multivariate approach to the problem of fish quality estimation. Food Chem. 87:531–535.
118. Dodd, T.H., Hale, S.A., Blanchard, S.M. (2004) Electronic nose analysis of tilapia storage. Trans. ASAE 47:135–140.
119. Haugen, J.E., Chanie, E., Westad, F., Jonsdottir, R., Bazzo, S., Labreche, S., Marcq, P., Lundb, F., Olafsdottir, G. (2006) Rapid control of smoked atlantic salmon (Salmo salar) quality by electronic nose: correlation with classical evaluation methods. Sens. Actuators B 116:72–77.
120. Olafsdottir, G., Chanie, E., Westad, F., Jonsdottir, R., Thalmann C.R., Bazzo, S., Labreche, S., Marcq, P., Lundby, F., Haugen, J.E. (2005) Prediction of microbial and sensory quality of cold smoked atlantic salmon (Salmo alar) by electronic nose. J. Food Sci. 70:S563–S574.
121. Korel, F., Luzuriaga, D.A., Balaban, M.O. (2001) Objective quality assessment of raw tilapia (Oreochromis niloticus) fillets using electronic nose and machine vision. J. Food Sci. 66:1018–1024.
122. Du, W.X., Lin, C.M., Huang, TS, Kim, J., Marshall, M.R., Wei, C.L. (2002) Potential application of the electronic nose for quality assessment of salmon fillets under various storage conditions. J. Food Sci. 67:307–313.
123. Olafsdottir, G., Nesvadba, P., Di natale, C., Careche, M., Oehlenschlager, J., Tryggvadottir, S.V., Schubring, R., Kroeger, M., Heia, K., Esaiassen, M., Macagnano, A., Jorgensen, B.M. (2004) Multisensor for fish quality determination. Trends Food Sci. Technol. 15:86–93.
124. Zhao, C.Z., Pan, Y.Z., Ma, L.Z., Tang, Z.N., Zhao, G.L., Wang, L.D. (2002) Assay of fish freshness using trimethylamine vapour probe based on a sensitive membrane on piezoelectric quartz crystal. Sens. Actuators B 81:218–222.
125. Anonymous (1996) Elektronische Nasen. Ernahrungsindustrie 6:54–55.
126. Mariaca, R. Bosset, J.O. (1997) Instrumental analysis of volatile (flavour) compounds in milk and dairy products. Lait 77:13–40.
127. Zannoni, M. (1995) Preliminary results of employ of an artificial nose for the evaluation of cheese. Sci. Tec. Lattiero Casearia 46:277–289.
128. Schaller, E., Bosset,, J.O., Escher, F. (1999) Practical experience with 'Electronic noses' systems for monitoring the quality of dairy products. Chimia 53:98–102.
129. Visser, F. R., Taylor, M. (1998) Improved performance of the AromaScan A32S electronic nose and its potential for detecting aroma differences in dairy products. J. Sens. Stud. 13:95–120.
130. van Ysacker, P., Ellen, G. (1998) Restricted possibilities for electronic nose applications in dairy industry. Voedingsmiddelentechnologie 31:11, 13–14.
131. Jou, K.D., Harper, W.J. (1998) Pattern recognition of Swiss cheese aroma compounds by SPMElGC and an electronic nose. Milchwissenschaft 53:259–263.
132. Harper, W.J., Sohn, S., Da Jou, K. (1996) The role of fatty acids in the aroma profiles of Swiss cheese as determined by an electronic nose. In: Olfaction and Electronic Nose, 3rd International Symposium, Toulouse.

133. Wijesundera, C., Walsh, T. (1998) Evaluation of an electronic nose equipped with metal oxide sensors for cheese grading. Aust. J. Dairy Technol 53:141.

134. Sberveglieri, G., Comini, E., Faglia, G., Niederjaufner, G., Benussi, G.P.,Contarini, G., Povolo, M. (1998) A novel electronic nose based on semiconductor thin films gas sensor to distinguish different heat treatments of milk. In Hurst, W.J. (ed) Seminars in Food Analysis. Chapman & Hall, New York, pp 3:67–76.

135. Sberveglieri, G., Benussi, G.P., Comini, E., Faglia, G., Niederjaufner, G., Contarini, G., Povolo, M. (1997) A novel electronic nose based on semiconductor films gas sensor to distinguish different types of milk. In: Authenticity and Adulteration of Food the Analytical Approach, Proceedings of Euro-FoodChem IX, Interlaken, Switzerland, 24–26 September 1997, pp 89–94.

136. Marsili, R.T. (1999) SPME-MS-MVA as an electronic nose for the study of off-flavors in milk. J. Agric. Food Chem. 47:648–654.

137. Korel, F., Luzuriaga, D.A. Balaban, M.O., (1999) Microbial, sensory and electronic nose evaluation of pasteurized whole milk. In: Hurst, W.J. (ed) Electronic noses and sensor array based systems. Lancaster, Basel, pp 154–161.

138. Trihaas, J. (2004) E-nose in Danish blue cheese production. Eur Dairy Mag 4:13–14.

139. O´Riordan, P.J., Delahunty, C.M. (2003) Characterisation of commercial cheddar cheese flavour. Part I and II. Int. Dairy J. 13: 355–370, 371–389.

140. Pillonel, L; Altieri, D; Tabacchi, R; Bosset, J.O. (2004) Comparison of efficiency and stability of two preconcentration techniques (SPME and INDEx) coupled to an MS-based 'electronic nose'. Mitt. Lebensmittelunters. Hyg. 95:85–98.

141. Drake, M.A., Gerard, P.D., Kleinhenz, J.P., Harper, W.J. (2003) Application of an electronic nose to correlate with descriptive sensory analysis of aged Cheddar cheese. Lebensm.-Wiss. Technol. 36:13–20.

142. Fenaille, F., Visani, P., Fumeaux, R., Milo, C., Guy, P.A. (2003) Comparison of mass spectrometry-based electronic nose and solid phase microextraction gas chromatography-mass spectrometry technique to assess infant formula oxidation. J. Agric. Food Chem. 51:2790–2796.

143. Ampuero, S., Zesiger, T., Gustafsson, V., Lunden, A., Bosset, J.O. (2002) Determination of trimethylamine in milk using an MS based electronic nose. Eur. Food Res. Technol. 214:163–167.

144. Marsili, R.T. (1999) SPME-MS-MVA as an electronic nose for the study of off-flavors in milk. J. Agric. Food Chem. 47:648–654.

145. Echeverria, G., Correa, E., Ruiz-Altisent, M., Graell, J., Puy, J., Lopez, L (2004) Characterization of Fuji apples from different harvest dates and storage conditions from measurements of volatiles by gas chromatography and electronic nose. J. Agric. Food Chem. 52:3069–3076.

146. Supriyadi, Shimizu, K., Suzuki, M., Yoshida, K., Muto, T., Fujita, A., Tomita, N., Watanabe, N. (2004) Maturity discrimination of snake fruit (Salacca edulis Reinw.) cv. Pondoh based on volatiles analysis using an electronic nose device equipped with a sensor array and fingerprint mass spectrometry. Flavour Fragrance J. 19:44–50.

147. Young, H., Rossiter, K., Wang, M., Miller, M. (1999) Characterization of Royal Gala apple aroma using electronic nose technology potential maturity indicator. J. Agric. Food Chem. 47:5173–5177.

148. Farnworth, E.R., McKellar, R.C., Chabot, D., Lapointe, S., Chicoine, M., Knight, K.P. (2002) Use of an electronic nose to study the contribution of volatiles to orange juice flavour. J. Food Qual. 25:569–576.

149. Nitz, S., Hanrieder, D. (2002) Möglichkeiten und Grenzen des Einsatzes von Gassensor-Arrays zur Qualitätsbeurteilung von Lebensmitteln. Adv. Food Sci. 24:154–169.
150. Shaw, P.E., Rouseff, R.L., Goodner, K.L., Bazemore, R., Nordby, H.E. Widmer, W.W. (2000) Comparison of headspace GC and electronic sensor techniques for classification of processed orange juices. Lebensm.-Wiss. Technol. 33:331–334.
151. McKellar, R.C., Rupasinghe, V., Lu, X., Knight, K.P. (2005) The electronic nose as a tool for the classification of fruit and grape wines from different Ontario wineries. J. Sci. Food Agric. 85:2391–2396.
152. Marti, M.P., Pino, J., Boque, R., Busto, O., Guasch, J. (2005) Determination of aging time of spirits in oak barrels using a headspace-mass spectrometry (HS-MS) electronic nose system and multivariate calibration. In: Analytical and Bioanalytical Chemistry 382 (2): The European Conference on Analytical Chemistry XIII, pp 440–443.
153. Marti, M.P., Busto, O., Guasch, J., Boque, R. (2005) Electronic noses in the quality control of alcoholic beverages. Trends Anal. Chem. 24:57–66.
154. Marti, M.P, Busto, O., Guasch, J. (2004) Application of a headspace mass spectrometry system to the differentiation and classification of wines according to their origin, variety and ageing. J. Chromatogr. A 1057:211–217.
155. Dittmann, B., Nitz, S. (2000) A new chemical sensor on a mass spectrometric basis-development and applications. In: Schieberle, P., Engel, K.H. (eds) Frontiers of Flavour Science, Proceedings of the 9th Weurman Flavour Research Symposium, Freising, Germany, 22–25 June 1999, pp 153–159.
156. Privat, E., Roussel, S., Grenier, P., Bellon-Maurel, V. (1998) Techniques for ethanol removal before discrimination of alcoholic drinks using electronic noses. Sci. Aliments 18:459–470.
157. Kojima, H., Araki, S., Kaneda, H., Takashio, M. (2005) Application of a new electronic nose with fingerprinting mass spectrometry to brewing. J. Am. Soc. Brew. Chem. 63:151–156.
158. McKellar, R.,C., Young, J.C., Johnston, A., Knight, K.P., Lu, X., Buttenham, S. (2002) Use of the electronic nose and gas chromatography-mass spectrometry to determine the optimum time for aging of beer. Tech. Q. Master Brew Assoc. Am. 39:99–105.
159. Tomlinson, J.B., Ormrod, I.H.L. and Sharpe, F.R. (1995) Electronic aroma detection in the brewery. J. Am. Soc. Brew. Chem. 53:167–173.
160. Bailey, T.P., Hammond, R.V. and Persaud, K.C. (1995) Application for an electronic aroma detector in the analysis of beer and raw materials. J. Am. Soc. Brew. Chem. 53:39–42.
161. Seregely, Z., Novak, I. (2005) Evaluation of the signal response of the electronic nose measured on oregano and lovage samples using diverent methods of multivariante analysis. Acta Aliment. 34:131–139.
162. Zhang, H., Balaban, M.O., Portier, K., Sims, C.A. (2005) Quantification of spice mixture compositions by electronic nose: Part II comparison with GC and sensory methods. J. Food Sci. 70:E259–E264 .
163. Baranauskiene, R., Venskutonis, P.R., Galdikas, A., Senuliene, D., Setkus, A. (2005) Testing of microencapsulated flavours by electronic nose and SPME-GC. Food Chem. 92:45–54.
164. Novak, I., Zambori-Nemeth, E., Horvath, H., Seregely, Z, Kaffka, K. (2003) Study of essential oil components in different origanum species by GC and sensory anlysis. Acta Aliment. 32:141–150.
165. Broda, S. Habegger, R., Hanke, A., Schnitzler, W.H. (2001) Characterization of parsley by chemosensory and other analytical methods. J. Appl. Bot. 75:201–206.

166. Dittmann, B., Zimmermann, B., Engelen, C., Jany, G., Nitz, S. (2000) Use of the MS-sensor to discriminate between different dosages of garlic flavoring in tomato sauce. J. Agric. Food Chem. 48:2887–2892.
167. Madsen, M.G., Grypa, R.D. (2000) Spices, flavour systems & the electronic nose. Food Technol. 54:44–46.
168. Lee, J.H., Sung, T.H., Lee, K.T., Kim, M.R. (2004) Effect of gamma-irridation on colour, pungency, and volatiles of Korean red pepper powder. J. Food Sci. 69:C585–C592.
169. Buratti, S., Benedetti, S., Cosio, M.S. (2005) An electronic nose to evaluate olive oil oxidation during storage. Ital. J. Food Sci. 17:203–210.
170. Garcia-Gonzalez, D.L., Barie, N. Rapp, M., Aparicio, R. (2004) Analysis of virgin olive oil volatiles by a novel electronic nose based on a miniaturized SAW sensor array coupled with SPME enhanced headspace enrichment. J. Agric. Food Chem. 52:7475–7479.
171. Garcia-Gonzalez, D.L, Aparicio, R. (2002) Detection of defective virgin olive oils by metal-oxide sensors. Eur. Food Res. Technol. 215:118–123.
172. Garcia-Gonzalez, D.L., Aparicio, R. (2002) Detection of vinegary defect in virgin olive oils by metal oxide sensors J. Agric. Food Chem. 50:1809–1814.
173. Aparicio, R., Rocha, S.M., Delgadillo, I., Morales, M.T. (2000) Detection of rancid defect in virgin olive oil by the electronic nose. J. Agric. Food Chem. 48:853–860.
174. Shiers, V., Adechy, M. (1998) Use of multi-sensor array devices to attempt to predict shelf-lives of edible oils. Semin. Food Anal. 3:43–52.
175. Cerrato Oliveros, C. Boggia, R., Casale, M., Armanino, C., Forina, M. (2005) Optimisation of a new headspace mass spectrometry instrument discrimination of different geographical origin olive oils. J. Chromatogr. A1076:7–15.
176. Aishima, T. (1991) Aroma discrimination by pattern recognition analysis of responses from semiconductor gas sensor array. J. Agric. Food Chem. 39:752–756.
177. Gardner, J.W., Shurmer, H.V., Tan, T.T. (1992) Application of an electronic nose to the discrimination of coffees. Sens. Actuators 6:71–75.
178. Marcone, M.F. (2004) Composition and properties of Indonesian palm civet coffee (Kopi Luwak) and Ethiopian civet coffee. Food Res. Int. 37:901–912.
179. Dorfner, R., Ferge, T., Yeretzian, C., Kettrup, A., Zimmermann, R. (2004) Laser mass spectrometry as on-line sensor for industrial process analysis: Process control of coffee roasting. Anal. Chem. 76:1386–1402.
180. Gretsch, C., Toury, A., Estebaranz, R., Liardon, R. (1998) Sensitivity of metal oxide sensors towards coffee aroma. Semin. Food Anal. 3:37–42.
181. van Deventer, D., Mallikarjunan, P. (2002) Comparative performance analysis of three electronic nose systems using different retained solvents of printed packaging. J. Food Sci. 67:3170–3183.
182. van Deventer, D., Mallikarjunan, P. (2002) Optimizing an electronic nose for analysis of volatiles from printing inks on assorted plastic films. Innov. Food Sci. Emerg. Technol. 3:93–99.
183. Heinio, R.L., Ahvenainen, R. (2002) Monitoring of taints related to printed solid boards with an electronic nose. Food Additives Contaminants 19(Suppl.):209–220.
184. Horner, G. (1999) Qualitative and quantitative evaluation methods for sensor arrays. In: Proceedings of the 6th International Symposium Olfaction and Electronic Nose, Tübingen, 20–22 September 1999.

185. James, D., Scott, S.M., Ali, Z., O'Hare, W.T. (2005) Review: Chemical sensors for electronic nose systems. Microchim. Acta 149:1–17.
186. Lindinger, W., Fall, R., Karl, T.G. (2001) Environmental, food and medical applications of proton-transfer-reaction mass spectrometry (PTR-MS). Adv. Gas Phase Ion Chem. 4:1–48.
187. van Ruth, S. M., Roozen, J.P. (2002) Delivery of flavors from food matrices. In: Taylor, A.J. (ed) Food Flavour Technology. Sheffield Academic Press, Sheffield, pp 167–184.
188. Taylor, A.J., Linforth, R.S.T., Harvey, B.A., Blake, B. (2000) Atmospheric pressure chemical ionisation mass spectrometry for in vivo analysis of volatile flavour release. Food Chem. 71:327–338.
189. Taylor, A.J., Sivasundaram, L.R., Linforth, R.S.T., Surawang, S. (2003) Time-resolved headspace analysis by proton-transfer-reaction mass-spectrometry. In: Deibler, K.D., Delwiche, J. (eds) Handbook of Flavor Characterization. Sensory Analysis, Chemistry and Physiology. Dekker, New York, pp 411–422.
190. Yeretzian, C., Jordan, A., Brevard, H., Lindinger, W. (2000) Identification of volatile compunds using combined gas chromotography electron impact atmospheric pressure ionization mass spectrometry. In: Taylor, A.J., Roberts, D.D. (eds) Flavour Release, ACS Symposium Series 763. American Chemical Society, Washington, pp 58–72.
191. Lindinger, W., Hansel, A., Jordan, A. (1998) Proton-transfer-reaction mass spectrometry (PTR–MS): on-line monitoring of volatile organic compounds at pptv levels. Chem. Soc. Rev. 27:347–354.
192. Fenaille, F., Visani, P., Fumeaux, R., Milo, C., Guy, P.A. (2003) Comparison of mass spectrometry-based electronic nose and solid phase microextraction gas chromatography-mass spectrometry technique to assess infant formula oxidation. J. Agric. Food Chem. 51:2790–2796.
193. Lindinger, W., Hansel, A., Jordan, A. (1998) Online monitoring of volatile organic compounds at pptv levels by means of proton-transfer-reaction mass spectrometry (PTR-MS). Medical applications, food control and environmental research. Int. J. Mass Spectrom. Ion Processes 173:191–241.
194. Lindinger, W., Hirber, J., Paretzke, H. (1993) An ion/molecule-reaction mass spectrometer used for online trace gas analysis. Int. J. Mass Spectrom. Ion Processes 129:79–88.
195. Hansel, A., Jordan, A., Holzinger, R., Prazeller, P., Vogel, W., Lindinger, W. (1995) Proton transfer reaction mass spectrometry: online trace gas analysis at the ppb level. Int. J. Mass Spectrom. Ion Processes 149/150:609–619.
196. Dorfner, R., Zimmermann, R., Kettrup, A., Yeretzian, C., Jordan, A., Lindinger, W. (1999) Vergleich zweier massenspektrometrischer Verfahren zur Direktanalyse in der Lebensmittelchemie. Lebensmittelchemie 53:32–34.
197. Gioumousis, G., Stevenson, D.P. (1958) Reactions of gaseous molecule ions with gaseous molecules. V. Theory. J. Chem. Phys. 29:294–299.
198. Yeretzian, C., Jordan, A., Lindinger, W. (2003) Analyzing the headspace of coffee by proton-transfer-reaction mass-spectrometry. Int. J. Mass Spectrom. 223–224:115–139.
199. Lindinger, C., Pollien, P., Ali, S., Yeretzian, C., Blank, I., Märk, T. (2005) Unambiguous identification of volatile organic compounds by proton-transfer-reaction mass-spectrometry (PTR-MS) coupled with GC-MS. Anal. Chem. 77:4117–4124.
200. Zimmermann, R., Heger, H.J., Yeretzian, C., Nagel, H., Boesl, U. (1996) Application of laser ionization mass spectrometry for online monitoring of volatiles in the headspace of food products: roasting and brewing of coffee. Rapid Commun. Mass Spectrom. 10:1975–1979.

201. Heger, H.J., Zimmermann, R., Dorfner, R., Beckmann, M., Griebel, H., Kettrup, A., Boesl, U. (1999) Online emission analysis of polycyclic aromatic hydrocarbons down to pptv concentration levels in the flue gas of an incineration pilot plant with a mobile resonance-enhanced multiphoton ionization time-of-flight mass spectrometer. Anal. Chem. 71:46–57.
202. Zimmerman, R., Heger, H.J., Kettrup, H.J., Boesl, U. (1997) A mobile resonance-enhanced multiphoton ionization time-of-flight mass spectrometry device for online analysis of aromatic pollutants in waste incinerator flue gases: first results. Rapid Commun. Mass Spectrom. 11:1095–1102.
203. Dorfner, R., Ferge, T., Kettrup, A., Zimmermann, R., Yeretzian, C. (2003) Real-time monitoring of 4-vinylguaiacol, guaiacol, and phenol during coffee roasting by resonant laser ionization time-of-flight mass spectrometry. J. Agric. Food Chem. 51:5768–5773.
204. Nijssen, L.M., Visscher, C.A., Maarse, H., Willemsens, L.C. Boelens, M.H. (1996) Volatile Compounds in Food, 7th edn. TNO Nutrition and Food Research Institute, Zeist.

16 Gas Chromatography–Olfactometry of Aroma Compounds

Werner Grosch
Deutsche Forschungsanstalt für Lebensmittelchemie,
Lichtenbergstraße 4, 85748 Garching, Germany

16.1 Introduction

The aroma of foods is caused by volatile compounds which are perceived by the human nose. Many studies (reviews in [1, 2]) have indicated that only a small fraction of the hundreds of volatiles occurring in a food sample contribute to its aroma. To detect these compounds, a method proposed by Fuller et al. [3] is used. In this procedure, which is designated gas chromatography–olfactometry (GC-O), the effluent from a gas chromatography column is sniffed by an expert who marks in the chromatogram each position at which an odour impression is perceived.

However, a single GC-O run only is usually insufficient to distinguish between the potent odorants that most likely contribute strongly to an aroma and those odorants that are only components of the background aroma. Therefore, to improve the results, two methods, combined hedonic aroma response measurements (CHARM) analysis [4] and aroma extract dilution analysis (AEDA) [5, 6] have been developed. As discussed in Sect. 16.4 in both methods serial dilutions of food extract are analysed by GC-O.

Reviews published by Acree and Teranishi [7], Blank [8], Grosch [1, 2, 9], Mistry et al. [10] and Schieberle [11] agree that GC-O was the starting point for the development of a systematic approach for the identification of the compounds causing food aromas. The aim of this chapter is to discuss the potential and the limitations of GC-O.

16.2 The GC-O Experiment

16.2.1 Introduction

The analysis of aroma compounds begins with the preparation of a concentrate containing the volatiles that smell like the starting material. However, as odorants are substances with a wide variety of functional groups, there is no ideal isolation procedure in aroma analysis. In consequence, the choice of the method is always a compromise. In general, mild conditions have to be used that allow the extraction of all of the important odorants and excludes the formation of artefacts, e.g. by the reactions listed in Table 16.1.

In bioactive materials, enzymatic reactions (nos. 1–3 in Table 16.1) are inhibited by homogenising the sample in the presence of calcium ions that precipitate the enzymes [12]. A lower pH value enhancing reactions 4–7 should be buffered and a higher temperature is avoided by distilling off the volatiles under vacuum. Samples containing hydroperoxides derived from unsaturated acyl lipids are sensitive to temperatures above 40 °C (no. 8).

16.2.2
Isolation of the Volatile Fraction

Recently, the procedures that are suitable to isolate the volatile fraction of a sample under mild conditions have been reviewed [1]. Three techniques—solvent extraction, distillation and solid-phase microextraction (SPME)—will be presented here.

16.2.2.1
Extraction

Solid samples are extracted with low-boiling solvents. As the polarity of the volatiles is different, a two-step extraction procedure is recommended, e.g. methylene chloride as the first solvent and diethyl ether as the second solvent [13]. The yield of the odorants is enhanced when the dry sample is soaked in water before the extraction procedure [14]. After filtration and drying, the extract is concentrated to approximately 50 mL and is then freed from the non-volatile material by using the solvent-assisted flavour evaporation (SAFE) method (Sect. 16.2.2.2).

16.2.2.2
Distillation

The compact distillation unit shown in Fig. 16.1 has been designed for the rapid and careful isolation of volatiles from the non-volatile food components [15]. This technique, denoted SAFE, is suitable for solvent extracts, aqueous samples, or matrices with high oil content.

The procedure is as follows. After application of high vacuum (approximately 5 mPa) to the apparatus, the distillation procedure is started by dropping aliquots of the sample into distillation flask no. 4 (Fig. 16.1). The volatiles, includ-

Table 16.1 Reactions leading to artefacts during isolation of volatiles

No.	Reaction
Enzymatic	
1	Hydrolysis of esters by esterases or lipases
2	Oxidative cleavage of unsaturated fatty acids by lipoxygenase and hydroperoxide lyase
3	Hydrogenation of aldehydes by alcohol dehydrogenases
Non-enzymatic	
4	Hydrolysis of glycosides and lactones
5	Formation of lactones from hydroxy acids
6	Cyclisation and rearrangement of *tert*-allylalcohols
7	Dehydration and rearrangement of *tert*-allylalcohols
8	Degradation of hydroperoxides

ing the solvent vapour, are transferred into distillation head no. 3. The distillate is condensed by liquid nitrogen in distillation flask no. 5.

16.2.2.3
SPME Extraction

This method is based on the partitioning of compounds between a sample and a coated fibre immersed in it [16–18]. The volatiles and other compounds are first adsorbed onto the fibre immersed in a liquid sample, an extract, or in the headspace above a sample for a certain period of time. After adsorption is complete, the compounds are thermally desorbed into a GC injector block for further analysis. Particularly in food applications, headspace SPME is preferred to avoid possible contamination of the headspace system by non-volatile food components [16].

An SPME unit consists of a piece of fused-silica fibre coated with a layer of a stationary phase such as non-polar poly(dimethylsiloxane) or polar polyacrylate or divinylbenzene/Carboxen/poly(dimethylsiloxane). The latter, for example, was suitable to trap the odorants (including sotolon) of soy sauce [19]. In the analytical procedure the fibre is exposed to the headspace of a food sample for 10–15 min. Then, the fibre is inserted into the injection port of a GC–mass spectrometry (MS) system. After desorption, the odorants are analysed. To improve the yields of the odorants, the fibre is placed in the effluent of a food sample purged with nitrogen [20].

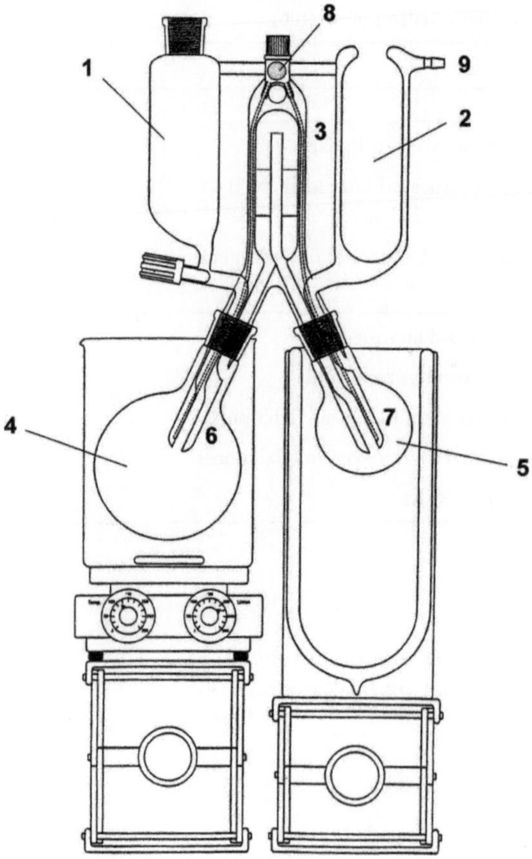

Fig. 16.1 Equipment for solvent-assisted flavour evaporation. *1* addition funnel, *2* cooling trap, *3* central head with thermostated water jacket, *4* distillation flask, *5* flask cooled with liquid nitrogen for distillate, *6*, *7* "legs" connected to funnel *1* and cooling trap *2*, *8* water inlet, *9* connection to the pump system. To ensure constant temperature during distillation, head *3* and "legs" *6* and *7* are connected by flexible polyethylene tubes that guide the water flask. [15]

16.2.3 Yield

Model experiments have been performed to show the yields of the odorants in the isolation procedure [21–23]. As an example, the values found for odorants from tomatoes by distillation with the SAFE method [23] are listed in Table 16.2. In agreement with other experiments, the result demonstrates that the losses of most of the odorants are high in the isolation procedure. In case of

Table 16.2 Yields of odorants from tomatoes obtained by distillation (solvent-assisted flavour evaporation)

Odorant	Yield (%)
3-Methylbutanal	24
1-Penten-3-one	37
Hexanal	39
(*Z*)-3-Hexenal	44
(*E*)-2-Hexenal	68
1-Octen-3-one	41
Methional	46
Phenyl acetaldehyde	26
3-Methylbutanoic acid	83
(*E*)-*β*-Damascenone	28
2-Phenylethanol	69
β-Ionone	18
4-Hydroxy-2,5-dimethyl-3(2*H*)-furanone	23
trans-4,5-Epoxy-(*E*)-2-decenal	27
Eugenol	53

[23]

labile odorants, further losses may occur during storage of the sample (cf. model experiment in [24]).

Owing to the limitations of the isolation procedures, it has to be examined sensorially whether the odour profiles of the concentrated extract and of the starting material agree (cf. discussion in [8]). In the SPME procedure this check demands an extraction of the odorants from the fibre as reported in [19].

16.3 Screening for Odorants by GC-O

After concentration of the extract by microdistillation [25] or by special procedures [26] to facilitate the identification of the odorants, an aliquot is separated by high-resolution GC and the effluent is split into a flame ionisation detector (FID) and a sniffing port [27]. The positions of the odorants in the gas chromatogram are assessed by sniffing the carrier gas as it flows from the port. This procedure is denoted GC-O.

16.4 Dilution Analysis

16.4.1 Introduction

In the majority of the studies on the composition of food aromas, AEDA is used for the determination of the relative odour potency of the compounds detected by GC-O (reviewed in [1]). The odour potency is proportional to the odour activity value (OAV) of the compound in air. The OAV is defined as the ratio of the concentration of a compound to its odour threshold [3].

16.4.2 Aroma Extract Dilution Analysis (AEDA)

An aliquot of the extract which was used for the first GC-O experiment is diluted with the solvent, usually as a series of 1+1 or 1+2 dilutions and each dilution is analysed by GC-O. This means that in each GC run the assessor records the retention time of each odour along with a descriptor of that odour. This procedure is continued until no odorants are perceivable. The highest dilution at which a compound can be smelled is defined as its flavour dilution (FD) factor. The FD factor is a relative measure, and is proportional to the OAV of the compound in air.

Dilution analyses rank the odorants present in an extract according to their relative OAV; the identification experiments are then focused on the odorants showing high FD factors.

It has been reported [28] that there may be a cross-adaptation between two odorants, causing a gap during sniffing of the dilution series. To avoid this phenomenon, AEDA should be performed within 2 days [11], e.g. GC-O of the concentrated extract and of the first dilutions 1:4, 1:16, 1:64, 1:256 and 1:1024 on the fist day, and the dilutions 1:2, 1:8, 1:32, 1:128 and 1:512 on the second day.

Some authors do not dilute the concentrated extract but dilute the sample before SPME and GC-O. Studies on soy sauce [9] and wine [29] are examples.

As an example of AEDA, Fig. 16.2 shows a plot of the FD factors of the odorants of parsley versus their retention indices; this plot is termed an FD chromatogram. As usual in dilution analyses, the result in Fig. 16.2 is not corrected for losses of odorants during the isolation and GC procedures; therefore, not only the odorants showing the highest FD factors (nos. 1, 2, 7 and 13 in Fig. 16.2) were identified but also all of the 14 odorants appearing in the FD-factor range of 4–512. The result is presented in the legend to Fig. 16.2.

The AEDA method has been applied to the volatile fractions of many foods (reviewed in [1]). Some recent studies which were not mentioned in [1] are listed in Table 16.3.

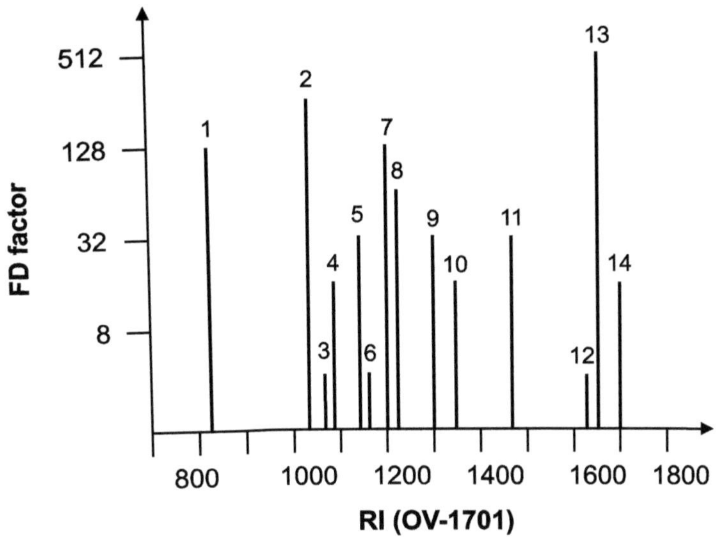

Fig. 16.2 Flavour dilution (*FD*) chromatogram obtained by application of aroma extract dilution analysis on an extract prepared from parsley leaves. The odorants were identified as *1* methyl 2-methylbutanoate, *2* myrcene, *3* 1-octen-3-one, *4* (*Z*)-1,5-octadien-3-one, *5* 2-isopropyl-3-methoxypyrazine, *6* *p*-mentha-1,3,8-triene, *7* linalool, *8* 2-*sec*-butyl-3-methoxypyrazine, *9* (*Z*)-6-decenal, *10* *β*-citronellol, *11* (*E*,*E*)-2,4-decadienal, *12* *β*-ionone, *13* myristicin, *14* unknown. *RI* retention index. [30, 31]

Odorants that cause aroma changes, e.g. off-flavours, may be detected by a comparative AEDA of fresh and deteriorated samples. Studies on storage defects of soybean oil [22, 51], buttermilk [52], boiled cod [53], dry parsley [54] and black and white pepper [55] are examples.

16.4.3 Aroma Extract Concentration Analysis

As reported in the previous section, AEDA is performed with a concentrated aroma extract. However, concentration of the volatile fraction might lead to losses of odorants, e.g. by evaporation and by enhanced side reactions in the concentrated extract. Consequently, the odour potency of these odorants can be underestimated in comparison to those whose levels are not reduced during concentration. To clarify this point, aroma extract concentration analysis (AECA) [56] should check the results of AEDA. AECA starts with GC-O of the original extract from which the non-volatile components have been removed. The extract is then concentrated stepwise by distilling off the solvent, and after each step an aliquot is analysed by GC-O [56].

Table 16.3 Some recent published applications of aroma extract dilution analysis (*AEDA*)

Material	Reference
Red pepper	[32]
Citrus flaviculpus Hort. ex Tanaka	[33]
Blue cheese	[34]
Apples (Elstar and Cox Orange)	[35]
Grenache rose wine	[36]
Coffee brew	[37]
Green tea	[38]
Black tea	[13]
Buckwheat honey	[39]
Brown rice	[40]
Yellow passion fruit	[41]
Non-fat dry milk	[42]
Soy sauce	[19]
Muskmelon	[43]
Laurus nobilis L. (leaves, buds, fruits)	[44]
Rose apple (*Syzygium jambos* Alston)	[45]
Chickasaw blackberry (*Rubus* L.)	[46]
Pinot Noir wine	[47]
Sake	[48]
Beer	[49]
Pineapple	[50]

In the case of boiled beef the results of AEDA were compared with those of AECA. Table 16.4 indicates that they agreed except in three cases. The odour potencies of 4-hydroxy-2,5-dimethyl-3(2*H*)-furanone, 3-mercapto-2-pentanone and methional were more than one dilution step higher in AECA than in AEDA [56]. Most likely, portions of these odorants had been lost during concentration of the extract for AEDA. AECA was also used in studies on the aroma of pepper [55], coffee [57] and Camembert cheese [58].

Table 16.4 Potent odorants of boiled beef—comparison of aroma extract concentration analysis (*AECA*) with AEDA [56]

Odorant	Extract volume (mL)[a,b]	
	AECA	AEDA
2-Furfurylthiol	100	50
4-Hydroxy-2,5-dimethyl-3(2*H*)-furanone	100	25
2-Methyl-3-furanthiol	50	50
1-Octen-3-one	12.5	6.25
(*E*)-2-Nonenal	12.5	6.25
3-Mercapto-2-pentanone	12.5	3.1
Methional	6.25	1.6
Butanoic acid	6.25	3.1
Guaiacol	6.25	3.1
3-Hydroxy-4,5-dimethyl-2(5*H*)-furanone	6.25	3.1
12-Methyltridecanal	6.25	3.1
Octanal	6.25	1.6
Nonanal	3.1	1.6
(*E,E*)-2,4-Decadienal	3.1	1.6

[a]The volume of the extract was adjusted to 200 mL and was then divided into halves that were subjected to AECA and AEDA, respectively.
[b]The extract volume at which the odorant was most (AECA) or least (AEDA) perceived by gas chromatography–olfactometry

16 4.4
GC-O of Static Headspace Samples

The highly volatile odorants are not detected or are underestimated when the screening method is applied to an aroma extract. These compounds are lost when the extract is concentrated or they are masked in the gas chromatogram by the solvent peak. To overcome this limitation, the screening has to be completed by GC-O of static headspace samples (GCOH; Fig. 16.3) [59–61].

In the sample of parsley (Table 16.5), the analysis was started with a headspace volume of 5 mL, in which GCOH revealed 15 odorants. Then, the headspace drawn from the sample was reduced in a series of steps to find the most potent odorants. GCOH of volumes of 2.5 and 1.25 mL indicated only seven and five odorants, respectively (Table 16.5); after reduction to 0.6 mL, only methanethiol, (*Z*)-3-hexenol and an unknown compound were the most potent, highly volatile odorants of parsley [31].

Table 16.5 Gas chromatography–olfactometry of static headspace samples of parsley leaves [31]

Odorant	Volume[a] (mL)	Flavour dilution factor[b]
Methanethiol	0.6	8.3
(*Z*)-3-Hexenal	0.6	8.3
Unknown	0.6	8.3
Myrcene	1.25	4
Myristicin	1.25	4
p-Methylacetophenone	2.5	2
(*Z*)-3-Hexenyl acetate	2.5	2
Unknown	5	1
2-*sec*-Butyl-3-methoxypyrazine	5	1
(*Z*)-3-Hexenol	5	1
1-Octen-3-one	5	1
(*Z*)-1,5-Octadien-3-one	5	1
ß-Phellandrene	5	1
1-Isopropenyl-4-methoxybenzene	5	1
p-Mentha-1,3,8-triene	5	1

[a]Lowest headspace volume required to perceive the odorant at the sniffing port
[b]The highest headspace volume was equated to a flavour dilution factor of 1. The flavour dilution factors of the other odorants were calculated on this basis. (Source [31])

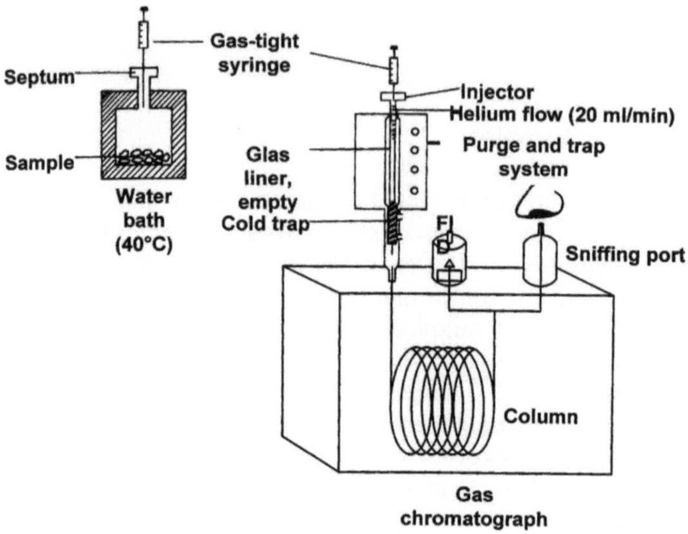

Fig. 16.3 Apparatus for gas chromatography–olfactometry of static headspace samples (from [60])

In most cases the concentrations of the compounds detected by GCOH are too small for the identification experiments; however, this disadvantage can be overcome when the odorants present in food are first detected in the extract by GC-O and then identified. Some of these odorants are also found by GCOH. As their odour quality, GC properties and chemical structures are known, they are easily identified in the headspace sample. In the case of parsley, a comparison of Fig. 16.2 with Table 16.5 indicates that odorant nos. 4, 6, 9, 11, 12 and 15 (Table 16.5) were known from AEDA. Further applications of GCOH are reviewed in [1].

16.4.5 Limitations of Extract Dilution Techniques

Besides the loss of odorants during extraction and concentration of the volatile fraction, the results of dilution experiments depend on:

- The sensitivity of the individual assessor to perceive odorants
- The chemical structure of the stationary phase used for GC-O

The influence of the sensitivity of the assessors on AEDA has been studied [11], with the result that the differences in the FD factors determined by a group of six panellists amount to not more than two dilution steps (e.g. 64 and 256), implying that the key odorants in a given extract will undoubtedly be detected. However, to avoid falsification of the result by anosmia, AEDA of a sample should be independently performed by at least two assessors. As detailed in [6], odour threshold values of odorants can be determined by AEDA using a "sensory" internal standard, e.g. (*E*)-2-decenal. However, as shown in Table 16.6 these odour threshold values may vary by several orders of magnitude [8] owing to different properties of the stationary phases. Consequently, such effects will also influence the results of dilution experiments. Indeed, different FD factors were determined for 2-methyl-3-furanthiol on the stationary phases SE-54 and FFAP: 2^{14} and 2^{6}, respectively. In contrast, 5-ethyl-3-hydroxy-4-methyl-2(5*H*)-furanone showed higher FD factors on FFAP than on SE-54: 2^{16} and 2^{5}, respectively. Consequently, FD factors should be determined on suitable GC capillaries [8]. However, the best method to overcome the limitations of GC-O and the dilution experiment is a sensory study of aroma models (Sect. 16.6.3).

16.5 Enrichment and Identification

In most cases only a few odorants selected for identification appear as clear peaks in the gas chromatogram. The majority of the odorants are concealed by peaks of the volatiles predominating in the extract. To enrich the odorants the extract is separated into the acid and the neutral/basic fractions and the latter is separated by chromatography on silica gel [21, 27]. If necessary, the fractions

Table 16.6 Odour threshold values (ng/L air) of some odorants as affected by the stationary phase of the gas chromatograph capillary [8]

Odorant	Stationary phase		
	SE-54	OV-1701	FFAP
2-Methyl-3-furanthiol	0.001–0.002	ND	5–10
5-Ethyl-3-hydroxy-4-methyl-2-(5*H*)-furanone (Abhexon)	2–4	ND	0.002–0.004
3-Hydroxy-4,5-dimethyl-2(5*H*)-furanone (sotolon)	ND	0.6–1.2	0.01–0.02
4-Hydroxy-2,5-dimethyl-3(2*H*)-furanone (furaneol)	ND	1–2	0.5–1.5
3,4-Dimethylcyclopentenolone	ND	1–2	0.05–0.1

ND not determined

obtained are further resolved by high-performance liquid chromatography [27, 62]. Thiols are enriched by reversible covalent chromatography [63, 64] or by a reaction with *p*-hydroxymercuribenzoic acid [65]. Finally, the analyte is purified by multidimensional GC (MDGC) [66, 67]. In MDGC the extract is separated on a polar precolumn, then a section of the effluent containing the analyte is cryofocused with liquid nitrogen and subsequently transferred to a non-polar main column that is combined with a mass spectrometer and a sniffing port.

In the identification experiments, the GC and MS data of the analytes have to be compared with those of corresponding authentic samples. However, as mentioned already, odorants are often concealed in the gas chromatogram by major volatile compounds; therefore, to avoid misidentification it is necessary to compare by GC-O the odour quality of the analyte with that of the authentic sample at approximately equal levels. The analyte, which has been perceived by GC-O in the volatile fraction, is only correctly identified if there is agreement in the sensorial properties, in addition to GC and MS data.

16.6 Aroma Model

Quantification of the odorants and calculation of their OAVs are the next steps to develop an aroma model.

16.6.1 Quantitative Analysis

As discussed in [1], precise quantitative results will be obtained when a stable isotope dilution assay (SIDA) is performed. In this procedure, stable isotopomers of the analytes are used as internal standards. Consequently, the major effort in the development of SIDA is the synthesis of the labelled standards since most of them are not commercially available.

The majority of the more than 100 odorants (reviewed in [1]) synthesised for use as internal standards are labelled with deuterium. However, during the quantification procedure some deuterated odorants might undergo deuterium–protium exchange, which would falsify the results. Examples are 4-hydroxy-2,5-dimethyl-3(2*H*)-furanone (furaneol) [68, 69] and 3-hydroxy-4,5-dimethyl-2(5*H*)-furanone (sotolon) [70], which are consequently labelled with ^{13}C.

The precision of SIDA has been checked in model experiments [22]. Although after cleanup the yields of some analytes were lower than 10 %, the results of quantification were correct as the internal standards showed equal losses.

16.6.2 Odour Activity Values

OAVs are calculated on the basis of odour threshold values which have been estimated in a medium that predominates in the food, e.g. water, oil or starch. As an example, the OAVs of the odorants of pineapples are listed in Table 16.7.

The highest OAVs were found for 4-hydroxy-2,5-dimethyl-3(2*H*)-furanone, followed by ethyl 2-methylpropanoate, ethyl 2-methylbutanoate, methyl 2-methylbutanoate and (*E,Z*)-1,3,5-undecatriene. It is assumed that these odorants contribute strongly to the aroma of pineapples [50]. However, FD factors and OAVs are functions of the odorants' concentrations in the extract, and are not psychophysical measures for perceived odour intensity [71, 72]. To take this criticism into account, aroma models are prepared on the basis of the results of the quantitative analysis (reviewed in [9]) and in addition omission experiments are performed [9].

16.6.3 Aroma Model

In the case of pineapples, the 12 odorants listed in Table 16.7 were dissolved in water in concentrations equal to those determined in the fruit [50]. Then the odour profile of this aroma model was evaluated by a sensory panel in comparison to fresh pineapple juice. The result was a high agreement in the two odour profiles. Fresh, fruity and pineapple-like odour notes scored almost the same intensities in the model as in the juice. Only the sweet aroma note was more intense in the model than in the original sample [50]. In further experiments, the contributions of the six odorants showing the highest OAV (Table 16.7) were evaluated by means of omission tests [9]. The results presented in Table 16.8 show that the omission of 4-hydroxy-2,5-dimethyl-3(2*H*)-furanone, ethyl 2-methylbutanoate or ethyl 2-methylpropanoate changed the odour so clearly that more than half of the assessors were able to perceive an odour difference between the reduced and the complete aroma model. Therefore, it was concluded that these compounds are the character-impact odorants of fresh pineapple juice.

Table 16.7 Potent odorants of fresh pineapple [50]

Odorant	Concentration[a] (μg/kg)	Threshold (μg/kg water)	Odour activity value[b]
Methyl 2-methylpropanoate	154	6.3	24
Ethyl 2-methylpropanoate	48.0	0.02	2,400
Methyl 2-methylbutanoate	1,190	2	595
Ethyl butanoate	75.2	1	75
Ethyl 2-methylbutanoate	157	0.15	1,050
Octanal	19.1	8	2
(*E,Z*)-1,3,5-Undecatriene	8.89	0.02	445
β-Damascenone	0.083	0.00075	111
δ-Octalactone	78.2	400	<1
4-Hydroxy-2,5-dimethyl-3(2*H*)-furanone	26,800	10	2,680
δ-Decalactone	32.7	160	<1
Vanillin	5.99	25	<1

[a]Quantitative analysis was performed using a stable isotope dilution assay.
[b]Odour activity values were calculated by dividing the concentrations of the odorants by their orthonasal odour thresholds in water

Table 16.8 Odour of the model for pineapple as affected by the absence of one compound[a] [50]

Odorant omitted from the aroma model	Number[b]
4-Hydroxy-2,5-dimethyl-3(2*H*)-furanone	11
Ethyl 2-methylbutanoate	9
Ethyl 2-methylpropanoate	8
(*E,Z*)-1,3,5-Undecatriene	7
ß-Damascenone	5
Methyl 2-methylbutanoate	4

[a]The aroma model contains the odorants listed in Table 16.7.
[b]Number of panellists (out of 15) detecting an odour difference between the reduced and the complete aroma model in a triangle test

Acknowledgements

The author wishes to thank R. Jauker for typing the manuscript and S. Bijewitz for preparing the drawings.

References

1. Grosch W (2004) In: Nollet LML (ed) Handbook of Food Analysis. Dekker, New York, p 717
2. Grosch W (2006) In: Ziegler G, Ziegler H (eds) Flavourings, 2nd edn. Wiley-VCH, Weinheim, p 695
3. Fuller GH, Steltenkamp GA, Tisserand GA (1964) Ann N Y Acad Sci 116:711
4. Acree TE, Barnard J, Cunningham DG (1984) Food Chem 14:273
5. Schmid W, Grosch W (1986) Z Lebensm Unters Forsch 182:407
6. Ullrich F, Grosch W (1987) Z Lebensm Unters Forsch 184:277
7. Acree TE, Teranishi R (1993) Flavor science. Sensible principles and techniques. ACS professional reference book. American Chemical Society, Washington, p 1
8. Blank I (1997) In: Marsili R (ed) Techniques for Analyzing Food Aroma. Dekker, New York, p 293
9. Grosch W (2001) Chem Senses 26:533
10. Mistry BS, Reineccius T, Olson LK (1997) In: Marsili R (ed) Techniques for Analyzing Food Aroma. Dekker, New York, p 265
11. Schieberle P (1995) In: Goankar AG (ed) Characterization of Food-Emerging Methods. Elsevier, Amsterdam, p 403
12. Buttery RG, Teranishi R, Ling LC (1987) J Agric Food Chem 35:540
13. Schuh C, Schieberle P (2006) J Agric Food Chem 54:916
14. Guth H, Grosch W (1993) Z Lebensm Unters Forsch 196:22
15. Engel W, Bahr W, Schieberle P (1999) Eur Food Res Technol 209:237
16. Blank I, Milo C, Lin J, Fay LB (1999) In Teranishi R, Wick EL, Hornstein I (eds) Flavor Chemistry. Thirty Years of Progress. Kluwer/Plenum, New York, p 63
17. Yang X, Peppard T (1994) J Agric Food Chem 42:1925
18. Jia M, Zhang QH, Min D (1998) J Agric Food Chem 46:2744
19. Baek HH, Kim HJ (2004) Food Sci Biotechnol 13:90
20. Grimm CC, Bergman C, Delgado JT, Bryant R (2001) J Agric Food Chem 49:245
21. Schieberle P, Grosch W (1987) J Agric Food Chem 35:252
22. Guth H, Grosch W (1990) Lebensm Wiss Technol 23:513
23. Mayer F, Takeoka G, Buttery R, Naim Y, Naim M, Bezman Y, Rabinowitch H (2003) In: Chadwallader KR, Weenen H (eds) Freshness and Shelf Life of Foods. ACS Symposium Series 836. American Chemical Society, Washington, p 144
24. Hofmann T, Schieberle P, Grosch W (1996) J Agric Food Chem 44:251
25. Bemelmans JMH (1979) In: Land DG, Nursten HE (eds) Progress in Flavour Research. Applied Science, Barking, p 79
26. Maarse H, Grosch W (1996) In Saxby MJ (ed) Food Taints and Off-Flavours. Blackie, London, p 72
27. Blank I, Sen A, Grosch W (1992) Z Lebensm Unters Forsch 195:239
28. Abbott N, Etievant P, Issanchou S, Danglois D (1993) J Agric Food Chem 41:1698
29. Marti MP, Mestres M, Sala C, Busto O, Guasch J (2003) J Agric Food Chem 51:7861
30. Jung HP, Sen A, Grosch W (1992) Lebensm Wiss Technol 26:55
31. Masanetz C, Grosch W (1998) Flavour Fragrance J 13:115
32. Jun H-R, Kim Y-S (2002) Food Sci Biotechnol 11:293
33. Choi HS, Sawamura M, Kondo Y (2002) J Food Sci 67:1713

34. Quian M, Nelson C, Bloomer S (2002) J Am Oil Chem Soc 79:663
35. Fuhrmann E, Grosch W (2002) Nahrung 46:187
36. Feirreira V, Ortin N, Escudero A, Lopez R, Cacho J (2002) J Agric Food Chem 50:4048
37. Sanz C, Czerny M, Cid C, Schieberle P (2002) Eur Food Res Technol 214:299
38. Kumazawa K, Masuda H (2002) J Agric Food Chem 50:5660
39. Zhou Q, Wintersteen CL, Cadwallader KR (2002) J Agric Food Chem 50:2016
40 Jezussek M, Bienvenido J, Schieberle P (2002) J Agric Food Chem 50:1101
41. Jordan MJ, Goodner K, Shaw PE (2002) J Agric Food Chem 50:1523
42. Karagul-Yuceer Y, Cadwallader KR, Drake MA (2002) J Agric Food Chem 50:305
43. Hayata Y, Sakamoto T, Maneerat C, Li X, Kozuka H, Sakamoto K (2003) J Agric Food Chem 51:3415
44. Kilic A, Hafizoglu H, Kollmannsberger H, Nitz S (2004) J Agric Food Chem 52:1601
45. Guedes C, Pinto A, Moreira R, De Maria C (2004) Eur Food Res Technol 219:460
46. Wang Y, Fin C, Quian MC (2005) J Agric Food Chem 53:3563
47. Fang Y, Quian M (2005) Flavour Fragrance J 20:22
48. Isogai A: Utsunomiya H, Kanada R, Iwata H (2005) J Agric Food Chem 53:4118
49. Fritsch H, Schieberle P (2005) J Agric Food Chem 53:7544
50. Tokitomo Y, Steinhaus M, Büttner A, Schieberle P (2005) Biosci Biotechnol Biochem 69:1323
51. Guth H, Grosch W (1990) Lebensm Wiss Technol 23:59
52. Heiler C, Schieberle P (1996) Lebensm Wiss Technol 29:460
53. Milo C, Grosch W (1995) J Agric Food Chem 43:459
54. Masanetz C, Grosch W (1998) Z Lebensm Unters Forsch 206:114
55. Jagella T, Grosch W (1999) Eur Food Res Technol 209:16; 22; 27
56. Kerscher R, Grosch W (1997) Z Lebensm Unters Forsch 204:3
57. Grosch W, Czerny M, Wagner R, Mayer F (1996) In: Taylor AJ, Mottram DC (eds) Flavour Science. Recent Developments. Royal Society of Chemistry, Cambridge, p 200
58. Kubickova J, Grosch W (1997) Int Dairy J 7:65
59. Holscher W, Steinhart H (1992) Z Lebensm Unters Forsch 195:33
60. Guth H, Grosch W (1993) Flavour Fragrance J 8:173
61. Semmelroch P, Grosch W (1995) Lebensm Wiss Technol 28:210
62. Czerny M, Wagner R, Grosch W (1996) J Agric Food Chem 44:3268
63. Full G, Scheier P (1994) Lebensmittelchemie 48:1
64. Semmelroch P, Grosch W (1996) J Agric Food Chem 44:537
65. Darriet P, Ominaga T, Lavigne V, Boidron JN, Dubourdieu D (1995) Flavour Fragrance J 10:385
66. Weber B, Maas B, Mosandl A (1995) J Agric Food Chem 52:2438
67. Reiners J, Grosch W (1999) Food Chem 64:45
68. Sen A, Schieberle P, Grosch W (1991) Lebensm Wiss Technol 24:364
69. Blank I, Fay LB: Lakner FJ, Schlosser M (1997) J Agric Food Chem 45:2642
70. Blank I, Schieberle P, Grosch W (1992) In: Schreier P, Winterhalter P (eds) Progress in Flavour Precursor Studies. Allured, Carol Stream, p 103
71. Fritjers JER (1978) Chem Senses 3:227
72. Audouin V, Bonnet F, Vickers ZM; Reineccius G (2001) In: Leland JV, Schieberle P, Buettner A, Acree TE (eds) Gas Chromatography-Olfactometry. The State of the Art. ACS Symposium Series 782. American Chemical Society, Washington, p 156

17 Enantioselective and Isotope Analysis—Key Steps to Flavour Authentication

A. Mosandl
Institut für Lebensmittelchemie,
Johann Wolfgang Goethe-Universität,
Max-von-Laue-Str. 9, 60438 Frankfurt am Main, Germany

17.1 Introduction

Authentication of genuine flavours is an important topic in view of quality assurance in the food industry and in consumer protection as well. Both *isotope discrimination* as well as *enantioselectivity* during biosynthesis may serve as inherent parameters of authenticity, provided that appropriate analytical methods and concise data from authentic samples are available.

Even if enantioselective capillary gas chromatography (enantio-cGC) and online isotope ratio mass spectrometry (IRMS) methods are highly efficient in the origin-specific analysis, analytical authentication remains a permanent challenge, owing to the complexity of natural product (food) matrices. At present, online coupling techniques are the methods of choice in the origin evaluation of flavour and fragrance compounds.

17.1.1 Isotope Discrimination

The reasons for isotope discrimination are isotope effects which are caused by both kinetic and thermodynamic factors. Especially the kinetic isotope effect during primary CO_2-fixation in photosynthesis is relevant for the source-specific discrimination of compounds from C_3 and C_4 plants.

Special techniques of mass spectrometry (MS) and of nuclear magnetic resonance (NMR) are employed for the assessment of isotope discrimination:

- IRMS: relations between stable isotopes ($^{13}C/^{12}C$; $^{2}H/^{1}H$; $^{18}O/^{16}O$;$^{15}N/^{14}N$)
- Site-specific natural isotope fractionation (SNIF) NMR (SNIF-NMR): quantitative ^{2}H-NMR measurements

17.1.2 Enantioselectivity

Enzyme-catalysed reactions usually proceed with high selectivity. Thus, high enantiomeric purity can be expected for chiral natural compounds. In the field of flavours and fragrances, enantio-cGC has proved to be highly efficient in origin-specific analysis. In order to obtain accurate information with respect to chirality, analytical procedures of the highest selectivity which employ chiral separation without racemisation must be utilised. In addition, references of definite chirality are essential.

17.2 Enantioselective Capillary Gas Chromatography

17.2.1 Scope

In the early 1980s, stereoanalysis of chiral flavour compounds was rather difficult, owing to the lack of suitable stationary GC phases.

A real breakthrough in this field occurred when enantio-cGC became more and more available. In particular, since 1988 selectively modified cyclodextrins have been synthesised, serving as chiral stationary phases in enantio-cGC, reported by Schurig and Novotny [1], König et al. [2, 3], Armstrong et al. [4], Dietrich et al. [5,6], Saturin et al. [7], and Bicchi et al. [8]. 6-O-silylated modified β-cyclodextrin and γ-cyclodextrin derivatives of well-defined structure and purity were synthesised and have proved to be chiral stationary phases of unique selectivity and versatility and, therefore, are successfully used in simultaneous enantio-cGC analysis [5,6]. Further derivatives were recently reported by Takahisa and Engel [9, 10], dealing with 2,3-di-*O*-methoxymethyl-6-*O*-*tert*-butyldimethylsilyl modified cyclodextrins as chiral stationary phases in enantio-cGC.

From our own experience, it should be emphasised that the enantioselectivity of modified cyclodextrin phases is considerably influenced by the polarity of the (non-chiral) polysiloxane solvents used.

Using a chiral column, coated with a definite modified cyclodextrin as the chiral stationary phase, the elution orders of furanoid and pyranoid linalool oxides are not comparable [11, 12]. Consistently, the chromatographic behaviour of diastereomers and/or enantiomers on modified cyclodextrins is not predictable (Fig. 17.1, Table 17.1). Even by changing the non-chiral polysiloxane part of the chiral stationary phase used, the order of elution may significantly be changed [13]. The reliable assignment of the elution order in enantio-cGC implies the coinjection of structurally well defined references [11–13].

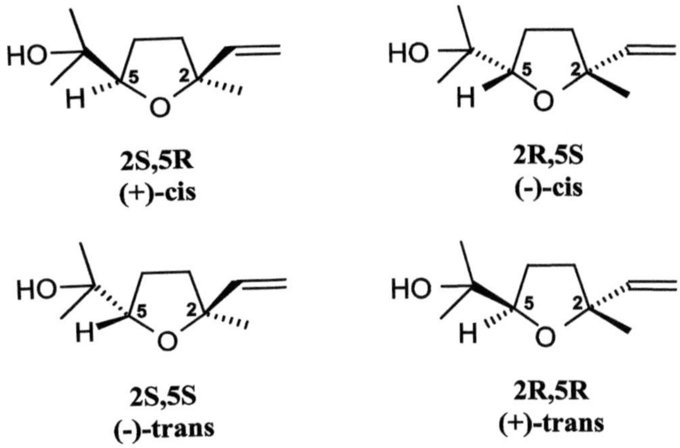

Fig. 17.1 Stereoisomers of linalool oxide [11]

Table 17.1 Elution order of the furanoid linalool oxides using different modified cyclodextrins (*CD*) as chiral stationary phases [11, 13]

Chiral selector	Solvent	I	II	III	IV
Permethyl-β-CD	OV-1701	*trans* (2*R*, 5*R*)	*trans* (2*S*, 5*S*)	*cis* (2*R*, 5*S*)	*cis* (2*S*, 5*R*)
Perethyl-β-CD	OV-1701	*trans* (2*R*, 5*R*)	*trans* (2*S*, 5*S*)	*cis* (2*R*, 5*S*)	*cis* (2*S*, 5*R*)
DIAC-6-TBDMS-β-CD	OV-1701	*trans* (2*S*, 5*S*)	*trans* (2*R*, 5*R*)	*cis* (2*R*, 5*S*)	*cis* (2*S*, 5*R*)
DIME-6-TBDMS-β-CD	OV-1701	*trans* (2*R*, 5*R*)	*cis* (2*R*, 5*S*)	*trans* (2*S*, 5*S*)	*cis* (2*S*, 5*R*)
DIME-6-TBDMS-β-CD	SE 52	*trans* (2*R*, 5*R*)	*cis* (2*R*, 5*S*)	*cis* (2*S*, 5*R*)	*trans* (2*S*, 5*S*)

DIAC heptakis(2,3-di-*O*-acetyl), *TBDMS* *tert*-butyldimethylsilyl, *DIME* heptakis(2,3-di-*O*-methyl)

17.2.2 Analytical Conditions

17.2.2.1 Stereodifferentiation of Enantiomers (Stereoisomers)

1. Evaluation of origin-specific enantiomeric ratios (of small ranges of variation), in correlation with their total amounts
2. Enantiomeric purity (ratio): measured ratio (expressed as a percentage) of the baseline-resolved enantiomers ($R_s \geq 1.5$)
3. Enantiomeric purity (ratio)—limitations: exact calculation of the enantiomeric ratio is defined by the given limits of detection and quantitation of the minor enantiomer (Fig. 17.2). Within this range, the minor enantiomer should be discussed as "detectable", but cannot be calculated exactly. Further details on the limits of detection and quantitation are given elsewhere [14].

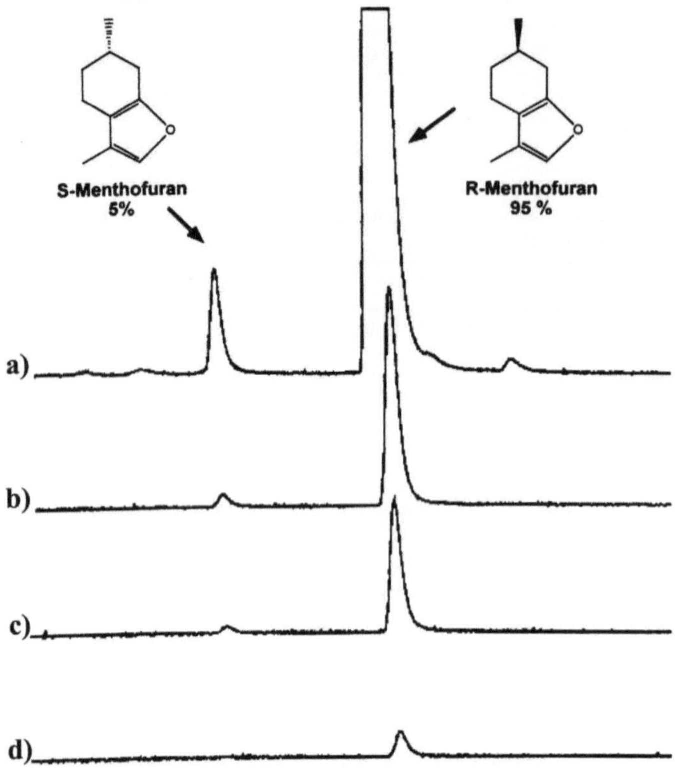

Fig. 17.2 Resolution of menthofuran enantiomers—quantitation of the minor enantiomer in relation to the concentration: quantitation accurate (*a*); approximate (*b*); impossible (*c*); analyte not detectable (*d*) [14]

17.2.2.2
Detection Limit

The limit of detection should be beneath the odour threshold. In this context one should keep in mind some special cases:

1. The odour threshold may be lower than the limit of analytical detection (e.g. sulphur compounds, pyrazines). In such cases authenticity assessment is definitely impossible.
2. Trace compounds without any sensorial relevance (odour activity value much less than 1) should not be evaluated in the sense of authenticity assessment, as the fraudulent addition of a sensorially ineffective compound makes no sense.
3. Legal assessment of trace amounts. In any case it depends on the expert witness to what extent sensorially irrelevant trace amounts, detected by (enantio)-cGC analysis, have to be classified as an avoidable contamination or have to be assessed as inevitable for technological reasons.

17.2.3
Enantioselective Multidimensional Gas Chromatography

Because of high complexity of natural flavours, essential oils or spice extracts, reliable chirality evaluation needs highly efficient sample cleanup procedures. The online GC-GC coupling, the so called enantioselective multidimensional gas chromatography (enantio-MDGC) system, has proved to be the method of choice. A schematic diagram of enantio-MDGC (Siemens Sichromat) is shown in Fig. 17.3 as a representative example. The multicolumn switching system (MCS2, GERSTEL) is the latest successful alternative (Fig. 17.12).

The design has been well proved in quality assurance and origin control of flavours and fragrances. A double-oven system is shown in the Fig. 17.3, with two independent temperature controls and two detectors (DM 1, DM 2). A "live switching" coupling piece is used to switch the effluent flow to either the first detector or the chiral column. With optimum pneumatic adjustment of the MDGC system, certain fractions are selectively transferred onto the chiral main column as they are eluted from the precolumn (heart-cutting technique) [15].

17.2.4
Detection Systems

If optimum chiral separation conditions and high-efficiency sample cleanup are properly employed, the first priorities in enantioselective analysis have been achieved. The ideal detector is universal yet selective, sensitive and structurally informative. MS currently provides the closest realisation to this ideal.

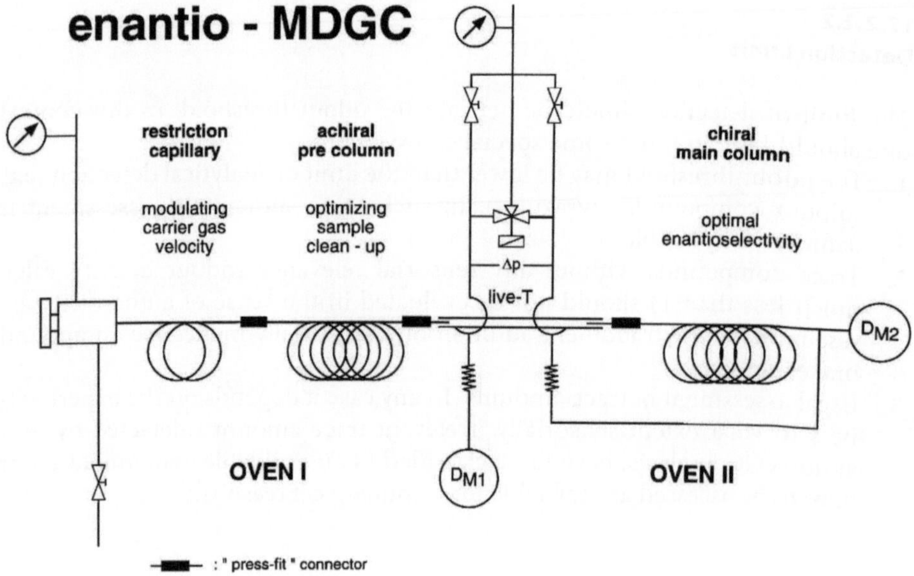

Fig. 17.3 Enantioselective multidimensional gas chromatography (*enantio-MDGC*), "Live-T" column switching, Siemens Sichromat [52]

The combination of enantio-MDGC with high-resolution MS or mass-selective detectors, both used in full scan or (at least) in the multiple ion monitoring (MIM) mode is currently the most potent analytical tool in enantioselective analysis of chiral compounds from complex mixtures.

17.2.4.1
Accuracy of Quantification

Internal standards of rather close relationship to the compounds analysed should be used, e.g. homologues (M+14) or isotopomers of analytes (^{2}H or ^{13}C labelling), owing to optimal identity of physical or chemical properties (e.g. Kovats index in GC).

17.2.4.2
Isotope Dilution Analysis

In combination with mass-selective detection (MIM mode), this technique may be ideal for quantitation of trace compounds from complex mixtures. But one should note that labelled internal standards may be discriminated by chemical and/or physical procedures (extraction, distillation, chromatography, derivatisation).

In particular, higher labelled isotopomers (e.g. CD_3 isotopomers and others) may (more or less) significantly differ from the corresponding unlabelled analytes.

17.2.4.3
Conclusion

Do not overestimate the use of labelled compounds as internal standards. In any case, proving the accuracy of sample cleanup by recovery experiments is imperative, no matter what kind of internal standard compound was used.

17.2.5
Limitations

Three types of limitations have to be accepted in enantio-cGC:

1. Racemates of natural origin, generated in some special cases [16,19–23]
2. Racemisation during processing or storage of foodstuffs, if structural features of chiral compounds are sensitive
3. Blending of natural and synthetic chiral compounds

17.2.5.1
Dihydroactinidiolide

In the flavour extract of apricots, racemic dihydroactinidiolide (DHA) was found as the first natural racemate detected by enantio-MDGC analysis [16]. The absolute configurations and the optical activities have been reported to be (*R*)-(-) and (*S*)-(+) enantiomers, respectively [17, 18].

Using amylose tris-3,5-dimethylphenylcarbamate as the chiral selector in enantioselective high-performance liquid chromatography, micropreparative resolution of the DHA racemate was achieved and the chromatographic behaviour in enantio-GC could be defined by coinjecting these references of definite chirality (Fig. 17.4) [13].

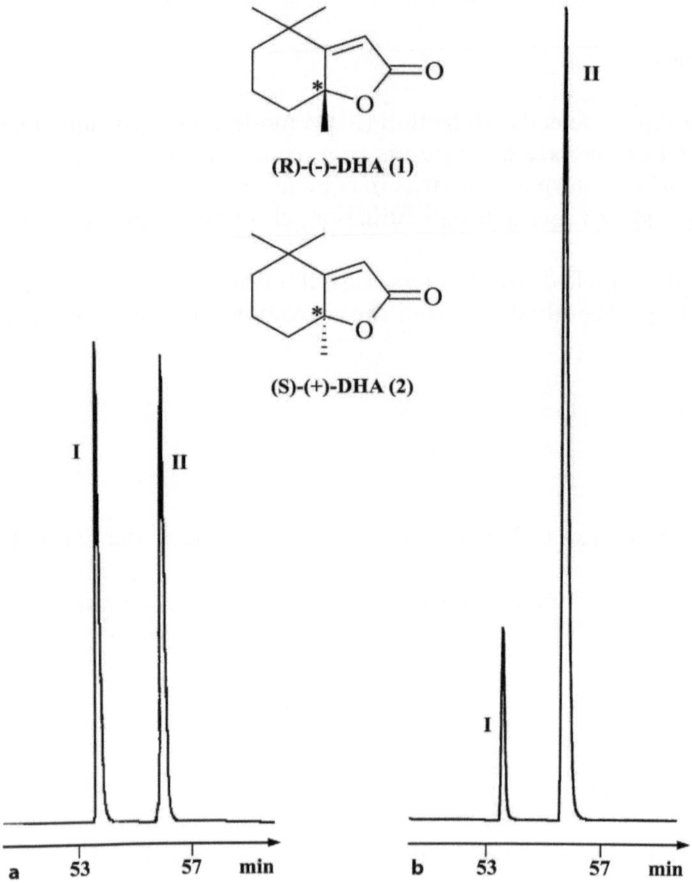

Fig. 17.4 Chromatographic behaviour of dihydroactinidiolide (*DHA*) enantiomers: synthetic racemate (**a**); DHA fractionation by enantioselective high-performance liquid chromatography (*HPLC*) (**b**). Chiral selectors used in enantio-GC: DIME-β-CD (30%) in SE 52; DIAC-β-CD (30%) in PS 268; DIAC-β-CD (50%) in OV 1701. Order of elution: *R* (*I*), *S* (*II*) in all cases [13]. *DIME* heptakis(2,3-di-*O*-methyl), *CD* cylclodextrin, *DIAC* heptakis(2,3-di-*O*-acetyl)

17.2.5.2 Germacrene D

The chiral hydrocarbon germacrene D is a widely spread plant constituent and is considered to be an important intermediate in the biosynthesis of many sesquiterpenes. Schmidt et al. [19, 20] have shown that the plant *Solidago canadensis* generates both optical antipodes of this compound by enzymatic cyclisation of farnesyl diphosphate using two different enantiospecific synthases. As to be seen in Fig. 17.5, the enantiomeric ratio of germacrene D in *Solidago canadensis* can vary from individual to individual [21].

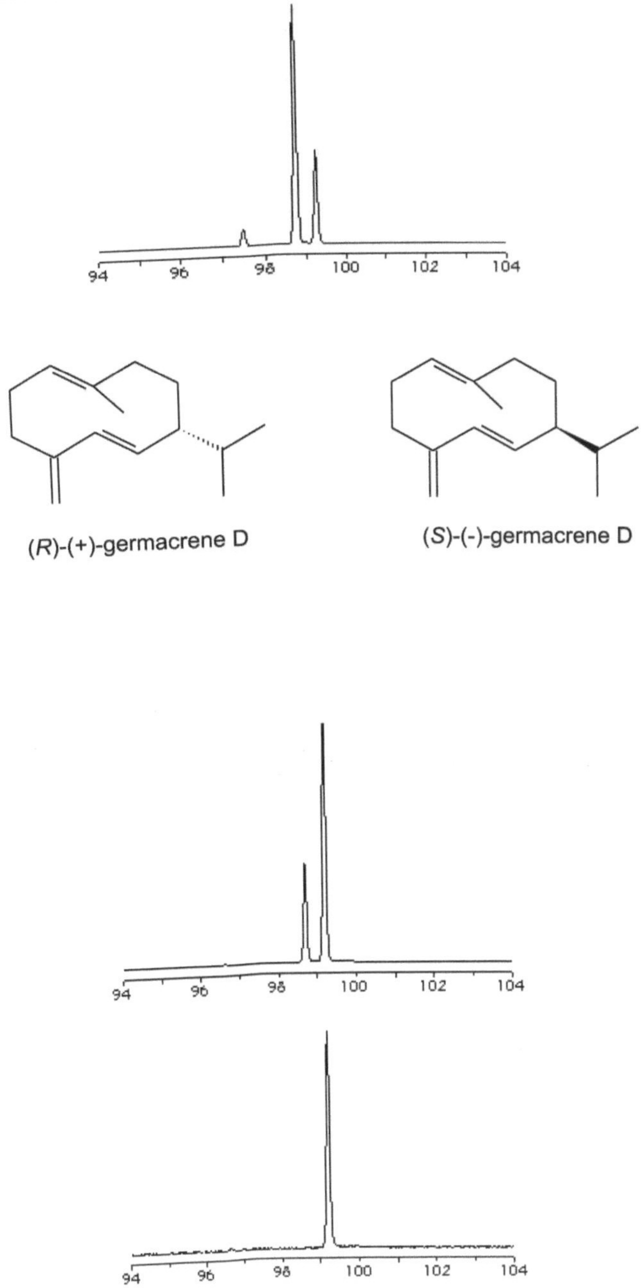

Fig. 17.5 Enantioselective analysis of germacrene D from the essential oil of different Solidago canadensis plants, using the enantio-MDGC–mass spectrometry (MS) technique [21]

17.2.5.3 Acid-Induced Keto/Enol Tautomerism

For a long time some important 2,5-dialkyl-4-hydroxy-3(2*H*)-furanones like Furaneol®, mesifuran or homofuraneol were successfully stereoanalysed on modified cyclodextrins as chiral stationary phases without any thermally induced racemisation during GC (Fig. 17.6). However, in view of authenticity assessment their stereodifferentiation remains useless, owing to the instability of dihydrofuranones in acidic media. This is the reason why these compounds were detected in strawberries, pineapples, grapes and wines as natural racemates [22,23].

In spite of these exceptional cases, the systematic evaluation of natural enantiomeric ratios has proved to be a valuable criterion for differentiating natural compounds from those of synthetic origin.

17.3 Results and Discussion

17.3.1 Chiral γ-Lactones and δ-Lactones

Owing to their pleasant odours many γ-lactones and δ-lactones are known to be important flavour compounds of fruits and contribute essentially to the characteristic and distinctive notes of strawberries, peaches, apricots and many other fruits [24]. Chiral aroma compounds from fruits and other natural sources are characterised by origin-specific enantiomeric ratios, as their biogenetic pathways normally are catalysed by enzymes.

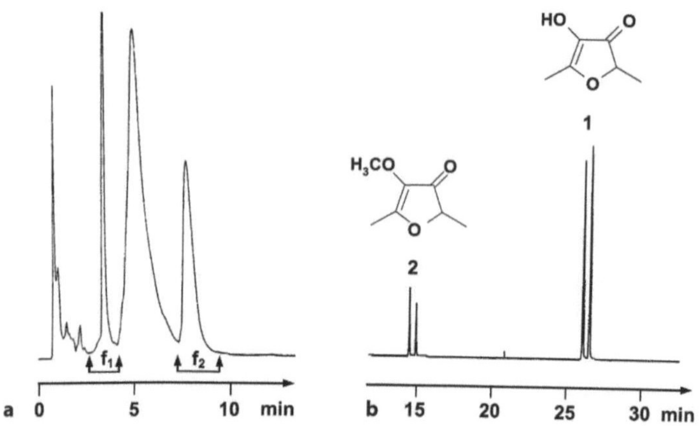

Fig. 17.6 Stereodifferentiation of Furaneol® (*1*) and mesifuran (*2*) from strawberries: **a** HPLC chromatogram of strawberry extract; mesifuran (fraction f_1), Furaneol® (fraction f_2); **b** HPLC fractions, analysed by enantioselective capillary GC [23]

Studies on the biosynthesis of lactones have shown that epoxidation of unsaturated fatty acids like, e.g., linoleic and linolenic acid may represent a common pathway to oxygenated derivatives of fatty acids. Epoxy fatty acid hydrolases were identified as key enzymes that exhibit high regioselectivity and enantioselectivity [25, 26].

Consequently, these intermediates are, in fruits, converted by β-oxidation steps to the corresponding even-numbered γ-lactones and δ-lactones.

The simultaneous stereoanalysis of γ-lactones and δ-lactones using enantio-MDGC has been reported (Fig. 17.7). This technique was applied to many fruits proving that enantiomeric ratios of γ-lactones and δ-lactones can be used as indicators of authenticity, as the genuine enantiomeric purities remain unaffected during fermentation and all other stages of fruit processing [27].

There are only few references on odd-numbered lactones in the literature. The first reports on the natural occurrence of γ-nonalactone and γ-undecalactone are known from the early flavour literature [28–30], long before sophisticated analytical techniques, such as enantio-cGC-MS, became available. These data have to be reevaluated, should the situation arise. Wörner et al. [31] provided the first report on γ-nonalactone among the volatile constituents of *Artemisia vulgaris* L. herb, revealing an amount between 1 and10 µg/kg and an enantiomeric distribution of (*R*)-γ-nonalactone to (*S*)-γ-nonalactone of 34:66 using enantio-MDGC, coupled online with MS.

Solid-phase extraction procedures and quantitative analysis of aliphatic lactones in wine were described by Ferreira et al. [32] dealing with, among others, the quantitation of γ-nonalactone and γ-undecalactone at trace levels.

However, it should be kept in mind that the origin and natural occurrence of odd-numbered γ-lactones is still not understood and their contribution to food flavour impression is rather limited or negligible, when trace amounts—far below their odour thresholds—are detected.

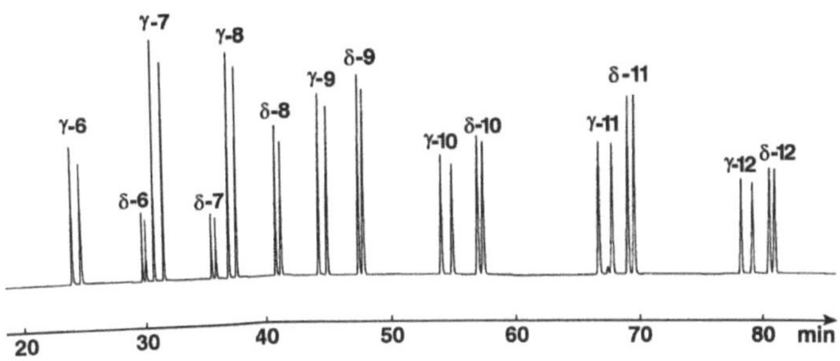

Fig. 17.7 Simultaneous stereoanalysis of γ-lactones and δ-lactones using enantio-MDGC (main column chromatogram of references). Elution order: γ-lactones: 4*R* (*I*), 4*S* (*II*); δ-lactones: δ-C_6, δ-C_7: 5*R* (*I*), 5*S* (*II*); δ-C_8–δ-C_{12}: 5*S* (*I*), 5*R* (*II*) [27]

17.3.2 2-Alkylbranched Acids (Esters)

From the analytical point of view, it is worth noting the biogenetic pathway of 2-methylbutanoic acid starting from isoleucine [(2*S*)-amino-(3*S*)-methylpentanoic acid]. The (*S*)-configuration of the precursor is expected to remain; but also enzymatic racemisation (by enolisation of the intermediate 2-oxo-3-methylpentanoic acid) is known from the literature. It is not surprising that in some cases 2-methylbutanoic acid is detected as an enantiomeric ratio more or less different from the expected homochiral *S* enantiomer (Table 17.2) [35–40].

Even (*R*)-2-methylbutanoic acid of high enantiomeric purity (more than 99%) has been reported as a natural compound in the extract of the steroid alkaloid containing drug *Veratrum album* L. [40].

Certainly, most of the data given in Table 17.2 are not qualified as indicators in authenticity assessment of food flavour, owing to their low and non-characteristic enantiomeric distributions, which could be simulated easily by calculated blending of the (*S*)-enantiomer (from biotechnological origin) with the synthetic racemate.

However, in the case of apples and many other fruits the (*S*)-enantiomer of ethyl 2-methylbutanoate, the impact flavour compound of apples, was identified with high enantiomeric purity, irrespective of the apple variety investigated and was unaffected by processing conditions (e.g. distillation, concentrating) or storage of apple juices.

Of course, during processing of fruit juices hydrolysis effects may occur, leading to decreased amounts of ethyl 2-methylbutanoate. However, its enantiomeric purity remains unchanged, whilst the corresponding 2-methylbutanoic acid is found as the (*S*)-enantiomer (99.5% or more) [33–37]. Consequently, the detection of racemic 2-methybutanoic acid (or the corresponding esters) definitely proves the addition of a synthetic (so called nature-identical) flavour compound.

In the context of EU food law, fruit juices must be genuine; in view of their aroma, only aroma concentrates of the fruit concerned are suitable for fruit juices from concentrates. Other natural flavourings (from other fruits or biotechnology) are not allowed.

17.4 Stir-Bar Sorptive Extraction–Enantioselective Multidimensional Gas Chromatography–Mass Spectrometry

A novel solventless simple technique for extraction of organic analytes from aqueous samples, stir-bar sorptive extraction (SBSE), was introduced by Baltussen et al. [41].

SBSE takes advantage of the high enrichment factors of sorptive beds, but with the application range and simplicity of solid-phase microextraction (SPME)

Table 17.2 Enantiomeric distribution of 2-methylbutanoic acid from different natural origins

	R (%)	*S* (%)	References
Fresh apples	<0.5	>99.5	[33–37]
Processed apples	<0.5	>99.5	[33–37]
Mutton tallow	25	75	[35]
Chamaemelum nobile L.	35	65	[35]
Theobroma cacao L.	30–25	70–75	[37]
Parmesan cheese	37–25	63–75	[38]
Rheum rhabarbarum L.	65	35	[39]
Veratrum album	>99	<1	[40]

[42]. The stir bar is coated with a thick film of poly(dimethylsiloxane) (PDMS), in which the aqueous sample extraction takes place during stirring for a predetermined time. After that time it is removed and placed into a glass tube, which is transferred into a thermal desorption system where the analytes are thermally recovered and evaluated online with a capillary MDGC-MS system (Fig. 17.8).

In addition to the extraction of organic analytes from aqueous samples, the PDMS stir bars are also suitable for headspace and in vivo headspace sampling. Headspace sampling is a technique widely used to characterise the volatile frac-

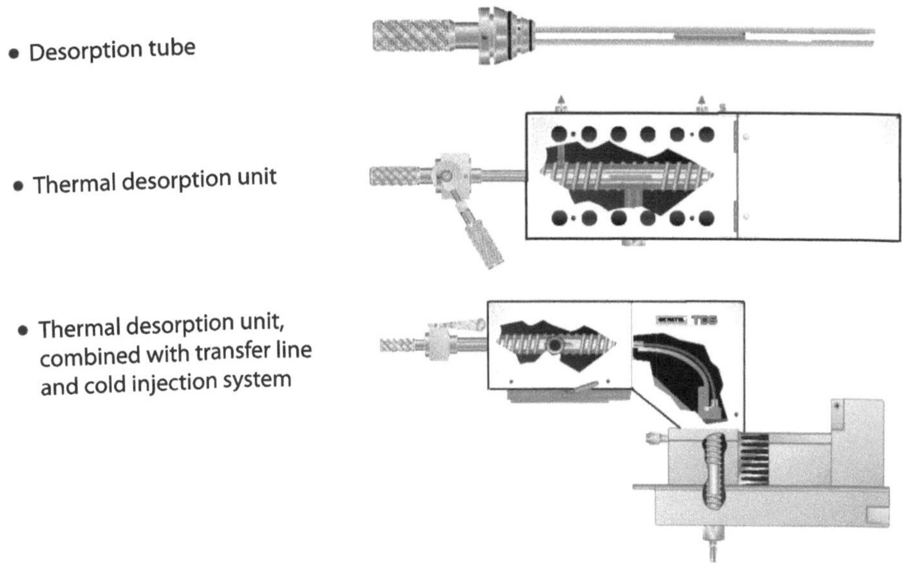

Fig. 17.8 Thermal desorption system (*TDS*), from GERSTEL, Mühlheim, Germany [52]

tion of several matrices, particularly aromatic and medicinal plants. SBSE has also been shown to be a successful technique for headspace sampling, since the PDMS stir bars enrich higher amounts of trapping material than SPME and therefore exhibit better extraction efficiency for analysing minor components [43].

This connection allows the combination of the high extraction efficiency of the stir bar (coated as a thick film of PDMS) with the high selectivity of the enantio-MDGC-MS system [44].

In this way, it is possible to determine the exact enantiomeric ratios of chiral compounds in complex natural materials such as food flavours or essential oils. Even headspace sampling and in vivo headspace sampling from living plants are successfully realised (Fig. 17.9).

17.4.1 Tea Tree Oils

The essential oils from *Melaleuca alternifolia* (Myrtaceae) are recommended for many medicinal and cosmetic purposes. More than 100 varieties of *Melaleuca* are known, having considerable differences in their essential oil composition (Fig. 17.10). In order to standardise the essential oil quality, minimum and maximum conditions are given by DAC (Deutscher Arzneimittel-Codex) and ISO 4730 (1996).

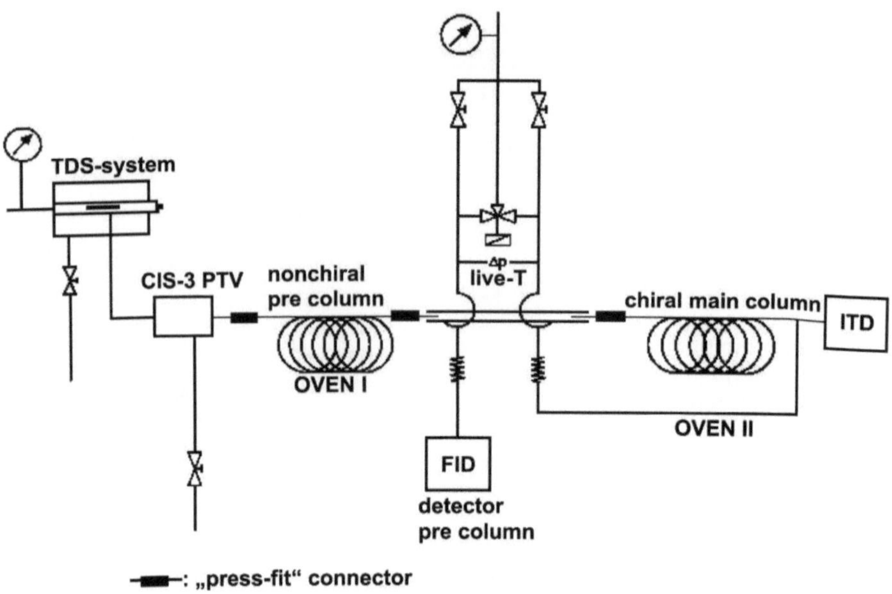

Fig. 17.9 TDS system [45]

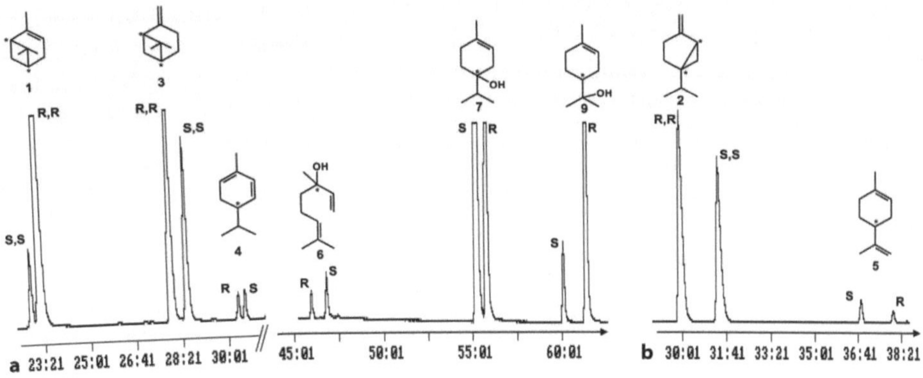

Fig. 17.10 Stir-bar sorptive extraction-*enantio*-MDGC-MS analysis of tea tree oil, main column separation [45]

Unfortunately, enantiomeric purities and total percentages of α-pinene, β-pinene, limonene and α-terpineol from tea tree oils more or less overlap with those of *Eucalyptus* oils (Table 17.3). Only enantiomeric purities and total percentages of terpinen-4-ol and α-phellandrene are significantly different, when *Melaleuca* and *Eucalyptus* oils are compared with regard to authenticity assessment [45].

Table 17.3 Monoterpene compounds from *Melaleuca* and *Eucalyptus* species [45]

	Chiral A	Non-chiral B	Tea tree oil		Eucalyptus oil	
			C	D	C	D
α-Pinene	√		*R*: 86–91	1.5 – 2.5	*R*: 93–99	2.0–8.0
β-Pinene	√		*R*: 58–65	0.1–1.0	*S*: 59–65	<0.5
α-Phellandrene	√		–	<0.1	–	<1.5
Limonene	√		*R*: 62–68	1.0–6.0	*R*: 64–72	4.0–12.0
1,8-Cineol		√	–	<15.0	–	>70.0
Camphor	√		–	–	–	<0.1
Terpinen-4-ol	√		*S*: 65–70	>30.0	*S*: 53–58	<1.0
α-Terpineol	√		*R*: 69–78	1.5–8.0	*R*: 66–72	<4.0

Tea tree oil: Melaleuca alternifolia Cheel, Melaleuca linariifolia Sm., Melaleuca dissitiflora Mueller; Eucalyptus oil: *Eucalyptus globulus* Labill., *Eucalyptus fructicetorum* F. v. Mueller ex Miquel, *Eucalyptus smithii* R. T. Baker
A Enantioselective multidimensional gas chromatography (*MDGC*)–mass spectrometry (*MS*), *B* gas chromatography (*GC*)–isotope ratio mass spectrometry (*IRMS*) multielement analysis ($\delta^{13}C$, $\delta^{2}H$, $\delta^{18}O$ values), *C* enantiomeric purity (%), *D* total percentage (%)

Enantio-cGC, however, fails in the case of non-chiral compounds, such as 1,8-cineol. In this special case 1,8-cineol may be attributed to high-level *Melaleuca* varieties or to the fraudulent addition of *Eucalyptus* oil. In order to get reliable results, enantio-MDGC-MS analysis and/or IRMS measurements (as far as possible) are necessary.

17.4.2
Isotope Discrimination

The natural cycles of the bioelements carbon, oxygen, hydrogen, nitrogen and sulphur) are subjected to various discrimination effects, such as thermodynamic isotope effects during water evaporation and condensation or isotope equilibration between water and CO_2. On the other hand, the processes of photosynthesis and secondary plant metabolism are characterised by kinetic isotope effects, caused by defined enzyme-catalysed reactions [46].

The highly precise measurement of isotope ratios has a long tradition in organic geochemistry. Nowadays, the elucidation of stable isotope distributions is highly desirable in view of fundamental studies in biochemistry, nutrition, drug research and also in the authenticity assessment of food ingredients.

In 1981 Martin and Martin [47] showed that the ^2H distribution in organic molecules does not follow a statistical pattern, but it is discriminated by isotopic effects, measurable by ^2H NMR and IRMS, respectively. Meanwhile, the systematics of $^{18}O/^2H$ patterns in natural plant products are being better and better understood and were reported by Schmidt et al. [48–50] as new and reliable tools for the elucidation of biosynthetic pathways and as helpful indicators in the authenticity assessment of natural compounds.

^2H SNIF-NMR and $^{18}O/^{16}O$ IRMS have been adopted as official methods by the Commission of the European Community for measurement of stable isotope ratios. These methods play a key role in detecting adulterations like addition of water and inadmissible wine sweetening or chaptalisation with beet or cane sugar [51].

17.5
Capillary Gas Chromatography–Isotope Ratio Mass Spectrometry Techniques

17.5.1
Fundamentals

IRMS has become more and more important in food authenticity assessment, since cGC, coupled online via a suitable combustion/pyrolysis interface with IRMS has been realised. The substances eluted from the cGC column are converted into the corresponding gas (carbon dioxide, nitrogen, hydrogen and car-

Table 17.4 IRMS online coupling techniques [52]

GC-combustion-IRMS	$\delta^{13}C$
GC-combustion/reduction-IRMS	$\delta^{15}N$
GC-pyrolysis-IRMS	$\delta^{18}O$, $\delta^{2}H$
Thermochemical conversion/element analyser	$\delta^{18}O$, $\delta^{2}H$

Table 17.5 Specifications for capillary GC-IRMS coupling techniques using DELTA plus XL, Thermo Electron, Bremen, Germany [52]

		On column		
Bioelement	Analysed gas	Need (mol)	Need (ng)	Precision (‰)
Carbon	CO_2	0.8 nmol C	10 ng C	0.2
Nitrogen	N_2	1.5 nmol N_2	42 ng N	0.5
Hydrogen	H_2	15 nmol H_2	30 ng H	3.0
Oxygen	CO	5 nmol O	80 ng O	0.8

bon monoxide, respectively) and are then directly analysed in the isotope mass spectrometer (Tables 17.4, 17.5). The spectrometer is adjusted for the simultaneous recording of the reactant gas isotopomers. Thus, the components can be detected in the nanomolar range with high precision.

17.5.2 Validation

Isotope ratios are given as deviations, in relation to a defined primary standard (zero point). The polyethylene foils CH 7 and NBS 22-oil are commercially available secondary standards, certificated and managed by the International Atomic Energy Agency. However, GC-IRMS systems cannot be calibrated without the aid of alternative peripheries like an elemental analyser (EA) or a dual inlet, owing to the lack of commonly accepted reference materials applicable in GC-IRMS techniques (Fig. 17.11).

In the course of a feasibility study, sponsored by the European Union, the components of the GC separation efficiency test, according to K. Grob, were tested for their usability as certificated tertiary standards. Seven compounds are now available as a ready-to-use mixture for testing the accuracy of the GC-IRMS measurements, and furthermore simultaneously provide important information about the actual quality status of the GC column system used [53].

The isotope ratio traces of the GC peaks exhibit a typical S shape. The heavier isotopic species of a compound are eluted more rapidly than the light species. Similar effects can be observed for all chromatographic processes, whereas the size of the isotope fractionation and the elution order of the isotopomers de-

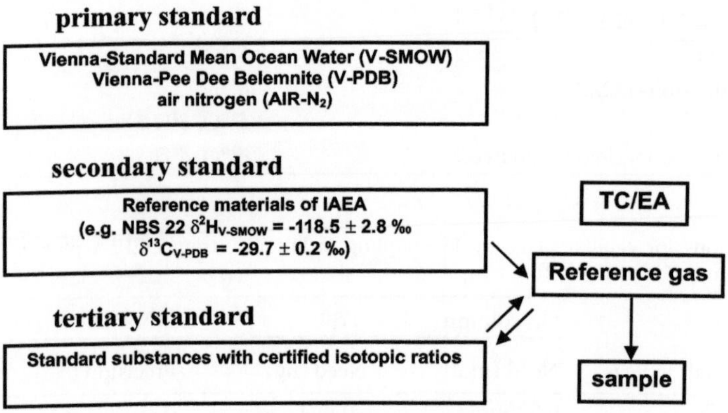

Fig. 17.11 GC–isotope ratio MS (*IRMS*)—calibration of the reference gas [53]

pend on (1) the chromatographic system applied, (2) the temperature and (3) the structural features of the compounds analysed. Care must be taken to integrate across the full width of the chromatographic peaks. Of course, reliable results on isotopic ratios from cGC-IRMS experiments can only be expected from very high resolution cGC ($R_s \geq 1.5$). Also, accurate sample cleanup procedures without any isotope fractionation must be guaranteed. Under the conditions of validated procedures and calibrated instruments, IRMS data are valuable indicators in the authenticity assessment of flavour and fragrance compounds.

The latest development is MDGC coupled online with IRMS. This coupling technique combines the advantages of both highly sophisticated techniques, to achieve the utmost accuracy of IRMS measurements [54].

Indeed, MDGC-IRMS is the method of choice for precise and accurate measurements of compounds from complex matrices, under the condition that the analyte is quantitatively transferred from the precolumn eluate to the main column.

17.6 Comprehensive Authenticity Assessment

17.6.1 (*E*)-*α*-Ionone and (*E*)-*β*-Ionone

The online determination of $\delta^2H_{V\text{-}SMOW}$ values using GC–pyrolysis–IRMS (GC-P-IRMS) was developed recently [55] and has proved to be a powerful tool to define the authenticity of natural compounds [56–61]. However, as fruit flavour extracts are rather complex, and the sample amount for hydrogen measurement has to be rather high owing to the low abundance of deuterium

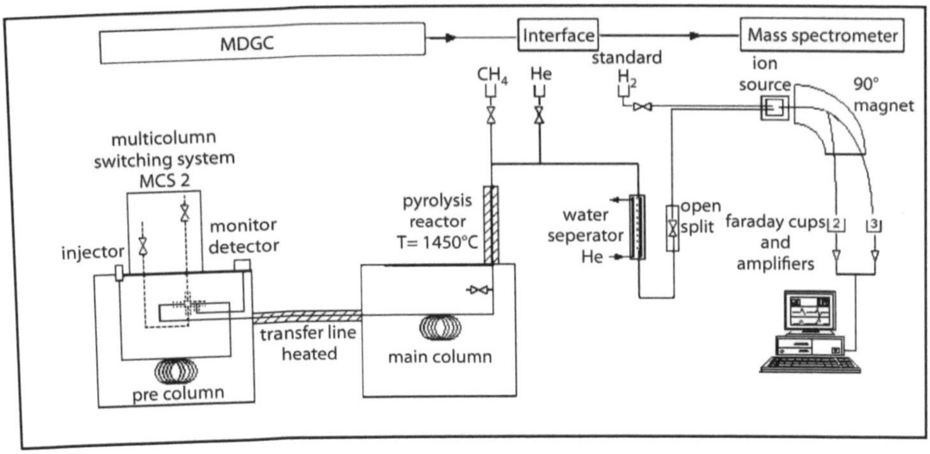

Fig. 17.12 MDGC–pyrolysis–IRMS; precolumn and main column are connected via the multi-column switching system MCS2 (GERSTEL). Cutting is realised by different gas flows through the MCS2 device [71]

isotopes, the use of single GC-IRMS is often not sufficient for the precise and accurate $\delta^2H_{V\text{-}SMOW}$ measurements of characteristic aroma components from fruit flavour extracts. The MDGC-IRMS technique was developed and introduced to the practice of authentication by Juchelka et al. [54] and Asche et al. [62] but until now this technique has been applicable only in the determination of $^{13}C/^{12}C$ ratios, for the following reason. As the carrier gas flow strongly depends on temperature, the classical pressure-controlled column-switching technique, which was introduced by Deans in 1968 and was realised in a modified version within the Siemens Sichromat MDGC system [63], is unsuitable for evaluating $^2H/^1H$ isotope ratios when temperature-programmed column switching becomes necessary.

The importance of a constant carrier gas flow for accurate $^2H/^1H$ isotope ratio measurements was demonstrated by Bilke and Mosandl [64]. A suitable residence time in the reactor is mandatory for a complete and subsequent quantitative pyrolysis, free of isotope discrimination. Furthermore, the amount of sample reaching the reactor is flow-dependent. With higher column temperature and constant gas pressure, the carrier gas flow decreases and less sample will pass the reactor in a certain time interval. This is why the constant-flow MDGC option was recognised as an essential prerequisite of reliable $\delta^2H_{V\text{-}SMOW}$ measurements. To meet these requirements, the multicolumn switching system MCS2 was used (Fig. 17.12). The accuracy and precision of this column-coupling technique is proved by comparative standard measuring using thermochemical conversion (TC)/EA-IRMS and MDGC-P-IRMS (Table 17.6).

By measuring standard compounds [5-nonanone, linalool, (-)-menthol, linalyl acetate, γ-decalactone, (E)-α-ionone, 1-octanol, dodecane, methyl decano-

Table 17.6 Comparison of $\delta^2H_{V\text{-}SMOW}$ values of tertiary standards, measured by thermochemical conversion/element analyser (*TC/EA*)–IRMS and MDGC–pyrolysis–IRMS (*MDGC-P-IMRS*)

	TC/EA-IRMS mean (‰)	MDGC-P-IRMS			
		Mean			Δ(MDGC-TC/EA) (‰)
		n[a]	(‰)	σ^b (‰)	
5-Nonanone	-89±3	30	-88	1.0	1
Linalool	-190±4	30	-190	2.3	0
(-)-Menthol	-242±3	30	-239	1.4	3
Linalyl acetate	-181±4	30	-184	2.0	-3
γ-Decalactone	-191±3	30	-191	1.2	0
(*E*)-*α*-Ionone	-197±3	30	-196	1.8	1
1-Octanol	-68±2	10	-72	1.4	-4
Dodecane	-128±3	10	-127	1.4	1
Methyl decanoate	-246±2	10	-247	0.8	-1
Methyl *N*-methylanthranilate	-133±4	10	-127	0.6	6
Methyl dodecanoate	-250±3	10	-249	1.8	1

[a]Number of measurements
[b]Standard deviation [71]

ate, methyl dodecanoate and methyl *N*-methylanthranilate], comparatively with TC/EA-IRMS and MDGC-P-IRMS, the accuracy of the new method was successfully demonstrated. As summarised in Table 17.6 all values determined via MDGC-P-IRMS comply with the TC/EA-IRMS values within the standard deviation range of 0–6‰. Thus, the direct and non-isotopic discriminating sample preparation via MDGC is proved.

From natural sources the (*R*)-enantiomer of (*E*)-*α*-ionone is detected with high enantiomeric purity (much more than 99%); hence, the authenticity of (*E*)-*α*-ionone is mostly proved via enantio-GC applications [27,65–67]. In the majority of cases synthetic ionones are produced via pseudoionone, prepared by base-catalysed condensation of citral with acetone. After acidic catalysis (using 85% phosphoric acid or concentrated sulphuric acid), this reaction yields racemic (*E*)-*α*-ionone and (*E*)-*β*-ionone [68].

With new upcoming techniques, such as simulated moving bed (SMB) chromatography [69], the production of large amounts of enantiopure (*R*)-configured (*E*)-*α*-ionone from the synthetic (*E*)-*α*-ionone racemate is conceivable, as reported by Zenoni et al. [70]. Consequently, enantioselective analysis is no longer sufficient for a comprehensive authenticity assessment of the named extracts [52] and, in general, the use of multielement/multicomponent IRMS analysis–in addition to enantio-cGC—is becoming more and more important. Constant-flow MDGC–combustion/pyrolysis–IRMS (MDGC-C/P-IRMS) and enantio-MDGC analysis have proved to be the most efficient online coupling techniques in the direct and comprehensive authenticity assessment of chiral and non-chiral analytes, such as (*E*)-*α*-ionone and (*E*)-*β*-ionone, from complex matrices without any risk of discrimination [71].

To point out the relevance of the new coupling technique, Fig. 17.13 shows a precolumn (Fig. 17.13a) and a main column (Fig. 17.13b) chromatogram of a raspberry extract (variety Rucami) measured by MDGC-P-IRMS. The concentrations of (*E*)-*α*-ionone and (*E*)-*β*-ionone were adjusted to the linearity range of the isotope ratio mass spectrometer (peak amplitude 4–7 V).

It is obvious that the precolumn separation of (*E*)-*α*-ionone is not sufficient for precise isotopic measurements. However, by cutting exclusively the precolumn section of (*E*)-*α*-ionone and (*E*)-*β*-ionone onto the main column, a sufficient chemical purification and adequate performance are achieved. To avoid isotopic discrimination during cutting, as reported by Juchelka et al. [54], the cut is chosen to be rather broad and both ionones are transferred by the same cut.

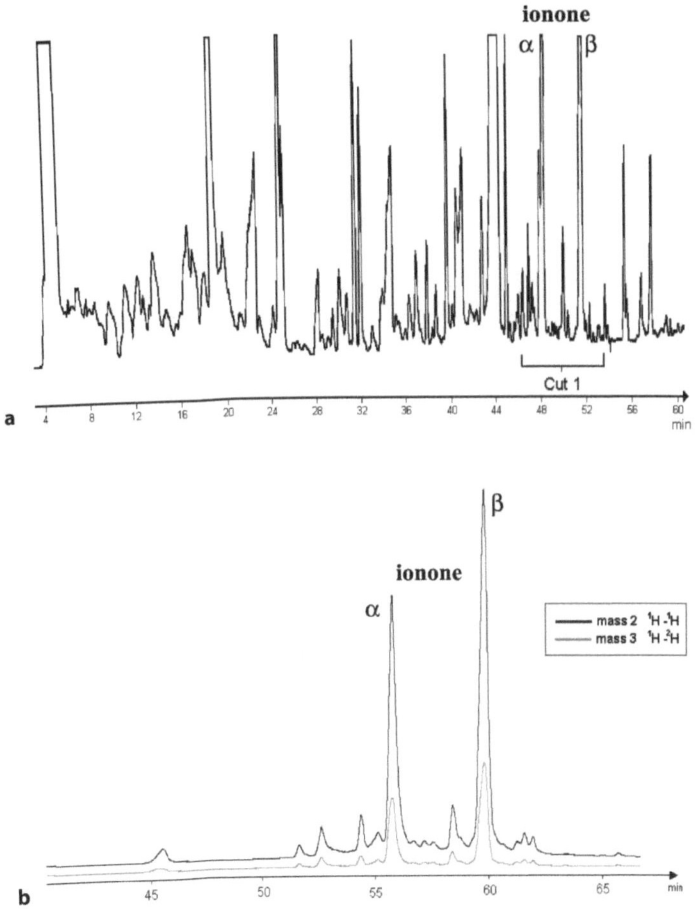

Fig. 17.13 Precolumn (**a,** flame ionisation detection) and main column (**b,** selected ion monitoring detection)–chromatograms of a raspberry extract [71]

17.6.2 Lavender Oil

For hundreds of years the essential oil of lavender has been well appreciated for perfumery purposes [72]. Lavender oil is obtained by steam distillation from the fresh-flowering tops of *Lavandula angustifolia* Miller (*Lavandula officinalis* Chaix) [73]. It is a colourless or pale yellow, clear liquid, with a fresh, sweet, floral, herbaceous odour on a woody balsamic base [73, 74]. According to the European Pharmacopoeia, characteristic components of lavender oils are limonene, cineol, 3-octanone, camphor, linalool, linalyl acetate, terpinen-4-ol, lavandulyl acetate, lavandulol and α-terpineol. Adulterations commonly include blends of lavender oils with lavandin oil or spike oil, and the addition of synthetic linalool and linalyl acetate. In contrast, genuine lavender oils contain as main constituents (*R*)-linalyl acetate and (*R*)-linalool of high enantiomeric purity (Fig. 17.14).

For that reason enantioselective analysis of linalool and linalyl acetate proved to be a powerful tool to detect adulterations with synthetic racemates of linalool and linalyl acetate, respectively [56, 75].

To conclude from the latest documentation of the European Directorate for the Quality of Medicines, European Pharmacopoeia Commission, the enantiomeric purity of linalool and linalyl acetate has been adopted into monograph no. 1338 *Lavender Oil of European Pharmacopoeia*. In accordance with this documentation, the percentage content of linalool (20.0–45.0%) and linalyl acetate (25.0–46.0%) in conjunction with the specification of (*S*)-linalool (maximum 12%) and (*S*)-linalyl acetate (maximum 1%) is now defined as a concept for the authenticity assessment of lavender oil [73]. However, by using upcoming techniques like SMB chromatography [69], the generation of large amounts of enantiopure (*R*)-linalool from synthetic racemate has become realistic. Consequently, enantioselective analysis may no longer be sufficient [52] and strategies for comprehensive authenticity assessment have been realised, including multielement/multicomponent GC/IRMS measurements as well as enantio-MDGC-MS analysis.

The determination of $\delta^{13}C_{V\text{-}PDB}$ and $\delta^{2}H_{V\text{-}SMOW}$ values of synthetic and natural linalool and linalyl acetate using IRMS has been reported by different authors [76–81]. With use of a pyrolysis interface, the determination of $^{18}O/^{16}O$ isotope ratios was proved to be a further important indicator in the authenticity assessment of lavender oils. In terms of authenticity assessment, three-dimensional plots of the $\delta^{18}O_{V\text{-}SMOW}$, $\delta^{2}H_{V\text{-}SMOW}$ and $\delta^{13}C_{V\text{-}PDB}$ values have been presented for both linalool and linalyl acetate [82].

A reliable authenticity assessment is concluded from the simultaneous consideration of multielement IRMS and enantioselective analysis. The differences of the stable isotope ratios of linalool and linalyl acetate are depicted as a three-dimensional plot of Δ values (δ values of linalool minus δ values of linalyl acetate for oxygen, hydrogen and carbon) (Fig. 17.15). This plot shows that the commercial samples S1–S5 are different from all the other samples investigated. Linalool and linalyl acetate of S1–S5 definitely are not genuine lavender oil compounds.

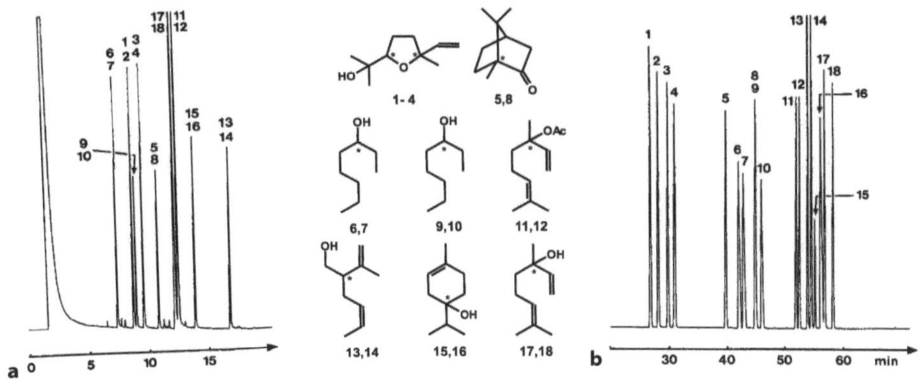

Fig. 17.14 Simultaneous stereoanalysis of *Lavandula* oil constituents, using enantio-MDGC (standard mixture). **a** Preseparation of racemic compounds; unresolved enantiomeric pairs of octan-3-ol (*6, 7*), *trans*-linalool oxide (*1, 2*), oct-1-en-3-ol (*9, 10*), *cis*-linalool oxide (*3, 4*), camphor (*5, 8*), linalool (*17, 18*), linalyl acetate (*11, 12*), terpinen-4-ol (*15, 16*) and lavandulol (*13, 14*). **b** Chiral resolution of enantiomeric pairs, transferred from the precolumn: *trans*-linalool oxide *1* (2*S*,5*S*), *2* (2*R*,5*R*); *cis*-linalool oxide *3* (2*R*,5*S*), *4* (2*S*,5*R*); camphor *5* (1*S*), *8* (1*R*); octan-3-ol *6 R*, *7 S*; oct-1-en-3-ol *9 S*, *10 R*; linalyl acetate *11 R*, *12 S*; lavandulol *13 R*, *14 S*; terpinen-4-ol *15 R*, *16 S*; linalool *17 R*, *18 S*. [75]

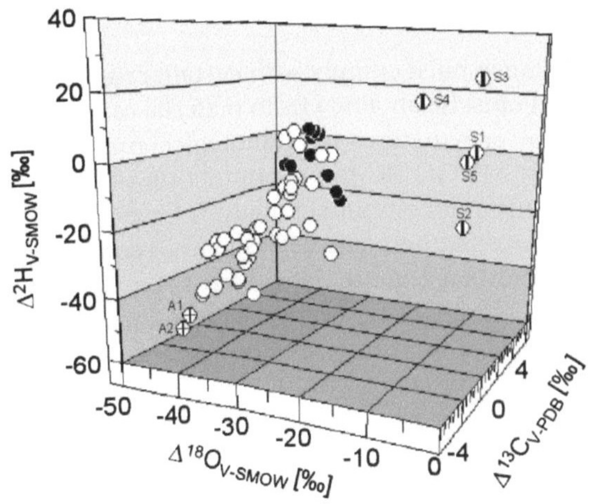

Fig. 17.15 Multielement IRMS analysis of lavender oil main compounds. Differential diagram ($\Delta = \delta_{\text{linalool}} - \delta_{\text{linalyl acetate}}$); authentic (*black circles*) and commercial (*white circles*) samples; commercial non-authentic (*circles with a line through*) and special aberrations (*circles with a cross*) [82]

Table 17.7 Enantiomeric ratios of linalool and linalyl acetate from non-authentic samples [82]

Sample	Linalool		Linalyl acetate	
	R	*S*	*R*	*S*
A1	96.2	3.8	>99.9	<0.1
A2	95.9	4.1	>99.9	<0.1
S1	70.7	29.3	52.9	47.1
S2	55.5	45.5	55.7	44.3
S3	62.0	38.0	51.8	48.2
S4	69.7	30.3	52.0	48.0
S5	60.8	39.2	53.3	46.7

According to Kreis and Mosandl [75] high enantiomeric purities of (*R*)-linalool (above 94 %) and (*R*)-linalyl acetate (above 98%) are known as characteristics of authentic lavender oils. In samples A1 and A2 the enantiomeric purity of (*R*)-linalyl acetate is better than ever detected in authentic lavender oils (above 99.9% (*R*)-enantiomer), whilst (*R*)-linalool is detected at the lower range of admissible purity of genuine linalool from lavender, produced under good manufacturing practice conditions [96.2 and 95.9% (*R*)-linalool, respectively]. Consequently, considering $\Delta\delta$ values in conjunction with enantio-MDGC-MS analysis leads to the conclusion that linalool and linalyl acetate from samples A1 and A2 and S1–S5 do not originate from lavender (Table 17.7, Fig. 17.15).

17.7 Concluding Remarks

In the legal sense a flavouring substance must comply with certain criteria if referred to as "natural". For example, it must be obtained from material of vegetable or animal origin, isolated by physical, enzymatic or microbiological processes.

Enantio-MDGC and/or (enantio)-MDGC-IRMS in conjunction with multi-element ($\delta^{13}C$, $\delta^{2}H$, $\delta^{18}O$, $\delta^{15}N$)- and multicomponent analysis have proved to be highly efficient in the comprehensive authenticity assessment of compounds originating from biogenesis, provided that concise data about (1) genuine enantiomers and/or isotopic ratios, (2) limits of natural variations and (3) their behaviour during processing or storage of foodstuffs are known.

All details about starting materials and central production features should be known, in order to define exactly the status "natural" of a flavouring substance. In principle, the burden of proof is the responsibility of the producer. If necessary, production documents should be disclosed, in order to get objective authenticity assessment by qualified and authorised experts. Constructive cooperation between food researchers, the food industry and authorities will be stimulating to quality assessment in the food industry and will enhance consumer confidence [83].

Acknowledgements

The presented work of the author's group was supported by the Deutsche Forschungsgemeinschaft (DFG) and the Forschungskreis der Ernährungsindustrie (FEI). Last, but not least, I would like to thank my co-workers for high motivation and enthusiastic support.

References

1. Schurig V, Novotny H-P (1988) Separation of enantiomers on diluted permethylated ß-cyclodextrin by high-resolution gas chromatography. J Chromatogr 441:155
2. König WA, Lutz S, Mischnick-Lübbecke P, Brassat B, Wenz G (1988) Cyclodextrins as chiral stationary phases in capillary gas chromatography. I. Pentylated α-cyclodextrin. J Chromatogr 447:193
3. König WA, Lutz S, Wenz G, van der Bey E (1988) Cyclodextrins as chiral stationary phases in capillary gas chromatography II. Heptakis(3-O-acetyl-2,6-di-O-pentyl)-β-cyclodextrin. J High Resolut Chromatogr Chromatogr Commun 11:506
4. Armstrong DW, Chang C-D, Li WY (1990) Relevance of enantiomeric separation in food and beverage analyses. J Agric Food Chem *38*:1674
5. Dietrich A, Maas B, Karl V, Kreis P, Lehmann D, Weber B, Mosandl A (1992) Stereoisomeric flavor compounds, part LV: Stereodifferentiation of some chiral volatiles on heptakis(2,3-di-O-acetyl-6-O-*tert*-butyl-dimethylsilyl)-ß-cyclodextrin. J High Resolut Chromatogr 15:176
6. Dietrich A, Maas B, Messer W, Bruche G, Karl V, Kaunzinger A, Mosandl A (1992), Stereoisomeric flavor compounds, part LVIII: The use of heptakis(2,3-di-O-methyl-6-O-*tert*-butyl-dimethylsilyl)-ß-cyclodextrin as a chiral stationary phase in flavor analysis. J High Resolut Chromatogr 15:590
7. Saturin C, Tabacchi R, Saxer A (1993) Gas chromatographic analysis of racemic mixtures on peralkylated cyclodextrins. Chimia 47:221
8. Bicchi C, D'Amato A, Manzin V, Galli A, Galli M (1996) Cyclodextrin derivatives in gas chromatographic separation of racemic mixtures of volatile compounds. X. (2,3-di-O-ethyl-6-O-*tert*-butyl-dimethylsilyl)-ß-and -γ-cyclodextrins. J Chromatogr A 742:161
9. Takahisa E, Engel K-H (2005) 2,3-Di-O-methoxymethyl-6-O-*tert*-butyl-dimethylsilyl-γ-cyclodextrin: a new class of cyclodextrin derivatives for gas chromatographic separation of enantiomers. J Chromatogr A 1063:181
10. Takahisa E, Engel K-H (2005) 2,3-Di-O-methoxymethyl-6-O-*tert*-butyldimethysilyl-β-cyclodextrin, a useful stationary phase for gas chromatographic separation of enantiomers. J Chromatogr A 1076:148
11. Kreis P, Dietrich A, Mosandl A (1996) Elution order of the furanoid linalool oxides on common gas chromatographic phases and modified cyclodextrin phases. J Essent Oil Res *8*:339
12. Weinert B, Wüst M, Mosandl A Hanssum H (1998) Stereoisomeric flavour compounds LXXVIII. Separation and structure elucidation of the pyranoid linalool oxide stereoisomers using common gas chromatographic phases, modified cyclodextrin phases and nuclear magnetic resonance spectroscopy. Phytochem Anal *9*:10
13. Bayer M (2006) Entwicklung neuer Trennphasen und -methoden für die enantioselektive Chromotographie. Dissertation, University of Frankfurt

14. Mosandl A, Hener U, Fuchs S (2000) Natürliche Duft-und Aromastoffe—Echtheitsbewertung mittels enantioselektiver Kapillar-GC und/oder Isotopenmassenspektrometrie. In: Analytiker-Taschenbuch B 21. Springer, Berlin Heidelberg New York, p 37
15. Mosandl A, Hener U, Hagenauer-Hener U, Kustermann A (1989) Stereoisomeric flavor compounds XXXII: Direct enantiomer separation of chiral γ-lactones from food and beverages by multidimensional gas chromatography. J High Resolut Chromatogr 12:532
16. Guichard E, Kustermann A, Mosandl A (1990) Chiral flavour compounds from apricots—distribution of γ-lactone enantiomers and stereodifferentiation of dihydroactinidiolide using multidimensional gas chromatography. J Chromatogr 498:396
17. Mori K, Khlebnikov V (1993) Carotenoids and degraded carotenoids VIII: Synthesis of (+)-dihydroactinidiolide, (+)- and (-)-actinidiolide, (+)- and (-)-loliolide as well as (+)- and (-)-epiloliolide. Liebigs Ann Chem 77
18. Yao S, Johannsen M, Hazell RA, Jörgensen KA (1998) Total synthesis of (R)-dihydroactinidiolide using asymmetric catalytic hetero-Diels-Alder methodology. J Org Chem 63:118
19. Schmidt CO, Bouwmeester HJ, de Kraker J-W, König WA (1998) Biosynthese von (+) und (-) Germacren D in *Solidago canadensis:* Isolierung und Charakterisierung zweier enantioselektiver Germacren-D-Synthasen. Angew Chem 110:1479
20. Schmidt CO, Bouwmeester HJ, Franke S, König WA (1999) Mechanisms of the biosynthesis of the sesquiterpene enantiomers (+) and (-) germacrene D in *Solidago canadensis.* Chirality 11:353
21. Steliopoulos P (2002) Biogenesestudien und Authenzitätsbewertung mittels stabiler Isotope. Dissertation, University of Frankfurt
22. Mosandl A, Bruche G, Askari C, Schmarr H-G (1990) Steroisomeric flavor compounds XLIV: Enantioselective analysis of some important flavor molecules. J High Resolut Chromatogr 13:660
23. Bruche G, Schmarr H-G, Bauer A, Mosandl A, Rapp A, Engel L (1991) Stereoisomere Aromastoffe LI: Stereodifferenzierung chiraler Furanone—Möglichkeiten und Grenzen der herkunftsspezifischen Aromastoff-Analyse. Z Lebensm Forsch 193:115
24. Maga JA (1976) Lactones in Foods. Crit Rev Food Sci Nutr *8*:1
25. Schöttler M, Boland W (1996) Biosynthesis of dodecano-4-lactone in ripening fruits: Cruial role of an epoxide-hydrolase in enantioselective generation of aroma components of the nectarine (*Prunus persica* var. *nucipersica*) and the strawberry (*Fragaria ananassa*). Helv Chim Acta 79:1488
26. Garbe L-A, Tressl R (2004) Metabolism of deuterated *threo*-dihydroxy fatty acids in *Saccharomyces cerevisiae:* Enantioselective formation and characterization of hydroxylactones and γ-lactones. Helv Chim Acta 87:180
27. Lehmann D, Dietrich A, Schmidt S, Dietrich H, Mosandl A (1993) Stereodifferenzierung von $\gamma(\delta)$-Lactonen und (E)-α-Ionon verschiedener Früchte und ihrer Verarbeitungsprodukte. Z Lebensm Unters Forsch 196:207
28. Fenaroli G (1975) Fenaroli's Handbook of Flavor Ingredients. CRC, Boca Raton, p 550
29. Opdyke DLJ (1975) Monographs on fragrance raw materials. γ-Nonalactone. Food and Cosmet Toxicol 13:889
30. Opdyke DLJ (1975) Monographs on fragrance raw materials. γ-Undecalactone. Food and Cosmet Toxicol 13:921

31. Wörner M, Pflaum M, Schreier P (1991) Additional volatile constituents of *Artemisia vulgaris* L. herb. Flavour Fragr J 6:257
32. Ferreira V, Jarauta I, Ortega L, Cacho J (2004) Simple strategy for the optimization of solid-phase extraction procedures through the use of solid-liquid distribution coefficients—application to the determination of aliphatic lactones in wine. J Chromatogr A 1025:147
33. Rettinger K, Karl V, Schmarr HG, Dettmar F, Hener U, Mosandl A (1991) Chirospecific analysis of 2-alkylbranched alcohols, -acids, and -esters; chirality evaluation of 2-methylbutanoates from apples and pineapples. Phytochem Anal 2:184
34. Karl V, Rettinger K, Dietrich H, Mosandl A (1992) 2-Alkylverzweigte Aromastoffe—Struktur, Geruch und chirospezifische Analyse. Dtsch Lebensm Rundsch 88:147
35. Karl V (1994) Chirale Aromastoffe—Alkylverzweigte Säuren, Ester und Alkohole—Analyse und Reindarstellung der Enantiomeren. Dissertation, University of Frankfurt
36. Schumacher K, Asche S, Heil M, Mittelstädt F, Dietrich H, Mosandl A (1998) Methyl branched flavor compounds in fresh and processed apples. J Agric Food Chem. 46:4496
37. Schumacher K (1999) Methoden zur Authentizitätskontrolle von Fruchtaromen. Dissertation, University of Frankfurt
38. Werkhoff P, Brennecke S, Bretschneider W, Güntert M, Hopp R, Surburg H (1993) Chirospecific analysis in essential oil, fragrance and flavor research. Z Lebensm Unters Forsch *196*:307
39. Dregus M, Schmarr H-G, Takahisa E, Engel K-H (2003) Enantioselective analysis of methyl-branched alcohols and acids in rhubarb (*Rheum rhabarbarum* L.) stalks. J Agric Food Chem 51:7086
40. Mosandl A, Rettinger K, Weber B, Henn D (1990) Untersuchungen zur Enantiomerenverteilung von 2-Methylbuttersäure in Früchten und anderen Lebensmitteln mittels multidimensionaler Gaschromatographie (MDGC). Dtsch Lebensm Rundsch 86:375
41. Baltussen E, Sandra P, David F, Cramers C (1999) Stir bar sorptive extraction (SBSE), a novel extraction technique for aqueous samples: theory and principles. J Microcolumn Sep 11:737
42. Pawliszyn J (ed) (1999) Application of the Solid Phase Microextraction. Royal Society of Chemistry: Cambridge
43. Bicchi C, Cordero C, Iori C, Rubiolo P, Sandra P (2000) Headspace sorptive extraction (HSSE) in the headspace analysis of aromatic and medicinal plants. J High Resol Chromatogr 23:539
44. Kreck M, Scharrer A, Bilke S, Mosandl A (2001) Stir bar sorptive extraction (SBSE)—enantio-MDGC-MS, a rapid method for the enantioselective analysis of chiral flavour compounds in strawberries. Eur Food Res Technol 213:389
45. Kreck M Scharrer A Bilke S Mosandl A (2002) Enantioselective analysis of monoterpene compounds in essential oils by stir bar sorptive extraction (SBSE)-enantio-MDGC-MS. Flavour Fragrance J 17:32
46. Schmidt H-L, Gleixner G (1998) Isotopic patterns in natural compounds origin and importance in authenticity analysis. In: SchreierP, Herderich M, Humpf H-U, Schwab W (eds) Natural Product Analysis. Vieweg, Braunschweig, p 271
47. Martin GJ, Martin ML (1981) Deuterium labeling at the natural abundance level as studied by high field quantitative ^{2}H NMR. Tetrahedron Lett 22:3525
48. Schmidt H-L, Werner RA, Eisenreich W (2003) Systematics of ^{2}H patterns in natural compounds and its importance for the elucidation of biosynthetic pathways. Phytochem Rev 2:61
49. Schmidt H-L, Werner RA, Roßmann A (2001) ^{18}O Pattern and biosynthesis of natural plant products. Phytochemistry 58:9

50. Schmidt H-L, Eisenreich W (2001) Systematic and regularities in the origin of ^{2}H patterns in natural compounds. Isotopes Environ Health Stud 37:253
51. Christoph N (2003) Possibilities and limitations of wine authentication using stable isotope and meteorological data, data banks and statistical tests. Part 1: Wines from Franconia and Lake Constance 1992 to 2001. Mitt. Klosterneuburg 53:23
52. Mosandl A (2004) Authenticity assessment—a permanent challenge in food flavor and essential oil analysis. J Chromatogr Sci 42:440
53. Mosandl A (2004) Authentizitätsbewertung von Aromastoffen mittels enantio-GC und Isotopen- MS. Mitt Lebensm Hyg 95:618
54. Juchelka D, Beck T, Hener U, Dettmar F, Mosandl A (1998) Multidimensional gas chromatography, online coupled with isotope ratio mass spectrometry (MDGC-IRMS): Progress in the analytical authentication of genuine flavor components. J High Resolut Chromatogr 21:145
55. Hilkert AW, Douthitt CB, Schlüter HJ, Brand WA (1999) Isotope ratio monitoring gas chromatography/mass spectrometry of D/H by high temperature conversion isotope ratio mass spectrometry. Rapid Commun Mass Spectrom 13:1226
56. Bilke S, Mosandl A (2002) Authenticity assessment of lavender oil using GC-P-IRMS: ^{2}H/^{1}H-ratios of linalool and linalyl acetate. Eur Food Res Technol 214:532
57. Bilke S, Mosandl A (2002) ^{2}H/^{1}H-and ^{13}C/^{12}C isotope ratios of trans-anethole using gas chromatography – isotope ratio mass spectrometry. J Agric Food Chem 50:3935
58. Preston C, Richling E, Elss S, Appel M, Heckel F, Hartlieb A, Schreier P (2003) On-line gas chromatography combustion/pyrolysis isotope ratio mass spectrometry (HRGC-C/P-IRMS) of pineapple (Ananas comosus L. Merr.) volatiles. J Agric Food Chem 51:8027
59. Fink K, Richling E, Heckel F, Schreier P (2004) Determination of ^{2}H/^{1}H and ^{13}C/^{12}C isotope ratios of (E)-methyl cinnamate from different sources using isotope ratio mass spectrometry. J Agric Food Chem 52:3065
60. Kahle K, Preston C, Richling E, Heckel F, Schreier P (2005) On-line gas chromatography combustion/pyrolysis isotope ratio mass spectrometry (HRGC-C/P-IRMS) of major volatiles from pear fruit (Pyrus communis) and pear products. Food Chem 91:449
61. Tamura H, Appel M, Richling E, Schreier P (2005) Authenticity assessment of γ- and δ-decalactone from Prunus fruits by gas chromatography combustion/pyrolysis isotope ratio mass spectrometry (GC-C/P-IRMS). J Agric Food Chem 53:5397
62. Asche S, Beck T, Hener U, Mosandl A (2000) Multidimensional gas chromatography, online coupled with isotope ratio mass spectrometry (MDGC-IRMS): a new technique for analytical authentication of genuine flavour components. In: Frontiers of Flavour Science. DFA, Garching
63. David F, Sandra P (1987) Capillary Gas Chromatography in Essential Oil Analysis. Hüthig, Heidelberg
64. Bilke S, Mosandl A (2002) Measurements by gas chromatography/pyrolysis/mass spectrometry: fundamental conditions in ^{2}H/^{1}H isotope ratio analysis. Rapid Commun Mass Spectrom 16:468
65. Braunsdorf R, Hener U, Lehmann D, Mosandl A (1991) Analytische Differenzierung zwischen natürlich gewachsenen, fermentativ erzeugten und synthetischen (naturidentischen) Aromastoffen I: Herkunftsspezifische Analyse des (E)-α(β)-Ionons. Dtsch Lebensm Rundsch 87:277

66. Werkhoff P, Bretschneider W, Güntert M, Hopp R, Surburg H (1991) Chirospecific analysis in flavor and essential oil chemistry. Part B. Direct enantiomer resolution of trans-α-ionone and trans-α-damascone by inclusion gas chromatography. Z Lebensm Unters Forsch 192:111
67. Larsen M, Poll L (1990) Odour thresholds of some important aroma compounds in raspberries. Z Lebensm Unters Forsch 191:129
68. Brenna E, Fuganti C, Serra S, Kraft P (2002) Optically active ionones and derivatives: preparation and olfactory properties. Eur J Org Chem 967
69. Juza M, Mazotti M, Morbidelli M (2000) Simulated moving-bed chromatography and its application to chirotechnology. Tibtech 18:108
70. Zenoni G, Quattrini F, Mazzotti M, Fuganti C, Morbidelli M (2002) Scale-up of analytical chromatography to the simulated moving bed separation of the enantiomers of the flavor norterpenoids α-ionone and α-damascone. Flavour Fragr J 17:195
71. Sewenig S, Bullinger D, Hener U, Mosandl A (2005) Comprehensive authentication of (E)-$\alpha(\beta)$-ionone from raspberries, using constant flow MDGC-C/P-IRMS and enantio-MDGC/MS. J Agric Food Chem 53:838
72. Roth l, Kormann K (1997) Duftpflanzen Pflanzendüfte. ecomed, Landsberg
73. European Pharmacopoeia Commission (2004) PA/PH/Exp. 13A/T (00) 40 DEF monograph no 1338
74. Bauer K, Garbe D, Surburg H (1990) Common Fragrance and Flavor Materials. VCH, Weinheim
75. Kreis P, Mosandl A (1992) Chiral compounds of essential oils XI: Simultaneous stereoanalysis of Lavandula oil constituents. Flavour Fragr J 7:187
76. Hener U, Braunsdorf R, Kreis P, Dietrich A, Maas B, Euler E, Schlag B, Mosandl A (1992) Chiral compounds of essential oils X: The role of linalool in the origin evaluation of essential oils. Chem Mikrobiol Technol Lebensm 14:129
77. Schmidt H-L, Werner RA, Eisenreich W (2003) Systematics of ^{2}H patterns in natural compounds and its importance for the elucidation of biosynthetic pathways. Phytochem Rev 2:61
78. Culp RA, Noakes JE (1992) Determination of synthetic components in flavor by deuterium/hydrogen isotopic ratios. J Agric Food Chem 40:1892
79. Hanneguelle S, Thibault J-N, Naulet N, Martin GJ (1992) Authentication of essential oils containing linalool and linalyl acetate by isotopic methods. J Agric Food Chem 40:81
80. Hör K, Ruff C, Weckerle B, König T, Schreier P (2000) Flavor authenticity studies by ^{2}H/^{1}H ratio determination using on-line gas chromatography pyrolysis isotope ratio mass spectrometry. J Agric Food Chem 49:21
81. Schmidt H-L, Rossmann A, Werner RA (1998) Flavourings. Wiley-VCH, Weinheim
82. Jung J, Sewenig S, Hener U, Mosandl A (2005) Comprehensive authenticity assessment of lavender oils using multielement/ multicomponent IRMS-analysis and enantioselective MDGC-MS. Eur Food Res Technol 220:232
83. Lebensmittelchemische Gesellschaft (2004) Authentizität von Aromastoffen. Lebensmittelchemie 58:54

18 Flavour-Isolation Techniques

Gary A. Reineccius
Department of Food Science and Nutrition,
University of Minnesota,
1334 Eckles Ave., Saint Paul, MN 55108 USA

18.1 Introduction

The challenges of isolating flavouring components from complex matrices is common in the analytical laboratory as well as in a manufacturing environment. In the laboratory, one has to isolate micro quantities of flavouring compounds to permit sophisticated gas chromatographic, mass spectrometric or liquid chromatographic methods to be applied in research or quality assurance settings. In manufacturing, the task is generally to isolate either a single chemical (perhaps a product of biotechnology) or a flavouring material from plant sources, reaction vessels, fermentation vats, or waste streams. While each of these tasks offers differences in matrices, target compounds, and/or production scale, there is a commonality in techniques applied. In this chapter we will discuss some of the basic methods for the isolation of flavourings and elaborate on their application in each of these settings.

18.2 Isolation of Flavour Compounds for Analysis

The task of isolating trace quantities of flavouring components from biological systems (plant or animal ingredients, or finished foods) for instrumental analysis is formidable. Most natural systems are composed of several hundred flavouring components that have an exceedingly broad range of chemical and physical properties. They are usually present in very low quantities (parts per million or parts per billion) in complex, natural matrices. This precludes ever truly producing a flavour isolate that is identical to the flavour profile in the food matrix itself but rather reflects the biases inherent to the method used to produce the flavour isolate. Unfortunately, there is no ideal method of isolating flavourings from a food so the researcher must pay great attention to the objectives of the study and ensure that the portion of flavour to be studied is truly reflected in the flavour isolate so produced. This is often done by smelling aroma isolates or tasting non-volatile flavour isolates to determine if the desired sensory properties are present in the isolate. If the isolate does not possess the desired sensory attribute, then an alternative flavour-isolation method must be applied.

With this introduction, we will present an overview of some of the more commonly used methods for aroma isolation. The reader is encouraged to obtain a comprehensive review of the topic for more detail and is referred to [1–3].

18.2.1 Absorption (Polymer Trapping, Solid-Phase Microextraction, Stir Bar, Solid-Phase Extraction)

Absorption methods (sorptive extraction) have become the method of choice for many researchers. They offer advantages of being rapid, solventless, automated, and reasonably sensitive and broad in isolation properties. However, they provide an aroma isolate that reflects the biases resulting from compound volatility and affinity for the absorbent matrix.

The earliest application of this method was the trapping of volatiles entrained in a gas stream on a granular polymer matrix, e.g. Tenax® (Fig. 18.1). The polymers chosen typically have little affinity for water but significant affinity for less polar analytes (e.g. aroma compounds). Thus, a food may be purged with an inert gas, and water (the major volatile compound in any food) passes through the trap, while most aroma compounds are absorbed on the polymer. The loaded polymer may be solvent-extracted, or thermally desorbed in the injection port of a gas chromatograph or an apparatus specifically designed for this purpose (e.g. a thermal desorber, [4]). Normally, one can use relatively large amounts of the absorbent, and thus trapping is quite efficient. This method is still very popular today.

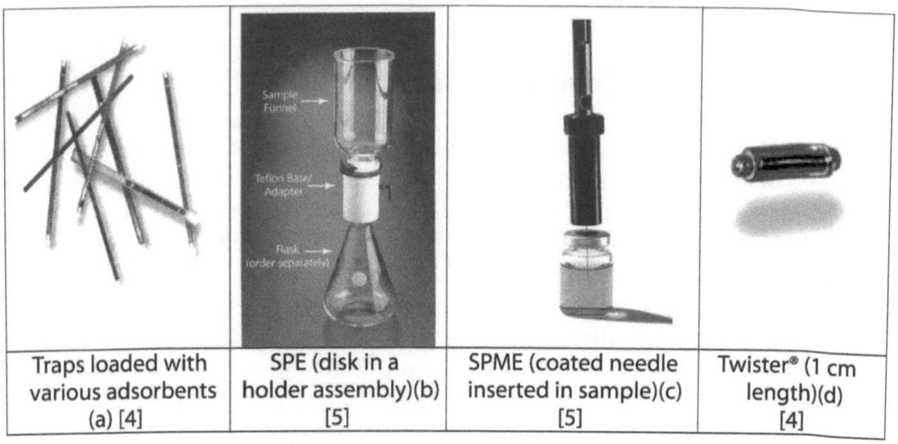

Fig. 18.1 Systems used to absorb aroma compounds from samples for analytical purposes. **a** Traps loaded with various adsorbents [4]. **b** Solid-phase extraction (disk in a holder assembly) [5]. **c** Solid-phase microextraction (coated needle inserted in sample) [5]. **d** Twister® (1-cm length) [4]. (Courtesy of GERSTEL GmbH and Co. KG)

A contemporary of the method just described is the use of an absorbent (e.g. C-18) bonded onto granular or disk-type supports (solid-phase extraction [5]). The granular material is used in cartridge form (typically less than 5 ml), while disk forms are placed in a funnel/holder such as shown in Fig. 18.1b. A liquid (e.g. water, milk, or juice) would be passed through the cartridge (or filter disk), the analytes absorbed in the stationary matrix, the absorbent washed with water, and then the analytes of interest eluted from the absorbent with an organic solvent. This method has found limited use in the isolation of volatiles from foods but continues to find significant application in the analytical field overall [6].

The most commonly used absorbent method in use today is solid-phase microextraction (SPME; Fig. 18.1c) [7–9]. In this method, an inert needle is coated with an absorbent (Table 18.1). The absorbent-coated needle may be placed above a food product, or in the food product. Depending upon the type of coating placed on the needle, volatiles with an affinity for the absorbent will migrate from the food matrix to the needle coating and be absorbed there.

The absorbed volatiles can be desorbed from the needle coating by placing the needle in a hot gas chromatographic (GC) injection port for several minutes. The volatiles are thermally desorbed and then analysed in the GC system. The virtues of this technique are widely acclaimed and include automation, solventless sample preparation, inexpensive, simple, and good recovery of volatiles [10]. Unfortunately, few reports mention the negatives of this method, which include competitive binding of volatiles, deterioration on use (may be used for 100 injections) may result in changing performance or fibre breakage, and very limited phase volume, which limits the technique to only the more abundant, non-polar volatiles [11, 12]. Also, if the fibre is placed in a fat-containing sample, the lipids will also be absorbed and create artefacts during thermal desorption.

A new version of this approach involves placing a coating on a small magnetic stir bar (Twister®, Gerstel, Baltimore, MD, USA; Fig. 18.1d). This configuration allows a significantly larger amount of absorbent phase to be used, and thus overcomes some of the disadvantages of SPME [13]. The bar may be placed

Table 18.1 Absorbents used in solid-phase microextraction [5]

Absorbent	Application
Poly(dimethylsiloxane)	Considered non-polar for non-polar analytes
Poly(dimethylsiloxane)/divinylbenzene	Ideal for many polar analytes, especially amines
Polyacrylate	Highly polar coating for general use, ideal for phenols
Carboxen/poly(dimethylsiloxane)	Ideal for gaseous/volatile analytes, high retention for trace analysis
Carbowax/divinylbenzene	For polar analytes, especially for alcohols
Carbowax/templated resin	Developed for high-performance liquid chromatograpy applications, e.g. surfactants
Divinylbenzene/Carboxen// poly(dimethylsiloxane)	Ideal for a broad range of analyte polarities, good for C3–C20 range

in the sample or in the sample headspace. Like SPME, thermal desorption of the bar can be automated [4].

18.2.2 Distillation (Simultaneous Distillation/Extraction, Vacuum Distillation)

One of the few properties all aroma compounds have in common is they must be volatile: if they are not volatile, they cannot make a contribution to olfaction. With this said, there is a very broad range in volatility across aroma-active compounds so one obtains a disproportionately large proportion of very volatile compounds and lesser amounts of low-volatility compounds in all aroma isolates obtained based on this property.

In terms of specificity in isolation, one will also isolate food constituents that are not aroma compounds (e.g. pesticides, herbicides, PCBs, plasticisers, and some antioxidants). Since these compounds are typically present in foods at very low levels, they generally present few complications. The primary volatile that complicates the application of this methodology is water. In all cases, one obtains an aroma isolate that consists of volatiles in an aqueous "solution". Thus, unless the amount of water is small and the subsequent analytical step is tolerant of some water, volatility-based techniques must include some water-removal process. This may be freeze-concentration, the addition of anhydrous salts, or solvent extraction. Distillation is often used to isolate aroma compounds from fat-containing foods. Since fat is not volatile (under isolation conditions), its presence does not prohibit the use of this methodology.

Volatility-based aroma-isolation techniques most commonly have involved either steam distillation or high-vacuum stripping. One of the oldest techniques falling in this category is simultaneous stream distillation/solvent extraction (Fig. 18.2, left [14]). In this methodology, the sample is dispersed in water which is heated to boiling. The steam that is generated carries volatiles with it into a section of the apparatus where the steam condenses in the presence of extracting solvent vapours. The co-condensation of volatile-laden steam and extracting solvent accomplishes an effective extraction of volatiles. As one can envision, this technique recovers aroma compounds on the basis of volatility *and* solubility in the extracting solvent (two biases). Yet it offers a relatively broad view of volatiles in foods with some loss of compounds that exhibit either extremes in volatility, or poor solubility in the extracting solvent. In addition, it provides an aroma isolate in a solvent that is reasonably concentrated: it can be used for several injections (GC/mass spectrometry, sniffing, etc.) as opposed to a single-use isolate. Artefact formation during distillation is problematic since it has traditionally been carried out at ambient pressure (100 °C). [Systems that operate under vacuum are available but their operation is problematic since the extracting solvent and water (steam) must be kept in the apparatus under this vacuum.] This technique is being used less today than in the past but still has great value.

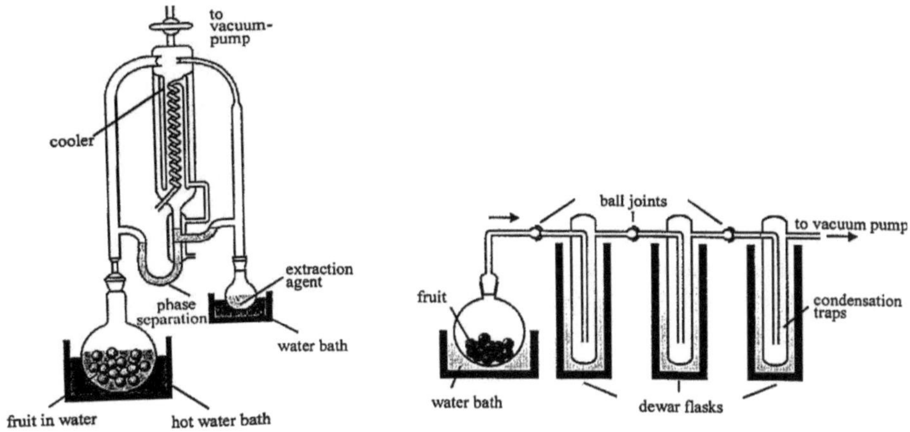

Fig. 18.2 Aroma-isolation techniques based on distillation. *Left* simultaneous distillation/extraction; *right* high-vacuum distillation with cryotrapping. (Reprinted with permission from [15]. Copyright 1998 American Chemical Society)

Alternatively, distillations may simply depend upon a vacuum to strip volatiles from a food. The volatiles stripped from a food may be condensed in a cold trap (Fig. 18.2, right) or passed through an absorbent trap (e.g. Tenax®) for collection. While these techniques have found substantial application in the past, in recent times they have seen less use.

18.2.3 Solvent Extraction

Solvent extraction is an excellent choice for aroma-compound isolation from foods when applicable. Unfortunately, many foods contain some lipid material, which limits the use of this technique since the lipid components would be extracted along with the aroma compounds. Alcohol-containing foods also present a problem in that the choice solvents (e.g. dichloromethane and diethyl ether) would both extract alcohol from the product, so one obtains a dilute solution of recovered volatiles in ethanol. Ethanol is problematic since it has a high boiling point (relative to the isolated aroma compounds), and in concentration for analysis, a significant proportion of aroma compounds would be lost with the ethanol. As one would expect, the recovery of aroma compounds by solvent extraction is dependent upon the solvent being used, the extraction technique (batch or continuous), and the time and temperature of extraction.

18.2.4 Combinations of Methods

It is very common to combine methods in obtaining aroma isolates. The simultaneous distillation/extraction method previously described is an example. Another popular combination method initially involves the solvent extraction of volatiles from a food and then high-vacuum distillation of the solvent/aroma extract to provide a fat-free aroma isolate. This technique is broadly used today to provide high-quality aroma extracts for numerous purposes. The apparatus used in solvent removal has been improved upon to reduce analysis time and efficiency: the modified method is termed solvent-assisted flavour extraction (SAFE) [16].

18.2.5 Comments on Aroma-Isolation Methods

It is important to remember that no method of obtaining an aroma isolate from a food gives a complete quantitative or qualitative picture of the aroma compounds actually present in the food. Every method used in aroma isolation has biases in isolation and the common need to combine methods (e.g. one based on volatility and then solubility in solvent extraction) introduces even more biases [17]. Thus, one has to use extreme care in choosing isolation methods and interpreting results obtained from them. It is also important to recognise that the scientific literature may be biased as well. Authors typically do not dwell on weaknesses of the methodology they are using but the positives. Thus, the task of choosing isolation methods must be approached in a thoughtful, knowledgeable manner.

18.3 Isolation of Flavour from Plant Materials for Commercial Use

In this section, we are interested in economically isolating flavouring materials indigenous to a plant source for commercial use. While the flavour industry continues to expand its production of synthetic chemicals and pure, natural chemicals made following the legal definition of "natural", the industry still depends very heavily upon flavouring materials isolated from plant sources. The citrus oils, mint oils, and vanilla are prime examples of flavouring materials derived from plants that are used in very large volumes by the industry.

It should be no surprise that the methods used to produce most flavourings from plant sources are based on similar principles as those used in the isolation of aroma compounds from foods. However, economics and scale play major roles in dictating methods. Additionally, the physical characteristics of the plant material, and concentrations and properties of flavouring materials also are considered. These constraints result in compromises in aroma recovery for it is not critical that all volatile compounds be quantitatively recovered from the starting plant materials. In fact it is not even required that the sensory properties of the flavour isolate obtained even resemble those of the starting plant material. Basically, it is only important that what is isolated from a plant source has commercial value. We will provide a brief overview of the methods used in the isolation of these flavouring materials. More comprehensive texts, for example those by Ziegler and Ziegler [18], Ashurt [19], or Reineccius [20], are recommend for detail.

18.3.1 Distillation (Essential Oils)

Distillation is very commonly used for the production of flavouring materials from plants. One would use differing types of distillation for this purpose depending upon the plant material used and the flavouring one wishes to recover. If one is using very fragile plant materials with a delicate flavour (e.g. flower petals), one would likely use a water distillation. The flower petals would be dispersed in water and then some portion of the water would be distilled from the distillation pot. Dry, rigid plant materials (e.g. leaves, twigs, bark, roots, or seeds) may require the use of low-pressure or high-pressure steam depending upon the plant material and the flavouring material desired. In most cases very hard plant materials (e.g. cinnamon bark or clove buds) would be ground to provide greater surface area for distillation. The distillate in all cases is collected and the oil and water are allowed to separate naturally owing to insolubility (gravity separation), or are solvent-extracted if the natural separation process is inefficient. Figure 18.3 shows the distillation equipment used to recover essential oils from plant materials [21].

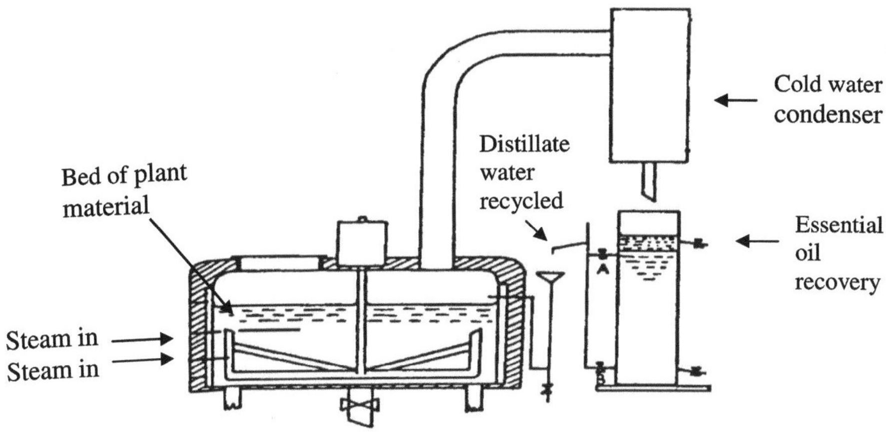

Fig. 18.3 Distillation equipment used to recover essential oils from plant materials [21]

In distillation processes, the non-volatile components (e.g. bite of pepper or ginger, and natural antioxidants) would remain with the spent plant material since they are non-volatile. Also, any water-soluble flavouring components would also be lost in the final distillate separation. The products so produced may only partially resemble the fresh starting plant material. Nevertheless, these materials are highly valued and are key ingredients for the flavour industry. Some are unstable to oxidation since the natural antioxidants remain with the spent plant materials.

18.3.2 Solvent Extraction (Oleoresins, Extracts, and Infusions)

Oils from some plants may be recovered either by distillation or by solvent extraction. It is readily understood that the flavour profile obtained by each process is unique: distillation yields only volatile flavouring components, while extraction yields extractable volatiles plus some non-volatile flavouring components. This difference is most evident when the plant material contains taste or chemesthetic components e.g. ginger. The essential oil of ginger has a very mild ginger aroma with no taste. It is broadly used in beverages and confectionary products. Ginger flavourings obtained by solvent extraction have a characteristic aroma and also the bite of the ginger root. This latter oil is used in most savoury products (although these delineations are becoming blurred).

As one would anticipate, the flavour character of the recovered oil is dependent upon the specific solvent used in extraction. For example, hexane extracts yield a different flavour character from that of those made using either supercritical CO_2 or acetone extraction. The increasing use of supercritical CO_2 has made some very true to character oils available to the industry, albeit at a higher price than traditional solvents.

18.3.3 Cold Pressing (Citrus Oils)

Some oils are sufficiently easily recovered that they can be pressed from the plant. This is generally limited to the citrus oils, where oil sacks are located near the surface of the peel. Citrus fruit is processed using equipment that simultaneously extracts the juice from the core of the fruit and oil from the peel. Lime oil may be produced either by distillation or pressing. Distilled lime oil is made from very small limes that are macerated and then distilled—no juice is recovered. The lime oil degrades greatly in the distillation pot owing to high temperatures and the low pH of the juice; however, the oil is very readily accepted since this is how most lime oil was originally produced and, thus, introduced to the market: the deteriorated flavour defined the product. More recently, lime oil has been recovered by pressing, and this oil is very fresh in character as opposed to the distilled oil. Oil yields by pressing are low and thus these products are typically expensive but yet they yield the true characteristic flavour of the citrus oils.

18.4 Isolation of Flavouring Materials from Waste Streams

The isolation of flavouring materials from waste streams is becoming of interest to the flavour industry. Historically, waste streams have been considered a disposal problem (cost) but the continuing demand for natural flavouring materials and the willingness to pay a premium for a natural flavouring makes the industry consider harvesting aroma compounds from these waste streams whenever possible. In some cases waste streams may provide natural chemicals (or flavouring mixtures) that are not available by alternative processes (e.g. biotechnology), giving them even greater value. For example, it is very difficult to produce natural forms of many of the Maillard products (heterocyclic compounds). They are uniquely formed from thermally induced reactions and thus are not found in plant sources or produced by enzymatic reactions (or fermentations). These types of compounds may be recovered from baking or roasting processes by condensing exhaust gases.

Flavour recovery from waste streams is also becoming of greater interest in the USA since the Environmental Protection Agency is starting to consider not only the liquid waste streams but also the exhaust gases of food-processing operations to be pollutants. When local manufacturing operations (e.g. bakeries) were producing baked goods for a small community, the aroma in the exhaust gases was quite dilute and considered pleasant. As commercial operations have grown in scale, the exhaust gases are more concentrated, may be less pleasant, and contribute considerably to air pollution. Thus, there may be a financial incentive to removing aroma chemicals from waste streams to reduce pollution-abatement costs.

The task of recovering aroma compounds produced in the industry by biotechnological processes is somewhat similar to flavour recovery from waste streams; however, biologically produced flavouring materials are generally somewhat easier to recover since the concentrations of volatiles are higher, and the volatiles produced are less complex in composition. In a biotechnological process, one aims for yields of target compounds in the grams per litre of fermentation broth range as opposed to the parts per million or parts per billion concentration ranges one might find in waste streams.

With the above introduction in mind, we will present an overview of the techniques used for the recovery of flavour compounds from waste streams. Pervaporation, a logical process for this application, is discussed in detail in Chap. 19, so it will not be discussed here.

The literature involving the further processing (e.g. fermentation, enzymatic or thermal processing) of a food waste stream to produce a flavouring is mentioned in two examples but discussion is limited.

18.4.1 Spinning Cone Concentrator

The use of a spinning-cone concentrator (SCC, Fig. 18.4) for the recovery of volatiles from food-processing streams has found considerable application [22, 23]. This technique may be carried out under vacuum (minimal heat damage) and is highly efficient owing to the thin film and the high surface areas provided by the design. The spinning column is made up of a central rotating shaft that has alternating rotating and stationary disks (Fig. 18.5).

A feed material is fed in at the top of the column and is allowed to flow through a series of disks. The first disk is fixed and thus the infeed flows down the disk by gravity to exit into the base of the first spinning disk. The centrifugal forces of each spinning disk result in the infeed liquid being forced up the disk in a thin film and then dropping onto the next fixed disk and flowing by gravity onto the next rotating disk. This flow pattern gives a very high, thin film flow pattern (Fig. 18.6). An inert gas (or steam, temperature dependent upon the operational vacuum) is drawn through the system countercurrent to the infeed flow. On exiting, the extracting gas passes through a condenser to recover the volatiles.

The efficiency of the SCC is illustrated in comparison with a traditional distillation-based essence recovery unit in Table 18.2. One can see that the aroma volatiles are preferentially stripped from the infeed in the SCC compared with the situation in a single-stage evaporator. Thus, highly concentrated aroma isolates can be produced. Flavortech [22] noted that essences of 1,500-fold may be produced from juices if a double pass is used (the infeed is the first pass and the acquired essence the second pass).

This process has found major application in the wine industry to control the alcohol content of wines (i.e. remove alcohol to the desired level). Wine is initially passed through the equipment at temperatures and pressures that primarily strip aroma components. The dearomatised wine is then passed a second time through the equipment at higher temperatures and vacuum to strip the desired amount of alcohol from the wine. The initially captured aroma fraction can then be added back to the reduced alcohol wine to produce the desired

Table 18.2 The amount of water (%) required to completely strip the aroma compounds from fruit juice [22]

Fruit	Stripping required for the total removal of fruit aroma (calculated using volatility relative to water)	
	Single-stage evaporator	Spinning-cone concentrator
Apple	10	0.5–1.0
Orange	20	1-2
Grape	42	2-3
Apricot	55	3-4
Strawberry	82	5-6

Fig. 18.4 A spinning-cone column apparatus [22]

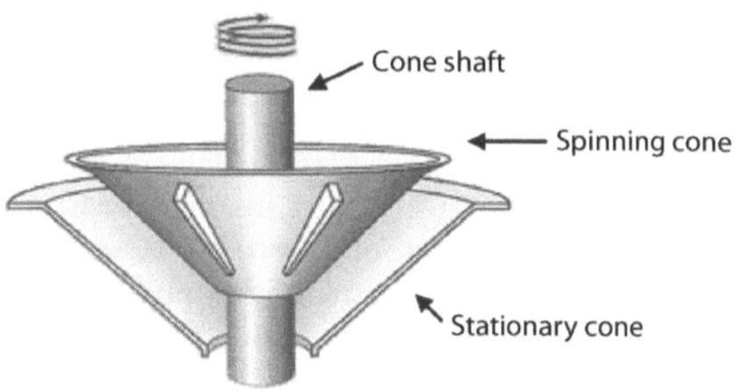

Fig. 18.5 One member of a spinning-cone concentrator (*SCC*) [22]

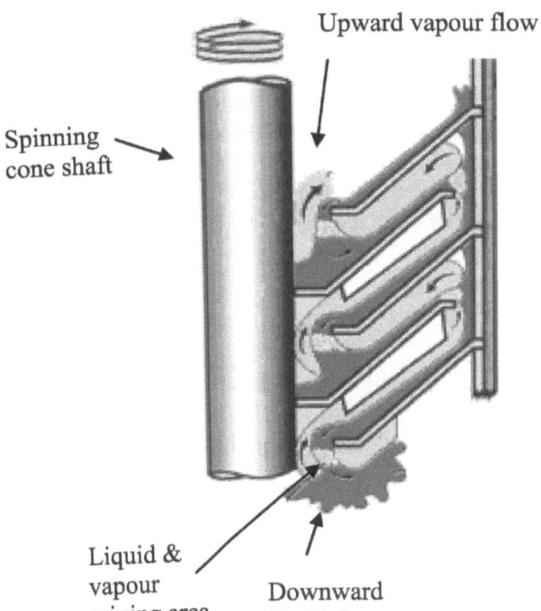

Fig. 18.6 The flow pattern of extracting gas and infeed liquid in a SCC [22]

product. This process may be used to produce alcohol-"free" wine (0.2% alcohol [24]) and more recently alcohol-"free" beer using the same process (0.1% alcohol [25]). This process is also used to obtain very highly concentrated, high-quality isolates from plant juices [23] and the recovery of volatiles from waste streams, notably apple or berry pumice, citrus and onion waste [22]. For example, the SCC is claimed to efficiently recover more than 90% of the citrus essential oils traditionally lost with the centrifuge waste [22].

18.4.2 Absorption/Adsorption

Adsorption (or absorption) involves passing an aroma-laden liquid (or gas) stream through a bed of adsorbent. Assuming that the adsorbent has a significant affinity for the aroma compounds of interest, they will be adsorbed onto the bed and concentrated. While for analytical purposes the bed is commonly thermally desorbed, it is more likely to be solvent-extracted in this application to recover the trapped volatiles.

One application of this process has been described for the recovery of aroma compounds from beer during fermentation [26]. During the manufacture of beer, large quantities of CO_2 are generated and liberated from the beer. The CO_2 carries along with it significant quantities of higher alcohols, esters, and hops compounds which have significant flavour value. Sanchez [26] devised a system whereby the fermentation gases were passed though a bed of adsorbent (not identified) and then desorbed by alcohol extraction of the column. Through the proper choice of adsorbent, a desirable aroma isolate was obtained that was low in short-chain esters and sulfite, both of which have a negative impact on beer flavour quality.

Adsorption methods have also been used in the recovery of flavourings created from the treatment of waste streams (e.g. spent coffee grounds). For example, selected volatiles generated in spent coffee grounds by thermal hydrolysis have been isolated using non-polar resins [27]. The gas stream emerging from the heated coffee grounds (220 °C) was passed through a resin (styrene or divinylbenzene, or activated carbon) until furfural breakthrough was noted. Then the adsorbent was desorbed, yielding an aroma isolate that was used to reinforce the aroma of soluble coffee. The primary components isolated in this process were acetaldehyde, diacetyl, acetone, 2-methylpropanal, 3-methylbutanal, 2,3-pentanedione, and small amounts of furfural.

While the first two examples of using adsorption methods to produce aroma isolates were from gas streams, Tan et al. [28] applied adsorption methods to the isolation of flavouring extracts from mushroom blanching water. Unfortunately, only an abstract was available of this work so it lacks detail. It appears that they evaluated the use of two different resins (not described) and ethanol, pentane, hexane, and other solvents for desorption. They claim to have had good success in obtaining a useful aroma isolate.

18.4.3 Extraction (from Gas or Liquid Streams) Using Cryogenic Traps or Solvents

As mentioned in the introduction to this section, there is the opportunity to recover aroma compounds from baking or roasting exhaust gases. The patent literature contains numerous references to the recovery of aroma compounds using this approach, most commonly from cocoa, coffee, or tea processing. Aroma compounds from the roaster exhaust gases are either condensed in cryogenic traps [29–32] or collected on absorbents (e.g. charcoal [33]) and then solvent-extracted to obtain a concentrated aroma extract. The concentrated extract may be used to aromatise a similar product (e.g. soluble coffee) or may be used to flavour other products (e.g. coffee-flavoured ice creams).

One of the earliest processes used charcoal traps to collect aroma from different stages of coffee processing (grinding, brewing, and concentration) [33]. Each processing step after roasting was carefully hooded so that all vapours from the coffee were passed through a charcoal bed. The charcoal bed was then extracted with either an organic solvent (ether, dichloromethane, or preferably dichloromonofluoromethane) or steam (121 °C). A minimum amount of steam (or solvent) was used to provide the most concentrated coffee essence. Extracts containing 20–40% coffee volatiles were prepared in this manner.

Numerous patents issued for the recovery of coffee exhaust gases since that time. The industry has generally chosen to cryogenically trap volatiles and the patent variations have largely been in the design of the vapour-trapping devices. One of the later versions of this process uses a series of cryogenic traps, each successive trap incorporating lower temperatures [29]. Since water is the most abundant volatile, the first trap is used to take the majority of water from the gas stream. Successively lowering the trap temperature effectively fractionates the volatiles, providing some control over the sensory properties of the collected fractions. A recent design of a cryogenic trap is shown in Fig. 18.7. In this process, liquid nitrogen is sprayed directly into the product vapour stream and a frost (organic volatiles) is formed. The cold gas stream with suspended frost is passed through a porous filter (Fig. 18.7, no. 20) where the frost is collected on the outside of the filter and the nitrogen and non-condensed volatiles pass out through the centre of the filter to enter the next colder trap. The traps are set up to periodically back-purge the filter rods with cold nitrogen to dislodge the captured frost. This keeps the traps free flowing. The dislodged frost is collected in the bottom of the vessel where it melts and is removed (Fig. 18.7, no. 27) on a continuous basis.

A similar apparatus has been used for recovery of aroma compounds from cacao during processing [34]. In this process, water and acetic acid are removed from the aroma-laden gas stream by the initial traps and then the gas is passed through traps of the same design as those described by Carns and Tuot [29]. The aroma isolate so provided is suggested to be useful for the flavouring of soluble cocoa beverages, cake mixes, and confectionery products.

While the bulk of the literature has focused on the recovery of coffee volatiles, there are a few publications describing the recovery of aroma compounds from other foods as well. One example is the recovery of hop aroma from kettle exhaust [35]. In this paper the authors described the condensation of vapours from a hop kettle, washing the condensate with NaClO, passing the wash through an active carbon column to absorb the hop oil, and finally solvent-extracting the carbon trap with an organic solvent to obtain a dilute hop oil extract. Concentration of this extract yielded a characteristic hop flavour which was reincorporated into the beer-making process.

If one is considering the recovery of aroma compounds from waste gas streams, one should investigate the pollution-control literature. There are a large number of patents and scientific articles that deal with this issue. The techniques used are generally aimed at the removal of trace volatiles in air streams and are potentially suited to aroma recovery. The primary consideration is whether the techniques yield an isolate safe for human consumption.

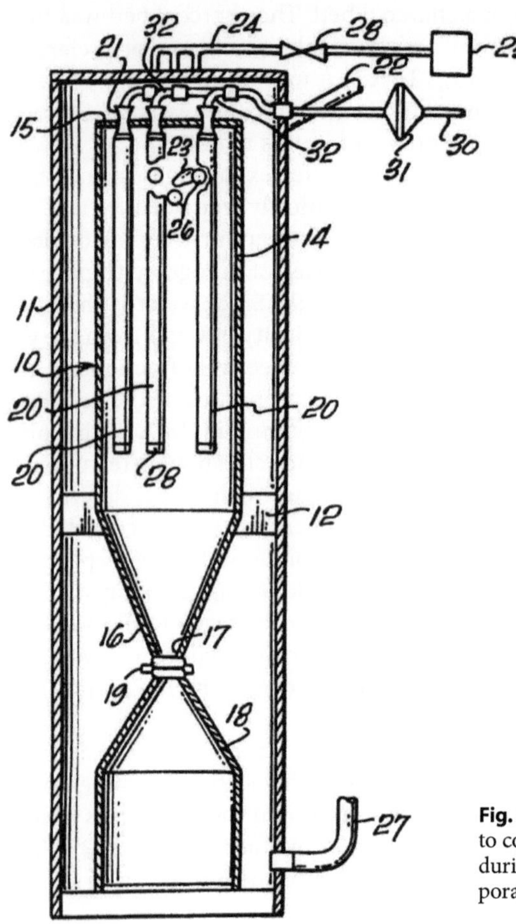

Fig. 18.7 A cryogenic trap used to collect coffee vapours lost during processing for reincorporation into soluble coffee [29]

18.4.4 Membranes

In some cases, the flavouring recovered from a waste stream is based on taste substances (non-volatiles). These materials are typically recovered by membrane methods. Membranes offer economy (high initial capital cost but less than 50% operating costs and 90% reduction in energy costs) and minimal heat damage compared with distillation processes. Chaing et al. [36] used ultrafiltration (UF) and reverse osmosis (RO) to recover mushroom flavouring materials from blanching water. They obtained concentrates of up to 20% solids, claiming 90% recovery of non-volatiles and 50% recovery of volatiles. They noted that the isolated fraction could not be differentiated from the starting material in sensory testing.

There has been considerable interest in recovering flavour components from seafood-processing operations. While solid waste from the cleaning operation may be used directly as base materials for the formation of seafood flavours based on process chemistry, the aqueous waste stream from cleaning is too dilute to be useful for that purpose. The cooking water, however, contains substantially higher concentrations of both volatile (aroma compounds characteristic of seafoods) and non-volatile [free amino acids (taurine, glutamic acid, glycine, etc.), peptides, nucleotides (purine derivatives), quaternary ammonium bases, organic acids (lactic acid), sugar (glucose, ribose) and inorganic salts (Na^+, K^+, Cl^-)] materials that have flavouring value [37]. Vandajon et al. [37] gave an example that a shrimp-processing line with a capacity of 2,000 t/year would generate about 15 t/year of organic material (potential flavouring materials) in the cooking water.

Vandanjon et al. [37] reported on using a combination of UF and RO, or UF and nanofiltration (NF) to remove both volatiles and non-volatiles from the cooking water of shrimp, buckies, and tuna. The preliminary use of UF is common in that it removes the larger materials that would clog the subsequent membrane steps. NF was not found to be as efficient in the recovery of volatiles as RO. Unfortunately, the authors did not evaluate the use of the recovered materials as flavourings.

More recently, Lin and Chaing [38] have reported on a membrane process to recover flavour compounds from salted shrimp processing wastewater. In many Asian countries, dry salted shrimp is a popular food item. This product is made by cooking shrimp in a 10% salt brine prior to shucking and drying; thus, the cooking water is high in salt, which poses a problem in flavour recovery. Lin and Chang [38] evaluated the use of loose RO membrane diafiltration or electrodialysis (ED) for desalting and flavour recovery. Using RO, they were able to remove about 93% of the salt but they recovered less than 50% of the free amino acids and nucleotides (target flavour compounds). Using ED, they removed less salt (85%) but improved their recovery of flavour compounds to more than 70%. On the basis of this work, they recommended the ED system for this purpose.

A group at North Caroline A&T has used combined fermentation and membrane processes to produce lactic acid from a cheese whey waste stream. Lactic acid is typically produced by the fermentation of glucose (starch source) and is recovered by neutralisation, filtration and reacidification. This process recovers the lactic acid but generates a significant waste stream. In work presented by Shahbazi et al. [39], lactic acid was produced by the fermentation of lactose and isolated using UF and NF membranes (Fig. 18.8). UF is used to retain cells and protein, while NF retains lactose while allowing lactic acid to pass through the membrane. The process does not produce a waste stream and offers economy in operation. This process is described in detail in other work [40].

Souchon et al. [41] used a combined membrane/solvent-extraction process to recover tomato volatiles from a model tomato aroma solution and an actual tomato-processing waste stream. This process uses a microporous membrane (polypropylene) to separate the waste stream from the extracting solvent [42]. This membrane is porous to the organic phase (hexane or Miglyol) and thus an overpressure must be applied to the aqueous stream in order to inhibit the flow of extracting solvent into the aqueous stream. The use of a membrane interface offers some advantages in that there are no issues with flooding, loading, or emulsification [41]. Furthermore, the pumping systems are low pressure and the separation of phases is not necessary (permits a wide range of solvents). Souchon et al. [41] noted that the primary disadvantage of this process is that the membrane provides a diffusional barrier to extraction, but the very high hollow fibre membrane surface area minimises this problem.

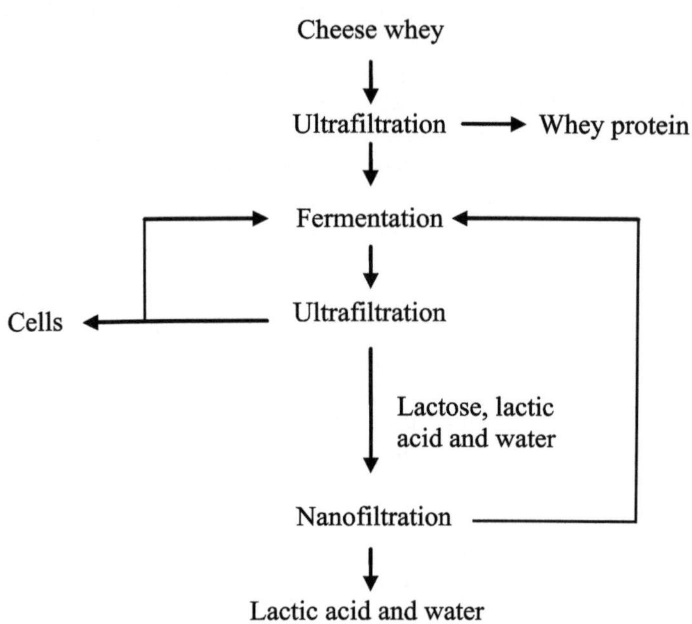

Fig. 18.8 The recovery of lactose from a whey waste stream [39]

Souchon et al. [41] found this process to be very efficient in the recovery of tomato volatiles both from the model system and from commercial tomato waste stream. They reported that using a 1.4-m^2 membrane surface, a 20-L/h waste stream flow, and hexane as the extracting solvent, they would recover approximately 95% of the *hydrophobic* tomato volatiles. Volatile recovery is dependent upon the type of volatile being extracted and the extracting solvent.

18.5 Summary

The recovery of aroma compounds from waste streams has been accomplished in only a few commercial applications, the best known being the recovery of coffee or cocoa volatiles during processing. The limitation in application is economics. The waste stream must have an adequate concentration of volatiles and the volatiles must be of high value. Few processing operations meet these economic criteria at this time. In the future, the high costs of recovering flavouring materials from waste streams will be partially offset by the saving in disposal costs associated with environmental issues.

References

1. Marsili R (2002) Flavor, fragrance, and odor analysis. Dekker, New York
2. Reineccius GA (2002) In: Taylor AJ (ed) Food flavour technology. Sheffield Academic Press, Sheffield, p 210
3. Reineccius GA (2006) Flavor chemistry and technology. Taylor & Francis, Boca Raton
4. Gerstel (2006) Thermal desorber catalogue. Gerstel, Baltimore. http://www.gerstel.com/tds-eng.pdf
5. Supleco (2006) SPME applications guide. Supleco, Bellefonte. http://www.sigmaaldrich.com/Graphics/Supelco/objects/8700/8652.pdf
6. Coulibaly K, Jeon IJ (1996) Food Rev Int 12:131
7. Marsili R (2002) Food Sci Technol 115:205
8. Braggins TJ, Grimm CC, Visser FR (1999) In: Hamilton NZ, Pawliszyn, J (eds) Applications of solid phase microextraction. Royal Society of Chemistry, Cambridge, p 407
9. Roberts DD, Pollien P (1998) Book of abstracts, 216th ACS national meeting, Boston, 23–27 August. American Chemical Society, Washington
10. Harmon AD (1997) In: Marsili R (ed) Techniques for analyzing food aroma. Dekker, New York, p 81
11. Nongonierma A, Cayot P, Quere JL, Springett M, Voilley A (2006) Food Rev Int 22:51
12. Roberts DD, Pollien P, Milo C (2000) J Agric Food Chem 48:2430
13. David FT, Tienpont B, Sandra P (2003) LC/GC 21:109
14. Chaintreau A (2001) Flavour Fragrance J 16:136
15. Guntert M, Krammer G, Sommer H, Werkhoff P (1998) In: Mussinan CJ, Morello MJ (eds) Flavor Analysis "Developments in Isolation and Characterization". American Chemical Society, Washington, p 40

16. Engel W, Bahr, W, Schieberle P (1999) Z Lebensm Unters Forsch 209:237
17. Reineccius GA (1993) In: Ho CT, Manley CJ (eds) Flavor measurement. Dekker, New York, p 61
18. Ziegler E, Ziegler H (1998) Flavourings: production, composition, applications, regulations. Wiley-VCH, New York
19. Ashurst PR (1995) Food flavorings. Blackie, New York
20. Reineccius G (1995) Source book of flavors, 2nd edn. Chapman Hall, New York
21. Heath HB, Reineccius GA (1986) Flavor chemistry and technology. Van Nostrand Reinhold, New York
22. Flavourtech Americas Inc (2006) http://www.flavourtech.com/index.htm
23. Sensus Flavors (2006) http://www.sensusflavors.com/index.php
24. Sutter Home Winery (2006) http://www.sutterhomefre.com/spinning_cone.html
25. Alpha Laval (2006) http://www.alfalaval.com/ecoreJava/WebObjects/ecoreJava.woa/wa/showNode?siteNodeID=6016&contentID=35744&languageID=1
26. Sanchez G (2001) Master Brew Assoc Am Q Bull 38:235
27. Cale KW, Imura N, Jasovsky GA, Katz SN (1990). US Patent 4,900,575
28. Tan CS, Fan YC, Lee TY, Chen YW, Lee S, Koo JT (1985) Proc Natl. Sci Counc Chin Part A Phys Sci Eng 9:258
29. Carns LG, Tuot J (1993) US Patent 5,323,623
30. Ghodsizadeh Y (1987) US Patent 5,030,473
31. Carns LG, Tuot J (1993) US Patent 5,182,926
32. Carns LG, Tuot J (1994) US Patent 5,323,623
33. Rooker W (1968) US Patent 3,418,134
34. Mazurek R, Temperini M, Barfuss D, Rushmore D (2000) US Patent 6,090,427
35. Kowaka M, Sakuma S, Nakayama K, Totsuka TT (1986) Tech Q Master Brew Assoc Am 23:57
36. Chiang BH, Chu CL, Hwang LS (1986) J Food Sci 51:608
37. Vandanjon L, Crosa S, Jaouen P, Quemeneur F, Bourseau P (2002) Desalination 144:379
38. Lin CY, Chaing BH (1993) Int J Food Sci Technol 28453
39. Shahbazi A, Li Y, Coulibaly S (2005) Lactic acid production from cheese whey, 1890 Joint Research and Extension Conference, New Orleans, 19–22 June. First Place Award of presentation competition of Association Research Director (ARD) and Association of Extension Administrator (AEA)
40. Shahbazi AM, Mims M, Li Y, Ibrahim SA, Vest S (2005) Appl Biochem Biotechnol 121–124:529
41. Souchon I, Pierre FX, Samblat S, Bes M, Marin M (2002) Desalination 148:87
42. Pierre FX, Souchon I, Marin M (2001) J Membr Sci 187:239

19 Aroma Recovery by Organophilic Pervaporation

Thomas Schäfer
Department of Chemistry and Industrial Chemistry,
University of Pisa, Via Risorgimento 26, 56126 Pisa, Italy

João G. Crespo
Requimte/CQFB, Department of Chemistry,
FCT/Universidade Nova de Lisboa,
Campus da Caparica, 2829-516 Caparica, Portugal

19.1 Membrane Processes in the Food Industry

Membranes are semipermeable barriers that permit the separation of two compartments of different composition or even condition, with the transport of components from one compartment to another being controlled by the membrane barrier. Ideally, this barrier is designed to let pass selectively only certain target compounds, while retaining all others—hence the denotation "semipermeable". Membrane separations are particularly suitable for food applications because (1) they do not require any extraction aids such as solvents, which avoids secondary contamination and, hence, the necessity for subsequent purification; (2) transfer of components from one matrix to another is possible without direct contact and the risk of cross-contamination; (3) membrane processes can, in general, be operated under smooth conditions and therefore maintaining in principle the properties and quality of delicate foodstuff.

Naturally, there exist a variety of membrane separation processes depending on the particular separation task [1]. The successful introduction of a membrane process into the production line therefore relies on understanding the basic separation principles as well as on the knowledge of the application limits. As is the case with any other unit operation, the optimum configuration needs to be found in view of the overall production process, and combination with other separation techniques (hybrid processes) often proves advantageous for large-scale applications.

Figure 19.1 gives an overview of some of the most common membrane separation techniques, their application range and their denotation. It should be pointed out that the terminology for membrane separation processes is partly traditional. The kind of membrane–solute interactions and the respective mass-transport phenomena can therefore not necessarily be derived from the designation of the membrane separation, and should always be evaluated for the individual application envisaged.

In general, but not as a rule, the smaller the target compounds to be separated, the denser should be the polymer network in order to give the most intense membrane–solute interactions during permeation. The driving force for the separation to take place should then act on the most significant difference

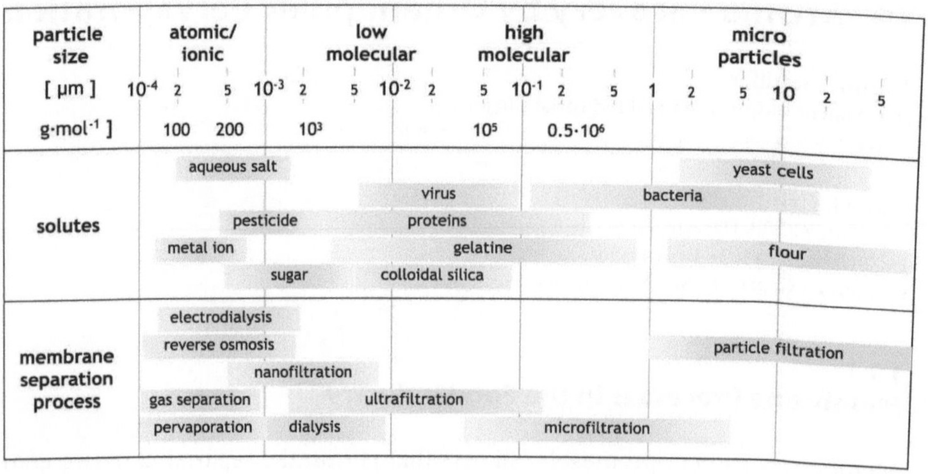

Fig. 19.1 Overview of some membrane separation processes and their application range (adapted from [2])

between the component(s) to be recovered and the bulk. Membranes can therefore be "porous" (Fig. 19.2a,d) or "non-porous" ("dense", Fig. 19.2c), they may be originally porous but confine a stationary phase of particular affinity within their pores (and hence become "dense", Fig. 19.2e), or they may be charged (Fig. 19.2f)—any configuration may be conceived depending on the particular separation task. A principle classification may be drawn between separations involving volatile and/or non-volatile components, as some processes involve mass-transport phenomena that rely on the volatility of the solute(s) to be recovered (Fig. 19.2c,d). As manifold as the membrane structures are the materials available: they can be polymers, inorganic matrices or composites [2, 3]. The wide range of possible materials underlines one of the strengths of membrane separations over evaporative techniques and solvent extraction: the possibility of tailoring and fine-tuning the separation barrier for an individual need, be it in its bulk properties or by suitable surface modifications [2, 4].

Porous (filtration) membranes separate primarily on the basis of size exclusion, with permeate fluxes being convective and relatively high (Fig. 19.2a). While it therefore appears that the choice of the membrane material is not primarily crucial for the separation, in practice it needs in fact to be made carefully in order to minimise undesired surface phenomena during operation (such as fouling [5]) that can strongly deteriorate and even govern the overall process performance. This also applies to membrane distillation and membrane osmosis, processes during which volatile compounds are evaporated through the pores of a membrane (Fig. 19.2d). The membrane serves in these cases more as a support structure rather than a selective barrier. The denser the membrane polymer network, the more is the intrinsic membrane transport diffusive and

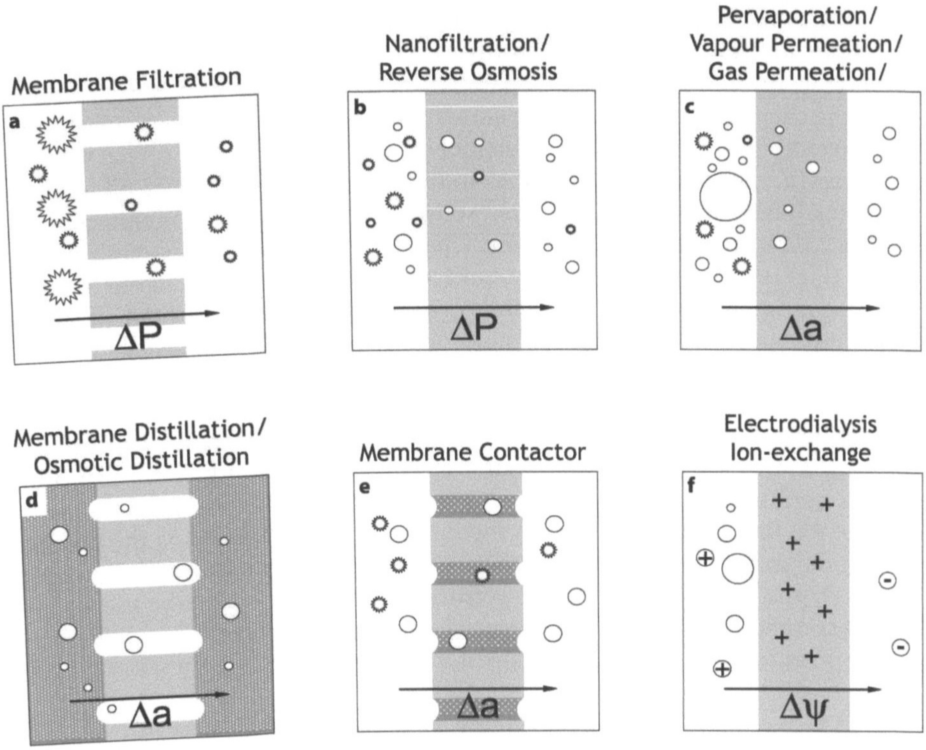

Fig. 19.2 The operation principle of the most common membrane separation processes, with membranes separating the feed (*left*) from the permeate phase (*right*). *Circles* and *stars* indicate volatile and non-volatile compounds, respectively. Driving forces acting upon solutes are indicated by *arrows* as gradients of pressure (P), activity (a) and electrostatic potential (ψ). It should be noted that all these driving forces are eventually based on a gradient of the chemical potential in its most general form

the lower become the intrinsic permeate fluxes. With the membrane–solute interactions being intense, the right choice of the membrane material becomes crucial for the separation process as it strongly determines the membrane selectivity. An example for such an application is the recovery of aroma compounds by pervaporation (Fig. 19.2c), which will be discussed in more detail next.

19.2 Recovery of Aromas and Aroma Profiles by Pervaporation

Pervaporation is a membrane separation process in which a dense, non-porous membrane separates a liquid feed solution from a vapour permeate (Fig. 19.2c). The transport across the membrane barrier is therefore based, generally, on a solution-diffusion mechanism with an intense solute–membrane interaction. It is a process that is employed when the target compounds are volatile and of low molecular size (Fig. 19.1), such as is the case with aroma compounds. The most appropriate driving force for the separation to take place is in this case a gradient in the chemical potential (or the activity) of the compound between the feed and the receiving (permeate) compartment.

Figure 19.3 schematically describes in more detail the transport phenomena occurring during pervaporation. First, solutes partition into the membrane material according to the thermodynamic equilibrium at the liquid–membrane interface (Fig. 19.3a), followed by diffusion across the membrane material owing to the concentration gradient (Fig. 19.3b). A vacuum or carrier gas stream promotes then continuous desorption of the molecules reaching the permeate side of the membrane (Fig. 19.3c), maintaining in this way a concentration gradient across the membrane and hence a continuous transmembrane flux of compounds.

From Fig. 19.3a–c, and as opposed to purely sorption controlled processes, it can be seen that during pervaporation both sorption and diffusion control the process performance because the membrane is a transport barrier. As a consequence, the flux J_i of solute i across the membrane is expressed as the product of both the sorption (partition) coefficient S_i and the membrane diffusion coefficient D_i, the so-called membrane permeability L_i, divided by the membrane thickness ℓ and times the driving force, which may be expressed as a gradient of partial pressures in place of chemical potentials [6]:

$$J_i = \frac{D_i S_i}{1}\left(p_{i,\mathrm{f}} - p_{i,\mathrm{p}}\right) = \frac{L_i}{1} H_i \left(x_{i,\mathrm{f}} - \frac{y_{i,\mathrm{p}} P_i^{\mathrm{o}}}{H_i} \right), \tag{19.1}$$

with $p_{i,f}$ and $p_{i,p}$ are the feed and permeate partial pressures of solute i, respectively, $x_{i,f}$ and $y_{i,p}$ are mole fractions of solute i in the feed and in the permeate, respectively, and P_i° is the saturated vapour pressure of solute i. It is pointed out that the use of Henry coefficients is strictly valid only for very dilute solutions. Whilst this is valid for most aroma-containing solutions, such as beverages, it always needs to be confirmed for the individual separation task.

When the driving force is maximum, i.e. when $p_{i,\mathrm{p}} \rightarrow 0$, then the pervaporative membrane selectivity α between two components can be expressed as

$$\alpha_i = \frac{\dfrac{J_i}{x_{i,\mathrm{f}}}}{\dfrac{J_j}{x_{j,\mathrm{f}}}} = \frac{D_i}{D_j}\frac{S_i}{S_j}\frac{H_i}{H_j} = \frac{L_i}{L_j}\frac{H_i}{H_j}, \tag{19.2}$$

with H_i/H_j being the selectivity of the vapour–liquid equilibrium, i.e. the selectivity of evaporative techniques. Compared with evaporative techniques, for example vacuum evaporation (spinning-cone column), whose driving force

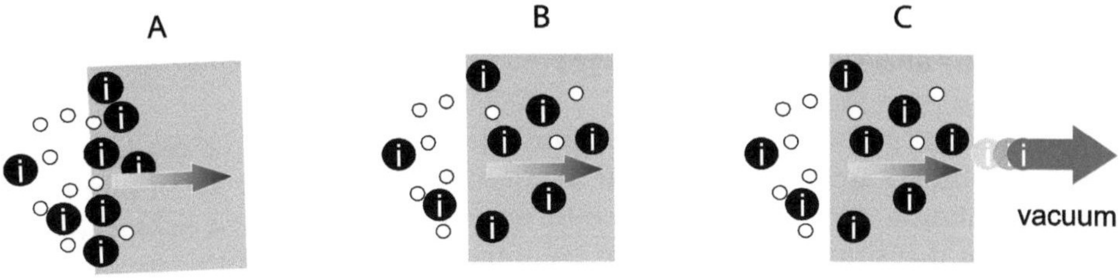

Fig. 19.3 The solution-diffusion transport model in pervaporation. **a** Solution of compounds from the feed phase into the membrane surface. **b** Diffusion across the membrane barrier. **c** Desorption from the membrane permeate (downstream) side into the permeate phase

is also a gradient in the chemical potential, pervaporation as a non-porous membrane separation technique has therefore the advantage of an additional selectivity stemming from the semipermeable membrane barrier between the feed liquid and the (permeate) vapour. Although this selectivity is theoretically gained at the cost of overall fluxes (the membrane is a transport resistance), this shortcoming can be compensated for in practice by a larger membrane area. Commercial membrane costs are competitively low, and in particular hollow-fibre membrane modules allow a very high membrane area-to-volume ratio and hence compact design of membrane modules.

In comparison with adsorptive/absorptive techniques for aroma recovery from bioconversions, the disadvantage of pervaporation is the fact that both sorption and diffusion determine the overall selectivity. While the sorption selectivity is very high (equal to that of adsorptive/absorption), the diffusion selectivity favours water owing to the simple fact that water is a smaller molecule than aroma compounds and thus sterically less hindered during diffusion (Table 19.1). The overall (perm)selectivity ($P=SD$) is therefore lower than in strictly sorption controlled processes, although it is still favourable compared with that for evaporation. This shortcoming compares, however, with operational advantages of pervaporation as outlined before.

Table 19.1 Selectivity of a pervaporation membrane (based on poly(octylmethylsiloxane) [7]) for ethyl hexanoate and isobutyl alcohol with respect to water

Compound	Sorption coefficient S ($mg \cdot mg^{-1}$)	Diffusion coefficient D ($m^2 \cdot s^{-1}$)	Sorption selectivity	Diffusion selectivity	Overall perm-selectivity
Water	5×10^{-4}	2.2×10^{-10}	1	1	1
Ethyl hexanoate	241	2.1×10^{-12}	482,000	0.0095	4,579
Isobutyl alcohol	1	5.4×10^{-12}	2,000	0.025	50

19.2.1
Limitations and Technical Challenges

While vapour permeation and hydrophilic pervaporation have readily found well-established areas for industrial application, in the case of organophilic pervaporation a clear industrial breakthrough has not yet been achieved. The reasons for this situation derive from the intrinsic character of this process and from the way some problems have been approached so far:

19.2.1.1
Membrane Selectivity

Although some membranes exhibit a high affinity towards aroma compounds—as a rule, pervaporative enrichment of alcohols is low, of aldehydes intermediate and of esters high [8]—the high diffusivity of water, even through "hydrophobic membranes", limits the degree of selectivity for aroma recovery from diluted aqueous media (Table 19.1). The membrane material of choice for organophilic pervaporation is poly(dimethylsiloxane), including chemically modified derivates through introduction of bulky side groups designed to reduce the partial water flux, e.g. poly(octylmethylsiloxane), and, additionally, other elastomeric materials such as polyether–polyamide block copolymers, ethylene–propylene–diene monomer elastomers and filler-type membranes [9]. Because one aims at employing selective membranes as thin as possible, in order to have a high sorption affinity and to minimise the relevance of diffusion selectivity, most membranes are composites consisting of a thin selective membrane and a macroporous support for mechanical stability.

19.2.1.2
Flux of Target Compounds

When using organophilic pervaporation for the recovery of aroma compounds, the partial fluxes (19.1) of the target aromas are lower than the corresponding fluxes when using evaporative techniques. This behaviour results from the fact that the membrane represents an additional barrier for mass transport (lower D_i), even if it exhibits a high sorption affinity (S_i) for the target aroma. This is the price to pay in order to achieve higher selectivities for the recovery of compounds of interest. Furthermore, in contrast to vapour permeation, where the feed stream can be compressed allowing the concentration of the target permeant solute to be increased and, hence, the driving force for transport, the minute concentration of aroma compounds to be recovered by organophilic pervaporation from aqueous streams leads typically to low transport rates and partial fluxes. This problem applies also to techniques for aroma recovery from aqueous streams based on liquid–vapour equilibrium, such as vacuum evaporation.

19.2.1.3
Module Design and Fluid Dynamics

Mass-transfer limitations due to poor hydrodynamic conditions in the feed-side–membrane interface are common in organophilic pervaporation (Fig. 19.4). This effect, usually known as feed-side concentration polarisation, may become particularly relevant for solutes with a high sorption affinity towards the membrane, which may lead to their depletion near the membrane interface if external mass-transfer conditions are not sufficiently good to guarantee their fast transport from the bulk feed to the interface. As a consequence of their depletion near the interface, the driving force for transport, and the resulting partial fluxes, becomes lower. This is not a membrane-intrinsic phenomenon, but stems from insufficient upstream flow conditions; in practice it may in fact not be overcome owing to module design limitations [10]. This problem is not relevant for hydrophilic pervaporation because water transport is mainly regulated by diffusion and not by selective sorption to the membrane. Better module design and new approaches for improved mass-transfer conditions, without dramatically increasing the energy input, are needed in this case; the recent work on the use of Dean vortices [11] and the assessment of full-scale vibrating pervaporation units [12] are examples of such effort.

Although less discussed in the technical and scientific literature, permeate-side concentration polarisation may also become a problem when using thin selective films that require macroporous supports for mechanical stability [13].

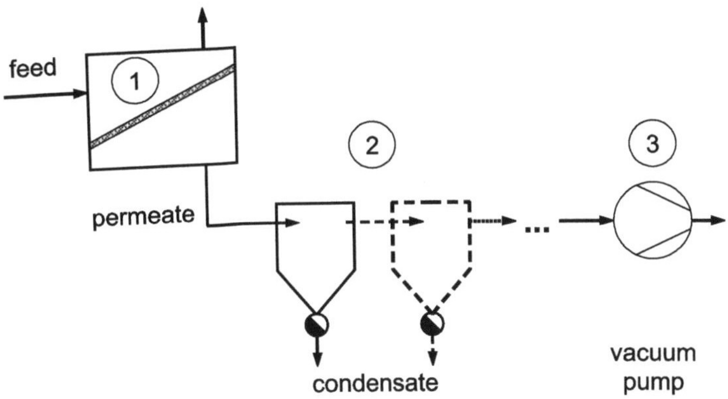

Fig. 19.4 Aspects of optimisation of the pervaporation process, apart from the membrane material: *1* module design for optimum upstream and downstream conditions; *2* condensation temperature(s) or aroma capture strategy; *3* vacuum applied and type of vacuum pump. All aspects of the optimisation are interdependent in pervaporation and therefore need to be tackled as a whole, rather than individually

These porous structures may hinder the transport of solutes away from the membrane downstream surface, causing a local increase of the solute partial pressure and hence a decrease of the driving force (19.1). Eventually, solute condensation may occur if the solute's local partial pressure surmounts its saturation vapour pressure. This problem becomes particularly relevant when dealing with high-boiling aroma compounds [14] and when pressure drop in the downstream circuit increases owing to poor module design.

19.2.1.4
Aroma-Capture Strategies

Stripping of the permeating compounds is achieved either by applying a vacuum in the downstream circuit or by using an inert sweeping gas stream. So far, vacuum pervaporation has deserved a higher degree of attention because it makes it easier to reach low solute partial pressures in the permeate circuit [15] and, consequently, higher driving forces, which are particularly relevant for the permeation of high-boiling compounds. The second main reason for not using sweeping gas pervaporation stems from the fact that recovery of target solutes (aroma compounds) by condensation becomes energetically inefficient owing to the cooling down of a large amount of non-condensable gases (sweep gas); under these circumstances the use of condensation approaches, namely fractionated condensation, becomes inadequate.

On the industrial scale, one aims at shutting off the vacuum pump once the appropriate vacuum level has been established in the downstream circuit in order to minimise energy costs and, ideally, let the condensation unit(s) alone maintain(s) the vacuum. Condensation can be carried out in a series of condensation stages, at different temperatures in order to achieve a permeate fractionation (fractionated condensation) and obtain different fractions enriched in target compounds (Fig. 19.4). The temperature of each condenser has to be adjusted according to the downstream pressure in the circuit and the character of the compounds to be separated and recovered [16, 17]. Capture of the target permeating compounds by condensation remains one of the main problems for competitive use of pervaporation systems, owing to the energy costs involved to keep an adequate downstream pressure and to cool down the permeating stream. This problem is not specific for pervaporation and applies also to the recovery of aroma compounds from vacuum streams originating, for example, during vacuum evaporation. The design and optimisation of adequate condensation strategies for aroma recovery is, for the same downstream pressure, independent of the technique used to generate such vapour. New ways of capturing the permeating vapours have to be developed in order to render this process competitive. This problem will be discussed later, in particular for situations where non-condensable gases permeate the membrane.

19.2.2 Market Opportunities

Recovery of aroma compounds from diluted aqueous streams (we are excluding from this discussion the recovery of aromas from vapour streams) may be of industrial interest under different circumstances: recovery of complex aroma profiles and/or target aroma compounds from active biocatalytic processes; recovery of complex aroma profiles and/or target aroma compounds from natural extracts and industrial process water (or effluent) streams.

Organophilic pervaporation allows for a selective aroma recovery from diluted aqueous streams with the advantage of leading to higher enrichment factors than evaporative techniques, owing to the additional selectivity introduced by the membrane. This selectivity translates into lower condensation energy needs when compared with techniques strictly based on a liquid–vapour equilibrium. Pervaporation offers a unique solution for the recovery of complex aroma profiles. An example for the recovery of complex aroma profiles faithful to their origin is the recovery of a muscatel aroma from an ongoing wine-must fermentation [7, 18].

Coupling pervaporation to active bioconversion processes is extremely interesting because it may allow for continuous removal of target compounds which, otherwise, may simultaneously exert an inhibitory effect over the biocatalysts (cells or enzymes) without detrimentally affecting the biological activity. Several examples have been discussed in the literature [19–22] referring the advantages of integrating bioconversion processes and pervaporation. However, not much has been discussed about the problem of production of non-condensable gases during biological processes (namely carbon dioxide), which permeate the membrane. The presence of non-condensable gases, as happens also during sweeping gas pervaporation, requires an additional energy input in order to keep the downstream pressure at desirable levels and leads to a decreasing energy efficiency of the condensation process (large amounts of energy are spent to cool down the non-condensable gases, lowering the energy efficiency of the process). Under these circumstances there is a well-identified need for development of new alternatives to conventional vacuum condensation for aroma capture, enabling continuous operation and reduced energy input.

One option involves the condensation of (or part of) the permeate under atmospheric instead of vacuum conditions. This requires the use of "dry-vacuum pumps", able to compress the permeate vapour from vacuum to atmospheric pressure, after which condensation is performed at a higher temperature [23]. In this case, the operating conditions have to be carefully monitored since these pumps may lead to unsuitable heating of the vapour and eventually aroma deterioration, despite the low residence time. Alternatively, the use of liquid ring vacuum pumps where the service liquid can take some of the aromas from the permeate stream has been proposed [24].

A second approach under investigation is the capture of the target aroma compounds by promoting their incorporation (i.e. by solubilisation) into a designed delivery system, which will be used directly in the food product. This is a very effective and elegant way to use the same system to, firstly, capture the aromas from the permeating vapour stream and, secondly, to deliver them into the final food product.

Most research on aroma recovery by organophilic pervaporation has been conducted using aqueous aroma model solutions [25–28], although in recent years significant interest has been devoted to the recovery of aroma compounds from natural complex streams, such as fruit juices [29-31], food industry effluents [32] and other natural matrixes [33]. The increasing demand for natural aroma compounds for food use, and their market value, opens a world of possibilities for a technique that allows for a benign recovery of these compounds without addition of any chemicals or temperature increase. However, in most situations, dedicated requests by industrialists are formulated in cooperation with marketing departments, which translate into the need for a correct public perception.

19.3
Concluding Remarks

The growing demand for nutritional food has had a positive impact on the demand for flavours, with consumers unwilling to compromise on taste. The global flavouring market was valued at $4.80 billion in 2005 and is likely to touch $6.22 billion in 2012. Beverages are the leading application segment for flavours and represented a consumption share of 31.1% in 2005 [34]. And with strong growth predicted in the low-fat and low-sugar foods and beverages market in 2006, the global demand for flavours can only grow.

The consistent development of new and innovative flavours is also driving the growth of the flavours market; hence, the development of new technologies and delivery systems that improve the application of flavourings in food products is likely to be crucial to the future development of this highly competitive market.

Pervaporation may certainly play an important role for replacement of evaporative techniques as well as aroma-recovery processes based on solvent extraction, in particular when the labelling "natural" is considered crucial. Some of the most relevant technical challenges discussed herein have to be addressed in order to render organophilic pervaporation a competitive process (Fig. 19.4). In particular, the way of capturing the target aromas from the permeate stream has to be reanalysed in terms of minimising energy consumption and labour-intensive operations.

References

1. Gekas V, Baralla G, Flores V (1998) Food Sci. Technol. Int. 4:311
2. Mulder M (1996) Basic Principles of Membrane Technology. Springer, Berlin Heidelberg New York
3. Baker RW (2004) Membrane Technology and Applications. Wiley, Berlin
4. Pinnau I, Freeman BD (2000) Membrane Formation and Modification, ACS Symposium Series. American Chemical Society, Washington
5. Belfort G, Davis RH, Zydney AL (1994) J. Membr. Sci. 96:1
6. Lonsdale HK, Merten U, Riley RL (1965) J. Appl. Polym. Sci. 9:1341
7. Schäfer T (2002) PhD thesis, Universidade Nova de Lisboa, Portugal
8. Böddeker KW (1994) In: Crespo JG, Böddeker KW (eds) Membrane Processes in Separation and Purification. Kluwer, Dordrecht, p 195
9. Rutherford SW, Kurtz RE, Smith MG, Honnell KG, Coons JE (2005) J. Membr. Sci. 263:57
10. Baker RW, Wijmans JG, Athayde AL, Daniels R, Ly JH, Le M (1997) J. Membr. Sci. 137:159
11. Moulin P, Veyret D, Charbit F (2001) J. Membr. Sci. 183:149
12. Vane LM, Alvarez FR (2002) J. Membr. Sci. 202:177
13. Lipnizki F, Olsson J, Wu P, Weis A, Tragardh G, Field RW (2002) Sep. Sci. Technol. 37:1747
14. Böddeker KW, Gatfield IL, Jahnig J, Schorm C (1997) J. Membr. Sci. 137:155
15. Vallières, C, Favre, E (2004) J. Membr. Sci. 244:17
16. Marin M, Hammami C, Beaumelle D (1996) J. Food Eng. 28:225
17. Brüschke HEA, Schneider W, Tusel GF (1989) Patent EP 0 332 738
18. Schäfer T, Bengtson G, Pingel H, Böddeker KW, Crespo JPSG (1999) Biotechnol. Bioeng. 62:412
19. Stefer B, Kunz B (2002) Chem. Ing. Tech. 74:1029
20. Bluemke W, Schrader J (2001) Biomol. Eng. 17:137
21. Bengtson G, Böddeker KW, Hanssen HP, Urbasch I (1992) Biotechnol. Tech. 6:23
22. Maume KA, Cheetham PSJ (1991) Biocatalysis 5:79
23. Willemsen JHA, Dijkink BH, Togtema A (2004) Membr. Technol. Feb:5
24. Jordt F, Ohlrogge K, Hapke J (1997) Proceedings of Euromembrane'97, p 329
25. Schäfer T, Vital J, Crespo JG (2004) J. Membr. Sci. 241:197
26. Baudot A, Marin M (1997) Food Bioprod. Process. 75:117
27. Borjesson J, Karlsson HOE, Tragardh G (1996) J. Membr. Sci. 119:229
28. Karlsson HOE, Tragardh G (1993) J. Membr. Sci. 76:121
29. Pereira CC, Rufino JRM, Habert AC, Nobrega R, Cabral LMC, Borges CP (2005) J. Food Eng. 66:77
30. Willemsen JHA (2003) Proc. Filtech Eur. 2:460
31. Alvarez S, Riera FA, Alvarez R, Coca J, Cuperus FP, Bouwer ST, Boswinkel C, van Gemert RW, Veldsink JW, Giorno L, Donato L, Todisco S, Drioli E, Olsson J, Tragardh G, Gaeta SN, Panyor L (2000) J. Food Eng. 46:109
32. Souchon I, Pierre FX, Athes-Dutour V, Marin A (2002) Desalination 148:79
33. Kattenberg HR, Willemsen JHA (1999) Patent WO 00/38540
34. http://www.foodnavigator.com

20 Encapsulation of Fragrances and Flavours: a Way to Control Odour and Aroma in Consumer Products

Jeroen J.G. van Soest
Kerkhuisstraat 13, 7037 DE Beek (Montferland), The Netherlands

20.1 Introduction

Fragrances or aroma chemicals are an essential additive in consumer products such as household detergent and laundry products [1–4]. They provide the control of odour. The search for attractive fragrances and making aromas durable on textiles is a long-time dream for textile chemists. Delivery of fragrances from detergents onto fabric is a challenge for the fabric-care industry. But, adsorption of fragrances to clothes is poorly understood [5].

Researchers are looking at controlled-release scents in order to extend fragrance longevity [6]. Encapsulation is a good route to control fragrance release and to make more durable fragrant finishing on textiles. However, the affinity between encapsulated aromas and fabrics is still a problem. Many washing products contain surfactants, which form micelles in water. As many fragrances are hydrophobic they tend to migrate to the micelles, rather than deposit on the substrate. A fixing agent can be applied with capsules on a fabric, but the fabric must pass a curing process to fix the capsules. Recently also various detergents were introduced on the market containing additives that absorb odours [7]. Presently, sustainability and making non-toxic and environmentally friendly products are a must in the laundry industry. Legislation and self-imposed industrial standards will provide the consumer with safe new products [8–10]. More efficient products are sought which reduce the amount of aroma chemicals which end up in the environment, for instance via the sewer. (Micro)encapsulation can be an important tool to protect unstable or non-substantive biodegradable fragrances from aggressive detergent components [11]. Also encapsulation, using natural products, could have a positive effect on reducing the frequency of perfume dermatitis in humans [12].

In this chapter, several (biopolymer-based) materials and encapsulation routes will be discussed in relation to their suitability for use as odour control in consumer and detergent products. The discussion of selected applications will illustrate current developments of delivery systems in perfumed laundry or home-care products.

20.2 Encapsulation

Encapsulation has been used in the pharmaceutical industry for many years, for controlled release and delivery of drugs [13]. Because of the additional high costs of early encapsulation techniques, the applicability of encapsulation has been limited. However, more cost-effective techniques and materials have been developed and production volumes are increasing; therefore, the application range has broadened, in particular in foods and consumer products [14–19]. One of the main application areas is encapsulation of aroma chemicals, flavour and fragrances. In the last decade the demand for fragranced products has been growing, and it is thought it will expand and diversify in the future. The following are examples of typical fragranced consumer products: air fresheners bath additives, candles, decorative cosmetics, deodorants, antiperspirants, perfumes, soaps, and hair-care, household, oral hygiene, personal-care, shaving, skin-care and laundry (detergents, softeners) products.

Detergent and laundry products, in general, have a fragrance level in the range 0.2–1%. Perfumes are added to fulfil three tasks:

1. To mask unpleasant odours of cleansing agents
2. To give the message of cleanness during storage and use
3. To impart a nice smell to the fabric

Encapsulation is an elegant way of improving the performance, such as substantivity, tenacity or endurance, of perfumes in washing powders, tablets or conditioners. The performance of fragrances tends to fade by evaporation, interactions with other components, oxidation and chemical degradation. Encapsulation can be the answer to various problems:

- Reduce the reactivity of the fragrance with the outside environment, for example oxygen, pH and water
- Decrease the evaporation rate of the fragrance, control the release rate and provide sustained release
- Promote the ease of handling of the fragrance
- Prevent lumping
- Improve the compatibility with other constituents
- Convert a gas or liquid to a solid form
- Promote easy mixing
- Dilute the core material to achieve uniform dispersion in the product
- Stabilise and protect the fragrance during storage
- Reduce the losses (of top notes) during repeated opening of the packages
- Increase use levels without affecting solubility and dispersing behaviour
- Reduce loss levels in washing water and sewers
- Extend shelf life
- Increase deposition and adhesion on textiles

20.2.1
Matrix or Coating Materials

There are three main types of encapsulated products based on size roughly divided into:

1. Macro-coated powders with sizes larger than 0.1 mm
2. Matrix microparticles or microcapsules with sizes in the range 0.1–100 μm
3. Nanoparticles or nanocapsules with sizes smaller than 0.1 μm

Macro-coating is used mainly to stabilise fragrances or transform them from liquid to free-flowing solid powder. Microencapsulation or nanoencapsulation is the process of enclosing a substance inside a miniature capsule. These capsules are referred to as microcapsules or nanocapsules. The substance inside the capsule can be a gas, liquid or solid. The capsule wall can consist of various materials, such a wax, plastic or biopolymers like proteins or polysaccharides.

In the literature a difference is made between "matrix" encapsulation and "true" encapsulation. In matrix encapsulation the resulting particles are more correctly described as aggregates of actives in a matrix material. A significant portion of the active is lying on the surface of the particles. True encapsulation is used for processes leading to core–shell-type products. However, this distinction of true and matrix is prone to argumentation.

The products can have a variety of shapes, such as spherical, oblong or irregular, can be monolithic or aggregates, and can have single or multiple walls. In Fig. 20.1 some typical morphologies of capsules are shown. The capsules consist of the coated or entrapped materials referred to as active, core material, fill, internal phase or payload (such as aroma chemicals). The coating or matrix material is called wall, membrane, carrier, shell or capsule.

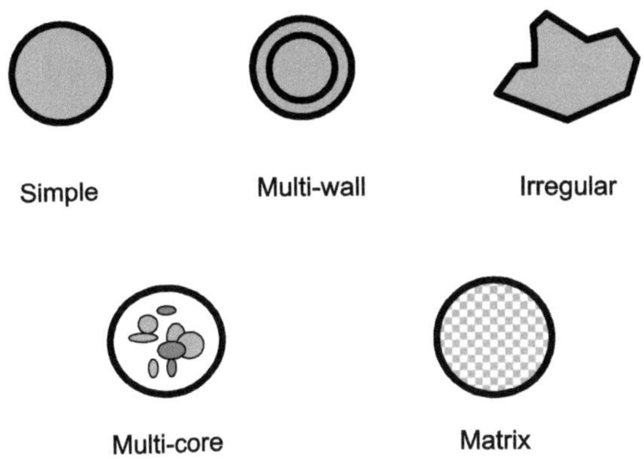

Fig. 20.1 Some typical forms of capsules

Amongst the most commonly used matrix materials are:

- Polysaccharides and sugars (gums, starches, celluloses, cyclodextrin, dextrose, etc.)
- Proteins (gelatin, casein, soy protein, etc.)
- Lipids (waxes, paraffin, oils, fats, etc.)
- Inorganics (silicates, clays, calcium sulphate, etc)
- Synthetics (acrylic polymers, poly(vinylpyrrolidone), etc.)

Biodegradable polymers, both synthetic and natural, have gained more attention as carriers because of their biocompatibility and biodegradability and therewith the low impact on the environment. Examples of biodegradable polymers are synthetic polymers, such as polyesters, poly(*ortho*-esters), polyanhydrides and polyphosphazenes, and natural polymers, like polysaccharides such as chitosan, hyaluronic acid and alginates.

20.2.2 Hydrophilic Matrices

Encapsulation of volatiles in glassy or crystalline matrices is used to extend the shelf life of aroma chemicals. Polysaccharides and glassy sugars, such as starch and maltodextrins, are very suitable for encapsulation of hydrophobic actives owing to the low solubility and low free volume in the glass available for diffusion [20, 21]. Furthermore, hydrophilic matrices have a low oxygen permeability, making them a protective environment for fragrances subject to oxidation.

20.2.3 Processing Routes

Various routes are available based on methods such as spray-drying, spray-cooling/chilling, spinning disk and centrifugal coextrusion, extrusion, fluidised bed, (complex) coacervation, alginate beads, liposomes, supercritical solution and inclusion encapsulation. For most techniques solvent evaporation (drying of water or evaporation of organic solvent in emulsions) plays an important role. Some typical examples are discussed inn the following subsections.

20.2.3.1 Spray-Drying

Spray-drying is an economical effective method widely used for flavour encapsulation [22–27]. The technology has been used in the food industry since the late 1950s to provide protection of aroma chemicals against oxidation or degradation and to convert liquids into free-flowing solids. The main limitations of the technology are that the process needs shell materials, which are soluble in water at acceptable levels and loss of significant amounts of actives. Typical shell materials are gum arabic, maltodextrins and modified starches. The usage of other polysaccharides and proteins is often very tedious and more expensive. The higher the water content in the feed, the higher the energy costs in evaporating the water during the process. Payloads of up to 50% have been achieved, while maintaining free-flowing properties. Double-layered microcapsules have been made using aqueous two-phase systems or multiple emulsions.

20.2.3.2 Spray-Cooling—Chilling

Spray-cooling or chilling is one of the least expensive methods, where the active is mixed with the carrier and atomised using cool air [14–16]. The matrix material is usually a regular, hydrogenated or fractionated vegetable oil. Spray-cooling is a matrix encapsulation method. A significant amount of the active is located at the surface, making the technique less efficient for volatile perfumes. Combinations of spray-drying and spray-cooling have also been described; however, the combined routes are more expensive and lead to low payloads.

20.2.3.3 Extrusion

Microencapsulation using extrusion is mainly described for glassy carbohydrate matrices [14–16, 28–29]. The glassy carbohydrates, such as starch and maltodextrins, are melted at elevated temperature and low water contents and are intensively mixed with the active in the extrusion barrel. Extrusion has been used for volatile and unstable flavours. The shelf life of flavour oils could be extended from several months to 5 years, compared with 1 year for spray-dried materials. The main drawbacks of the technology are the high investments costs and the formation of rather large particles (500–1,000 μm).

20.2.3.4 Rotational Suspension Separation

This is a relatively new technology involving mixing of the core and wall material and a rotational or centrifugal step [14–16]. Typical and similar processes are spinning disk and centrifugal coextrusion. The techniques are industrial alternatives for other traditional encapsulation methods using conventional devices to atomise suspensions or emulsions such as spray-draying or spray-cooling. Spinning-disk technology is an interesting route because of the high throughput and similar processing costs as spray-drying and spray-cooling. The continuous process can take place within seconds to minutes. Solids, liquids or suspensions of 30–200 μm can coated with a layer of 1–200 μm of matrix material. Typical matrix materials are meltable hydrophobic substances such as fats and poly(ethylene glycol). Centrifugal extrusion has been performed using various biopolymer coatings, such as alginates, gums and caseins, giving spherical microcapsules. The technique is more prone to clogging than spray-drying.

20.2.3.5
Air Suspension or Spray-Coating

Air suspension coating is done by suspending a solid core material in a fluid bed of heated or cooled air and spraying the solid with a molten or dissolved matrix material [14–16]. Fluidised-bed technology can be used to apply a uniform layer of almost any kind of material (polysaccharides, proteins, fats, etc.) [30]. The technology is limited to solids or frozen products with minimal particle sizes of approximately 100 μm, making it not so suited for most fragrances. An agglomeration or granulation step can be an integral part of this technology, leading to perfume materials with controlled-release features of fragrances in wash liquors.

20.2.3.6
Coacervation

Coacervation [14–16] consists of the following steps:

1. Disperse the oil (active) in n solution of a surface-active hydrocolloid.
2. Precipitate the hydrocolloid onto the oil by lowering the solubility of the hydrocolloid (add a non-solvent or change pH or temperature).
3. Induce the formation of a polymer–polymer complex by addition of a second complexing hydrocolloid.
4. Optionally, add a cross-linker to stabilise or improve barrier properties of the microcapsules.
5. Dry the material to form microparticles with sizes of 10–250 μm.

Simple or complex coacervation is still not commonly used to encapsulated flavours or fragrances. The technique is complicated and expensive to use. In particular for food ingredients, there are only a few food-grade coating polymers available, such as gum arabic and gelatin. For gelatin systems, additional cross-linking of the shell is done using glutaraldehyde, making it less "label"-friendly. Eventually harmful cross-linkers could be replaced by enzymatic treatments, although industrially viable enzymes are presently not available. It is said that the processing costs can be reduced by optimisation of the drying step. By replacing the usual isolation-drying step (filtration followed by fluidised bed or freeze drying) with a spray-drying step, costs can be reduced significantly. The advantage of coacervation is the efficiency in encapsulation of the actives making high payloads possible of more than 90%. The technique is used for encapsulation of essential oils and fish oil.

20.2.3.7
Emulsion and Interfacial Polymerisation

Microcapsules can be made using oil-in-water or water-in-oil emulsions (or multiple emulsions) [14–16, 31]. The actives are trapped inside a monomer or polymer matrix, which can be polymerised and cross-linked. After breaking the emulsions, the microcapsules can be dried by solvent evaporation or other drying methods. Interfacial polymerisation occurs with monomers or polymers with surface-active properties or which are rendered insoluble by the polymerisation or cross-linking reactions [32]. Polymerisation takes place at the water–oil interface. The use of these methods is limited since the preferred matrix or coating materials are non-renewable or non-food grade, such as polyesters, polyamides, polyurethanes, polyacrylates or polyureas, often leaving traces of toxic monomers. More recently also polysaccharide-based systems have been described using food-grade cross-linkers.

20.2.3.8
Miscellaneous Routes

Various routes are described in the literature which are based on very specific interactions of actives with a specific polymer or coating molecule or specific processing techniques [14–16]. Some of them are mentioned next.

20.2.3.8.1
Liposomes

Liposome entrapment [14–16] is mainly used in pharmaceutical and cosmetic applications. Liposomes (the most common being phospholipids) can form membrane-like vesicles, with diameters in the range 25 nm–10 μm, which show selective permeability for small molecules. Both hydrophobic and hydrophilic ingredients can be entrapped. Application of liposome entrapment is still limited in food (flavour) or fragranced household products because of the high price of phospholipids and difficulties in scaling up the process at acceptable cost in use and creating a good delivery form. Research is progressing in finding cheap alternatives for phospholipids based on hydrophobic emulsifiers and using microfluidisation as a cost-effective continuous processing method.

20.2.3.8.2
Inclusion Complexation

Inclusion complexation or molecular encapsulation is based on the molecular inclusion of an active inside the cavity of another molecule. The most well-known systems are based on cyclodextrins [33]. Cyclodextrins are used to protect heat-, light- or oxygen-sensitive ingredients. They are used to increase the solubility of hydrophobic substances and to reduce the volatility of aroma chemicals. The central cavity of the cyclodextrin is hydrophobic, making it attractive for hydrophobic substances to occupy it. To obtain complexation, guest molecules are coprecipitated or cocrystallised from aqueous solution. To obtain high loadings from hydrophobic actives with low solubility, the method is expensive because of the high drying costs and the high price of cyclodextrin.

Although in principle amylose can also be used to form inclusion complexes, its use is not widespread because of the low solubility and high price of pure amylose and the low specificity of high-amylose containing starches [34].

20.2.3.8.3
Alginate Beads

Gelling gum based beads can be produced very easily on a laboratory scale [16]. The technique is well described in scientific literature for the preparation of alginate-based microcapsules [35]. Scaling up of the small batch process to an economically viable process is difficult, although recently several methods have been described to facilitate scaling up. Furthermore the beads are very porous, making them not very suitable for aroma chemicals, where extended shelf life or sustained release is wanted. Most attention has been given to alginates (being easy to use and renewable). However other gelling agents are being used already in various fragranced consumer products, such as gellan and carrageenan.

20.2.3.8.4
Cocrystallisation

Cocrystallisation is mainly done from supersaturated sugar solutions [15]. Aggregated particles (of 3–30 μm) of sugar crystals are formed which entrap guest molecules. The sugars form an oxygen barrier, thereby extending the shelf life of aroma chemicals. The procedure is simple and inexpensive, because relatively cheap encapsulation matrices can be used, such as sucrose.

20.2.3.8.5
Supercritical Solutions

Supercritical solutions can be regarded as dense solvating gasses or low-viscous low-density liquids. The most well-known and probably most interesting candidate is based on carbon dioxide. Supercritical carbon dioxide can be regarded as an organic solvent. Various concepts have been developed using supercritical fluids, such as a method similar to conventional spray-drying and rapid expansion of supercritical solutions [16]. The early methods were restricted to shell materials which could dissolve in the supercritical fluid; however, a slight adaptation of the process broadened the applicability to matrix material, which can swell in supercritical fluids, such as proteins and polysaccharides. The use of supercritical carbon dioxide renders the use of organic solvents obsolete and makes the technology environmentally interesting and interesting for food applications.

20.2.4 Recent Developments and Trends

There are various reasons for applying encapsulation. Numerous patents are filed every year dealing with new microencapsulation techniques. Some of these new technologies and processes have currently no industrial relevance, because of high cost in use, difficult scale-up and narrow range of applicability. For fragranced consumer products, controlling costs is even more important than in applications found in pharmacy or even foods. The market is very competitive and therefore additional costs should be considered. But, some old or new technologies look promising for the near future. Old technologies have become more efficient and scaling-up processes have improved. Also environmental issues related to raw material use, energy and waste control are more important. Designing fragrances with better biodegradability has led to fragrances with too low stability towards oxygen and water, making them unsuitable for most applications. Encapsulation could be a good tool in protecting these fragrances.

20.2.4.1 New Technologies

Development of new encapsulation methods is time- and effort-consuming, requiring a multidisciplinary approach. In contrast with foods, materials used for fragrance encapsulation are not subject to the extensive legislation that applies to food approval. This makes the use of new materials as matrix materials easier. Some new developments with potential for the near future are discussed next.

20.2.4.1.1 Nanotechnology

Nanotechnology is hot in the world of science [36]. Research focuses on properties, which arise from scaling down structural features of materials to the nanometre range. Two strategies are used to make nanostructured materials:

1. Top down—break down larger structures
2. Bottom up–build from individual atoms or molecules capable of self-assemblage

The most important materials developed are nanocomposites and nanotubes. Fabrication of the first nanocomposites was inspired by nature (biomineralisation). Nanocomposites based on nanoclays and plastics are seen as ideal materials for improved barrier properties against oxygen, water, carbon dioxide and volatiles [37]. This makes them in particular suitable for retaining flavours in foods. The technology is rather straightforward using commercially available nanoclays and extrusion processing.

Even newer generations of nanomaterials are based on carbon nanotubes using the bottom-up approach. The materials are still very expensive, but the technology is evolving rapidly. Another type of nanotube has been prepared based on self-assembly of specific molecules such as chitosan-based nanoparticles of polypeptides, DNA or synthetic polymers. Phospholipids or dendrimer-coated particles are suitable for the entrapment of actives in very small vesicles. The current materials are still lacking in selectivity and yield (costs).

For delivery systems to be effective, the encapsulated active compounds need to be delivered to the appropriate locations, without losing activity. In particular in textile washing, fragrances need to be delivered to the cloth during washing, without losing activity during storage or without losing fragrances in the wash water. In particular, nanoparticles or nanospheres are said to have improved encapsulation and release characteristics [38]. Also the small sizes make them more suitable for adjusting the adhesion properties to various textile fibres. Although the preparation of nanoparticles is more developed for synthetics and inorganics, also biopolymer-based technologies are being developed. Examples are given in the literature of polysaccharide-based materials [39, 40].

Manipulation of materials at the nanometre level opens to door to improved functionality of aroma chemicals. However, nanotechnology needs to be made more economically viable to have lower cost-in-use.

20.2.4.1.2 Colloidosomes

A very recent development is encapsulation of actives in colloidosomes [16, 41]. The method is analogous to liposome entrapment. Selectively permeable capsules are formed by surface-tension-driven deposition of solid colloidal particles onto the surface of an inner phase or active ingredient in a water-in-oil or an oil-in-water emulsion composed of colloidal particles. Initially synthetic polymer microparticles were used but more recently a natural alternative has been described based on small starch particles. After spray-drying, redispersible emulsions can be formed.

20.2.4.1.3 High-Pressure Gelation

High-pressure gelation could be an interesting new approach [16]. It has been shown that native starches can be gelatinised using high-pressure treatments

[42]. It is possible to control the degree of granule disintegration much better than with low-pressure gelatinisation. Also proteins can be gelled at high pressure and the microencapsulating capability of this process has been shown [16].

20.2.4.1.4
Sol–Gel Processing

The sol–gel process originates from the ceramic industry [16, 43]. It can be regarded as the inorganic analogue of interfacial polymerisation encapsulation. During the sol–gel encapsulation, an inorganic gel network is formed by gelation of a sol (a colloidal suspension). The most commonly used precursors are metal alkoxides, which can react and undergo the sol–gel transition in aqueous environment.

20.2.4.2
Recent New Materials

Besides new technologies also new materials are being found. Several (patent) overviews of the art of the encapsulation of various materials, such as flavours and fragrances, can be found in the literature [44, 45]. This section highlights some typical more recent new patented carrier materials used for improvement of fragrance performance in detergents using encapsulation methods.

New synthetic-based matrices are being developed. Enhanced deposition of fragrances to fabrics is obtained using a fragrance-containing acrylate-based gel capable of being mixed with a detergent composition [46]. Enhanced longevity is also claimed. In another invention, microcapsules are described based on free-radical polymerisation [47]. A more scientific development is based on the grafting of temperature-sensitive hydrogels to fabrics. Environment-sensitive deodorant fibres and delivery fabrics can be made on the basis of these hydrogels [48].

In particular, matrices based on polysaccharides or other biopolymers are of interest using well-known technologies such as spray-drying and extrusion. Mixtures of various carriers can be used to tailor properties such as release, deposition and substantivity [49]. A particular example of a new material used for fragrance encapsulation is the use of polysaccharide esters such as starch acetate [50]. The methods described are very easy and can be scaled up. The patent addresses the issues of fragrance stability during storage and the loss of most of the fragrances in the wash water [51]. Also matrices containing inorganic materials have been developed that are suitable for transforming fragrances into free-flowing powder with improved deposition properties for laundry application [52].

20.3 Performance of Fragrances in Consumer Products

To be able to bring the "message of freshness, cleanness, newness" of fragrances to the consumer, it is of great importance to understand the way fragrances work in various applications and environments [3, 4, 53–58]. Research has focused on the interactions with the complex chemical compositions of the various products and the targeted substrates. The way the consumer perceives the odours depends not only on the fragrance composition but also on the other components and on the type of substrate, such as fabric, floor, hair and skin. A classification of perfumed consumer products can be made depending on how the product is transferred to the substrate:

- Direct application of the perfumed product to the substrate (e.g. deodorant, cream)
- Transfer of the product via a wash or rinse step (e.g. detergent, softener, shampoo)

The consumer experiences various stages in perceiving the odour sensation:

- Odour of the product
- Wet odour impact
- Dry odour (*tenacity*) or initial dry odour impact (*perceived substantivity*)
- Odour during use (*long-lasting*)

It is clear that encapsulation has a strong effect on all of these properties. In using encapsulation for the design of new fragranced consumer products, the effects have to be taken into account in the reformulation procedures.

Measurements of fragrance performance involve the following aspects:

- Control of experimental settings
- Dynamics if processing (e.g. drying rate and temperature)
- Interpretation in terms of olfactory dose–response characteristics

Without a doubt the process is influenced by factors such as water solubility and hydrophobicity of the fragrance constituents and the presence of surfactants and cosurfactants, and many more. To help the development of effective fabric-care products, it is important to develop a better understanding of the factors that influence retention of aroma chemicals on textiles and their release. New methodologies, also used to study the distribution of chemical finishing agents and soils on fibres, can be helpful to study the distribution of unsaturated aroma chemicals on textiles, in order to gain a deeper understanding of the mechanisms of their deposition, adsorption and retention on fabrics. One of the main tools in measuring odour characteristics is quantitative gas chromatography–headspace analysis [53–57]. The aroma chemical distribution on cotton, Lyocell and polyester fibres was studied using backscattered electron microscopy and X-ray microanalysis [58]. Various parameters have been identified to determine perceptible odour, such as vapour pressure, water solubility, temperature, logP(oil/water) or hydrophobicity, volatility, and concentration of fragrance and other ingredients. It is obvious that most parameters will be strongly affected by the encapsulation matrix material and the method used, as illustrated in Fig. 20.2.

The selection of a specific encapsulation route or delivery system depends on the nature of the product where the delivery system will be used, on which property one wants to improve (process retention, protection, deposition or release mechanism). Fragranced consumer products can make use of a broader selection of matrices than foods and pharmaceuticals; however, there are other strict constraints. Fragrances are commodity ingredients in consumer products and additional costs owing to the encapsulation process should be low. In order to meet consumer acceptance, the products should meet the olfactory requirements. Another selection criterion is the presence of water in the product and the humidity during storage. At low water levels or in dry products, hydrophilic matrices can be used. Water is a plasticiser for hydrophilic matrices such as starches. The lower glass-transition temperature, swelling and dissolving effect of water will have a negative effect on storage stability and retention. In liquid aqueous (wet) products, hydrophobic matrices are to be used. Hydrophobic matrices are worse oxygen barriers and are less effective as a barrier for other hydrophobic ingredients. The next selection is based on the release characteristics required, such as the mechanism and kinetics (sustained versus triggered). Examples of release triggers giving burst-like release are water, heat, mechanical stress, enzymes, ion strength and pH. Additional criteria are toxicity, compatibility and biodegradability.

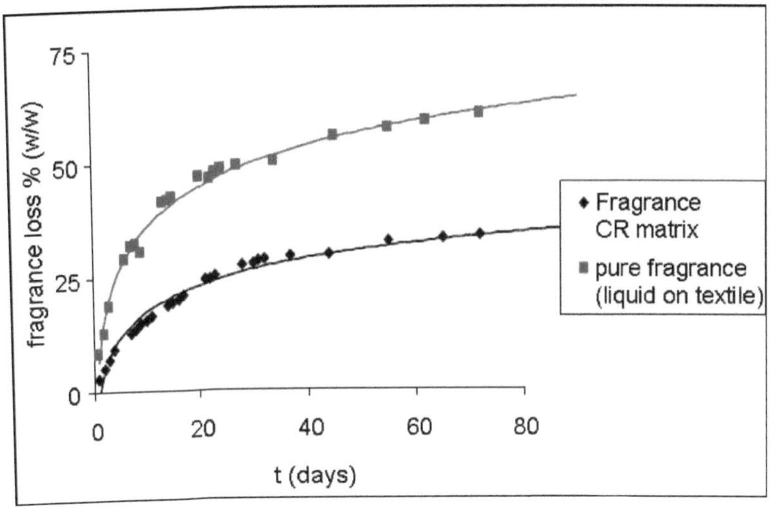

Fig. 20.2 Release profile of an encapsulated fragrance compared with that of the pure fragrance. *CR* controlled release

20.4 Market Developments and Products

There are two reasons why it is difficult to give examples of already marketed products based on encapsulation used in detergents or even household and consumer products. One is the fact that the technology is still not always applicable in a cost-effective way. The other reason is that it is not always known that an encapsulated material is used, because companies like to keep the know-how in-house. Still some technologies are being used in daily life already [1, 2, 59].

One of the widely known novelties of using microencapsulation technologies in a consumer product is the InstaScent™ (scratch and sniff) or Snap&Burst™ scented overprint varnishes. Tiny glass-based capsules contain a liquid scent and are glued onto paper. This product manufactured by Lipo Technologies is a cost-effective way of presenting fragrances to customers. When the paper is scratched, some of the capsules are ruptured and the scent is released. Another technology making use of fragrance microcapsules, which make use of triggered release by breaking the capsules, was developed by Bayer-Lanxess (Euderm® and Bayscent®). The microcapsules are prepared using interfacial polymerisation and are applied to leather or textiles by spraying.

A well-known example of the use of cyclodextrins is found in Fébrèze from Proctor & Gamble as odour control. Fébrèze, a spray used for eliminating bad odours on fabrics, has been adapted for use in fabric softener (Lenor Stayfresh). An alternative is provided by Henkel's Neutralin technology, which combines the odour-reducing zinc ricinoleate (Tegosorb™ from Degussa) with a fragrance. The material is claimed to perform better in water, making it suitable for detergent applications.

An example of a fragranced consumer product is Crayola® Magic Scent (from Binney & Smith) food-scented crayons containing gelatin-encapsulated aromas like orange, cherry, chocolate, strawberry, peach, blueberry, liquorice, lime, bubble gum, banana, lemon, coconut and grape. In personal care, Kleenex® Cold-Care facial tissues from Kimberly-Clark make use of the same type of capsules to protect volatile menthol fragrance. The Breathe Right® family of products (from CNS) and Vicks® (from Proctor & Gamble) were developed to make it easier for more people to breathe freely using encapsulated mentholated vapours.

New technologies are being developed or adapted for household cleaning and detergents. Examples are Microflex (a microemulsion delivery system for fragrances from International Speciality Products) and Hallcrest's microcapsules based on coacervation and liquid crystals. Henkel developed a new technology making it possible to selectively deposit a chemically linked fragrance compound on a fabric. Slow release is then triggered by air humidity.

Various companies are expressing there efforts in the areas of innovative delivery technologies for the soap and detergent market, such as Alco (part of National Starch and ICI), ISP, Rhodia, Cognis and Ciba. Alco has access to the flavour encapsulating starch technologies from National Starch and acquired Salvona delivery technologies and is adapting them with the focus on detergents and fabric softeners.

For powdered detergents Givaudan launched Granuscent® encapsulation technology. Protective granules are made by spray-drying a fragrance emulsion, forming a glassy hydrophilic matrix. Similar efforts are being made by Symrise (formerly Dragoco and Haarmann & Reimer). They are exploring the use of the starch-based InCap and poly(vinyl alcohol) PolyCap technology for dry products. Their urea resin or gum-based SymCap system is directed towards liquid systems. Ciba is pursuing fragrance delivery form the point of extending scent longevity on the shelf (using the excited state quencher, ESQ™, technology developed for increased colour stability).

20.5 Conclusions

Various encapsulation techniques are available for improving the efficiency of aroma chemicals in fragranced consumer products. Encapsulation techniques are still being optimised in terms of fragrance performance, scaling up and costs. Environmental aspects are becoming more important, putting constraints on the use of non-biodegradable fragrances, which are used in large excess and end up in the environment. Encapsulation could be the tool to make more efficient use of fragrances as slow or controlled delivery systems. Encapsulation opens the way in using biodegradable fragrances which could not be used before because of the too low chemical stability (during processing, storage or usage). A straightforward route to develop microencapsulated fragrance materials is to adapt existing methods developed for pharmacy, foods, agriculture or cosmetics. However, industrial constraints (cost in use) should be taken into account in a cost-competitive market area such as consumer products.

Acknowledgements

PFW Aroma Chemicals B.V. (Barneveld, The Netherlands) is highly appreciated for their willingness to make results presented in Fig. 20.2 available for presentation.

References

1. McCoy M (2006) Chem Eng News 84:13
2. McCoy M (2005) Chem Eng News 83:15
3. Stora T, Escher S, Morris A (2001) Chimia 55:406
4. Quellet C, Schudel M, Einggenberg R (2001) Chimia 55:421
5. Reutenauer S, Thielmann F (2003) J Mater Sci 38:10
6. Nelson G (2002) Int J Pharm 242:55
7. Szejtli J (2003) Starch 55:191
8. Tas JW, Balk RA, Ford E, van de Plassche EJ (1997) Chemosphere 35:2973

9. Scientific Committee on Cosmetology (1997) Notes of Guidance for Testing of Cosmetic Ingredients for Their Safety Evaluation; 2nd revision 1997, Ann 9. http://europa.eu.int/documents/comm/dg24/health/sc/sccp/out07_en.html, http://europa.eu.int/documents/comm/dg24/health/sc/sccp/out08_en.html.
10. Regulation (EC) No. 648/2004 of the European Parliament and Council on Detergents (2004) Off J Eur Union L 104
11. Anderson K (2006) [TC]^2. http://www.techexchange.com/thelibrary/innovateor.html
12. Jackson EM (1998) Am J Contact Dermatitis 9:193
13. Kosaraju SL (2005) Crit Rev Food Sci Nutr 45:251
14. Gibbs B, Kermasha S, Alli I, Mulligan CN (1999) 50:213
15. Jackson LS, Lee K (1991) Lebensm Wiss Technol 24:289
16. Gouin S (2004) Trends Food Sci Technol 15:330
17. Reineccius GA (1989) Food Rev Int 5:147
18. Zeller BL, Salleb FZ, Ludescher RD (1999) Trends Food Sci Techol 9:389
19. Pothakamury UR, Barbosa-Cánovas GV (1995) Trends Food Sci Techol 6:397
20. Korus J, Tomasik P, Lii CY (2003) J Microencapsulation 20:47
21. Qi ZH, Xu A (1999) Cereal Food World 44:460
22. Sheu TY, Rosenberg M (1995) J Food Sci 60:98
23. Rosenberg M, Kopelman IJ, Talmon Y (1990) J Agric Food Chem 38:1288
24. Soottitantawat A, Bigeard F, Yoshii H, Furuta T, Ohkawara M, Linko P (2005) Innov Food Sci Emerg Technol 6:107
25. Yoshii H, Soottitantawat A, Liu X-D, Atarashi T, Furuta T, Aishima S, Ohgawara M, Linko P (2001) Innov Food Sci Emerg Technol 2:55
26. Buffo R, Reineccius G (2000) Perfum Flavor 25:37
27. Bayram ÖA, Bayram M, Tekin AR (2005) J Food Eng 69:253
28. Yilmaz G, Jongboom ROJ, van Soest JJG, Feil H (1999) Carbohydr Pol 38:33
29. Doane WM (1993) Ind Crops Prod 1:83
30. Dimantov A, Greenberg M, Kesselman E, Shimoni E (2004) Innov Food Sci Emerg Technol 5:93
31. van Soest JJG, van Schijndel RJG, Gotlieb KF (2002) US Patent Appl 6,340,527
32. Park JH, Ye M, Park K (2005) Molecules 10:146
33. Szente L, Szejtli J (2004) Trends Food Sci Technol 15:137
34. Wulff G, Avgenaki G, Guzmann MSP (2005) J Cereal Sci 41:239
35. Smidsrod O, Skjak-Braek G (1990) Trends Biotechnol 8:71
36. Moraru CI, Panchapakesan CP, Huang Q, Takhistov P, Liu S, Kokini JL (2003) Food Technol 57:24
37. van Soest JJG (2006) ACS Symp Ser 921:111
38. Kumar MNVR (2000) J Pharm Pharma Sci 3:234
39. van Soest JJG, Dziechciarek Y, Philipse AP (2002) In: Yuryev V, Cesaro A, Bergthaller W (eds) Starch and Starch Containing Origins—Structure, Properties and New Technologies. Nova, New York
40. van Soest JJG, van Schijndel RJG, Stappers FJM, Gotlieb KF, Feil H (2004) US Patent Appl 6,755,915
41. Dinsmore AD, Ming FH, Nikolaides MG, Marquez M, Bausch AR, Weitz DA (2002) Science 298:1006

42. Douzals JP, Marechal PA, Coquille JC, Gervis PJ (1996) Agric Food Chem 44:1405
43. Kickelbrick G (1996) Prog Polym Sci 28:83
44. Risch SJ (1995) ACS Symp Ser 590:196
45. Porzio MA, Popplewell LM (2001) US Patent Appl 6,187,351
46. Pashkovski EE,. Farooq A, Heibel M, Mehreteab A, Miller L, Mastrull J, Theiler R (2002) Patent Appl WO 02/77150
47. Jahns E, Boeckh D, Bertleff W, Neumann P (2003) US Patent Appl 2003/0125222
48. Liu B, Hu J (2005) Fibres Textiles 13:45
49. Lou WC, Popplewell LM (2003) US Patent Appl 2003/0077378
50. Vedantam VK, Yong TT (2004) Patent Appl WO 04/083356
51. van Alst M, van Tuil R, van Soest JJG (2003) Biopolymers: Health, Food and Cosmetic Appl, Proceedings of the European Confrence Polymerix 2003, Rennes, France, 21–22 May 2003
52. van Schijndel RJG, Westerweele E, Sivasligil DS, van Soest JJG, Lenselink W (2002) EP Patent Appl 1,229,789
53. Yven C, Guichard E, Giboreau A, Roberts DD (1998) J Agric Food Chem 46:1510
54. Verma M, Borse BB, Sulochanamma G, Raghavan B (2005) Flavour Fragrance J 20:122
55. Seuvre A-M, Philippe E, Rochard S, Voiley A (2006) Food Chem 96:104
56. Lethuaut L, Brossard C, Meynier A, Rouseau F, Llamas G, Bousseau B, Genot C (2005) Int Diary J 15:485
57. van Ruth SM, King C (2003) Flavour Fragrance J 18:407
58. Liu HQ, Obendorf SK, Young TJ, Incorvia MJ (2004) J Appl Pol Sci 91:3557
59. McCoy M (2004) Chem Eng News 82:23

21 Creation and Production of Liquid and Dry Flavours

Rainer Barnekow, Sylvia Muche, Jakob Ley, Christopher Sabater,
Jens-Michael Hilmer, Gerhard Krammer
Symrise GmbH & Co. KG,
Mühlenfeldstraße 1, 37603 Holzminden, Germany

21.1 Modern Flavour Creation

21.1.1 The Roots of Flavour Work

Among the oldest known and documented formulae, biblical anointment oils represent an interesting combination of spices such as cinnamon and fragrance materials, for example myrrh [1]. In those days, culinary developments and fragrance creations were heavily influenced by religious ceremonies. In the middle ages both the gastronomic and the fragrance aspects were influenced by new technologies like beer brewing or baking technology as well as the distillation of essential oils. The age of enlightenment and the curiosity of researchers led to the so-called great cycle of the aroma and fragrance industry, which generated numerous aroma chemical and fragrance materials which were all based on the combination of analytical identification of a chemical structure, synthesis, scale up and subsequent production.

In 1874 Wilhelm Haarmann started to produce the first synthetic aroma chemical, vanillin [2–4]. Since then the flavour, fragrance and aroma chemical industry has shown rapid progress. In the beginning, perfumers created the first flavour formula with synthetic aroma chemicals. Over the years many different parameters, like the availability of natural products, the development of food industry and changes in consumers' lifestyles, have led to a broad range of widely accepted flavourings.

In parallel, the fragrance industry has grown to meet consumer preferences with regard to the use of perfumes and also other aspects, such as personal identity, human odours, mood preferences, emotions and psychology [5].

The introduction and rapid development of highly effective analytical instrumentation like the combination of gas chromatography with mass spectrometry (GC-MS) facilitated a significant increase of known aroma chemicals from around 600 in the early 1960s to around 15,000 nowadays. In parallel, the flavourist work profile received strong impulses from the food industry with regard to flavour stability, dosage and flavour application, which finally initiated the development of a sophisticated flavour technology portfolio which comprises liquid and dry blending, plating, spray-drying, emulsions and various encapsulation techniques.

Additionally, in the last 5–10 years an emerging number of low-volatile taste-modifying molecules were found using sophisticated analytical methods based on liquid chromatography (LC), namely taste dilution analysis, LC-MS or LC-NMR methods.

21.1.2 Raw Materials—the Foundation of Every Creation

The world of aroma compounds is becoming more and more complex. In the early days people used aromatic products like fruit juices or fruit juice concentrates which were relatively weak and still close to the related foodstuff. Later, with more knowledge of separation techniques, infusions, extracts, oleoresins and absolutes ranging from weak to strong impact were used to impart aroma. Essential oils such as spice oils already had a very strong impact. Modern analytical technologies allowed the evaluation of the chemical compositions of extracts and essential oils, so that isolates either as powerful mixtures or even as single compounds could be obtained.

The route from cinnamon via the extract, the resin to the cinnamon bark oil and finally to cinnamic aldehyde stands only as one example of the increase in the number of natural aroma compounds. Later, the availability of nature-identical, synthetic aroma chemicals opened great opportunities for flavouristic creativity. In the future, with the completion of the EU positive list and based on the existing FEMA list, the modern raw material portfolio will provide a range of selected aroma chemicals with a defined safety standard. At the same time, well-established aroma chemicals such as estragol have to be omitted because of toxicological considerations. Another growing area a flavourist has to be aware of is the field of non-volatile taste compounds, since a modern flavour solution in the future will comprise the aroma and also a taste part or a taste-modifier part (i.e. umami enhancement, sweet enhancement, bitter masking).

The right choice of raw materials is crucial for creative development (Table 21.1). The final application, the market for which the flavouring will be developed, legislative and ethnic implications, and customer requirements all have to be considered by flavourists when choosing their starting materials.

21.1.2.1 Natural Raw Materials

The field of natural raw materials is dominated by plant derivatives. Important representatives of naturals are the botanical extracts. Extracts can be obtained by water or alcohol–water extraction. Onion extract, for example, is produced by squeezing the washed and ground onion bulbs in large filter presses. The resulting onion juice can then be concentrated to give a stable raw onion extract with superior flavour properties. A valuable by-product is the onion oil which

Table 21.1 Definitions and examples for common raw materials (not necessarily legal definitions)

Flavouring type	Definition/production	Example
Food with flavouring properties	Food or processed food with strong flavours	Vanilla beans, spices
Resins	Evaporated extracts	Pepper oleoresin
Essential oil	Steam distillation	Spearmint oil, lemon oil, pepper oil
Extracts	Alcoholic extracts	Vanilla extract
Natural aroma chemicals	Isolation and purification via physical processes	Citral from lemon grass oil, eugenol from cloves, menthol from *Mentha* species
Natural aroma chemicals	Production via fermentation or enzymatic treatment	Oxidation of 2-methylbutanol to 2-methylbutyric acid via *Acetobacter* species
Nature-identical aroma chemicals	Occur in nature and obtained via synthesis	Vanillin produced from lignin or catechol
Artificial	Do not occur in nature	Ethyl vanillin
Reaction flavours	Thermal treatment of amino acids and reducing carbohydrates based on Maillard reaction	Caramel and malt flavours
Smoke flavours	Preparation based on smoke, produced via defined processes	Smoky ham note

can be obtained via solvent extraction from the condensates of the concentration process.

These extracts are produced from a large variety of plants, like herbs and spices, with or without prior enzyme treatment for the hydrolysis of the cell walls. Extracts from plant or animal material can be generated by solvent extraction but also by complete enzyme hydrolysis of plant derivatives (wheat gluten, soy, etc.) or of real meat, filtration and subsequent concentration of the liquid extract (hydrolysed vegetable/animal protein). These hydrolysed vegetable proteins or hydrolysed animal proteins generate savoury-like notes and also contain a natural content of flavour enhancers such as monosodium glutamate (MSG), inosine 5′-monophosphate and guanosine 5′-monophosphate. The extracts can be used as such or can be further heat-treated as so-called thermally treated extracts or process flavours.

One of the most important and popular extracts from a market-potential perspective is certainly vanilla extract. North America is the largest market, followed by the European market, with ice cream being the largest single application. The predominant vanilla species is *Vanilla planifolia*, which is the basis for a large volume of available extracts. Madagascar is still the largest producer of high-quality vanilla extract, followed by Indonesia.

Vanilla is an excellent example in which a flavourist has to understand the market the flavour is to be created for. The Americans, for example, prefer the more prominent, phenolic and smoky notes, while French consumers are more interested in anisic notes.

In Germany buttery, creamy and balsamic profiles have a long tradition, while vanillin itself represents the key driver for Scandinavia.

The cured vanilla bean consists approximately of 98% of water, fats, waxes, sugar, cellulose, etc. Only some 2% is flavour compounds, the main constituent (approximately 90%) of these being vanillin. Roughly 9% are *p*-hydroxybenzaldehyde, vanillic acid and *p*-hydroxybenzoic acid, which do not contribute much to the overall flavour profile. The remaining approximately 1% of the constituents of the flavour compounds reveals the most significant flavour properties. This part itself comprises more than 400 chemicals giving an extract its specific sensorial "fingerprint" (Fig. 21.1). There are significant differences in the chemical compositions and therefore also in the sensory profiles of vanilla extracts as a result of the geographical origin, the soil, the climate and the processing conditions. Depending on the type of flavour that needs to be developed, a flavourist can start with a specific vanilla extract already supporting the desired flavour profile or with a more neutral extract which is typified with specific qualities.

Another important group is the essential oils which are manufactured mainly from herbs and spices mostly by steam distillation. The advantage of steam distillation is the fact that a clean and powerful oil can be isolated after the distillation step without waxes and other non-volatile compounds but with an odour

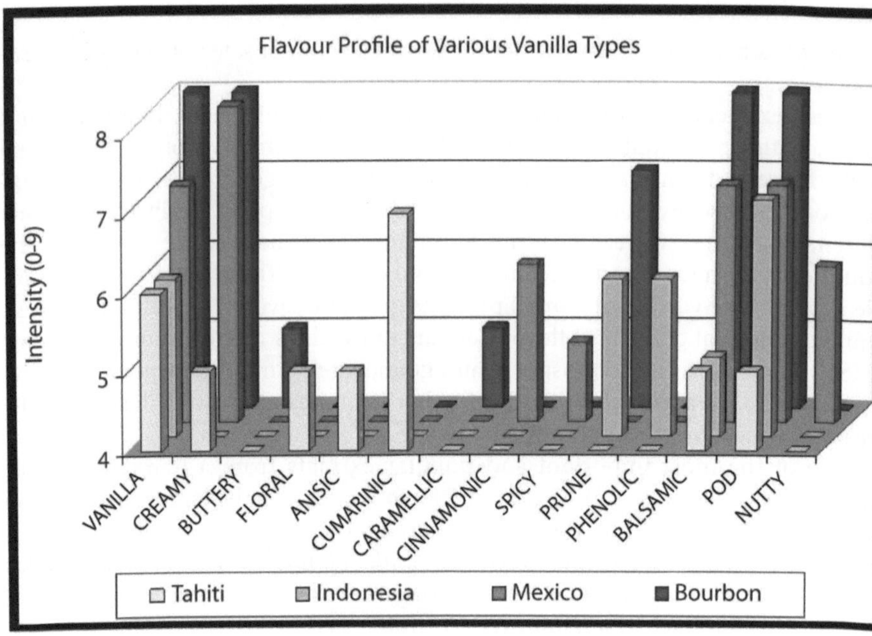

Fig. 21.1 Flavour profiles of different vanilla types

that strongly resembles the odour of the original spice or plant. The disadvantage of losses of highly volatile compounds and non-volatiles as well has to be taken into account. In addition possible chemical reactions and thermal degradation during the process affect the original flavour profile. In this way the essential oil loses some of the freshness and authenticity compared with the original material. Moreover, taste-sensation materials remain to a large extent in the botanical residue. While pepper extracts still have all the pungent spiciness of real pepper, the compounds actually responsible are not present in the pepper oil.

The essential oils from citrus fruits are often obtained through a cold-pressing step from the peel (e.g. orange peel oil). For this purpose several technologies are in use. The most prominent examples are *sfumatura* ("slow folding") with a superficial grazing or total abrasion of the whole fruit and the so-called *pelatrice speciale* method with a constant amount of water for the extraction of the oil. Both processes are very gentle and give very authentic essential oils. The problem of these citrus oils, mainly orange oils, is the presence of high amounts (80–95%) of the non-oxygenated terpenes, limonene being predominant. These terpenes, which do not contribute much to the aroma, can be oxidised when exposed to air and can generate off-flavours. The insolubility of these terpenes, for example, in clear beverage applications remains another disadvantage if citrus oils are used as such. A variety of processes like distillation, solvent extraction and washing can be used to remove the non-oxygenated terpenes to a large extent and to enrich the desired oxygenated terpenes. These processes lead to powerful multiconcentrated oils with a higher solubility in aqueous applications.

Other useful enriched natural materials such as paprika extract are predominantly produced through solvent-extraction methods using solvents or supercritical fluids like CO_2.

Single natural aroma compounds like natural vanillin are obtained through physical separation techniques from edible materials or through natural fermentative processes.

Natural raw materials are of high importance in flavour development. Natural isolates (Table 21.2) serve as a basis for most natural flavourings which can be blended with single natural aroma chemicals. The performance of nature-identical flavourings will be supported by using extracts and oils as they significantly enhance the complexity of flavourings and increase their authenticity.

21.1.2.2
Nature-Identical Raw Materials

Aroma chemicals which are found in natural sources or food preparations but are synthesised by normal chemical procedures are defined by the status "nature-identical". Most of them were discovered and developed during the nineteenth and twentieth centuries. The most important single aroma chemicals produced in very large amounts are vanillin, menthol, citral and anethol. They are used not only by flavour producers but also in large amounts in fragrance

Table 21.2 Important isolates from natural sources

Name	Lead compound(s)	Application
Anise oil	*trans*-Anethol	Alcoholic beverages, oral care
Bitter almond oil	Benzaldehyde	Pistachio flavours
Buchu	(+)-*trans*-8-Mercapto-*p*-menthan-3-one	Black currant flavours
Caraway oil	(+)-Carvone	Savoury flavours
Cardamom oil	1,8-Cineol, α-terpinyl acetate	Baked products
Cinnamon oil	*trans*-Cinnamic aldehyde	Flavours for confectionery products
Grapefruit oil	Nootkatone	Beverage flavours
Clove oil	Eugenol	Oral-care flavours, savoury flavours
Sweet fennel oil	Anethol	Beverage flavours
Ginger oil	β-Sesquiphellandrene	Beverage flavours
Ginger oleoresin	gingerols, shogaols	Hotness, savoury food, confectionery
Juniper berry oil	α-Pinene	Alcoholic drinks
Laurel leaf oil	1,8-Cineol	
Marjoram oils	1-Terpinen-4-ol, *cis*-sabinenhydrate	Savoury flavours
Cornmint	(-)-Menthol	Chewing gum, oral care
Spearmint	(-)-Carvone	Chewing gum, oral care
Origanum oils	γ-Terpinene, *p*-cymene, thymol, carvacrol	Savoury flavours
Star anise oil	*trans*-Anethol	Beverage and confectionery flavours
Thyme oil	Thymol	

applications. On the other hand, there are a lot of so-called high-impact flavour chemicals, which are produced only in very small amounts as a result of their low threshold levels, for example acetyl thiazoline and 1-menthen-3-thiol. Most of these materials are produced in high purity and therefore provide highly standardised sensorial properties for top note creation.

21.1.2.3
Ethical Requirements

Flavourings created for the US market or Israel normally have to follow requirements for kosher status, whereas markets as the Near and Middle East and parts of Asia (e.g. Indonesia, Philippines) have a strong need for halal flavourings. As the flavour market is becoming more and more global, even the European companies in the flavour industry have to be certified by the respective certifying authorities. In general these requirements result in a reduced number of raw materials and in specific cases also carrier materials (e.g. omission of ethanol for halal flavours) for the daily project work of a flavourist.

21.1.3
Process Flavours

Process flavours or process flavours play a key role in those food products which have been exposed thermal treatment during processing and final preparation, where heating steps during preparation are applied.

Since process flavours are generated by the interaction of raw materials like protein derivatives (amino acids) and reducing sugars (Maillard reaction), it is obvious that a large number of prepared food products are affected:

- Meat products, e.g. beef, chicken, pork, lamb
- Vegetables, e.g. onions, potatoes, garlic
- Roasted products, e.g. coffee, cocoa, roasted nuts, popcorn
- Cereal products, e.g. biscuits, bread, extrudates
- Beverages, e.g. beer, wine, whiskey

The generation of non-volatile components plays an important role because important attributes like umami, mouthfeel, texture, etc. can be given to the final products.

Besides the well-known Maillard reaction, additional reactions like sugar degradation, fat oxidation and interaction of Maillard intermediates are major sources for powerful flavour materials.

21.1.3.1
Process-Flavour Creation

For the production of process flavours, heat has to be applied to the raw materials for the thermal processing. This offers different possibilities for the production process:

- Heating in a (non-sealed) reactor (without pressure)
- Heating in an autoclave (with increased internal pressure)
- Heating in an extruder (continuous process)

All these methods provide different possibilities regarding throughput, temperature, pressure, etc. While non-sealed reactors have the simplest production setup, they are limited in the possible reaction temperature to the boiling point of the solvent, e.g. 100 °C for water. In contrast, autoclaves allow much higher reaction temperatures. An extruder system is especially useful for products with higher viscosity and provides all the advantages of a continuous production process rather than batchwise manufacturing.

The maximum permitted temperature for the production of process flavours is around 180 °C, as defined by legal regulations, but in general, the temperatures actually used are much lower in order to be able to reach a broad variety of different flavour profiles, such as cooked, boiled, fried, roasted and shallow-fried notes. The pressure during the reaction is usually below 10 bar (10,000 hPa).

21.1.3.2
Process-Flavour Stability

In order to achieve a long shelf life for the products, drying techniques like spray-drying, vacuum-drying or evaporation can be applied to produce dry powders or paste products. Conventional carrier materials are, for example, sugars (i.e. glucose or lactose) or high molecular weight products like gum arabic or malto-dextrin.

Dry products can not only improve the microbiological stability but can also improve the sensorial stability, since chemical and physical interactions and degradations are limited to a minimum.

In the modern production of process flavours, the following topics are coming more and more into focus:

- Generation of inexpensive high-impact products (low dose)
- Generation of flavourful food preparations, based on the reaction of foodstuffs
- Generation of allergen-free products

21.1.4
Taste Modifiers

In the past the most common tastants used were sweet carbohydrates, inorganic salts (mainly sodium chloride, but also buffering salts), amino acids, especially MSG, fruit acids and phosphoric acid and to some extent bitter components such as caffeine, but also low-volatile chemosensates such as capsaicin and cooling compounds such as menthyl lactate. Most people are adapted to the classical ingredients which show a very high positive hedonic score and now are in several cases disfavoured for health-related reasons: the extensive consumption of (saturated) fats, sucrose, glucose and high-fructose corn syrup (HFCS) may cause obesity under certain circumstances, and as a consequence, some related diseases such as type II diabetes and cardiovascular disease and sodium chloride can cause hypertension, especially in persons who are prone to this effect [6]. MSG—one of the world's most important umami compounds—is found in many food preparations introduced by basic ingredients, for example vegetables such as tomatoes and various meat selections or seafood materials. At the same time, MSG is discussed in the context of adverse effects like the so-called Chinese restaurant syndrome. As a consequence, added MSG is disfavoured in some countries.

As a direct result of this so-called food-minus trend some of the ingredients have to replaced or reduced. Owing to their role as preference drivers in food consumption, it is important to retain the whole flavour and taste profile of the original product, which can be done in most cases using a mixture of flavours, tastants, taste modifiers and texturants.

Another problem arising from modern food trends is the off-taste generated by fortification. The fortification with healthy polyphenols, for example from grape seed or green tea, causes pronounced astringency or bitterness. The addition of selected fats such as fish oil causes strong metallic and rancid off-flavours. A very special problem arises in the growing use of soy products, which in several cases cause astringent, "beany" and bitter off-notes.

21.1.4.1
Masking Technologies

The challenge of masking is to suppress unpleasant tastes in functional or lightened-up products without negatively affecting the sensory profile, mouthfeel or the effectiveness of functional ingredients in a label-friendly way. Masking is a very complex challenge since there is no off-taste blocker system which can be applied in a universal way. In fact, rather a detailed analysis of the product formulation and a tailor-made application of appropriate masking strategies are required. There are several ways to fight against off-tastes:

- Identification and elimination of compounds causing off-tastes. Elimination is often process-optimisation oriented and in general not a matter of flavour development or optimisation.
- Retardation of release of functional ingredients. Since most of the biofunctionals express their functionality after passing the oral cavity, emulsion and encapsulation techniques are a perfect delivery system for food applications. Important aspects of this approach are stability and the beneficial combination with flavour systems.
- Masking via enhancing positive sensory drivers can lead to the suppression of negative sensory drivers. This can be done by addition of sweeteners/acidulants, congruent flavours with the ability of suppression and round-off, and real taste-masking and flavour-modifying components.

The masking topic is especially difficult. Off-taste generating ingredients can act very differently in the receptor landscape in the mouth. Especially for bitter taste, roughly 24–30 different receptors are known, which show a certain binding pattern to bitter molecules [7]. In addition, the transduction mechanisms of taste signals in bitter and sweet taste cells are very similar; therefore, it is difficult to develop a "universal" bitter blocker. Several molecules such as adenosine 5′-monophosphate (**1**) [8], neodiosmine (**2**) [9], homoeriodictyol (**3**) [10], γ-aminobutyric acid (**4**) [11] and some Maillard products originating from β-alanine or γ-aminobutyric acid (e.g. **5**) [12] were described as bitter-masking compounds (cf. Figure 21.2) and some of them were approved as generally recognised as safe (GRAS) (e.g. adenosine 5′-monophosphate) or Flavour and Extract Manufacturer's Association (FEMA) GRAS such as homoeriodictyol.

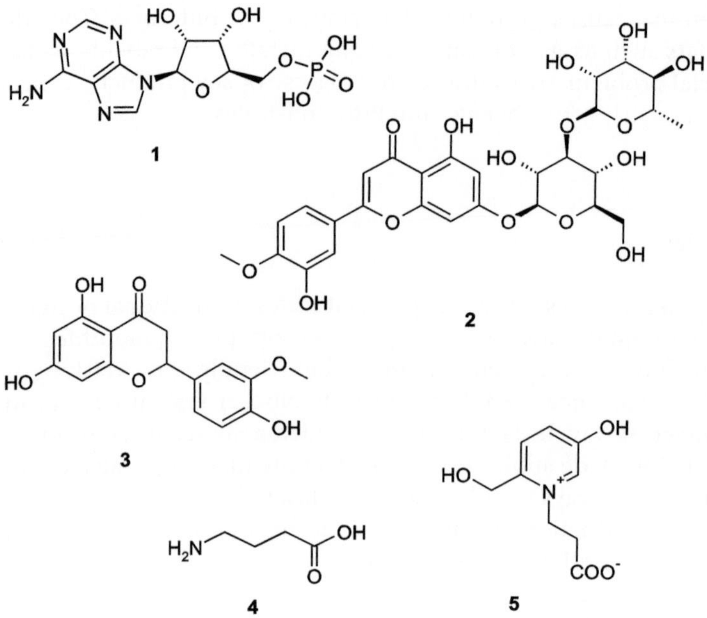

Fig. 21.2 Masking molecules towards bitter taste

21.1.4.2
Sweet Optimisation

Foods which have a high sugar content (primarily sucrose, lactose, glucose or fructose or mixtures thereof) are usually strongly preferred by consumers owing to their sweetness. On the other hand, it is generally known that a high content of easily metabolised carbohydrates allows the blood sugar level to increase greatly. This leads to the formation of fatty deposits and can ultimately lead to health problems.

As a consequence, in most cases, mixtures of low-calorie sweeteners are used to address this issue. At the same time, numerous sensory and consumer tests have shown major differences between low-calorie sweeteners and sucrose or HFCS with regard to body and aftertaste.

In these cases, masking flavours can be used together with a rebalancing of the flavour profile to cover the changes in perception. The use of sweet inhibitors such as lactisol (**6**, Fig. 21.3) can help to reduce the lingering aftertaste in some cases.

An improvement in the taste properties, in particular the aftertaste problem of non-nutritive, highly intensive sweeteners, can be achieved by the use of tannic acid [13, 14].

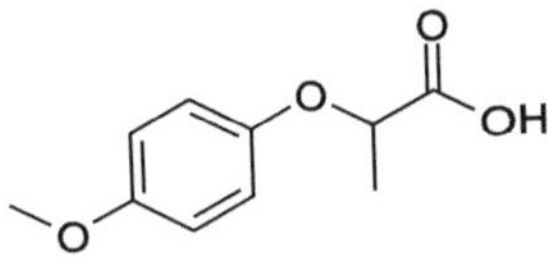

6

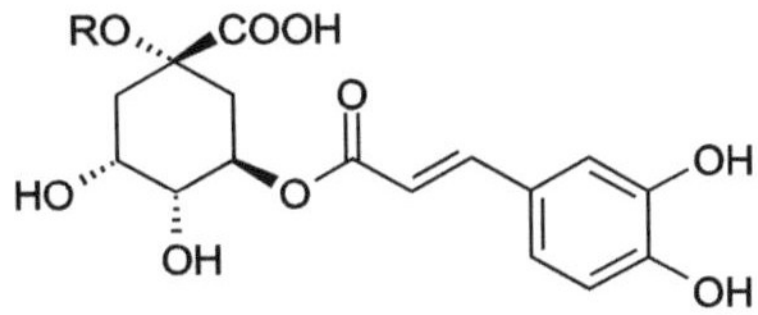

7: R = H
8: R = caffeoyl

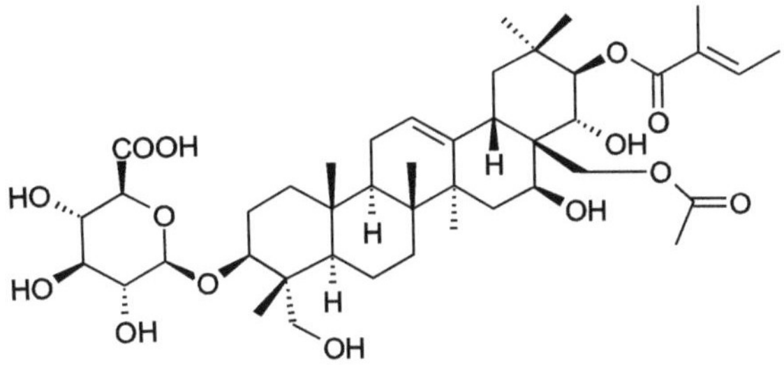

9

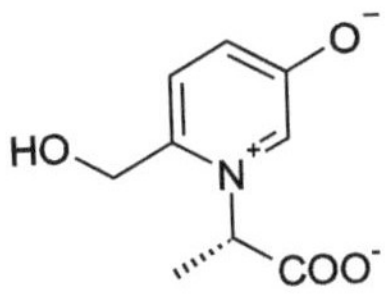

10

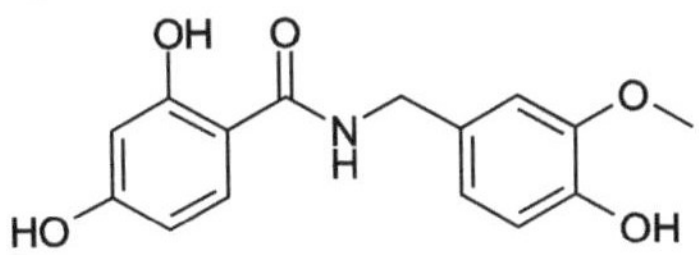

11

Fig. 21.3 Sweet-taste modifiers

Some other sweet taste modifiers have been described in the literature. Chlorogenic acid (**7**) and 1,5-dicaffeoyl quinic acid (cyanarin, **8**) from artichoke [15] can induce a sweet water taste, i.e. a sweet impression of water which is applied to the tongue after rinsing the mouth with the solution of the caffeic acid derivatives (Fig. 21.3). The proteins miraculin from miracle fruit [16] and neoculin and curculin occurring in the fruit of *Curculigo latifolia* [17] are able to induce sweet taste by using acidic solutions. Unfortunately, in both cases the effects are only perceived in the consecutive sequence and therefore the effects cannot really be used in food. Some triterpenoids such as gymnemic acids (e.g. **9**) occurring in *Gymnema sylvestre* are able to inhibit sweet perception similar to lactisol [18].

For some applications, it is of great interest to increase sweetness of sugar or HFCS-reduced products without using sweeteners. In several cases, it is possible to use supportive flavours [19] or to optimise the flavour to improve the overall profile [20–22] and to retain the preference. Significant importance is attributed to compounds which may increase the sweet sensation without showing significant intrinsic sweetness and strong flavour profile. Just recently, interesting compounds such as alapyraidine (**10**) [23], a general taste enhancer, were studied. In addition, sweetness-enhancing hydroxybenzoic acid amides of vanillylamine (e.g. **11**) were reported [20].

21.1.4.3
Savoury Enhancement

Owing to the unfavourable effects of high levels of ingested sodium ions on blood pressure, lowering the sodium content of food is one of the hottest topics in food development. On the other hand, salt taste, which is mainly caused by sodium chloride, is a main preference driver for most consumers. In liking tests, the sample containing higher amounts of sodium is often preferred. Additionally, sodium chloride or other sodium salts such as sodium gluconate have been used as maskers for bitter off-tastes [24] and as a common taste enhancer since ancient times. Therefore, the sodium problem is correlated not only to savoury products but also to sweet foods or beverages.

Reduction of the sodium chloride level can result in taste problems and flavour shifts. There are several approaches to maintain salt taste. Most often, potassium chloride is used, because it shows the most prominent salty taste of those applicable inorganic salts. Lithium chloride is the most salty salt but cannot be used for toxicological reasons. Most consumers, however, complain about the bitter, chalky taste of KCl-containing formulations. Development of sodium-reduced products using mineral salts is a challenge and the whole product formula has often to be adapted [25]. Therefore, the main focus of the research was the search for masking compounds or technologies to cover the bad taste of KCl, e.g. phenolic acids and derivatives [26] and lactisol [27].

The salty and savoury character of salt-reduced food can be maintained by using glutamic acid salts, mainly MSG, but also the corresponding potassium and calcium salts [28, 29]. This strategy does not find wide acceptance because of the previously mentioned reasons.

In some cases, yeast preparations which contain a high amount of nucleotides can be used to increase saltiness in combination with masking off-notes of KCl [30]. Additionally, use of low amounts of fruit acids may reduce the bad taste of KCl-containing food preparations [31]. Usage of low amounts of sweeteners such as thaumatin [32] or neotame [33] was described to mask the off-taste of KCl.

Salty taste enhancing preparations or compounds besides KCl were described. For example, a mixture of certain amino acids based on L-lysine were used to increase the saltiness of a NaCl-reduced preparation [34]; γ-aminobutyric acid (**4**) was also used as a salty taste enhancer [35]. Some dipeptides such as *N*-L-ornithyl taurine hydrochloride or *N*-L-lysinyl taurine hydrochloride were described as very salty with a clean salt taste [36]. Additionally, choline chloride was suggested as a salt enhancer [37].

More recently, some new savoury-enhancing molecules have been described by Soldo and Hofmann [12] (alapyraidine **10**, Fig. 21.3). In addition, it was found that potentially pungent compounds such as midchain unsaturated alkamides **12** and **13** can enhance the salty or savoury taste [38, 39]. Most of the new savoury enhancers which were found via high-throughput screening based on the umami receptor [40] are not based on traditional natural product chemistry, e.g. oxalamides **14–16** and benzoic acid amide **17**. Combinations of pungent chemosensates such as cetylpyridinium chloride in combination with amino acids such as arginine were described as salt-taste enhancers [41].

Other components described in the literature which are able to enhance saltiness or umami taste are umami-tasting glutamate glycoconjugates (e.g. **18** or **19**) [42], (*S*)-malic acid 1-*O*-D-glucopyranoside (morelid **20**) [43], theogalline (**21**) [44], *N*-lactoyl ethanolamine (**22**) [45] and *N*-gluconyl ethanolamines (**23**) [46], α-keto acids derived from amino acids (e.g. **24**) [47] and some *N*-succinoyl derivatives of aspartic acid or glutamic acid (e.g. **25** and **26**) [48].

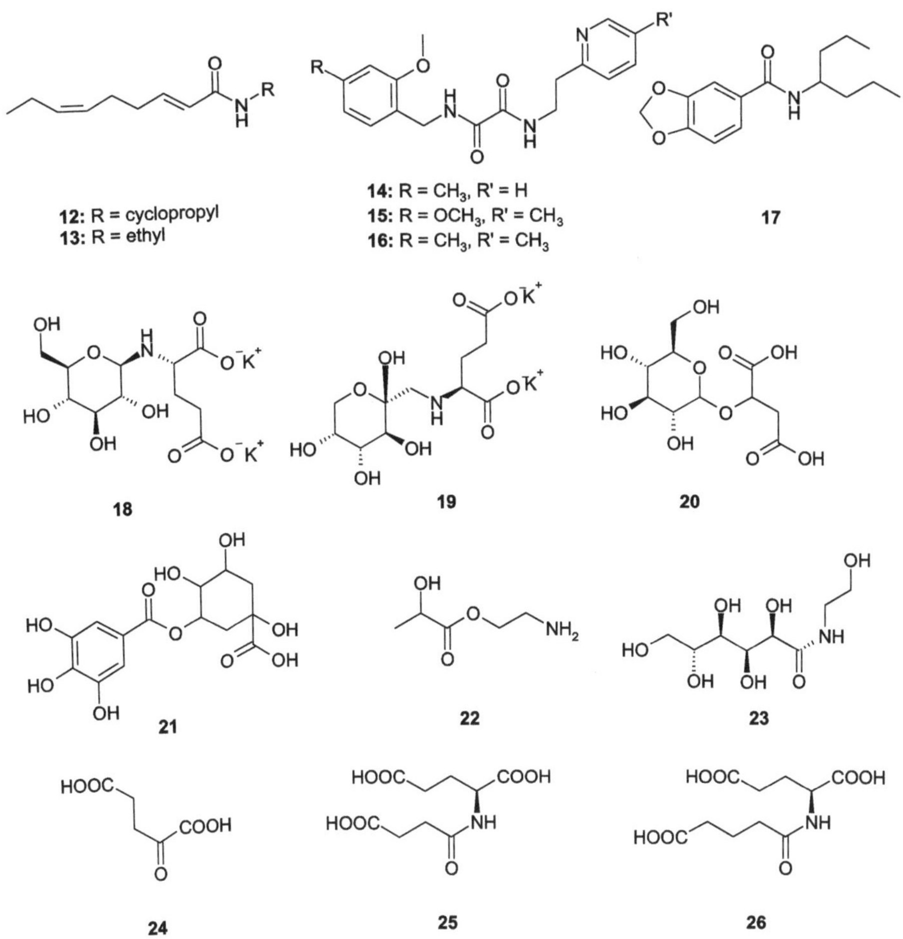

Fig. 21.4 New savoury-taste enhancers

21.1.5 Chemosensates

Chemosensates play a tremendous role in flavour creation, especially for exotic and spicy foods or beverages. Chemosensates can be roughly classified into pungent/hot, tingling, cooling and astringent compounds. In contrast to the volatile flavours, which mainly act on the olfactory epithelium and bind to the olfactory receptors, and the tastants, which act on the taste buds and bind to the different taste receptors, the chemosensates act directly on the afferent nerve endings of the trigeminal ganglion in the face or mouth or on the afferent nerve endings of the dorsal root ganglions (DRG) in the skin. There are different types of trigeminal/DRG fibres: some are sensitive to heat, heat and mechanical stress, coldness, cold/heat and mechanical stress, etc. It is now known that temperature changes, the absolute temperature and mechanical stress can increase the signalling frequency of the neurons, which results in more or less severe pain feelings and reactions. The chemosensates are able to change the sensation threshold of the fibres or of the temperature-sensing receptors and can therefore initiate signalling at body temperature. As a consequence, chemosensates can induce a temperature feeling without changing the physical temperature [49]. Astringency is sometimes referred to as being a trigeminal sensation and is caused in most cases by precipitation of proline-rich proteins in the saliva by astringents such as catechins or gallated carbohydrates; there are indications that astringent polyphenols such as quercetin glycosides be directly act on the receptor level [50].

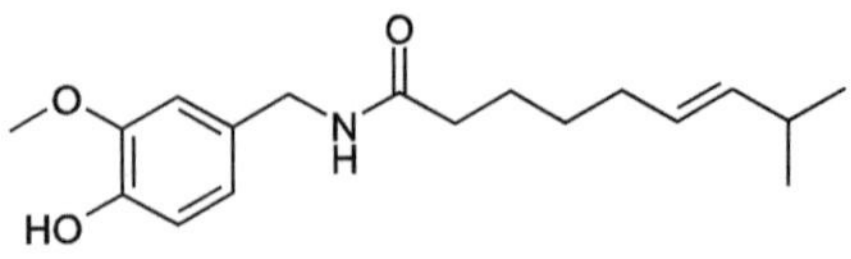

27

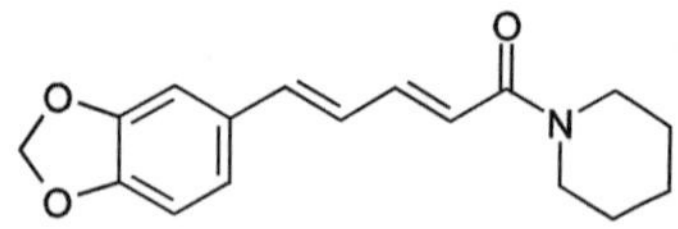

28

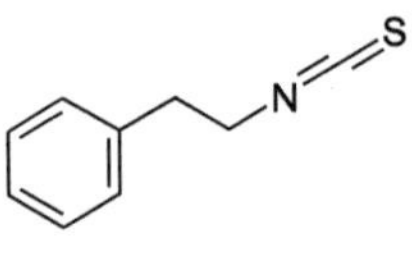

29

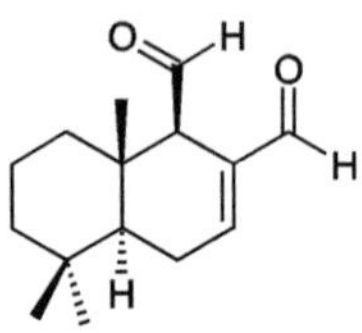

30

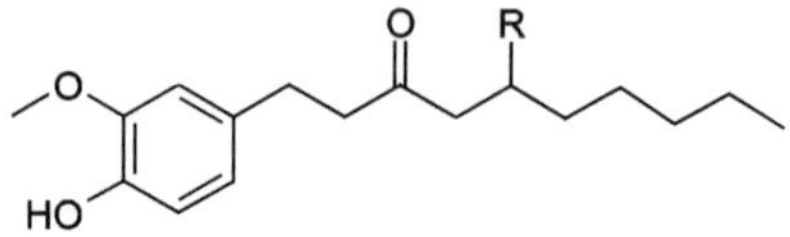

31: R = H
32: R = OH

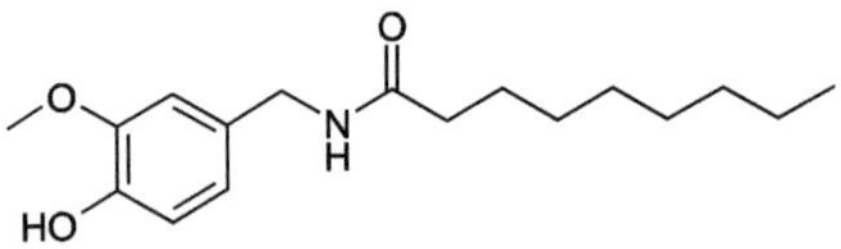

33

Fig. 21.5 Hot and pungent chemosensates

21.1.5.1
Pungent/Hot Compounds

A huge number of compounds causing a pungent or hot sensation are known from natural sources. In Fig. 21.5, some of the most important examples are illustrated: capsaicin (**27**) from chili pepper, piperine (**28**) from pepper, isothiocyanates (e.g. **29**) from Brassicaceae ssp., some dialdehydic drimanes (e.g. **30**) occurring in Tasmanian or water pepper, paradols (e.g. 6-paradol **31**) or gingerols (e.g. 6-gingerol **32**).

These compounds are all used in the form of ingredients occurring in the aforementioned oleoresins or extracts obtained from the corresponding plants. In some cases (like capsaicine and piperine) single compounds are used in flavourings. Capsaicine is restricted in the EU owing to its genotoxic potential [51]. Another capsaicinoid is nonivamide (**33**), which is characterised by a saturated side chain.

Pungent and hot compounds are used especially in savoury formulations and seasonings but can also be formulated in low amounts for other applications.

21.1.5.2
Tingling Compounds

Tingling sensation is an unusual and polarising sensation. The effect can be described by the perceived sensation after administering a 9-V battery to the tongue. There are assumptions that the effect is caused by a simultaneous activation of thermal and mechanosensitive fibres [52]. Important compounds inducing a tingling sensation are unsaturated medium-chain alkamides such as spilanthol (**34**) from *Spilanthes* ssp. or *Heliopsis longipes* [53], sanshools (e.g. α-hydroxysanshool, **35**) or pellitorine (**36**) from Roman pellitory root or *Piper* ssp. (Fig. 21.6).

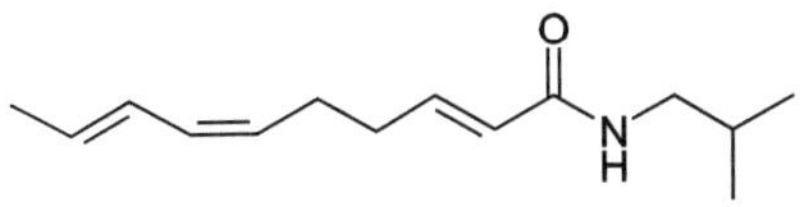

34

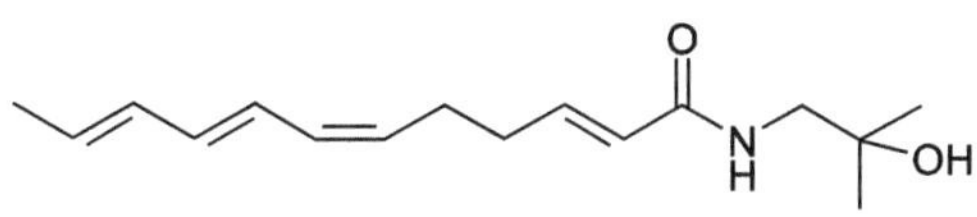

35

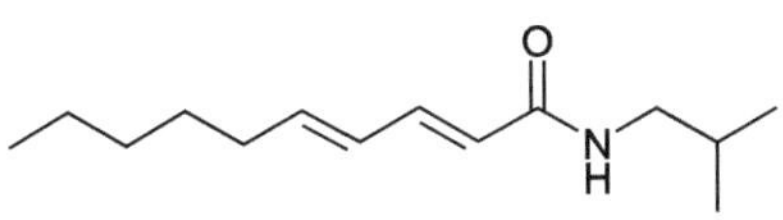

Fig. 21.6 Tingling chemosensates

Tingling compounds are mostly used in oral-care products and chewing gums. In low amounts, they can be used for spicy formulations. Most of the tingling compounds can trigger salivation to a certain extent [54].

21.1.5.3
Cooling Compounds

Besides the widespread usage in oral-care products and chewing gums, strongly cooling compounds are very rarely found in nature. Among the compounds which occur in nature, the odour-active L-menthol (**37**) and some direct derivatives such as the more or less tasteless L-menthyl succinate (**38**) and L-menthyl glutarate (**39**) [55] and the L-menthyl lactates (**40**) are known [56]. In addition, cubebol (**41**) was described as a moderately effective cooling compound [57]. On the other hand, a lot of artificial cooling compounds were developed in the past and the most important are shown in Fig. 21.7. The menthane carboxylic acid amides WS 3 (**42**) and WS 5 (**43**) are the most active cooling compounds known so far. Furthermore WS 23 (**44**), the l-menthone ketal of glycerol **45**, L-menthyl glyceryl ether **46**, and the L-menthyl carbonates of propylene glycol and ethylene glycol (**47** and **48**) are used as artificial compounds.

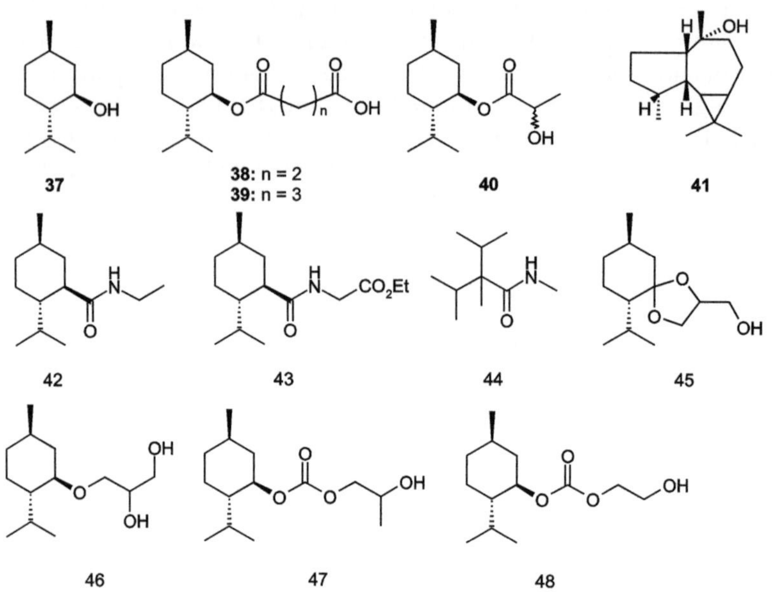

Fig. 21.7 Cooling chemosensates

21.1.6 Modern Tools for Flavour Development—Flavour Creation

Flavour work is characterised by basically two different views: the perspective of the application segment and the types of raw materials which are used. Complex raw materials like spices for seasoning blends and citrus oils for beverage flavours always bring advantages with regard to the body and the completeness of the final composition. Single molecules and purified fractions of, for example, essential oils usually offer the possibility to do a complete flavour design without any limitations arising from complex raw materials such as extracts and distillates from plants, roots, barks and fermented foodstuffs. The art of the flavour work comes into play when the flavourist starts to construct quantitative and qualitative skeletons for his composition. The main pillars of the first formula differ between the different application segments. In the field of non-alcoholic beverage development, juice, juice-derived materials, extracts, distillates and essential oils are of key importance. This is also true for a variety of vanilla flavourings. For many sweet and confectionery flavours, single aroma chemicals such as ethyl butyrate for fruity notes and sulphur compounds like methyl thioethyl propionate for tropical notes are essential for the final flavouring. In particular savoury flavours are based on important aroma chemicals such as 2-methyl furanthiol.

Young flavourists create the first formula in a trial-and-error approach. Experienced flavourists know how to create the first blueprint based on a qualitative and quantitative skeleton in a focussed and efficient way, because the roles and the interactions of single compounds have been learned over the years.

21.1.6.1 Modular Flavour Concept

A successful flavour is based on different elements. A very important part is the selection of the most potent volatile aroma compounds. All we eat, every fruit, every meal, every drink, every cup of fresh coffee contains a combination of various volatile flavour ingredients.

A strawberry, for example, has a total content of 10-ppm volatile flavour compounds. This amount of flavour is made up of at least 300 single raw materials. Each of these single ingredients has a special flavour character. Some of them are very strong, others are quite weak. It is very important to know the character and the strength of each individual ingredient. Only then is it possible to know the exact influence of a single ingredient in the total flavour. It is also important to know that all components have a more or less different flavour character. Some of them are more fruity, others are more floral and some of them have a very fresh leafy character, like freshly cut grass. Some of them are quite similar. It is possible to distinguish between all of these characters and select those which have a similar flavour impression. On the basis of this selection it is possible to develop a so-called building-block concept, which comprises different groups of important flavour characteristics. The example of modular strawberry flavour illustrates the blueprint and the strategy behind this flavour-creation concept.

A fresh strawberry flavour contains a group of flavour materials which are described as "green", like fresh leaves or similar to freshly cut grass. The main representatives are chemicals like (*E*)-2-hexenal or (*Z*)-3-hexenol. On the basis of a detailed description of all ingredients, it is possible to select all raw materials which are described as "green". Finally all those similar green components are put together into one basic mixture—a so-called green base.

Using the same approach, we can combine all other important volatile ingredients of a strawberry to form specific bases. All fruity esters like ethyl butyrate, ethyl caproate or ethyl isobutyrate can be combined to a "fruity base", and all caramel-like ingredients to a "caramel base", and so on. Finally, all major flavour characters of a strawberry can be obtained by appropriate combination of only a few bases.

With these building blocks it is possible to reassemble various types of strawberry flavours with fruity, spicy or ripe flavour character without dealing with the complexity of more than 300 single raw materials. This approach speeds up the combinatorial elements of the flavour work and provides an excellent learning platform for young flavourists.

21.1.6.2
Odour- and Taste-Activity Concept

Modern scientific tools like the so-called odour-activity value (OAV) concept were developed to unravel the quantitative and qualitative pattern of individual chemicals. Quantitation is achieved in an extremely accurate way by means of isotope dilution analysis.

In combination with the threshold of a flavour compound in a given matrix, it is possible to calculate the OAV value using the following formula:

$$\text{OAV} = \frac{\text{concentration of a flavour compound in a food}}{\text{threshold of a flavour compound in the corresponding food matrix}}$$

The complete OAV analysis of a food target is time-consuming and requires an excellent analytical setup. In many cases the well-established combination of GC combined with olfactometry provides an excellent insight with regard to the composition of aroma compounds in a mixture and the role of individual chemicals.

In recent years, non-volatile taste compounds have been becoming more important in the area of modern flavour development. Therefore, the principal approach of the OAV has been adapted for the taste side in the form of the so-called taste-activity value. In order to facilitate the search for taste-active materials and for a better understanding of the "taste dimension" of foodstuffs, a new instrumental setup called LC-Taste® has been developed [58].

Non-volatile ingredients play an important role in the overall flavour character of fruits and other foodstuffs. Many non-volatiles have a strong effect on the sensorial properties such as mouthfeel, creaminess and juiciness. Experienced flavourists know how to combine both the volatile and the non-volatile worlds of raw materials for delicious flavours.

21.1.6.3 The Sensorial Relevance of Ripening Effects

Apple flavour is an excellent example for chemical reactions which are responsible for the so-called ripening effect of juices, distillates and purees.

Starting from only one single ingredient, e.g. (*Z*)-3-hexenal, many other ingredients are formed during treatment and ageing of a fruit (Table 21.3). This effect has a strong contribution to the strength and character of the flavour.

During the production of recovery flavours, apple wines or brandies, the interaction with ethanol, acetaldehyde and acetic acid represents the next level of interactions. The reaction products contain compounds which result from esterification and acetal formation reactions, which are summarised in Table 21.4.

Table 21.3 Reaction products from isomerisation and degradation of (*Z*)-3-hexenal

Compound	Intensity
(*Z*)-3-Hexenal	Very strong
(*Z*)-3-Hexenol	Medium strong
(*E*)-2-Hexenal	Very strong
(*E*)-2-Hexenol	Medium strong
(*E*)-2-Hexenoic acid	Weak
3-Hydroxy hexanoic acid	Weak

Table 21.4 Reaction products originating from (*Z*)-3-hexenal, ethanol, acetaldehyde and acetic acid

Compound	Intensity
(*Z*)-3-Hexenyl acetate	Strong
(*E*)-2-Hexenal diethyl acetal	Medium strong
(*E*)-2-Hexenyl acetate	Strong
Ethyl-(*E*)-2-hexenoate	Medium strong
Ethyl 3-hydroxyhexanoate	Medium strong
(*E*)-2-Hexenal di-(*E*)-2-hexenyl acetal	Weak
(*E*)-2-Hexenyl-(*E*)-2-hexenoate	Weak
(*E*)-2-Hexenyl-3-hydroxyhexanoate	Weak

21.1.7 The Specifics of Flavour Application

The field of flavour application is basically driven by three main influencing factors. The market for flavoured foodstuffs and the technology which is needed for the flavour formulation have always played an important role in the flavour industry. In recent years, the relevance of the corresponding food technology has grown significantly. Among the most important fields of application, the beverage market represents a key area. Soft drinks, fruit-juice-containing beverages as well as alcoholic beverages and instant drinks are produced in an enormous variety of flavour and packaging materials. In the UK the market showed an increase of soft drinks and fruit juices of some 7% in 2003 over 2002. In the 10 years between 1993 and 2003 this same market grew from a consumption of around 9 billion litres to around 14 billion litres in a full year.

This growth rate is accompanied by an ever-increasing range of flavours and ingredients. Products are more and more being designed for lifestyle and for age groups. Meanwhile, a remarkable portfolio of product types has been established in the market (Table 21.5).

The aforementioned products are commercialised at different concentration levels, for example ready to drink or dilute to taste in the case of syrups.

The matrix for flavour applications in the beverage industry illustrates the tremendous variation for the required flavour systems with regard to sensory and

Table 21.5 The variety of product types of flavoured beverages (modified after Ashurst [65])

Soft drinks	Other drinks	Alcoholic beverages	Dry beverages
Definition	Tea	Flavoured beer	Instant beverages
Sweetened Water-based Balancing acidity	Flavoured ice tea	Shandy-type products	Fruit beverages
Juice and/or pulp content	Coffee-type Beverages	Wine coolers	Flavoured teas
No juice Low juice High juice	With flavours With dairy raw materials	Low flavoured High flavoured	Liquid flavour Dry flavour
Carbonation	Milk beverages	Liqueurs	Dry coffee type beverages
Carbonated Non-carbonated	No juice Low juice	Flavoured Non-flavoured	With creamer Without creamer
Alcohol content	Soy beverages	Spirits	Vending machines
Alcohol-free Low alcohol content	Masking flavours Cream flavours	Extracts Distillates	Flavoured pads

regulatory background. In addition, the wide range from liquid to dry applications emphasises the broad scope from emulsion technologies to spray-drying.

21.2 From Formula to Product

The formula of a liquid or dry flavouring represents the blueprint for a final product. At this stage various parameters of influence have to be considered. Besides the compounding or mixing instructions with impact on the solubility of compounds, the chemical interaction of formula constituents is one of the most important parameters.

The formula of a liquid flavouring usually comprises the list of ingredients and a short summary of the corresponding blending instructions. At this stage the solubility of the ingredients in the carrier system is of high importance. Additional criteria can be summarised under the headline of shelf-life stability.

21.2.1 Shelf-Life Stability

The modern food industry is confronted with the consumers' increased requirements concerning their final products. Apart from product safety, further crucial characteristics such as colour, flavour, texture and sensory stability are important issues for the evaluation by the consumer. Thus, the requirements for the flavour of the final product are increasing. This requires extensive knowledge in various fields.

- Processing properties of the flavour
- Behaviour of the flavour during the production of the food
- Behaviour of the flavour during the shelf-life of the food
- Interactions of flavour and foodstuff

Apart from simple handling and easy dosage, a constant high product quality and a guaranteed shelf life are absolutely essential. That means that the taste of the food products has to be constant during the entire product life [59].

The term "stability" in the following text will represent the sensorial quality of a food product during its whole life span under suitable and defined conditions. The stability of a food product is affected by a multiplicity of factors, like process temperature and time, residual moisture, degree of browning and the physical and chemical constitution of the ingredients. A further factor of crucial importance is the type and the quality of the packaging material [60]. The execution of real-time stability tests is essential for control purposes in the food industry. Storage conditions have to be adapted as realistically as possible: the products should be kept in their original packaging, temperature and light conditions should be realistically adjusted and varying conditions, e.g. during transportation, should also be included in the test parameters. During the development of food products real-time tests have a crucial disadvantage. They are very time-consuming.

21.2.2 Accelerated Shelf-Life Testing

There is considerable evidence in the literature that temperature plays a major role in causing changes in food quality during storage. Higher storage temperatures generally lead to increased quality deterioration. In the past, there have been several attempts to use mathematical techniques and models to describe changes in food quality as influenced by storage and temperature [61]. Not all ageing processes are influenced in the same way. So, for instance, a storage temperature increase of 10 °C will speed up Maillard reactions by a factor of 2.5 but will speed up oxidation reactions only by a factor of 1.5. This also depends on the total composition of a product. In the past, accelerated stability testing methodologies based on time, temperature and modified oxygen atmosphere were developed in order to accelerate shelf-life stability under defined conditions [62]. A new rapid method was developed which is even more adaptable to different food product types.

In general, a dry or liquid ingredient is brought into a high-pressure container without any sample preparation (Fig. 21.8). After closing this vessel, the product is exposed to oxygen pressure and heat. After a defined storage period, the sample can be used for further testing. In the subsequent sensory panel work, the samples are tested in a triangle test by comparing the nonstressed sample with the stressed one. If the panel is not able to detect a significant difference between the two samples, the sample can be considered as "stable" for the simulated time frame.

In case there is a significant difference between the stressed and the nonstressed sample, off-flavours or strong sensorial deviations are monitored. As a next step, further analytical evaluations, e.g. GC-MS, are performed for the identification of relevant degradation products.

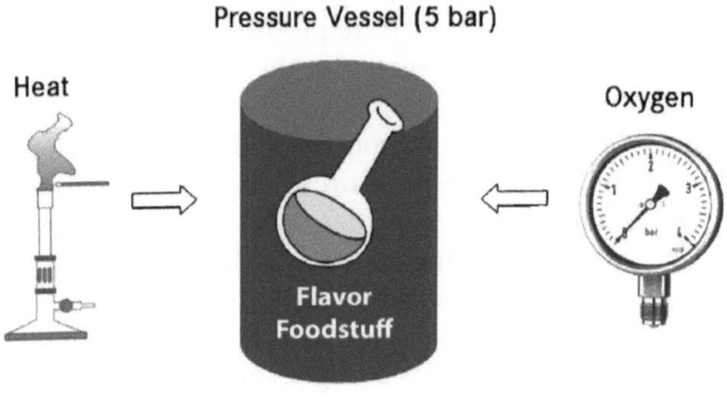

Fig. 21.8 Principle of accelerated stability testing [63]

Figure 21.9 illustrates the correlation between temperature and testing time for real-time conditions and accelerated test conditions.

The kinetics of degradation processes in foodstuffs are not simulated by the accelerated shelf-life testing method in the same way. Different parameters like a_w value, pH value, acidity levels, sulphur compounds, etc. affect flavour degradation in the foodstuff very individually; therefore, a careful transfer of results is absolutely essential. The tested products have to be clustered depending on their ingredients and parameters for the accelerated testing have to be adapted (Fig. 21.10).

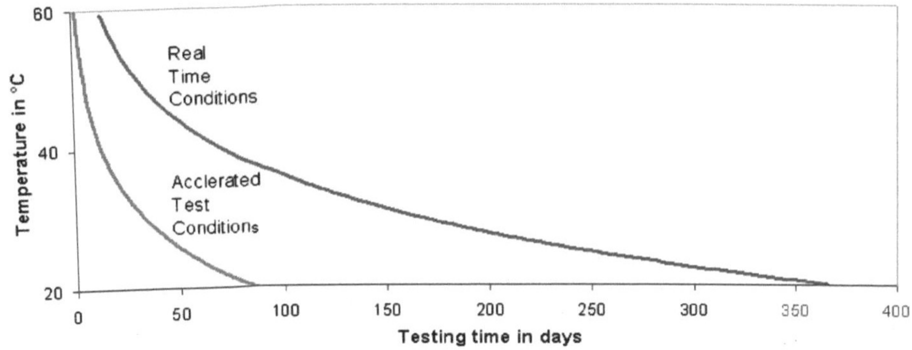

Fig. 21.9 Correlation of conditions in accelerated stability testing with real-time testing (1 year at room temperature) [63]

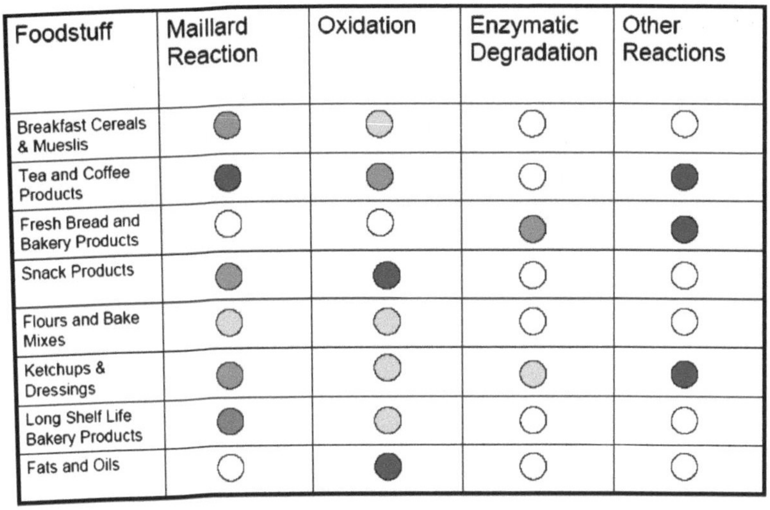

Fig. 21.10 Degradation hazards in different foodstuffs, white: low, black: high [63]

21.2.3
Chemical Interactions

The stability of the flavour in the food is an enormously complex issue. In order to come to a reliable prediction, the reactivity of flavour compounds and the embedding in the corresponding food matrix have to be considered. The interactions of flavours and foodstuff can be clustered into two main groups:

1. Degradation of flavour ingredients and subsequent flavour loss
2. Formation of new flavouring substances and, as a direct result, off-flavour formation

The typical flavour profile of many foodstuffs is not only characterised by so-called character-impact compounds like cinnamic aldehyde for cinnamon, vanillin for vanilla and eugenol for cloves. Flavour changes during processing and storage and the corresponding flavour stability are based on numerous chemical interactions which are directly linked to organoleptic properties. Table 21.6 summarises the main chemical reactions which are responsible for causing flavour changes.

Table 21.6 Overview of chemical interactions responsible for flavour changes

Reaction type	Main examples
Acetal formation	

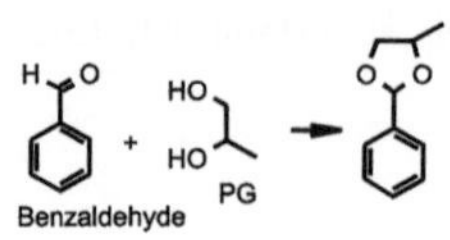

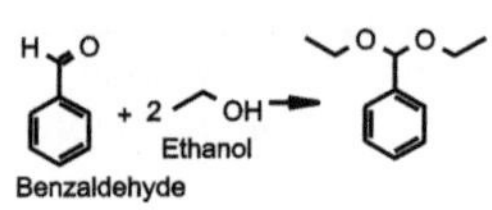

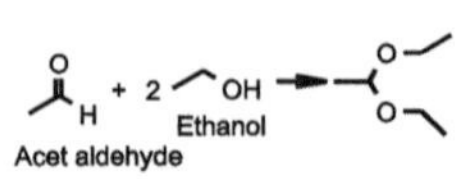

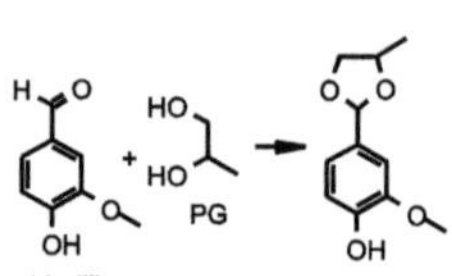

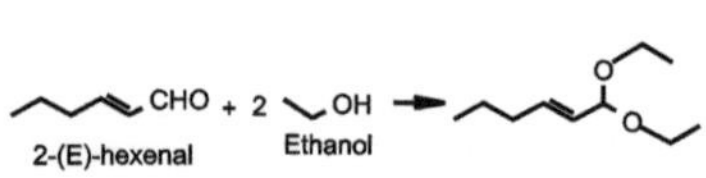

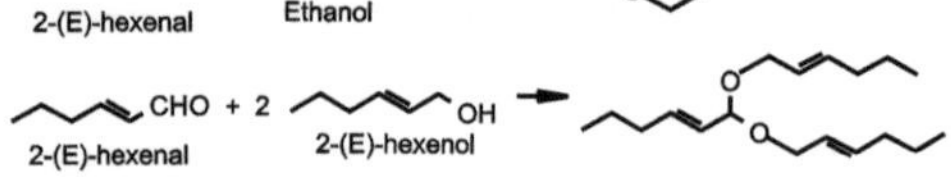

Table 21.6 (*Continued*) Overview of chemical interactions responsible for flavour changes

Reaction type	Main examples
Mercaptals and hemimercaptals	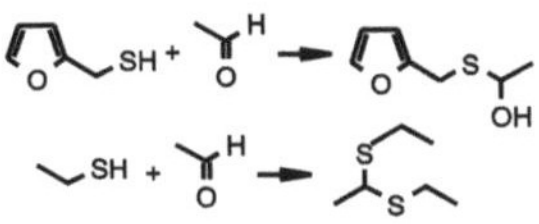
Aldol formation	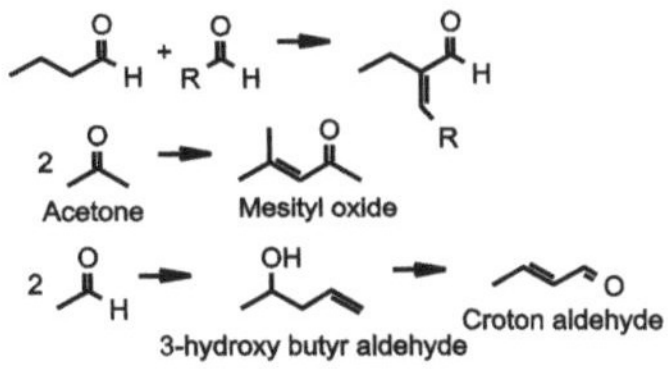
Schiff base formation	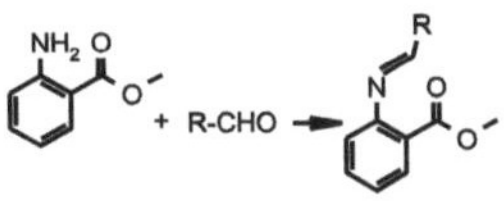
Decarboxylation and deamination	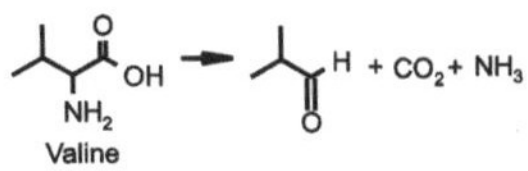
Esterification	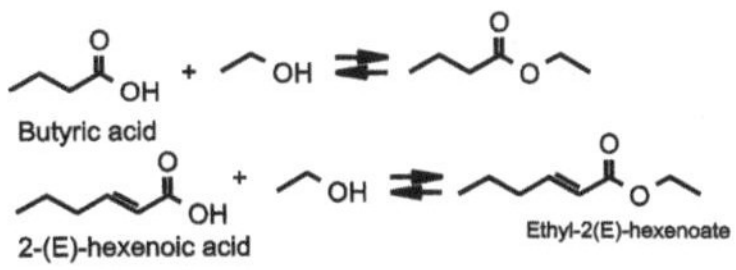
Ring opening	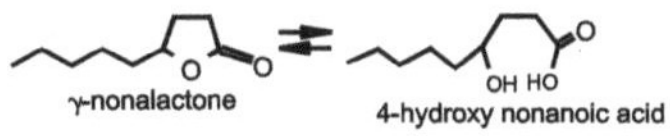
Isomerisation	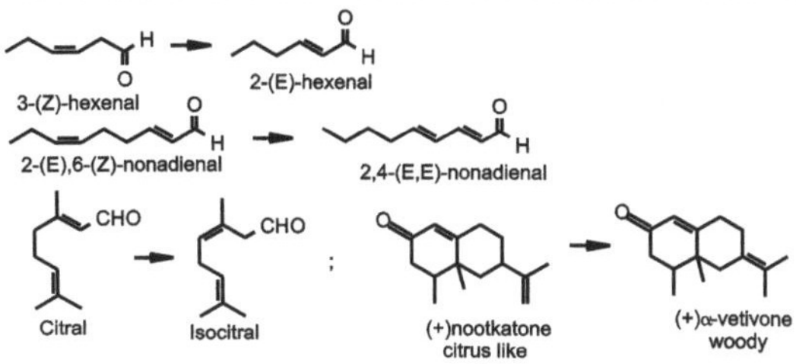
Oxidation	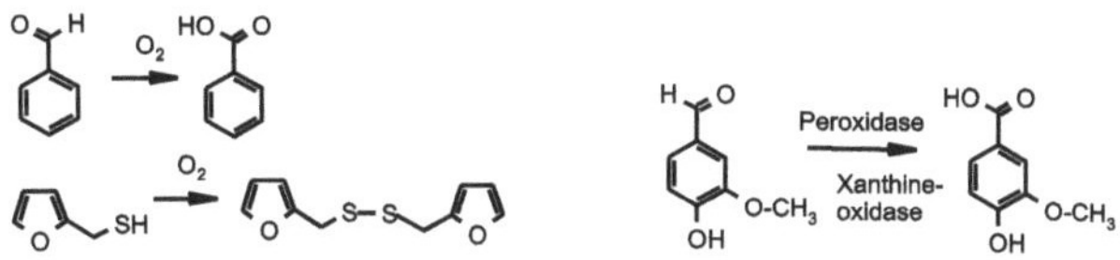

As soon as the cause of the instability is known, means can be found to improve the shelf-life of the product. So, in case of oxidation effects, antioxidants could be added, and the oxidative risk could be minimised by reducing oxidising substances, varying production parameters or optimising the packaging of a product.

21.3 Flavour Production

From its nature, a flavour is defined as a multicomponent blend of volatiles, non-volatiles and complex raw materials which is responsible for the final product properties. In flavour production, the volume-dominated operation units are mixing processes of liquids and dry blends.

Under technical production aspects, the manufacturing of flavourings can be divided into:

- The production of valuable aroma-active components
- The refining, blending and transformation of the flavour in the final physical, potentially encapsulated product

Properties such as viscosity are very important for the production process.

21.3.1 Liquid Flavours

Liquid flavours can be divided into low-viscous liquids, medium-viscous liquids, emulsions, pastes and suspensions. The main processing of liquid flavour production is basically liquid blending. The most popular carriers for flavours for aqueous systems are ethanol, propylene glycol or glycerol. For fat-soluble flavours, triacetin or vegetable oils are the most important carriers.

Long process times are required when single raw materials, compounds or natural extracts exhibit high viscosity. In general, in the flavour industry two approaches are used to reduce viscosity:

1. dilution of the highly viscous extract with a solvent
2. thermal treatment (heating) of the highly viscous extract

The disadvantage of the dilution of extracts is that the flavour concentration is lower and the flavour dose has to be increased. Significant heat treatment can influence the flavour stability in a negative way because oxidation and Maillard reactions are enhanced and the flavour might be less stable. Owing to the fact that flavours are mainly complex mixtures, their rheological properties, particularly in presence of hydrocolloids, fibres or other macromolecules, are often not Newtonian. This can cause problems in the production process. In order to simplify and optimise the dosage and the blending of micro components it is essential to work with pre-mixes. The mixing sequence is of elementary significance and determines product quality. In general, macro components, inert materials and low volatiles are added in the first production step. In a second step, this premix will incorporate the high volatiles or partly reacting components. For the production of liquid flavourings, a definition of flavour additives is needed such as antioxidants or preservatives in order to maintain the required shelf-life stability.

21.3.2
Dry Flavours

There are advantageous criteria for the application of dry flavours in food products. In numerous food products only dry flavours can be utilised owing to their physical properties. The physical form and the properties of a dry flavour are of fundamental importance for the successful processing of a food product. For instance, for a dry tea flavour blending and the filling process, the flavour has to fulfil several properties, e.g. defined particle size and shape, and a given hygroscopicity and flowing behaviour. Another reason for the application of dry flavours in the food industry is the beneficial application of encapsulated flavours in food products. The advantages of these flavour types are primarily an improved flavour stability and controlled flavour-release mechanisms. Meanwhile a broad range of technologies exist for flavour encapsulation. The most commonly used processes are spray-drying, spray-chilling, encapsulation, melt extrusion, coacervation, and β-cyclodextrin complexation. In order to select a suitable, specific encapsulation technology the final application of the flavour has to be known.

21.3.2.1
Plated Flavours

One of the oldest production methods for the production of dry flavours is the plating of a liquid flavour or extract onto a solid carrier. Carriers of main importance for the food industry are salt, lactose, starch and maltodextrin [64].

In the plating process it is essential to guarantee a homogenous blending besides homogenous addition/distribution of the liquid material. The liquid feeding can be carried out by inlet lances, injectors or nozzles, while agglomeration of the plated powder material has to be suppressed by a chopper or cutter.

Essential advantages of this flavour technology are the low production and investment costs. The fundamental disadvantage of this technology, however, is a far lower flavour stability owing to the fact that the specific surface of the flavour has been enlarged considerably and this results in a much higher sensitivity of the flavour towards oxidative degradation reactions.

21.3.2.2
Spray-Dried Flavours

The most common method to simultaneously dry and encapsulate flavours is the spray-drying technique (Fig. 21.11). For this technology, carrier materials like maltodextrin, starch and gum arabic are dissolved in water. As a next step, the liquid flavour raw material is emulsified in this slurry. Also non-volatile flavour components can be added. The slurry is "atomised" and dried in a spray-drying facility.

Spray-drying consists of four separate process steps:

1. Slurry preparation
2. "Atomisation" of a slurry
3. Drying and encapsulation of the flavour molecules
4. Separation of the dried flavour from the exhaust air

The typical flavour load of a spray-dried product amounts to 18–25%.

Besides the drying process, the flavour components are also encapsulated in the carrier matrix. After the slurry has been "atomised", all volatile components, including water, which are located at the surface of the droplet are immediately evaporated. Thereby the remaining carrier substance forms a membrane around the droplet. This membrane is semipermeable and inhibits further evaporation of flavour molecules. This production step is controlled by diffusion mechanisms. Water as a molecule with a small molecular size can pass through the membrane, while the larger flavour molecules are not able to permeate it.

An optimal dehumidified spray-dried product consists of small, round particles, whose size is almost similar to their former droplet size. They are hollow and the encapsulated flavour molecules are situated in the outer shell.

The advantages of spray-dried flavours are the high flavour load and the fast release. The process is very economical. A disadvantages of the flavour powder is the physical demixing in dry blends with sugar, tea, cereals or granulates.

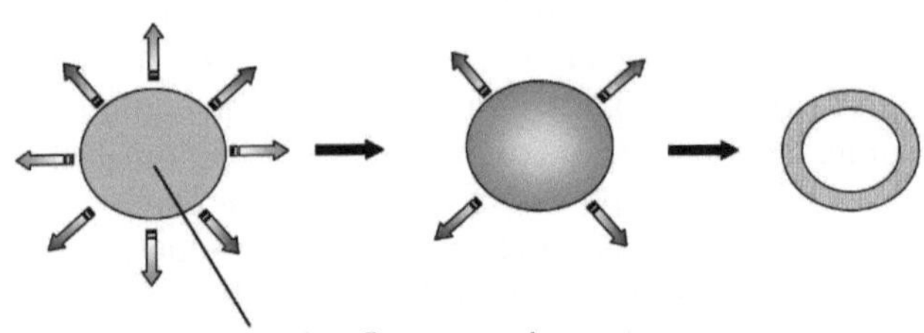

Fig. 21.11 The mechanism of encapsulation during spray-drying

21.3.2.3 Compacted Flavours

Compacted flavours are granules of size between 0.5 and 5 mm. The main application of the compacted flavour granules in the food industry is tea leaf flavouring for tea bags. Powdered flavours are not suitable because of demixing of the leaves and the powder during the blending process.

The spray-dried flavours or powder blends are processed by a roller compactor into lumps (Fig. 21.12). These lumps are crushed into granules. This process cannot be categorised as a direct encapsulating technique, since the flavour-encapsulating effect of compacted flavours is based on the use of spray-dried raw material.

A particular advantage of the compacted granules is the flexibility of the particle size. Each size between 0.5 and 5.0 mm is adjustable. A further advantage is shaped particles, colour and a combination of spray-dried flavour and additives (vitamins, minerals, functional food ingredients) can be combined in the granulated matrix.

In recent years, specific requirements with regard to shelf-life stability and tailor-made release behaviour led to the development of a range of specific encapsulation technologies such as glass-encapsulated flavours or seamless capsules with liquid cores.

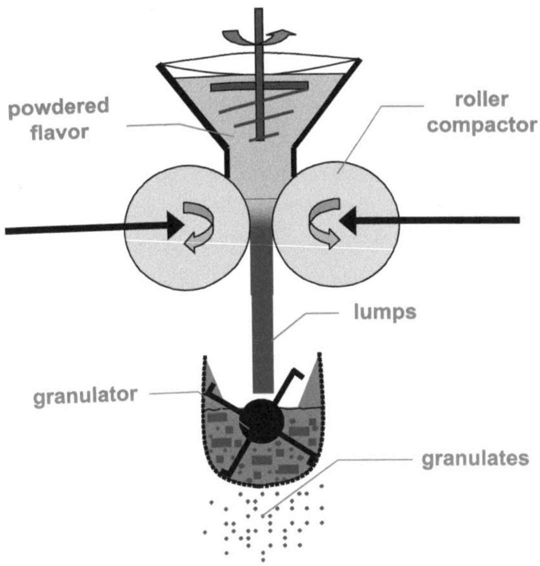

Fig. 21.12 Roller compaction

21.4 Conclusion

While in ancient times, the sensorial properties of a flavour for foodstuffs were of major importance, modern flavours have to perform like multifunctional systems. Physical form, chemical and mechanical stability and controlled release mechanisms are meanwhile essential criteria for the flavour quality. All these properties have to be addressed by a flavourist in close cooperation with technologists. Therefore, knowledge about food product properties must lead to a careful and intelligent evaluation of the flavour system as an important driver for the success of the final product.

Acknowledgement

The authors would like to thank Ian Gatfield for reviewing the manuscript, as well as Regina Albrecht for her skilled assistance

References

1. The Bible, Exodus 23
2. Haarmann W (1877) Kaiserliches Patentamt no 576
3. Rabenhorst J, Hopp R (1991) DE 3 920 039
4. Rabenhorst J, Hopp R (1997) EP 0 761 817
5. Shaikh Y (2002) Specialty Aroma Chemicals in Flavors and Fragrances. Allured, Carol Stream, p vii
6. Murray N (2004) Food Process May 21
7. Meyerhof W (2005) Rev. Physiol. Biochem. Pharmacol. 154:37
8. Gravina SA, McGregor RA, Nossoughi R, Kherlopian J, Hofmann T (2003) In: Hofmann T, Ho C-T, Pickenhagen W (eds) Challenges in Taste Chemistry and Biology. ACS Symposium Series 867. American Chemical Society, Washington, p 91
9. Gentili B, Guadagni DG (1979) US Patent 4,154,862
10. Ley JP, Krammer G, Reinders G, Gatfield IL, Bertram H-J (2005) J. Agric. Food Chem. 53:6061
11. Rotzoll N, Dunkel A, Hofmann T (2006) J. Agric. Food Chem. 54:2705
12. Soldo T, Hofmann T (2005) J. Agric. Food Chem. 53:9165
13. Syed S (1998) WO 98 20 753
14. Lee CH, Scarpellino RJ, Murtagh MM (1975) US Patent 3,924,017
15. Bartoshuk LM, Leyy C-H, Scarpellino R (1972) Science 178:988
16. Gibbs BF, Alli I, Mulligan C (1996) Nutr. Res. 16:1619
17. Shirasuka Y, Nakajima K, Asakura T, Yamashita H, Yamamoto A, Hata S, Nagata S, Abo M, Sorimachi H, Abe K (2004) Biosci. Biotechnol. Biochem. 68:1403
18. Suttisri R, Lee IS, Kinghorn AD (1995) J. Ethnopharmacol. 47:9
19. Djordjevic J, Zatorre RJ, Jones-Gotman M (2004) Chem. Sens. 2004, 29:199

20. Ley JP, Paetz S, Kindel G, Freiherr K, Krammer GE (2006), Abstracts of Papers, 231st ACS National Meeting, Atlanta, GA, 26–30 March, AGFD-142
21. Ley J, Kindel G, Bertram HJ, Krammer G (2006) WO 2006 024 587
22. King BM, Arents P, Bouter N, Duineveld CAA, Meyners M, Schroff SI, Soekhai ST (2006) J Agric. Food Chem. 54:2671
23. Soldo T, Blank I, Hofmann T (2003) Chem. Sens. 28:371
24. Keast RSJ, Breslin PAS, Beauchamp GK (2001) Chimia 55:441
25. Bala R, Kaur A, Bakshi AK (2004) J. Food Sci. Technol. 41:668
26. Fuller W, Kurtz RJ (1998) US Patent 5,637,618
27. Roy G (1997) Modifying Bitterness. Technomic, Basel
28. Okiyama A, Beauchamp GK (1998) Physiol. Behav. 65:177
29. Ball P, Woodward D, Beard T, Shoobridge A, Ferrier M (2002) Eur. J. Clin. Nutr. 56:519
30. Shackelford JR (1981) US Patent 4,297,375
31. Lontz J, Alonso CD (1974) US Patent 3,782,974
32. Takahiro N (1988) JP 63 137 657
33. Bakal A (2006) US 2006 024 422
34. Lee T (1992) US Patent 5 145 707
35. Wada T (2004) JP 2004 275 098
36. Belitz HD, Grosch W, Schieberle P (2001) Lehrbuch der Lebensmittelchemie. Springer, Berlin Heidelberg New York, p 32
37. Locke KW, Fielding S (1994) Physiol. Behav. 55:1039
38. Pei T, Conklin G, Janczuk A, Smith CM, Dewis ML (2006) US 2006 057 268
39. Dewis ML, John TV, Eckert MA, Colstee JH, Da Costa NC, Pei T (2005) US 2005 075 368
40. Tachdjian C, Patron AP, Qi M, Adamski-Werner SL, Tang XQ, Chen Q, Darmohusodo V, Lebl-Rinnova M, Priest C (2006) US 2006 045 953
41. DeSimone JA, Heck GL (1992) US Patent 4,997 672
42. Beksan E, Schieberle P, Robert F, Blank I, Fay LB, Schlichtherle-Cerny H, Hofmann T (2003) J. Agric. Food Chem. 51:5428
43. Rotzoll N, Dunkel A, Hofmann T (2005) J. Agric. Food Chem. 53:4149
44. Kaneko S, Kumazawa K, Masuda H, Henze A, Hofmann T (2006) J. Agric. Food Chem. 54:2688
45. de Rijke E, Ruisch BJ, Bouter N, König T (2006) Mol. Nutr. Food Res. 49:351
46. Visser J, Renes H, Winkel C, Noomen S, Visser J (2006) WO 2006 009 425
47. Van Den Ouweland G, Benzi F, Van Beem N, Vanrietvelde C (2001) US Patent 6,287,620
48. Frerot E, Benzi F (2004) WO 2004 075 663
49. Jordt SE, McKermy DD, Julius D (2003) Curr. Opin. Neurobiol. 13:487
50. Scharbert S, Hofmann T (2005) J. Agric. Food Chem. 53:5377
51. Scientific Committee on Food (2002) Opinion of the Scientific Committee on Food on Capsaicin SCF/CS/FLAV/FLAVOUR/8 ADD1 Final, European Comission, Brussels, Belgium
52. Bryant BP, Mezine I (1999) Brain Res. 842:452
53. Molina-Torres J, Salazar-Cabrera CJ, Armenta-Salinas C Ramírez-Chávez E (2004) J. Agric. Food Chem. 52:4700
54. Ley JP, Krammer G, Looft J, Reinders G, Bertram HJ (2006) In: Bredie WLP, Petersen MA (eds) Flavour Science: Recent Advances and Trends. Elsevier, Amsterdam, p 21
55. Hiserodt RD, Adedeji J, John TV, Dewis ML (2004) J. Agric. Food Chem. 52:3536

56. Gassenmeier K (2006) Flavour Fragrance J. 21:725
57. Velazco MI, Wuensche L, Deladoey P (2005) EP 1,541,039
58. Krammer G, Sabater C, Brennecke S, Liebig M, Freiherr K, Ott F, Ley JP, Weber B, Stöckigt D, Roloff M, Schmidt CO, Gatfield I, Bertram HJ(2006) In: Bredie WLP, Petersen MA (eds) Flavour Science: Recent Advances and Trends, Elsevier, Amsterdam, p 169
59. Winkel C (2005) In: Rowe DJ (ed) Flavors and Fragrances. Blackwell, Oxford, p 244–259
60. Stöllman U, Johansson F, Leufvén A (1994) In: Man CMD, Jones AA (eds) Shelf Life Evaluation of Foods, 1st edn. Chapman & Hall, London
61. Singh RP (1994) In: Man CMD, Jones AA (eds) Shelf Life Evaluation of Foods, 1st edn. Chapman & Hall, London
62. Ellis MJ (1994) In: Man CMD, Jones AA (eds) Shelf Life Evaluation of Foods, 1st edn. Chapman & Hall, London
63. Muche S (2006) Lebensmitteltechnik 2006 1–2:38
64. Reineccius GA (2006) In: Flavor chemistry and technology, 2nd edn. Taylor & Francis, Boca Raton, pp 351–389
65. Ashurst PR (2005) Chemistry and Technology of Soft Drinks and Juices Blackwell, Oxford

22 Enzymes and Flavour Biotechnology

M. Menzel, P. Schreier
Lehrstuhl für Lebensmittelchemie,
Universität Würzburg,
Am Hubland, 97074 Würzburg, Germany

22.1 Introduction

There are about 25,000 enzymes present in nature and about 400 have been commercialised mainly for stereoselective organic synthesis and also for the biotechnological production of flavour compounds. The worldwide market for enzymes is more than US $1 billion.

The majority of enzymes in food biotechnology comprise hydrolytic enzymes, transferases, oxireductases and lyases.

Microbial enzymes play the greatest role in production of flavour compounds; they can also be expressed in recombinant microorganisms.

This chapter is not only an update of our review of 1997 [1] but also an overview of the latest development in enzyme-based flavour technology. Some aspects of the present chapter are based on the previous review [1].

22.2 Hydrolytic Enzymes

22.2.1 Lipases (EC 3.1.1.X)

Lipases are serine hydrolases that catalyse the hydrolysis of lipids to fatty acids and glycerol [2]. In contrast to esterases, they work at the lipid–water interface and show only little activity in aqueous solutions. Studies of the X-ray structures of human lipase [3,4] and *Mucor miehei* lipase [5,6] revealed a change in conformation at the lipid–water interface, which explains the increase of activity.

Both of the aforementioned lipases contain Asp-His-Ser triades; different catalytic triades can be found, e.g., in *Geotrichum candidum* (Glu-His-Ser) [7] or in *Humicola lanuginosa* (Asp-His-Tyr) [8].

Lipases play an important role in organic synthesis and also in flavour biotechnology. Pig pancreatic extract and especially many microbial lipases are used for ester hydrolysis, esterification (alcohol and acid), transesterification (ester and alcohol), interesterification (ester and acid) and transfer of acyl groups from esters to other nucleophiles like amines or thiols [1].

Some criteria of selectivity are important for these catalysed reactions: substrate selectivity [9], regioselectivity [10], stereoselectivity (*endo*/*exo* [11] and *Z*/*E* [12] differentiation), enantioselectivity [13], *meso* differentiation [14] and prochiral recognition [15].

In many cases, the stereoselectivity of the enzyme used in water and in an organic solvent is the same [16, 17]; thus, complementary stereoisomers can be produced. If an enzyme prefers the *R* enantiomer of a chiral ester over the *S* ester, the *R* alcohol and the *S* ester can be obtained after a hydrolytic reaction. As the enzyme's stereochemical preference remains the same, transesterification in organic solvents will produce the *S* alcohol and *R* ester.

Theoretically, both reactions will stop at 50% conversion and will give both enantiomers with 100% enantiomeric excess [$ee=(R-S)/(R+S)\times 100$ for $R>S$], if the enzyme has an absolute stereoselectivity.

22.2.1.1
Lipolysis

Lipolysed milk fat was one of the first flavours produced with the help of enzymes. The original process was based on the controlled lipase-catalysed hydrolysis of cream [18]. For instance, *Mucor miehei* lipase possesses a high selectivity towards flavour-active short-chain fatty acids. Additionally, lipases that prefer long-chain fatty acids or lipases without particular preferences can be found. The free fatty acids produced can be isolated by steam distillation and further purified. Thus, it is possible to obtain pure short-chain fatty acids like butanoic, hexanoic, octanoic and decanoic acid.

Lipolysed milk fat products can serve as cream-like/butter-like flavouring agents [19].

22.2.1.2
Kinetic Resolution of Racemates

Stereoselectivity of lipases is often used to yield pure optically active flavour compounds from racemic precursors. This fact is important if one isomer of a molecule has more desirable properties than the other one.

For instance, (-)-menthol (*p*-menthan-3-ol) is one of the most important flavouring agents and is the major compound in natural peppermint oil. The characteristic peppermint odour and the typical cooling effect is limited to (-)-menthol. The other isomers do not show this refreshing effect. A racemic mixture of menthol holds an intermediate position: the cooling effect is still perceptible.

There are several biochemical and chemical processes for the resolution of a racemic mixture of menthol. Many microbiological lipases hydrolyse menthyl esters and prefer the (-)-menthyl esters, whereas (+)-menthyl esters are not hydrolysed at all. This asymmetric hydrolysis of menthyl esters can be performed with lipases from *Penicillium*, *Rhizopus*, *Trichoderma* and various bacteria [20].

The enantioselective hydrolysis of racemic menthyl benzoate (industrially key compound) by recombinant *Candida rugosa* lipase LIP1 leads to optically pure l-(-)-menthol; ee>99% [21]. This pathway is part of a menthol synthesis developed by the flavour industry.

The resolution of the commercially available racemic *trans*-jasmonate to (-)-*trans*-jasmonate by microbial lipase has been described by Serra et al. [22].

Nozaki et al. [23] characterised the production of (+)-mesifuran [2,5-dimethyl-4-methoxy-3(2*H*)-furanone], an important flavour compound in arctic bramble, but which also occurs in strawberry and pineapple. After lipase-catalysed (*Candida antarctica*) enantioface-differentiating hydrolysis of the enol acetate, the pure optically active (+)-mesifuran could be obtained.

Kinetic resolution of branched-chain fatty acids has been reported recently by Franssen et al. [24]. With the help of immobilised *Candida antarctica* lipase B, racemic 4-methyloctanoic acid (responsible for sheep-like and goat-like flavours in sheep and goat milk and cheese, respectively) was esterified with ethanol. Only the *R* ester could be obtained, whereas (*S*)-4-methyloctanoic acid was not converted (Scheme 22.1).

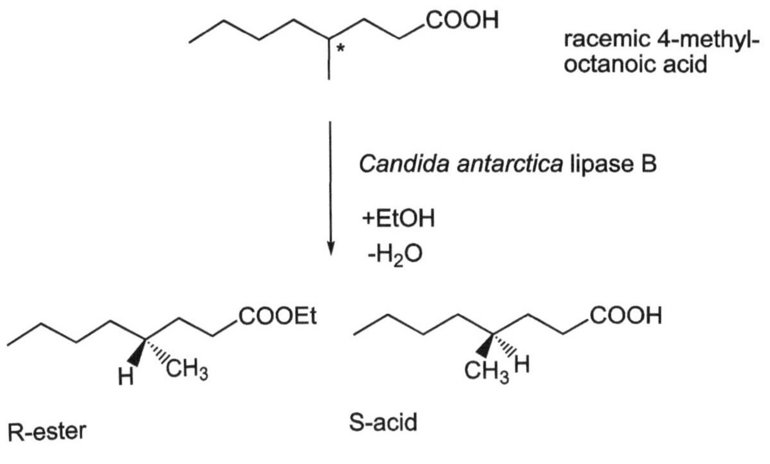

Scheme 22.1 Kinetic resolution of racemic 4-methyloctanoic acid with *Candida antarctica* lipase B [24]

22.2.1.3 Catalysis in Organic Media

Lipase-catalysed esterification and transesterification reactions have a wide range of applications in the synthesis of aroma compounds.

The reaction conditions have a great influence on the enzyme-catalysed reactions in organic media and determine the reaction's yield and selectivity.

Enzymes require only a monomolecular water phase for their activity in organic solvents [25]; the pH of the water phase [26], temperature [27], type of solvent [28] and immobilisation techniques [29] will influence the reaction too.

Of course, the selection of the appropriate enzyme is fundamental because yield and selectivity of the enzymes vary extremely. For instance, *Candida rugosa* lipase will give high yields but has a low selectivity. In contrast, lipase from *Aspergillus niger* exhibits high selectivity [13].

The biotechnological production of flavour compounds is particularly focused on esters and lactones. Lipase from *Mucor miehei* is the most widely studied fungal lipase [30–35]. Esters of acids from acetic acid to hexanoic acid and alcohols from methanol to hexanol, geraniol and citronellol have been synthesised using lipases from *Mucor miehei, Aspergillus sp., Candida rugosa, Rhizopus arrhizus* and *Trichosporum fermentans* [32-37].

Methylbutanoates and methylbutyl esters are essential flavour compounds in fruit flavours; they can be produced biotechnologically as mentioned before. Chowdary et al. [33] have described the production of a fruit-like flavour: isoamyl isovalerate by direct esterification of isoamyl alcohol and isovaleric acid in hexane with the help of *Mucor miehei* lipase immobilised on a weak anion-exchange resin.

Synthesis of short-chain geranyl esters catalysed by esterase from *Fusarium oxysporum* in an organic solvent was reported by Stamatis et al. [39].

Large-scale synthesis of (*Z*)-3-hexenyl acetate in hexane with lipase, (*Z*)-3-hexenol and acetic acid was described by several authors [40–42]. (*Z*)-3-Hexenyl acetate has a fruity odour and shows a significant green note flavour. It can be produced using lipase from *Candida antarctica* immobilised on an acrylic resin [40, 41] or using immobilised lipase from *Mucor miehei* [42]. The conversion was reported to be about 90%.

An optimised enzymatic synthesis of methyl benzoate in an organic medium was reported by Leszczak and Tran-Minh [43]. Methyl benzoate is part of the aroma of some exotic fruits and berries. The ester has been produced by direct esterification of benzoic acid with methanol in hexane/toluene catalysed by lipase from *Candida rugosa*.

Gatfield et al. [44] reported in 2001 a method to produce natural ethyl (*E,Z*)-2,4-decadienoate, the impact compound of pear. Immobilised lipase from *Candida antarctica* is capable of transesterifying *Stillingia* oil in the presence of ethanol. By this process, a complex mixture of ethyl esters is generated. By fractional distillation, the ethyl ester of (*E,Z*)-2,4-decadienoate can be isolated from the mixture in a total yield of about 5% and with a high degree of purity. As only natural precursors, physical and biological processes were used, the aroma compound obtained can be labelled as natural according to the legislation of the European Union.

In 2004, Ley et al. [45] showed a stereoselective enzymatic synthesis of *cis*-pellitorine [*N*-isobutyldeca-(2*E*,4*Z*)-dienamide], a taste-active alkamide naturally occurring in tarragon. The reactants were ethyl (*E*,*Z*)-2,4-decadienoate—the pear ester described before—and isobutyl amine. The reaction is catalysed by lipase type B from *Candida antarctica* (commercially available), which shows a remarkable selectivity towards the 2*E*,4*Z* ester. The yield was about 80%.

The biotechnological synthesis of lactones has reached a high standard. Besides microbial production, lactones can also be enzymatically produced. For instance, a lipase-catalysed intramolecular transesterification of 4-hydroxycarboxylic esters leads enantioselectively (ee>80%) to (*S*)-γ-lactones; the chain length may vary from C5 to C11 [13]. γ-Butyrolactone can be produced in that way with lipase from *Mucor miehei* [30].

The preparation of optically active δ-lactones is more difficult because of the lack of selectivity of most lipases.

22.2.2 Glycosidases (EC 3.2.1.X)

It is well-known that in plant tissues certain amounts of flavour compounds are bound as non-volatile sugar conjugates. Most of these glycosides are β-glucosides, but there are other glycones like pentoses, hexoses, disaccharides and trisaccharides too [46]. Acylated glycosides and phosphate esters have also been reported [47, 48]. Information about the analysis of glycosides can be found in the work of Herderich et al. [49].

Besides the structural elucidation of glycosides, research is focused on the application of glycosidases to liberate the aroma-active aglycons from their bound forms. The development of a continuous process of enzymatic treatment (simultaneous enzyme catalysis extraction) [50] opened the doors for the industrial large-scale production of aroma compounds from their non-volatile conjugates.

Major interest has been directed to wine. During winemaking, the grape's β-glucosidase is rapidly inactivated. Glucosidases from *Saccharomyces cerevisiae* and *Candida molischiana* have been suggested to solve this problem [51]. Nonetheless, many fungal glycosidases will not work properly, because they are inhibited by glucose, fructose, ethanol and the relatively low pH of wine. Some glycosidases from *Aspergillus* sp. (e.g. some β-apiosidases, α-arabinosidases and α-rhamnosidases) do not have these disadvantages. The formation of these enzymes can be induced by the presence of the respective glycoside; their use has been patented for application to grape must [52]. Cabaroglu et al. [53] have given a comprehensive overview of wine flavour enhancement by the use of fungal glycosidases and have shown that enzymatically treated wine was preferred in sensory analyses.

Sensory quality of food can be improved by synergistic action of monoglycanases, oligoglycanases and polyglycanases. A process for the production of vanilla extracts involving the treatment of crushed green vanilla beans with enzymatic preparations that degrade plant cell walls and the glucosidic precursor together has been patented [54]. Similarly, a cellulase possessing glucosidase side activity has been reported to liberate benzaldehyde from its bound form during the processing of peach [55].

Raspberry ketone [4-(4′-hydroxyphenyl)-butan-2-one], the impact compound found in raspberries, can be obtained by enzymatic reactions: The first step is the β-glucosidase-catalysed hydrolysis of the naturally occurring betuloside to betuligenol. The latter can be transformed into raspberry ketone by microbial alcohol dehydrogenase (Scheme 22.2) [56].

Conjugates of flavour compounds were also found in milk: phenols can be liberated by β-glucuronidase, arylsulfatase and acid phosphatase from their respective precursors [57].

Besides the liberation of bound flavour compounds, the creation of these conjugates is becoming more and more important, especially for convenience food. A bound, non-volatile aroma compound allows the slow liberation of the flavour upon heating. These slow-release compounds are produced with the help of glucosidases in a reversed hydrolysis reaction. For instance, the production of geranyl glucoside was described by de Roode et al. [58] and Franssen et al. [24]. Glycosyl transferases are also able to produce glycosides, but they are more complicated to handle than glucosidases [24].

There are synthetic acetal derivatives of flavour-active aldehydes like benzaldehyde and cinnamaldehyde [59].

As the chemical synthesis of glycosides is cumbersome, biotechnological transglycosidation using glycosidases is attracting more and more attention [60].

22.2.3 Flavorzyme®

Flavorzyme® is a commercially available proteolytic enzyme preparation by Novo Nordisk Bioindustrials. It can be used to obtain a meat-like process flavouring from defatted soybean meal. With the help of aroma extract dilution analysis, Wu and Cadwallader [61] showed in their study of 2002 the presence of key aroma compounds of roasty, meat-like aroma in the enzymatically hydrolysed and heated hydrolysed protein, e.g. maltol, furaneol, methanethiol and furanthiol derivatives.

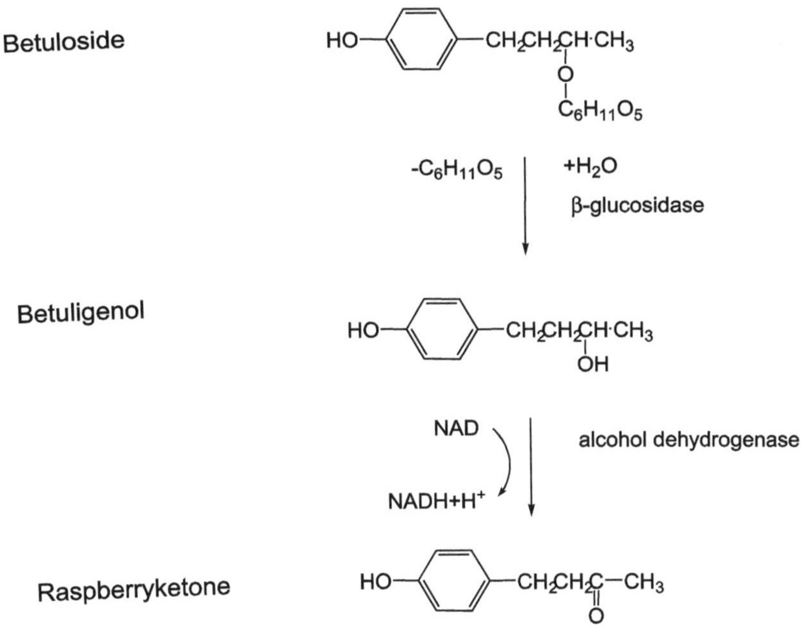

Scheme 22.2 Enzymatic production of raspberry ketone from betuloside [56]

22.3 Oxireductases

Many enzyme-catalysed redox processes include the transfer of the equivalent of two electrons by one two-electron step or two one-electron steps. The latter is considered as a radical process involving the use of cofactors like flavin, quinoid coenzymes or transition metals.

The two-electron process is either a hydride transfer or a proton abstraction followed by two-electron transfer.

22.3.1 Horse Liver Alcohol Dehydrogenase (EC 1.1.1.1)

Horse liver alcohol dehydrogenase is able to oxidise primary alcohols—except methanol—and to reduce a large number of aldehydes. Aqueous solution or organic solvents can be used [62]. As there are no new developments concerning this enzyme, the reader is referred to the review of Schreier [1].

22.3.2 Lipoxygenase (EC 1.13.11.12)

Lipoxygenase (LOX) is a non-haem, iron-containing dioxygenase that catalyses the regioselective and enantioselective dioxygenation of unsaturated fatty acids containing at least one (*Z,Z*)-1,4-pentadienoic system. For instance, LOX from soy converts linoleic acid to the (*S*)-13-hydroperoxide [1].

It is supposed that the catalytic mechanism proceeds through a free-radical intermediate which reacts directly with oxygen or an organic iron intermediate [63]. The three-dimensional protein structure of the native form of LOX isoenzyme L-1 from soybean has already been described [64, 65].

LOX is an important factor in the large-scale use of plant enzymes for the production of natural "green note" aroma compounds, a group of isomeric C6 aldehydes and alcohols [66].

In nature, the green notes are produced after the destruction of the plant's tissue (leaves, fruits or vegetables). Destruction of the cell wall leads to a cascade of enzyme-catalysed reactions; polyunsaturated fatty acids with the diene system described before are converted into hydroperoxides by LOX catalysis. The hydroperoxide lyase cleaves the hydroperoxides; in the whole cascade, oxireductases are involved too. The biotechnological large-scale production of natural green notes follows the natural pathway.

A patented process for the production of green notes applying baker's yeast for in situ reduction of enzymatically produced aldehydes [67, 68] has been called into question regarding the effective production of (*Z*)-3-hexenol. According to Gatfield's report [69] the isomerisation of (*Z*)-3-hexenol to (*E*)-2-hexenal is a very fast process. The latter undergoes facile conversion to hexanol. Beside this, baker's yeast can add activated acetaldehyde to (*E*)-2-hexenal, forming 4-octen-2,3-diol.

At present, there are some patents concerning the production of green notes by recombinant guava 13-hydroperoxide lyase expressed in *Escherichia coli* [70, 71] and *Cucumis melo* hydroperoxide lyase; the latter yields a mixture of C6 and C9 compounds [72].

Fungal LOXs exhibit different regioselectivity from LOX from higher plants; they catalyse the formation of 10-hydroperoxides from linoleic and linolenic acid by dioxygenation. Hydroperoxide lyase and subsequent enzymes in the damaged fungal cells are able to form the typical volatile mushroom aroma substances, including the impact compound (*R*)-1-octen-3-ol. The latter can be produced industrially by feeding the mycelia with linoleic acid [73, 74].

It is a well-known fact that soybean LOX is able to cooxidise plant pigments, such as carotenoids and chlorophyll in the presence of linoleic acid. The hypothesis of a free-radical mechanism has been supported by stereochemical studies of the unselective formation of epoxides during LOX-catalysed cooxidation [75].

A pathway for the production of α-ionone and β-ionone by LOX-catalysed cooxidation of carotenes has been described [76].

22.3.3 Peroxidases (EC 1.11.1.X)

22.3.3.1 Soybean Peroxidase

The production of methyl anthranilate, which has a fruity odour, by enzymatic N-demethylation of methyl *N*-methyl anthranilate (Scheme 22.3.) has been reported by van Haandel et al. [77]. Self-prepared soybean peroxidase (haem-based enzyme) preparation and H_2O_2 were used.

The reaction product can be labelled as natural if the methyl *N*-methyl anthranilate used has a natural origin, e.g. methyl *N*-methyl anthranilate extracted from citrus leaves.

An alternative method for the production of methyl anthranilate with the help of *Bacillus megaterium* was recently reported by Taupp et al. [78]; the latter pathway resulted in higher yields of methyl anthranilate.

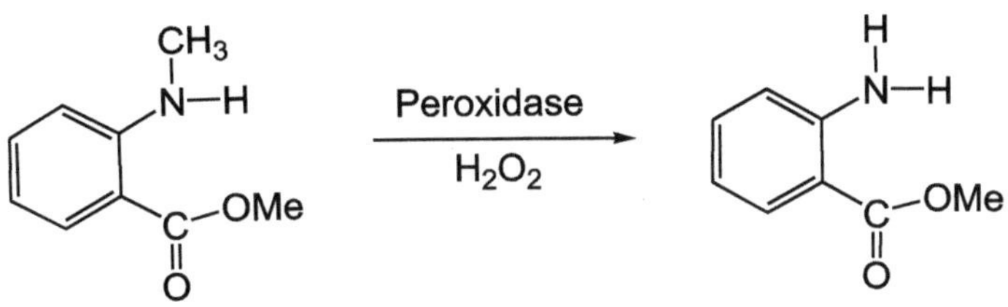

Scheme 22.3 Production of methyl anthranilate by enzymatic N-demethylation of methyl *N*-methyl anthranilate [77]

22.3.3.2 Horseradish Peroxidase (EC 1.11.1.7)

The haem peroxidases are a superfamily of enzymes which oxidise a broad range of structurally diverse substrates by using hydroperoxides as oxidants. For example, chloroperoxidase catalyses the regioselective and stereoselective halogenation of glycals, the enantioselective epoxidation of distributed alkenes and the stereoselective sulfoxidation of prochiral thioethers by racemic arylethyl hydroperoxides [62]. The latter reaction ends in (*R*)-sulfoxides, (*S*)-hydroperoxides and the corresponding (*R*)-alcohol, all in optically active forms.

Horseradish peroxidase catalysed kinetic resolution of racemic secondary hydroperoxides has been described by Adam et al. [79]. The reaction yields (*R*)-hydroperoxides up to ee>99% and (*S*)-alcohols up to ee>97%. Optically active hydroperoxides as potential stereoselective oxidants can be obtained by this process.

As horseradish peroxidase is relatively expensive and possesses only little thermostability, the industrial application of horseradish peroxidase is limited [77].

22.3.3.3
Lepista irina Peroxidase

In 2003, Zorn et al. [80] discovered a fungal peroxidase from *Lepista irina*—a valued edible fungus—that cleaved *β,β*-carotene to flavour-active compounds. According to the authors, the cleavage of *β,β*-carotene to aroma compounds by a fungal peroxidase had not been reported before.

It was found that extracellular liquid of the fungus can degrade *β,β*-carotene to *β*-cyclocitral, dihydroactinidiolide, 2-hydroxy-2,6,6-trimethylcyclohexanone, *β*-apo-10′-carotenal and *β*-ionone; the last two compounds are the main prod-

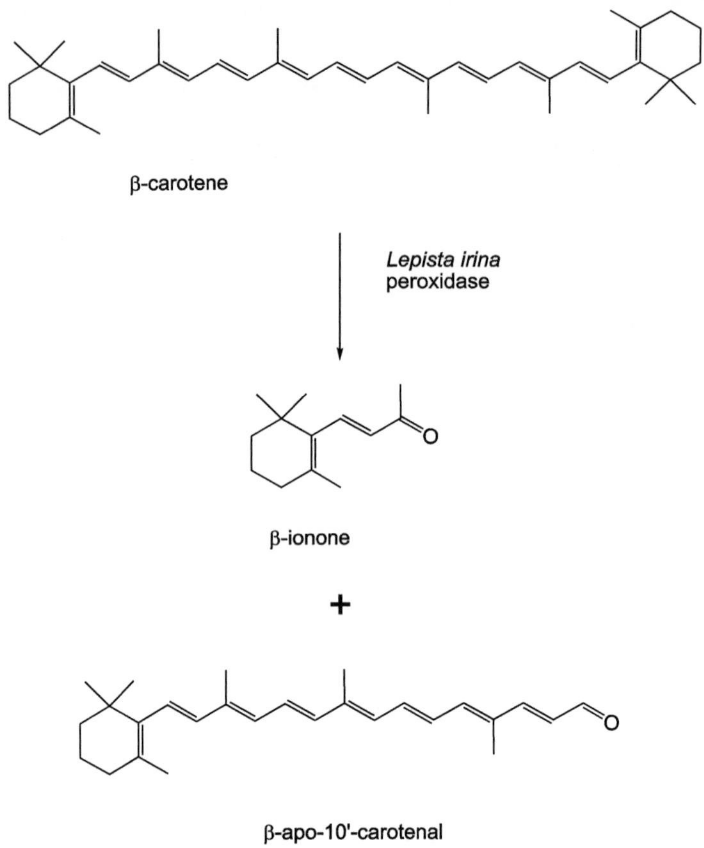

Scheme 22.4 Cleavage of *β*-carotene by *Lepista irina* peroxidase [80]

ucts (Scheme 22.4). The key enzyme catalysing the oxidative cleavage was isolated and characterised.

As there is great interest from the detergent, food and perfume industry in the potent aroma compounds formed by carotenoid breakdown, and as the β-ionone obtained can be labelled as natural aroma—if natural carotenoids are used—this cleavage reaction might have a high potential.

22.3.4 Laccase (EC 1.10.3.2)/Germacrene A Hydroxylase

Laccase, a group of multi-copper proteins of low specificity, acting on both *o*-quinols and *p*-quinols and often on aminophenols and phenylenediamine, is used for the biotechnological production of nootkatone, the impact compound of grapefruit. Huang et al. [81] described a process for the laccase-catalysed oxidation of valencene to nootkatone; they used whole microorganisms with laccase activity—such as from *Botrytis cinera*—but they reported a process with isolated laccase too. The first step of the reaction is the formation of valencene hydroperoxide, which undergoes a non-enzymatic degradation to nootkatone. The yield was about 60%.

Franssen et al. [24] pointed out an alternative method of production of nootkatone from valencene catalysed by (+)-germacrene A hydroxylase, an enzyme of the cytochrome P450 monooxygenase type that was isolated from chicory roots. In general, this enzyme appeared to accept a broad range of sesquiterpenes and hydroxylates exclusively at the side-chain's isopropenyl group. Valencene is an exception: it was not hydroxylated at the side chain, but β-nootkatol was formed in the first step (Scheme 22.5); it is not yet clear if the second step is enzyme-catalysed.

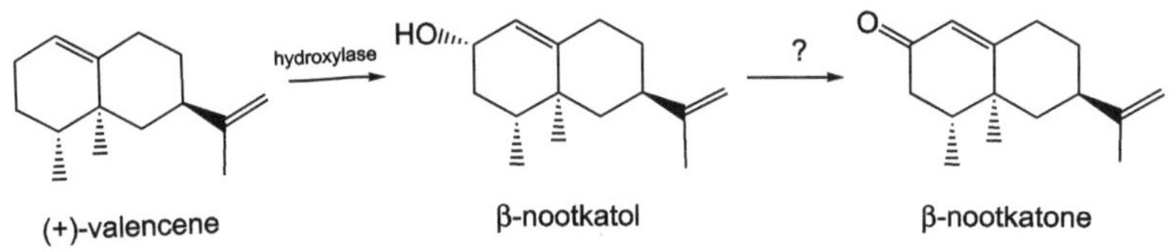

Scheme 22.5 Production of nootkatone from valencene catalysed by (+)-germacrene A hydroxylase [81]

22.3.5 Microbial Amine Oxidases (EC 1.4.3.X)

Amine oxidase from *Aspergillus niger* and monoamine oxidase from *Escherichia coli* can be used for the oxidative deamination of amines, forming the corresponding aldehydes, hydrogen peroxide and ammonia. Using these enzymes, Yoshida et al. [82] described a pathway for the production of vanillin (4-hydroxy-3-methoxy-benzaldehyde).

Vanillylamine [(4-hydroxy-3-methoxy-phenyl)methylamine] is the substrate of choice for the formation of vanillin with the help of amine oxidase. It can be obtained by cleavage of capsaicin (*N*-[(4-hydroxy-3-methoxy-phenyl)methyl]-8-methyl-6-nonenamide) isolated from pepper and capsicum [83]. As natural vanillin extracted from beans of *Vanilla planifolia* is rare and extremely expensive, this pathway for the production of natural vanillin is regarded to have a great potential. The vanillin obtained by the process can be labelled as natural if the cleavage of capsaicin is performed enzymatically.

22.3.6
Vanillyl Alcohol Oxidase (EC 1.1.3.38)

Vanillyl alcohol oxidase (VAO) is a flavoenzyme from the ascomycete *Penicillium simplicissimum* that converts a broad range of 4-hydroxybenzyl alcohols and 4-hydroxybenzylamines into the corresponding aldehydes. This large substrate specificity makes it possible to obtain vanillin from two major pathways.

As VAO is able to perform an oxidative deamination of capsaicin-derived vanillyl amine, vanillin can be produced by the pathway described in the previous subsection. Van den Heuvel et al. [83] pointed out this biocatalytic route of synthesis in 2001 using penicillin G acylase to obtain vanillyl alcohol from natural capsaicin (Scheme 22.6). As the vanillin obtained can be labelled as natural,

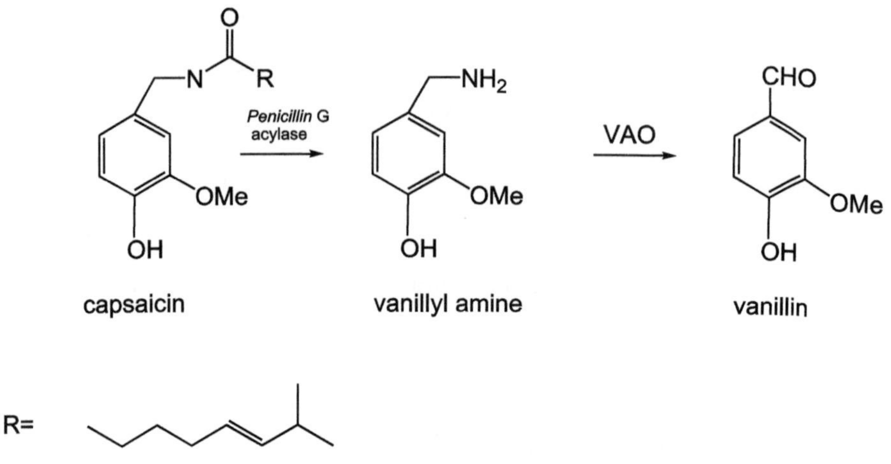

Scheme 22.6 Oxidative deamination of capsaicin-derived vanillyl amine and formation of vanillin [83] *VAO* vanillyl alcohol oxidase

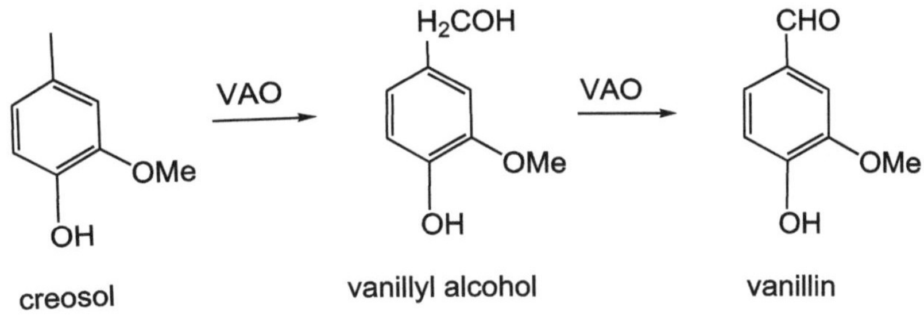

Scheme 22.7 Production of vanillin from creosol by two enzymatic reactions [83]

the enzymes used do not require expensive cofactors and the enzymes can be produced on a large scale, this bi-enzymatic process could be promising.

The second pathway using VAO reported by van den Heuvel et al. [83] is the VAO-catalysed oxidation of vanillyl alcohol to vanillin. Vanillyl alcohol is not very abundant in nature but can be generated by VAO-catalysed conversion of creosol (2-methoxy-*p*-cresol). As creosol can be found in creosote obtained from heating wood or coal tar, the feedstock for this pathway is very abundant.

The process comprises two steps: the conversion of creosol to vanillyl alcohol and the oxidation of the alcohol to vanillin (Scheme 22.7). Interestingly, these two steps are catalysed by the same enzyme, i.e.VAO.

In 2004, van den Heuvel et al. [84] described in another study the characteristics of VAO and pointed out details of the reaction's mechanism.

22.4 Transferases

22.4.1 Cyclodextrin Glucanotransferase (EC 2.4.1.19)

In 2002, Do et al. [85] proposed a pathway for the enzymatic synthesis of (-)-menthyl α-maltoside and α-maltooligosides from (-)-menthyl α-glucoside using cyclodextrin glucanotransferase obtained from *Bacillus macerans*. The reaction can be performed in a reactor containing (-)-menthyl α-glucoside, the enzyme and soluble starch; the yield was about 80%:15% (-)-menthyl α-maltoside and 65% (-)-menthyl α-maltooligosides, respectively. Treatment of the starch with α-amylase can raise the proportion of (-)-menthyl α-maltoside.

At first, (-)-menthyl α-maltoside has a bitter and sweet taste, but after a few minutes, the refreshing flavour occurs. It has the potential to become a slow-release aroma compound in foods or cigarettes because it possesses higher solubility in water and has a lower tendency to sublimate.

22.5 Lyases

22.5.1 D-Fructose-1,6-biphosphate Aldolase (EC 4.1.2.13)

The formation of C–C bonds by aldol condensation is a very useful method in synthesis. Besides the chemical synthesis, aldolases can be used to perform this reaction. The reaction yields a stereoselective condensation of an aldehyde with a ketone donor.

In nature, four complementary aldolases can be found in the carbohydrate metabolism. They show different stereoselectivity and this broad range of enzymes makes it possible to fulfil a large variety of synthetic tasks. In biotechnology, Furaneol® (2,5-dimethyl-4-hydroxy-2*H*-furan-3-one) can be produced from fructose-1,6-biphosphate with the help of a three-step enzymatic process involving fructose-1,6-bisphosphate aldolase (rabbit muscle aldolase). The first step is the aldolase-catalysed cleavage of the sugar biphosphate to dihydroxyacetone phosphate and glyceraldehyde phosphate. The latter is isomerised by a coimmobilised triose phosphate isomerase to obtain dihydroxyacetone phosphate, which is the substrate for the aldolase-catalysed aldol condensation with d-lactaldehyde. The condensation's product, 6-deoxyfructose phosphate, can be easily converted to Furaneol® [86].

In spite of the intensive effort regarding the biosynthesis of Furaneol® (including the detection of some important enzymes), the biosynthesis in plants is still not fully understood [87].

22.5.2 Sesquiterpene Synthase (EC 4.2.3.9)

In the last few years, sesquiterpene synthase from different plants has raised attention. In 2004, Schalk and Clark [88] described a process (patented by Firmenich, Switzerland) that makes it possible to obtain sesquiterpene synthase and to produce various aliphatic and oxygenated sesquiterpenes from farnesyl diphosphate. For instance, valencene can be obtained in this way.

One year later, Schalk [89] described a process for cloning sesquiterpene synthases from patchouli plants (*Pogostemon cablin*) and the enzyme-catalysed terpenoid production. Various sesquiterpenes can be obtained by this method, for instance patchoulol and other germacrene-type sesquiterpenes.

22.6 Conclusion

Thanks to the intense research during the last 20 years, flavour biotechnology is an integrated part of industrial aroma production, in which enzyme-catalysed reactions are an alternative to microbial-based processes. Besides the production of aroma, enzymes can also be used to refine aroma of cheese or wine.

The great advantage of enzymes is their stereoselectivity and the ability to produce "natural" aroma if "natural" substrates are available.

References

1. Schreier P (1997) In: Berger RG (ed) Biotechnology of aroma compounds. Springer, Berlin Heidelberg New York, p 51
2. Desnuelle P (1972) In: Boyer PO (ed) The enzymes, vol 7. Academic, New York, p 575
3. Winkler FK, D'Arey, Hunziker W (1990) Nature 343:771
4. Van Tilbeurgh H, Sarda L, Verger R, Cambillau C (1992) Nature 359:159
5. Brzozowski AM, Derewenda U, Derewenda ZS, Dodson GG, Lawson DM, Turkenburg JP, Bjorkling F, Huge-Jensen B, Patkar SA, Thim L (1991) Nature 351:491
6. Derewenda U, Brzozowski AM, Lawson DM, Derewenda ZS (1992) Biochemistry 31:1532
7. Schrag JD, Li Y, Wu S, Cygler M (1991) Nature 351:761
8. Liu WH, Beppa T, Arima K (1977) Agric Biol Chem 41:131
9. Francalani F, Cesti P, Cabri D, Bianchi T, Martinego T, Foa M (1987) J Org Chem 52:5079
10. Poppe L, Novak M, Kajtar-Peredy M, Szantay C (1993) Tetrahedron Asymmetry 4:2211
11. Meltz M, Saccomano NA (1992) Tetrahedron Lett 33:1201
12. Lutz D, Güldner A, Thums R, Schreier P (1990) Tetrahedron Asymmetry 1:783
13. Lutz D, Huffer M, Gerlach D, Schreier P (1992) In: Teranishi R, Takeoka GR, Güntert M (eds) ACS symposium series 490. American Chemical Society, Washington, p 32
14. Tanaka M, Yoshioka M, Sakai K (1993) Tetrahedron Asymmetry 4:981
15. Guanti G, Banfi L, Riva R (1994) Tetrahedron Asymmetry 5:1
16. Kirchner G, Scollar MP, Klibanov AM (1985) J Am Soc 107:7072
17. Adler U, Breitgoff D, Klein P, Laumen K, Schneider M (1989) Tetrahedron Lett 30:1793
18. Kilare A (1985) Process Biochem 20:35
19. Chen JP, Yang BK (1992) J Food Sci 57:781
20. Fukui S (1981) Appl Microbiol Biotechnol 22:199
21. Vorlová S, Bornscheuer U, Gatfield I, Hilmer JM, Bertram HJ, Schmid R (2002) Adv Synth Catal 344:1152
22. Serra S, Fuganti C, Brenna E (2005) Trends Biotechnol 23:193
23. Nozaki M, Suzuki N, Tsuruta H (2000) In: Schieberle P, Engel KH (eds) Frontiers of flavour science. Deutsche Forschungsanstalt für Lebensmittelchemie, Garching, p 426
24. Franssen M, Alessandrini L, Terraneo G (2002) Pure Appl Chem 77:273
25. Valivety RH, Halling PJ, Macrae AR (1992) Biochem Biophys Acta 218:222
26. Welsh FW, Williams RE (1990) Enzyme Microb Technol 12:743
27. Phillips RS (1992) Enzyme Microb Technol 14:417
28. Tawaki S, Klibanov AM (1992) J Am Chem Soc 114:1882
29. Cao SG, Liu ZB, Feng Y, Ma L, Ding ZT, Cheng YH (1992) Appl Biochem Biotechnol 32:1
30. Gatfield I (1992) In: Patterson RLS, Charlwood BV, MacLeod G, Williams AA (eds) Biotransformation of flavours. Royal Society of Chemistry, Cambridge, p 171
31. Laboret F, Perraud R (1995) In: Etiévant P, Schreier P (eds) Bioflavour '95. INRA, Paris, p 355
32. Laboret F, Perraud R (1999) Appl Biochem Biotechnol 82:185

33. Chowdary GV, Ramesh MN, Prapulla SG (2000) Process Biochem 36:331
34. Bruno L, De Lima Filho J, De M. Melo E, De Castro H (2004) Appl Biochem Biotechnol 113–116:189
35. De BK, Chatterjee T, Bhattacharyya DK (1999) J Am Oil Chem Soc 76:1501
36. Langrand G, Rondot N, Triantaphylides C, Baratti J (1990) Biotechnol Lett 12:581
37. Nakagawa H, Watanabe S, Shimura S, Kirimura K, Usami S (1998) World J Microb Biot 14:219
38. Larios A, García H, Oliart RM, Valerio-Alfaro G (2004) Appl Microbiol Biotechnol 65:373
39. Stamatis H, Christakopoulos P, Kekos D, Macris BJ, Kolisis FN (1998) J Mol Catal B 4:229
40. Bourg-Garros S, Razafindramboa N, Pavia A (1998) Biotechnol Bioeng 59:495
41. Bourg-Garros S, Razafindramboa N, Pavia A (1998) Enzyme Microb Technol 22:240
42. Shieh CJ, Chang SW (2001) J Agric Food Chem 49:1203
43. Leszczak JP, Tran-Minh C (1998) Biotechnol Bioeng 60:356
44. Gatfield I, Hilmer JM, Bertram HJ (2001) Chimia 55:397
45. Ley J, Hilmer JM, Weber B, Krammer G, Gatfield I, Bertram HJ (2004) Eur J Org Chem 24:5135
46. Winterhalter P, Schreier P (1994) Flavour Fragrance J 9:281
47. Roscher R, Steffen JP, Herderich M, Schwab W, Schreier P (1996) J Agric Food Chem 44:1626
48. Ney I, Jäger E, Herderich M, Schreier P, Schwab W (1996) Phytochem Anal 7:233
49. Herderich M, Roscher R, Schreier P (1996) In: Takeoka G, Teranishi R (eds) ACS symposium series 637. American Chemical Society, Washington, p 261
50. Schwab W, Schreier P (1988) J Agric Food Chem 36:1238
51. Vasserot Y, Arnaud A, Galzy P (1993) Bioresour Technol 43:269
52. Baumes R, Brillouet JM, Tapiero C, Bayonove C, Cordonnier (1991) EP 416713 A1
53. Cabaroglu T, Selli S, Canbas A, Lepoutre JP, Günata Z (2003) Enzyme Microb Technol 33:581
54. Brunerie P (1993) WO 9304597
55. Di Cesare LF, Nani R, Forni E, Scotto P (1993) Flüss Obst 60:254
56. Hugueny P, Dumont B, Ropert F, Belin JM (1995) In: Etiévant P, Schreier P (eds) Bioflavour '95 INRA, Paris, p 269
57. Lopez V, Lindsay RC (1993) J Agric Food Chem 41:446
58. de Roode M, Oliehoek L, van der Padt A, Franssen M, Boom R (2001) Biotechnol Prog 17:881
59. Anderson DA (1993) CHEMTECH 23:45
60. Davies G, Gloster T, Henrissat (2005) Curr Opin Struct Biol 15:637
61. Wu YF, Cadwallader K (2002) J Agric Food Chem 50:2900
62. Wong CH, Whitesides GM (1994) Enzymes in synthetic organic chemistry. Pergamon, Oxford
63. Corey EJ (1987) In: Bartmann W, Sharpless KB (eds) Stereochemistry of organic and bioorganic transformations. VCH, Weinheim, p 1
64. Boyington JC, Gaffney BJ, Amzel M (1993) Science 260:1482
65. Minor W, Steczko J, Bolin JT, Otwinowski Z, Axelrod B (1993) Biochemistry 160:24
66. Muller BL, Dean C, Whitehead IM (1995) In: Etiévant P, Schreier P (eds) Bioflavour '95. INRA, Paris, p 339
67. Brunerie P (1989) Fr Patent 2,652,587

68. Muller, Gauthier A, Dean C, Kuhn JC (1993) WO Patent 9324-644
69. Gatfield I (1995) Perfumer Flavorist 20:5
70. Whitehead IM, Slusarenko AJ, Waspi U, Gaskin DJ, Brash AR, Tijet N (1999) WO Patent 99-IB807
71. Whitehead IM, Slusarenko AJ, Waspi U, Gaskin DJ, Brash AR, Tijet N (2002) US Patent 2002-42991
72. Brash AR, Tijet N, Whitehead IM (2001) US Patent 2000-537357
73. Berger RG, Drawert F, Tiefel P (1992) Naturally-occurring flavours from fungi, yeasts, and bacteria. In: Patterson RLS, Charlwood BV, MacLeod G, Williams AA (eds) Bioformation of flavours. Royal Society of Chemistry, Cambridge, p 21
74. Kermasha S, Perraud X, Bisakowski B, Husson F (2002) J Mol Catal B 19–20:479
75. Waldmann D, Schreier P (1995) J Agric Food Chem 43:626
76. Ropert F, Dumont B, Belin JM (1995) In: Etiévant P, Schreier P (eds) Bioflavour '95. INRA, Paris, p 275
77. van Haandel MJH, Sarabèr FCE, Boersma MG, Laane C, Fleming Y, Weenen H, Rietjens IMCM (2000) J Agric Food Chem 48:1949
78. Taupp M, Harmsen D, Heckel F, Schreier P (2005) J Agric Food Chem 53:9586
79. Adam W, Hoch U, Lazarus M, Saha-Möller CR, Schreier P (1996) J Am Chem Soc 117:11898
80. Zorn H, Langhoff S, Scheibner M, Nimtz M, Berger RG (2003) Biol Chem 384:1049
81. Huang R, Christenson P, Labuda IM (2001) US Patent 6,200,786
82. Yoshida A, Takenaka Y, Tamaki H, Frébort I, Adachi O, Kumaga H (1997) J Ferment Bioeng 84:603
83. van den Heuvel RHH, Fraaije MW, Laane C, van Berkel W (2001) J Agric Food Chem 49:2954
84. van den Heuvel RHH, van den Berg WAM, Rovida S, van Berkel W (2004) J Biol Chem 32:33492
85. Do H, Sato T, Kirimura K, Kino K, Usami S (2002) J Biosci Bioeng 94:119
86. Kreye L, Lüke HJ, Scheper T (1995) Proceedings of the 7th European congress on biotechnology, Nice, poster MAP 201
87. Raab T (2004) PhD thesis, University of Wuerzburg, Germany
88. Schalk M, Clark A (2004) WO Patent 2003-IB5072
89. Schalk M (2005) WO Patent 2003-IB6459

23 Microbial Flavour Production

Jens Schrader
DECHEMA e.V., Karl-Winnacker-Institut,
Biochemical Engineering,
Theodor-Heuss-Allee 25, 60486 Frankfurt, Germany

23.1 Introduction and Scope

For thousands of years microbial processes have accompanied mankind playing the decisive but unrecognised role of producing more flavourful foods and beverages such as bread, cheese, beer, wine and soy sauce. It was in 1923 that the first scientific review on microbial flavours appeared [1]. With the dynamic development of modern analytical techniques in the middle of the twentieth century when isolation, chromatographic separation and structural identification of volatiles became routine, the basis for a more systematic elucidation of microbial flavour generation was given. Research in the last decades has led to a tremendous increase in knowledge of microbial and enzymatic flavour generation which has been frequently reviewed [2–8] and was reviewed in several multiauthor works dealing with this topic [9, 10] and one comprehensive book exclusively dedicated to aroma biotechnology published in 1995 [11].

Nowadays, biotechnological production of flavour compounds is a mature discipline in the chemical industry, with an estimated 100 molecules in the market produced by enzymatic or microbial processes [7]. The predominant driving force was and still is the fact that flavour compounds produced from natural raw materials by microbial or enzymatic methods can be labelled 'natural' in accordance with European and US legislation, thereby satisfying the unbroken consumer trend towards all 'bio' or 'natural' products in the food sector. By contrast, the involvement of chemical means leads to the less appreciated labels 'nature-identical' (EC Flavour Directive 88/388/EEC) or 'artificial' (US Code of Federal Regulations 21 CFR 101.22) for flavours not occurring in nature. This from the scientist's viewpoint rather surprising situation paved the way for 'self-sufficient' research on biocatalytic and fermentative flavour production, which started several decades ago. These research activities steadily expanded to almost all natural key flavour compounds which cannot be economically provided by classic isolation from their natural sources, e.g. by extraction or distillation, owing to too low concentrations. This happened although many of the target compounds could and still can be produced in a more efficient and less expensive way by chemical syntheses because the natural flavours achieve significantly higher market prices of up to 2 orders of magnitude. For 2005 the total worldwide flavour and fragrance market was estimated to be about US $16.0 billion, with a growth in local currencies of about 3% in the same year (http://www.leffingwell.com/top_10.htm). In 2001 the percentage of natural flavours of all added flavours amounted to 90% (EU) and 80% (USA) in beverages, to 80% (EU and USA) in savoury foods, and to 50% (EU) and 75% (USA) in dairy foods [6].

Nevertheless, enhanced competitive pressure and a less distinguishing food labelling legislation ('natural flavouring' vs. 'flavouring' in the EU) cause companies to increasingly evaluate natural flavours by their production costs in comparison with the costs of their chemically synthesised counterparts and in most cases do not leave room for high extra charges for the naturals anymore. Instead, three characteristics of most biotechnological processes are increasingly influencing academic as well as industrial considerations: biocatalytic reactions usually (1) are highly selective (chemo, regio, stereo), (2) start from natural raw materials/renewable resources and (3) are environmentally friendly and sustainable (Table 23.1). Especially the fact that evolution has optimised biological systems on the basis of metabolism of natural organic molecules makes biotechnology an outstanding technology for the development of sustainable production processes—outdoing classic syntheses starting from petrochemicals—which will have an increasing share in the chemical industry of the twenty-first century. The nowadays much-cited discipline 'white biotechnology' as a synonym for industrial biotechnology which bundles lots of economic and ecological hopes has already blossomed into numerous examples of efficient bioprocesses in the area of microbial flavour synthesis owing to the very special situation of a virtual non-competitiveness against chemical synthesis. On the basis of the long and sound research tradition in aroma biotechnology, novel approaches combining the emerging opportunities given by modern molecular biology including '-omics' and metabolic engineering technologies, and advanced bioprocess engineering, e.g. in situ product removal strategies, will definitely lead to even more biotechnologically produced flavours in the future.

The scope of this chapter is to give a comprehensive overview of microbial processes used in industry or microbial strategies investigated in application-oriented research for the production of single flavour compounds. The chapter is subdivided by structural substance classes and exclusively focuses on compounds produced by fermentation or whole-cell biocatalysis. Biotransformation with isolated enzymes, a mature discipline of aroma biotechnology, is excluded since a separate chapter of this book is dedicated to this topic. It was the intention to primarily treat those examples within each substance class discussed which are already being industrially applied or where significant product concentrations have been reported and, thus, the research results justify the assumption of a short-term to medium-term technical realisation. Traditional non-volatile flavour compounds are included, because some of them, e.g. monosodium glutamate (MSG) or citric acid, are industrial bulk products with market volumes exceeding 1,000,000 t $year^{-1}$. These examples illustrate extremely well the beneficial impact of biotechnology on the chemical industry as commodities can be produced from renewable resources based on a sustainable technology. This chapter does not cover academic research activities in the field of biocatalytic flavour and fragrance synthesis, which are predominantly carried out on an

Table 23.1 Driving forces to use biotechnological methods for flavour production (adapted from [270])

'Market pull'	'Technical push'
Increasing consumers' demand for 'organic', 'bio', 'healthy', 'natural'	High chemo-, regio- and stereo-selectivities of biocatalytic systems
Industrial dependence on distant (frequently overseas) raw materials, undesired/limited raw materials	Sustainability of bioprocesses
Search for natural character-impact compounds	Improved biocatalysts by evolutionary and rational enzyme engineering and metabolic engineering
Search for natural flavour compounds with additional functionalities (e.g. antimicrobial properties)	Improved downstream processing, especially in situ product-recovery techniques

analytical scale to elucidate sensory properties of enantiomerically pure compounds not or hardly achievable by chemical catalysis [12]. Also, genetic engineering and biochemical elucidation of microbial flavour generation pathways are only discussed if necessary for understanding or if not already covered in Chap. 26 by Schwab. Finally, as only microbially produced single-flavour molecules are discussed, traditional food fermentation processes and the impact of selecting and engineering starter cultures and/or fermentation conditions (e.g. alcoholic beverages, dairy, meat and bakery products) are not covered. The interested reader is referred to respective reviews in this field [13–16].

23.2 Characteristics of Microbial Flavour Production

Although for a multitude of microorganisms the metabolic potential for de novo flavour biosynthesis is immense and a wide variety of valuable products can be detected in microbial culture media or their headspaces, the concentrations found in nature are usually too low for commercial applications. Furthermore, metabolic diversity often leads to a broad product spectrum of closely related compounds. Exceptions to the rule can be found where the flavour compounds are derived from primary metabolism as is the case for some of the non-volatiles (e.g. glutamic acid, citric acid). Therefore, the biocatalytic conversion of a structurally related precursor molecule is often a superior strategy which allows the accumulation of a desired flavour product to be significantly enhanced. As a prerequisite for this strategy, the precursor must be present in nature and its isolation in sufficient amounts from the natural source must be easily feasible in an economically viable fashion. Additionally, if product and precursor are closely related with respect to their physicochemical properties, a selective product recovery during downstream processing becomes a major issue for the bioprocess development. Many of the industrially relevant microbial flavour production processes follow this 'precursor approach' (e.g. vanillin from ferulic acid or eugenol, 4-decanolide from ricinoleic acid, 2-phenylethanol from L-phenylalanine). Besides the problems arising from metabolic diversity, the cytotoxicity of the flavour products and often also of their precursors is another big hurdle during bioprocess development. Here, very often in situ product recovery or sequential feeding of small amounts of precursor becomes essential to improve the overall performance of a bioprocess and to render it economically viable. Owing to their hydrophobicity, flavour compounds preferentially partition to lipid structures, which makes cellular membranes the main target for product accumulation during microbial processes. The flavour molecules negatively influence the cell physiology by enhancing the membrane fluidity, eventually leading to collapsing transmembrane gradients and, consequently, to the loss of cell viability [17]. An empirical correlation was found between the $\log P_{\text{octanol/water}}$ value of an organic solvent and its corresponding partition coefficient for the cell membrane–water system ($\log P_{\text{membrane/water}}$) [18]:

$$\log P_{\text{membrane/water}}=0.97\log P_{\text{octanol/water}}-0.64.$$

With this equation the actual membrane concentration of a hydrophobic compound can be estimated if its concentration in the water phase is known. For instance, limonene, a hydrophobic precursor in many biotransformations to produce monoterpenoid flavour compounds ($\log P_{\text{octanol/water}}=4.5$), would accumulate within membranes in concentrations of up to 530 mM if it is present in the water phase at saturation concentration of only 0.1 mM [19]. This concentration would clearly not allow conventional microorganisms to survive.

The design of recombinant microorganisms for the improved production of natural flavour molecules is being intensively investigated in academic and industrial research since, of course, it can provide the same economic benefits as in other industrial applications of modern biotechnological production processes, e.g. in the pharmaceutical industry. Although genetic engineering in food-related applications has been the subject of a controversial public discussion for quite some time, the fact that in aroma biotechnology genetically modified organisms are used as biocatalysts which are completely separated from the volatiles during the product-recovery step raises hope that this technique will also be applicable in industrial flavour production processes in the future. Further improvements will certainly be triggered by the enormous progress currently being made in the field of total genome sequencing. The time needed to determine complete microbial genomes has dramatically decreased during the last few years. Among the microorganisms already sequenced, several bacteria and fungi can be found which are valuable candidates with respect to food and flavour applications, e.g. *Bacillus subtilis*, *Brevibacterium linens*, *Clostridium acetobutylicum*, *Corynebacterium glutamicum*, *Gluconobacter oxydans*, *Lactococcus lactis*, *Pseudomonas putida*, *Streptococcus thermophilus*, *Saccharomyces cerevisiae*, *Yarrowia lipolytica* and *Aspergillus niger*.

Table 23.2 summarises the main issues of microbial flavour production and how they may be addressed by biotechnological methods.

Table 23.2 Main drawbacks during microbial flavour production and biotechnological strategies to address them

Characteristics	Biotechnological strategy	Exemplary product
Formation of unwanted by-products owing to complex metabolic pathways	Overexpression of key genes of the synthetic pathways	3-Methylbutyl acetate [86] (Sect. 23.4.1.4)
	Heterologous gene expression/use of engineered enzymes	Cinnamyl alcohol [136] (Sect. 23.4.2), verbenol [187, 188] (Sect. 23.4.3.2)
	Knockouts of genes involved in product degradation	Vanillin [89] (Sect. 23.4.2)
	'Precursor approach' instead of de novo biosynthesis	2-Phenylethanol [120] (Sect. 23.4.2), raspberry ketone [103, 106] (Sect. 23.4.2)
	Screening; enrichment cultures	Perillyl alcohol [184] (Sect. 23.4.3.2), 10-hydroxypatchoulol [207] (Sect. 23.4.3.3)
	Subsequent biotransformation converting a by-product to the desired product	4-Decanolide [222] (Sect. 23.4.4)
Toxic properties of the flavour compounds produced	In situ product recovery by:	
	Adsorption, e.g. on XAD resins	6-Pentyl-α-pyrone [241] (4.4)
	Stripping and adsorption	C2–C5 alkyl esters [85] (Sect. 23.4.1.4), furfurylthiol [256] (Sect. 23.4.5)
	Extraction (two-phase bioprocess)	2-Phenylethanol [120] (Sect. 23.4.2), phenylacetaldehyde [132] (Sect. 23.4.2)
	Membrane-based processes	Isonovalal [181] (Sect. 23.4.3.2), 2-phenylethyl acetate [119] (Sect. 23.4.2)
	Resting cells instead of growing ones	Acetaldehyde [53] (Sect. 23.4.1.2)
	Product-tolerant strains	Vanillin [90] (Sect. 23.4.2)
Toxic properties of the precursor molecules	Sequential precursor feeding	4-Octanolide [234] (Sect. 23.4.4), 3-methylbutyl acetate [48] (Sect. 23.4.1.4), carboxylic acids [39] (Sect. 23.4.1.1)
	On line monitoring of precursor/bioactivity	Limonene transformation products [108, 142] (Sect. 23.4.3.1)
	Immobilisation of microorganisms	Propanoic acid [41] (Sect. 23.4.1.1), phenylacetic acid [133] (Sect. 23.4.2)
	Two-phase bioprocess with an organic solvent as the precursor reservoir	5-Decanolide [231] (Sect. 23.4.4), 4-hexanolide [236] (Sect. 23.4.4)
	Resting cells instead of growing ones	Carvone [170, 172] (Sect. 23.4.3.2)
	Precursor-tolerant (solvent-tolerant) strains	Perillic acid [156] (Sect. 23.4.3.2), carvone [171, 172] (Sect. 23.4.3.2)
	Fungal spores instead of mycelia	Methylketones [69, 70] (Sect. 23.4.1.3)

In Fig. 23.1 the main biotechnological routes to flavour molecules are summarised, illustrating that a wide range of microbial (and enzymatic) processes have been developed exclusively relying on the bioconversion of natural precursors and/or the cultivation on non-fossil carbon sources, such as glucose and fatty acids. These characteristics make aroma bioprocesses a prime example of a sustainable industrial technology based on renewable resources.

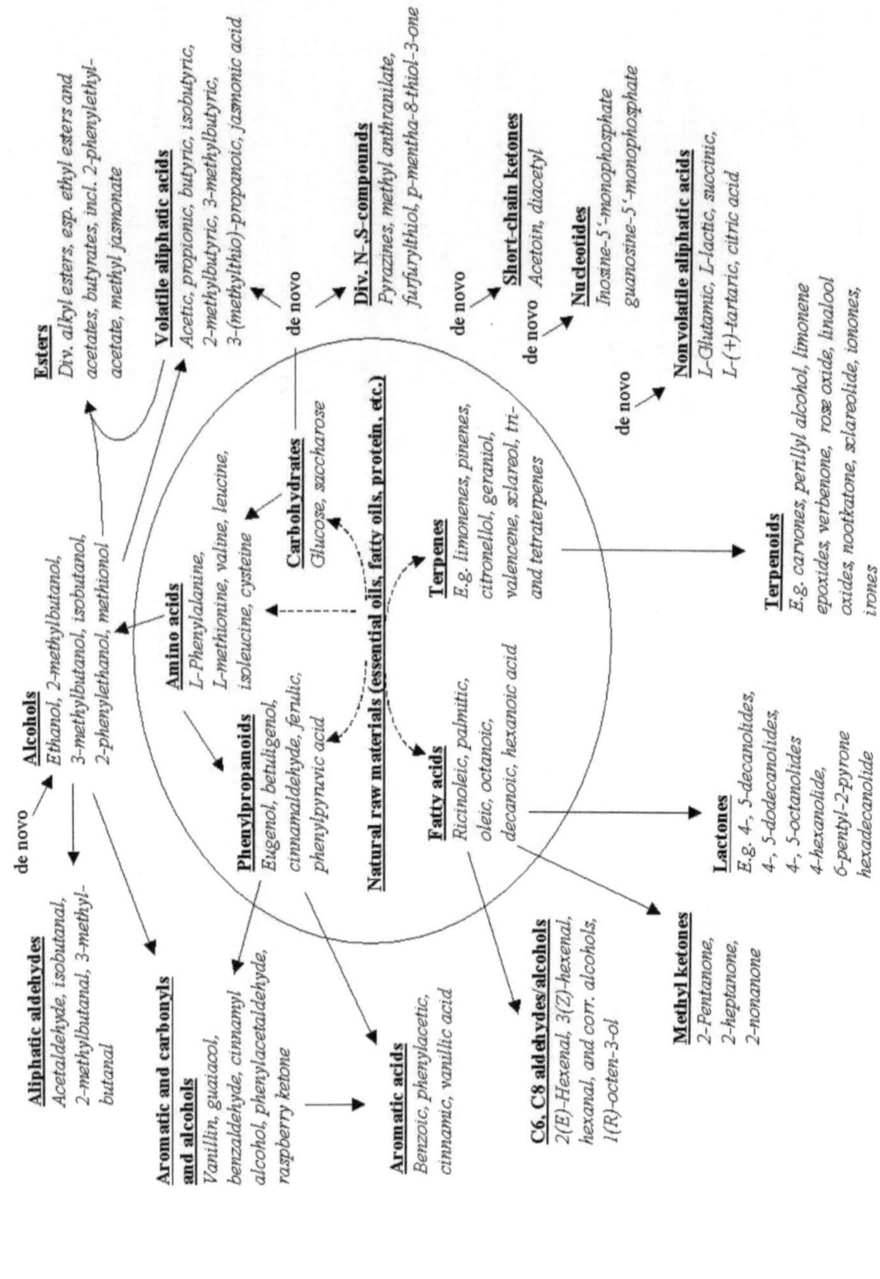

Among the natural flavour molecules produced with microorganisms are some real bulk products, such as L-glutamic acid and citric acid manufactured on the million-ton scale, but the majority of the target compounds are produced for highly specific applications and thus are rather niche products with market volumes below 1 t $year^{-1}$. Here, industry avoids costly research and development effort to establish more sophisticated processes owing to the limited market volume of these products. Nevertheless some natural flavours which have a broader application are produced in amounts of around one to several tons per year, such as vanillin, 2-phenylethanol and 4-decanolide. These flavour compounds have an increasing market share owing to steadily improved bioprocesses: for instance, the price for the peach-like 4-decanolide dropped from about US $20,000 per kilogram in the 1980s to about US $300 per kilogram in 2004 [8, 20]. Table 23.3 summarises some natural flavour compounds currently being produced by microbial processes in industry. The microbial production of these and other flavour compounds is discussed in more detail in the following sections. When a flavour compound is mentioned for the first time within the respective section it is written in bold letters.

23.3 Non-volatile Flavour Compounds

Non-volatile flavour compounds in the sense of this chapter are defined as molecules which cause one of the sensory impressions sweet, bitter, salty, sour or umami (Fig. 23.2).

By market volume the most important flavour molecule is **L-glutamic acid.** In 2004, the worldwide annual MSG production was estimated to be amount 1,500,000 t [21]. The amino acid is extensively used as taste enhancer, frequently in conjunction with nucleotides, a flavour impression which is also referred to as 'umami', a term derived from the Japanese meaning deliciousness or a savoury or palatable taste. MSG is manufactured by aerobic cultivation of *Corynebacterium glutamicum* on starch hydrolysates or molasses media in large-scale bioreactors (up to 500 m^3). Production strains with modified metabolic flux profiles and highly permeable cell walls for an improved product secretion are

◂ **Fig. 23.1** Microbial routes from natural raw materials to and between natural flavour compounds (*solid arrows*). Natural raw materials are depicted within the *ellipse*. Raw material fractions are derived from their natural sources by conventional means, such as extraction and hydrolysis (*dotted arrows*). *De novo* indicates flavour compounds which arise from microbial cultures by de novo biosynthesis (e.g. on glucose or other carbon sources) and not by biotransformation of an externally added precursor. It should be noted that there are many more flavour compounds accessible by biocatalysis using free enzymes which are not described in this chapter, especially flavour esters by esterification of natural alcohols (e.g. aliphatic or terpene alcohols) with natural acids by free lipases. For the sake of completeness, the C6 aldehydes are also shown although only the formation of the corresponding alcohols involves microbial cells as catalysts. The list of flavour compounds shown is not intended to be all-embracing but focuses on the examples discussed in this chapter

Table 23.3 Some microbially produced flavour compounds and corresponding bioprocess features

Product	Precursor	Microorganism	Process data	Remarks	References
L-Glutamic acid	–	*Corynebacterium glutamicum*	150 g L^{-1}, 60 h, 1.500,000 t $year^{-1}$	Aerobic cultivation; up to 500-m^3 scale; mutants with highly permeable cell walls	[21, 22]
Citric acid	–	*Aspergillus niger*	>200 g L^{-1}, 9–12 days, 1,000,000 t $year^{-1}$; yield >95%	Downstream processing by precipitation as calcium citrate	[21, 22, 25]
Acetic acid	Ethanol	*Acetobacter, Gluconobacter*	'Vinegar' with 10 to >20 %, >190,000 t $year^{-1}$; yield ~98%	Aerobic cultivation at 100-m^3 scale; Frings aerator for high oxygen transfer rates	[21, 22, 38]
L-Lactic acid	-	*Lactobacillus*	210 g L^{-1}, 140,000 t $year^{-1}$; yield >90%	More than 100-m^3 scale; recovery of lactic acid by salt splitting technology	[21, 22, 31]
Vanillin	Ferulic acid	*Amycolatopsis, Streptomyces*	Up to 18 g L^{-1}, 50 h, 1–10 t $year^{-1}$	In situ product recovery by crystallisation at more than 10 g L^{-1} possible	[8, 90, 91]
(*Z*)-3-Hexenol ('leaf alcohol')	Linolenic acid	Soy lipoxygenase + plant hydroperoxide lyase + baker's yeast	4 g kg^{-1}, 5–10 t $year^{-1}$ (also by isolation from plant oils)	Addition of baker's yeast to obtain the alcohol; without yeast the aldehyde is the major product	[60, 66]
4-Decanolide (γ-decalactone)	Ricinoleic acid	*Yarrowia lipolytica*	11 g L^{-1}, 55 h, several tons per year	Final acidification and temperature increase effect cyclisation of all 4-hydroxydecanoic acid to the corresponding lactone	[222, 224, 228]
2-Phenylethanol	L-Phenylalanine	Diverse yeasts; e.g. *Saccharomyces* and *Kluyveromyces*	>10 g L^{-1}, 30 h, 0.5–1 t $year^{-1}$	Fed-batch cultivation; in situ product recovery by two-phase system with more than 25 g L^{-1} in the organic phase possible	[109, 120]
Short-chain carboxylic acids, e.g. 2-, and 3-methylbutyrate	Fusel alcohols	*Gluconobacter, Acetobacter*	Up to 95 g L^{-1}, 72 h	Two-step cultivation: biomass + bioconversion period; used as flavour acids but also for ester syntheses	[39]

used in a controlled bioprocess and up to 150 g L^{-1} L-glutamic acid is obtained in 60 h [22].

Citric acid is another prominent biotechnological bulk product of the chemical industry. About 1,000,000 t was produced in 2004 [21]. Citric acid is used in different industrial areas: in the food sector, the odourless acid is used for its pleasant acid taste and as a preservative [22, 23]. For industrial production, submerged fed-batch cultivation of *Aspergillus niger* on starch hydrolysates or cheap sucrose sources, e.g. molasses, in large-scale stirred or tower fermenters (50–1,000 m^3) is preferred and final concentrations of more than 200 g L^{-1} can be achieved after 9–12 days [22, 24]. Special cultivation conditions, such as maintaining a high concentration of a rapidly consumable carbon source (more than 50 g L^{-1}), excess aeration, suboptimal phosphate concentration, Mn^{2+} limitation and a decreasing pH value falling below 3 lead to high yields of up to 95 kg of citric acid per 100 kg of supplied sugar [25]. The downstreaming includes the processing steps filtration of the cell mass, precipitation of calcium citrate by adding $Ca(OH)_2$, redissolving with sulfuric acid, filtration of $CaSO_4$ and crystallisation of citric acid.

Although citric acid is by far the most important fruit acid, the more expensive natural **L-(+)-tartaric acid** is also routinely used in beverage products for its milder sourness and as an antioxidant. Although different enzymatic approaches have been investigated, natural tartaric acid is still obtained by conventional purification from residues of wine fermentation and constitutes an estimated world market of 50,000–70,000 t $year^{-1}$ [26]. Two product categories exist: low-grade grape debris and yeast cell material containing 15–25% potassium tartrate (dregs, wine lees, *Weinhefegeläger*, *Weinhefe*) and high-grade material (cream of tartar, *Fassweinstein*) with 60–70% potassium tartrate. The latter can be purified by heating (140–150 °C) and neutralising with lime milk containing calcium sulfate. Pure L-(+)-tartaric acid is obtained after treating the calcium salt with sulfuric acid, filtering off the calcium sulfate, evaporation and crystallisation.

Lactic acid can be produced as a racemic mixture from lactonitrile, but this chemical synthesis has diminished since most applications require enantiopure **L-lactic acid** [27]. Up to 290,000 t $year^{-1}$ of L-lactic acid is currently produced biotechnologically of which 150,000 t is used for the production of polylactate, a biodegradable polymer, and 140,000 t is used in the fields textiles, leather and food [21]. In foods and beverages, lactic acid is used for its pleasant mild sour taste, e.g. as an additive, preservative or acidulant in fruit juices, syrups, jellies, or for the preparation of sourdoughs [22, 28]; therefore, lactic acid is discussed in this chapter although it shows a low volatility causing a slightly sour odour note [23]. Depending on the carbon source used, different *Lactobacillus* species are exploited, such as *L. delbrueckii*, *L. leichmannii*, *L. bulgaricus* and *L. lactis* [22]. In industrial bioreactors of more than 100 m^3, productivities of up to 3 kg m^{-3} h^{-1} are obtained. In (fed-)batch fermentations with homofermentative lactic acid bacteria, final concentrations of more than 200 g L^{-1} are obtained and high yields of more than 90% can be achieved [29–31]. The conventional purification protocol uses lime [$Ca(OH)_2$] or chalk ($CaCO_3$) as a neutralising agents (pH 5.0–6.8 depending on the respective strain used), causing the formation of calcium lactate during fermentation. Free lactic acid is formed by adding sulfuric acid, leading to gypsum as a by-product. Advanced technical concepts are under investigation, e.g. in situ product recovery by ion-exchange resins and organic solvents to overcome end-product inhibition caused by the undissociated form of lactic acid, or electrodialysis for selective product recovery from fermentations in cheap but complex raw materials, such as molasses and whey, which cause purification problems [32]. Significant progress in downstream processing has been made by the so-called salt-splitting technology, i.e. the lactic acid is extracted from a lactate salt concentrate by an amine-based organic solvent (forming a trialkylamine lactate in the organic phase) from which it can be recovered by re-extraction with water; another salt-splitting technology is based on a water-splitting electrodialysis with bipolar membranes [27].

Succinic acid has found some use in flavour compositions to introduce tartness which is not achievable by means of conventional acids [33]. It is manufactured chemically by hydrogenation of fumaric or maleic acid [26] but its use in food applications would probably increase significantly once an economically viable biotechnological process for natural succinic acid production from glucose has been established. Currently, research activities in this field are stimulated by the fact that succinic acid can serve as an ideal platform building block for a variety of commodity and specialty chemicals, and is thus a perfect candidate for the biorefinery concept based on renewable resources, e.g. starch hydrolysates [34]. With anaerobic bacteria, especially *Actinobacillus succinogenes* and *Anaerobiospirulum succiniciproducens*, which tolerate high product concentrations, 80 to more than 100 g L^{-1} succinate can be produced [34, 35]. Strains of *Actinobacillus succinogenes* with enhanced product tolerance were selected via screening on fluoroacetate media. For *Anaerobiospirulum succiniciproducens*, a continuous bioprocess was described using ultrafiltration for cell recycling and a monopolar electrodialysis unit for concentrating the permeate solution.

The most important of the aforementioned flavour-enhancing nucleotides are **inosine 5′-monophosphate** (IMP) and **guanosine 5′-monophosphate** (GMP), of which about 2,000 and 1,000 t $year^{-1}$ are produced by biotechnological processes worldwide [22] and which are used as their disodium salts. The nucleotides contribute to the flavour-enhancing effect brought into food by yeast hydrolysates. Different biotechnological strategies have been developed for the production of pure nucleotides:

1. *Candida utilis* is grown to high biomass concentrations and the extracted RNA is subsequently hydrolysed into the four 5′ nucleotides adenosine 5′-monophosphate (AMP), GMP, cytidine and uridine 5′-monophosphate by crude nuclease P1 from *Penicillium*; the desired nucleotides are isolated by ion-exchange chromatography and AMP is converted to IMP by adenyl deaminase from *Aspergillus* [22, 36].
2. The desired nucleotides are produced directly by fermentation in concentrations above 30 g L^{-1}: IMP or inosine, which can be chemically converted into

IMP, is synthesised with mutants of *Bacillus subtilis* or *Corynebacterium ammoniagenes*. Xanthosine 5′-monophosphate is produced with *Corynebacterium* or *Bacillus* and subsequently converted into GMP by *Bacillus* and other strains [6]. Alternatively, another related compound, 5′-amino-4-imidazole carboxamide-1-riboside-5′-phosphate, is produced by *Bacillus megaterium* and chemically converted into GMP [22, 36].

Certain polypeptides resulting from protease digestion of proteins contribute to the typical taste of savoury foods. The DNA sequence coding for an octapeptide known as **beefy meaty peptide** was cloned into yeast as a fusion with the yeast α factor to be secreted as free octapeptide into the medium which facilitated its recovery [37]. Alternatively, intracellular expression of tasty peptides

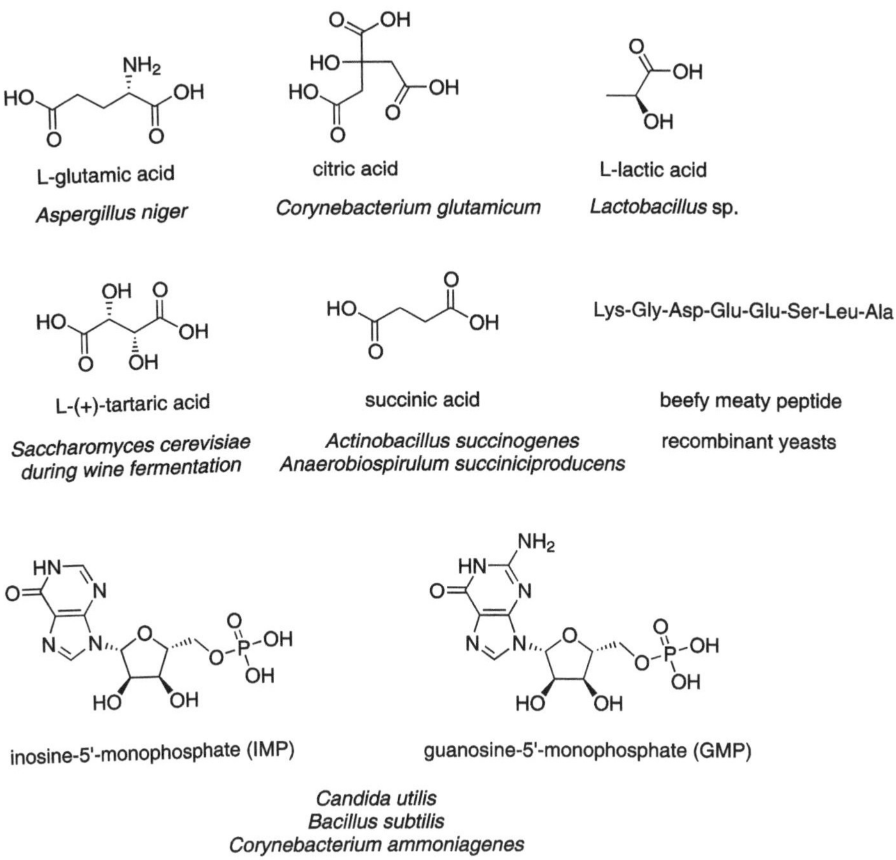

Fig. 23.2 Some non-volatile flavour compounds produced with microorganisms

cloned into yeasts may lead to yeast extracts with improved flavouring characteristics.

23.4 Volatile Flavour Compounds

23.4.1 Aliphatic Compounds

23.4.1.1 Carboxylic Acids

Fermentatively produced **acetic acid**, which is used as **'vinegar'** in food applications for its freshness, sourness and preservative properties, amounts to 190,000 t annually worldwide [21]. Today, the strictly aerobic cultivation of acetic acid bacteria, i.e. *Acetobacter* or *Gluconobacter* strains, in a mash containing ethanol, initial acetic acid and nutrients is mainly performed by submerged cultivation in specially designed stirred-tank reactors of about 100 m^3 (Frings Acetator®) [38]. A key feature is a Frings aerator, which is a self-aspirating rotor–stator system leading to high oxygen transfer rates at low aeration rates (approximately 0.1 vvm), thereby reducing the loss of the volatile ethanol and acetic acid via the exhaust air. The basic bioprocess leads to final acetic acid concentrations of approximately 10%. Repeated fed-batch processes have been developed for product concentrations of more than 15% and have to follow a carefully designed protocol to maintain optimum conditions concerning the oxygen, ethanol and acetic acid concentrations during cultivation. Final yields of up to 98% are common. In a two-stage process scheme, a portion of a first fermenter, which is replenished by new mash, is transferred into a second fermenter, where an almost complete ethanol oxidation is achieved—a final concentration of more than 20% (corresponding to a productivity of up to 50 L m^{-3} day^{-1} [22]) can be obtained, which is necessary for the canning industry. The biochemical principle of this primary metabolism is the stepwise oxidation of ethanol via acetaldehyde to acetic acid by the action of two pyrolloquinoline quinone dependent enzymes bound to the cytoplasmic membrane, alcohol dehydrogenase and aldehyde dehydrogenase, which feeds electrons into the respiratory chain of the organisms.

The strong oxidative capabilities of acetic acid bacteria are also harnessed for the production of other flavour acids from their corresponding alcohols, such as **propanoic acid, butanoic acid, 2-methylpropanoic acid, 2-methylbutanoic acid** and **3-methylbutanoic acid** (Scheme 23.1).

These natural acids synthesised from natural alcohols have market prices of less than €100 per kilogram and are of great importance to the flavour industry either because of their intense smell and sour taste or as substrates for enzymatic syntheses of flavour esters [39]. The industrial process based on *Gluconobacter oxydans* DSM 12884 established by the German company Haarmann & Reimer (now Symrise) obtains molar yields above 90% and final product concentrations of more than 80 g L^{-1} within 65–92 h (Table 23.3) [39]. The submerged cultivation process at 0.5 vvm aeration, 30 °C and 500 rpm usually starts with a batch fermentation period of 16–22 h for biomass growth and is followed by a bioconversion period where the alcohol is fed continuously into the reactor. By this means toxic effects of the alcohol and its loss via the exhaust air are minimised.

A microbial resolution of racemic 2-methylbutanoic acid was performed with a novel *Pseudomonas* sp. strain isolated from soil [40]. The strain was selected by screening on a medium containing racemic 2-methylbutanoic acid as the sole carbon source. The strain preferentially catabolised the fruity (*S*)-2-methylbutanoic acid, thereby yielding optically pure **(*R*)-2-methylbutanoic acid** which has a distinct odour described as being cheesy, sweaty and sharp.

It should be noted here that **propanoic** and **butanoic acid** can also be efficiently synthesised as metabolic end products of classic anaerobic fermentations on different sugars with various microorganisms, such as *Clostridium*, *Butyribacterium*, *Propiobacterium* and *Lactococcus*, which was investigated decades ago as a spin-off of acetone–butanol fermentation research [2]. Bioprocess and genetic engineering methods, e.g. in situ product removal, cell immobilisation and targeted gene inactivation, can help to significantly improve productivities and final product concentrations. Recently, immobilisation of *Propionibacterium acidipropionici* ATCC 4875 in a fibrous-bed bioreactor running under fed-batch conditions led to 72 g L^{-1} propanoic acid [41]. Knocking out the *ack* gene (acetate kinase) decreased unwanted acetic acid formation by 14% [42]. The same reactor type and gene interruption strategy were successfully applied for butanoic acid production by *Clostridium tyrobutyricum* ATCC 25755 to yield about 40 g L^{-1} from both xylose or glucose as the carbon source (corresponding to 0.43 g g^{-1}) [43, 44]. Coupling the reactor to an external hollow-fibre membrane module containing Alamine 336 in oleyl alcohol for in situ product extraction dramatically enhanced product concentration and reactor productivity [45]. The extractant was continuously regenerated by stripping with NaOH in a second membrane contactor. Thus, an impressive final butanoic acid concentration of more than 300 g L^{-1} and a productivity of 7.37 g L^{-1} h^{-1} were achieved.

In plants the 13-hydroperoxide produced from linolenic acid by lipoxygenase (Sect. 23.4.1.2) can be converted to the allene oxide by allene oxide synthase followed by cyclisation, reduction and β-oxidation to form **jasmonic acid**, an important plant growth factor; the corresponding **methyl jasmonate** is a valuable flavour and fragrance compound that imparts a sweet-floral, jasmine-like note [46]. Recently, a patent described the use of *Diplodia gossypina* ATCC 10936 for the production of natural jasmonic acid [47]. With submerged cultures, up to 1.5 g L^{-1} jasmonic acid was obtained after 11 days of incubation; the addition of 10-oxo-8-*trans*-decenoic acid, a hormone stimulating mycelial growth, proved to be advantageous; methyl jasmonate was obtained by autoclaving the

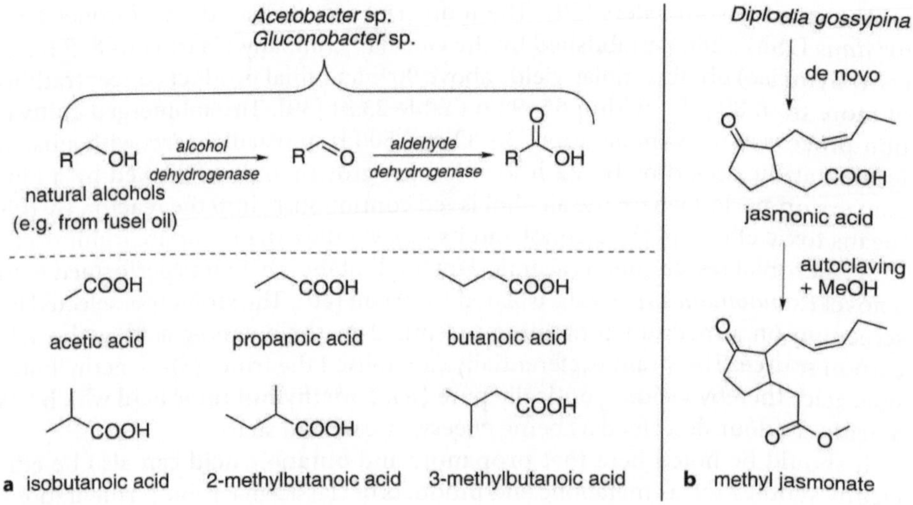

Scheme 23.1 **a** Short-chain flavour acid production from natural alcohols by acetic acid bacteria. **b** Jasmonic acid and methyl jasmonate production with *Diploida gossypina*

jasmonic acid extracted from the fermentation broth in the presence of methanol (Scheme 23.1).

23.4.1.2 Alcohols and Aldehydes

By volume, **ethanol** can be viewed as the most prominent flavour-active or flavour-enhancing compound produced by biotechnology, although in the scientific literature it is usually not categorised among the flavour compounds. With respect to the focus of this chapter on bioprocesses which aim at the synthesis of single-flavour compounds and not on traditional food and beverage fermentation processes, it should only be noted here that by traditional fermentation processes about 140,000,000 t of beer and 27,000,000 t of wine are produced worldwide annually [21]. This corresponds to roughly 5,000,000 and 2,000,000 t ethanol, respectively, making it a real 'bulk' chemical among the alcohols used in the food sector. Natural raw materials, such as starch hydrolysates or molasses, are fermented with yeasts to convert these cheap sugars to ethanol, a process which is currently gaining new public attention for its promising perspectives to provide an ecologically sound fuel from renewable resources (more than 18,500,000 t $year^{-1}$ 'bioethanol' [21]). Important flavour alcohols are derived from these ethanol-producing fermentation processes. During distillation of bioethanol or spirits, a cheap by-product of the yeast metabolism can be recovered, 'fusel oil'. This fraction usually contains **2-methylbutanol, 3-methyl-**

butanol (isoamyl alcohol) and **2-methyl-1-propanol** as the main constituents in concentrations of 10–40 vol% [48], which are used directly in fruit flavour compositions or as starting materials for the biotechnological synthesis of natural flavour acids (Sect. 23.4.1.1), aldehydes or esters (Sect. 23.4.1.3). The fusel alcohols are products of the yeast's Ehrlich pathway, a three-enzyme cascade by which amino acids, here valine, leucine and isoleucine, are converted into their corresponding alcohols. This pathway, which is ubiquitous among yeasts, is described in more detail in Sect. 23.4.2.

Short-chain aliphatic aldehydes, such as **acetaldehyde**, **2-methyl-1-propanal**, **2-methylbutanal** and **3-methylbutanal (isovaleraldehyde)**, impart fruity and roast characters to flavour compositions [49]. Natural acetaldehyde is an important compound naturally occurring in a broad range of fruit flavours, essential oils and distillates; it augments fruit flavours and, for instance, it decisively contributes to the 'freshness' and 'juiciness' of foods and beverages, such as citrus juices [23, 50].

The aforementioned aldehydes can be efficiently produced by oxidation of the corresponding alcohols with alcohol oxidase (AOX) or alcohol dehydrogenase expressing microorganisms. The methylotrophic yeast *Pichia pastoris* can grow on methanol as the sole carbon and energy source using its strong alcohol oxidase (AOX) which is induced by methanol. The flavin-containing AOX naturally oxidises methanol to formaldehyde by reducing molecular oxygen to H_2O_2. This toxic intermediate is immediately cleaved into water and oxygen by the action of catalase, which co-acts with the AOX within special cell compartments, the peroxisomes. AOX has a low substrate specificity and also accepts alcohols other than methanol. Since the subsequent enzyme of the methanol degradation pathway, formaldehyde dehydrogenase, is highly specific, other alcohols are only converted into their corresponding aldehydes [51] (Scheme 23.2).

This makes *Pichia pastoris* an interesting biocatalyst for aldehyde production from alcohols in general. High product yields of *Pichia pastoris* catalysed oxidation of different short-chain alcohols have been described [51, 52]. In the case of acetaldehyde, a final concentration of 35 g L^{-1}, corresponding to an acetaldehyde productivity of 1.38 g g_{cdw}^{-1} h^{-1}, has been reported [52], although only in small-scale analytical experiments. Resting cells were used as biocatalysts in a tris(hydroxymethyl)aminomethane (Tris)–HCl buffer to alleviate product inhibition by chelating the produced aldehyde with Tris. Performing the reaction at 5 °C instead of 30 °C and using a high Tris-HCl concentration of 3 M eliminated catabolite inactivation and product inhibition effects, respectively [53]. In a semi-continuous closed-loop pressurised bioreactor, high yields of up to 130 g L^{-1} were obtained within 4 h. Stripping the volatile product via the exhaust air (where it can be recovered by cold or chemical trapping) was a simple alternative to maintain the acetaldehyde concentration below 0.2 g L^{-1} in a 10-L airlift bioreactor [54]. If biphasic reaction systems are used, also more hydrophobic long-chain aliphatic, C6–C11, and aromatic alcohols, such as benzyl alcohol, 2-phenylethanol and 3-phenyl-1-propanol, are converted [51, 55] (Sect. 23. 4.2). With another methylotrophic yeast, *Candida boidinii*, an effective alcohol oxidation process on a preparative scale was established [56]: induced yeast cells, grown in a methanol-fed prefermenter, were used as a biomass suspension of 33 g L^{-1} in 15 L phosphate buffer pH 7.5 to convert isoamyl alcohol to isovaleraldehyde with a yield of 44% and a final concentration of about 40 g L^{-1} within 7 h.

A process for aldehyde production using two bioreactors continuously operating in series was patented [57]. The first reactor was used for yeast production (e.g. *Candida boidinii, Pichia pastoris, Hansenula polymorpha, Torulopsis* sp.) on methanol, the effluent of which was directed into the second alcohol-fed reactor where the transformation to the aldehyde at a rate of 1.72 g L^{-1} h^{-1} occurred. Another method to produce aldehydes is alcohol dehydrogenation with acetic acid bacteria. In this case special mutants having low aldehyde dehydrogenase activities are needed to accumulate the aldehydes; otherwise overoxidation to the carboxylic acids dominates (Sect. 4.1.1). Such a mutant strain of *Gluconobacter oxydans* was isolated from beer and exploited to produce different short-chain aldehydes, such as acetaldehyde, **propanal**, **butanal** and isovaleraldehyde [58]. Especially the isoamyl alcohol oxidation worked very efficiently and showed a molar conversion of more than 90% and a productivity of about 1.5 g L^{-1} h^{-1} in 8 h. An integrated bioprocess with a hollow-fibre membrane contactor coupled to the bioreactor for liquid–liquid extraction with isooctane as the organic solvent was chosen to recover isovaleraldehyde continuously and thereby reduce toxic effects [59]. By this means the final product concentration was increased to 35 g L^{-1} after 16 h.

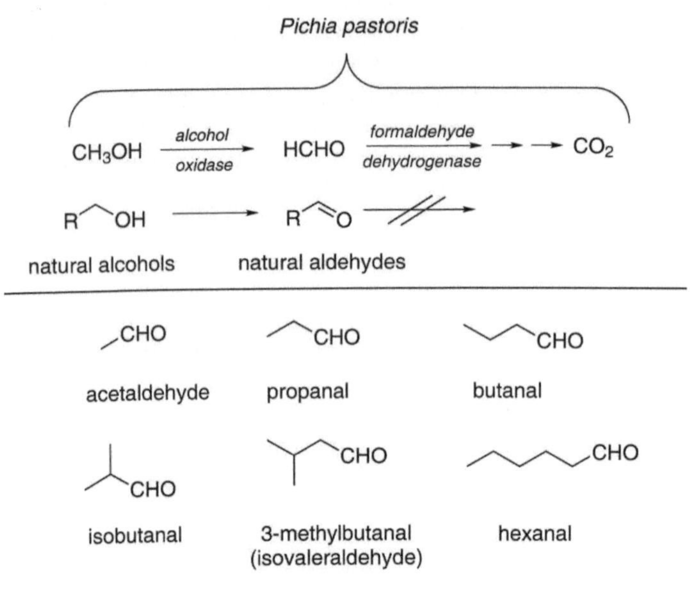

Scheme 23.2 Production of aliphatic flavour aldehydes from natural alcohols using alcohol oxidase activity of *Pichia pastoris* cells

Among the aliphatic alcohols and aldehydes, a group of structurally related C6 compounds, comprising **(*Z*)-3-hexenal**, **(*E*)-2-hexenal**, **hexanal** and their corresponding alcohols, are of great importance to the flavour industry since they are responsible for a 'green' organoleptic sensation ('green notes'). In 1995 the market for natural green notes was estimated at 5–10 t year^{-1} and US $3,000 per kilogram [60]. In nature these and also higher aliphatic aldehydes, such as C8 and C9 compounds, are derived from hydroperoxidation and cleavage of linoleic and linolenic acid by the sequential action of lipoxygenase and hydroperoxide lyase. Alcohol dehydrogenases synthesise the corresponding alcohols. The biochemistry of this reaction sequence as well as recent genetic engineering developments in this field are comprehensively described in Chap. 26. A series of quite similar biocatalytic strategies have been described based on the aforementioned biochemical principle during the last two decades [61–64]; by these methods, e.g., natural **(*Z*)-3-hexenol** is produced competitively to its isolation from peppermint oil distillation fractions [65]. A bioprocess patented by Firmenich [66] mimics the plant biochemistry starting from linoleic and linolenic acid and exploiting crude plant enzyme preparations of lipoxygenase (soya flour) and hydroperoxide lyase (e.g. guava fruit homogenate) to produce the desired aldehydes. Additionally, whole microbial cells, baker's yeast, are used as a reducing catalyst to convert the aldehydes into their corresponding alcohols, if desired (Scheme 23.3).

It is worth mentioning that the authors claimed the possibility to direct the bioprocess to each single target compound by variation of the process protocols. For instance, to obtain the desired 'leaf alcohol', (*Z*)-3-hexenol, linolenic acid is activated with lipoxygenase in the first step, but in the second step, hydroperoxide lyase and baker's yeast are added simultaneously to avoid chemical conversion of the aldehyde into its more stable isomer (*E*)-2-hexenal ('leaf aldehyde'). On the other hand, instead of adding the baker's yeast the pH is decreased to 6.5 and the temperature elevated to 50 °C to enhance leaf aldehyde formation. By this means about 4 g kg^{-1} (*Z*)-3-hexenol and 1.5 g kg^{-1} (*E*)-2-hexenal were obtained, indicating that the yields are still relatively low. Significant improvements of this process can be expected by heterologous expression of the respective enzymes, thereby enhancing and/or combining the activities within one host organism [37, 67] (Chap. 26).

In fungi a homologous reaction sequence leads to the formation of aliphatic C8 compounds, among which **(*R*)-1-octen-3-ol** is the most important one with an intensive mushroom-like odour (Scheme 23.3). In plants, the biosynthesis of the C6 volatiles is initiated after damage of the cells contacting the enzymes and the substrates which are located in different compartments and allowing molecular oxygen to penetrate into the tissue ('freshly cut green grass'). This principle has been transferred to a production process for natural mushroom flavour: after submerged fermentation of edible fungi, such as *Pleurotus* sp. or *Morchella* sp., the fungal mycelium suspension is fed into a homogeniser to break the cells, thereby inducing the lipoxygenase-catalysed reaction sequence followed by an agitation/aeration vessel to enable a high oxygen supply [68]. The biomass suspension is recirculated several times before it is harvested and freeze-dried to give a mushroom powder containing approximately 1.2 g kg^{-1} (*R*)-1-octen-3-ol,

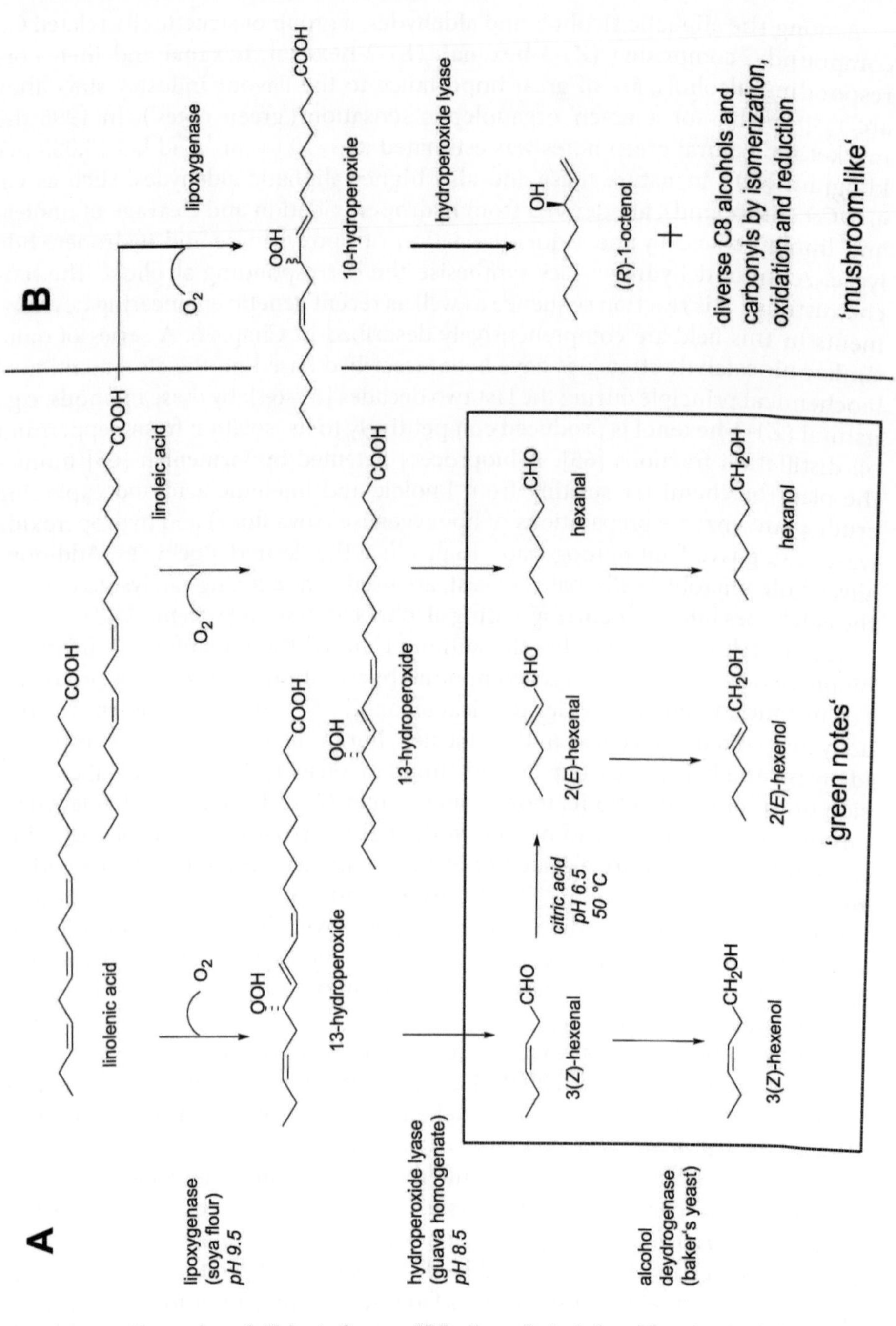

Scheme 23.3 Formation of aliphatic flavour aldehydes and alcohols. **a** Biotechnological reaction sequence mimicking plant biosynthesis of C6 compounds ('green notes'). **b** Homologous reaction sequence in fungi leading to mushroom-like C8 compounds. The stoichiometric formation of ω-oxo-carboxylic acids during hydroperoxide lyase cleavage is not depicted

besides other C8 alcohols and carbonyls, for flavouring purposes. In another industrial-scale process wasted mushroom stems are used as enzyme-containing raw material mixed with linoleic acid as a precursor [46].

23.4.1.3 Ketones

The odd-numbered **methylketones** have characteristic nutty cheese-like notes and are used in cheese flavour compositions [49]. The distinct taste of Roquefort cheese is substantially due to **2-heptanone** and **2-nonanone**. Methylketone formation is an aerobic process which is strongly favoured when the fungal growth is restricted and which does not occur with long-chain fatty acids. The fatty acid degradation pathway involves a key component, 3-oxoacylcoenzyme A (3-oxoacyl-CoA), which can be converted either into methylketone, by hydrolysis through thiohydrolase action followed by decarboxylation, or into CO_2, by thiolase followed by the citric acid cycle (β-oxidation of fatty acids) (Scheme 23.4). The bioformation of the methylketones results from an overflow of the β-oxidation cycle, where an excess of 3-oxoacyl-CoA ester is accumulated. One industrial process for the production of C5–C9 methylketones from the corresponding C6–C10 fatty acids uses *Penicillium roquefortii* grown by solid-state fermentation on buckwheat seeds [69, 70]. The whole sporulation medium is used for bioconversion without discarding the grains. This lowers the viscosity of the culture liquid, permitting a higher oxygen input than in a typical filamentous culture. A two-phase water/tetradecane system is used for in situ extraction of the product; there are no toxic effects of the organic solvent on the fungal spores: $\log P_{\mathrm{octanol/water}}$(tetradecane)=7. Different product yields are reported: depending on the respective fatty acid used as the starting material 20 g L^{-1} **2-pentanone**, 75 g L^{-1} **2-heptanone** and 60 g L^{-1} **2-nonanone** were produced.

3-Hydroxy-2-butanone (**acetoin**) is a characteristic constituent of butter flavour used for flavouring margarine and can be obtained as a by-product of molasses-based and lactic acid fermentations [49, 71]. The closely related **2,3-butanedione** (**diacetyl**) has a much lower organoleptic threshold than acetoin and is an important strongly butter-like flavour compound in butter and other dairy products [72]; in buttermilk, for instance, the diacetyl concentration is only about 2–4 mg L^{-1} [73]. α-Acetolactate (α-AL) is an intermediate of lactic acid bacteria mainly produced from pyruvate by α-acetolactate synthase. In most lactic acid bacteria, α-AL is decarboxylated to the metabolic end product acetoin by α-AL decarboxylase (ALDB) [71] (Scheme 23.5).

Special flavour-active strains, however, which do not contain ALDB, accumulate α-AL and, as a result of its chemical oxidative decarboxylation, generate high diacetyl levels in dairy products. Consequently, several processes have been patented for the production of natural diacetyl in the past few decades which usually involve a chemically enhanced conversion of α-AL into diacetyl or aim at α-AL itself as the biological product, which can serve as a less-volatile

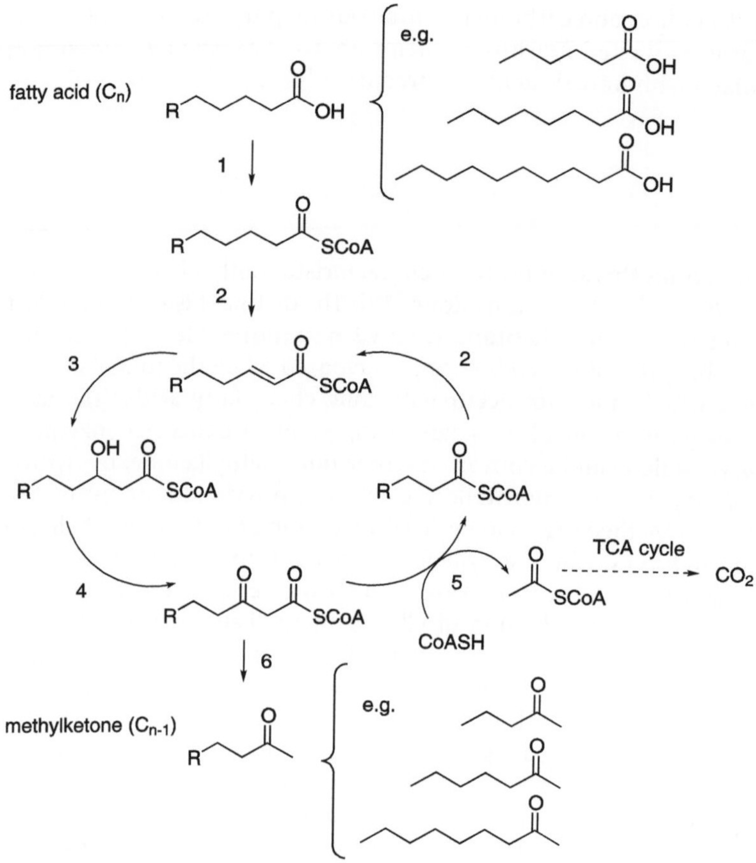

Scheme 23.4 Production of methylketones from fatty acids by *Penicillium roqueforti*. *1* ATP-dependent acylcoenzyme A (*acyl-CoA*) synthase; *2* flavin adenine dinucleotidedependent acyl-CoA dehydrogenase; *3* enoyl-CoA hydratase; *4* NAD-dependent 3-hydroxyacyl-CoA dehydrogenase; *5* 3-oxoacyl-CoA thiolase; *6* 3-oxoacyl-CoA thiolester hydrolase and 3-oxoacid decarboxylase. (Adapted from [46])

diacetyl precursor in food applications [74–76]. High diacetyl concentrations of up to 14 g L^{-1} have been described for a patented process based on *Streptococcus cremoris* and *S. diacetylactis* in a milk or whey medium supplemented with citric acid as a precursor [70]. A characteristic feature of this process was the need for an oxidising reagent during steam distillation, e.g. ferric chloride. Even higher concentrations of acetoin plus diacetyl of 35 g L^{-1} in total were described for the cultivation of *Enterobacter cloacae* ATCC 27613 in a complex nutrient broth followed by chemical conversion of the microbially produced acetoin, resulting in an overall yield of 60% diacetyl calculated on the basis of sugar consumed [77]. In recent years, detailed knowledge of the metabolism of lactic acid bac-

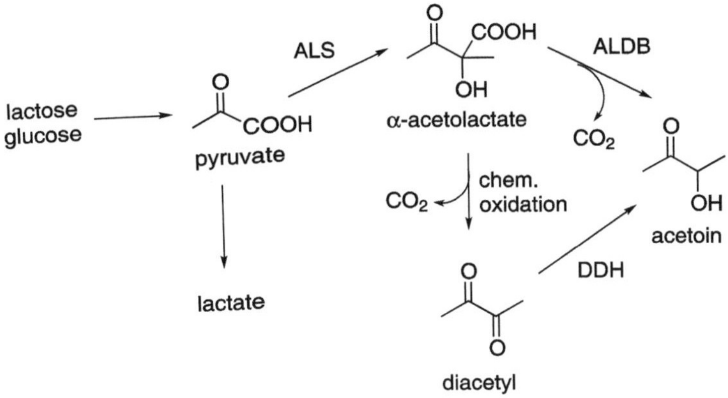

Scheme 23.5 Metabolic pathways of lactic acid bacteria leading from pyruvate to α-acetolactate and acetoin and chemical diacetyl formation. *ALS* α-acetolactate synthase, *ALDB* α-acetolactate decarboxylase, *DDH* diacetyl dehydrogenase. (Adapted from [72])

teria has led to innovative strategies for engineering *Lactococcus lactis* strains to enhance α-AL diacetyl, and acetoin production [72]. With *Lactococcus lactis* overexpressing *Streptococcus mutans* NADH oxidase (to redirect the pyruvate pool from lactate production to NADH-independent pathways) and having an inactivated ALDB, no more lactic acid production was observed and the strain converted glucose into α-AL, diacetyl and acetoin with yields of 57, 16 and 5%, respectively, as well as into acetate and CO_2 as by-products without the need for any other precursor. Nevertheless, to improve natural diacetyl production, here 137 mg L^{-1}, the physicochemical reaction conditions are to be adjusted to enhance the chemical oxidative decarboxylation of α-AL, e.g. by extending the aeration time preferentially at a lower pH than used during fermentation [72].

23.4.1.4 Esters

Esters are widespread in fruits and especially those with a relatively low molecular weight usually impart a characteristic fruity note to many foods, e.g. fermented beverages [49]. From the industrial viewpoint, esterases and lipases play an important role in synthetic chemistry, especially for stereoselective ester formations and kinetic resolutions of racemic alcohols [78]. These enzymes are very often easily available as cheap bulk reagents and usually remain active in organic reaction media. Therefore they are the preferred biocatalysts for the production of natural flavour esters, e.g. from short-chain aliphatic and terpenyl alcohols [7, 8], but also to provide enantiopure novel flavour and fragrance compounds for analytical and sensory evaluation purposes [12]. Enantioselectivity is an important factor as many flavour esters often have a very different sensory profile depending on their enantioforms [79]. Since a separate chapter of this book is devoted to the use of isolated enzymes in flavour science and technology (Chap. 22 by Menzel and Schreier), the focus here is on ester formation strategies based on whole microbial cells which also yield high product concentrations.

One interesting approach takes advantage of the high esterase activity of some fungi which can be harnessed without isolating the enzymes: dried fungal mycelium, especially from *Rhizopus oryzae*, can be used as an effective ester synthesising biocatalyst in organic solvents [80–82]. After growth on different Tweens (20, 40, 60 and 80) as the main carbon source the fungus show significant carboxylesterase activity. This strategy alleviates any costly enzyme preparation; moreover, the endogenous enzyme system is stabilised by the cellular structures, such as membranes, and the lyophilised biomass can be used as a self-immobilised catalyst for efficient flavour synthesis, e.g. for the direct esterification of 2-methylbutanol, 3-methylbutanol and hexanol with acetic acid or butanoic acid [81]. With butanoate, almost quantitative conversion of 65 mM of the respective alcohol was achieved after 24 h with *Rhizopus oryzae* cells in *n*-heptane. Up to 30 g L^{-1} **isopentylhexanoate** per gram of acetone-dried mycelium of *Rhizopus arrhizus* was achieved in a column reactor [83]. The esterification of a racemic mixture of **2-octanol** and **butanoic acid** proceeded with more than 97% enantiomeric excess for the *R* ester [82]. Aromatic acids are also substrates suitable for this approach using mycelium-bound carboxylesterases (Sect. 23. 3.2). High yields of short-chain fatty acid ethyl esters (C2–C8) were obtained with lyophilised *Rhizopus chinesis* cells, e.g. 98.5% for **ethylhexanoate** after 80 h using 0.6 M hexanoic acid in heptane and an acid-to-alcohol ratio of 1:1.3 [84].The initial water activity turned out to be an important parameter and a_w values ranging from 0.66 to 0.97 led to higher yields. The whole-cell lipase approach contributed to a long-lasting operational stability of the biocatalyst with half lives of up to 981 h as determined by its multiple reuse in consecutive batches. In 1987 an elegant bioprocess was patented based on living microorganisms to produce C2–C5 alkyl esters useful as natural fruit-like flavours [85]. *Geotrichum fragrans* was the preferred yeast for this synthesis, which needs two kinds of precursors, C5–C6 amino acids, i.e. leucine, isoleucine or valine, and natural aliphatic alcohols, such as those described in Sect. 23.4.1.2. The key steps of this bioconversion pathway are the initial oxidative deamination of the amino acids followed by decarboxylation/CoA ester formation by α-ketoacid dehydrogenase—a multienzyme complex—resulting in activated C4–C5 carboxylic acids (Fig. 23.3). Externally added alcohol, such as ethanol, causes interesterification of the CoA esters to the desired flavour esters. When mixtures of amino acids were used, complex fruity ester compositions were obtained. For commercially feasible yields of esters, continuous sweeping of the volatile products into the air stream was necessary. The product recovery included adsorption to activated charcoal from the fermentation exhaust air stream, extraction of the loaded charcoal by an organic solvent and distillation.

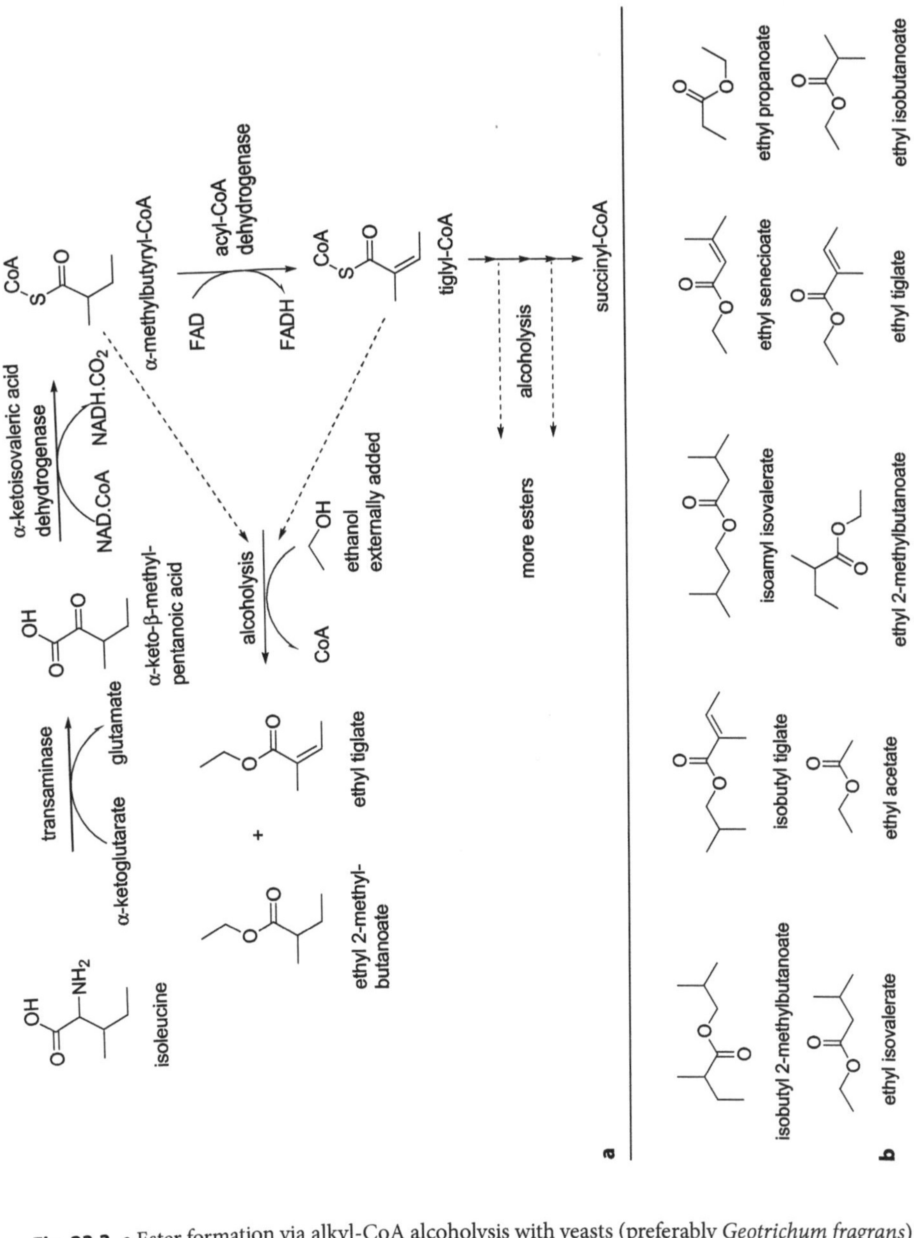

Fig. 23.3 **a** Ester formation via alkyl-CoA alcoholysis with yeasts (preferably *Geotrichum fragrans*) according to [85], exemplarily shown for ethyl-2-methylbutanoate and ethyl tiglate. **b** Some possible flavour esters producible depending on amino acid and alcohol used as substrates

The membrane-bound alcohol acetyl transferase is the key enzyme for another yeast-based ester synthesis: natural **ethyl acetate**, the most common ester in fruits and which is used for fruit and brandy flavours [49], can be produced in high concentrations with *Candida utilis* [83]. Under iron-limiting conditions, the tricarboxylic acid cycle cycle is inhibited and the intracellular pool of acetyl-CoA increases. In a fed-batch process the concentration of ethanol produced was maintained at a high level, thereby yielding 10–15 g L^{-1} ethyl acetate. Owing to the Crabtree effect ('aerobic fermentation'), *Candida utilis* converts most of the added glucose to ethanol under aerobic conditions, thus providing the substrate ethanol for the desired ester formation by alcoholysis of the acetyl-CoA. This is postulated to be a protective mechanism of the yeast against the toxic ethanol by producing the less toxic and more volatile ethyl acetate. Another yeast, *Williopsis saturnus* var. *mrakii*, shows a remarkably high de novo activity to produce fruity esters owing to its strong alcohol acetyl transferase which converts branched alcohols from amino acid metabolism into the corresponding acetates; this ester formation can be drastically enhanced by addition of the alcohols, e.g. those isolated from fusel oil, to the culture [48]. Owing to toxic effects, the fusel oil is added at low levels after the growth phase during the stationary bioconversion phase. The esters are recovered from the exhaust air by sorption on activated charcoal followed by organic solvent extraction. 3-Methylbutanol was the preferred alcohol which was converted into **3-methylbutyl acetate**, the character-impact compound of banana aroma, in a high yield (approximately 90%). Recently, a recombinant sake yeast overexpressing the *ATF1* gene coding for alcohol acetyl transferase was successfully engineered and produced up to the fivefold 3-methylbutyl acetate concentration compared with the wild type [86]; owing to a self-cloning strategy, this strain is not treated as a genetically modified organism in Japan.

23.4.2 Aromatic Compounds

In this section microbially produced benzene derivatives which are important as natural flavour compounds are discussed without further subdividing this section according to the functional groups, such as alcohols, aldehydes and acids. Metabolic pathways leading to the desired targets usually start from aromatic amino acids and/or phenylpropanoids, such as cinnamic acid, ferulic acid, eugenol and phenylpyruvic acid [6]. White-rot fungi, especially basidiomycetes found on living or dead wood, are capable of degrading lignin, a polymer of substituted *p*-hydroxycinnamyl alcohols. These fungi are the preferred microorganisms for studying aromatic flavour generation owing to their versatile enzyme machinery which has emerged during evolution [11, 87]. Nevertheless, characteristic disadvantages of these filamentous fungi, e.g. slow growth, difficult technical handling of the mycelia-forming organisms, a vast number of concurrently formed flavour-active products and low yields of the target compounds, impede their routine application in industrial processes. Thus, although the flavour compounds discussed in this section can all be produced by higher fungi, too, bacteria and yeasts are preferred, as they have a faster metabolism and, given that an appropriate production strain can be found (e.g. by enrichment cultures or mutagenesis/selection approaches), they usually result in narrower product spectra and higher yields in precursor-supplemented media. For the production of natural aromatic flavours it is of great benefit that L-phenylalanine has been made available as a natural precursor by microbial fermentation from the industrial L-aspartame process. Other phenylpropanoid precursors, such as eugenol or ferulic acid, can be found abundantly in nature.

Vanillin is undoubtedly the most important flavour compound with respect to both market volume and market value; a separate chapter of this book is dedicated to this flavour compound (Chap. 9 by Verpoorte and Korthout). It is the main aroma compound of the cured pods of *Vanilla* sp. [88]. Annually more than 10,000 t is produced, mainly by chemical synthesis. Whereas chemically produced vanillin is a cheap 'bulk flavour compound' available for about US $10 per kilogram, natural vanillin derived from *Vanilla* is only available in very low amounts and is therefore limited to a few select premium food applications. This opens an attractive market niche for biotechnology: natural vanillin from microbial processes currently costs up to US $1,000 per kilogram [8]. The annual world market volume of biotechnologically produced vanillin can be estimated to be 1–10 t and to be expanding steadily. This illustrates the increasing popularity of natural biotech vanillin although its discrimination from natural vanillin 'ex *Vanilla*' by isotope analysis is possible [88]. Different bioprocess strategies have been investigated based on bioconversion of ferulic acid, phenolic stilbenes, isoeugenol or eugenol and on de novo biosynthesis, applying bacteria, fungi, plant cells or engineered microorganisms [89]. The current industrial processes are based on the bioconversion of ferulic acid by different bacteria, which obviously have an outstanding tolerance against vanillin, which is cytotoxic at higher concentrations: a process patented by Haarmann & Reimer [90] uses the actinomycete *Amycolatopsis* sp. HR 167, with which a product concentration around 12 g L^{-1} can be obtained in a fed-batch process (Scheme 23.6). A similar approach using *Streptomyces setonii* as a production strain was patented by Givaudan, leading to final concentrations of up to 16 g L^{-1} after 50 h [91]. **Guaiacol**, missing the aldehyde group of vanillin, is a valuable by-product of this bioconversion (up to 0.4 g L^{-1}) because it significantly contributes to the characteristic flavour of *Vanilla* extracts and is often used together with vanillin in flavour compositions (Scheme. 23.6). In another patent, an *Amycolatopsis* mutant essentially free of by-product formation was described together with the downstream processing including precipitation of the vanillin by concentrating and cooling and further purification of the solid vanillin using a liquefied gas, preferentially CO_2 [92]. This strain converted 32 g L^{-1} ferulic acid to almost 18 g L^{-1} within approximately 50 h. The precursor currently used in all industrial biotech vanillin production processes, ferulic acid, can be obtained by microbial bioconversion of eugenol—abundantly available from the essential

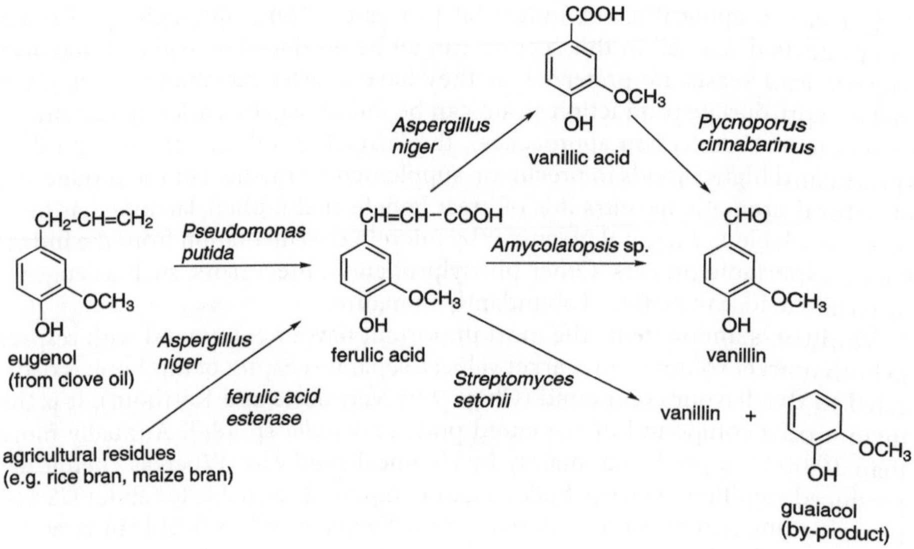

Scheme 23.6 Microbial strategies for the production of natural vanillin

oil of the clove tree *Eugenia cariophyllus*—with a eugenol-tolerant *Pseudomonas putida* [93] under sequential precursor-feeding conditions or by direct isolation from plant materials, e.g. rice bran, using ferulic acid esterase. Genetic engineering has been successfully applied to produce vanillin by direct bioconversion of the cheaper precursor eugenol instead of ferulic acid using metabolically engineered *Pseudomonas* or *Rhodococcus* strains [89, 94]. Nevertheless this route is currently not feasible for commercial production in Europe owing to the negative public perception of any food-related use of genetically modified microorganisms. Finally, a strategy using two filamentous fungi in succession for direct bioconversion of maize bran into vanillin is worth mentioning [95]. *A. niger* is exploited to release ferulic acid from the natural raw material by its feruloyl esterase activity and to subsequently metabolically convert ferulic acid into vanillic acid, which is further transformed into vanillin by *Pycnoporus cinnabarinus*. Under theses conditions, 584 mg L^{-1} vanillin was produced directly from ferulic acid containing raw material in a 'one-pot' approach (Scheme 23.6).

Benzaldehyde, with its bitter almond flavour, is the second-most important flavour compound, with a world market of approximately 7,000 t $year^{-1}$ [96]. Whereas by far the majority is chemically synthesised, there is, nevertheless, a growing market for the natural flavour compound, accounting for approximately 100 t $year^{-1}$ [87]. But, about 80% of this natural benzaldehyde represents a grey zone as it cannot be officially regarded as 'natural' according to EU legislation since a chemical hydrolysis is involved in its preparation from cassia oil. A biocatalytic route starting from amygdalin, a glycoside present in fruit kernels, based on the consecutive use of β-glucosidase (to release mandelonitrile) and mandelonitrile lyase has been used in the industry but owing to safety problems associated with the generation of equimolar amounts of hydrogen cyanide alternative strategies are needed [70]. With different basidiomycetes, such as *Trametes, Ischnoderma, Polyporus* and *Bjerkandera* species, benzaldehyde concentrations in the range of several hundred milligrams per litre up to about almost 1 g L^{-1} can be achieved in media supplemented with L-phenylalanine [97–100]. Benzyl alcohol was usually produced concomitantly during the same processes. In situ product recovery techniques, such as adsorption to a styrene/divinylbenzene copolymer resin selective for aromatic compounds, or organophilic pervaporation with poly(dimethylsiloxane) (PDMS) membranes have been successfully applied to improve productivities and final product concentrations [97, 101, 102] (Fig. 23.4). These effects were attributed to the circumvention of both product inhibition by benzaldehyde and its further conversion to the corresponding alcohol. Nevertheless, a disadvantage of these processes based on basidiomycetes is the long cultivation time needed for growth and bioconversion which usually amounts to more than 10 days.

Natural **raspberry ketone**, 4-(4′-hydroxyphenyl)-butan-2-one, is the character-impact compound of the aroma of raspberries. Although the chemically synthesised ketone only costs about US $10 per kilogram the flavour industry would prefer the natural ketone for many food applications. Unfortunately, its recovery from the natural source is impractical owing to the very low concentrations found in the berries (less than 4 mg kg^{-1} [48]). This situation has stimulated various attempts at a biotechnological production. Nevertheless up to now no economically viable biotechnological production has been described although the target substance may achieve a market price of more than US $1,000 per kilogram as a 'natural' flavour compound. Up to now, mainly two biotechnological strategies have been proposed (Scheme 23.7). The de novo synthesis with the basidiomycete *Nidula niveo-tomentosa* can be significantly enhanced by adding natural amino acid precursors, L-tyrosine or L-phenylalanine, to the medium. In an optimised medium, this basidiomycete produced raspberry ketone and its corresponding alcohol betuligenol with a total product yield (raspberry ketone and betuligenol) of approximately 200 mg L^{-1} after 22 days, a 50-fold increase compared with the non-optimised system [103]. Nevertheless, the less-flavour-active alcohol is the primary product and the overall productivity is still far too low for a commercial application. On the other hand, this approach illustrates that raspberry ketone production starting from the cheap precursor L-phenylalanine is, in principle, possible, justifying further elucidation of the respective pathway, which differs from that found in raspberry as evidenced by stable isotope labelled precursor feeding studies [104]. The second strategy to produce natural raspberry ketone is a biocatalytic two-step conversion involving the β-glucosidase-catalysed hydrolysis of the naturally occurring betuloside, a 2-glycoside of 4-(4-hydroxyphenyl)-2-butanol (glucoside, mannoside). This precursor occurs in different plants: the bark of the European white birch (*Betula alba*), rhododendron (*Rhododendron* spp.), maple (*Acer* spp.), fir (*Ab-*

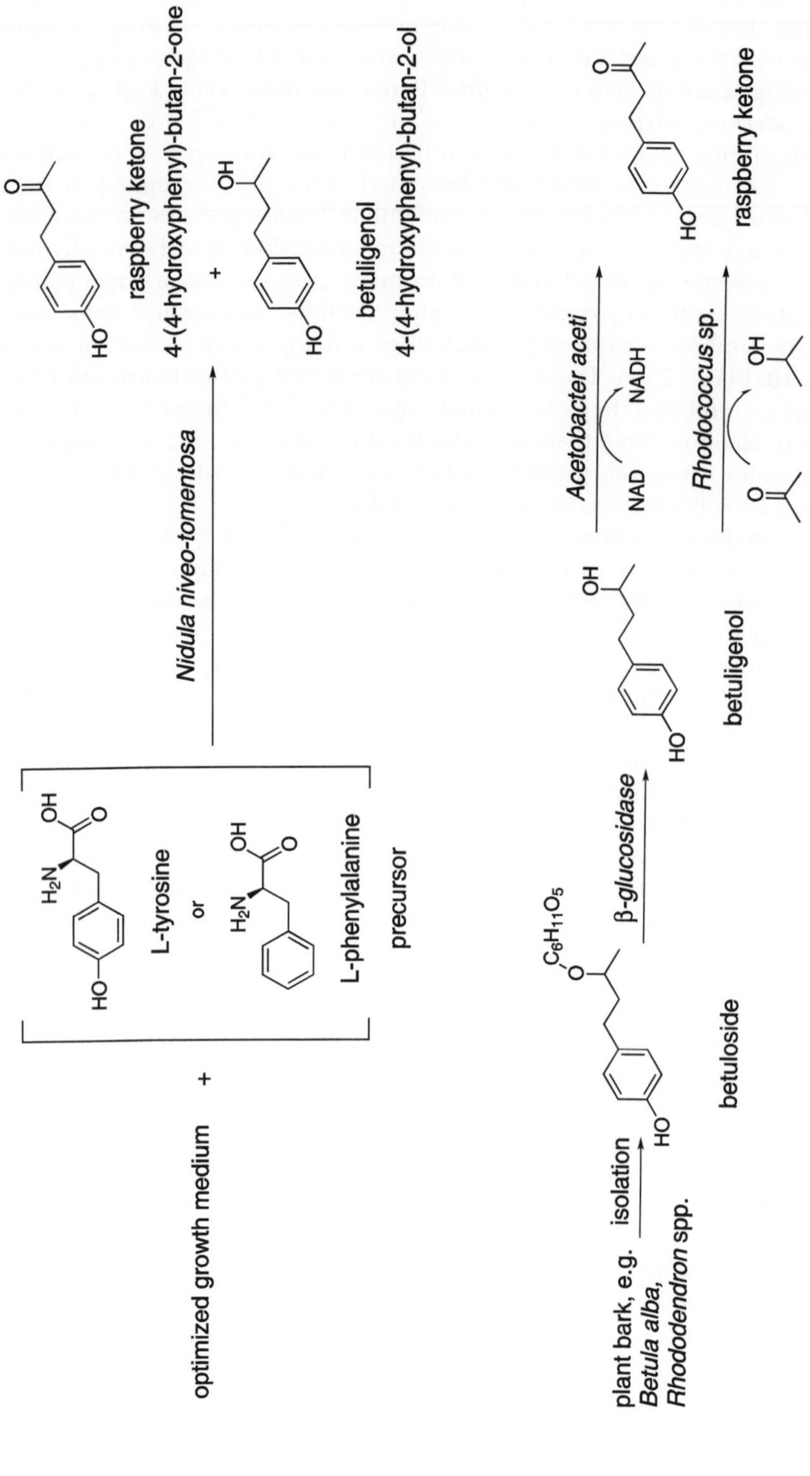

Scheme 23.7 Biotechnological strategies for the production of natural raspberry ketone

ieta spp.) and yews (*Taxus* spp.). By hydrolysis, betuligenol is released, and is transformed by a microorganism containing a secondary alcohol dehydrogenase such as *Acetobacter aceti* into the corresponding raspberry ketone [105]. In a recent publication, the oxidation was performed with lyophilised *Rhodococcus* cells in phosphate buffer containing 10% v/v acetone as a hydrogen acceptor [106]. This biocatalytic oxidation shows a high yield of 89% and can be performed at precursor concentrations of up to 500 g L^{-1}. Thus, the main bottleneck still preventing an industrial application is now obviously the lack of an economically viable supply of the natural precursor betuligenol rather than the biooxidation process itself.

2-Phenylethanol has a rose-like odour and makes the chemically produced compound the most used fragrance chemical in perfume and cosmetics, with a world market of about 7,000 t $year^{-1}$ [107, 108]. 2-Phenylethanol is also found in many foods as a characteristic flavour compound rounding off the overall aroma, especially in foods obtained by fermentation, such as wine, beer, cheese, tea leaves, cocoa, coffee, bread, cider and soy sauce [109]. In food applications, natural 2-phenylethanol is preferred rather than its nature-identical counterpart from chemical synthesis and it has a market volume of 0.5–1 t $year^{-1}$. This product is sold at market prices of up to US $1,000 per kiklogram and is mainly produced by yeast-based bioprocesses since its isolation from natural sources, e.g. rose oil, would be too costly [109].

Although 2-phenylethanol can be synthesised by normal microbial metabolism, the final concentrations in the culture broth of selected microorganisms generally remain very low [110, 111]; therefore, de novo synthesis cannot be a strategy for an economically viable bioprocesses. Nevertheless, the microbial production of 2-phenylethanol can be greatly increased by adding the amino acid L-phenylalanine to the medium. The commonly accepted route from L-phenylalanine to 2-phenylethanol in yeasts is by transamination of the amino acid to phenylpyruvate, decarboxylation to phenylacetaldehyde and reduction to the alcohol, first described by Ehrlich [112] and named after him (Scheme 23.8).

During the last few decades a series of microbiological and technical approaches have been published aiming at improving growth-associated 2-phenylethanol formation based on Ehrlich bioconversion [113–118]. *Kluyveromyces* and *Saccharomyces* species have been shown to be efficient biocatalysts leading to molar conversion yields of more than 90%. In situ product removal is essential for high-performance processes by alleviating product inhibition which can already significantly impair growth at a 2-phenylethanol concentration of about 0.3 g L^{-1} strain-dependently [119]. Coupling an organophilic pervaporation membrane to a bioreactor cultivation of the thermotolerant yeast *Kluyveromyces marxianus* CBS 600 at 40 °C resulted in volumetric productivities of up to 5.2 mmol L^{-1} h^{-1} [119] (Fig. 23.4).

The flavour product from L-phenylalanine included 2-phenylethanol as the main product and **2-phenylethyl acetate** as a side product, which is also a valuable rose-like flavour compound with a more fruity note and which is formed

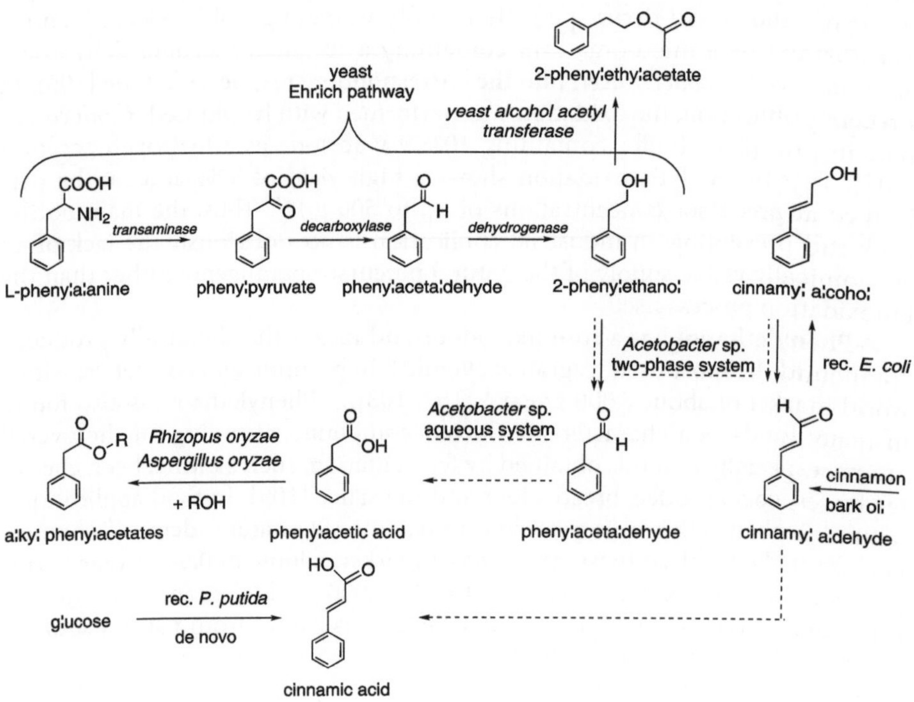

Scheme 23.8 Some microbial pathways and biotransformations leading to aromatic flavour molecules

within the yeast metabolism by the action of alcohol acetyl transferase. Performing the integrated bioprocess at 45 °C yielded only a slightly lower overall product concentration but with the acetate as the major product. Recently an aqueous–organic two-liquid-phase bioprocess has been described reporting the highest 2-phenylethanol and 2-phenylethyl acetate space-time yields and final concentrations so far [120]. Again, *Kluyveromyces marxianus* CBS 600 was used in a fed-batch emulsion system with poly(propylene glycol) 1200 as a non-volatile and biocompatible organic solvent efficiently extracting the flavour compounds from the aqueous culture medium (phase ratio of approximately 1:1) within the bioreactor (Fig. 23.5). Space-time yields of 0.33 and 0.08 g L^{-1} h^{-1} were obtained for 2-phenylethanol and 2-phenylethyl acetate, respectively, corresponding to final concentrations of 26.5 and 6.1 g L^{-1} in the organic phase after 30 h. The amino acid was provided as the sole nitrogen source in high excess (above its solubility threshold), making complicated feeding strategies unnecessary.

The elucidation of the genetics and regulations of the Ehrlich pathway leading from amino acids to alcohols and the corresponding acids and esters—a pivotal metabolic route to flavours generated by traditional food fermentation processes—has attracted much research interest in the past. More recent inves-

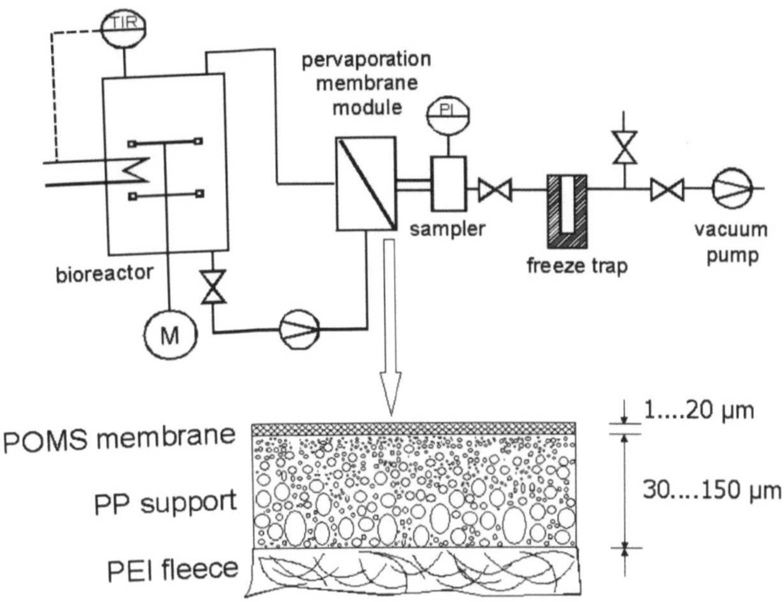

Fig. 23.4 Organophilic pervaporation (*PV*) for in situ recovery of volatile flavour compounds from bioreactors. The principle of PV can be viewed as a vacuum distillation across a polymeric barrier (membrane) dividing the liquid feed phase from the gaseous permeate phase. A highly aroma enriched permeate is recovered by freezing the target compounds out of the gas stream. As a typical silicone membrane, an asymmetric poly(octylsiloxane) (*POMS*) membrane is exemplarily depicted. Here, the selective barrier is a thin POMS layer on a polypropylene (*PP*)/poly(ether imide) (*PEI*) support material. Several investigations of PV for the recovery of different microbially produced flavours, e.g. 2-phenylethanol [119], benzaldehyde [264], 6-pentyl-α-pyrone [239], acetone/butanol/ethanol [265] and citronellol/geraniol/short-chain esters [266], have been published

tigations reveal a surplus of isogenes responsible for each enzymatic transformation step and indicate complex regulation principles on both transcriptional and posttranscriptional levels [121–129]. The accumulated knowledge should lead to improved production strains for this type of amino acid derived flavour compound by genetic engineering in the near future.

Nevertheless, for the production of the flavour-active aromatic alcohol derivatives, such as the corresponding aldehydes and acids, metabolic engineering approaches have to compete with conventional oxidative biocatalysis starting from the natural alcohol as a substrate. For instance, the whole-cell oxidation system based on *Pichia pastoris* AOX already described in Sect. 23.4.1.2 can also be used to convert benzyl alcohol to benzaldehyde in aqueous media although product inhibition restricted the final product concentration to about 5 g L^{-1}, indicating the need for aqueous–organic two-phase reaction media [51]. **Phenylacetalde-**

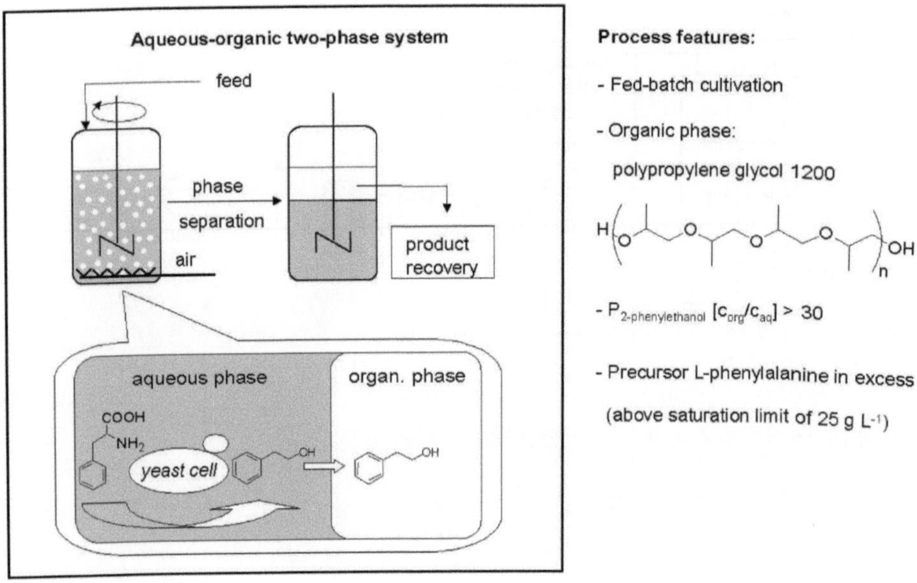

Fig. 23.5 Aqueous–organic two-liquid-phase system for microbial production of flavour compounds. Here the formation of 2-phenylethanol from L-phenylalanine is exemplarily shown [120]. The organic solvent used for in situ extraction has to be carefully selected on the basis of multiple criteria, such as biocompatibility, non-flammability and legislative regulations. For a more detailed description of flavour production in two-phase systems, see Chap. 24 by Larroche et al.

hyde can be efficiently synthesised with acetic acid bacteria making use of their strong oxidative capacity provided by the dehydrogenase system (Sect. 23.4.1.2) [130]. An *Acetobacter* sp. strain immobilised in alginate beads produced 1.92 g L^{-1} phenylacetaldehyde from 4 g L^{-1} 2-phenylethanol and showed higher production rates than non-immobilised cells, which was explained by protection from toxic effects caused by the product and/or the precursor. In situ product recovery by a two-liquid-phase system consisting of isooctane–water (1/1 v/v) was successfully performed and yielded 9 g L^{-1} phenylacetaldehyde recovered in the organic phase from 10 g L^{-1} 2-phenylethanol within 4 h using another *Acetobacter* strain [131]. The composition of the medium in this biotransformation can be exploited as a control mechanism to direct the oxidation of aromatic alcohols to either the aldehyde in the presence or the acid in the absence of the organic phase [132] (Scheme 23.8). By this means 2-phenylethanol and cinnamyl alcohol were transformed to the corresponding acids, **phenylacetic acid** and **cinnamic acid**, in water with yields of more than 97% within 3 and 8 h, respectively; phenylacetaldehyde and **cinnamyl aldehyde** were produced from the alcohols within only 45 min in water–isooctane with yields of 90 and 77%, respectively. The process for the production of acids from aliphatic alcohols with *Gluconobacter oxydans* DSM 12884 described in Sect. 23.4.1.1 was also successfully applied to aromatic alcohols: benzyl alcohol, 2-phenylethanol, and cinnamyl alcohol were converted into **benzoic acid**, phenylacetic acid and cinnamic acid [39].

Immobilisation of an *Acetobacter aceti* strain in calcium alginate resulted in improvement of the operational stability, substrate tolerance and specific activity of the cells and 23 g L^{-1} phenylacetic acid was produced within 9 days of fed-batch cultivation in an airlift bioreactor [133]. Lyophilised mycelia of *Aspergillus oryzae* and *Rhizopus oryzae* have been shown to efficiently catalyse ester formation with phenylacetic acid and phenylpropanoic acid and different short-chain alkanols in organic solvent media owing to their carboxylesterase activities [134, 135] (Scheme 23.8). For instance, in *n*-heptane with 35 mM acid and 70 mM alcohol, the formation of ethyl acetate and propylphenyl acetate was less effective (60 and 65% conversion yield) than if alcohols with increased chain lengths were used (1-butanol 85%, 3-methyl-1-butanol 86%, 1-pentanol 91%, 1-hexanol 100%). This effect was explained by a higher chemical affinity of the longer-chain alcohols, which are more hydrophobic, to the solvent.

Since cinnamyl aldehyde is the main component of cassia oil (approximately 90%) and Sri Lanka cinnamon bark oil (approximately 75%) [49], it is industrially more important to generate **cinnamyl alcohol**, which is less abundantly available from nature but is important as cinnamon flavour, by biotransformation of natural cinnamyl aldehyde than vice versa. Recently, a whole-cell reduction of cinnamyl aldehyde with a conversion yield of 98% at very high precursor concentrations of up to 166 g L^{-1} was described [136]. *Escherichia coli* DSM 14459 expressing a NADPH-dependent *R* alcohol dehydrogenase from *Lactobacillus kefir* and a glucose dehydrogenase from *Thermoplasma acidophilum* for intracellular cofactor regeneration was applied as the biocatalyst (Scheme 23.8).

The microbial production of significant amounts of cinnamic acid from glucose by cloning phenylalanine ammonia lyase (PAL) from the yeast *Rhodosporidium toruloides* into a solvent-tolerant *Pseudomonas putida* strain was described for the first time [137] (Scheme 23.8). Random mutagenesis and selection on the toxic antimetabolite *m*-fluorophenylalanine followed by high-throughput screening led to the isolation of a mutant strain with improved de novo phenylalanine biosynthesis and consequently a higher cinnamic acid production. In a nitrogen-limited fed-batch fermentation on glycerol, 750 mg L^{-1} cinnamic acid formed within 50 h, corresponding to a conversion yield of 6.7% based on the carbon consumed. Higher productivities are being aimed at by integrating in situ product-recovery techniques owing to the inhibition of PAL by cinnamic acid and by further enhancing the intracellular phenylalanine level by proteomics and transcriptomics methodologies.

23.4.3
Terpenes

23.4.3.1
General Considerations

With some estimated 20,000 to more than 40,000 different molecules known, terpenes (isoprenoids) are the largest family of natural compounds in nature [138–140]. Whereas oxyfunctionalised monoterpenes and sesquiterpenes are extensively applied in industry as flavour and fragrance compounds, their precursors, the terpene hydrocarbons, are usually separated from their natural sources, essential oils, as they contribute little to flavour and fragrance and may also cause undesirable off-flavours and precipitations. The essential oil content of plants is, however, low, with concentrations of less than 0.1 to 5% and the commercial extraction of minor compounds is only in rare cases economically viable. As many terpene hydrocarbons are abundantly available in nature, e.g. (+)-limonene and the pinenes, which are the main components of citrus and turpentine oils, respectively, e.g. more than 90% (+)-limonene in orange oil, they represent an ideal starting material for biocatalytic oxyfunctionalisations leading to natural terpenoid flavour and fragrance compounds. This research area has therefore been the target of many research groups for decades focussing on individual types of bacteria that degrade terpenes, e.g. *Pseudomonas*, *Rhodococcus* and *Bacillus*, on deuteromycetes, e.g. aspergilli and penicillia, and especially on all the higher fungi of the ascomycetes and basidiomycetes, which have a marked capacity for terpene de novo biosynthesis and biotransformation. In terpene transformation a manageable number of enzyme reactions are frequently found owing to the uniform basic terpene structure which derives from the general biosynthesis principle based on five-carbon (isoprene) units [141]. Most of the monooxyfunctionalisation reactions are believed to be catalysed by cytochrome P450 monooxygenases. In contrast to most of the chemical oxidation processes, which often suffer from harsh reaction conditions and the need for hazardous reagents, e.g. toxic heavy metals, showing a low discrimination of the carbon atoms in terpene hydrocarbons, these enzymes are able to regioselectively transform multifunctional terpenoid substrates at specific sites under mild conditions. Even non-activated chemically inert carbon atoms can be functionalised by enzymatic reactions. However, only a few terpenes are produced biotechnologically on an industrial scale despite their often unique organoleptic properties, the growing demand and the unstable supply situation from the traditional (frequently overseas) sources. The main reasons stem from the physicochemical properties of terpenes, such as low water solubility, high volatility and cytotoxicity of the terpenoid precursor and the product, which impede conventional bioprocesses (Sect. 23.2). In fungal terpene biotransformations, monitoring the oxygen uptake rate [142] or the terpene concentration in the exhaust air [108] was shown to be helpful for feeding the toxic precursor in a biocompatible way; nevertheless, the engineering of terpene biotransformations definitely needs further impetus by combining tailored process modifications, e.g controlled precursor feeding with in situ product recovery, to obtain higher product yields and to establish economically viable processes. Moreover, owing to microbial metabolic versatility, one precursor is usually converted into a variety of derivatives, partly representing undesired or not readily separable by-products which should be addressed by appropriate measures as shown in Table 23.2. Comprehensive reviews covering the research until 2001, in which the interested reader is guided through a wealth of microbial terpene biotransformation strategies, have been published [143–145]. Therefore, this chapter only gives a few examples of microbial transformations that stand out from the vast number of reported biotransformations with respect to the product concentrations achieved. Furthermore, the focus is also on more recent publications which describe the formation of highly valuable terpenoid flavour compounds (albeit usually still at low concentrations) and on approaches which may point the way ahead towards more sophisticated bioprocesses in the future by targeting modern molecular biological or biochemical engineering aspects.

23.4.3.2 Monoterpenes

The unsaturated monterpene-triene β-myrcene, frequently found in the terpene hydrocarbon fraction of many essential oils, has been shown to be a suitable precursor for biotransformations with basidiomycetes, although usually a multitude of metabolic derivatives in low concentrations appeared [146]. In contrast, the transformation of β-myrcene by a mutant obtained by transposon mutagenesis from a parental β-myrcene degrading *Pseudomonas* strain yielded **myrcen-8-ol** as the main product [147] (Scheme 23.9). Citronellol, an acyclic monoterpene alcohol, which can be isolated, e.g., from *Boronia*, *Eucalyptus* and geranium and rose oils in high concentrations [49] was converted by the basidiomycete *Cystoderma carcharias* to 3,7-dimethyl-1,6,7-octane-triol as the main product, but ***cis*-rose oxide/*trans*-rose oxide**, an industrially important fragrance compound, arose as one of the minor products. In a 2-L bioreactor fed-batch process, rose oxide was trapped out of the exhaust air by adsorption, albeit only in the milligram range [148]. A screening based on solid-phase microextraction revealed that sporulated surface cultures of *Aspergillus* and *Penicillium* species can also produce rose oxides from citronellol, although here again only as minor metabolic by-products [149]. Submerged shaking cultures of *Aspergillus niger* ATCC 9142 have been used to transform linalool into the furanoid and pyranoid ***cis*-linalool/*trans*-linalool oxides**, which are of interest for lavender notes in perfumery [150]. An industrial-scale bioprocess for the transformation of monoterpenes to the corresponding acids, which are important as building blocks for natural flavour esters, was patented [151]. Under aerobic conditions and at alkaline pH, geraniol was enantioselectively oxidised to **(*E*)-geranic acid** (85%) and (+)-**citronellic acid** (15%) using commercial baker's yeast. Geranic acid reached a maximum concentration of 3.6 g L^{-1} after 48 h. A NADH-depen-

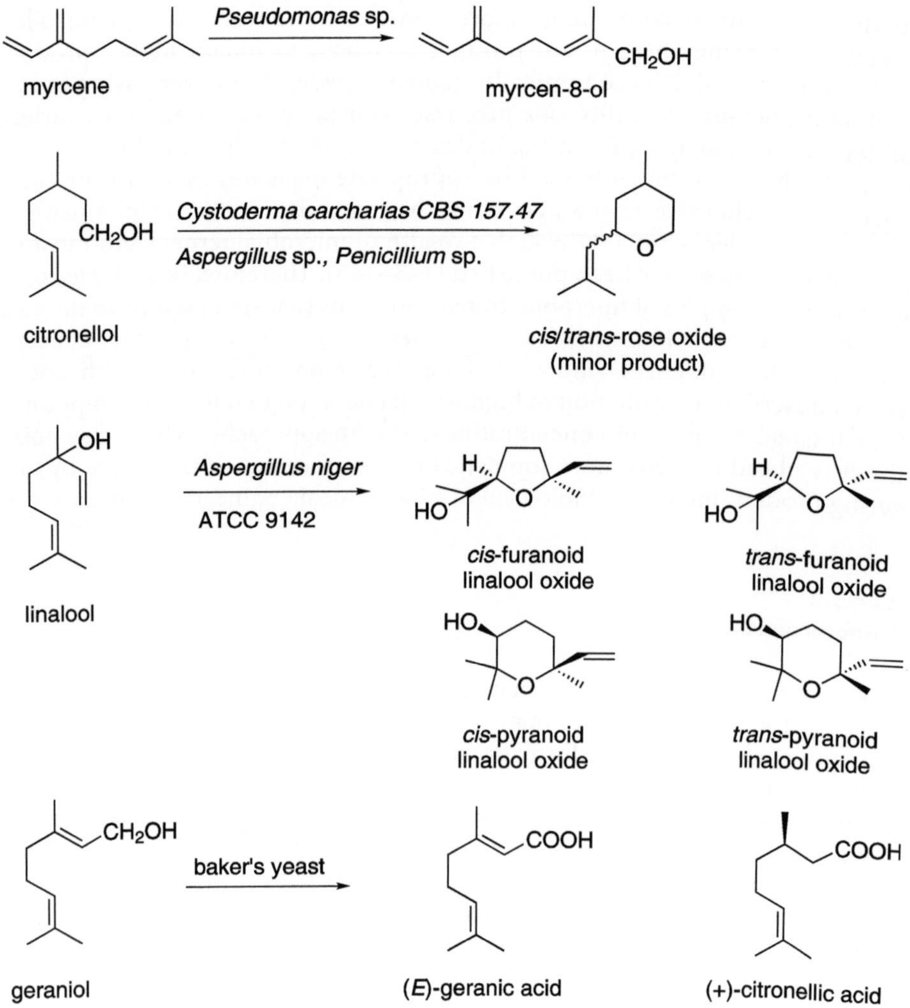

Scheme 23.9 Some microbial monoterpene transformations leading to interesting flavour molecules

dent pathway from geraniol via geranial, neral and citronellal to citronellol was proposed branching off from geranial to geranic acid and from citronellal to citronellic acid. Geranic acid was shown to be the sole product formed during biotransformation of geraniol with a *Rhodococcus* sp. strain termed GR3 isolated from soil [152]. After 3 days of growth in a conventional medium containing minerals, dextrose and peptone, the bioconversion was started by adding geraniol to the medium at 1% v/v and geranic acid formed with saturation kinetics leading to a final yield of 50% after 96 h. Higher precursor concentrations caused lower conversion yields obviously owing to toxic effects of geraniol.

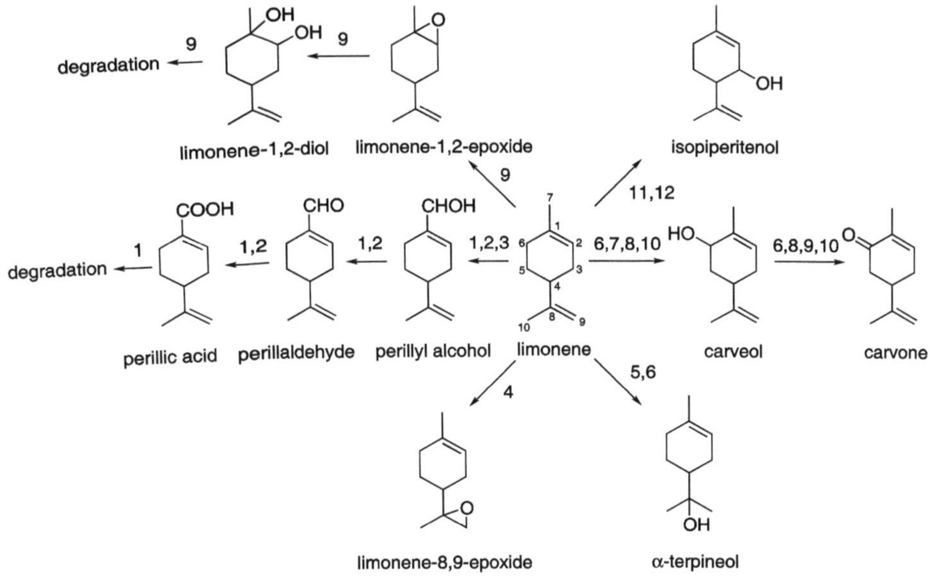

Scheme 23.10 Microbial limonene transformation routes. *1* Frequently found in bacilli and pseudomonads; *2 Pseudomonas putida* DSM12264; *3 Bacillus stearothermophilus*, recombinant *Pseudomonas putida* expressing a P450 monooxygenase from *Mycobacterium* sp.; *4 Xanthobacter* sp. C20; *5 Penicillium digitatum* NRLL 1202 and DSM 62840; *6 Pseudomonas aeruginosa*; *7 Rhodococcus opacus* PWD4, *Fusarium proliferatum*; *8 Rhodococcus globerulus* PWD8; *9 Rhodococcus erythropolis* DCL14; *10 Pleurotus sapidus*; *11 Hormonema* sp. UOFS-Y-0067; *12* recombinant *Escherichia coli* expressing an evolved mutant of P450cam from *Pseudomonas putida*. For information about the stereochemistry of the biotransformations, see the text

Probably the most intensively studied monoterpene precursor is (+)-limonene, the main constituent of citrus essential oils, which accumulates in amounts of more than 50,000 t year^{-1} as a cheap by-product of the citrus processing industry [153]. The current status of microbial and plant biotransformation of limonene was recently reviewed and at least six different molecule sites for initial limonene oxyfunctionalisation were reported [154]. Microbial transformations of limonene are summarised in Scheme 23.10. It is assumed that limonene-degrading *Pseudomonas* and *Bacillus* strains gain most of their carbon and energy by initially attacking the 7-position in a rather regiospecific way leading to perillyl alcohol followed by further progressive oxidation via perillaldehyde to **perillic acid**, which is mineralised by a β-oxidation-like mechanism [154]. *Pseudomonas putida* DSM 12264 growing on *p*-cymene as the sole carbon source can be used to convert limonene to perillic acid at high yields because the three enzymes, *p*-cymene monooxygenase, *p*-cumic alcohol dehydrogenase and *p*-cumic aldehyde dehydrogenase, show significant side activities towards limonene and its analogue derivatives **perillyl alcohol** and perillaldehyde [155]. Further degradation of perillic acid, the only product of limonene biotransformation, is obviously hampered by the high substrate specificity of the respective catabolic enzymes. The advantage of this *Pseudomonas* strain is its solvent tolerance, allowing for growth in the presence of high limonene concentrations (e.g. 150 mM) dramatically exceeding its saturation concentration (approximately 0.1 mM), causing a separate limonene phase finely dispersed in the aqueous phase upon stirring. In an optimised mineral medium with (+)-limonene and glycerol as the cosubstrate up to 3.0 g L^{-1} (+)-perillic acid formed after 120 h. In a 3-L bioreactor under fed-batch conditions with non-limiting amounts of glycerol, ammonia and limonene, the final product concentration was increased to 11 g L^{-1} (+)-perillic acid after 7 days, obviously limited by product inhibition [156]. Although these are the highest product concentrations reported for a microbial limonene oxyfunctionalisation, from the flavourist's viewpoint, the more volatile precursors of the acid, perillaldehyde and perillyl alcohol, are the industrially more interesting targets.

A *Bacillus stearothermophilus* strain with a temperature optimum near 55 °C isolated from orange peel by enrichment on (+)-limonene as the sole carbon source was able to convert the monoterpene cometabolically to perillyl alcohol as the major product (200 mg L^{-1}), whereas **α-terpineol** and perillaldehyde were by-products [157]. The respective (+)-limonene conversion pathways encoded on a 9.6-kb chromosomal fragment was cloned into *Escherichia coli* where the lilac-like fragrance α-terpineol became the major product [158]. By subcloning of smaller fragments, the product diversity was further narrowed and 235 mg L^{-1} α-terpineol was produced after 3 days using (+)-limonene as a neat substrate phase supplying the precursor and for extractive in situ product recovery; **carvone** formed as by-product [159]. The bioconversion was carried out at 50–60 °C to repress undesired metabolic side activities of the host by processing the bioconversion at the optimum temperature of the donor strain. With use of *Penicillium digitatum* NRLL 1202, racemic limonene was converted to (+)-α-terpineol since only (+)-limonene was accepted as a substrate enantiospecifically. Bioconversion activity increased up to 12-fold after sequential substrate induction and a yield of about 3.2 g L^{-1} α-terpineol was achieved after 96 h, albeit on a 5-mL analytical scale [160]. The limonene hydroxylation at the 8-position which corresponds to a hydration of the 8,9 double bond is probably catalysed by a P450 monooxygenase leading to the epoxide as an intermediate rather than by a hydratase but it has not been isolated and identified so far [160]. Immobilisation in calcium alginate and the use of organic cosolvents were proposed to improve the product yields of this system [161, 162]. Recently a fed-batch 3-L bioprocess with *Penicillium digitatum* DSM 62840 was reported yielding 0.5 g L^{-1} within 7 days using a two-step approach comprising a biomass growth period followed by a biotransformation period in the same reactor [163]. A 3-L bioreactor with a closed gas loop and terpene-saturated process air was described to alleviate any loss of terpenes via the exhaust air [164]. *P. digitatum* DSM 62480 was exploited in this reactor system to produce more than 1 g L^{-1} α-terpineol after about 120 h of biotransformation. In contrast, **limonene-8,9-epoxide** has been shown to be the main product of *Xanthobacter* sp. C20 catalysed conversion of both limonene enantiomers with a pro-8*R* stereospecificity [165]. The strain was pregrown on cyclohexane as the sole carbon and energy source and used as resting cells for biocatalysis in phosphate buffer yielding up to 0.8 g L^{-1} of the epoxide.

Carvone is an important monoterpene ketone, of which the (+)-isomer represents the character-impact compound of caraway flavour (up to 60% in caraway oil), whereas the (-)-isomer has a typical spearmint note (70–80% in spearmint oil). A *Pseudomonas aeruginosa* strain was described that was capable of converting (+)-limonene into carvone and α-terpineol as the main products at 37 °C in 200-mL shake flasks after 13 days, and final concentrations of up to 0.63 and 0.24 g L^{-1}, respectively, were achieved [166]. The toluene-degrading strain *Rhodococcus opacus* PWD4 was found to oxyfunctionalise (+)-limonene exclusively at the 6-position, yielding enantiomerically pure ***trans*-(+)-carveol**, whereas **(+)-carvone** formed as a by-product (1.3% of the amount of ***trans*-carveol**) [167]. The initial specific activity for carveol formation was 14.7 U g_{cdw}^{-1} accompanied by a molar yield of 94–97%. One of the enzymes from the toluene-degradation pathway has to be responsible for the (+)-limonene hydroxylation since, on the one hand, cells pregrown on glucose did not convert limonene at all and, on the other hand, toluene proved to be a strong competitive inhibitor. Another toluene degrader, *Rhodococcus globerulus* PWD8, showed a lower specific activity of 3 U g_{cdw}^{-1} and slowly overoxidised most *trans*-(+)-carveol to 0.29 mM (+)-carvone, the more valuable terpene ketone, from 1.2 mM (+)-limonene after 27 h under small-scale (2.5-mL) assay conditions. *Rhodococcus erythropolis* DCL14, which grows on limonene as the sole carbon source, starts metabolising limonene by a rather uncommon epoxidation at the 1,2 double bond, forming **limonene-1,2-epoxide** [168], while further mineralisation proceeds via limonene-1,2-diol and 1-hydroxy-2-oxolimonene, pointing to a β-oxidation degradation. The same strain also contains several carveol dehydrogenases enabling it to convert carveol stereospecifically to carvone [169]. This enzyme activity was exploited to produce **(-)-carvone** from *cis*-(-)-carveol/*trans*-(-)-carveol corresponding to a diastereomeric excess of more than 98% and a yield of 0.68% w/w [170]. The cells were used in an aqueous–organic two-liquid-phase air-driven column reactor containing *n*-dodecane as a protecting organic phase, but product inhibition above 50 mM carvone impeded higher product concentrations. These limitations were overcome by adapting *Rhodococcus erythropolis* cells in mineral medium to carveol and carvone dissolved in *n*-dodecane prior to biotransformation [171]. The air-driven column reactor was used after an adaptation period of 197 h and an 8.3-fold increase in carvone production rate compared with non-adapted cells was achieved. The highest final concentration was achieved with cells adapted for 268 h which produced 1.03 M carvone after 167 h at room temperature. The cellular adaptation mechanism was explained by a dose-dependent increase in the degree of saturation of the membrane phospholipids [172]. The basidiomycete *Pleurotus sapidus* was shown to regiospecifically oxyfunctionalise the same limonene 6-position: by means of preculitivation in the presence of small amounts of the precursor fed via the gas phase, the concentrations of ***cis*-carveol**/*trans*-carveol and carvone increased to yield a total product concentration of more than 0.1 g L^{-1} [173]. In contrast to *Pleurotus sapidus*, the ascomycete *Fusarium proliferatum* did not form measurable amounts of carvone, but converted both limonene enantiomers: (+)-limonene to *cis*-(+)-carveol, and (-)-limonene predominantly to *trans*-(-)-carveol, which could be further oxidised to (-)-carvone, again by *Pleurotus sapidus* [163]. These examples illustrate that by combining two microbial strains with hydroxylase and dehydrogenase activity a biotechnological stereospecific production of both carvone enantiomers from (+)-limonene and (-)-limonene may become possible in the future.

The black yeast *Hormonema* sp. UOFS Y-0067 isolated from monoterpene-rich environments transformed (+)-limonene into ***trans*-isopiperitenol** by regioselectively attacking the 3-position, a biotransformation reported for the first time [174]. A product concentration of 0.5 g L^{-1} was achieved after 12 h of incubation in shake flasks (31% molar conversion), obviously limited by product and/or precursor toxicity. Unfortunately, the product concentration of *trans*-isopiperitenol, which requires only a catalytic hydrogenation step to yield the desired (-)-menthol, was not always reproducible owing to morphological mutability and, thus, this microorganism was obviously not well suited for further process development.

The pinenes are a cheap natural starting material abundantly available as main constituents of turpentine oil (up to 75–90%) with up to 160,000 t α-pinene and 26,000 t β-pinene per year and are also found in relevant amounts in the essential oils of non-coniferous plants, e.g. up to 12% in citrus oils [175]. As for most terpenes, the microbial metabolism of the pinenes often leads to diverse degradation pathways and therefore to a wide variety of products [176]. This effect is even more marked in the case of the structurally more complex bicyclic monoterpenes compared with the aforementioned acyclic and monocyclic monoterpenes [177]. Some interesting terpenoid compounds accessible by microbial conversion of α-pinene are illustrated in Scheme 23.11. From the biochemical engineer's viewpoint, microbial pinene transformations yielding only one or a few main products are of special interest as they are close to commercialisation or may serve as the starting point for further improvements by process and/or genetic engineering approaches. **Verbenone**, an impact compound of rosemary oil, and its precursor ***trans*-verbenol** were described to be the main products of an α-pinene conversion using the self-isolated black yeast *Hormonema* sp., which has already been mentioned for the limonene transformations. Although yielding extraordinarily high concentrations with respect to pinene bioconversions (0.3 g L^{-1} verbenone and 0.4 g L^{-1} *trans*-verbenol after 96 h), the unwanted morphological characteristics of the microorganism restricted further process development, as mentioned above. Verbenone was also produced as the main product using *Aspergillus niger* in a two-step approach [178]: with resting cells, pregrown until the late-exponential phase in potato–dextrose broth with 6% (w/v) glucose, 200 mg L^{-1} α-pinene was converted into 32.8 mg L^{-1} verbenone after 6 h of incubation. The yield of **verbenol**, itself a valuable flavour compound with a fresh pine, ozone odour, was improved compared with that of UV mutant strains described earlier [179] by generating an intergeneric hybrid strain from an *Aspergillus niger* strain showing high product yields and *Penicillium digitatum* showing high biomass yields [180]. By this means (-)-*cis*-α-pinene was transformed into **(-)-*cis*-verbenol** at a yield of 60%; nevertheless, the corresponding product concentration of 1.08 mg g^{-1} biomass is obviously still too low for industrial application. One example where an outstanding productivity and product concentration was achieved is the formation of **(Z)-2-methyl-5-isopropyl-2,5-hexadienal (isonovalal)**, a non-plant fragrance compound with a citrus-like note for potential use in perfume formulations. It was produced from **α-pinene oxide** in concentrations of up to 400 g L^{-1} within 2.5 h using 25 g L^{-1} precultivated *Pseudomonas rhodesiae* CIP 107491 biomass; the cells had been permeabilised by freeze-thawing and organic solvent treatment prior to use. The bioprocess was performed with in situ product recovery using hexadecane in a biphasic medium and by sequential feeding of biomass and precursor to compensate the irreversible biocatalyst inactivation by the product. Recently, a bioreactor coupled to an external membrane module for in situ product removal was reported for the same biotransformation reaction with *Pseudomonas fluorescens* NCIMB 11671 [181]. A dense silicone membrane comprising 70% PDMS and 30% fumed silica coiled into a hexadecane reservoir was used to separate the aqueous fermentation broth, which was recirculated inside the membrane tubing from the organic phase. With an optimised continuous feeding of the precursor α-pinene oxide directly into the fermentation broth containing about 60 g L^{-1} biomass, a stable process for over 400 h was achieved and more than 100 g L^{-1} isonovalal in the organic phase was produced despite the limitations of the membrane area used.

The general concept of heterologous expression of terpene functionalising enzymes in tailored host microorganisms has been extensively pursued in recent years not only for the purpose of biochemical characterisation of novel enzymes, e.g. those from plant terpene biosynthesis (which is not the focus of this review; for examples see [182, 183]), but also for designing more efficient whole-cell biocatalysts. A recent example is the production of **perillyl alcohol** from limonene by using a recombinant cytochrome P450 alkane hydroxylase expressed in *Pseudomonas putida* [184]. The monooxygenase together with a ferredoxin reductase and a ferredoxin—a typical bacterial class 1 cytochrome P450 system—was encoded by an operon found in a novel *Mycobacterium* sp. strain termed HXN-1500. This strain had been selected from 1,800 mainly hydrocarbon degrading bacteria screened for their ability to hydroxylate limonene at the 7-position. With a recombinant *Pseudomonas putida* host strain which allowed selection for growth on alkanes if alkane hydroxylase is functionally expressed, a two-liquid-phase biotransformation in a 2-L bioreactor was performed after the cells had been pregrown to a cell density of approximately 10 g L^{-1} on *n*-octane. Bis(2-ethylhexyl)phthalate served as the organic phase for in situ precursor supply and product recovery from the aqueous phase, and 2.3 g L^{-1} perillyl alcohol was produced after 75 h under fed-batch conditions (feeding of *n*-octane as the carbon source), calculated for the total liquid content of the reactor. Although still showing a threefold lower enzyme activity than the wild-type *Mycobacterium* strain, which is categorised as safety class 2, this recombinant

Pseudomonas putida strain is a safety class 1 microorganism and, furthermore, it still harbours great potential for optimisation (e.g. substrate uptake, product export). Recently, with P450cam-catalysed camphor to 5-*exo*-hydroxycamphor transformation as a model reaction, a tenfold increase of the activity was generated by coexpression of the *Pseudomonas* P450cam system and glycerol dehydrogenase for enhanced cofactor (NADH) regeneration within recombinant *E. coli* cells [185].

The active site of P450cam was remodelled by site-directed mutagenesis and the most active double mutant Y96F-V247L showed completely different substrate and product spectra [186]: (-)-limonene and (+)-α-pinene were transformed with high regioselectivities and stereoselectivities to **(-)-*trans*-isopiperitenol** and **(+)-*cis*-verbenol** as main products (about 70% of all products formed), respectively. The triple mutant F87W-Y96F-V247L was less active but even more

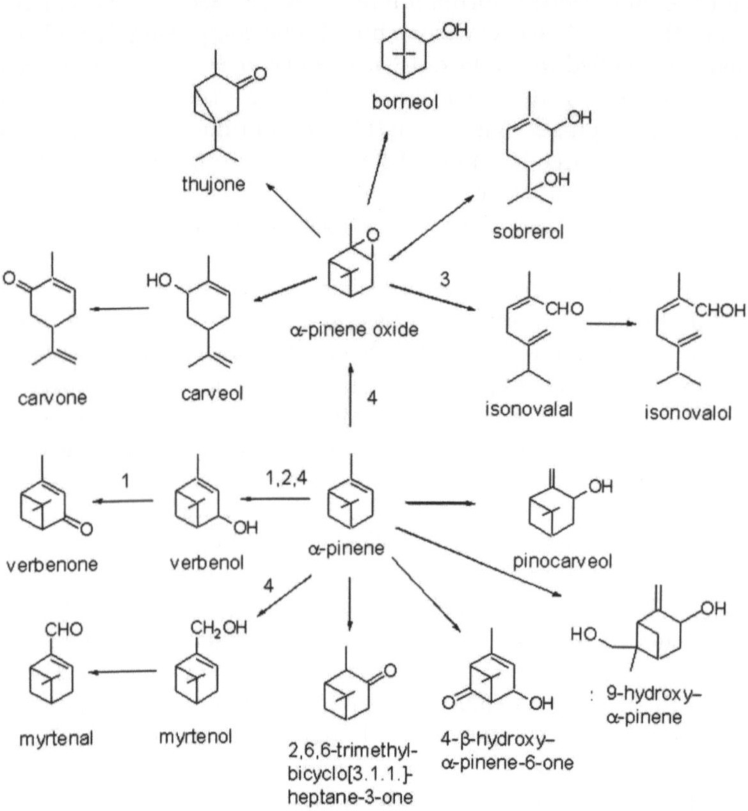

Scheme 23.11 Microbial transformations of α-pinene [176, 267–269]. *1 Hormonema* sp. UOFS-Y-0067, *Aspergillus niger*, recombinant *Escherichia coli* expressing an evolved mutant of P450cam from *Pseudomonas putida*; *2 Aspergillus niger + Penicillium digitatum* fusant strain; *3 Pseudomonas rhodesiae* CIP107491, *Pseudomonas fluorescens* NCIMB11671; *4* recombinant *Escherichia coli* expressing an evolved mutant of P450 BM3 from *Bacillus megaterium*

selective giving 85% (+)-*cis*-verbenol. By crystal-structure-based engineering of the active site another selective triple mutant (F87W-Y96F-L244A) was created which gave 86% (+)-*cis*-verbenol and 5% **(+)-verbenone**, while Y96F-L244A-V247L gave 55 and 32%, respectively [187, 188]. A triple mutant of P450 BM3 from *Bacillus megaterium* (F87V-L188Q-A74G) designed by rational evolution [189] capable of oxyfunctionalising hydrophobic aliphatic and aromatic substrates was exploited to convert α-pinene [164]. A recombinant *E. coli* strain was used as whole-cell biocatalyst in an aqueous–organic two-liquid-phase bioprocess to produce verbenol, **myrtenol** and α-pinene oxide in a total concentration of several hundreds of milligrams per litre after about 8 h of bioconversion.

23.4.3.3 Sesquiterpenes and Diterpenes, Norisoprenoids

Sesquiterpenes, biosynthetically derived from the trimeric precursor farnesyl diphosphate, constitute the structurally most diverse class of terpenoids and play key roles in food flavours and fragrances as well as pharmacologically active compounds. Their difficult total synthesis coupled with the abundance of non-functionalised, economically less important sesquiterpene hydrocarbons in many essential oils have stimulated much research during the last few decades dealing with sesquiterpene substrates from the over 70 subclasses [190]. The following discussion will be limited to only a few examples involving biotechnology which are of industrial relevance either because of the key character of the target flavour compounds or because of the progress in process development which has already been made.

The biotransformation of (+)-valencene, a sesquiterpene hydrocarbon found in orange oil, to **(+)-nootkatone** has attracted much research activity during the last few decades. (+)-Nootkatone possesses a citrus/grapefruit-like aroma and a bitter taste; it has a very low odour threshold (1 ppb) [175] and as a character-impact constituent of citrus aromas it is a very sought after natural flavour compound for foods and beverages. Recently, it has also been described to lower the somatic fat ratio, making it a natural product demanded by the cosmetic and fibre industries [191]. Although enzymatic cooxidation in the presence of lipoxygenase or laccase [192, 193] and bacterial valencene biotransformation with a *Rhodoccocus* strain [194] have been patented, it is doubtful that these processes will ever be applied owing to low specificities and/or activities. Recently, a relatively high selectivity was described for *Gynostemma pentaphyllum* cell cultures which converted valencene to nootkatone with 72% conversion yield corresponding to 650 mg L^{-1} nootkatone after 20 days [195]. **α-Nootkatol** (11%) and **β-nootkatol** (5%) were minor products; they are the direct metabolic precursors of nootkatone produced by an initial hydroxylation of valencene which, upon dehydrogenase-catalysed oxidation, are transformed into the target product (Scheme 23.12). Although this plant cell culture is obviously still too inefficient and too costly for commercial application, it is rather productive compared with other plant cell culture based biotransformations. Microsomal

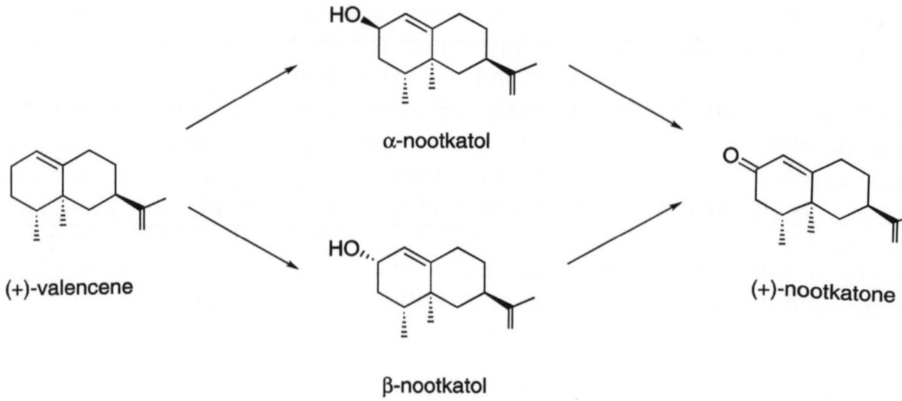

Scheme 23.12 Biotransformation of (+)-valencene to (+)-nootkatone via α-nootkatol and/or β-nootkatol

enzyme preparations from chicory (*Cichorium intybus* L.) roots have also been shown to catalyse the reaction from valencene to nootkatone as the main product with only negligible by-product formation [196]. Here, β-nootkatol turned out to be the only intermediate. Different higher fungi, such as *Mucor* species, *Botryosphaeria dothidea* and *Botryodiplodia theobromae*, and, interestingly, also green algae *Chlorella* species are also promising valencene-to-nootkatone biocatalysts [191]. For instance, *Chlorella pyrenoidosa* converted 89% of (+)-valencene added to the culture after 7 days of precultivation (20 mg in 50 mL) into (+)-nootkatone within a further 12 days, while *Chlorella vulgaris* even showed a conversion yield of 100% under the same conditions; with the fungus *Mucor* sp. a comparable yield of 82% was obtained after 7 days of precultivation followed by 7 days of biotransformation. During investigations with submerged cultures of the ascomycete *Chaetomium globosum*, it was found that the biotransformation proceeded via α-nootkatol as the intermediate and that major parts of the valencene and its monooxyfunctionalisation products accumulated within the cells, while dioxygenated and polyoxygenated products were found in the medium [197]. The bioprocessing limitations associated with the hindered mass transfer of terpenes across microbial cell membranes, especially the inefficient export of the transformation products out of the cells, may be overcome by an alternative cell preparation which has been described in a patent application very recently [198]. It is claimed to treat filamentous fungi known for their versatile terpene catabolism by lyophilising the mycelia prior to biotransformation which was preferentially carried out in an aqueous–organic two-phase system with *n*-decane as the organic phase. The authors claimed a better availability of the terpenes to the membrane enzymes after lyophilisation, leading to a more efficient biotransformation system; nevertheless, no yields have been reported.

Recently, an industrial process development for nootkatone production from valencene by microbial transformation (bacteria, fungi) was mentioned [199, 200]. Although no details were given, the author claimed the development of an in situ product-removal technique by which an extremely selective recovery of nootkatone from the reaction mixture and the excess precursor during the proceeding production was achieved and which was said to be essential for an economically viable bioprocess.

The same rational P450cam mutants which have already been described for limonene and pinene oxyfunctionalisations were also successfully applied to valencene. In whole-cell biotransformations *β*-nootkatol and **nootkatone** formed as main products with up to 25% overall yield, corresponding to activities of up to 9.9 nmol (nmol P450)$^{-1}$ min^{-1} [201]. Higher activities (up to 43 min^{-1}) but lower selectivities than those with P450cam were obtained with mutants derived from *Bacillus megaterium* P450 BM3.

The sesquiterpene aldehydes ***α*-sinensal** and ***β*-sinensal** contribute particularly to the special sweet orange aroma and also occur in other citrus oils; the former has a very low odour threshold of 0.05 ppb [175]. The sesquiterpene hydrocarbon farnesene may serve as closely related starting material and, consequently, farnesene isomers were used in biotransformations with *Arthrobacter, Bacillus, Nocardia*, and *Pseudomonas* with the aim to produce precursors of sinensal, but only little conversion was achieved when using the more stable farnesene sulfones [202]. Another strategy to produce *α*-sinensal starts from *trans*-nerolidol and aims at microbial ω-hydroxylation with fungi or bacteria, such as *Aspergillus* and *Rhodococcus* species, to produce 12-hydroxy-*trans*-nerolidol, which itself serves as precursor for the chemical conversion to the desired product [203–205] (Scheme 23.13). Certain self-isolated *Aspergillus* strains were shown to be very regioselective (74% of total product formed) [204]. The physiological state of an *Aspergillus* culture before nerolidol addition—monitored by on-line quantification of titrant addition in pH control—had a major impact on the biotransformation efficiency [205]. The maximal conversion yield of 8–9% was obtained by addition of a (±)-*cis*-nerolidol/(±)-*trans*-nerolidol mixture to the culture in the postexponential phase at high dissolved oxygen pressure (above 50% air saturation) in minimal and complex media after 25 and 14 h, respectively.

Patchouli alcohol (patchoulol) is a major constituent (30–45%) in steam distillates of dried leaves of *Pogostemon cablin* (Blanco) Benth; around 1,000 t of essential oil is produced worldwide per annum, primarily in Indonesia [49, 206]. Patchouli oil is very tenacious and is used in perfumery for oriental and masculine notes. The primary fragrance molecule in the essential oil is the sesquiterpene alcohol **norpatchoulenol**, which is present at 0.35–1.0% or less. In 1981, a combined biocatalytic and chemocatalytic method for the preparation of norpatchoulenol from patchoulol was published [207] (Scheme 23.14). The first step involved a microbial process to convert patchoulol to **10-hydroxypatchoulol**. *Pithomyces* species, filamentous fungi isolated from soil samples by enrichment on patchoulol as sole the carbon source, catalysed the regioselective

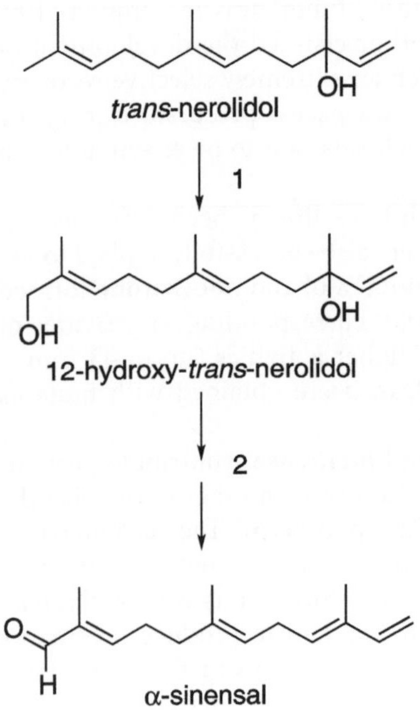

Scheme 23.13 Biocatalytic–chemocatalytic reaction sequence to produce α-sinensal from *trans*-nerolidol. *1 Aspergillus niger* sp., *Aspergillus niger* ATCC 9142, *Rhodococcus rubropertinctus* DSM 43197; *2* chemical conversion steps

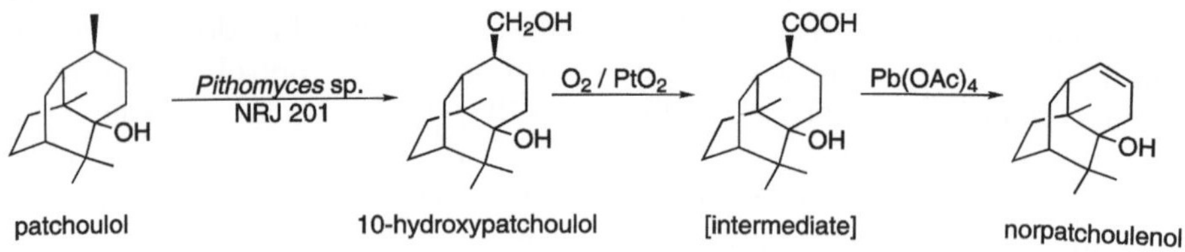

Scheme 23.14 Regioselective biohydroxylation of patchoulol and the following chemical steps to produce norpatchoulenol according to [207]

hydroxylation reaction with yields of up to 45%, corresponding to a maximum product concentration of 1.1 g L^{-1}. Maximum yields were achieved after biotransformation periods of 3–7 days which were carried out in 1–5-L bioreactors with fungal cultures pregrown in complex media for about 3 days. The 10-hydroxy compound was subsequently converted chemically via a two-step process to norpatchoulenol.

(-)-**Ambergris oxide** (Ambrox®) is one of the most important ambergris fragrance compounds and is a key compound of ambra, a secretion product of the sperm or cachalot whale, possibly resulting from pathological conditions [23, 49]. A novel microorganism, classified as *Hyphozyma roseoniger* CBC 214.83 (ATCC 20624), which can exist in both yeast-like and filamentous forms, was isolated and was capable of forming a diol from the diterpene alcohol sclareol found in the leaf oil from *Salvia sclarea* L.; the conversion proceeded in one microbial step via a cascade of reactions in high yields of more than 75%, but only after around 12 days of incubation [208]. Subsequently, other suitable microbial strains have been found by continued screening; e.g. the yeast *Cryptococcus albidus* ATCC 20918 which can metabolise sclareol even further, producing the ketone lactone sclareolide at high yields of more than 100 g L^{-1} [209]. The sclareolide is then chemically converted back to the diol and further to ambergris oxide (Ambrox®) (Scheme 23.15).

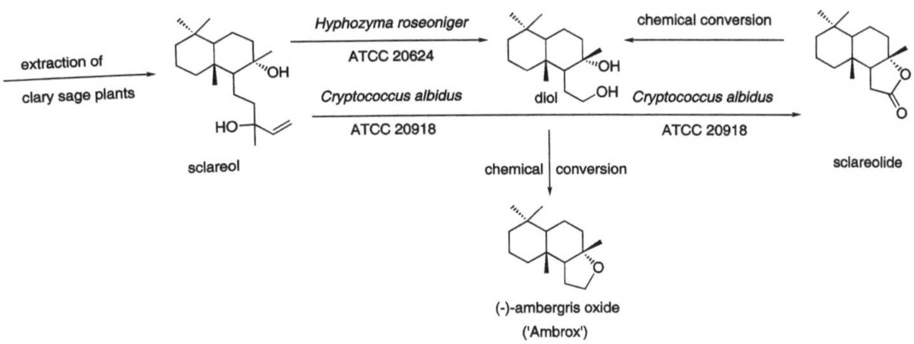

Scheme 23.15 Biocatalytic–chemocatalytic synthesis of Ambrox® (adapted from [270])

Although certain microorganisms, especially higher fungi, show a remarkable capability for de novo biosynthesis of terpenoid flavours, product titers of single terpenoid flavour molecules rarely exceed 100 mg L^{-1} and are, thus, too low for commercial processes. This situation may change dramatically in the near future owing to the great progress currently being made by metabolic engineering of microbial terpene biosynthesis and by heterologous expression of key enyzmes catalysing plant terpene functionalisation reactions in tailored host microorganisms. Recently, the total biosynthesis of terpenoids by engineering the mevalonate-dependent isoprenoid (MEV) pathway from *Saccharomyces cerevisiae* in *Escherichia coli* thereby alleviating the bacterial 1-deoxy-d-xylulose-5-phosphate (DXP) pathway has been reported [210]. By this means, the sesquiterpene amorphadiene, a precursor of the antimalaria drug artemisinin, was produced by successfully cloning a sequence comprising nine genes leading from acetyl-CoA via the universal C5 units isopentenyl pyrophosphate (IPP) and dimethylallyl pyrophosphate (DMAPP) to the target compound. Because IPP and DMAPP are the ultimate precursors for all terpenoids, such a strain, after further enhancing its metabolic terpene flux, may serve as a platform cell factory for de novo biosynthesis of any terpenoid for which the biosynthetic genes are available, i.e. flavour and fragrance compounds included. The same group also showed that engineering both *Escherichia coli*'s DXP pathway as well as *Saccharomyces cerevisiae*'s MEV pathway in the respective native hosts can also serve as a promising alternative strategy to design high-performing terpenoid producer strains [211, 212].

Closely related to terpenes and thus generally considered a subclass are the C13 norisoprenoids, e.g. **α-ionone and β-ionone**, volatile ketones generated by oxidative degradation of carotenoids, and **irones**, e.g. **α-irone and γ-irone**, C14 ketones derived from the triterpenes of the iridal type (Scheme 23.16). The best biotechnological strategy for the production of the most important natural ionone, β-ionone, a violet-like flavour and fragrance compound (threshold in water 0.007 ppb), still relies on the cooxidative cleavage of carotenoid-rich raw materials using enzymatic oxidation systems, e.g. lipoxygenase or xanthine oxidase in the presence of unsaturated fatty acids. These enzymes initially oxidise the unsaturated fatty acids, e.g. linoleic or linolenic acid, to free-radical species which themselves attack the conjugated double bonds of carotenoids, resulting in a non-specific cleavage pattern and thus broad product spectra [61, 213].

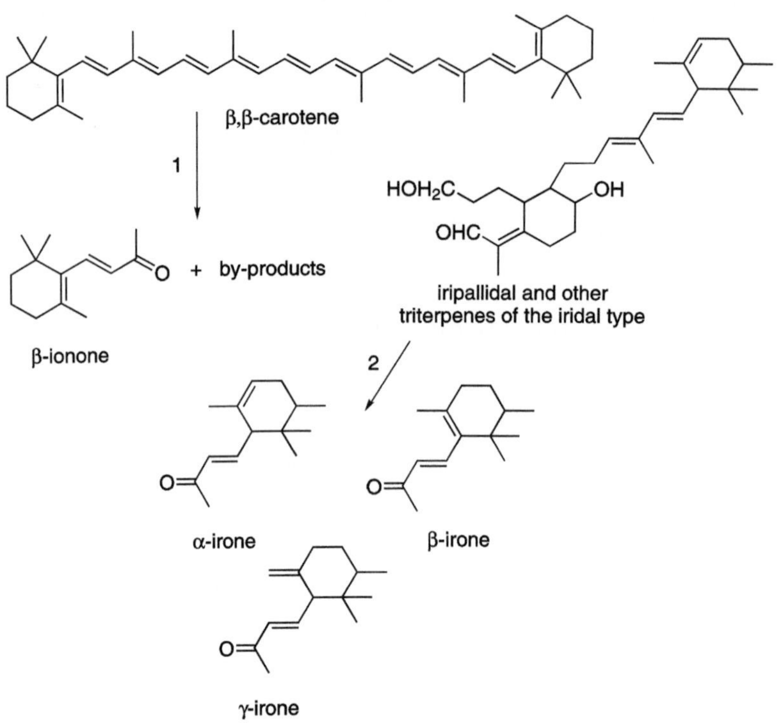

Scheme 23.16 Microbial pathways from triterpene and tetraterpene (carotenoid) precursors to valuable flavour and fragrance compounds. *1* Carotenoid-cleaving peroxidase-containing supernatant of certain fungal cultures, e.g. *Lepista irina*; *2 Serratia liquefaciens*, *Botrytis* sp.

A direct cleavage mechanism has been proposed for novel carotenoid-cleaving peroxidases found in the basidiomycete *Lepista irina* [214] and other fungi [215]. Whereas submerged cultures did not accumulate significant amounts of volatile degradation products (probably owing to total catabolism), mycelium-free supernatants yielded β-ionone and further degradation products. Microbial whole-cell approaches are not yet used in industry for the production of natural ionones but, here as well, metabolic engineering might grant access to the desired target compounds in a superior way, i.e. highly regioselective reaction, in the future. On the one hand, carotenoid-producing *E. coli* strains have successfully been designed during the past decade [216] and, on the other hand, the native plant enzymes responsible for the regioselective cleavage of carotenoids to produce C13 norisoprenoids and other highly desired flavour-active apocarotenoids, the carotenoid-cleaving dioxygenases, have recently been identified and functionally expressed in *E. coli* for the first time [217]. If the technical potential of such engineered organisms is evaluated, in situ product recovery strategies will certainly be needed to circumvent catabolism of the volatiles by their rapid removal from the cells.

Conventional approaches to produce irones, also valuable fragrance compounds with a typical violet-like (orris) odour, were published in the 1980s. It is known that during storage of *Iris* rhizomes the content of the desired irones increases slowly, probably by chemical oxidative degradation of the triterpenes initially present in the rhizomes; therefore, prior to the production of the orris root oil by steam distillation the rhizomes are stored for several years [49]. Bacteria, especially *Serratia liquefaciens*, isolated from *Iris palladia* rhizomes, were used to convert rhizome preparations naturally containing the triterpenoid iridales to the desired target compounds [218, 219]. By this means, the four natural isomeric irones, ***trans*-α-irone, *cis*-α-irone, *cis*-γ-irone** and **β-irone** formed in similar proportions as in the traditionally processed rhizomes and a maximum irone content of 1.2 g kg^{-1} was obtained after 3–4 days of cultivation. Owing to an early discontinuance of microbial growth which already occurred at day 1 (probably because of toxic effects of the products), the irone formation was supposed to be a combined biocatalytic–chemocatalytic reaction sequence initialised by microbial activity. *Botrytis* species were also claimed as biocatalysts for the same purpose [220]. Dried organic solvent extracts of the rhizomes were added as an emulsion in water–acetone–Tween 80 after 2 days to the fungal culture in a corn steep liquor medium. Up to 2.3 g kg^{-1} irones was produced after a further 48 h and the irones were isolated by steam distillation.

23.4.4 Lactones

Saturated and unsaturated γ-lactones and δ-lactones which are synthesised from the corresponding acyclic hydroxy fatty acids by intramolecular esterification are important flavour compounds found ubiquitously in fruits and also in milk and fermentation products in parts-per-million concentrations. The natural lactones belong to the most desired targets for aroma biotechnology. It seems as if almost every big flavour company has claimed a preparation method starting from natural fatty acids, hydroxy fatty acids or unsaturated lactones as precursors during the last two decades. Scheme 4.17 summarises different strategies for the production of natural *γ*-lactones and *δ*-lactones.

4-Decanolide (γ-decalactone), which imparts a powerful fruity, especially peach-like aroma has a market volume of several hundred tons per year. In the early 1980s, natural 4-decanolide was an extremely expensive, rare natural flavour (price in excess of US $10,000 per kilogram). The subsequent introduction and optimisation of its biotechnological production has resulted in a steady decrease of the price to approximately US $300 per kilogram and an increase of the market volume to several tons per year [8].

Most of the commercial processes for the formation of 4-decanolide are based on the natural hydroxy fatty acid ricinoleic acid [(*R*)-12-hydroxy-(*Z*)-9-octadecenoic acid], the main fatty acid of castor oil, or esters thereof as substrates and fatty acid degrading yeasts or higher fungi as biocatalysts [221]. Ricinoleic acid is degraded by four cycles of β-oxidation and one double-bond hydrogenation into 4-hydroxydecanoic acid, which lactonises at lower pH to 4-decanolide, resulting in the same *R* configuration of the lactone as is found in peaches and other fruits [222]. Many processes for which high product concentrations have been reported are based on strains of *Yarrowia lipolytica*, a yeast which is particularly well adapted to hydrophobic environments and which was patented for 4-decanolide production for the first time in 1983 [223].

In a process established by Haarmann & Reimer, up to 11 g L^{-1} 4-decanolide was obtained in 55 h with a wild-type strain and with raw castor oil as the substrate [224]. Metabolic engineering of *Yarrowia lipolytica* aims at optimising the flux along the complex β-oxidation pathway and decreasing the formation of unwanted by-products, such as 3-hydroxy-4-decanolide, dec-2-en-4-olide and dec-3-en-4-olide [225, 226]. More genetic engineering approaches with *Yarrowia lipolytica* can be expected in the future since its total genome has been sequenced recently [227]. An elegant conventional method to improve the overall yield of 4-decanolide uses baker's yeast for reduction of the double bond of the decenolides produced as by-products [70, 222].

Another process patented by Givaudan uses *Mucor circinelloides* as a biocatalyst for the production of 4-decanolide [228]. Here the natural substrate is the ethyl ester of decanoic acid which is isolated from coconut oil. The key microbial activity harnessed in this process is the stereoselective and regioselective hydroxylation of the fatty acid in the γ-position, which is followed by spontaneous lactonisation of the hydroxy fatty acid under acidic conditions and results in yields of up to 10.5 g L^{-1} 4-decanolide after 60 h.

The closely related **5-decanolide (*δ*-decalactone)**, not only found in many fruits but also found in dairy products, exhibits a creamy-coconut, peach-like aroma [49] and can be synthesised from the corresponding α,β-unsaturated lactone 2-decen-5-olide found in concentrations of up to 80% in *Massoi* bark oil using basidiomycetes or baker's yeast [229]. After about 16 h of fermentation, 1.2 g L^{-1} 5-decanolide was obtained. At the same time, the minor lactone in

Massoi bark oil, 2-dodecen-5-olide (7%), is converted to **5-dodecanolide**, which is also a desired fruity lactone. Different bacteria were used for the same *Massoi* lactone conversion in a medium containing natural oils as cosolvents for dissolving the precursor [230]. From 30-L culture volume, 195 g 5-decanolide was isolated after 48-h aerobic biotransformation with *Pseudomonas putida* ATCC 33015, corresponding to a conversion yield of 99.1%. More recently, growing *Saccharomyces cerevisiae* cells were claimed to be used in a two-phase bioprocess with triglycerides or high molecular weight hydrocarbons, e.g. Neobee® (C8–C10 fatty acid triglyceride), olive oil or hexadecane, as the organic phase containing the *Massoi* lactones as precursors [231]. With use of this two-phase system, toxic effects of the precursors and the products on the cells were avoided and further downstream processing was facilitated. Feeding dextrose to adjust a low operational concentration (preferably at 0.03–0.07 g L^{-1}) and maintaining a sufficiently high oxygen supply (more than 10% dissolved oxygen pressure) yielded maximum 5-decanolide and 5-dodecanolide concentrations of up to 7.45 and 1.7 g L^{-1} after 60–70 h, respectively. Other strategies for 5-decanolide production start from other natural precursors, such as 11-hydroxypalmitic acid (sweet potato, Jalap resin) and coriolic acid (13-hydroxyoctadeca-9,11-dienoic acid) (*Coriana nepalensis* seed oil) and use *Saccharomyces cerevisiae* and *Cladosporium suaveolens* as biocatalysts [222].

Oxidation of oleic acid to 10-hydroxyoctadecanoic acid by a gram-positive bacterium was described with a transformation yield of 65% at a concentration of 50 g L^{-1} oleic acid after 72 h in a medium containing Tween 80 [232]. The hydroxy fatty acid can be converted to **4-dodecanolide**, an important coconut-fruity like lactone, by β-oxidation with yeasts, affording a total lactone yield of about 20% from oleic acid [222, 232].

The bioconversion of native oils, e.g. sunflower, castor oils and especially coconut oil, which is rich in octanoic acid, with fungal catalysts, such as *Cladosporium suaveolens*, *Aspergillus niger* or *Pichia etchellsii*, yields about 1 g L^{-1} **4-octanolide**, which is also a desired lactone-type flavour compound with a sweet herbaceous coconut-like odour [233]. Even higher concentrations of up to 7.56 g L^{-1} were obtained in a bioreactor with octanoic acid or its ethyl ester as a substrate and *Mortiella isabellina* as a biocatalyst [234]. The bioconversion was carried out in a complex nutrient broth with 0.05% Tween 80 as cosolvent and 0.5 vvm aeration at pH 6 and 27 °C. After 5 h the bioconversion was started by feeding ethyl octanoate (or octanoic acid) and after 77 h the reaction was completed by acidifying the culture broth to pH 2–3 and heating it to 121 °C for 15 min (lactonisation). The same precursor was converted to 11.2 g L^{-1} 4-octanolide *by Mucor circinelloides* within 47 h [235]. **4-Hexanolide** can be produced by a homologous strategy from natural hexanoic acid found in palm, milk and coconut fats using *Aspergillus oryzae* or *Mortiella isabellina* as biocatalysts in a two-phase system (e.g. Primol® as an organic phase containing hexanoic acid) with sufficient oxygen supply [236]; final product concentrations of up to 19.4 g L^{-1} 4-hexanolide were obtained, while more than 16 g L^{-1} **2-pentanone** formed as an additional, valuable flavour-active product during the same cultivation. The methylketone was recovered from the exhaust air by trapping it on charcoal. **5-Octanolide**, naturally found in meat, cheese, fermented beverages and fruits, can be produced biotechnologically as a by-product besides 5-decanolide when a mixture of 11-hydroxypalmitic acid and 3,11-dihydroxymyristic acid from Jalap resin is converted by *Saccharomyces cerevisiae* [237].

The twofold unsaturated short-chain lactone **6-pentyl-α-pyrone** imparts a strong coconut-like odour and, interestingly, it was found to be the major volatile product from de novo biosynthesis of the fungus *Trichoderma*, with concentrations of up to 200 mg L^{-1}, which was described in the early 1970s [238] (Scheme 4.18). After an extended cultivation of 27 days, the harvested fermentation broth was processed by organophilic pervaporation and about 1 g L^{-1} calculated on the basis of culture volume was recovered; the efficiency of coupling organophilic pervaporation to the bioreactor for continuous product removal was limited by too low feed concentrations of the aroma compound [239]. Other in situ product-removal techniques, such as adsorption to XAD resins and aqueous–organic two-liquid-phase fermentation [240, 241], have also been tried to enhance overall yields by circumventing product inhibition effects which already occur at low 6-pentyl-α-pyrone concentrations (100 mg L^{-1}). The combination of in situ product removal by extractive bioconversion and cofermenting *Rhizoctonia solani* as an elicitor strain showed a significantly positive effect on 6-pentyl-α-pyrone production with *Trichoderma harzianum* [242]. The presence of non-viable mycelium of the phytopathogenic fungus *Rhizoctonia solani* led to an increase of product concentration from 147 to 474 mg L^{-1} and a decrease of process time from 192 to 96 h. A surface culture of *Trichoderma harzianum* was shown to be superior to a submerged culture which produced 455 mg L^{-1} 6-pentyl-α-pyrone after 96 h and 167 mg L^{-1} after 48 h, respectively, under the same bioreactor conditions [243].

Musks are important ingredients of fragrance formulations, but almost all the musks used are polycyclic aromatics produced chemically from petrochemically derived raw materials. Naturally occurring musks include the macrocyclic lactones found in some plants, such as ambrette seedoil and galbanum, and the keto musks produced by some animals, such as musk deer and civet cats.

The macrocyclic lactones are preferred to traditionally synthesised nitromusk compounds owing to their better skin compatibility and natural degradation [222]. **Hexadecanolide** is efficiently produced by a combined biosynthetic and chemosynthetic reaction sequence: the yeast *Torulopsis bombicola* converts palmitic acid, its ester (or even hexadecane), by ω-hydroxylation and ω-1-hydroxylation in very high yields of up to 40% [244] (Scheme 23.18). Owing to a concurrent glycosyl transfer, up to 300 g L^{-1} sophorolipids can be produced by this fermentative approach. Subsequent acid hydrolysis and lactonisation yielded hexadecanolide and **methylcyclopentadecanolide** in a 93:7 mixture, a difficult reaction as at high concentrations ω-hydroxypalmitic acid tends to polymerise. In a comparable process patented by a Japanese company, tridecane and pentadecane were converted by *Candida tropicalis* into the corresponding terminal dicarboxylic acids, which, upon chemical conversion and polymerisation steps, yielded the musk fragrance macrocycles **ethylene brassylate** and **cy-**

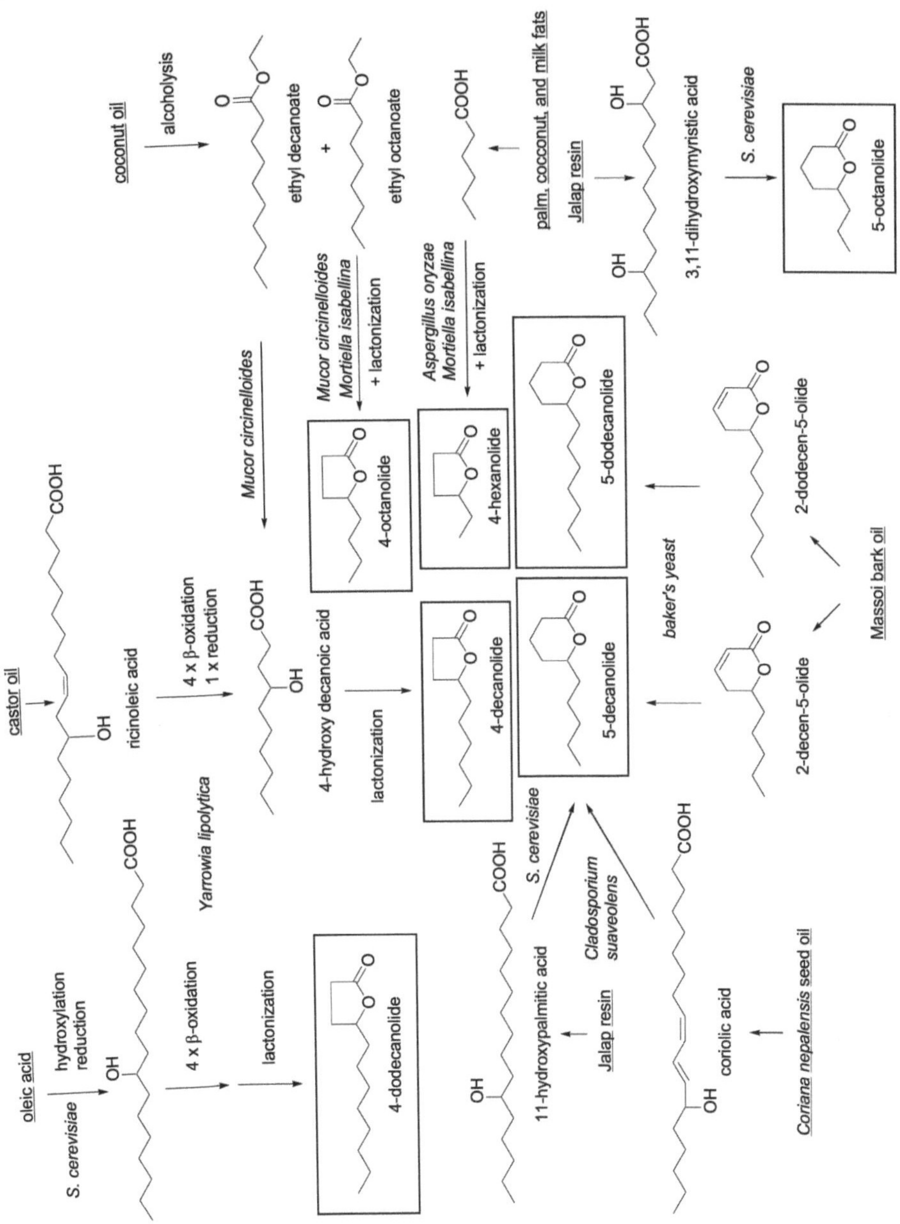

Scheme 23.17 Microbial processes for the production of natural flavour-active lactones

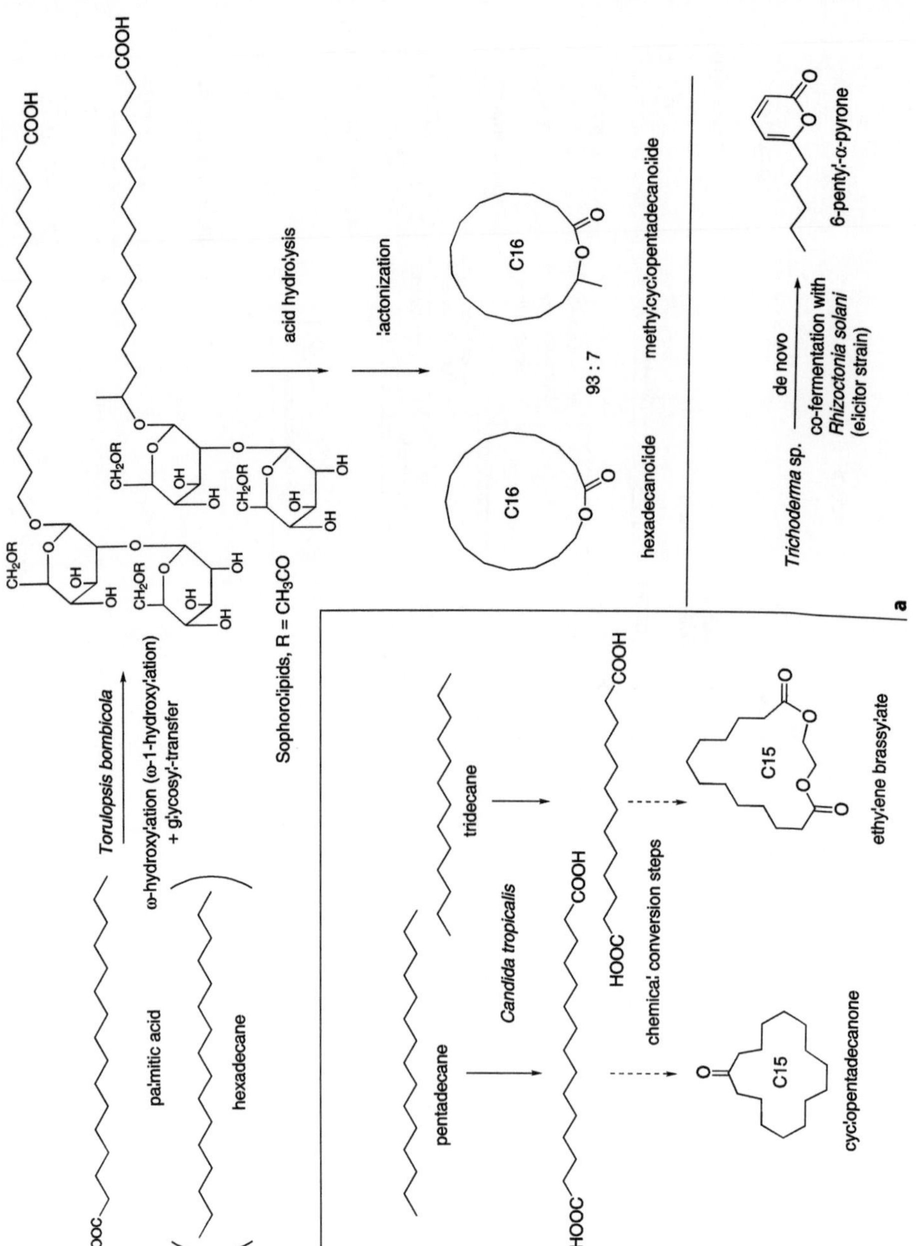

Scheme 23.18 **a** De novo biosynthesis of coconut-like 6-pentyl-α-pyrone by *Trichoderma* sp. **b** Production of macrocyclic musk-like lactones by a combination of microbial ω-hydroxylations and ω-1-hydroxylations and subsequent chemical conversion steps. **c** Production of macrocyclic musk fragrances initiated by terminal oxidation of hydrocarbons with *Candida tropicalis*

clopentadecanone, respectively [3] (Scheme 23.18). Final concentrations of up to 120 g L^{-1} and a final product purity of 94% at 20-m^3 scale were reported.

23.4.5
O-Heterocycles, S- and N-Containing Compounds

Besides the aforementioned lactones, **Furaneol®, 2,5-dimethyl-4-hydroxy-2*H*-furan-3-one** (DMHF), is another very important O-heterocyclic compound where biotechnology is involved in its manufacture. It is a key-impact compound of pineapple and strawberry aroma and is also found in savoury foods. It exhibits a pineapple, strawberry-like odour in dilute solutions and a caramel-like one in concentrates and is synthesised by heating rhamnose with an amine source, preferentially proline.

Biocatalysis can be used for the generation of rhamnose (6-deoxymannose) by the selective enzymatic hydrolysis of plant-derived flavanoid glycosides containing rhamnose in the terminal position, such as naringin or rutin (Scheme 23.19). The yield of the subsequent flavour-development step by heating is reduced by even small traces of glucose, which cause an off-taste. This problem can also be solved biocatalytically by eliminating the glucose via selective conversion of the glucose using immobilised *Saccharomyces cerevisiae* (to ethanol and CO_2) or using *Gluconobacter suboxydans* to (to 5-ketogluconic acid, which is precipitated as the calciumsalt) [70]. Pure rhamnose for DMHF production may also be produced by cultivating *Pseudomonas aeruginosa*, which synthesises large amounts of rhamnolipids in oil-containing media (50 to more than 130 g L^{-1}) [245, 246], and subsequent selective rhamnolipid hydrolysis and purification [247]. Recently, a nonpathogenic species, *Pseudomonas chlororaphis*, was described to produce about 1 g L^{-1} rhamnolipids on glucose, an amount comparable to the levels reported for the pathogenic *Pseudomonas aeruginosa* under the same conditions, which might give access to a food-grade strategy in the future [248].

The production of a closely related furanone starts with natural 5-oxo-gluconic acid production from glucose with *Gluconobacter suboxydans;* the acid is recovered by precipitation as the calcium salt; for flavour applications, it is converted by heating to **4-hydroxy-5-methyl-2*H*-furan-3-one**, a typical savoury reaction flavour with a meat-like taste [70] (Scheme 23.19).

Owing to very low thresholds, volatile sulfur compounds (VSCs) usually have prime impact on food aromas; they are found in lots of natural sources, including fermented foods (e.g. wine, beer, cheese), and act as both flavours and off-flavours [249, 250]. Although their biogenetic formation has been elucidated in detail, only few biotechnological processes with potential for commercial application have been reported. The sulfur-containing amino acids L-methionine and L-cysteine are the natural precursors of a wide variety of VSCs. **Methanethiol** is the most frequently found VSC in cheese and can be readily oxidised to other VSCs, such as **dimethyl sulfide** and **dimethyl disulfide**, or

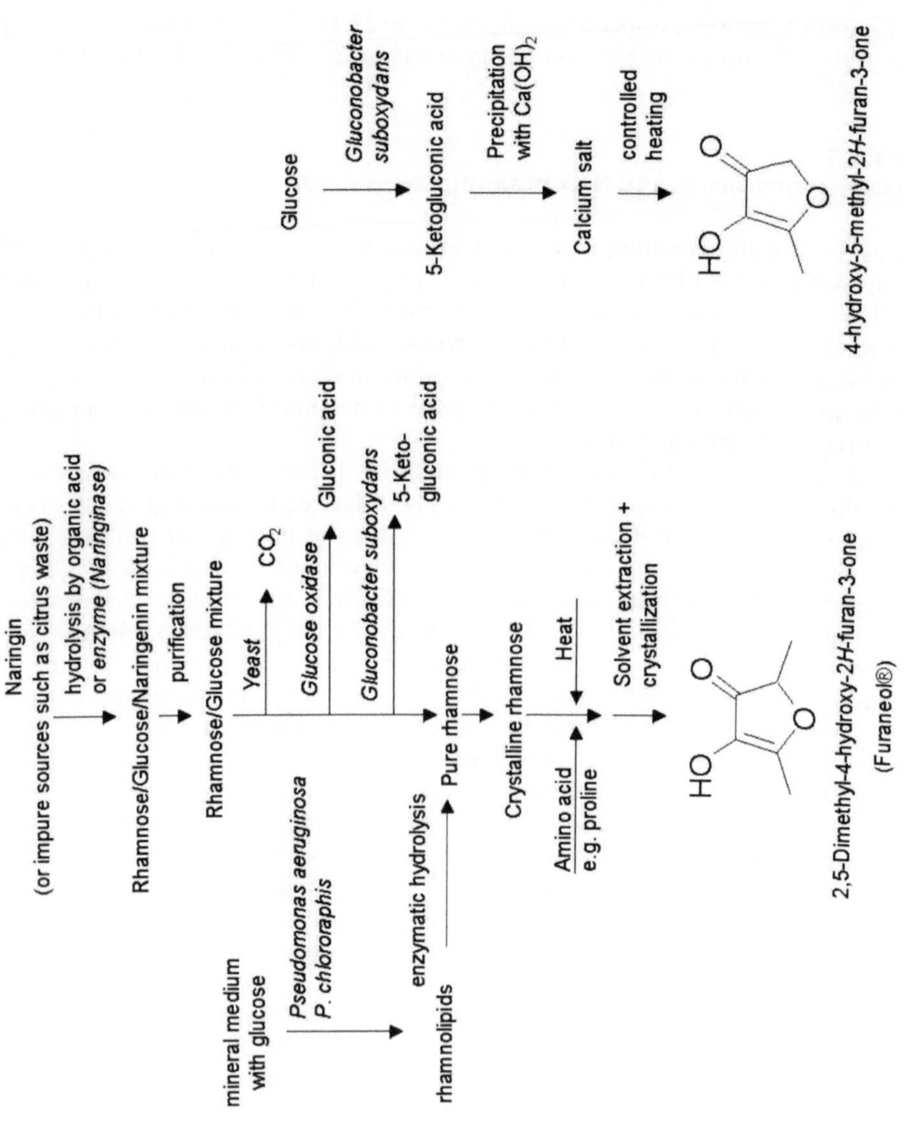

Scheme 23.19 Furanone production schemes involving biocatalytic steps (*italicised*) (adapted from [270, 271])

esterified to S-methylthioesters, e.g. ***S*-methyl acetate** [249] (Scheme 23.20). For instance, the cheese-ripening yeast *Geotrichum candidum* produces methanethiol and *S*-methyl acetate in the lower parts-per-million range [249, 251]. Recently, the gene encoding L-methionine-γ-lyase (MGL) was cloned from the cheese-ripening bacterium *Brevibacterium linens* [252]. MGL converts L-methionine to methanethiol, α-ketobutyrate and ammonia. The potato-like **me-**

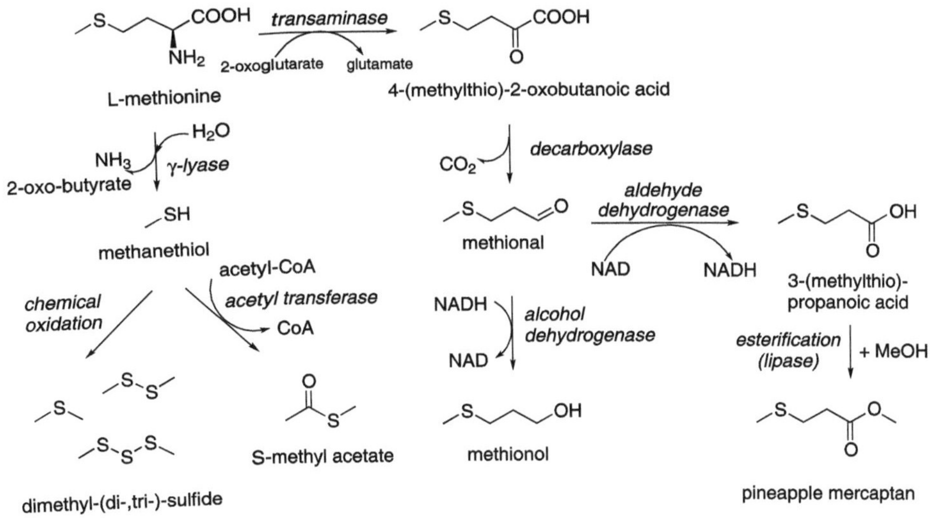

Scheme 23.20 Some sulfur-containing flavour compounds generated from L-methionine by microbial metabolism plus chemical or enzymatic transformations

thylthiopropanal (**methional**) was shown to be produced from L-methionine in concentrations of up to 62 mg L^{-1} by *Lactococcus lactis* under assay conditions [253]. The corresponding alcohol, **3-(methylthio)propan-1-ol**, known as **methionol**, which has also a potato-like odour, can be formed by yeast-based bioconversion of L-methionine following the Ehrlich pathway as already described for the other flavour-active alcohols (Sect. 23.4.2). Depending on the redox status of yeast cells, methional is reduced by alcohol dehydrogenase to methionol or is oxidised by aldehyde dehydrogenase to the corresponding acid, **3-(methylthio)propanoic acid** [129] (Scheme 23.20). Both products were obtained in total yields of up to 55% and concentrations of up to 11.2 g L^{-1} with different yeasts, especially with *Saccharomyces* and *Hansenula*, in a fermentation with a high biomass loading and glucose and precursor feeding [105, 254]. The acid can also be synthesised from L-methionine by oxidation with acetic acid bacteria as described for aliphatic acids in Sect. 23.4.1.1 [39]. The methyl ester of the acid, the important flavour compound **pineapple mercaptan**, can be obtained by subsequent lipase-catalysed esterification. Nevertheless, for labelling the products as 'natural', a natural source of L-methionine must be available, which is not the case so far, although fermentative L-methionine production has been improved during the last few years [255].

Furfurylthiol is a key flavour especially for coffee, beef and roast-like food aromas. It was synthesised in concentrations of up to 1 g L^{-1} using β-lyase activity of whole bacterial cells, e.g. *Enterobacter cloacae* or *Eubacterium limosum* [256] (Scheme 23.21). Resting cells were used to cleave the sulfur–carbon bond of a furfural–cysteine conjugate and an XAD-4 resin connected to the gas outlet

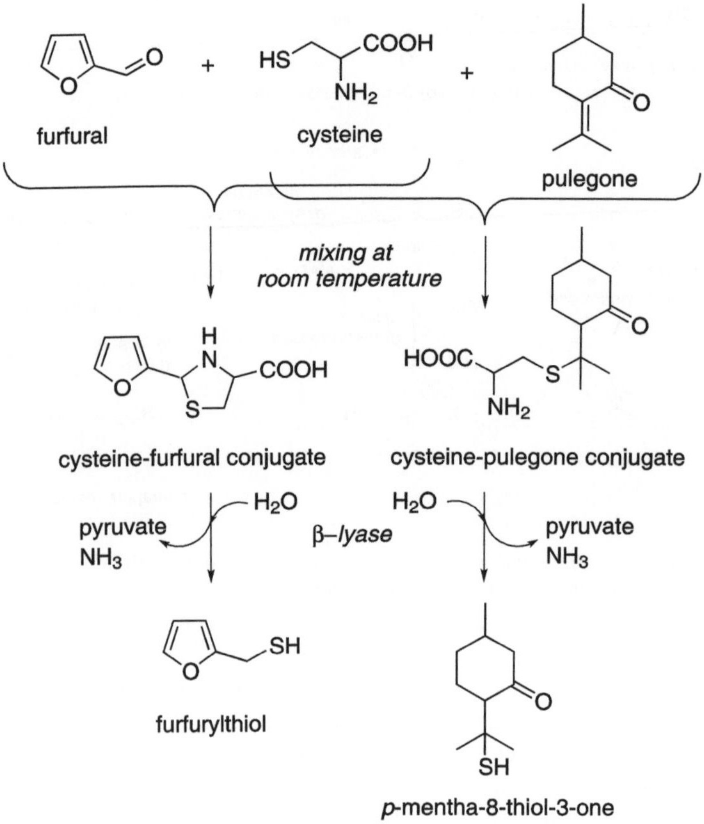

Scheme 23.21 Syntheses of valuable sulfur-containing flavour compounds involving β-lyase activity of *Enterobacter cloacae* or *Eubacterium limosum* cells

served for in situ product removal by periodical nitrogen flushing. The same β-lyase based strategy was pursued earlier to produce ***p*-mentha-8-thiol-3-one**, a very potent, extremely low-threshold blackcurrant flavour compound, from a pulegone–cysteine conjugate using *Eubacterium limosum* ATCC 10825 [257] (Scheme 23.21). After synthesis of the *S*-cysteinyl-pulegone simply by mixing at room temperature and allowing for precipitation over 3 days, the filtered conjugate was converted by resting cells, pregrown anaerobically on a complex medium, in a buffer system.

As for S-containing heterocycles, many N-containing heterocycles are also found in heat-treated foods as secondary flavours as a result of Maillard-type reactions between reducing sugars and amino acids. Pyrazines are N-heterocycles important contributors to the taste and aroma of roasted and toasted foods as well as vegetables and fermented foodstuffs. In cultures of *Pseudomonas perolens* ATCC 10757, amino acids such as valine, glycine and methionine were shown to

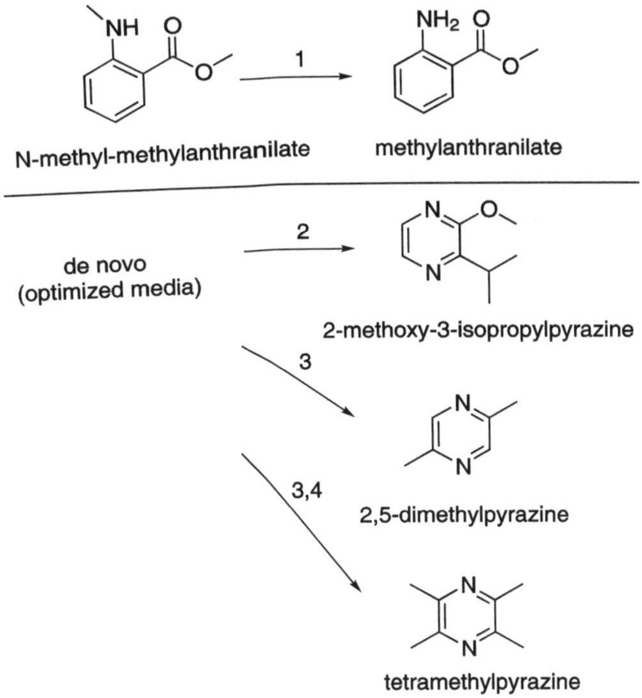

Scheme 23.22 Some nitrogen-containing flavour compounds produced by microorganisms. **a** Methylanthranilate formation from *N*-methyl methylanthranilate: *1 Trametes* sp., *Polyporus* sp. **b** Different pyrazines produced with microorganisms in optimised media: *2* mutant strain from *Pseudomonas perolens* ATCC 10757; *3 Bacillus subtilis, Brevibacterium linens*; *4* mutant strain of *Corynebacterium glutamicum*

be precursors for the production of the musty, potato-like **2-methoxy-3-isopropylpyrazine** [258], with lactate and pyruvate being important components to enhance pyrazine formation [259] (Scheme 23.22). Nevertheless, the final product concentrations remained low although medium optimisation and selection of a mutant strain auxotrophic for leucine increased 2-methoxy-3-isopropylpyrazine production from around 20 μg L^{-1} to 15.7 mg L^{-1}. *Bacillus subtilis* and *Brevibacterium linens* were shown to naturally produce pyrazines under solid substrate conditions from ground soy beans or soy flour enriched with L-threonine, acetoin and ammonia as precursors fed separately to the medium [69, 70, 260]. By this means, up to 4 g L^{-1} **2,5-dimethylpyrazine** and 1 g L^{-1} **tetramethylpyrazine** were produced. A mutant of *Corynebacterium glutamicum* deficient in a single enzyme of the isoleucine–valine pathway was described to produce 3 g L^{-1} tetramethylpyrazine after 5 days, which crystallised upon cooling of the culture broth; thiamine was found to be essential for product formation [261].

Methylanthranilate, which occurs in small quantities (0.5–3 mg L^{-1}) [87] in a large number of blossom essential oils, grapes and citrus oils is mainly used as a fragrance compound but also finds application as a flavour compound in grape and citrus compositions [49]. A patent describes the formation of the natural compound by microbial N-demethylation of *N*-methyl methylanthranilate that is abundant in 'petit grain' mandarin leaf oil [262] (Scheme 23.22). Higher fungi, preferentially *Trametes* and *Polyporus* strains, tolerating elevated concentrations of the toxic precursor were used; for instance, after biomass growth *T. versicolor* was supplemented with 0.2% (w/w) *N*-methyl methylanthranilate and converted it into methylanthranilate with a yield of about 30% after 3 days of incubation. *N*-formyl methylanthranilate formed as an undesired by-product but could be transformed to the target compound by heat or base-catalysed decarboxylation. Nevertheless, compared with this microbial demethylation approach, the use of free lipase to esterify natural anthranilate, found in protein hydrolysates from food processing, with methanol in biphasic media may be the more efficient strategy to yield the natural flavour compound [263].

References

1. Omelianski VL (1923) J Bacteriol 8:393
2. Welsh F, Murray WD, Williams RE (1989) Crit Rev Biotechnol 9:105
3. Janssens L, De Pooter HL, Schamp NM, Vandamme EJ (1992) Process Biochem 27:195
4. Hagedorn S, Kaphammer B (1994) Annu Rev Microbiol 48:773
5. Schreier P (1989) Food Rev Int 5:289
6. Berger RG, Krings U, Zorn H (2001) In: Taylor A (ed) Food flavour technology. Sheffield Academic Press, Sheffield, p 60
7. Gatfield IL (1999) In: Teranishi R, Wick EL, Hornstein I (eds) Flavor chemistry. Thirty years of progress. Kluwer/Plenum, New York, p 211
8. Schrader J, Etschmann MMW, Sell D, Hilmer JM, Rabenhorst J (2004) Biotechnol Lett 26:463
9. Gabelman A (1994) (ed) Bioprocess production of flavor, fragrance, and color ingredients. Wiley, New York
10. Berger RG (1997) (ed) Biotechnology of aroma compounds. Advances in biochemical engineering biotechnology. Springer, Berlin Heidelberg New York
11. Berger RG (1995) Aroma biotechnology. Springer, Berlin Heidelberg New York
12. Serra S, Fuganti C, Brenna E (2005) Trends Biotechnol 23:193
13. Hammes WP, Hertel C (1998) Meat Sci 49:125
14. Pretorius IS (2000) Yeast 16:675
15. Mollet B (1999) Int Dairy J 9:11
16. Randez-Gil F, Pascual S, Prieto JA (1999) Trends Biotechnol 17:237
17. Sikkema J, Bont de JAM, Poolman B (1995) Microbiol Rev 59:201
18. Sikkema J, Bont de JAM, Poolman B (1994) J Biol Chem 269:8022
19. Berger RG, De Bont JA, Eggink G, Da Fonseca M, Gehrke M, Gros JB, Van Keulen F, Krings U, Larroche C, Leak DJ, Van der Werf MJ (1999) In: Swift KAD (ed) Current topics in flavours and fragrances. Kluwer, Dordrecht, p 139
20. Bonnarme P, Feron G, Durand A (1996) Eur Food Drink Rev 67

21. DECHEMA (2004) (ed) White biotechnology: opportunities for Germany. Position paper DECHEMA, Frankfurt
22. Schmid RD (2003) Pocket guide to biotechnology and genetic engineering. Wiley-VCH, Weinheim
23. Burdock GA (2004) Fenaroli's handbook of flavor ingredients, 5th edn. CRC, Washington
24. Roehr M, Kubicek CP, Kominek J (1996) In: Rehm H-J, Reed G (eds) Biotechnology, vol 6. VCH, Weinheim, p 307
25. Karaffa L, Kubicek CP (2003) Appl Microbiol Biotechnol 61:189
26. Roehr M, Kubicek CP (1996) In: Rehm H-J, Reed G (eds) Biotechnology, vol 6. VCH, Weinheim, p 363
27. Gruber P, Henton DE, Starr J (2006) In: Kamm B, Gruber PR, Kamm M (eds) Biorefineries—industrial processes and products, vol 2. Wiley-VCH, Weinheim, p 381
28. Kascak JS, Kominek J, Roehr M (1996) In: Rehm H-J, Reed G (eds) Biotechnology, vol 6. VCH, Weinheim, p 293
29. Kwon S, Yoo I-K, Lee WG, Chang HN, Chang YK (2001) Biotechnol Bioeng 73:25
30. Jain MK, Zeikus JG (1999) In: Flickinger MC, Drew SW (eds) Encyclopedia of bioprocess technology: fermentation, biocatalysis, and bioseparation, vol 1. Wiley, p 150
31. Bai D-M, Wei Q, Yan Z-H, Zhao X-M, Li X-G, Xu S-M (2003) Biotechnol Lett 25:1833
32. Narayanan N, Roychoudhury PK, Srivastava A (2004) Electron J Biotechnol 7:167
33. Arctander S (1969) Perfume and flavor chemicals. Aroma chemicals. Arctander, Montclair
34. Werpy T, Frye J, Holladay J (2005) In: Kamm B, Gruber P, Kamm M (eds) Biorefineries—industrial processes and products, vol 2. Wiley-VCH, Weinheim, p 367
35. Taillade P, Lapadatescu C, Lavie-Cambot C (2003) In: Le Quéré J-L, Étiévant PX (eds) Flavour research at the dawn of the twenty-first century. Intercept, London, p 437
36. Kuninaka A (1996) In: Rehm H-J, Reed G (eds) Biotechnology, vol 6. VCH, Weinheim, p 561
37. Muheim A, Häusler A, Schilling B, Lerch K (1998) Perfum Flavor 23:21
38. Ebner H, Sellmer S, Follmann H (1996) In: Rehm H-J, Reed G (eds) Biotechnology, vol 6. VCH, Weinheim, p 381
39. Rabenhorst J, Gatfield IL, Hilmer J-M (2001) EP 1081229
40. Tachihara T, Hashimoto T, Ishizaki S, Komai T, Fujita A, Ishikawa M, Kitahara T (2006) In: Bredie WLP, Petersen MA (eds) Flavour science: recent advances and trends. Developments in food science, vol 43. Elsevier, Amsterdam, p 97
41. Suwannakham S, Yang ST (2005) Biotechnol Bioeng 91:325
42. Suwannakham S, Huang Y, Yang ST (2006) Biotechnol Bioeng 94:383
43. Zhu Y, Wu Z, Yang S-T (2002) Proc Biochem 38:657
44. Liu X, Zhu Y, Yang S-T (2006) Enzyme Microb Technol 38:521
45. Wu Z, Yang ST (2003) Biotechnol Bioeng 82:93
46. Häusler A, Münch T (1997) ASM News 63:551
47. Farbood MI, Blocker RW, McLean LB, Sprecker MA, McLean MP, Kossiakoff N, Kim AY, Hagedorn M (2002) US Patent 6,458,569
48. Vandamme EJ, Soetaert W (2002) J Chem Technol Biotechnol 77:1323
49. Surburg H, Panten J (2006) Common fragrance and flavor materials—preparation, properties and uses, 5th edn. Wiley-VCH, Weinheim
50. Goodrich RM, Braddock RJ, Parish ME, Sims CA (1998) J Food Sci 63:445
51. Duff SJB, Murray WD (1989) Biotechnol Bioeng 34:153

52. Duff SJB, Murray WD (1988) Biotechnol Bioeng 31:44
53. Duff SJB, Murray WD (1992) In: Charalambous G (ed) Food science and human nutrition. Elsevier, Amsterdam, p 1
54. Duff SJB, Murray WD (1988) Ann N Y Acad Sci 542:428
55. Murray WD, Duff JB (1990) Appl Microbiol Biotechnol 33:202
56. Nozaki M, Washizu Y, Suzuki N, Kanisawa T (1995) In: Étiévant P, Schreier P (eds) Bioflavour 95. INRA, Versailles, p 255
57. Gabelman A, Luzio GA (1998) US Patent 5,783,429
58. Molinari F, Villa R, Manzoni M, Aragozzini F (1995) Appl Microbiol Biotechnol 43:989
59. Molinari F, Aragozzini F, Cabral JMS, Prazeres DMF (1997) Enzyme Microb Technol 20:604
60. Muller BL, Dean C, Whitehead IM (1995) In: Étiévant P, Schreier P (eds) Bioflavour 95. INRA, Versailles, p 339
61. Belin JM, Dumont B, Ropert F (1998) US Patent 5,705,372
62. Goers SK, Ghossi P, Patterson JT, Young CL (1989) US Patent 4,806,379
63. Brunerie P, Koziet Y (1997) US Patent 5,620,879
64. Holtz RB, McCulloch MJ, Garger SJ, Teague RK, Phillips HF (2001) US Patent 6,274,358
65. Muheim A, Häusler A, Münch T (1999) BioWorld 1:2
66. Muller B, Gautier A, Dean C, Kuhn JC (1995) US Patent 5,464,761
67. Häusler A, Lerch K, Muheim A, Silke N (2001) US Patent 6,238,898
68. Schindler F, Seipenbusch R (1990) Food Biotechnol 4:77
69. Larroche C, Gros JB (1997) In: Berger RG (ed) Biotechnology of aroma compounds. Advances in biochemical engineering biotechnology. Springer, Berlin Heidelberg New York, p 179
70. Cheetham PSJ (1999) In: Flickinger MC, Drew SW (eds) Encyclopedia of bioprocess technology: fermentation, biocatalysis, and bioseparation. Wiley, New York, p 1004
71. Jyoti BD, Suresh AK, Venkatesh KV (2003) World J Microbiol Biotechnol 19:509
72. Hugenholtz J, Kleerebezem M, Starrenburg M, Delcour J, De Vos W, Hols P (2000) Appl Environ Microbiol 66:4112
73. Rattray FP, Myling-Petersen D, Larsen D, Nilsson D (2003) Appl Environ Microbiol 69: 304
74. Shukunobe Y, Takato S (1989) JP 01168256A
75. Okonogi S, Tomita M, Shimamura S, Toxama K, Miyagawa H, Fujimoto M (1992) JP 4169166
76. Verhue WMM, Tjan SB, Verrips CT, Van Schie BJ (1992) CA 2048977
77. Gupta KG, Yadav NK, Dhawan S (1978) Biotechnol Bioeng 20:1895
78. Bornscheuer UT, Kazlauskas RJ (1999) Hydrolases in organic synthesis. Wiley-VCH, Weinheim
79. Brenna E, Fuganti C, Serra S (2003) Tetrahedron Asymmetry 14:1
80. Molinari F, Villa R, Aragozzini F (1998) Biotechnol Lett 20:41
81. Molinari F, Marianelli G, Aragozzini F (1995) Appl Microbiol Biotechnol 43:967
82. Gandolfi R, Converti A, Pirozzi, D, Molinari F (2001) J Biotechnol 92:21
83. Armstrong DW, Brown LA (1994) In: Gabelman A (ed) Bioprocess production of flavor, fragrance and color ingredients. Wiley, New York, p 41
84. Xu Y, Dong W, Mu XQ, Zhao GA, Zhang KC (2002) J Mol Catal B 18:29
85. Farbood MI, Morris JA, Seitz EW (1987) US Patent 4,686,307
86. Hirosawa I, Aritomi K, Hoshida H, Kashiwagi S, Nishizawa Y, Akada R (2004) Appl Microbiol Biotechnol 65:68

87. Lomascolo A, Stentelaire C, Asther M, Lesage-Meessen L (1999) Trends Biotechnol 17:282
88. Desmurs JR, Giannotta D, Gelo-Pujic M, Role C, Lancelin P (2004) Perfum Flavor 29:32
89. Priefert H, Rabenhorst J, Steinbüchel A (2001) Appl Microbiol Biotechnol 56:296
90. Rabenhorst J, Hopp R (2000) US Patent 6,133,003
91. Muheim A, Müller B, Münche T, Wetli M (1998) EP 0885968A1
92. Heald S, Myers S, Robbins K, Walford T, Hill CA (2004) WO 2004085663
93. Hopp R, Rabenhorst J (1994) US Patent 5,371,013
94. Plaggenborg R, Overhage J, Loos A, Archer JA, Lessard P, Sinskey AJ, Steinbuchel A, Priefert H (2006) Appl Microbiol Biotechnol 72:745
95. Lesage-Meessen L, Lomascolo A, Bonnin E, Thibault JF, Buleon A, Roller M, Asther M, Record E, Ceccaldi BC, Asther M (2002) Appl Biochem Biotechnol 102–103:141
96. Clarke GS (1995) Perfum Flavor 20:53
97. Lomascolo A, Asther M, Navarro D, Antona C, Delattre M, Lesage-Meessen L (2001) Lett Appl Microbiol 32:262
98. Berger RG, Böker A, Fischer M, Taubert J (1999) In: Teranishi R, Wick EL, Hornstein I (eds) Flavor chemistry: 30 years of progress. Kluwer/Plenum, New York, p 229
99. Lapadatescu C, Feron G, Vergoignan C, Djian A, Durand A, Bonnarme P (1997) Appl Microbiol Biotechnol 47:708
100. Kawabe T, Morita H (1994) J Agric Food Chem 42:2556
101. Lamer T, Spinnler HE, Souchon I, Voilley A (1996) Process Biochem 31:533
102. Lomascolo A, Lesage-Meessen L, Labat M, Navarro D, Delattre M, Asther M (1999) Can J Micriobiol 45:653
103. Boker A, Fischer M, Berger RG (2001) Biotechnol Prog 17:568
104. Zorn H, Fischer-Zorn M, Berger RG (2003) Appl Environ Microbiol 69:367
105. Whitehead IM (1998) Food Technol 52:40
106. Kosjek B, Stampfer W, van Deursen R, Faber K, Kroutil W (2003) Tetrahedron 59:9517
107. Fabre CE, Blanc PJ, Goma G (1998) Perfum Flavor 23:43
108. Abraham WR, Hoffmann HMR, Kieslich K, Reng G, Stumpf B (1985) In: Porter R, Clark S (eds) Enzymes in organic synthesis. Ciba Foundation symposium 111, London, 15–17 May 1984. Pitman, London, p 146
109. Etschmann MM, Bluemke W, Sell D, Schrader J (2002) Appl Microbiol Biotechnol 59:1
110. Fabre CE, Blanc PJ, Goma G (1996) Sci Aliments 16:61
111. Janssens L, de Pooter HL, Vandamme EJ, Schamp NM (1988) In: Schreier P (ed) Bioflavour '87. de Gruyter, Berlin, p 453
112. Ehrlich F (1907) Ber Dtsch Chem Ges 40:1027
113. Akita O, Ida T, Obata T, Hara S (1990) J Ferment Bioeng 69:125
114. Albertazzi E, Cardillo R, Servi S, Zucchi G (1994) Biotechnol Lett 16:491
115. Fabre CE, Blanc PJ, Marty A, Goma G (1996) Perfum Flavor 21:27
116. Fabre CE, Blanc PJ, Goma G (1998) Biotechnol Prog 14:270
117. Etschmann MMW, Sell D, Schrader J (2003) Biotechnol Lett 531
118. Stark D, Münch T, Sonnleitner B, Marison IW, von Stockar U (2002) Biotechnol Prog 18:514
119. Etschmann MM, Sell D, Schrader J (2005) Biotechnol Bioeng 92:624
120. Etschmann MM, Schrader J (2006) Appl Microbiol Biotechnol, 71:440
121. Fujiwara D, Yoshimoto H, Sone H, Harashima S, Tamai Y (1998) Yeast 14:711
122. Mason AB, Dufour JP (2000) Yeast 16:1287
123. Dickinson JR, Harrison SJ, Hewlins MJE (1998) J Biol Chem 273:25751

124. Dickinson JR, Harrison SJ, Dickinson JA, Hewlins MJE (2000) J Biol Chem 275:10937
125. Dickinson JR, Salgado LEJ, Hewlins MJE (2003) J Biol Chem 278:8028
126. Wittmann C, Hans M, Bluemke W (2002) Yeast 19:1351
127. Verstrepen KJ, Derdelinckx G, Dufour JP, Winderickx J, Pretorius IS, Thevelein JM, Delvaux FR (2003) FEMS Yeast Res 4:285
128. Verstrepen KJ, Van Laere SDM, Vanderhaegen BMP, Derdelinckx G, Dufour JPPIS, Winderickx J, Thevelein JM, Delvaux FR (2003) Appl Environ Microbiol 69:5228
129. Vuralhan Z, Luttik MA, Tai SL, Boer VM, Morais MA, Schipper D, Almering MJ, Kotter P, Dickinson JR, Daran JM, Pronk JT (2005) Appl Environ Microbiol 71:3276
130. Manzoni M, Molinari F, Tirelli A, Aragozzini F (1993) Biotechnol Lett 15:341
131. Molinari F, Gandolfi R, Aragozzini F, Leon R, Prazeres DMF (1999) Enzyme Microb Technol 25:729
132. Gandolfi R, Ferrara N, Molinari F (2001) Tetrahedron Lett 42:513
133. Gandolfi R, Cavenago K, Gualandris R, Sinisterra Gago JV, Molinari F (2004) Process Biochem 39:747
134. Converti A, Gandolfi R, Zilli M, Molinari F, Binaghi L, Perego P, Del Borghi M (2005) Appl Microbiol Biotechnol 67:637
135. Romano A, Gandolfi R, Molinari F, Converit A, Zilli M, Del Borghi M (2005) Enzyme Microb Technol 36:432
136. F. Chamoleau, C. Hagedorn, O. May, H. Gröger, Flavour & Fragrance Journal, 2007, in press.
137. Nijkamp K, van Luijk N, de Bont JA, Wery J (2005) Appl Microbiol Biotechnol 69:170
138. Hill RA (1993) In: Thomson RH (ed) The chemistry of natural products. Blackie, London, p 106
139. McCaskill D, Croteau R (1997) In: Berger RG (ed) Biotechnology of aroma compounds. Advances in biochemical engineering biotechnology. Springer, Berlin Heidelberg New York, p 107
140. Aharoni A, Jongsma MA, Bouwmeester HJ (2005) Trends Plant Sci 10:594
141. Little DB, Croteau RB (1999) In: Teranishi R, Wick EL, Hornstein I (eds) Flavor chemistry. Thirty years of progress. Kluwer/Plenum, New York, p 239
142. Schäfer S, Schrader J, Sell D (2004) Process Biochem 39:2221
143. Schrader J, Berger RG (2001) In: Rehm HJ, Reed G (eds) Biotechnology, vol 10. Wiley-VCH, Weinheim, p 373
144. Demyttenaere JCR (2001) In: Atta-ur-Rahman (ed) Bioactive natural products (part F). Studies in natural products chemistry, vol 25. Elsevier, Amsterdam, p 125
145. van der Werf MJ, Bont de JAM, Leak DJ (1997) In: Berger RG (ed) Biotechnology of aroma compounds. Advances in biochemical engineering biotechnology. Springer, Berlin Heidelberg New York,p 147
146. Busmann D, Berger RG (1994) J Biotechnol 37:39
147. Iurescia S, Marconi AM, Tofani D, Gambacorta A, Paterno A, Devirgiliis C, van der Werf MJ, Zennaro E (1999) Appl Environ Microbiol 65:2871
148. Onken J, Berger RG (1999) Appl Microbiol Biotechnol 51:158
149. Demyttenaere JC, Vanoverschelde J, De Kimpe N (2004) J Chromatogr A 1027:137
150. Demyttenaere JCR, Willemen HM (1998) Phytochem 47:1029
151. Whitehead IM, Ohleyer E (1997) US Patent 5,599,700
152. Chatterjee T (2004) Biotechnol Appl Biochem 39:303

153. Braddock RJ (1999) Handbook of citrus by-products and processing technology. Wiley, New York
154. Duetz WA, Bouwnmeester H, van Beilen JB, Witholt B (2003) Appl Microbiol Biotechnol 61:269
155. Speelmans G, Bijlsma A, Eggink G (1998) Appl Microbiol Biotechnol 50:538
156. Mars AE, Gorissen JPL, Beld Ivd, Eggink G (2001) Appl Microbiol Biotechnol 56:101
157. Chang HC, Oriel P (1994) J Food Sci 59:660
158. Chang HC, Gage DA, Oriel PJ (1995) J Food Sci 60:551
159. Savithiry N, Cheong TK, Oriel P (1997) Appl Biochem Biotechnol 63–65:213
160. Tan Q, Day DF, Cadwallader KR (1998) Process Biochem 33:29
161. Tan Q, Day DF (1998) Process Biochem 33:755
162. Tan Q, Day DF (1998) Appl Microbiol Biotechnol 49:96
163. Kaspera R, Krings U, Pescheck M, Sell D, Schrader J, Berger RG (2005) Z Naturforsch C 60:459
164. Schewe H, Pescheck M, Sell D, Schrader J (2006) In: Bredie WLP, Petersen MA (eds) Flavour science: recent developments. Developments in food science, vol 43. Elsevier, Amsterdam, p 45
165. van der Werf MJ, Keijzer PM, van der Schaft PH (2000) J Biotechnol 84:133
166. Acosta M, Mazas N, Mejías E, Pino J (1996) Alimentaria 272:73
167. Duetz WA, Fjällman AHM, Ren S, Jourdat C, Witholt B (2001) Appl Environ Microbiol 67:2829
168. van der Werf MJ, Swarts HJ, de Bont JAM (1999) Appl Environ Microbiol 65:2092
169. van der Werf MJ, Boot AM (2000) Microbiology 146:1129
170. de Carvalho CCR, da Fonseca MMR (2002) J Mol Catal B 19–20:377
171. de Carvalho CC, Poretti A, da Fonseca MM (2005) Appl Microbiol Biotechnol 69:268
172. de Carvalho CC, Parreno-Marchante B, Neumann G, da Fonseca MM, Heipieper HJ (2005) Appl Microbiol Biotechnol 67:383
173. Onken J, Berger RG (1999) J Biotechnol 69:163
174. van Dyk MS, van Rensburg E, Moleleki N (1998) Biotechnol Lett 20:431
175. Ohloff G (1994) Scent and fragrances. Springer, Berlin Heidelberg New York
176. Trudgill PW (1994) In: Ratledge C (ed) Biochemistry of microbial degradation. Kluwer, London, p 33
177. Mikami Y (1988) In: Russell GE (ed) Biotechnology and genetic engineering reviews, vol 6. Intercept, Newcastle, p 271
178. Agrawal R, Joseph R (2000) Appl Microbiol Biotechnol 53:335
179. Agrawal R, Nazhath-Ul-Ainn D, Joseph R (1999) Biotechnol Bioeng 63:249
180. Rao SCV, Rao RR, Agrawal R (2003) Biotechnol Appl Biochem 37:145
181. Boontawan A, Stuckey DC (2006) Appl Microbiol Biotechnol 69:643
182. Wüst M, Little DB, Schalk M, Croteau R (2001) Arch Biochem Biophys 387:125
183. Wüst M, Croteau RB (2002) Biochemistry 41:1820
184. van Beilen JB, Holtackers R, Lüscher D, Bauer U, Witholt B, Duetz WA (2005) Appl Environ Microbiol 71:1737
185. Mouri T, Michizoe J, Ichinose H, Kamiya N, Goto M (2006) Appl Microbiol Biotechnol 72:514
186. Bell SG, Sowden RJ, Wong LL (2001) Chem Commun 635

187. Bell SG, Chen X, Xu F, Rao Z, Wong LL (2003) Biochem Soc Trans 31:558
188. Bell SG, Chen X, Sowden RJ, Xu F, Williams JN, Wong LL, Rao Z (2003) J Am Chem Soc 125:705
189. Appel D, Lutz-Wahl S, Fischer P, Schwaneberg U, Schmid RD (2001) J Biotechnol 88:167
190. Lamare V, Furstoss R (1990) Tetrahedron 46:4109
191. Furusawa M, Hashimoto T, Noma Y, Asakawa Y (2005) Chem Pharm Bull 53:1513
192. Muller B, Dean C, Schmidt C, Kuhn JC (1998) US Patent 5,847,226
193. Huang R, Christenson PA, Labuda IM (2001) US Patent 6,200,786
194. Okuda H, Sonohara H, Takigawa H, Tajima K, Ito S (1994) JP 06303967
195. Sakamaki H, Itoh K-I, Taniai T, Kitanaka S, Takagi Y, Chai W, Horiuchi CA (2005) J Mol Catal B 32:103
196. de Kraker J-W, Schurink M, Franssen MCR, König WA, de Groot A, Bouwmeester HJ (2003) Tetrahedron 59:409
197. Kaspera R, Krings U, Nanzad T, Berger RG (2005) Appl Microbiol Biotechnol 67:477
198. Mueller M, Dirlam K, Henning WH, Berger RG, Krings U, Kaspera R (2005) WO 2005078110
199. Sime JT (1997) Spec Publ R Soc Chem 200:190
200. Sime J (1998) Chem Br May 26
201. Sowden RJ, Yasmin S, Rees NH, Bell SG, Wong LL (2005) Org Biomol Chem 3:57
202. Abraham WR, Arfmann HA, Giersch W (1992) Z Naturforsch C 47:851
203. Arfmann H-A, Abraham W-R, Kieslich K (1988) Biocatalysis 2:59
204. Madyastha KM, Gururaja TL (1993) Indian J Chem 32B:609
205. Hrdlicka PJ, Sorensen AB, Poulsen BR, Ruijter GJR, Visser J, Iversen JJL (2004) Biotechnol Prog 20:368
206. Seitz EW (1994) In: Gabelman A (ed) Bioprocess production of flavor, fragrance and color ingredients. Wiley, New York, p 95
207. Suhara Y, Itoh S, Ogawa M, Yokose K, Sawada T, Sano T, Ninomiya R, Maruyama HB (1981) Appl Environ Microbiol 42:187
208. Farbood MI, Willis BJ (1989) US Patent 4,798,799
209. Farbood MI, Morris JA, Downey AE (1990) US Patent 4,970,163
210. Martin VJJ, Pitera DJ, Withers ST, Newman JD, Keasling JD (2003) Nat Biotechnol 21:796
211. Reiling KK, Yoshikuni Y, Martin VJJ, Newman J, Bohlmann J, Keasling JD (2004) Biotechnol Bioeng 87:200
212. Ro D-K, Paradise EM, Ouellet M, Fisher KJ, Newman KL, Ndungu JM, Ho KA, Eachus RA, Ham TS, Kirby J, Chang MCY, Withers ST, Shiba Y, Sarpong R, Keasling JD (2006) Nature 440:940
213. Ropert F, Dumont B, Belin JM (1995) In: Etievant P, Schreier P (eds) Bioflavour 95. INRA, Versailles, p 275
214. Zorn H, Langhoff S, Schreibner M, Nimtz M, Berger RG (2003) Biol Chem 384:1049
215. Zorn H, Langhoff S, Scheibner M, Berger RG (2003) Appl Microbiol Biotechnol 26:331
216. Sandmann G, Albrecht M, Schnurr G, Knörzer O, Böger P (1999) Trends Biotechnol 17:233
217. Schwartz SH, Qin X, Zeevart JAD (2001) J Biol Chem 276:25208
218. Belcour B, Courtois D, Ehret C (1990) US Patent 4,963,480
219. Belcour B, Courtois D, Ehret C, Petiard V (1993) Phytochemistry 34:1313

220. Gil G, Le Petit J (1992) US Patent 5,106,737
221. Krings U, Berger RG (1998) Appl Microbiol Biotechnol 49:1
222. Gatfield IL (1997) In: Berger RG (ed) Biotechnology of aroma compounds. Advances in biochemical engineering biotechnology. Springer, Berlin Heidelberg New York, p 221
223. Farbood MI, Willis BJ (1983) WO 8301072
224. Rabenhorst J, Gatfield IL (2000) WO 0024920
225. Wache Y, Aguedo M, LeDall M-T, Nicaud JM, Belin JM (2002) J Mol Catal B 19–20:347
226. Wache Y, Aguedo M, Nicaud JM, Belin JM (2003) Appl Microbiol Biotechnol 61:393
227. Dujon B, Sherman D, Fischer G, Durrens P, Casaregola S, Lafontaine I, De Montigny J, Marck C, Neuveglise C, Talla E, Goffard N, Frangeul L, Aigle M, Anthouard V, Babour A, Barbe V, Barnay S, Blanchin S, Beckerich JM, Beyne E, Bleykasten C, Boisrame A, Boyer J, Cattolico L, Confanioleri F, De Daruvar A, Despons L, Fabre E, Fairhead C, Ferry-Dumazet H, Groppi A, Hantraye F, Hennequin C, Jauniaux N, Joyet P, Kachouri R, Kerrest A, Koszul R, Lemaire M, Lesur I, Ma L, Muller H, Nicaud JM, Nikolski M, Oztas S, Ozier-Kalogeropoulos O, Pellenz S, Potier S, Richard GF, Straub ML, Suleau A, Swennen D, Tekaia F, Wesolowski-Louvel M, Westhof E, Wirth B, Zeniou-Meyer M, Zivanovic I, Bolotin-Fukuhara M, Thierry A, Bouchier C, Caudron B, Scarpelli C, Gaillardin C, Weissenbach J, Wincker P, Souciet JL (2004) Nature 430:35
228. Münch T, Kümin B (1997) EP 0795607
229. van der Schaft PH, Burg ter N, van den Bosch S, Cohen AM (1992) Appl Microbiol Biotechnol 36:712
230. Gocho S, Rumi K, Tsuyoshi K (1998) US Patent 5,763,233
231. Farbood MI, Kizer LE, Morris J, Harris G, McLean LB (2001) US Patent 6,187,741
232. Gocho S, Tabogami N, Inagaki M, Kawabata C, Komai T (1995) Biosci Biotechnol Biochem 59:1571
233. Cardillo R, Fuganti C, Sacerdote G, Barbeni M, Cabella P, Squarcia F (1990) US Patent 4,950,607
234. Farbood MI, McLean LB, Morris JA, Bondarovich HA (1992) US Patent 5,112,803
235. Page GV, Eilerman RG (1991) US Patent 5,032,513
236. He F, Farbood MI, Kizer LE (2000) US Patent 6,110,520
237. Boog A, Peters A, Roos R (1993) US Patent 5,215,901
238. Collins R, Halim A (1972) J Agric Food Chem 20:437
239. Bengtson G, Böddeker KW, Hanssen HPUI (1992) Biotechnol Tech 6:23
240. Prapulla SG, Karanth NG, Engel KH, Tressl R (1992) Flavour Fragrance J 7:231
241. Serrano-Carreon L, Balderas-Ruiz K, Galindo E, Rito-Palomares M (2002) Appl Microbiol Biotechnol 58:170
242. Serrano-Carreon L, Flores C, Rodriguez B, Galindo E (2004) Biotechnol Lett 26:1403
243. Kalyani A, Prapulla SG, Karanth NG (2000) Appl Microbiol Biotechnol 53:610
244. Jeffcoat R, Willis BJ (1988) In: Lawrence BM, Mookherjee BD, Willis BJ (eds) Developments in food science, vol 18. Elsevier, Amsterdam, p 743
245. Lacy D, Linhardt RJ, Bryan BA, Mayerl F, Pickenhagen W (1988) EP 0282942
246. Giani C, Wullbrandt D, Rothert R, Meiwes J (1997) US Patent 5,658,793
247. Mixich J, Rapp KM, Vogel M (1992) WO 9205182
248. Gunther NW, Nunez A, W. F, Solaiman DKY (2005) Appl Environ Microbiol 71:2288

249. Arfi K, Spinnler HE, Tache R, Bonnarme P (2002) Appl Microbiol Biotechnol 58:503
250. Blank I (2002) In: Reineccius GA, Reineccius TA (eds) Heteroatomic aroma compounds. ACS symposium series 826. American Chemical Society, Washington, p 25
251. Berger C, Khan JA, Molimard P, Martin N, Spinnler HE (1999) Appl Environ Microbiol 65:5510
252. Amarita F, Yvon M, Nardi M, Chambellon E, Delettre J, Bonnarme P (2004) Appl Environ Microbiol 70:7348
253. Amarita F, Fernandez-Espla D, Requena T, Pelaez C (2001) FEMS Microbiol Lett 204:189
254. Whitehead IM, Ohleyer E (1993) WO 9308293
255. Kumar D, Gomes J (2005) Biotechnol Adv 23:41
256. van der Schaft P, van Geel I, de Jong G, ter Burg N (1994) In: Maarse H, van der Heij DG (eds) Trends in flavour research. Elsevier, Amsterdam, p 437
257. Kerkenaar A, Schmedding DJM, Berg J (1988) US Patent 5,182,194
258. Cheng T-B, Reineccius GA (1991) Appl Microbiol Biotechnol 36:304
259. McIver RC, Reineccius GA (1986) In: Parliment TH, Croteau R (eds) Biogeneration of aromas. ACS symposium series 317. American Chemical Society, Washington, p 266
260. Larroche C, Besson I, Gros J-B (1999) Proc Biochem 34:667
261. Demain AL, Jackson M, Trenner NR (1967) J Bacteriol 94:323
262. Page G, Scire B, Farbood M (1993) US Patent 5,200,330
263. Kittleson JR, Pantaleone DP (1995) US Patent 5,437,991
264. Souchon I, Lamer T, Spinnler HE, Voilley A (1995) In: Étiévant P, Schreier P (eds) Bioflavour 95. INRA, Versailles, p 409
265. Bengtson G, Pingel H, Böddeker KW (1991) In: Bakish R (ed) Proceedings of the 5th international conference on pervaporation processes in the chemical industry, Heidelberg, 11–15 March 1991. Bakish Material Corporation, Englewood
266. Bluemke W, Schrader J (2001) Biomol Eng 17:137
267. Busmann D, Berger RG (1994) Z Naturforsch C 37:39
268. Berger RG, Latza E, Neuser F, Onken J (1999) In: Schieberle P, Engel KH (eds) Frontiers of flavour science. Deutsche Forschungsanstalt für Lebensmittelchemie, Garching, p 394
269. Farooq A, Tahara S, Choudhary MI, Atta-ur-Rahman, Ahmed Z, Baser KHC, Demirci F (2002) Z Naturforsch C 57:303
270. Cheetham PSJ (1997) In: Berger RG (ed) Biotechnology of aroma compounds. Advances in biochemical engineering biotechnology. Springer, Berlin Heidelberg New York, p 1
271. Cheetham PSJ (1991) In: Moses V, Cape RE (eds) Biotechnology, the science and the business. Harwood, Chur, p 481

24 Microbial Processes

C. Larroche, J.-B. Gros, P. Fontanille
Laboratoire de Génie Chimique et Biochimique,
Université Blaise Pascal, Polytech' Chermont-Ferrand,
24 avenue des Landais, BP 206, 63174 Aubière cedex, France

24.1 Introduction: General Concepts on Biotransformation Multiphase Systems

Biotransformation, also called microbial transformation or bioconversion, can be defined as a process dealing with the conversion of a compound, often called a precursor, into a structurally related compound(s) by a biocatalyst in a limited number of enzymatic steps.

This process is often preferred to chemical ones when high specificity is required, to attack a specific site on the substrate and to prepare a single isomer of the product. While chemical methods usually lead to the formation of a mixture of isomers and by-products, biotechnological methods are suited to achieve this type of transformation, as enzymes generally show a pronounced regioselectivity and stereoselectivity which leads to single enantiomeric products with regulatory requisites for pharmaceutical, food and agricultural use. The biocatalysts are whole cells, spores, crude enzymes or purified enzymes.

For a long time, applications of biotransformations to synthetic routes have been limited owing to the general idea that biocatalysts are only active in aqueous solutions and under mild conditions. More recently, it has become clear that biocatalysts are not as sensitive as expected. They can be active under harsh conditions, such as extreme pH, temperature and pressure, high salt concentrations or in the presence of other additives. They were also found to be active in all sorts of non-conventional media, such as organic solvents, aqueous two-phase systems, solid media, gases and supercritical fluids [1]. These acknowledgements allowed a drastic increase of the applicability of biocatalysts in organic synthesis. The advantages and drawbacks of selected multiphasic systems are described in Table 24.1 and are detailed in the following sections.

24.1.1 Supercritical Fluids

Enzymes can express activity in supercritical and near-supercritical fluids, such as carbon dioxide, freons (CHF_3), hydrocarbons (ethane, ethylene, propane) or inorganic compounds (SF_6, N_2O). The choice of supercritical fluids is often

Table 24.1 Main advantages and drawbacks of some multiphasic systems used in biotransformations

Process used	Main advantages	Main drawbacks
Supercritical fluid systems	Low surface tension and viscosity High mass-transfer rates Easy separation of reaction products	Problem of stability and activity of enzyme High energy and equipment costs owing to the use of high pressures
Aqueous–organic reaction systems		
Water and water-miscible solvent biphasic systems	Enhanced enzyme activity and stability at low concentration	Inhibitory effect on the biocatalyst at high concentration
Aqueous–organic biphasic systems	High substrate and product solubilities Improved volumetric productivity of the reaction Reduction in substrate and product inhibition Facilitated recovery of product and biocatalyst High gas solubility in organic solvents Shift of reaction equilibrium	Biocatalyst denaturation and/or inhibition by organic solvent Increasing complexity of the reaction
Microheterogeneous systems	High mass transfer Avoid loss of activity Reduce product inhibition or toxicity	Recovery of reaction product
Very low water systems	Reduction of water-dependent unwanted side reactions Improved thermostability of enzyme Manipulation of the enantioselectivity	Interactions between solvents and enzymes
Solid-state fermentation	Less effluent generation Low capital investment	Heat and mass-transfer limitations Difficulties in process control and scaling up

limited to compounds having a critical temperature between 0 and 60 °C since higher temperature could affect enzyme stability. The most universal system is supercritical carbon dioxide, which is probably explained by the fact that its critical point of 73.8 bar and 31.1 °C makes equipment design and reaction setup relatively simple. Moreover, it is non-toxic, non-flammable and safe for human beings and can be removed easily after the reaction [2].

In some aspects, supercritical fluids, which represent a state between the gaseous and liquid phases, have properties resembling those of non-polar solvents in being adequate for biotransformations of hydrophobic compounds. Although the use of supercritical fluids is not restricted to hydrolases, the use of this class of enzymes, especially lipases, dominates [3, 4]. Esters represent the main flavour compounds produced by this process [5].

Small changes in the temperature or pressure of a supercritical fluid may result in great changes in its viscosity and in the diffusivity and solubility of compounds dissolved within it. In such systems, the bioconversion rate is increased thanks to the high diffusion rates which facilitate transport phenomena. In some cases a high diffusion rate can also facilitate product separation.

The major drawback of this reaction system is the high energy and equipment costs due to the use of high pressures. In addition, the use of supercritical carbon dioxide can have adverse effects on enzymes, for example, by decreasing the pH of the microenvironment of the enzyme, by the formation of carbamates owing to covalent modification of free amino groups at the surface of the protein and by deactivation during pressurisation–depressurisation cycles [4].

24.1.2 Ionic Liquids

Over the past 5 years, the most exciting development in biocatalysis in multiphasic media was the use of ionic liquids to improve the activity, stability and selectivity of enzymes [4]. An ionic liquid is a salt in which the ions are poorly coordinated, which results in these solvents being liquid below 100 °C, or even at room temperature. At least one ion has a delocalised charge and one component is organic, which prevents the formation of a stable crystal lattice. These solvents have many useful properties, among them very low vapour pressure and excellent chemical and thermal stabilities (up to 400 °C), which make them environmentally friendly. Additionally, their physical properties, such as density, viscosity, melting point, polarity and miscibility with water or organic solvents, can be fine-tuned by changing either the anion or the substituents in the cation or both [6–8]. This is important because by manipulating the solvent properties, one is allowed to design an ionic liquid for specific reaction conditions. It has been shown that room-temperature ionic liquids are being used more and more in biotransformation processes as a substitute for the organic solvents in chemical reactions. Nevertheless, the use of such systems for flavour synthesis is still in its infancy [4, 9–11]. The most common ones are imidazolium (Fig.

24.1) and pyridinium derivatives, but also phosphonium or tetralkylammonium compounds can be used. Lately, an environmental friendly halogen-free ionic liquid has been tested.

A variety of enzymes, particularly those that tolerate conventional organic solvents, show excellent performance in ionic liquids. Activities and stabilities are generally comparable with or higher than those observed in conventional organic solvents. Even if ionic liquids will clearly not provide advantages in all systems, improvements in reactivity or selectivity are notably observed for certain biotransformations of highly polar substrates, such as (poly)saccharides, amino acids and nucleotides, which cannot be performed in water owing to equilibrium limitations, when the appropriate combination of cation and anion is selected. The subject is attracting and many interesting reviews have reported the most recent progress on the topic [4, 7, 12].

However, one of the recurrent problems that occur when ionic liquid techniques are used for organic synthesis is the recovery of non-volatile or low-volatility products. Extraction with solvents that are immiscible with the ionic liquids, giving biphasic systems, is one of the simplest methodologies to separate the products from the ionic liquid phase. Drawbacks are the extraction of small amounts of the ionic liquid and eventually of the catalyst, if a catalytic reaction is involved. Also, the partitioning of the solute between the phases limits the extent of solute extraction, and obviously the use of volatile organic solvents is a breakdown point for the integral green design of the process [6]. Pervaporation, nanofiltration [13] or extraction with supercritical carbon dioxide in an ionic liquid [6] have also been proposed and could represent interesting green industrial processes. Finally, there is a major difficulty to be overcome: ionic liquids are today about 800 times more expensive than organic solvents [14], rendering them economically viable mainly when the product is of high added value [4]. However, it is necessary to keep in mind that these compounds should normally be recycled, which should attenuate the preceding affirmation.

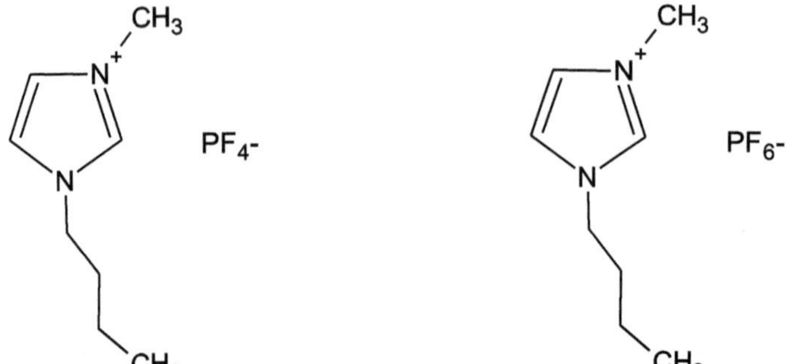

Fig. 24.1 Chemical structure of common ionic liquids in biotransformation: 1-butyl-3-methyl-imidazolium tetrafluoroborate and 1-butyl-3-methyl-imidazolium hexafluoroborate

24.1.3
Organic–Aqueous Reaction Systems

Organic–aqueous media offer important advantages for the industrial applications of enzymatic and whole-cell catalysis when substrates are poorly soluble in water [15–17]. This is the case for most of the flavour compounds like terpenes [18, 19]. Use of an organic phase in the aqueous reaction system has become a current way to improve biotransformation processes. Some of them are described next.

24.1.3.1
Water-Miscible Organic Solvents

Water-miscible organic solvents such as dimethyl sulfoxide, acetone and ethanol are often added to reaction mixtures to increase the aqueous solubility of poorly water soluble reactants. At low concentrations this strategy can be effective without adversely affecting enzyme activity and stability. It has the advantage of generally not presenting mass-transfer limitations as they are homogeneous systems. At higher cosolvent concentrations, however, biocatalyst inhibition or inactivation may become prohibitive, thus limiting the maximum cosolvent concentration which may be used. Another disadvantage of this approach is that the use of miscible cosolvents does not automatically simplify downstream recovery of the biocatalyst or product separation [20].

24.1.3.2
Water-Immiscible Organic Solvents

In such systems, biotransformations are generally carried out in a reaction medium composed of an aqueous phase containing the biocatalyst and a water-immiscible organic solvent which may be the substrate itself to be converted [21] or may serve as a reservoir for substrates and products [22] (Fig. 24.2). In these conditions, a constant substrate feeding in the aqueous phase is obtained owing to the partition coefficient. The substrate is used by the biocatalyst to be converted into the product of interest, which is then continuously extracted into the organic phase.

The introduction of an organic solvent in the reaction system has several advantages (Table 24.1). It allows a relatively high solubility of many poorly water soluble or insoluble compounds, may improve conversion rates and generally simplifies the conversion process. Another important advantage is that the equilibrium of a hydrolytic reaction can be shifted in favour of the product, this being extracted into the organic phase; therefore, biocatalyst and product recovery will be facilitated. Water–organic biphasic systems diminish undesirable side reactions in organic media as well as substrate and product inhibition; thus, high product yields may be achieved.

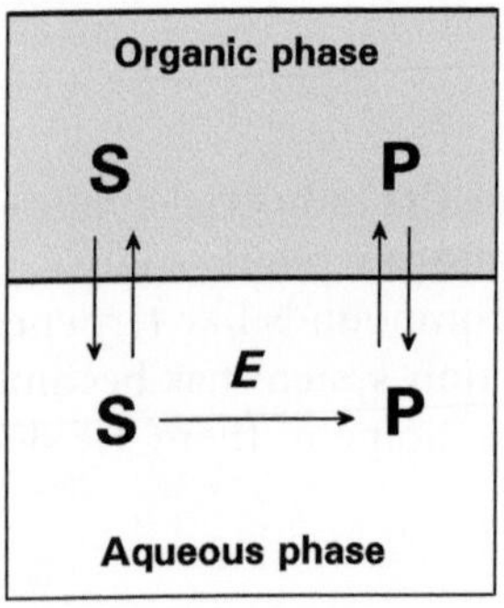

Fig. 24.2 Enzymatic conversion in a two-phase system. *S* substrate, *P* product, *E* enzyme

In spite of numerous advantages of aqueous–organic biphasic systems, drawbacks also exist. The biocatalyst may be denaturated by the organic solvent, and in addition the introduction of an organic phase leads to an increasing of the reaction complexity. The selection of appropriate solvents represents an efficient way to avoid these phenomena.

24.2 Solvent Selection in Organic–Aqueous Systems

The choice of an organic solvent for a given reaction can be determined by three main factors [23]:

1. The effect of the solvent on biocatalyst stability.
2. The effect of the solvent on the reaction, including the solubility of the substrates and products, the effect on equilibrium yields, kinetics and enzyme specificity.
3. The safety of the solvent, which is important for food and pharmaceutical-based processes, where compliance with safety and solvent-disposal legislation will be a major consideration.

Other characteristics such as chemical and thermal stability, a low tendency for forming highly stable emulsions with water media, non-biodegradability, a non hazardous nature and low market price have to be taken into account too.

The toxic effect on biocatalytic activity and stability in two-phase reaction system media can be divided into two effects. The first one, called the molecular-toxicity effect, is a direct toxic effect of the solvent molecules, which are dissolved in the aqueous phase and interact with the biocatalyst, particularly with whole cells. The second one, which is created by the presence of an interface between the aqueous and the organic solvent phase, is called the phase-toxicity effect [2, 24].

Inactivation of the biocatalyst owing to these effects can be a significant limitation for industrial application of enzymatic and whole-cell biotransformation. For more than 20 years, many attempts have been made to associate the toxicity of different solvents with some of their physicochemical properties and to explain the influence of the two-phase system composition on bioconversion efficiency.

24.2.1
Molecular Toxicity of the Solvent

The requirement of biocompatibility is a restrictive criterion, in particular for whole-cell biocatalysis [24]. Solvents are known to partition into and disrupt the bacterial cell membrane, thus affecting the structural and functional integrity of the cell [25, 26]. The most popular parameter used to classify organic solvent toxicity is $\log K_{ow}$, also often referred to as $\log P$, which is defined as the decimal logarithm of the partition coefficient of the given solvent in a mixture of 1-octanol and water at a given temperature, generally 25 °C. This value can be determined experimentally or can be calculated (see later). Laane et al. [27] observed that a correlation exists between the $\log K_{ow}$ value and the toxicity of the solvent. The greater the polarity, the lower the $\log K_{ow}$ value and the greater the toxicity of the solvent. When plotting cellular activity retention against $\log K_{ow}$, a sigmoïdal curve is obtained (Fig. 24.3).

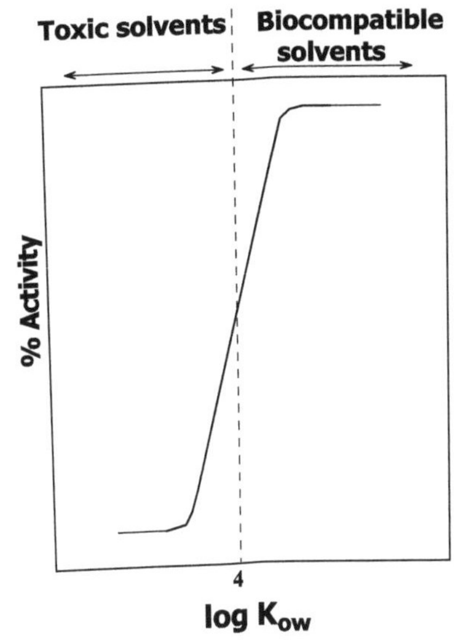

Fig. 24.3 Relationship between the activity retained by cells exposed to an organic solvent and the $\log K_{ow}$ value of the solvent

Biocatalysis in organic solvents is generally low in polar solvents having a $\log K_{ow}$ lower than 2, is moderate in solvents having a $\log K_{ow}$ between 2 and 4 and is high in apolar solvents having a $\log K_{ow}$ higher than 4. Several authors have demonstrated that the inflection point of these curves depends on the microorganism studied. The characteristics of the cell membrane could influence the solvent tolerance of the microorganisms [25-28]. Gram-negative bacteria are in general more tolerant than Gram-positive bacteria probably because of the presence of the outer membrane.

Recently, organic solvent tolerant bacteria, a novel group of extremophilic microorganisms that combat these destructive effects and thrive in the presence of high concentrations of organic solvents as a result of various adaptations, are being explored for their potential in industrial and environmental biotechnology [29].

The preceding discussion dealt with membrane-bearing systems. In the case of acellular systems, i.e. with crude cell extracts or purified enzymes, the validity of $\log K_{ow}$ as a criterion for biocompatibility is questionable. As a result, other parameters, such as interfacial tension, have been suggested to predict the effect of solvents on enzyme stability [30].

24.2.2 Phase-Toxicity Effect

When water-immiscible liquids are used, three quite different classes of inactivation mechanism must be distinguished. First, in some cases inactivation is related to removal of water from the molecular environment of the enzyme rather than any direct effect of the solvent itself. A second possibility is that individual molecules of the organic species dissolved in an aqueous phase around the enzyme may interact with it. Third, contact of the enzyme molecules with the bulk organic liquid at the phase interface may be involved. There is evidence that in many cases interfacial effects provide the dominant mechanism.

In order to avoid mass transfer of apolar reactant towards the aqueous phase being rate-limiting, the interfacial area has to be increased by a high agitation level. This may also increase interfacial effects [31]. This inactivation was found proportional to the area of the organic solvent exposed [32, 33]. Figure 24.4 illustrates a mechanism of enzyme inactivation at the aqueous–organic interface which takes place by unfolding of enzyme molecules adsorbed at the interface, followed by enzyme aggregation and finally precipitation from solution. Baldascini and Janssen [34] have recently shown this inactivation for an epoxide hydrolase at the octane–water interface. A high stirring rate increases here again the rate of interfacial inactivation. This effect can be due to an increase in the rate of desorption of inactivated enzyme molecules from the interface, which then allows active enzyme in solution to become adsorbed and inactivated in turn.

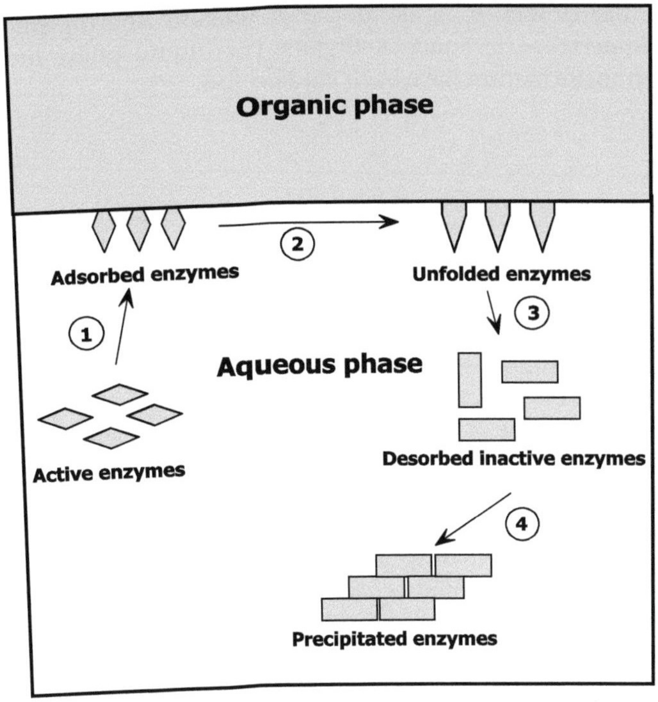

Fig. 24.4 Mechanism of enzyme inactivation at an aqueous–organic interface. Step *1*: reversible enzyme adsorption to the interface and concomitant enzyme structural rearrangement at the interface. Step *2*: unfolding of enzyme molecule at the interface. Step *3*: desorption of inactivated/unfolded enzyme molecules from the interface. Step *4*: irreversible aggregation and precipitation of inactivated enzyme. (From [34])

By comparing interfacial inactivation rates in a stirred-cell (low and controlled area of exchange) and an emulsion system (high interfacial area), these authors have shown that the use of an emulsion system can be exploited to obtain high solute interphase mass-transfer rates since the rate of specific interfacial inactivation remains low. However, in this system, the presence of an epoxide substrate at high concentration in the organic phase increases the rate of interfacial inactivation. Addition of a sacrificial protein to the system, which can prevent adsorption of the catalytic enzyme at the interface, could provide a method to reduce the rate of interfacial inactivation.

The composition of the aqueous phase plays a critical role in interfacial reactions too. Sah and Bahl [35] showed that critical factors such as pH, buffer type and concentration affected the destabilisation of β-lactoglobulin towards emulsification. In particular, pHs away from the pI and low buffer/salt concentrations are beneficial for minimising the interfacial inactivation.

The effects of many other factors, such as interfacial tension, stirring rate, phase volume ratio or temperature, in aqueous–organic two-liquid-phase media on the stability of biotransformation have been studied [36, 37].

24.3 Engineering Aspects

In multiphase systems, biological reactions are always carried out in the presence of water. This is true even if the presence of water is almost negligible. The biocatalyst may be present as a solid phase, for example as immobilised enzymes or cells, or as an individual cell; the substrate may also constitute a solid phase. When necessary, gas is sparged into reactors to supply oxygen or a gaseous substrate and to remove carbon dioxide. Thus, heterogeneous systems with four phases involved are very typical cases.

Bioreactors are operated in discontinuous mode, with a sequential or continuous feed of the substrate (fed-batch operation) or in continuous mode. The choice of the operating mode depends mainly on the reaction characteristics:

- Any reaction exhibiting substrate inhibition should not be carried out in batch since it results in a longer residence time; the high concentration of the substrate at the beginning lowers the reaction rate. A continuously operated stirred tank is preferred. At laboratory scale, fed-batch operation enables a low substrate concentration in the reactor and a higher reaction rate.
- If product inhibition occurs, either a stirred-tank reactor in batch or a plug-flow reactor should be used. In these two reactors, the product concentration increases with time. Alternatively a reactor with integrated product separation (membrane, solvent, etc.) is preferable.

Most reactors used for biotransformation are mechanically stirred tank reactors, aerated or not. At a first approximate, stirred tanks behave as perfectly mixed reactors; this is nearly always the case at the laboratory scale. The term "perfectly mixed" applies to the liquid phase only. In an aerobic culture or biotransformation, it can be advantageous to contact the perfectly mixed liquid phase with a gas phase that goes through the reactor in plug flow since it will give the highest rate of mass transfer of a gaseous substrate to the liquid; this contributes to an optimal utilisation of the gas phase. Note that some processes are operated with an external recycle of fluids. When the recycle ratio is high, the reactor is nothing more than a well-mixed reactor.

Hereafter we will treat only perfectly mixed stirred-tank reactors, which are considered, and rightly so, as the reference reactors. We consider a rather general case of biotransformation processes involving aerated systems comprising both water and hydrophobic compounds. These last components are often volatile, as in the case of aroma. As a result, losses by gas stripping can be important.

The partitioning of compounds among the different phases leads to a stripping of substrates or products, creating emission of volatile organic compounds

(VOCs), and to a decrease in production yields. Quantitation of these phenomena and determination of material balances and conversion yields remain the bases for process analysis and optimisation. Two kinds of parameters are required. The first is of thermodynamic nature, i.e. phase equilibrium, which requires the vapour pressure of each pure compound involved in the system, and its activity. The second is mass-transfer coefficients related to exchanges between all phases (gas and liquids) existing in the reaction process.

24.3.1 Vapour Pressure

The (saturation) vapour pressure of a chemical is the pressure its vapour exerts in equilibrium with its liquid or solid phase. It is thus a property of a given pure compound, i.e. its molecular structure, which is temperature-dependent and usually referred to as P^{o}, P^{s} or P^{vp}.

The values reported for vapour pressure of chemicals at ordinary temperatures (-40 to 40 °C) range from 760 to less than 1×10^{-6} mmHg, i.e. from 101.325 to 1.3×10^{-5} kPa. It must be emphasised that lower values are difficult to measure and, in practice, they are often estimated by calculation using models [38]. A very interesting source of data is the PhysProp database, available at http://www.syrres.com/esc/physdemo.htm, which contains data for about 25,000 compounds which can be accessed, for the on-line version, through the CAS number. ChemFinder (http://www.chemfinder.com) is also an efficient source of data. Most databases available on the Internet, free or not, are given on the Website of *Links for Chemists* (http://www.liv.ac.uk/Chemistry/Links/links.html).

The fundamental relationship that allows the determination of the vapour pressure P^{o} of a pure condensed phase as a function of temperature is the Clapeyron equation [38, 39]. The simplest equation that can result from its integration is [40]

$$\ln P^{o} = A - \frac{B}{T}, \tag{24.1}$$

where A and B are compound-specific constants. This equation is valid if temperature variation is confined in a rather narrow range. Values for A and B can be found for various organic compounds [41]. For larger variations, the so-called Antoine equation [42], where A, B and C are empirical constants, and T is expressed in degrees Celsius, is more accurate (24.2):

$$\ln P^{o} = A - \frac{B}{T + C}. \tag{24.2}$$

The values of A, B and C can be found in textbooks such as Reid et al [43]. Grain [40] and Sage and Sage [44] suggested a method to estimate the constants which allows a fully predictive calculation. It should be emphasised that sophisticated expressions are often prone to errors when calculations are carried out by hand. As a result, if a series of calculations have to be made, one can recommend the use of commercial software such as MPBPVP from Syracuse Research Corporation (http://www.syrres.com/esc/mpbpvp.htm), ACD Boiling Point (Advanced Chemistry Development, http://www.acdlabs.com/products/phys_chem_lab/), prediction software from Pirika (http://www.pirika.com/) or the VLEcalc program, which predicts both boiling point and vapour pressure (http://www.vlecalc.org/). Using a set of 185 compounds, representative of classes of aroma molecules (alkanes, alkenes, alkynes, alcohols, ethers, ketones, esters, terpenoids), we found that high values of vapour pressures could be quite accurately predicted by available procedures. Values lower than 100 mmHg were more difficult to obtain, and it appeared that the method of Grain, used in the MPBPVP program, was a good method for these kinds of molecules.

24.3.2
Phase Equilibrium. Activity Coefficients

24.3.2.1
Gas–Liquid Equilibrium

As already stated, one of the important pieces of data for biotransformation processes is knowledge of phase equilibrium and the activity of solutes involved. Hence, assuming that gas and liquid phases are at thermodynamic equilibrium, we can write

$$y = \gamma x \frac{P^{\circ}}{P}, \tag{24.3}$$

where y is the mole fraction of a solute in the gas, γ is the activitycoefficient of the solute in a liquid phase, x is its mole fraction in a liquid phase, P° is the vapour pressure and P is the total pressure in the system. The product γx is the so-called activity of the solute, which is equal in all phases at thermodynamic equilibrium.

Equation 24.3 is useful for estimating the losses of compounds by gas stripping. Indeed Gy, where G is the molar gas flow rate, gives the molar loss rate of a given compound in the environment, solvent included.

24.3.2.2
Liquid–Liquid Equilibrium. Aqueous Solubility

Several liquid phases coexist in a system when the solvents are not completely miscible. Liquid–liquid equilibrium properties are very useful in solvent extraction and in biotransformation or enzymatic syntheses in two-solvent systems. One speaks about liquid–liquid equilibrium in two cases: (1) if the two solvents are not completely miscible, it is said that there is partial miscibility of the two solvents; (2) if there is distribution of a compound in the two non-miscible solvents.

At equilibrium, equality of chemical potentials of a component in two liquid phases L_1 and L_2 leads to

$$\gamma_{L_1} x_{L_1} = \gamma_{L_2} x_{L_2} . \tag{24.4}$$

One can define a liquid–liquid equilibrium coefficient K_{ll}:

$$K_{LL} = \frac{x_{L_2}}{x_{L_1}} = \frac{\gamma_{L_1}}{\gamma_{L_2}} . \tag{24.5}$$

The octanol–water partition coefficient of a solute, defined as its concentration in the octanol-rich phase over its concentration in the water-rich phase at infinite dilution, is one interesting example of a liquid–liquid equilibrium coefficient.

Knowledge of K_{LL} and solving for the system made of the MES equations (component material balances, equilibrium relationships, and sum equations i.e. mole-fraction constraint) make it possible to calculate the composition of each phase at equilibrium, which is a a realistic assumption in most processes.

For compounds exhibiting a low solubility in a solvent, Eq. 24.4 simplifies [45] to

$$\gamma^{\infty} = \frac{1}{x_s} , \tag{24.6}$$

with γ^{∞} being the activity coefficient at infinite dilution and x_s the mole fraction in the solvent at saturation. Equation 24.6 is valid for many organic compounds in water as a solvent and means that, in those cases, the determination of the activity coefficient and that of the aqueous solubility are in fact the same question.

24.3.2.3
Prediction of Aqueous Solubility

Large databases on aqueous solubility exist, such as AQUASOL dATAbASE (http://www.pharmacy.arizona.edu/outreach/aquasol/), which contains almost 20,000 solubility records for almost 6,000 compounds, or the already mentioned PhysProp. However, not all situations are covered and the ability to predict this property is still useful. This remark has favoured the development of numerous mathematical models and much prediction software [46].

A comprehensive review of this field is beyond the scope of this chapter. As for vapour pressure, the aim is to give easy-to-use tools for non-specialists in the field. The general solubility equation, initially introduced by Yalkowski and Valvany [47], then revised by the same team [48], is probably the simplest:

$$\log S_{\mathrm{w}} = 0.5 - \log K_{\mathrm{ow}} - 0.01(\mathrm{MP} - 25), \tag{24.7}$$

where K_{ow} is the octanol–water partition coefficient of the solute and MP its melting point in degrees Celsius. If the solute melts below 25 °C, MP is set at 25 °C and the last term vanishes. S_{w} is the aqueous solubility in moles per litre. The only input data here is K_{ow}, which can be found in several sources, such as PhysProp or LOGKOW, a free-access database provided by Sangster Research Laboratories (Quebec, Canada) and available at http://logkow.cisti.nrc.ca/logkow/index.jsp. The most extensive hard-copy database is given by Hansch et al [49].

If needed, this parameter can also be estimated, and an abundant literature exists in this topics. Many free-access software applications are available on-line (e.g. http://www.pirika.com, http://www.syrres.com/esc/est_kowdemo.htm and http://www.daylight.com/daycgi/clogp). Other software applications are commercial packages (e.g. http://www.chemsilico.com/CS_prLogP/LPhome.html, http://www.ap-algorithms.com/articles.htm, http://www.acdlabs.com/products/phys_chem_lab/logp/, and http://www.itscb.com/newsitetest/services/asg/physicalproperties.shtml#logP). Other methods can be found in Sangster [50], Baum [38] or Leo [51]. A hand calculation can be made using the equation of Meylan and Howard [52], which is the basis for the LOGKOW program. Using the same set of flavouring compounds as for vapour pressure estimation, we found that ClogP was a good free tool for these kinds of compounds.

The water solubility itself can also, of course, be obtained with more refined (and generally more accurate) models. A model usable for hand calculations is that proposed by Meylan et al [53]:

$$\log S_{\mathrm{w}} = 0.796 - 0.854 \log K_{\mathrm{ow}} - 0.00728(\mathrm{MW}) + \sum f_i, \tag{24.8}$$

where S_w is the solubility in moles per litre and Σf_i the summation of all correction factors applicable to a given compound. Values of f_i are given in the original paper.

Most of the Web addresses given above also give predictions for S_w. One of the most popular models for the activity coefficient is the UNIQUAC functional group activity coefficient (UNIFAC), which derives from UNIQUAC [54] and for which two major modifications exist, UNIFAC Dortmund [55, 56] and UNIFAC Lyngby [57]. It is a group-contribution approach which is continuously updated, and a consortium, headed by Jürgen Gmehling (University of Oldenburg, Germany) exists (http://www.unifac.org). However, this algorithm generally gives inaccurate values for hydrophobic compounds in water. A way to circumvent this problem is to estimate first the activity coefficient of a solute in 1-octanol with UNIFAC, then to deduce the water solubility by (24.9) [58]:

$$\log S_w = -\log K_{ow} - \log \gamma_o^\infty + 3.916, \tag{24.9}$$

with γ_o^∞ being the activity coefficient of the solute at infinite dilution in 1-octanol and S_w the solubility in water in millimoles per litre. The activity coefficients can be easily calculated with standard UNIFAC software freely available for download, such as Unifacal (http://www.eng.auburn.edu/users/guptarb/classes/chen7200/?S=A) or the software available from http://www.che.udel.edu/thermo/basicprograms.htm#UNIFAC. These programs are based only on the original version of UNIFAC [54].

The already mentioned data set used by us for solubility estimation showed that both (24.8) and (24.9), combined with ClogP for K_{ow} prediction, were suited for classic aroma compounds.

24.3.3 Contact Between Phases. Mass Transfer

In many cases, the transport of substrates to the cells and that of metabolites from the surface of the cells to the culture medium are carried out at rates characterised by time constants of the same order of magnitude as those of the biological reactions. Transport or transfer of matter must thus be included in an analysis of the behaviour of a bioreactor as well as the kinetic rates [59, 60].

24.3.3.1 Gas–Liquid Dispersion

Transfer from a gas phase to a microorganism occurs according to the following mechanisms: (1) transport by convection in the gas bubble; (2) diffusion through the gas boundary layer in the vicinity of the gas–liquid interface; (3)

transport across the gas–liquid interface; (4) diffusion in the liquid boundary layer in the vicinity of the gas–liquid interface; (5) transport by convection in the liquid phase; (6) diffusion in the liquid layer in the vicinity of the microorganism.

Normally transport by convection is sufficiently fast that the concentrations are homogeneous, and the crossing of the gas–liquid interface is instantaneous; there is no resistance to transfer. There is thus thermodynamic equilibrium at the interface. This assumption is questionable in culture media containing proteins, inorganic ions, etc.

In most processes, steps 1, 3, 5 and 6 are in pseudo steady state and the mass transfer is governed by diffusion through the gas–liquid layers (steps 2 and 4). An additional step can appear if one deals with aggregates of cells (pellets), but we will not examine this case.

The mass transfer rate Q of a component from one phase to another is given by a relation of the type

$$Q = K_l A(C^* - C), \tag{24.10}$$

with

$$y = HC^*. \tag{24.11}$$

K_l is the overall mass-transfer coefficient based on the liquid phase. A is the total interfacial area in the gas–liquid dispersion. C is the concentration in the liquid phase. C^* thus corresponds to equilibrium with the gas phase of composition y. H is the Henry coefficient for the gas. In the case of oxygen or a sparingly soluble compound, H is large and resistance to mass transfer is located in the liquid phase.

In a bioreactor, one is interested in the transfer per unit of volume of reactor, called $K_l a$ or the volumetric mass-transfer coefficient. a is the interfacial surface area per unit of volume of liquid. In a perfectly mixed tank, C has identical values at any point and C^* depends on the conditions in the gas phase at the outlet of the reactor. Several authors [60] consider that a better estimate of the driving force is given by the logarithmic mean concentration difference between the entry and the exit of gas.

Many correlations of experimental results have been published [61, 62]. Most of these correlations are written in the form

$$K_l a_d = k\, u_s^a \left(\frac{P_d}{V_l}\right)^b . \tag{24.12}$$

Coefficients k, a and b depend on the reactor design. u_s is the surface velocity of the gas, P_d is the power consumption and V_l is the liquid volume. It is observed that the volumetric mass-transfer coefficient for a non-coalescing medium is higher by about a factor of 2 than that measured for a coalescing medium under the same operating conditions. Fermentation media are in general non-coalescent, but biotransformation media, even very simple ones, can be coalescent.

Note that in the definition of K_la, the interfacial surface area is in general based on the liquid volume. This definition is consistent with the material balances in the reactor and in particular the gas-phase balances. However, in correlations published for K_la values, most authors use a specific area a_d based on the total volume of the gas–liquid dispersion (24.12). a_d and a are connected via the gas holdup ε:

$$a_d = (1-\varepsilon)a. \tag{24.13}$$

In fact none of these models of the mass-transfer coefficient are of much use for the calculation of K_la values in small scale reactor conditions and we have to obtain them by experiments. However, these models can be used as a guide to estimate the influence of the physical properties of the medium. They also make it possible to consider relative values of K_la for compounds for which in experiments the value of K_la is not measurable as easily as for gases such as oxygen.

The interfacial gas–liquid area a is a function of the size of the gas bubble dispersion:

$$a = \frac{6\varepsilon}{(1-\varepsilon)d_b}, \tag{24.14}$$

where d_b is the average bubble diameter, often taken as the Sauter diameter d_{32}. There are correlations making it possible to estimate ε and d_b as functions of the viscosity of the medium, its surface tension, its density and the characteristics of the gas injector, surface gas velocity and power dissipated by mixing [60]. If we remember the complex composition of culture media where the presence of inorganic ions, proteins, etc. strongly affects the gas–liquid interface and thus coalescence, they can be used only to detect tendencies.

If one needs accurate values, the gas-balance method, operated during the course of a culture or a biotransformation, is the only one that gives a K_la value averaged over the whole reactor [63].

24.3.3.2
Liquid–Liquid Dispersion. Drop Diameter

In liquid–liquid reacting systems, one of the important parameters is the surface area per unit volume, a, in the dispersion, which can be related to the Sauter mean drop diameter d_{32}. In some processes, the drop size distribution and especially the minimum drop size or the maximum stable drop diameter are also important factors in analysing the process results.

For dilute dispersions with a non-viscous dispersed phase where the viscous energy within a drop is negligible compared to the surface energy, the maximum stable drop diameter d_{max} is given by

$$\frac{d_{max}}{D} = cWe_{st}^{-0.6}, \tag{24.15}$$

where $We_{st} = \frac{\rho_c N^2 D^3}{\sigma}$ is the Weber number in the stirred tank; ρ_c is the density of the continuous phase, N is the impeller speed rate and D is the impeller diameter. c lies between 0.05 and 0.06 for six-bladed Rushton turbines depending on published correlations [64]. When the volume fraction φ of the dispersed phase becomes important, (24.15) is multiplied by the correction factor $(1+4\varphi)$.

It was Sprow [65] who first assumed that the Sauter mean diameter is proportional to the maximum stable drop diameter, i.e.

$$d_{32}=cd_{max}, \tag{24.16}$$

and then verified the relationship with experimental data. c lies between 0.42 and 0.69 and decreases with an increase in N, but seems independent of the geometry of tanks and impellers [64]. When N is high (greater than 20 s^{-1}), c approaches a constant value and c=0.5 can be considered as a design value.

24.3.3.3
Gas–Liquid–Liquid Dispersion

The addition of a dispersed liquid phase (immiscible organic solvent) changes the rate of transfer of the solute gas across the boundary layer. Physical properties (density, viscosity, gas solubility and gas diffusivity) of the liquid mixture are changed and the gas–liquid characteristics (possible pathway for mass transfer and gas–liquid interfacial area) can be changed owing to the interfacial properties of the dispersed liquid. Three distinct approaches have been reported in the literature to explain the change in mass transfer in a gas–liquid–liquid system [66–68]:

1. Direct gas–liquid (dispersed phase) contact forming a "gas–organic complex".
2. Shuttle effect of droplets carrying a gas solute from the gas–liquid interface to the liquid bulk.
3. Dynamic interaction of the solvent droplets with the concentration boundary layer causing increased turbulence or mixing in this layer.

In order to understand the mechanisms governing the mass transfer in three-phase systems, the distribution of organic and water phases near the gas–liquid interface has been estimated using various possible mechanisms for mass transfer.

Basically two possible pathways exist:

1. Transfer in series. There is a gas-to-water mass transfer into the liquid and no direct gas-to-organic phase contact is possible.
2. Transfer in parallel. Gas-to-solvent contact is possible and the gas-to-water mass transfer as well as the gas-to-organic compound mass transfer occur.

It is necessary to note that some investigators have highlighted the importance of the interfacial properties of the three-phase systems on these possible pathways. Indeed, according to these authors, the interfacial properties of the organic phase–water system expressed through the notion of the spreading coefficient S_{ow} could have a strong influence on the pathways for mass transfer.

The spreading coefficient, S_{ow}, of a solvent on water is defined as

$$S_{ow} = \sigma_{wg} - (\sigma_{og} + \sigma_{ow}), \tag{24.17}$$

where σ_{ij} is the surface tension between phase i and phase j. As defined by (24.17), S_{ow} quantifies the ability of an organic phase to either bead up (form a droplet) or spread out (form a film) when contacting an aqueous phase. Compounds with a negative spreading coefficient ($S_{ow}<0$) tend to form discrete droplets, whereas those with a positive spreading coefficient ($S_{ow}>0$) tend to spread as a thin film over the bubbles.

For beading oils, the most probable pathway is mass transfer in series. Assuming a "shuttle effect" of the oil phase, investigators consider that the solute absorbed in the oil droplets near the gas–aqueous phase interface is given up to the water phase outside the boundary layer.

For spreading oils, the most probable pathway is the transfer in parallel. Brilman et al. [68] have suggested the possible direct gas–solvent contact through

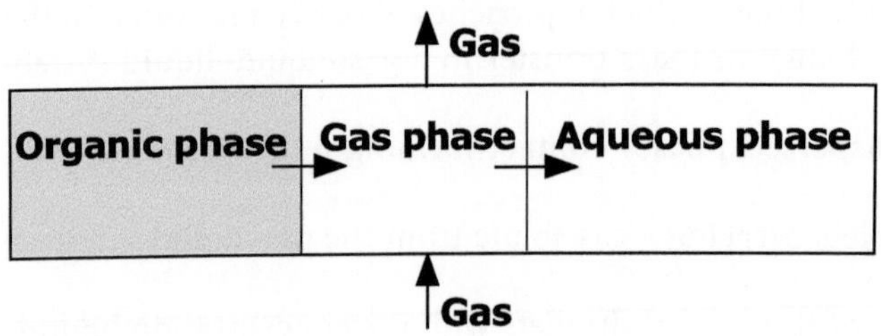

Fig. 24.5 Transfer between phases in the two-phase system described by Larroche et al. [71, 72]

the formation of complexes of gas–organic drops which is dependent not only on the spreading coefficient but also on the bubble and droplet size. Triglyceride oils with medium-chain to long-chain fatty acids, such as soybean, sunflower or liquid butter oil, have $S>0$ and can therefore spread at the air–water interface. Oil spreading is inhibited by the presence of an adsorbed protein layer [69].

Let us examine a biotransformation involving a gas–liquid–liquid system. Mikami et al. [70] have reported the biotransformation of β-ionone by *Aspergillus niger* JTS 191 into a mixture very effective for tobacco flavouring. Larroche et al. [71] have also reported a similar process, which involved fed-batch biotransformation of β-ionone by *Aspergillus niger* IFO 8541 entrapped in calcium alginate beads operated in an aerated, stirred bioreactor. In all cases, the apparent reaction yield was very far from 100%, but no convincing argument was provided to explain this behaviour. The biotransformation process was carried out at a low stirring rate, 5 s^{-1}, in order to avoid particle damage. Preliminary experiments showed that the rates of the phenomena appearing in an abiotic system operated under these conditions were lowered, indicating phase-transfer limitation. The culture medium was made of three phases, i.e., two liquid phases and the gas, because the precursor was present at a concentration higher than its water-solubility limit. The liquid phases corresponded to an organic one, made of pure β-ionone, and an aqueous solution. Experimental data enabled the phase-transfer fluxes to be calculated. The results demonstrated that the main β-ionone transfer took place from the organic phase to the gas phase and that the aqueous phase was mainly fed by the gas (Fig. 24.5). It could thus be considered that the triphasic system mainly involved serial transport in the direction organic phase to gas to aqueous solution, which is an unusual situation.

The amounts of β-ionone stripped and of the chemical products synthesised during a biotransformation experiment were calculated using the above mass-transfer rate model. The amount of residual β-ionone available for biotransformation was then deduced and compared to the amount of biological metabolites obtained. The results demonstrated that the true biotransformation yield was very close to 100%. Further analysis of β-ionone behaviour demonstrated that stripping was the main process, which involved up to 65% of total precursor consumption, while the biotransformation part accounted for about 30% [72, 73].

From a process point of view, this very high loss by stripping is a strong drawback. Strategies to reduce this phenomenon could be the use of an apolar organic solvent in order to reduce the solute activity in the system, a tight control of the aeration rate or even a partial recycling of the gas.

24.4 Conclusion

Engineering approaches for a chemical or biochemical process focus on the various ways to improve the economics of the overall system. This leads, among others, to formove an optimisation of reaction rates and yields. The question which arises now is to see if these objectives are compatible with or can improve the "sustainability" of the process.

Improvement of rates is mainly the result of biocatalyst engineering, while improvement of yields result from the biocatalyst selectivity and from mass transport between phases. This last phenomenon is also a key feature for environmental aspects. Hence, most of the impacts of a biological process deal with carbon release in the environment. This release takes place in the form of VOCs, including CO_2. If it is difficult to avoid CO_2 production when microorganisms are involved (it is still the same with enzymes because they were preliminary produced by cell cultivation), care can be taken for other organic compounds.

This chapter shows how a biphasic medium can help in reducing loss of volatile compounds in a gaseous phase exiting from a bioreactor, in comparison with pure aqueous systems. It also emphasises the usefulness of solvents having low vapour pressure (heavy organic solvents or ionic liquids) in the reduction of the release of compounds into the environment. There are, from this point of view, common interests between engineering needs and environmental concerns in the flavouring industry.

References

1. Tramper J, Vermüe M, Beeftink HH, von Stockar U (eds) (1992) Biocatalysis in non-conventional media Elsevier, Amsterdam
2. Vermüe M, Tramper J (1995) Pure Appl Chem 64:345
3. Cabral JMS (2001) Basic biotechnology, 2nd edn. Cambridge University Press, Cambridge
4. Krieger N, Bhatnagar T, Baratti JC, Baron AM, de Lima V, Mitchell D (2004) Food Technol Biotechnol 42:279
5. Kumar R, Modak J, Madras G (2005) Biochem Eng J 23:199
6. Lozano P, de Diego T, Gmouh S, Vaultier M, Iborra JL (2004) Biotechnol Prog 20:661
7. Van Rantwijk F, Madeira RL, Sheldon RA (2003) Trends Biotechnol 21:131
8. Brennecke JF, Maginn EJ (2001) AIChE J 47:2384
9. Lozano P, De Diego T, Guegan JP, Vaultier M, Iborra JL (2001) Biotechnol Bioeng 75:563
10. Howarth J, James P, Dai J (2001) Tetrahedron Lett 42:7517
11. Erbeldinger M, Mesiano AJ, Russell AJ (2000) Biotechnol Prog 16:1129

12. Yang Z, Pan W (2005) Enzyme Microb Technol 37:19
13. Kragl U, Eckstein M, Kaftzik N (2002) Curr Opin Biotechnol 13:565
14. Park S, Kazlauskas RJ (2003) Curr Opin Biotechnol 14:432
15. Schrader J, Etschmann MMW, Sell D, Hilmer JM, Rabenhorst J (2004) Biotechnol Lett 26:463
16. Ishige T, Honda K, Shimizu S (2005) Curr Opin Chem Biol 9:174
17. Rasor JP, Voss E (2001) Appl Catal A 22:145
18. Schrader J, Berger RG (2001) In: Rehm H-J, Reed G (eds) Biotechnology, vol 10, 2nd edn. Wiley-VCH, Weinheim, p 373
19. De Carvalho CCCR, Da Fonseca MMR (2006) Biotechnol Adv 24:134
20. Hatti-Kaul R (2001) Mol Biotechnol 19:269
21. Grivel F, Larroche C (2001) Biochem Eng J 7:27
22. Fontanille P, Larroche C (2003) Appl Microb Biotechnol 60:534
23. Bell G, Halling PJ, Moore BD, Partridge J, Rees DG (1995) TIBTECH 13:468
24. León R, Fernandes P, Pinheiro HM, Cabral JMS (1998) Enzyme Microb Technol 23:483
25. Sardessai Y, Bhosle S (2002) Res Microb 153:263
26. De Carvalho CCCR, Da Fonseca MMR (2004) Bioprocess Biosyst Eng 26:361
27. Laane C, Boeren S, Vos K, Veeger C (1987) Biotechnol Bioeng 30:81
28. Inoue A, Horikoshi K (1991) J Ferment Bioeng 7:194
29. Sardessai Y, Bhosle S (2004) Biotechnol Prog 20:655
30. Ross AC, Bell G, Halling PJ (2000) J Mol Cat B 8:183
31. Halling PJ, Ross AC, Bell G (1998) Prog Biotechnol 15:365
32. Ghatorae AS, Guerra MJ, Bell G, Halling PJ (1994) Biotechnol Bioeng 44:1355
33. Ross AC, Bell G, Halling PJ (2000) Biotechnol Bioeng 67:498
34. Baldascini H, Janssen DB (2005) Enzyme Microb Technol 36:285
35. Sah H, Bahl Y (2005) J Controlled Release 106:51
36. Fan KK, Ouyang P, Wu X, Lu Z (2001) Enzyme Microb Technol 28:3
37. Cruz A, Fernandes P, Cabral JMS, Pinheiro HM (2002) J Mol Cat B 19:371
38. Baum EJ (1998) Chemical property estimation. Theory and application, CRC, Boca Ratton
39. de Swaan Arons J, de Loos TW (1994) In: Sandler SI (ed) Models for thermodynamic and phase equilibria calculations. Dekker, New York, p 363
40. Grain CF (1990) In: Lyman WJ, Reehl WF, Rosenblatt DH (eds) Handbook of chemical property estimation methods. American Chemical Society, Washington, p 14.1
41. Schlessinger GG (1972) In: Weast RC (ed) Handbook of chemistry and physics, 53rd edn. CRC, Cleveland, p D151
42. Antoine C (1888) C R Acad Sci, 107:681
43. Reid RC, Prausnitz JM, Poling BR (1987) The properties of gases and liquids, 3rd edn. McGraw-Hill, New York.
44. Sage ML, Sage GW (2000) In: Boethling RS, Mackay D (eds) Handbook of property estimation methods for chemicals: environmental and health sciences. CRC Boca Raton, p 53
45. Sherman SR, Trampe DB, Bush DM, Schiller M, Eckert CA, Dallas AJ, Li J, Carr PW (1996) Ind Eng Chem Res 35:1044
46. Delaney JS (2005) Drug Discov Today 10:289
47. Yalkowsky YSH, Valvany SC (1980) Pharm Sci, 69:912
48. Ran Y, Jain N, Yalkowsky SH (2001) J Chem Inf Comp Sci 41:1208

49. Hansch C, Leo A, Hoekman D (1995) Exploring QSAR. Hydrophobic, electronic, and steric constants. American Chemical Society, Washington
50. Sangster J (1997) Octanol-water partition coefficients: fundamentals and physical chemistry. Wiley series in solution chemistry, vol 2. Wiley, Chichester
51. Leo A (2000) In: Boethling RS, Mackay D (eds) Handbook of property estimation methods for chemicals: environmental and health sciences. CRC, Boca Raton, p 89
52. Meylan WM, Howard PH (1995) J Pharm Sci 84:83
53. Meylan WM, Howard PH, Boethling RS (1996) Environ Toxicol Chem 15:100
54. Fredenslund Aa., Jones RL, Prausnitz JM (1975) AIChE J 21:1086
55. Gmehling J, Li J, Schiller MA (1993) Ind Eng Chem Res 32:178
56. Lohmann J, Joh R, Gmehling J (2001) Ind Eng Chem Res 40:957
57. Larsen BL, Rasmussen P, Fredenslund A (1987) Ind Eng Chem Res, 26:2274
58. Gros JB, Larroche C (2005) In: Pandey A, Webb C, Soccol CR, Larroche C (eds) Enzyme technology. Asiatech, New Delhi, p 479
59. Scragg AH (1991) Bioreactors in biotechnology. A practical approach, Horwood, New York
60. Nielsen J, Villadsen J, Liden G (2003) Bioreaction engineering principles, 2nd edn. Kluwer/ Plenum, New York
61. Van't Riet K (1983) Mass transfer in fermentation. Trends Biotechnol 1:113
62. Perry RH, Green DW (1997) Perry's chemical handbook, 7th edn. McGraw-Hill, New York
63. Poughon L, Duchez D, Cornet JF, Dussap CG (2003) Bioprocess Biosystem Eng 25:341
64. Zhou G, Kresta SM (1998) Chem Eng Sci 53:2063
65. Sprow FB (1967) Chem Eng Sci 22:435
66. Césario MT, de Wit HL, Tramper J, Beeftink HH (1997) Biotechnol Prog 13:399
67. Dumont E, Delmas H (2003) Chem Eng Process 42:419
68. Brilman DWF, Goldschmidt MJV, Versteeg GF, van Swaaij WPM (2000) Chem Eng Sci 55:2793
69. Hotrum NE, Cohen Stuart MA, Van Vliet T, Avino SF, van Aken GA (2005) Colloids Surface A 260:71
70. Mikami Y, Fukunaga Y, Arita M, Obi Y, Kisaki T (1981) Agric Biol Chem 43:791
71. Larroche C, Grivel F, Creuly C, Gros JB (1995) In: Etievant P, Schreier P (eds) Bioflavour 95. INRA, Paris, p 309
72. Grivel F, Larroche C, Gros JB (1999) Biotechnol Prog 15:697
73. Grivel F, Larroche C (2001) Biochem Eng J 7:27

25 The Production of Flavours by Plant Cell Cultures

A.H. Scragg
Centre for Environmental Sciences,
Faculty of Applied Sciences,
University of the West of England,
Bristol BS16 1QY, UK

25.1 Introduction

The interest in sustainable industrialisation or development was probably launched in 1987 by a report to the World Commission on Environment and Development (The Brundtland Commission) [1], although concern had been voiced prior to this. The Brundtland Commission report was confirmed at the UN Earth Summit in Rio de Janeiro in 1992. The objective was to achieve agricultural and industrial production and energy generation where environmental and economic systems are in balance. Sustainable development was defined as "strategies and actions that have the objective of meeting the needs and aspirations of the present without compromising the ability to meet those of the future". Another definition for sustainable development was "to prolong the productive use of our natural resources over time, while at the same time retaining the integrity of their bases, thereby enabling their continuity" [2]. It is self-evident that the exploitation of non-renewable resources cannot continue unchecked and that a balance has to be achieved between their consumption and the use of renewable resources.

In terms of energy, the International Energy Agency [3] has predicted that the supply of crude oil will peak around 2014 and then decline, and that coal will last until 2200 [4].

Fossil fuels are not only used to produce energy; they are the raw material for a very wide range of industries producing both bulk and fine chemicals. These chemicals include plastics, paints, antifreeze, insecticides, vitamins, adhesives, detergents, butyl rubber, resins, dyes and flavours [5]. In the past many of these chemicals would have been obtained from plants. Plants have been used for thousands of years as a source of fuel, food, construction materials, textiles, bulk and fine chemical, medicines, oils, dyes, poisons, rubber, resins, gums, fragrances and flavours. For example, soybeans were not only used to produce oil for cooking and meal for animal foods, but in the 1920s soybeans were used to produce adhesives, plastics, insulating foams, paints, textiles, lubricants, emulsifiers and binders [6]. Since then these have been replaced by chemicals produced from fossil fuels.

The chemical synthesis of natural flavours started some time ago with the synthesis of coumarin in 1868 and vanillin in 1874 [7]. The development of the petrochemical industry and the availability of cheap oil has meant that most of the plant-derived products are now synthesised from crude oil. In addition, flavours can now be produced using microbial cultures. Thus, to achieve sustainable development plants will have to provide many of the products currently produced from petrochemicals, including flavours. In this chapter the possible use of plant tissue culture techniques and processes in the sustainable production of flavours is outlined and discussed.

25.2 Flavours

The market for flavours and aromas is large and was worth $16 billion in 2003 [7, 8]. There are about 6,500 flavours known but of these only 300 are commonly used. At present 50–100 are produced by microbial fermentation, and many of the rest are chemically synthesised. In many cases, flavours and aromas are very complex mixtures extracted from pulp, bark, peel, leaf, bud, berry and flowers of fruit, vegetables, spices and other plants. The particular flavour or aroma will depend on the balance of these compounds, although a number are due to a single compound.

Many of the single-flavour compounds have been chemically synthesised and are therefore available cheaply and in large quantities. However, recently there has been a move towards natural colours and flavours, distinct from sustainability, which often carries a premium price. For example, vanillin, the characteristic flavour component of cured vanilla pods, has an annual consumption in food of 6,000 t. The main method of vanillin production is chemical synthesis from guaiacol and lignin, but the price of this "nature-identical" vanillin is very low ($15 per kilogram) compared with the "natural" vanillin extracted from cured vanilla pods ($1,200–4,000 per kilogram) [7, 9]. The reasons for the high price of natural vanillin are the limited supply of pods owing to climate affecting yields, economic and political problems, and the labour required for harvesting and curing the pods. Despite the higher cost there is a customer-led demand for natural flavours developing alongside the demand for organically grown food. "Natural" flavours have been defined in the USA and Europe as flavours only prepared either by extraction from natural sources or by transformation of natural precursors using enzymes or microbial cultures. Any chemically synthesised flavours must therefore be regarded as nature-identical [10, 11]. Figure 25.1 shows the pathways that are available for the production of natural flavours as defined by these regulations. In the supply of natural flavours or flavour precursors there are three options: collection from the wild plant population, agricultural cultivation, and plant tissue culture (Fig. 25.2).

Collection from the wild is perhaps the easiest and has been used with many flavour-producing plants but overcollection has endangered the stocks in many

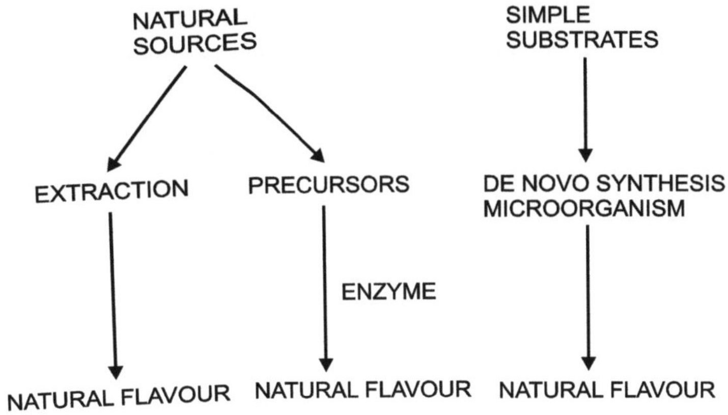

Fig. 25.1 The three pathways for the preparation of 'natural' flavours. The first two involve the extraction of the flavour or precursors from natural sources. The precursors can then be converted to the natural flavour by enzymes extracted from plants or microorganisms. The last method is the de novo synthesis of the flavour by microorganisms growing on simple substrates such as glucose and sucrose

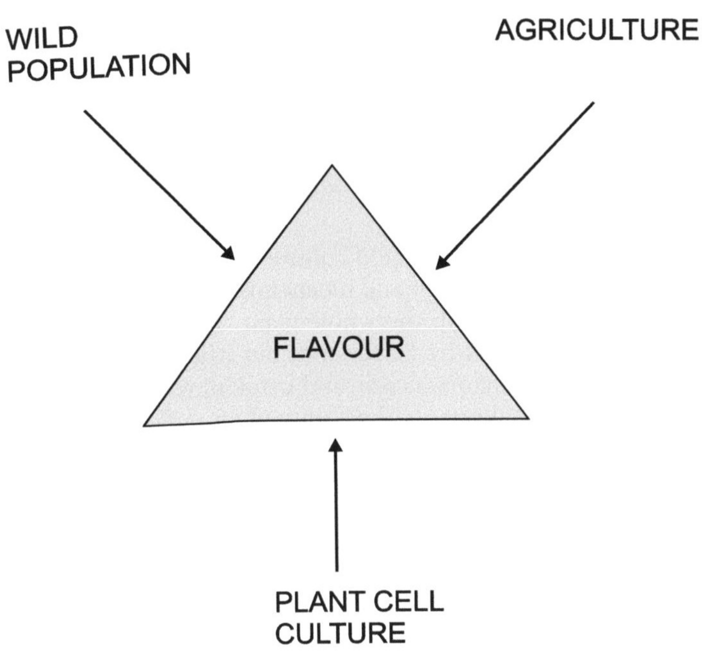

Fig. 25.2 Three possible sources of natural flavours

cases. The supply can be supplemented by agricultural cultivation but in some cases the wild populations require specific growth conditions which cannot be reproduced elsewhere. Propagation may also be difficult, so agricultural cultivation may not be possible. If agriculture is not a viable option, other methods have to be found to preserve and maintain the wild population while providing the material for flavour extraction. In this case plant tissue techniques may be suitable for the multiplication of the plant and/or its conservation. The agricultural cultivation of the plant or related species is clearly the most economic solution to the flavour supply. However, the crop may suffer from pests and diseases, and adverse climatic conditions which can affect yield and quality. In some cases political factors can also affect supplies from some countries and regions. The plant may also be difficult to propagate and may require exacting condition for growth as found with vanilla pods. Under these conditions the techniques of plant cell culture may help to alleviate the pressure on the supply of natural flavours in a sustainable manner by helping with the propagation of the particular plant or the de novo production of the flavour itself.

25.3 Plant Cell and Tissue Culture

The techniques of plant tissue culture offer a number of options in the quest for the sustainable production of natural flavours. These are as follows:

- Micropropagation: the provision of plants difficult to propagate using normal methods or those of endangered species
- The de novo production of the flavours using callus and suspension cultures of the source plant
- The use of whole cells or extracted enzymes to carry out biotransformations of precursors to the flavour compound

The culture of plant cells on solid or in liquid culture was developed as a research tool in order to study the physiology and biochemistry of plants without the complications of having to deal with the whole plant. The idea of culturing plants cells was proposed in 1904 by Haberlandt but it was not until the discovery of the plant growth regulators auxins and cytokinins in 1943–1960 that plant cells could be reliably cultivated. The ability of an individual cell to grow and divide in a self-regulating manner is referred to as totipotency. Thus, a totipotent cell should be able to regenerate a whole plant from a single cell. A distinction should be made between organ and tissue culture. Root and shoot cultures are examples of organ culture where the plant material maintains its morphological identity. Tissue culture is the culture of non-differentiated cells in liquid or on a solid medium and examples are cell suspension and callus. Figure 25.3 outlines the development of both types of culture [12–14]. Plant cell cultures are normally grown under sterile conditions so that any part taken from the plant, known as an explant, will be surface-sterilised. Once sterile, the explant is placed on a solid medium containing major and minor salts, a carbon and energy source, normally sucrose, and the growth regulators auxins and cytokinins. The major component of the medium is sucrose, which is a renewable resource. It is the growth regulators that direct the growth, elongation and differentiation of the cells in the explants. It is the balance of these two regulators that controls whether shoots, roots or a mass of undifferentiated cells, a callus, is formed. It is the callus material which is added to a liquid medium to form the suspension cultures. Suspension cultures generally have a faster growth rate, are more homogeneous than callus material and thus can be cultivated on a large scale in bioreactors. This is an important feature when developing an industrial process.

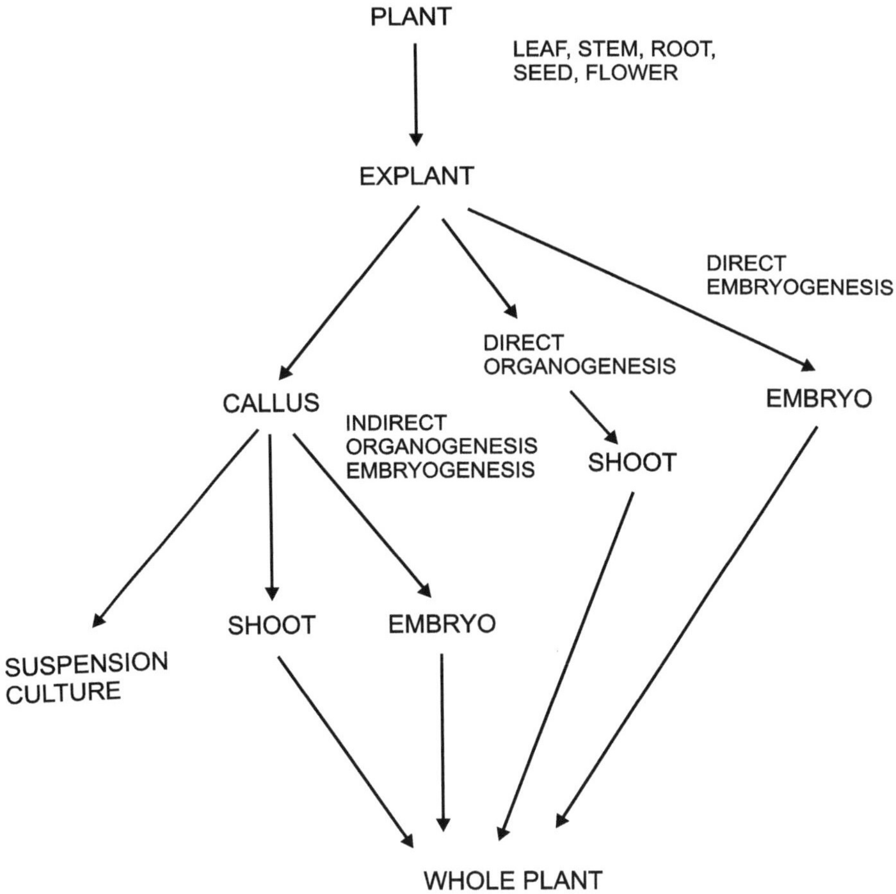

Fig. 25.3 The pathways that can be taken in the development of plant tissue and cell cultures starting from a part of a plant (explant). The explant can via direct embryogenesis or oganogenesis form embryos or shoot and roots, respectively, which can be converted into plants. In another path, the explant can form a callus, which can then be used to form a suspension culture. In addition, indirect organogenesis or embryogenesis of the callus can lead to plant formation

25.3.1
Micropropagation

The ability of the growth regulator balance to stimulate shoot and root formation means that a single explant can be used to form a large number of plants in a process known as micropropagation. Micropropagation is now a commercially efficient technology producing over 500 million plants annually [15]. The advantages are:

- Production of a very large number of cloned plants in a short time
- Production of disease-free plant material
- Production of a large stock of true-to-type propagation material
- Easy transportation of plant material
- Production of a large number of plants from elite or difficult-to-grow and slow-to-grow plants, bringing new plants to the market rapidly
- Conservation of plant genetic resources, preserving those plants threatened in the wild

There are a number of books and reviews on the micropropagation of plants, and a large number of plants have been micropropagated, including flavour-producing plants [14–16]; therefore, there are in the literature many methods for the micropropagation of flavour-producing plants, and some recent examples are *Theobroma cacao* [17, 18], *Mentha arvensis* [19], onion and garlic [20].

25.3.2
Plant Cell Suspensions

In the second option, plant cell cultures can be used to produce flavours de novo.

The potential for plant cell cultures to produce compounds of value has concentrated on pharmaceuticals, and a wide range of compounds have been detected in cell suspensions, some at high concentrations [21–23]. Because of the high value of the pharmaceuticals, these have dominated the research and potential commercialisation of tissue cultures and this has limited the investigated of flavour production. The production of flavours using plant cell cultures has a number of advantages and disadvantages when compared with production by cultivation of the plant or chemical synthesis. The advantages are as follows:

- Not affected by weather or pests and disease
- Flavours can be produced in situ, in every country
- Uses sustainable resources, mainly in the medium
- Flavours are not derived from petrochemicals
- Saves agricultural land
- A defined production system giving consistent quality and quantity
- Free from embargos and political interference
- Higher production than for whole plants especially when only found in very small quantities

- Saves stocks of rare or slow-growing plants
- Only source because the complex nature of the flavours means chemical synthesis is not possible and is too expensive

The disadvantages are low yields of the product and high costs of the process. The production of flavour compounds using plant cell cultures offers a process which uses a sustainable carbon source, sucrose, which is the major component of the medium. Production using this method may be used to supply only part of the material required, taking pressure off the wild stocks of the plant. However, if plant cell cultures are to be used on an industrial scale a number of conditions need to be achieved:

- High yields of the compound or compounds
- High growth rate of cells
- Ability to scale up the process
- Will the product be accepted as natural?

The high growth rate and scale-up can only be achieved with suspension cultures grown in large bioreactors. Until a flavour is produced from plant cell culture on a large scale no decision can be made as to its natural origin. A very wide range of compounds have been detected in plant cell cultures and many of these are pharmaceuticals. However, flavour compounds have been detected and some examples of these are shown in Table 25.1. The flavours detected have mainly been restricted to the single characteristic flavour compounds and there are a number of reviews on the subject [24–27]. High yields of compounds have been obtained for a number of secondary metabolites, mainly pharmaceuticals. Some of the yields obtained for these compounds are given in Table 25.2, including values for two microbial products, penicillin and citric acid. Some of the secondary products have reached yields and productivities approaching that of penicillin, which is acceptable since penicillin has been under continuous development since the 1940s [28]. However, many of the high yielding compounds are of no industrial value, such as rosmarinic acid, in contrast to the low yields of the high-value drug taxol. In contrast, the yield of flavour compounds is generally low, in part because some of the compounds are volatile, in some cases high levels can be toxic, and accumulation often occurs in specialised cells.

A number of strategies have been adopted to increase yields. The first is to screen and select for high yields and rapid growth rates. This is not as easy to carry out with plant cells compared with microbes as plants are difficult to grow as single cells, and the detection of the compounds in the single cell or clump also poses considerable problems. The compounds of interest are often produced after growth has ceased, which makes selection difficult. Some of the high-yielding cultures (Table 25.2) were isolated without screening and selection, which is perhaps the luck factor. Secondly, considerable effort has been placed on the manipulation of cultural conditions to stimulate secondary product accumulation. These have been reviewed in a number of reports [23, 27] and the conditions are briefly:

- Quality and quantity of carbon source
- Nitrate levels
- Phosphate levels
- Growth regulators
- Addition of precursors
- Changing conditions such as temperature, light, pH, agitation or aeration

Manipulation of the culture environment has proved successful in many cases, stimulating the accumulation of secondary products in plant cell cultures, but each treatment will not always be successful with every culture [21, 23, 28, 29]. A range of treatments may have to be tried for each individual culture. All the changes in growth conditions and medium have perhaps a common feature in that they all cause some form of stress. Stress is known to trigger changes in cells and this may stimulate the accumulation of secondary products (Fig. 25.4).

Other methods used to increase secondary product accumulation are elicitation, permeabilisation, product removal, immobilisation and differentiation. Elicitation is the triggering of plant defence mechanisms by the addition of abiotic and biotic elicitors. Elicitor refers to chemicals which can trigger physiologi-

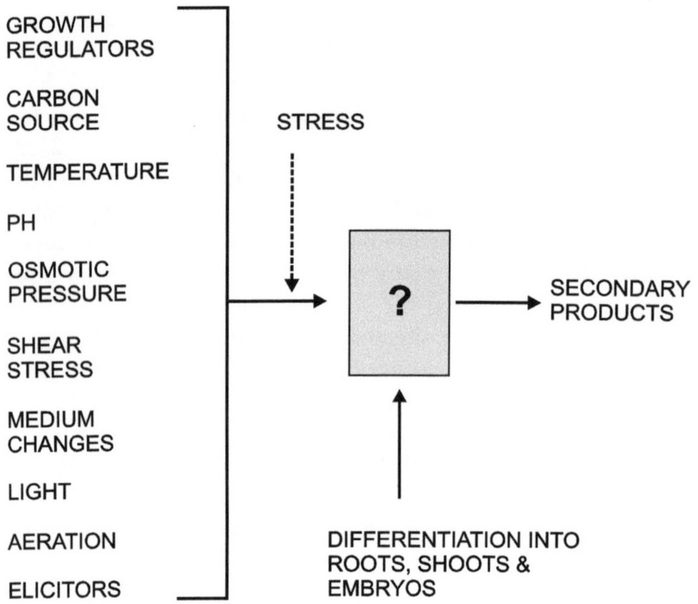

Fig. 25.4 A possible link between the number of changes in cultural conditions that have been used to increase secondary product accumulation through the response of the cells to stress. The other route is to induce some form of differentiation

Table 25.1 Flavour compound found in plant cell and tissue cultures

Flavour	Plant	Metabolite	Culture type	Amount accumulated	References
Aniseed	*Pimpinella anisum*	Anethole/chavicol	Callus	0.37%	[51]
Basmati rice	*Oryza sativa*	2-Acetyl-pyroline	Callus	ND	[52]
Shilli	*Capsicum annuum*	Capsaicin	Suspension, immobilised	0.4%	[53–55]
Cocoa	*Theobroma cacao*	Complex mixture	Callus	ND	[56]
Garlic	*Allium sativum*	Diallyl disulphide	Callus	14% of explant	[57, 58]
Ginseng	*Panax ginseng*	Ginsenosides	Suspension	1.57 g/l	[49, 59, 60]
Grape	*Vitis vinifera*	*β*-Damascenone	Callus	8 ng/g	[61]
Guava	*Psidium guajava*	Mixture	Callus	ND	[62]
Hop	*Humulus lupulus*	α acids	Suspension	ND	[63]
Liquorice	*Glycrrhiza glabra*	Glycyrrhizin	Callus, suspension	ND	[64]
Onion	*Allium cepa*	Dipropyl disulphide	Callus, organs[a]	ND	[65, 66]
Peppermint	*Mentha piperita*	Menthol	Suspension	0.0012%	[67]
Saffron	*Crocus sativus*	Crocin, safranol	Callus, organs	ND	[68–70]
Stevioside	*Stevia rebandiana*	Stevioside	Plantlets	ND	[71, 72]
Tarragon	*Artemisia dracunulus*	Methyl chavicol	Callus, suspension	ND	[73]
Thaumatin	*Thaumatococcus daniellii*	Thaumatin	Callus	ND	[74]
Vanilla	*Vanilla planifolia*	Vanillin	Callus, organs	0.099	[75–77]
Vermouth	*Artemisia absinthum*	Pinene, thujyl alcohol	Callus	ND	[24]

ND not determined
[a]Refers to root or shoot development

Table 25.2 Yields and productivity of secondary products in plant cell suspensions

Culture	Product	Biomass (g/l dry wt)	Time (days)	Yield (% dry wt)	Yield (g/l)	Productivity (g/l/day)	References
Coleus blumei	Rosmarinic acid	25.7	6	21.4	5.5	0.91	[78]
Dioscorea sp.	Diosgenin	11.3	16	3.8	0.43	0.028	[79]
Coptis japonica	Berberine	70	6	5	3.5	0.6	[80]
Perilla frutescens	Anthocyanins	13.5	10	8.89	1.2	0.12	[81]
Anchusa officinalis	Rosmarinic acid	35	25	11.4	4	0.16	[82]
Papaver somniferum	Sanguinarine	12.1	9	2.5	0.3	0.025	[83]
Salvia officinalis	Rosmarinic acid	17.8	30	36	6.4	0.22	[84]
Panax notoginseng	Saponins	35	28	4.48	1.57	0.055	[60]
Taxus chinensis	Taxol	–	12	–	0.027	0.00225	[86]
Panax ginseng	Saponins	10	12	7.5	0.75	0.0625	[85]
Lavandula vera	Rosmarinic acid	29.2	12	1.7	0.507	0.0423	[87]
Taxus chinensis	Taxanes	22.7	21	1.3	0.278	0.013	[88]
Taxus chinensis	Taxanes	18.5	20	14.2	0.229	0.0135	[89]
Taxus chinensis	Taxanes	27	23	10.2	0.274	0.0093	
Penicillium sp.	Penicillin	–	–	–	–	1.4–2.1	[21]
Aspergillus niger	Citric acid	–	–	–	–	30–38	[21]

cal and morphological responses which lead to secondary product accumulation. Abiotic elicitors include metal ions and inorganic compounds and examples of biotic elicitors are cell wall extracts from yeast, fungi and bacteria [30].

Many secondary products are stored in the cell's vacuole and the release of these compounds from the vacuole can be achieved by permeabilising the vacuole and cell membranes. A variety of permeabilising agents have been used, including organic solvents such as dimethyl sulphoxide and 2-propanol, and polysaccharides like chitosan. Other methods have included ultrasonication, electroporation and ionophoretic treatment [31]. The objective is to permeabilise the cells for a short time to allow the release of the product while maintaining the cell's viability. In this case the cell can continue to grow and accumulate the secondary product. Low levels of product accumulation have also been attributed to feedback inhibition, degradation of the product in the medium or its volatility causing losses [23, 32]. Permeabilisation should reduce the feedback effect. In situ product collection will reduce degradation and can be carried out by adding solid or non-miscible liquid phases which will accumulate any compound released into the medium [32–36].

Secondary product accumulation appears to be stimulated by cell-to-cell contact which appears to occur when plant cells are immobilised [37–41]. The advantages of immobilisation are the cells are easily recovered or retained, allowing a continuous process to be used, the product is easily separated from the cells and the cells show increased longevity and are protected from shear forces. In addition, the cell-to-cell contact may induce cytodifferentiation, which may stimulate secondary product formation. The main disadvantage is extracting the product if it is retained within the cell. Differentiation of the culture into roots or shoots [23] can initiate the accumulation of secondary products. Normal root and shoot cultures grow very slowly but *Agrobacterium*-infected roots grow rapidly. Hairy roots are induced by infection of plants by *Agrobacterium rhizogenes*. Hairy roots have the advantages that they grow rapidly compared with normal roots and do not require growth regulators. The main problem with organised cultures such as roots and shoots is growth in bioreactors. They are sensitive to shear because of their large size and shear is high in the standard stirred-tank bioreactor. Therefore, to cultivate such cultures alternative bioreactor designs are required.

25.3.3 Biotransformation

Biotransformation is the conversion of a compound into the product using living plant cells or enzymes extracted from plants [42, 43]. This is the third option for using plant cell cultures for flavour production. Some examples are given in Table 25.3, but at present the yields are still low.

Table 25.3 Biotransformations carried out by plant cell and organ cultures

Plant	Precursor	Product	References
Capsicum frutescens (immobilised)	Ferulic acid, vanillyamine	Capsaicin, vanillin	[42]
Capsicum frutescens (immobilised)	Protocatechuic aldehyde, caffeic acid	Vanillin, capsaicin	[23]
Coffea arabica	Vanillin	Vanillin-β-D-glucoside	[91]
Crocus sativus	Crocetin	Crocetin dineapolitanosyl	[92]
Mentha sp. (immobilised)	(-)-Menthone	(+)-Neomenthol	[93]
Mentha piperita and *Mentha canadensis*	Menthyl acetate	Menthol	[94]
Solanum aviculare and *Dioscorea deltoidea*	(*S*)-(-)-Limonene, (*R*)-(+)-limonene	Carvone	[95]
Peganum harmala	Menthyl acetate	Menthol	[96]

25.3.4 Scale-Up

If an industrial scale is to be achieved, the plant culture of whatever type will have to be cultivated in bioreactors of up to 75,000 l in size. Plant cell suspensions, shoot and root cultures pose very different problems in bioreactors compared with microbial cultures. There are a number of reviews on the subject of bioreactor growth and scale-up [21, 44–48]. Briefly, plant cells grow slowly, the cells are large and generally form clumps which make them more sensitive to shear associated with agitation and give long processing times. Organ cultures are far more sensitive to shear and apart from having hairy roots grow very slowly. These characteristics mean that alternative impeller and bioreactor designs to the normally used stirred-tank system have been investigated. The main design feature is to avoid or reduce shear within the bioreactor.

25.4 Discussion

The plant cell culture technique of micropropagation of flavour-producing plants will be able to help with their agricultural cultivation and will relieve the pressure on the wild populations. Micropropagation will be able to propagate those plants where conventional propagation is difficult or will be to multiply elite stock. This may be required if demand for natural flavours continues to increase.

In the second option the development of a plant cell culture process for the production of flavours requires a high yield, fast growth rate, high biomass, and the ability to grow in bioreactors. It is clear that the yields of many of the flavours compounds detected (Table 25.1) are low. The area will require continued research if high yields are to be obtained. Perhaps one exception is ginseng, which is probably owing to its use as a medicine and health tonic rather than a flavour. Although ginseng is used in China, Japan and Korea, there has been considerable interest worldwide and sales have a value of $1 billion [49]. Ginseng was originally collected from the wild but it is now farmed, but farming is time-consuming, labour-intensive and takes 5–7 years from sowing to harvest. Considerable effort has been made to increase the yield of the active ingredients in plant cell suspensions and a fed-batch system has yielded a biomass level of 35 g/l with a productivity of 1.57 g/l in 20 days (0.078 g/l/day). The level of productivity compares quite well with that of microbial processes such as for citric acid and penicillin (Table 25.2).

The use of plant cells or enzymes extracted from the cells for the biotransformation of exogenous substances offers another method for producing flavours. This would be useful when compounds are not found in cell suspensions. Many studies have been carried out on the biotransformation of xenobiotics or pharmaceuticals by plant cell culture [43, 50].

References

1. World Commission on Environment and Development (1987) Our common future. Oxford University Press, Oxford
2. De Paula GO, Cavalcanti RN (2000) J Clean Product 8:109
3. International Energy Authority (2002) Renewables in global energy supply. International Energy Authority, Paris
4. Evans J (2000) Chem Br August 30
5. Evans J (1999) Chem Br August 38
6. Kumar R, Choudhary V, Mishra S, Varma IK, Mattiason B (2002) Ind Crops Prod 16:155
7. Serra S, Fuganti C, Brenna E (2005) Trends Biotechnol 23:193
8. Priefert H, Rabenhorst J, Steinbuchel A (2001) Appl Microbiol Biotechnol 56:296
9. Walton NJ, Mayer MJ, Narbad A (2003) Phytochemistry 63:505
10. US Code of Federal Regulations (1985) 21, 101.22a.3. FDA, Washington
11. The Council of Europe Communities (1988) Council directive 88/388/EEC of 22 June 1988
12. Dixon RA (ed) (1985) Plant cell culture. A practical approach. IRL, Oxford
13. Stafford A, Warren G (eds) (1991) Plant cell and tissue culture. Open University Press, Milton Keynes
14. Hall RD (ed) (1999) Plant cell culture protocols. Humana, Totowa
15. Altman A (ed) (1998) Agricultural biotechnology. Dekker, New York
16. Bajaj YPS (ed) (1991) Biotechnology in agriculture and forestry, vols 17–21. Springer, Berlin Heidelberg New York
17. Tan CL, Furtek DB (2003) Plant Sci 164:407
18. Benett AB (2003) Trends Plant Sci 8:561

19. Phatak SV, Heble MR (2002) Fitoterapia 73:32
20. Mukhopadhyay MJ, Sengupta P, Mukhopadhyay S, Sen S (2005) Sci Hortic 104:1
21. Scragg AH (1995) Plant Cell Tissue Organ Cult 43:163
22. Verpoorte R, van der Heijden R, ten Hoopen HJG, Memelink J (1999) Biotechnol Lett 21:467
23. Ramachandra Rao S, Ravishankar GA (2002) Biotechnol Adv 20:210
24. Balandrin MF, Klocke JA (1988) In: Bajaj YPS (ed) Biotechnology in agriculture and forestry, vol 4. Springer, Berlin Heidelberg New York, p 3
25. Mulder-Krieger T, Verpoorte R, Svendse A, Scheffer J (1988) Plant Cell Tissue Organ Cult 13:85
26. Collin HA (1988) In: Constabel F, Vasil IK (eds) Cell culture and somatic cell genetics of plants, vol 5. Academic, New York, p 569
27. Scragg AH (1997) In: Berger RG (ed) Biotechnology of aroma compounds. Advances in biochemical engineering biotechnology, vol 55. Springer, Berlin Heidelberg New York, p 239
28. Verpoorte R, van der Heijden R, Scripresma J, Hoge JHC, ten Hoopen HJG (1993) J Nat Prod 56:186
29. Zenk MH, El-Shagi H, Schulte U (1975) Planta Med Suppl 79
30. Zhao J, Davis LC, Verpoorte R (2005) Biotechnol Adv 23:283
31. Brodelius P (1988) Appl Microbiol Biotechnol 27:561
32. Biederbeck R, Knoop B (1988) In: Bajaj YPS (ed) Biotechnology in agriculture and forestry, vol 4. Springer, Berlin Heidelberg New York, p 123
33. Hooker BS, Lee JM (1990) Plant Cell Rep 8:546
34. Buitelaar RM, Langenhoff AAM, Tramper J (1991) Enzyme Microb Technol 13:487
35. Bassetti L, Tramper J (1995) Enzyme Microb Technol 17:353
36. Chang HN, Sim SJ (1995) Curr Opin Biotechnol 6:209
37. Williams PD, Mavituna F (1992) In: Fowler MW, Warren GS (eds) Plant biotechnology. Pergamon, Oxford, p 701
38. Brodelius P, Deus B, Mosbach K, Zenk MH (1979) FEBS Lett 103:93
39. Li C, Moon K, Honda H, Kobayashi T (2000) Biochem Eng J 4:169
40. Kargi F, Ganapathi B (1991) Enzyme Microb Technol 13:643
41. Rhodes MJC, Smith JI, Robins RJ (1987)Appl Micro Biotechnol 26:28
42. Johnson TS, Ravishankar GA, Venkataraman LV (1996) Plant Cell Tissue Organ Cult 44:117
43. Giri A, Dhingra V, Giri CC, Singh A. Ward OP, Narasu ML (2001) Biotechnol Adv 19:175
44. Taticek RA, Lee CWT, Shuler ML (1994) Curr Opin Biotechnol 5:165
45. Kieran PM, MacLoughlin PF, Malone DM (1997) J Biotechnol 59:39
46. Doran PM (1997) Hairy roots: culture and applications. Harwood, Amsterdam
47. Tescione LD, Ramakrishnan D, Curtis WR (1997) Enzyme Microb Technol 20:207
48. Scragg AH (1991) In: Stafford A, Warren G (eds) Plant cell and tissue culture. Open University Press, Milton Keynes, p221
49. Wu J, Zhong JJ (1999) J Biotechnol 68:89
50. Ishihara K, Hamada H, Hirata T, Nakajima N (2003 J Mol Catal B 23:145
51. Ernst D (1989) In: Bajaj YPS (ed) Biotechnology in agriculture and forestry, vol 7 Medical and aromatic plants II. Springer, Berlin Heidelberg New York, p 381
52. Suvarnalatha G, Narayan MS, Ravishankar GA, Venkataraman LV (1994) J Sci Food Agric 66:439
53. Lindsey K, Yeoman MM (1984) Planta 162:595

54. Ochoa-Alejo N, Gomez-Peralta JE (1993) J Plant Physiol 141:147
55. Williams PD (1993) In: Bajaj YPS (ed) Biotechnology in agriculture and forestry, vol 21. Springer, Berlin Heidelberg New York, p51
56. Townsley PM (1972) J Inst Can Sci Technol Aliment 7:76
57. Madhavi DL, Prabha TN, Singh NS, Patwardhan MV (1991) J Sci Food Agric 56:15
58. Ohsumi C, Hayashi T, Sano K (1993) Phytochemicals 33:107
59. Zhong JJ, Chen F, Hu WW (1999) Process Biochem 35:491
60. Zhang YH, Zhong JJ (1997) Enzyme Microb Technol 21:59
61. Shure KB, Acree TE (1994) Plant Cell Rep 13:477
62. Prabha TN, Narayanan MS, Patwardhan MV (1990) J Sci Food Agric 50:105
63. Fowler MW, Stafford A, Martin PA, Stuart GG, Kember RMJ (1987) European Biotechnology Congress 313
64. Hayashi H, Fukui H, Tabata M (1988) Plant Cell Rep 7:508
65. Collin HA, Musker D (1988) In: Constabel F, Vasil I (eds) Cell culture and somatic cell genetics of plants, vol 5. Academic, San Diego, p 475
66. Hughes J, Tregova A, Tomsett AB, Jones MG, Cosstick R, Collin HA (2005) Phytochemistry 66:187
67. Chung IS, Kang YM, Oh JM, Kim T, Lee HJ, Chae YA (1994) Biotechnol Tech 8:789
68. Sarma KS, Sharada K, Maesato K, Hara T, Sonoda Y (1991)Plant Cell Tissue Organ Cult 26:11
69. Sujata V, Ravishankar GA, Venkataraman LV (1990) Biotechnol Appl Biochem 12:336
70. Chen S, Zhao B, Wang X, Yuan X, Wang Y (2004) Biotechnol Lett 26:27
71. Geuns JMC (2003) Phytochemistry 64:913
72. Swanson SM, Mahady GB, Beecher CWW (1992) Plant Cell Tissue Organ Cult 28:151
73. Gramshaw JW, Cotton CM, Evans LV (1987) In: Schreier T (ed) Bioflavour '87. de Gruyter, Berlin, p 341
74. van de Wel H, Ledeboer AM (1989) Thaumatins. In: Stumpf PK, Conn EE (eds) The biochemistry of plants—a comprehensive treatise, vol 15. Academic, San Diego, p 379
75. Knorr D, Caster C, Dornburg H, Graf S, Havin-Frenkel D, Podstolski A, Werrman U (1993) Food Technol 47:57
76. Westcott RJ, Cheetham PSJ, Barraclough AJ (1994) Phytochemistry 35:135
77. Knuth ME, Sahai OP (1991) US Patent WO8900820
78. Ulbrich B, Wiesner W, Arens H (1985) In: Deus-Neumann B, Barz W, Reinhard E (eds) Secondary metabolism of plant cell cultures. Springer, Berlin Heidelberg New York, p 293
79. Sahai O, Knuth M 91985) Biotechnol Prog 1:1
80. Fujita Y (1988) In: Ciba Foundation symposium 137. Applications of plant cell and tissue culture. Wiley, Chichester, p 228
81. Zhong JJ, Konstantinov KB, Yoshida T (1994) J Ferment Bioeng 77:445
82. Su WW, Humphrey AE (1993) Biotechnol Bioeng 42:884
83. Park JM, Yoon SY, Giles KL, Songstad DD, Eppstein D, Novakovski D, Friesen L, Roewer I (1992) J Ferment Bioeng 74:292
84. Hippolyte I, Marin B, Baccou JC, Jonard R (1992) Plant Cell Rep 11:109
85. Wu J, Lin L (2002) Appl Micro Biotechnol 59:51
86. Yu LJ, Lan WZ, Qin WM, Xu HB (2002) Process Biochem 38:201
87. Ilieva M, Pavlov A (1997) Appl Micro Biotechnol 47:683

88. Dong HD, Zhong JJ (2002) Enzyme Microb Technol 31:116
89. Pan ZW, Wang HQ, Zhong JJ (2000) Enzyme Microb Technol 27:714
90. Wang HQ, Yu JT, Zhong JJ (1999) Process Biochem 35:479
91. Kometani T, Tanimoto H, Nishimura T, Okada S (1993) Biosci Biotechnol Biochem 57:1290
92. Dufresne C, Cormier F, Dorion S, Niggli UA, Pfister S, Pfander H (1999) Enzyme Microb Technol 24:453
93. Galun E, Aviv D, Dantes A, Freeman A (1985) Plant Med 51:511
94. Werrmann U, Knorr D (1993) J Agric Food Chem 41:517
95. Vanek T, Valterova I, Vaisar T (1999) Phytochemistry 50:1347
96. Zhu W, Asghari G, Lockwood GB (2000) Fitoterrapia 71:501

26 Genetic Engineering of Plants and Microbial Cells for Flavour Production

Wilfried Schwab
Biomolecular Food Technology,
Technical University Munich,
Lise-Meitner-Str. 34, 85354 Freising, Germany

26.1
Genetic Engineering

Genetic engineering is defined as the laboratory technique used to change the DNA of living organisms. Changes in the genetic constitution of cells (apart from selective breeding) result from the introduction or elimination of specific genes through modern molecular biology techniques. Usually this technology is based on the use of a vector for transferring useful genetic information from a donor organism into a cell or organism that did not previously possess it. If the acceptor organism receives an additional structural gene coding for a functional polypeptide, the corresponding protein can be isolated and used as a biocatalyst in industrial applications. Alternatively, cells can be fitted with genes that create new biosynthetic pathways, allowing the overproduction of aroma substances and other desired compounds. This technique is known as metabolic engineering. Advances in genome sequencing enable access to an incredible number of genes from microorganisms and, more recently, from plants through *in silico* screening for putative functions in flavour formation.

Both plants and microorganisms are also suitable hosts for cloning vectors [1]. Bacteria like *Escherichia coli* and *Bacillus subtilis* offer the advantages of simple physiology, short generation times and high protein yields. With *B. subtilis* and some others, it is possible to induce secretion of a gene product into the surrounding medium, thereby facilitating biocatalyst isolation. However, eukaryotic recombinant proteins in bacterial cells often do not fold properly or are toxic to the cells, preventing cell cultures from reaching high densities. In addition, bacteria lack enzymes required for posttranslational modifications. For this reason, simple eukaryotes such as yeasts are used, which not only perform posttranslational modifications, but can also be induced to secrete certain proteins into the growth medium for harvesting.

Transgenic plants generated by *Agrobacterium*-mediated transformation or direct gene transfer are increasingly considered to be economically competitive systems for the production of foreign proteins (Fig. 26.1). Metabolic engineering in plants is also feasible nowadays, but it requires extensive knowledge of the relevant biosynthetic pathways [2]. Different vectors are available that enable the expression of the gene product in the cytosol or even in plastids [3]. How-

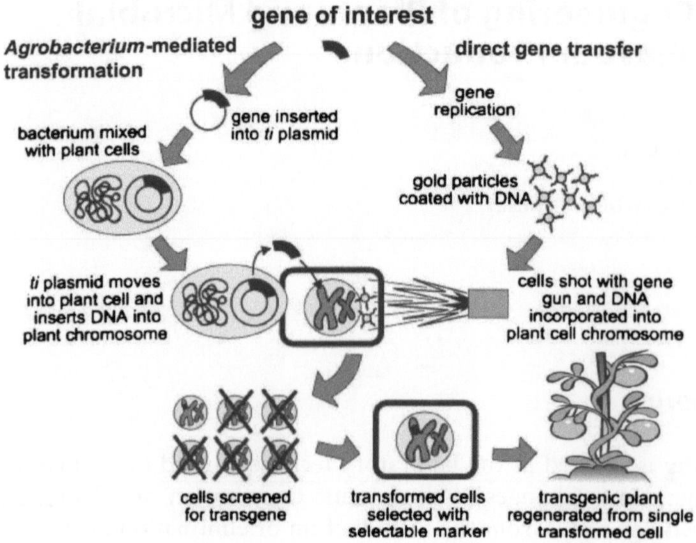

Fig. 26.1 Transgenic plants generated by *Agrobacterium*-mediated transformation or direct gene transfer

ever, all currently available plant expression methods suffer from limitations, such as the long time frame necessary for stable transformation and the low yield obtained with transient expression systems. Recently, an efficient transient plant expression system that is based on the *in planta* assembly of functional viral vectors from separate provector modules was developed to address these difficulties. The process is very fast and provides very high protein yield (up to 80% of total soluble protein) [4].

This review summarises the recent advances in engineering microbial and plant cells for flavour production. Examples from a number of compound classes will be presented to illustrate the use of transgenic organisms as sources for biocatalysts utilised in fermentation processes, as well as the application of metabolic engineering to produce a specific desired compound.

26.2 Terpenoids

Terpenoids are synthesised by the condensation of a series of isoprene (2-methylbuta-1,3-diene) units, followed by enzymatic cyclisation by a terpene cyclase, and subsequent modification such as hydroxylation, and are grouped on the basis of their carbon chain length. *Monoterpenes* and *sesquiterpenes* consisting of ten and 15 carbon atoms, respectively, are ranked among the most important aroma compounds. Despite their diversity, all terpenoids are synthesised from the common precursors dimethylallyl diphosphate and isopentenyl diphosphate. This occurs through two distinct pathways, the *mevalonate* and the *deoxyxylulose phosphate* (DXP) pathways, both of which have been targets for metabolic engineering [5]. Although microorganisms produce some terpenoids, most economically significant terpene products are found in plants, and extraction of these compounds from plant sources often involves expensive, low-yield processes.

Much attention has been paid to the last step of the formation of *monoterpenes* and *sesquiterpenes*, which is catalysed by terpenoid synthases. Over 30 complementary DNAs (cDNAs) encoding plant terpenoid synthases involved in the primary and secondary metabolism have been cloned, characterised, and the proteins heterologously expressed [6]. However, because geranyl diphosphate and farnesyl diphosphate are not readily available substrates, their biotransformation by terpenoid synthases is not economically viable. As a result, considerable effort has been put into engineering the total plant terpenoid biosynthetic pathway in recombinant microorganisms.

In the past, economical production of plant terpenes in recombinant *E. coli* has been limited by two major obstacles—low precursor supply and low plant enzyme expression levels or activity. The first was overcome by engineering the mevalonate isoprenoid pathway from *Saccharomyces cerevisae* into recombinant *E. coli*, thereby bypassing the microbial DXP pathway for isoprenoid biosynthesis and achieving high yields of artemisin precursor amorphadiene [7]. Poor expression of the plant amorphadiene synthase enzyme in *E. coli* was overcome by synthesising the gene from oligonucleotides incorporating *E. coli* codon bias. These innovative engineering efforts provide an excellent platform for further development of recombinant terpenoid production.

Another example of successful engineering of terpene biosynthesis is the constitutive overexpression of the gene encoding the first-step enzyme 1-deoxy-D-xylulose-5-phosphate synthase (DXPS) in the DXP pathway in bacteria and *Arabidopsis*. In both cases, increased enzyme activity caused increased accumulation of downstream terpenoids, suggesting that DXPS is rate-limiting [8].

The DXP pathway is also thought to be responsible for the synthesis of essential oil monoterpenes, which accumulate in glandular trichomes in mint and other plants [9]. In one of the first successful genetic modifications of a plant terpenoid pathway, Mahmound and Croteau [10] reported increasing flux through the monoterpene pathway in mint plants, resulting in an increased essential oil yield. They also improved the quality of the oil by expressing an antisense derivative of the menthofuran synthase gene to downregulate synthesis of the undesirable constituent menthofuran. Their work builds on a recent major revision in understanding the plant terpenoid metabolism and represents a useful example of the current state and future directions of metabolic engineering. It constitutes the first successful yield increase in an essential oil crop.

Not all attempts at metabolic engineering deliver the expected results. For example, Lücker et al. [11] transformed petunia (*Petunia hybrida*) with the (*S*)-linalool synthase (LIS) gene from *Clarkia breweri* (Scheme 26.1) , but despite correct expression of the foreign gene in the transgenic lines, the primary product of the enzyme was not observed. Analysis of the non-volatile metabolites revealed that the tertiary terpene alcohol was being immediately metabolised to the (*S*)-linalyl-*β*-D-glucopyranoside. In a different study, attempts to modify floral scent in carnation by the constitutive expression of the *Clarkia breweri LIS* gene resulted in an unsuccessful olfactory outcome although (*S*)-linalool biosynthesis was achieved in the transgenic lines [12]. In this case, the amount of the terpenol was either below the threshold for human perception or masked by other volatiles, even though (*S*)-linalool and its derivatives (*cis*-linalool oxide and *trans*-linalool oxide) constituted almost 10% of the bouquet. In yet another study, the same gene was introduced into tomato (*Lycopersicon esculentum*), under the control of a fruit-specific promoter [13]. In this case the accumulation of (*S*)-linalool and 8-hydroxy-(*S*)-linalool was observed in tomato fruit and the changes in fruit aroma volatiles were successfully discriminated by humans.

The transformation of *Arabidopsis thaliana* with a cDNA from strawberry fruits encoding a dual (*S*)-linalool/(*S*)-nerolidol synthase also led to the production of both (*S*)-linalool and its glycosylated and hydroxylated derivatives in the leaves [14]. Surprisingly, the formation and emission of (*S*)-nerolidol was detected as well, suggesting that a small pool of its precursor farnesyl diphosphate is present in the plastids. The newly emitted (*S*)-linalool and (*S*)-nerolidol showed the same diurnal emission pattern as the pristine volatiles.

By genetically modifying tobacco (*Nicotiana tabacum*) using three different monoterpene synthases from lemon (*Citrus limon*) and the subsequent combination of these three into one plant by crossing, Lücker et al. [15] showed that it is possible to increase the amount and alter the composition of the blend of monoterpenes produced in tobacco plants. The results demonstrated that there is a sufficiently high level of substrate accessible for the introduced enzymes. The transgenic tobacco line containing the three *Citrus limon* monoterpene synthases produced (+)-limonene, *γ*-terpinene, and (-)-*β*-pinene as their main products and was transformed with a fourth gene, a limonene-3-hydroxylase cDNA, isolated from *Mentha spicata* [16]. The targeting sequences of these synthases indicate that they are probably localised in the plastids, whereas the sequence information of the P450 hydroxylase implicated transport to the endoplasmatic

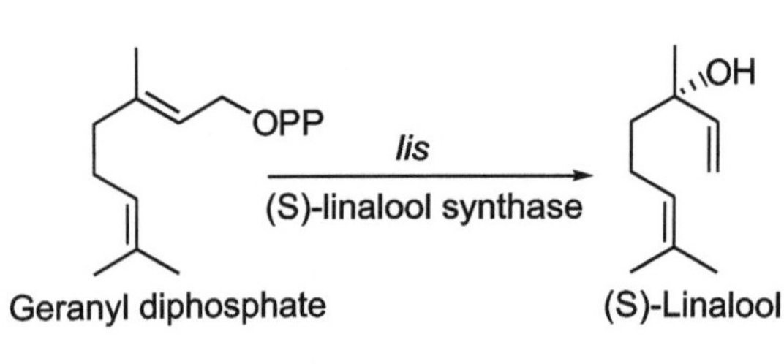

Scheme 26.1 Catalytic formation of (*S*)-linalool from geranyl diphosphate. *OPP* denotes the diphosphate moiety

reticulum. Despite the different locations of the enzymes, the introduced P450 hydroxylase proved to be functional in the transgenic plants as it hydroxylated (+)-limonene, resulting in the emission of (+)-*trans*-isopiperitenol.

The last examples demonstrate that attention has shifted away from single-gene engineering strategies and towards more complex approaches involving the simultaneous overexpression and/or suppression of multiple genes. The use of regulatory factors to control the abundance or activity of several enzymes is also becoming more widespread.

26.3 Hexenals

The cloning, characterisation and expression of many lipoxygenase (LOX) [17] and hydroperoxide lyase (HPL) [18] genes has led researchers to propose new processes for the production of "green note" flavours. HPL specifically produces the highly demanded compound *cis*-3-hexenal from the 13-hydroperoxide of linolenic acid and hexanal from the hydroperoxide of linoleic acid, both of which are formed by LOXs (Scheme 26.2).

Since soybean (*Glycine max* L.) seeds are a rich source for LOXs, genetic engineering approaches focus on the production of recombinant HPL, the limiting enzyme for biocatalytic processes. Recombinant HPL from alfalfa (*Medicago sativa* L.) expressed in *E. coli* forms *cis*-3-hexenal and its isomerisation product *trans*-2-hexenal from linolenic hydroperoxide [19–20]. Recently, the cloning of a *HPL* gene from watermelon (*Citrullus lanatus*) leaves and the overexpression of the corresponding protein in *Nicotiana tabacum* leaves have been reported [21]. These examples show that recombinant expression is an excellent way to increase the availability of HPL used in biotechnological processes.

However, expression in a microbial cell is not always straightforward. For example, recombinant enzyme activity may be different from that of the native enzyme. When incubated in a mixture of hydroperoxides, a HPL from green bell pepper (*Capsicum annuum* L.) that was expressed in *Yarrowia lipolytica* favours the production of hexanal although the native enzyme produces the unsaturated aldehyde *cis*-3-hexenal, both within the green bell pepper itself and when expressed in *E. coli* [22].

In soybean seeds, three distinct isoforms of LOXs have been described, on the basis of differences in pH optima, substrate specificity, and their product formation. Soybean isoenzyme LOX3 not only produces less hydroperoxide but also converts them to ketodiene products, which are not substrates for HPLs. Thus, elimination of LOX3 facilitates greater production of hexenals [23]. The use of LOX2 alone yielded the highest hexenal production, while a two-step conversion was required for LOX1 to produce hexenals at high levels owing to the different pH optima of the enzymes involved. Consequently, the utilisation of pure recombinant LOX2 in combination with recombinant HPL in a biocatalytic process has great potential.

Often attempts to modify aroma profiles end up revealing unforeseen complexities. To modify the flavour properties of tomato (*Lycopersicon esculentum* Mill.) fruits, cucumber (*Cucumis sativus*) HPL, which acts on 9-hydroperoxides of fatty acids to form *cis*-3-nonenal and *cis,cis*-3,6-nonadienal, was introduced to tomato plants [24]. However, the composition of volatile short-chain aldehydes and alcohols in the transgenic fruits was minimally modified although enzyme assays demonstrated high HPL activity. When linoleic acid was added to a crude homogenate prepared from the transgenic tomato fruits, a large amount of C9 aldehyde was formed, but C6 aldehyde levels were almost equivalent to those in control tomatoes. It has been revealed that 13-hydroperoxides of fatty acids are preferably formed from endogenous substrates, but 9-hydroperoxides are formed from exogenous fatty acids.

Five tomato *LOX* genes have been shown to be expressed during fruit ripening, *TomloxA* to *TomloxE*. Antisense-suppression of TomloxA and TomloxB in tomato fruit causes no significant changes in the production of the known tomato flavour volatiles [25]. However, the specific depletion of TomloxC by co-suppression or antisense inhibition leads to major decreases in the flavour volatiles in both fruit and leaves. This suggests that TomloxC is specifically involved in the generation of C6 aldehydes and alcohols, while the functions of the other *LOX* genes remain unknown [26].

Similarly, in potato (*Solanum tuberosum*), silencing *LOX-H1* caused a severe decrease in the amount of volatiles produced by the leaves and in the intensity of their aroma, while the depletion of HPL increased the content of C5 (2-pentenal, pentanal, 1-penten-3-ol and *cis*-2-pentenol) volatiles [27]. These examples clearly demonstrate that the fatty acid metabolism involved in aroma biosynthesis is not as simple as initially supposed.

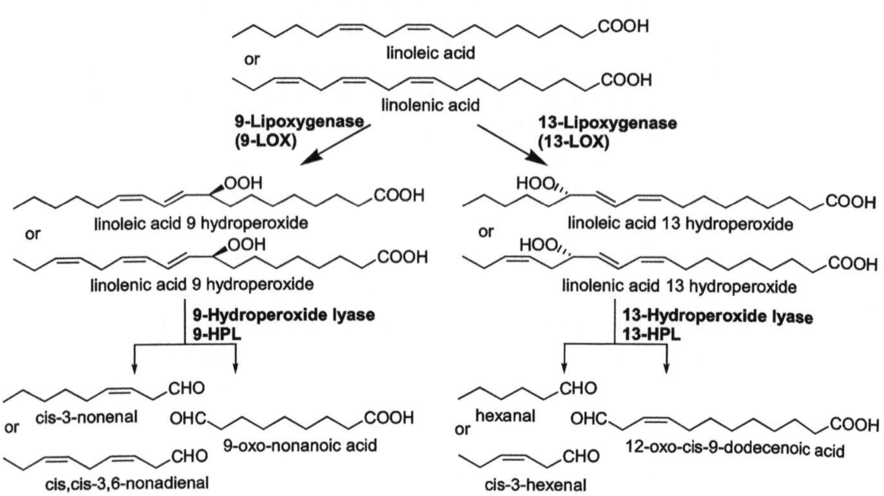

Scheme 26.2 Short-chain aldehyde formation by lipoxygenases and hydroperoxide lyases

26.4 Esters

The biosynthesis of volatile esters, which are important flavour and fragrance components, from the condensation of a coenzyme A (CoA) bound acid component with an alcohol has been shown in a variety of species (Scheme 26.3). In general, the availability of precursors for the enzyme catalysing the reaction determines the quantity and quality of desirable products synthesised in transgenic plants. This principle was applied to *Petunia hybrida* plants transformed with the strawberry alcohol acyltransferase (SAAT) enzyme [28]. Here, the lack of substrates resulted in an unaltered volatile profile, even though both *SAAT* expression and the corresponding enzyme activity were detected in transgenics and inherited in the T2 generation [29]. The feeding of isoamyl alcohol to explants of transgenic lines resulted in the formation of the corresponding acetyl ester, showing that the availability of alcohol substrates is an important parameter to consider when engineering volatile ester production in plants.

Increases in substrate levels that were accidentally produced by metabolic engineering also resulted in an olfaction-detectable increase in the methyl benzoate emission in transgenic carnation [30]. The metabolic flux from the anthocyanin pathway was redirected towards benzoic acid, the methyl ester precursor, by antisense suppression of the flavanone 3-hydroxylase.

In *Saccharomyces cerevisiae* the expression levels of *ATF1* and *ATF2* greatly affect the production of ethyl acetate and isoamyl acetate. But the corresponding proteins are also responsible for the formation of propyl acetate, isobutyl acetate, pentyl acetate, hexyl acetate, heptyl acetate, octyl acetate and phenylethyl acetate [31]. Because double-deletion strains still produced considerable amounts of certain esters, it was assumed that the yeast proteome contains additional as-yet-unknown ester synthases. Genetically engineered *E. coli* cells expressing the *ATF2* gene produced isoamyl acetate from intracellular acetyl-CoA pools, when isoamyl alcohol was added externally to the cell culture medium [32]. Inactivation of the acetate production pathway enhances the production of isoamyl acetate, since it competes with the ester production pathway for the common intracellular metabolite acetyl-CoA. Isoamyl acetate production can further be enhanced by overexpressing the pantothenate kinase gene *panK*, creating an increase in intracellular CoA/acetyl-CoA [33]. Cofactor manipulation is thus an additional tool to achieve metabolic engineering objectives, including increased metabolite production.

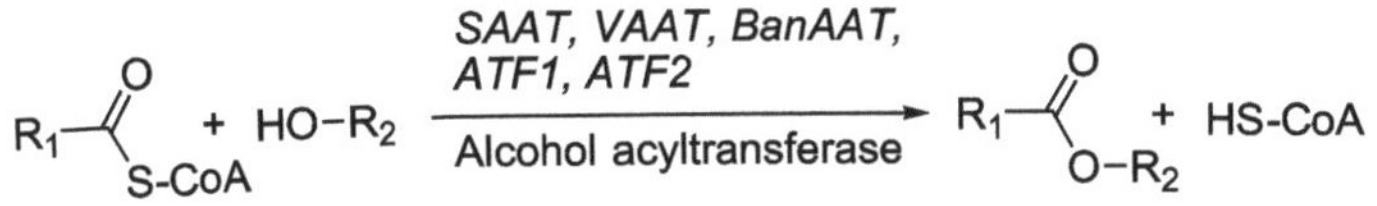

Scheme 26.3 Biosynthesis of volatile esters catalysed by alcohol acyltransferases. *SAAT* strawberry alcohol acyltransferase, *VAAT Fragaria vesca* alcohol acyltransferase, *ATF1, AFT2 Saccharomyces cerevisiae* alcohol acetyltransferases, *BanAAT* banana alcohol acyltransferase

26.5 Vanillin

Vanillin is the most universally accepted aroma chemical used in processed foods, pharmaceuticals and perfumeries [34]. Pods of *Vanilla planifolia* or *Vanilla tahitiensis* are the major natural sources for vanillin. The beans are largely produced in Madagascar and Indonesia and contain 2–3% by weight of vanillin in the cured pod [35]. Of the 12,000-t world consumption, only 20 t is extracted from the *Vanilla* beans, the overwhelming deficit being filled by synthetic vanillin. The price of pure natural vanillin ranges from €1,000 to €3,500 per kilogram, owing mostly to tedious cultivation practices, while the synthetic equivalent costs about €10 per kilogram. The search for inexpensive "biovanillin" led to the application of biotransformation processes to microbial and plant cell cultures, enabling the production of vanillin from cheap substrates such as eugenol and ferulic acid [34].

The bioconversion of eugenol and ferulic acid to vanillin was first characterised in *Pseudomonas fluorescens* (Scheme 26.4) [36, 37]. However, an enzyme of the pathway, vanillin:NAD^+ oxidoreductase, catalysed the removal of vanillin from the medium through the formation of vanillic acid [38]. Deletion of the oxidoreductase was, however, only partially successful, largely because vanillin is also the substrate of coniferyl aldehyde dehydrogenase, an enzyme of the eugenol degradative pathway present in *Pseudomonas* sp. [39].

The expression of genes of the biotransformation pathway in host organisms is the most innovative approach. In *E. coli* cells that were transformed with the genes coding for 4-hydroxycinnamate:CoA ligase and 4-hydroxycinnamoyl-CoA hydratase/lyase (HCHL) vanillin formation at millimolar levels has been observed in resting cells supplied with ferulic acid [40]. Later, two genetically modified *E. coli* strains were used in a two-step biotransformation system designed to produce vanillin [41]. In the first step, resting cells of the first strain, which had been transformed with the vanillyl alcohol oxidase gene from *Penicillium simplicissimum* and the coniferyl alcohol and aldehyde dehydrogenase genes from *Pseudomonas* sp., produced up to 14.7 g of ferulic acid per litre from eugenol with a molar yield of 93%. In the second step, the second strain converted this ferulic acid to form vanillin. The entire process resulted in 0.3 g of vanillin per litre, along with 0.1 g of vanillyl alcohol per litre and 4.6 g of ferulic acid per litre.

In another study, the generation of vanillin from glucose was attempted via the shikimate pathway using genetically engineered *E. coli* in a fed-batch fermentation process [42]. The engineered strain carried a mutated shikimate dehydrogenase locus, a 3-dehydroquinate synthase and a 3-dehydroshikimate dehydratase gene together with a catechol-*O*-methyltransferase and a 3-deoxy-D-*arabino*-heptulosonic acid 7-phosphate synthase gene that was insensitive to feedback inhibition (Scheme 26.5). The organism showed an increased capacity to generate 3-dehydroshikimate but was blocked in the further conversion to shikimate. The introduced 3-dehydroshikimate dehydratase produced protocatechuate, which was methylated by catechol-*O*-methyltransferase to produce vanillin. The final conversion to vanillin was performed extracellularly by an aryl dehydrogenase partially purified from *Neurospora crassa*. Supplementation with L-methionine increased the level of vanillate, suggesting a limiting supply of *S*-adenosylmethionine for the *O*-methyltransferase.

Plants contain only negligible amounts of vanillin. Even in *Vanilla* beans, vanillin is only released from its β-D-glucoside during the postharvest curing of the pods. At this time, no reports of the genetic engineering of plants for higher vanillin yields are available. Because 4-hydroxycinnamoyl-CoA thioesters are intermediates of the plant's central phenylpropanoid and lignin pathway, the expression of HCHL in plants is of particular interest. Although HCHL has been successfully expressed in plant systems, no vanillin or vanillin-β-D-glucoside has been detected thus far (A Mitra, MJ Mayer, personal communication).

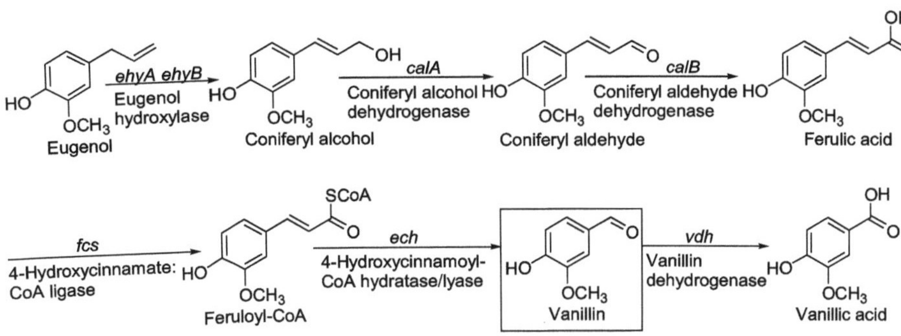

Scheme 26.4 Cloned genes and characterised enzymes involved in the conversion of eugenol and ferulic acid to vanillin in *Pseudomonas* sp. (adapted from [35])

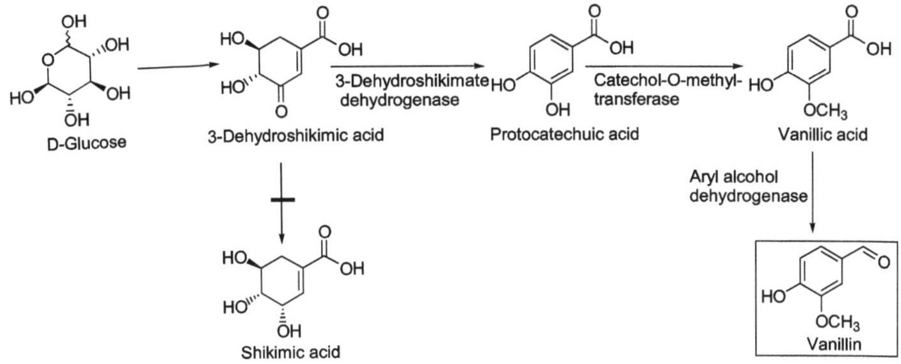

Scheme 26.5 Biocatalytic transformation of glucose to vanillin (adapted from [35])

26.6 Miscellaneous

Metabolic engineering of aroma has also been applied to tomato, potato, milk products and alcoholic beverages. In tomato, at least four attempts have been made to modify flavour constituents in ripening fruit. The expression of a yeast Δ-9 desaturase gene in tomato changed the fatty acid composition in tomato fruits, successfully modifying their flavour profile [43]. Another group transformed tomato plants with a gene construct containing a tomato alcohol dehydrogenase (*ADH*) cDNA in the sense orientation, producing fruits with increased levels of hexanol and *cis*-3-hexenol, whereby the concentrations of the respective aldehydes were generally unaltered—alterations that have been associated with a more intense "ripe fruit" flavour [44]. One other study found that antisense suppression of LOX [25] and overexpression of a 9-HPL [24] in tomatoes resulted in remarkably lower LOX and higher HPL activity, but these changes did not affect the flavour profiles.

The characteristic flavour compound responsible for the particular aroma of baked potatoes is methional formed by Strecker degradation from L-methionine, the loss of which is a major problem associated with potato processing. For economic reasons, neither methional nor its precursor L-methionine is added back to the potato following processing. A solution to this problem is to increase the level of soluble L-methionine by introducing to the potato an *A. thaliana* cystathionine γ-synthase gene, a key gene regulating L-methionine biosynthesis in plants [45]. This results in an up to sixfold enhancement of L-methionine levels in the leaves, roots and tubers of transgenic potato plants compared with those of control potato plants, and as high as 4.4-fold enhancements of methional levels in baked tubers of field-grown transgenic potato lines.

Enzymatic degradation of amino acids also plays an important role in the development of cheese flavour. Usually, branched-chain amino acids are precursors of cheesy aroma compounds, such as isovalerate and isobutyrate, whereas aromatic amino acids are precursors of floral or phenolic aroma compounds. The limiting factor for the transamination reaction of amino acids to aroma compounds is the level of available α-keto acids in cheese. Hence, the glutamate dehydrogenase gene (*GDH*) from *Peptostreptococcus asaccharolyticus* was introduced into *Lactococcus lactis* so that this organism could produce α-ketoglutarate from glutamate [46]. The *GDH*-transformed strain produced higher proportions of volatile carboxylic acids than the control strains and therefore has potential value for cheese ripening.

In addition, cofactor engineering has been used to deliberately modify the intracellular NADH/NAD^+ ratio that plays a predominant role in controlling the *Lactococcus lactis* fermentation pattern. The introduction of the *nox* gene, which codes for a NADH oxidase (NOX) that converts molecular oxygen to water at the expense of NADH, to a strain with an inactivated copy of the *aldB* gene for α-acetolactate decarboxylase led to the efficient metabolism of the native pyruvate to α-acetolactate and diacetyl. The resulting strain could convert up to 80% of the available pyruvate into the butter aroma compound and its precursor [47].

Considerable effort has been given to the reduction of ethanol in alcoholic beverages, since they are of such great commercial value. For example, the glycerol-3-phosphate dehydrogenase gene (*GPD1*) was overexpressed in an industrial lager brewing yeast (*Saccharomyces cerevisiae carlsbergensis*) to reduce the content of ethanol in beer [48]. The amount of glycerol produced by the *GPD1*-overexpressing strain was increased up to sixfold and the amount of ethanol was decreased by 18% compared with the production in the wild type. Only minor changes in the concentration of higher alcohols, esters and fatty acids were observed but the levels of acetoin, diacetyl and acetaldehyde were considerably increased.

Finally, the yeast *Yarrowia lipolytica* is able to transform ricinoleic acid (12-hydroxy oleic acid) into γ-decalactone, a desirable fruity and creamy aroma compound; however, the biotransformation pathway involves ß-oxidation and requires the lactonisation at the C10 level. The first step of ß-oxidation in *Y. lipolytica* is catalysed by five acyl-CoA oxidases (Aox), some of which are long-chain-specific, whereas the short-chain-specific enzymes are also involved in the degradation of the lactone. Genetic constructions have been made to remove these lactone-degrading activities from the yeast strain [49, 50]. A strain displaying only Aox2p activity produced 10 times more lactone than the wild type in 48 h but still showed the same growth behaviour as the wild type.

26.7 Conclusion

The introduction of new genes into microorganisms and plants has become more or less a routine. But as long as the regulation mechanisms of biosynthetic pathways are not thoroughly understood, increased levels of desired metabolites will only be achieved randomly by metabolic engineering. Although genome sequencing projects constantly provide a huge number of new genes, their primary functions remain unknown, even when they show high similarity with already characterised structural genes. Detailed biochemical analyses of the recombinant proteins and studies with transgenic lines, where the gene has been downregulated and/or upregulated, are essential. Nevertheless, the few examples presented here already demonstrate that genetic and metabolic engineering has been quite successful for the production of flavours and has great potential.

Acknowledgement

Heather A. Coiner is thanked for revising the manuscript.

References

1. Watts KT, Mijts BN, Schmidt-Dannert C (2005) Advanced Synthesis and Catalysis 347:927–940
2. Verpoorte R, van der Heijden R, Memelink J (2000) Transgenic Research 9:323–343
3. Klaus SMJ, Huang F-C, Golds TJ, Koop H-U (2004) Nature Biotechnology 22:225–229
4. Marillonnet S, Giritch A, Gils M, Kandzia R, Klimyuk V, Gleba Y (2004) Proceedings of the National Academy of Sciences of the United States of America 101:6852–6857
5. Broun P, Somerville C (2001) Proceedings of the National Academy of Sciences of the United States of America 98:8925–8927
6. Trapp S, Croteau RB (2001) Genetics 158:811–832
7. Martin VJJ, Pitera DJ, Withers ST, Newman JD, Keasling JD (2003) Nature Biotechnology 21:796–802
8. Estévez JM, Cantero A, Reindl A, Reichler S, León P (2001) Journal of Biological Chemistry 276:22901–22909
9. McCaskill D, Croteau R (1995) Planta 197:49–56
10. Mahmound SS, Croteau R (2001) Proceedings of the National Academy of Sciences of the United States of America 98:8915–8920
11. Lücker J, Bouwmeester HJ, Schwab W, Blaas J, van der Plas LH, Verhoeven HA (2001) Plant Journal 27:315–324
12. Lavy M, Zuker A, Lewinsohn E, Larkov O, Ravid U, Vainstein A, Weiss D (2002) Molecular Breeding 9:103–111
13. Lewinsohn E, Schalechet F, Wilkinson J, Matsui K, Tadmor Y, Nam KH, Amar O, Lastochkin E, Larkov O, Ravid U, Hiatt W, Gepstein S, Pichersky E (2001) Plant Physiology 127:1256–1265
14. Aharoni A, Giri AP, Deuerlein S, Griepink F, de Kogel W-J, Verstappen FWA, Verhoeven HA, Jongsma MA, Schwab W, Bouwmeester HJ (2003) Plant Cell 15:2866–2884
15. Lücker J, Schwab W, van Hautum B, Blass J, van der Plas LHW, Bouwmeester HJ, Verhoeven HA (2004) Plant Physiology 134:510–519
16. Lücker J, Schwab W, Franssen MCR, van der Plas LHW, Bouwmeester HJ, Verhoeven HA (2004) Plant Journal 39:135–145
17. Casey R, West SI, Hardy D, Robinson DS, Wu Z, Hughes RK (1999) Trends in Food Science and Technology 10:297–302
18. Noordermeer MA, Veldink GA, Vliegenthart JFG (2001) Chembiochem 2:494–504
19. Noordermeer MA, van Dijken AJH, Smeekens SCM, Veldink GA, Vliegenthart FG (2000) European Journal of Biochemistry 267:2473–2482
20. Noordermeer MA, van der Goot W, van Kooij AJ, Veldsink JW, Veldink GA, Vliegenthart JFG (2002) Journal of Agricultural and Food Chemistry 50:4270–4274
21. Fukushige H, Hildebrand DF (2005) Journal of Agricultural and Food Chemistry 53:2046–2051
22. Bourel G, Nicaud J-M, Nthangeni B, Santiago-Gomez P, Belin J-M, Husson F (2004) Enzyme and Microbial Technology 35:293–299
23. Fukushige H, Hildebrand DF (2005) Journal of Agricultural and Food Chemistry 53:6877–6882

24. Matsui K, Fukutomi S, Wilkinson J, Hiatt B, Knauf V, Kajwara T (2001) Journal of Agricultural and Food Chemistry 49:5418–5424
25. Griffiths A, Prestage S, Linforth R, Zhang J, Taylor A, Grierson D (1999) Postharvest Biology and Technology 17:163–173
26. Chen G, Hackett R, Walker D, Taylor A, Lin Z, Grierson D (2004) Plant Physiology 136:2641–2651
27. Salas JJ, Sanchez C, Garcia-Gonzalez DL, Aparicio R (2005) Journal of Agricultural and Food Chemistry 53:1648–1655
28. Aharoni A, Keizer LC, Bouwmeester HJ, Sun Z, Alvarez-Huerta M, Verhoeven HA, Blass J, van Houwelingen AM, De Vos RC, van der Voet, Jansen H, Guis M, Mol J, Davis RW, Schena M, van Tunen AJ, O'Connell AP (2000) Plant Cell 12:647–662
29. Beekwilder J, Alvarez-Huerta M, Neef E, Verstappen FWA, Bouwmeester HJ, Aharoni A (2004) Plant Physiology 135:1865–1878
30. Zuker A, Tzfira T, Ben-Meir H, Ovadis M, Shklarman E, Itzhaki H, Forkmann G, Martens S, Neta-Sharir I, Weiss D, Vainstein A (2002) Molecular Breeding 9:33–41
31. Verstrepen KJ, Van Laere SDM, Vanderhaegen BMP, Derdelinckx G, Dufour J-P, Pretorius IS, Winderickx J, Thevelein JM, Delvaux FR (2003) Applied and Environmental Microbiology 69:5228–5237
32. Vadali RV, Bennett GN, San K-Y (2004) Applied Microbiology and Biotechnology 63:698–704
33. Vadali RV, Bennett GN, San K-Y (2004) Metabolic Engineering 6:294–299
34. Ramachandra Rao S, Ravishankar GA (2002) Biotechnology Advances 20:1001–153
35. Walton NJ, Narbad A, Faulds CB, Williamson G (2000) Current Opinion in Biotechnology 11:490–496
36. Gasson MJ, Kitamura Y, McLauchlan WR, Narbad A, Parr AJ, Parsons ELH, Payne J, Rhodes MJC, Walton NJ (1998) Journal of Biological Chemistry 273:4163–4170
37. Mitra A, Kitamura Y, Gasson MJ, Narbad A, Parr AJ, Payne J, Rhodes MJC, Seweter C, Walton NJ (1999) Archives of Biochemistry Biophysics 365:10–1
38. Okeke BC, Venturi V (1999) Journal of Bioscience and Bioengineering 88:103–106
39. Overhage J, Priefert H, Rabenhorst J, Steinbüchel A (1999) Applied Microbiology and Biotechnology 52:820–828
40. Overhage J, Priefert H, Steinbüchel A (1999) Applied and Environmental Microbiology 65:4837–4847
41. Overhage J, Steinbüchel A, Priefert H (2003) Applied and Environmental Microbiology 69:6569–6576
42. Li K, Frost JW (1998) Journal of the American Chemical Society 120:10545–10546
43. Wang C, Chin C-K, Ho C-T, Hwang C-F, Polashock JJ, Martin CE (1996) Journal of Agricultural and Food Chemistry 44:3399–3402
44. Speirs J, Lee E, Holt K, Yong-Duk K, Scott NS, Loveys B, Schuch W (1998) Plant Physiology 117:1047–1058
45. Di R, Kim J, Martin MN, Leustek T, Jhoo J, Ho C-T, Tumer NE (2003) Journal of Agricultural and Food Chemistry 51:5695–5702
46. Rijnen L, Courtin P, Gripon J-C, Yvon M (2000) Applied and Environmental Microbiology 66:1354–1359

47. Hugenholtz J, Kleerebezem M, Starrenburg M, Delcour J, De Vos W, Hols P (2000) Applied and Environmental Microbiology 66:4112–4114
48. Nevoigt E, Pilger R, Mast-Gerlach E, Schmidt U, Freihammer S, Eschenbrenner M, Garbe L, Stahl U (2002) FEMS Yeast Research 2:225–232
49. Waché Y, Laroche C, Bergmark K, Møller-Andersen C, Aguedo M, Le Dall M-T, Wang H, Nicaud J-M, Belin J-M (2000) Applied and Environmental Microbiology 66:1233–1236
50. Groguenin A, Waché Y, Garcia EE, Aguedo M, Husson F, LeDall M-T, Nicaud J-M, Belin J-M (2004) Journal of Molecular Catalysis B: Enzymatic 28:75–79

Subject Index

C

D

H

S